DRY SCRUBBING TECHNOLOGIES FOR FLUE GAS DESULFURIZATION

DRY SCRUBBING TECHNOLOGIES FOR FLUE GAS DESULFURIZATION

Sponsored by

Ohio Coal Development Office
Ohio Department of Development
Jacqueline F. Bird, Director

KLUWER ACADEMIC PUBLISHERS
Boston / Dordrecht / London

Distributors for North, Central and South America:
Kluwer Academic Publishers
101 Philip Drive
Assinippi Park
Norwell, Massachusetts 02061 USA
Telephone (781) 871-6600
Fax (781) 871-6528
E-Mail <kluwer@wkap.com>

Distributors for all other countries:
Kluwer Academic Publishers Group
Distribution Centre
Post Office Box 322
3300 AH Dordrecht, THE NETHERLANDS
Telephone 31 78 6392 392
Fax 31 78 6546 474
E-Mail <services@wkap.nl>

 Electronic Services <http://www.wkap.nl>

Library of Congress Cataloging-in-Publication Data

A C.I.P. Catalogue record for this book is available
from the Library of Congress.

Copyright © 1998 by Kluwer Academic Publishers

All rights reserved. No part of this publication may be reproduced, stored in a retrieval system or transmitted in any form or by any means, mechanical, photo-copying, recording, or otherwise, without the prior written permission of the publisher, Kluwer Academic Publishers, 101 Philip Drive, Assinippi Park, Norwell, Massachusetts 02061

Printed on acid-free paper.

Printed in the United States of America

Table of Contents

Table of Figures xvii
Table of Tables xxix
Preface xxxiii
Acknowledgments xxxix

CHAPTER 1   Flue Gas Desulfurization for Acid Rain Control 1

J.R. Kadambi, R.J. Adler,
Case Western Reserve University
M.E. Prudich, Ohio University
L.S. Fan, K. Raghunathan,
The Ohio State University
S.J. Khang, T.C. Keener,
University of Cincinnati

Abstract 1
Background 2
General Discussion of Current FGD Technology 4
   Overview 4
Process Research Studies 6
   Wet/Wet Systems 6
   Dry/Dry Systems 6
   Wet/Dry Systems 8
In-Furnace Injection 10
Fundamentals 12
   Sulfur Evolution 12
Modeling of Sorbent Reaction 22
Technology 23
Scope 27
Economizer Zone Injection, 900-1200°F 28
Injection of Hydrated Lime Downstream of the Air Preheater 29
Fundamental Studies 29
   Dry/Dry Systems 29
   Wet/Dry Systems 31
Spray Drying Absorption 34
   Introduction 34
Description of the Spray Drying FGD Process 37
   Reagent Preparation 37
   Dryer Configuration 39
   Waste Disposal 44
Process Chemistry 45
Particulate Control 50
Electrostatic Precipitator (ESP) 51
   Fundamentals and Operating Parameters 52

TABLE OF CONTENTS

Operating Conditions 56
ESP Design Considerations 58
Fabric Filters (Baghouse) 60
Wet Scrubber 64
Mechanical Particulate Control Devices 65
Comparison of SO_2 Removal in Fabric Filter (Baghouse) and ESPs 67
Scope 70
Sorbent Forms and Additives 71
    Calcium-Based Sorbents 71
    Chemical Form 72
    Calcium/Magnesium Ratio 72
Morphology 74
    Preparation by Aqueous Hydration 77
The Use of Additives to Enhance Reactivity 77
    Deliquescents, Buffers, and Sodium Additives 78
    Alcohol and Sucrose Hydration 82
    Silicate Additives 82
Miscellaneous Additives - High Temperature Applications 83
Contacting Flow Patterns and Multiphase Flows 87
    Contacting Flow Patterns and Some Novel Techniques 87
    Novel Techniques with FGD Applications 88
    Circulating Fluidized Bed Absorber 90
    Limestone Emission Control (LEC) System 91
    Novel Impeller Fluidizer 92
    Acoustic and Other Techniques 95
Fundamental Fluid Mechanics, Mass and Heat Transfer Consideration in
FGD Processes Involving Multiphase Flows 96
Transport Properties of Dry Sorbent and Sorbent Slurries 96
Conclusions 97
    In-Furnace Injection 97
    Economizer Zone Injection 97
    Injection of Hydrated Lime Downstream of the Air Preheater 97
    Spray Drying Absorption 98
    Particulate Control 98
Future Research Needs 99
    Reaction Mechanisms and Sorbent Material 99
    Transport Effects and Modeling 100
    Process Development 100
References 102

**CHAPTER 2 New Calcium-Based Sorbents for Flue Gas
 Desulfurization** 115

Ray-Kuang Chiang, Gwanghoon Kwag and Malcolm E. Kenney
Department of Chemistry
Case Western Reserve University

- Abstract 115
- Introduction 116
  - Properties of SO_2 116
  - Properties of CO_2 116
  - Properties Required in Sorbents 117
  - Previously Considered Sorbents 117
- Techniques and Apparatus 118
  - Infrared Spectroscopy 118
  - X-Ray Powder Diffractometry 119
  - BET Nitrogen Adsorption Surface Area Determinations 119
  - Scanning Electron Microscopy 119
  - SO_2 Uptake Measurements 120
  - Milling 123
  - Shaking 123
- Survey of Calcium Silicates 123
  - Introduction 123
  - Tabulation of Known Calcium Silicates 124
  - Further Description of C-S-H 124
  - Conclusions 139
- Pure Calcium Silicate Sorbents 139
  - Introduction 139
  - Experimental 140
  - Results and Discussion 142
  - Conclusions 151
- SiO_2-Modified $Ca(OH)_2$ Sorbents 151
  - Introduction 151
  - Experimental 155
  - Results and Discussion 157
  - Conclusions 167
- Hydrated Ca_3SiO_5 AND β-Ca_2SiO_4 Sorbents 167
  - Introduction 167
  - Experimental 172
  - Results And Discussion 175
  - Conclusions 181
- Hydrated Portland Cement Sorbents 182
  - Introduction 182
  - Experimental 189
  - Results and Discussion 191
  - Conclusions 197

TABLE OF CONTENTS

Summary 198
    List of Symbols and Abbreviations 199
Notes and References... 200

CHAPTER 3   Fundamental Studies Concerning Calcium-Based Sorbents 207

D. Mandal, R. Venkataramakrishnan, K.J. Sampson, M.E. Prudich
Department of Chemical Engineering
Ohio University

    Abstract 207
    Literature 208
        Background 208
        Sorbent Preparation 208
    Limestone/Lime Characterization Study 221
        Limestone/Lime Samples................................. 221
        Calcine Preparation 222
        Hydrate Preparation 223
        Sorbent Characterization 224
    Effect of Chemical Additives on Duct Injection/Spray Drying Performance 229
        Introduction/General Description......................... 229
        Chemical Effects of Additives 234
        Model Comparisons 249
        Conclusions 249
        Recommendations 250
    Nomenclature... 251
    References 253

CHAPTER 4   Sorbent Transport and Dispersion 255

L.-S. Fan, E. Abou-Zeida,
S.-C. Liang, X. Luo
The Department of Chemical Engineering
The Ohio State University

    Abstract 255
    Introduction 256
    Powder Characterization 257
        General Theory of Interparticle Forces 257
        Results of Powder Characterization....................... 263
        Mechanical Properties 296
    Powder Dispersion.. 303
        Introduction 303
        Experimental... 303
        Simulation of the Gas Flow Field in the Entrances 311
        Results and Discussion.................................. 311
    Modeling 317

A Stochastic Model for Attrition of Sorbent Particles. 317
Integral Model for Powder Dispersion. 329
Nomenclature. 336
References 339

CHAPTER 5 Transport Processes Involved in FGD 343

J.R. Kadambi, P., Chinnapalaniandi
C.U.Yurteri, V.P. Kadaba, M.A. Assar
Mechanical and Aerospace Engineering
Case Western Reserve University

Abstract 343
Introduction 344
   Issues Regarding Transfer Processes for FGD Processes 344
Literature Review. 346
   Single Phase Flow Confined Ducted Jets 346
   Free Coaxial Jets 347
   Confined Jet Flow with Sudden Expansion. 347
   Particle-Laden Flows, Turbulent Free Jet 348
   Concentric Jet Flows 352
   Pipe Flow 353
   Modeling 353
   Conclusions 354
   The Objectives of Investigation of Transport Processes in FGD. 355
Design Consideration for the Sorbent Injection Facility 355
   Geometric Similarity 356
   Particle-Laden Jet-to-Stream Momentum Flux Ratio 356
   Reynolds Number 357
   Stokes Number. 357
   Froude Number 358
   Particle Loading and Particle-to-Fluid Density Ratio 358
Sorbent Injection Facility. 358
   Sorbent Injection (Particle-Laden Jet Test Facility) 358
   Characteristics of Particle Phase (Glass Particles, Lime) 363
   Glass Particles 363
   Lime Particles. 364
   Laser Based Optical Measurement systems. 364
   Calibration Tests 368
Test Results and Discussion. 375
   Axial Velocity Profile Across the Jet. 376
   Particle-Laden Flow Measurements. 379
   Axial Development Along the Centerline 379
   Axial Velocity Profile Across the Jet. 384
   Jet Spreading Rate and Entrainment 394
   Measurement of Mean Particle Diameter and Particle Concentration. 396

TABLE OF CONTENTS

Conclusions 399
    Single-Phase Flow 400
    Particle-Laden Flows 400
Investigation of Spray-Concurrent Flow 401
    Experimental Setup 402
    Experimental Procedure 403
    Experimental Results 405
    Experimental Velocity Profiles and Prediction 405
    Flow Reversal 407
    Development of Criterion for Prediction of Flow Reversal .. 409
Conclusions 412
    Particle-Laden Jet Reversal in Cocurrent Flow 412
Summary 413
Nomenclature 415
References 417

CHAPTER 6 High Temperature Desulfurization of Flue Gas Using Calcium-Based Sorbents 421

L.-S. Fan, A. Ghosh-Dastidar, S. Mahuli, R. Agnihotri
Department of Chemical Engineering
The Ohio State University

Abstract 421
Introduction 422
    Furnace Sorbent Injection (FSI) 424
    High Temperature Phenomena: Calcination, Sintering and Sulfation 424
    Role of Surface Area and Porosity of the Calcines 428
    Role of Particle Size 429
    Role of Sorbent Type 429
    Role of Temperature 430
High-Temperature Entrained Flow Reactor Setup 430
    Introduction 430
    Entrained Flow Reactor System 431
    The Powder Feed System 433
    The High-Temperature Reactor 435
    The Probe System 438
    Particle Collection/Classification System 451
    Data Acquisition System 452
Experimental Approach 452
    Operating Conditions 452
    On-Line Estimation of Particle Residence Time 453
    Post-Reaction Analyses of Data on Reaction Kinetics 455
Ca-Based Sorbents Used in High-Temperature Flue Gas Desulfurization 458
Experimental Results and Discussion: Calcination of Calcium Based Sorbents .. 460
    Introduction 460
    Effect of Temperature on Calcination 460

    Internal Surface Area Development with Calcination................... 462
    Effect of Particle Size on Calcination 468
    Development of Porosity and Pore Size Distribution of Calcined
       Sorbent Particles 469
    Experimental Results and Discussion: Sulfation of Calcium-Based Sorbents ... 476
       Introduction 476
       Effect of Temperature on the Sulfation of CaO and $Ca(OH)_2$............. 477
       Effect of Initial Surface Area and Porosity on CaO Sulfation............. 481
       Effect of Particle Size on Sulfation 483
       Development of Internal Structural Properties During Sulfation 484
    Reaction Modeling for High-Temperature SO_2 Sorption 488
       Introduction 488
       Calcination Modeling... 493
       Sintering Modeling 495
       Comprehensive Sulfation Modeling 497
       Comparison of Experimental Data and Model Predictions 501
    Modified Calcium-Based Sorbent 512
       Introduction 512
       Surfactant Modifiers 513
       Lignosulfonate Modified Calcium Hydroxide........................ 513
       Experimental Results and Discussion: Comparison Between
          Modified and Pure Calcium Hydroxide........................... 513
    Conclusions 519
    Nomenclature 523
       Greek Letters 524
    Reference... 525

CHAPTER 7   **Kinetic Studies on the Medium Temperature $Ca(OH)_2$ Sorbent Injection FGD Process........ 529**

Soon-Jai Khang, Timothy C. Keener*
Anbo Wang, Zhenwei Wang
Department of Chemical Engineering
*Department of Civil &
Environmental Engineering
University of Cincinnati

    Abstract 529
    Introduction 530
       Background 530
    Entrained Flow Reactor.. 536
       Introduction 536
       Entrained Flow Reactor Construction and Operation 537
       Entrained Flow Reactor II 539
       Particle Heat Transfer Rate Determination 543
       Particle Residence Time Determination 547

Dehydration Reaction of $Ca(OH)_2$ 547
   Literature Review 547
   Equilibrium of the Dehydration Reaction 550
   Dehydration Kinetics in the Entrained Flow Reactor 552
   Dehydration Model 555
   The Pseudohomogeneous Model 557
   Conclusions 560
Carbonation Reaction 561
   Literature Review 561
   Carbonation of $Ca(OH)_2$ in Entrained Flow Reactors 561
   Reactivity of CaO with CO_2 567
   First-Order Deactivation Model 568
   Third-Order Deactivation Model 569
   General dth-Order Deactivation Model 572
   Model Application and Results 573
   Conclusions 576
Sulfation Reaction 577
   Introduction 577
   Sulfation in an Entrained Flow Reactor 577
   Reactivity of CaO with SO2 578
   Sulfation Model 579
   Model Application and Results 581
   Conclusions 584
Simultaneous Sulfation and Carbonation Reaction 585
   Literature Review 585
   Simultaneous Dehydration, Sulfation and Carbonation in the
     Entrained Flow Reactor 587
   Kinetic Model for Simultaneous Dehydration, Sulfation and Carbonation ... 593
   Reaction with Nascent $CaCO_3$ 595
   Activity 597
   Model Application 598
   Conclusions 601
   Recommendations 602
References 604

CHAPTER 8   **Advances in Spray Drying Desulfurization for High-Sulfur Coals** 607

Tim C. Keener*, Jun Wang*, and Soon-Jai Khang**
*The Department of Civil and Environmental Engineering
**The Department of Chemical Engineering
The University of Cincinnati

   Abstract 607
   Introduction 608
   Literature Review 611
      General Spray Dryer Operations for Desulfurization 611

Reagent Properties and Preparations	611
Spray Drying Desulfurization	613
Particulate Collection	615
Recycle, Solid Waste Treatment and Disposal	616
Methods of Increasing SO_2 Removal and Sorbent Utilization	617
Previous Lime and Limestone Dissolution Rate Studies	619
Spray Dryer FGD Models	622
Lime Dissolution Rate Studies	623
Introduction	623
Experimental Approach and Limestone Test Results	623
Rotating Disk Technique and Lime Dissolution Rate Experimental Studies	626
Pilot Spray Dryer Experimental Approach and Results	641
Spray Dryer and Sampling System	641
Baseline Test and Parameters that Affect Spray Dryer Desulfurization Processes	644
Additives to Increase $Ca(OH)_2$ Utilization	647
Hydration of Fly Ash with $Ca(OH)_2$ to Increase Sorbent Reactivity and Utilization	658
Spray Dryer Modeling	664
Introduction to "SPRAYMOD" and its Application to Baseline Test Results	664
SO_2 Absorption Process	668
Modification of the Model by Hygroscopic Additive Effects: The Mechanism of Hygroscopicity	670
Conclusions	680
Nomenclature	682
Greek Letters	684
References	685

CHAPTER 9   Low Temperature Dry Scrubbing/LEC Process Support ... 691

K.W. Appell, M.J. Visneski, S. Reddy, M. Maldei, K.J. Sampson
M.E. Prudich
Department of Chemical Engineering
Ohio University

Abstract	691
Literature Review	692
General Background	692
Fixed-Bed Limestone Emission Control	705
Moving-Bed Limestone Emission Control	709
Limestone Solubilities and Dissolution Rates	713
Process Theory and Model Development	716
Assumptions	718

TABLE OF CONTENTS

    Removal Rate of Sulfur Dioxide 719
    Interfacial Concentrations 720
    Reaction Site and Determination of Control 721
Fixed-Bed Process Model 723
    Mass and Enthalpy Balances 723
    Model Simulations 730
Moving-Bed Reactor Process Model 733
    Model Development 733
    Model Simulations 734
Limestone Solubilities and Rates of Solubilization 737
    Theory 737
    Experimental 751
    Results and Discussion 759
    Conclusions and Recommendations: Solubilities and Rates
       of Dissolution 767
Nomenclature 769
References 771

CHAPTER 10   Simulation and Optimization of a Granular Limestone Flue Gas Desulfurization Process 775

D.W. Duespohl, K.J. Sampson, S. Chattopadhyay, M.E. Prudich
Department of Chemical Engineering
Ohio University

    Abstract 775
    Introduction 775
    Literature Review 778
       Transport and Reaction 778
       Economics 780
    Process Model 780
       Reactor Model 781
       Sorbent Blinding 781
       Dry Capture Model 786
       Differential Equations 787
       Solution Algorithm for Partial Differential Equations 790
       Calculation Speed Enhancement 792
       Cost Model 794
       Plant Simulation Model 795
       Optimization Algorithm 798
       Design Cases 801
       Computer Code 803
    Results 803
       Discussion of Results 806
    Conclusions 807
    Nomenclature 808

References 816

Appendix A–Governing Differential Equations 818
Material and Enthalpy Balances 818
Gas Phase Material Balances....................................... 820
Solid Phase Material Balances..................................... 822
Liquid Phase Material Balances.................................... 824
Detailed Enthalpy Balances.. 824

Appendix B Transport and Physical Property Correlations 827
Sulfur (Dioxide) Transport Rate (gS) 827
Water Transport Rate (gW) 833
Gas/Liquid Heat Transport Rate (gg)............................... 833
Gas/Liquid Heat Transfer Coefficient (hg) 833
Calcium Pseudo-Mass Transfer Coefficient (kmC).................... 834
Water Vapor Mass Transfer Coefficient (kmS)....................... 835
Colburn j Factor for Heat and Mass Transfer in the Gas Phase (Jg) 835

Appendix C - Plant Simulation and Cost Model.................... 852
Mass Balance Equations 852
Equipment Size Parameters.. 853

Dry Scrubbing Technologies for Flue Gas Desulfurization

Table of Figures

Chapter 1   Flue Gas Desulfurization for Acid Rain Control............ 1

Figure 1.1	Typical injection locations for calcium-based sorbents............ 5
Figure 1.2	Differential reactor setup used by Borgwardt [1985].............. 14
Figure 1.3	Equilibrium diagram for CaO, CaS and $CaSO_4$ with gas species [77]....................................... 21
Figure 1.4	A schematic of the LIMB process [91]........................ 24
Figure 1.5	Direct desulfurization process proposed by Chughtai (197)......... 26
Figure 1.6	Spray dryer absorber [112].................................. 39
Figure 1.7	Inlet of a spray dryer using rotary atomization [112].............. 42
Figure 1.8	Bypass gas options for temperature control [110]................ 43
Figure 1.9	Classic drying curve for slurries in spray dryers.................. 45
Figure 1.10	Dissociation constants of water and dissolved SO_2.............. 48
Figure 1.11	Equilibrium constants for the dissociation of $Ca(OH)_2$ and the precipitation of $CaSO_3 \cdot 1/2H_2O$....................... 49
Figure 1.12	Typical temperature-resistivity relationship [126]................ 54
Figure 1.13	Comparison of ESP with bag filter for SO_2 removal [126]......... 69
Figure 1.14	Comparison of ESP with bag filter for SO_2 removal [126]......... 70
Figure 1.15	SO_2 capture reactivity versus the surface area of the calcined hydrate [161]................................. 75
Figure 1.16	Effect of sorbent surface area on low temperature SO_2 capture performance: (a) 150°F, 60% relative humidity, 1000 ppm SO_2, 60 minutes, Yoon et al. [17]. (b) 150°F, 20 vol % H_2O, 1500 ppm SO_2, Borgwardt and Bruce [162]. 76
Figure 1.17	Effect of additive concentration on sorbent utilization Kirchgessner and Lorrain [181].85
Figure 1.18	Scheme of experimental set-up for circulating fluidized bed absorber [192]............................... 91
Figure 1.19	Single impeller experiment................................. 93
Figure 1.20	Effect of rotation rate, single impeller. Impeller speeds are marked in rpm. 94
Figure 1.21	Cocurrent recycling scheme utilizing the impeller fluidizer......... 95

Chapter 2   New Calcium-Based Sorbents for Flue Gas Desulfurization 115

Figure 2.1	Custom-built sand-bed SO_2 sorption system................... 121
Figure 2.2	(a) Seven-coordination of Ca in CaO_2 core of calcium silicate layer in 11-Å tobermorite. (b) CaO_2 core of this layer (Hamid's data [74])....................................... 135

TABLE OF FIGURES

Figure 2.3 (a) Part of an oligomeric $(Si_3O_9H)n$ chain of calcium silicate layer in 11-Å tobermorite. (b) $(Si_3O_9H)n$ chains oriented as in layer (Hamid's data [74]). 136

Figure 2.4 Structure of 11-Å tobermorite (Hamid's data [74]). 137

Figure 2.5 (a) Colloidal-particle model of C-S-H, (b) extended-sheet model of C-S-H. 139

Figure 2.6 X-ray powder patterns of xonotlite, 11Å-tobermorite and C-S-H. ... 147

Figure 2.7 X-ray powder patterns of hillebrandite and a-dicalcium silicate hydrate. 148

Figure 2.8 Structure of $Ca(OH)_2$: (a) a single layer projected along <001>; (b) edge view of two layers (Petch's data [98]). 154

Figure 2.9 X-ray powder patterns of products of runs 2 through 7, Table 2.12. 160

Figure 2.10 Infrared spectra of products of runs 2 through 7. Table 2.12. 161

Figure 2.11 Schematic representation of the $Ca(OH)_2$-fumed SiO_2 reaction. 162

Figure 2.12 X-ray powder patterns of products of runs 8 through 11, Table 2.13. 166

Figure 2.13 Infrared spectras of products of runs 12 and 13, Table 2.14. 168

Figure 2.14 Calorimetric curve for hydration of Ca_3SiO_5 at a 1:2 water-to-solids ratio during (a) early part of hydration reaction and (b) intermediate part of hydration reaction (Jennings [120]). 169

Figure 2.15 Schematic representation of Ca_3SiO_5 hydration mechanism. 170

Figure 2.16 X-ray powder patterns of products of runs 1 and 4, Table 2.15. 178

Figure 2.17 X-ray powder patterns of intermediate and final products of run 3, Table 2.15. 178

Figure 2.18 X-ray powder patterns of intermediate and final products of run 8, Table 2.15. 179

Figure 2.19 (a) Schematic structure of $[Ca_2Al(OH)_6](OH).6H_2O$ and $[Ca_2Al(OH)_6][Al(OH)_4].3H_2O$ showing $[Ca_2Al(OH)_6]+$ layers and OH- or $Al(OH)_4$- ions between layers. (b) Structure of $[Ca_2Al(OH)_6]+$ layers (Taylor's data [143]). ... 184

Figure 2.20 Portion of unit cell of $Ca_3Al2(O_4H_4)_3$, hydro-garnet, showing $Al(OH)_6$ and $(OH)4$ units in structure (Foreman's data [144]). 184

Figure 2.21 (a) Structure of $[Ca_3Al(OH)_6.12H_2O]2(SO_4)_3.2H_2O$, trisulfoaluminate hydrate, projected on the <001> showing the columns and channels present. (b) Structure of $[Ca_3Al(OH)_6.12H_2O]3+$ columns (Taylor's data [145]). 186

Figure 2.22 Structure of $[Ca_2Al(OH)_6]SO_4.6H_2O$, monosulfo-aluminate hydrate, showing layers of $[Ca_2Al(OH)_6.]+$ and H_2O and SO_4^{2-} ions between layers (Allmann's data [146]). 186

Chapter 3 Fundamental Studies Concerning Calcium-Based Sorbents 207

Figure 3.1	SO_2 capture reactivity versus surface area of the calcined hydrate [11].	211
Figure 3.2	Effect of sorbent surface area on low temperature SO_2 capture performance: (1) 150°F, 60% relative humidity, 1000 ppm SO_2, 60 minutes [12]; (b) 150oF, 20 vol% H_2O, 1500 ppm SO_2 [13].	213
Figure 3.3	Schematic of the bench-scale calcination reactor.	223
Figure 3.4	Schematic of the bench-scale slaking reactor.	224
Figure 3.5	Comparison of the BET surface areas of -80+100M limestone samples.	225
Figure 3.6	Comparison of the BET Surface area -80z+100 limestone samples and their calcines.	226
Figure 3.7	Comparison of BET surface area of -80+100M limestone samples, their calcines, and their hydrates.	227
Figure 3.8	CaO solubility versus additive concentration of 70°C.	228
Figure 3.9	Effect of solubility additives on hydrate BET surface area.	229
Figure 3.10	Illustration of the accumulation of sorbent and product at the droplet surface as evaporation shrinks the surface [40].	232
Figure 3.11	SO_2 capture efficiency versus additive concentration (vapor pressure depression effect).	236
Figure 3.12	Percentage sulfur dioxide capture versus dissolution rate constant for lime (slurry injection case).	239
Figure 3.13	Reduction in dissolution rate constant as a function of NaOH concentration.	241
Figure 3.14	Percentage sulfur dioxide capture versus equilibrium solubility of lime at various lime dissolution rates constants.	242
Figure 3.15	Percentage SO_2 capture versus the SO_2 absorption enhancement factor.	243
Figure 3.16	Percentage SO_2 capture efficiency versus Henry's law constant.	244
Figure 3.17	Sulfur dioxide capture efficiency as a function of solid-phase alkalinity.	245
Figure 3.18	Percentage SO_2 capture efficiency versus additive concentration.	246
Figure 3.19	Percentage SO_2 capture efficiency versus droplet size.	248

Chapter 4   Sorbent Transport and Dispersion 255

Figure 4.1	Sorbent contact geometries.	262
Figure 4.2	SEM micrographs of four sorbent powders.	265
Figure 4.3	Sedigraph analyses of four sorbents.	266
Figure 4.4	Theoretical force versus particle diameter for hydrated lime.	268
Figure 4.5	Comparison between calcite (sphere-flat plate) calcite (sphere-sphere) and hydrated lime (sphere-sphere) van der Waals forces.	269
Figure 4.6	Experimental apparatus used in transport experiments.	270

Dry Scrubbing Technologies for Flue Gas Desulfurization

TABLE OF FIGURES

Figure 4.7	Comparison of pre- and post- transport volume distributions.	272
Figure 4.8	Comparison of Black River Hydrate size distributions from the furnace at 21°C (~) and 825°C (O).	275
Figure 4.9	Comparison of 1.3% lignosulfonated hydrate size distributions from the furnace at 30°C (ª) and 839°C (~).	276
Figure 4.10	SEM micrograph of pure, as-hydrated $Ca(OH)_2$ sample. The hydrate is composed of numerous ill-defined solid crystals.	278
Figure 4.11	SEM micrograph of 0.38 wt% lignosulfonate-modified sample. Large, hexagonal single crystals are clearly observed.	278
Figure 4.12	SEM micrograph of pure hydrate. The ill-defined, tiny crystals appear to form large agglomerates after drying.	280
Figure 4.13	SEM micrograph of 1.5 wt% lignosulfonate-modified sample. This breakage of agglomerates is not observed in the pure, unmodified hydrate sample.	281
Figure 4.14	Surface area versus mass % lignosulfonate added to final product hydrate.	282
Figure 4.15	% conversion and surface area versus mass % lignosulfonate for the unfiltered hydrates.	283
Figure 4.16	Schematic of the experimental setup.	284
Figure 4.17	Variation of charge-to-mass ratio with powder mass flow rate for hydrated lime.	286
Figure 4.18	Effect of gas velocity on the charge-to-mass ratio for hydrated lime.	289
Figure 4.19	Electric field at the wall surface as a function of powder mass flow rate.	291
Figure 4.20	Comparison of experimental charge-to-mass ratio with maximum values based on the breakdown field strength of the air.	292
Figure 4.21	Effect of four tube materials on the charge-to-mass ratio for hydrated lime.	294
Figure 4.22	UPS spectrum of Cu at 25°C.	295
Figure 4.23	Variation of charge accumulation with time at three low gas velocities.	296
Figure 4.24	Hausner ratio and compressibility.	300
Figure 4.25	Dispersibility and cohesion number.	301
Figure 4.26	Schematic diagram of powder dispersion system.	304
Figure 4.27	Detailed configuration of different powder injection nozzles.	305
Figure 4.28	Schematic diagram of the powder dispersion visualization system.	306
Figure 4.29	Samples of the original images and the processed images: (a) original images, (b) processed images.	307
Figure 4.30	Radial distribution of solid concentration at two axial locations in the simple expansion nozzle.	311
Figure 4.31	Comparison of agglomerate size distribution in different nozzles	313

Figure 4.32	Radial profiles of the agglomerate size at two axial locations in the expansion nozzle with two booster jets. 314
Figure 4.33	Radial profiles of solid concentration at two axial positions in the expansion nozzle with two booster jets. 314
Figure 4.34	Flow patterns in different nozzels: (a) the simple expansion nozzle, (b) the confined coaxial jets nozzle, (c) the expansion nozzle with two booster jets................ 316
Figure 4.35	Effects of total air flow rate on powder dispersion. 317
Figure 4.36	Size distribution of attrited particles for different values of parameters λ5. 328
Figure 4.37	Dynamics of agglomerate size variation....................... 329
Figure 4.38	Axial variation of cross-sectionally averaged turbulent energy dissipation rate. 330
Figure 4.39	Comparison of the simulation results with experimental data. 336

Chapter 5   Transport Processes Invloved in FGD 343

Figure 5.1	Schematic of particle-laden jet flow test facility................ 360
Figure 5.2	Particle feeder. 361
Figure 5.3	Schematic diagram of seeding generator...................... 362
Figure 5.4	Particle size distribution from the Sedigraph and PDA (m=10). 364
Figure 5.5	PDI/LDA principles...................................... 366
Figure 5.6	The wire disk assembly and the LDV probe volume. 369
Figure 5.7	Histogram for two rotating wire velocities..................... 370
Figure 5.8	Histogram of the two rotating wire transit time. The larger value is for the 1150 micron wire.......................... 370
Figure 5.9	Histogram of product vt for the two rotating wires. The larger value is for the 1150 micron wire................ 371
Figure 5.10	The effect of measurement volume size and number of fringes crossed on the TTLDV measurements................. 373
Figure 5.11	Glass particle size distribution obtained from PDA and TTLDV (160 mm lens, TTLD mean = 41.0 μm, PDA mean =43.2 μm)........... 374
Figure 5.12	Comparison of the lime particle size distribution obtained from TTLDV, PDA, and Microtrac results.................... 374
Figure 5.13	a: Intensity of axial turbulence at axial locations of x/d=5, 10, 15 and 20 (case 1). b: Intensity of radial turbulence at axial locations of x/d=5, 10, 15 and 20 (case 1)................... 378
Figure 5.14	a: Mean particle phase and gas-phase velocities for case 1, m=2.5. b: Mean particle phase-velocity along the centerline for case 1, m=2.5, m=10 380
Figure 5.15	Effect of wake due to jet tube wall thickness at (a)x/d=0.1, (b)x/d=3, (c)x/d=5 from U0=35, Udf=20,m=5................ 382
Figure 5.16	Axial fluctuating velocity along the centerline for case 1 at m=2.5 and m=10...................................... 383

Dry Scrubbing Technologies for Flue Gas Desulfurization

TABLE OF FIGURES

Figure 5.17 Radial fluctuating velocity along the centerline for case 1 at m=2.5 and m=10. 384
Figure 5.18 Velocity profile at the exit of the jet for x/d=0.1 (case 1). 385
Figure 5.19 Gas-phase and particle-phase velocity profile at the exit of the jet, x/d=0.1 (case 1, m=2.5 and m=10). 387
Figure 5.20 a: Particle-phase velocity and particle number concentration profile for case 1, m=2.5, axial location x/d=20. b: Particle-phase velocity and particle number concentration profile for case 1, m=5, axial location x/d=20. 388
Figure 5.21 Mean axial velocity and concentration profiles at x/d=20 for case 1, m=5. 390
Figure 5.22 Effect of mass loading on particle-phase and gas-phase centerline velocity at axial location x/d=15 (case 1). 391
Figure 5.23 a: Intensity of axial velocity fluctuation, x/d=20 (case Hx m=5). b: Intensity of axial velocity fluctuation, x/d=20 (case 1, m=5).392
Figure 5.24 Turbulent shear stress at axial location x/d=20, case 1. 394
Figure 5.25 Entrainment in single and particle-laden jets (m=5, case 3). 396
Figure 5.26 Particle number concentration profile for case 1. 398
Figure 5.27 Particle volumetric concentration profile, x/d=15, case 3. 399
Figure 5.28 Velocity profiles showing axisymmetric spray. 403
Figure 5.29 Comparison of experimental velocity profile with the cosine model for case 4: uo=681.48 m/s, u=6m/s. 406
Figure 5.30 Map of flow field showing flow reversal. 407
Figure 5.31 Spray parameters. 409
Figure 5.32 Comparison of the experimental velocity profile with the modified cosine law (test number 3, x/d=15, U=6 m/s). 413

Chapter 6 High Temperature Desulfurization of Flue Gas Using Calcium-Based Sorbents 421

Figure 6.1 (a) Inward growth where reactants move from gas/solid interface to solid/solid interface; (b) outward growth where reactants move from solid/solid interface to gas/solid interface (Hsia et al. [14]). 427
Figure 6.2 Schematic of the entrained flow reactor system. 432
Figure 6.3 Schematic of the continuous powder feed system. 434
Figure 6.4 Details of the reactor entrance block. 437
Figure 6.5 Typical axial temperature profile inside the reactor with a nominal reactor temperature of 1323 K. 438
Figure 6.6 Schematic of the collection probe. 440
Figure 6.7 Front view of the collection probe head. 441
Figure 6.8 Side view of the collection probe head. 442
Figure 6.9 Design details of the collection probe assembly. 443
Figure 6.10 Front view of the injection probe head. 444

Figure 6.11	Side view of the injection probe head.	445
Figure 6.12	Design details of the injection probe assembly.	446
Figure 6.13	Design of optical guide.	447
Figure 6.14	Theoretically calculated particle heat-up rate assuming the Nusselt number to be equal to 2.0.	450
Figure 6.15	Injection and collection probe signals and their cross-correlation. Calculated residence time = 4 ms.	456
Figure 6.16	Injection and collection probe signals and their cross-correlation. Calculated residence time = 476 ms.	457
Figure 6.17	Effect of temperature on calcination of 3.6 mm Linwood $Ca(OH)_2$.	461
Figure 6.18	Arrhenius plots to estimate activation energies of calcination and sintering.	463
Figure 6.19	Effect of temperature on surface area evolution during calcination of 3.6 µm Linwood $Ca(OH)_2$.	464
Figure 6.20	Surface area development with calcination of $Ca(OH)_2$ at 1050°C.	466
Figure 6.21	Influence of particle size on surface area development during calcination of $Ca(OH)_2$ at 1050°C.	467
Figure 6.22	SEM photomicrographs of partially calcined $Ca(OH)_2$ particles. (a) 42% calcined, 1.5 µm particles after 24 ms at 1223 K, surface area of 24 m2/g. (b) 90% calcined, 1.5 µm particles after 235 ms at 1223 K, surface area of 13 m2/g.	468
Figure 6.23	Influence of particle size on calcination of Linwood $Ca(OH)_2$ at 1323 K.	470
Figure 6.24	Effect of particle size on surface area evolution during calcination of Linwood $Ca(OH)_2$ at 950°C.	471
Figure 6.25	$Ca(OH)_2$ adsorption-desorption isotherm.	472
Figure 6.26	Porosity development with calcination of $Ca(OH)_2$ at 1000°C.	474
Figure 6.27	Pore size distribution during calcination of $Ca(OH)_2$ at 1000°C.	475
Figure 6.28	Pore size distribution during calcination of $Ca(OH)_2$ at 1080°C.	476
Figure 6.29	Effect of temperature on sulfation of pre-sintered h-CaO (3.9 mm) with an initial surface area of 3.9 m2/gm.	479
Figure 6.30	Effect of temperature on sulfation of 3.9 mm Linwood $Ca(OH)_2$.	480
Figure 6.31	Effect of initial surface area of pre-sintered CaO (3.9 mm) on sulfation reaction at 1308 K.	482
Figure 6.32	Effect of particle size on sulfation of Linwood $Ca(OH)_2$ at 1388 K.	483
Figure 6.33	Surface area development during sulfation of $Ca(OH)_2$ at 1353 K.	485
Figure 6.34	Surface area development during sulfation and during calcination only of $Ca(OH)_2$ at 1353 K.	486
Figure 6.35	Porosity evolution during sulfation and during calcination only of $Ca(OH)_2$ at 1353 K.	487
Figure 6.36	Pore size distribution of partially sulfated $Ca(OH)_2$ at 1080°C.	488

TABLE OF FIGURES

Figure 6.37	Schematic illustration of model. (a) Single, spherical $Ca(OH)_2$ grain. (b) Partially calcined grain with inner $Ca(OH)_2$ core and sintering CaO micrograins.	491
Figure 6.38	Schematic of the overlapping pore structure of the random pore model. (a) Early stage, showing product layer around each pore. (b) Intermediate stage, showing some overlapping reaction surfaces (after Bhatia and Perlmutter [30]).	498
Figure 6.39	Effect of temperature on calcination of 3.6 μm Linwood $Ca(OH)_2$. Experimental data and model predictions.	502
Figure 6.40	Effect of temperature on surface area evolution during calcination of 3.6 μm Linwood $Ca(OH)_2$. Experimental data and model predictions.	504
Figure 6.41	Ratio of interface H_2O partial pressure to the dissociation pressure with time. Model simulation.	505
Figure 6.42	Comparison of comprehensive model predictions with experimental sulfation data for 3.9 μm Linwood $Ca(OH)_2$.	507
Figure 6.43	Variation of CaO porosity as predicted by the calcination and sintering model and used in the sulfation modeling.	508
Figure 6.44	Variation of CaO surface area as predicted by the calcination and sintering model and used in the sulfation modeling.	509
Figure 6.45	Variation of the structural parameter ψ with time; model simulation.	510
Figure 6.46	Variation of the parameter β with time. Model simulation.	511
Figure 6.47	Arrhenius plot to estimate activation energies of surface reaction and product layer diffusion.	512
Figure 6.48	Comparison of pore size distribution for pure $Ca(OH)_2$ and 1.5% ligno-$Ca(OH)_2$.	515
Figure 6.49	Primary particle size distribution from Sedigraph analysis.	516
Figure 6.50	Sulfation of 3.9 μm pure hydrate and 1.5% lignohydrate at 1080°C.	517
Figure 6.51	Pore volume distribution after 20 ms of calcination at 1080°C. Particle size: 3.9 μm.	518
Figure 6.52	Comparison of pore volumes determined experimentally after 20 ms of sulfation and calcination at 1080°C.	519
Figure 6.53	Pore volume distribution after 20 ms of sulfation at 1080°C. Particle size: 3.9 μm.	520

Chapter 7   Kinetic Studies on the Medium Temperature $Ca(OH)_2$ Sorbent Injection FGD Process 529

Figure 7.1	Dry SO_2 removal processes.	531
Figure 7.2	Schematic diagram of entrained flow reactor I.	537
Figure 7.3	Detail structure of the sorbent feeder for entrained flow reactor I.	538
Figure 7.4	Temperature in the reaction chamber of entrained flow reactor I.	539
Figure 7.5	Schematic diagram of the entrained flow reactor II.	540
Figure 7.6	Detailed structure of the entrance block design.	541
Figure 7.7	Schematic diagram of the sorbent feeder for entrained flow reactor II.	543

Figure 7.8	Temperature increase of a sorbent particle injected into a furnace at 800°F (427°C).	545
Figure 7.9	Temperature increase of a sorbent particle injected into a furnace at 1000°F (538°C).	546
Figure 7.10	TGA setup at preheated stage.	551
Figure 7.11	TGA setup during test.	552
Figure 7.12	Dehydrated reaction of $Ca(OH)_2$ under the medium temperature range (residence time 50 ms).	554
Figure 7.13	Dehydration reaction of $Ca(OH)_2$ in the residence time from 35 to 75 ms (reaction temperature 900°F).	555
Figure 7.14	Arrhenius plot of dehydration data.	559
Figure 7.15	Effects of residence time on carbonation conversion at various temperatures.	564
Figure 7.16	Effect of temperature on the carbonation reaction.	565
Figure 7.17	BET surface area versus carbonation temperature (residence time 50 ms).	566
Figure 7.18	BET surface area versus reaction time.	566
Figure 7.19	Flat geometry of the reactant core and the product layer for the carbonation reaction.	570
Figure 7.20	Model validation for the carbonation data using the first-order deactivation model.	574
Figure 7.21	Arrhenius equation fit for Kc.	575
Figure 7.22	Arrhenius equation fit for KdC.	576
Figure 7.23	Validation of the sulfation model.	582
Figure 7.24	ln KA versus 1/T plot for determination of the activation energy for the sulfation reaction.	583
Figure 7.25	ln KdA versus 1/T plot for determination of the activation energy for the sulfation deactivation.	583
Figure 7.26	Sulfation and carbonation conversions at 600°F in 14% CO_2 and 3000 ppm SO_2.	590
Figure 7.27	Sulfation and carbonation conversions at 900°F in 14% CO_2 and 3000 ppm SO_2.	591
Figure 7.28	Sulfation and carbonation at 1100°F in 14% CO_2 and 3000 ppm SO_2.	591
Figure 7.29	Influence of SO_2 on carbonation conversion at 600°F.	592
Figure 7.30	Influence of SO_2 on carbonation conversion at 900°F.	592
Figure 7.31	Influence of SO_2 on carbonation conversion at 1100°F.	593
Figure 7.32	Overall kinetic scheme of the simultaneous sulfation and carbonation of $Ca(OH)_2$.	594
Figure 7.33	Model validation for carbonation conversion with the simultaneous reaction.	599
Figure 7.34	Model validation for sulfation conversion with the simultaneous reaction.	600

Dry Scrubbing Technologies for Flue Gas Desulfurization

TABLE OF FIGURES

Figure 7.35   In KCO₃ versus 1/T. 601

Chapter 8   Advances in Spray Drying Desulfurization for High-Sulfur Coals 607

Figure 8.1   Solubility of lime expressed as $Ca(OH)_2$ versus temperature [2]. ... 612
Figure 8.2   Schematic diagram of experimental setup. 624
Figure 8.3   Acid HCl weight in the reservoir versus time at pH upper limit 5.6 in the Fisher reagent limestone dissolution test. 625
Figure 8.4   Acid HCl weight in the reservoir versus time: pH upper limit 5.2; Fisher reagent limestone dissolution test. 625
Figure 8.5   TGA dehydration test for lime powder. 628
Figure 8.6   TGA dehydration test for 24 MPa pressed. 629
Figure 8.7   TGA dehydration test for 48 MPa pressed lime. 630
Figure 8.8   Lime dissolution rate versus disk rotation speed. 631
Figure 8.9   Lime dissolution rate versus solution pH. 632
Figure 8.10 Solution temperature versus time for lime dissolution study. 633
Figure 8.11 Acid HCl weight in reservoir versus time for lime dissolution study. 633
Figure 8.12 $Ca(OH)_2$ solubility in sugar solution [2]. 634
Figure 8.13 Lime dissolution rate for different additive solutions. 635
Figure 8.14 Lime dissolution rate versus temperature as derived from activation energy formula. 637
Figure 8.15 Effect on lime dissolution rate by ammonium compounds. 638
Figure 8.16 Schematic diagram of pilot spray dryer system. 641
Figure 8.17 Schematic diagram of spray dryer nozzles. 643
Figure 8.18 $NaHCO_3$ baseline SO_2 removal in pilot spray dryer. 644
Figure 8.19 $Ca(OH)_2$ baseline SO_2 removal in pilot spray dryer. 645
Figure 8.20 SO_2 removal efficiency versus approach to saturation temperature. . 647
Figure 8.21 The pH of saturated $Ca(OH)_2$ solution after additivie addition. 648
Figure 8.22 SO_2 removal efficiency versus additive concentration. 649
Figure 8.23 Calcium sulfite and calcium sulfate solubilities versus pH. 653
Figure 8.24 SO_2 Removal for $Ca(OH)_2$ with H_2O_2 additive. 655
Figure 8.25 SO_2 removal for $Ca(OH)_2$ with sugar additive. 656
Figure 8.26 SO_2 Removal for $Ca(OH)_2$ with benzoic acid. 657
Figure 8.27 SO_2 Removal for $Ca(OH)_2$ with formic acid. 657
Figure 8.28 Surface area development in MF7 sorbents (n) and MF8 sorbents (n).659
Figure 8.29 Surface area development in MF8 sorbents heated at 80°C (■) and 95°C(■). 659
Figure 8.30 SO_2 removal for $MF7/Ca(OH)_2$ (■) and $MF8/Ca(OH)_2$ (∗). Calcium ultilization for $MF7/Ca(OH)_2$ (■) and $MF8/Ca(OH)_2$ (∗). 660
Figure 8.31 SO_2 removal for $MF8/Ca(OH)_2$: heated (▲) and unheated (●). Calcium utilization: Heated (▲) and Unheated (●). 661

Figure 8.32	SO_2 removal for baseline $Ca(OH)_2$ slurry (-), 1:4 MF8/CaO mixture (▲), 1:1 MF8/CaO mixture (■), and 4:1 MF8/CaO mixture (★).	661
Figure 8.33	Sulfur capture in solution from heated MF7/Ca(OH)$_2$ mixture (●) and heated MF8/Ca(OH)$_2$ mixture (■).	663
Figure 8.34	Schematic diagram for SO_2 absorption into a slurry drop.	669
Figure 8.35	Water vapor partial pressure over NaOH solution plane surface.	672
Figure 8.36	Ratio of pressure of droplet surface to pure water plane surface versus droplet size.	672
Figure 8.37	SO_2 removal versus Ca/S molar ratio for baseline $Ca(OH)_2$.	673
Figure 8.38	$Ca(OH)_2$ particle size distribution (measured by Coulter Counter).	674
Figure 8.39	Baseline SO_2 removal versus approach to saturation temperature.	674
Figure 8.40	Comparison of model prediction and experimental data on SO_2 removal versus approach to saturation temperature.	676
Figure 8.41	Comparison of model prediction and experimental data on SO_2 removal versus approach to saturation temperature.	677
Figure 8.42	Schematic drawing for derivation of spray dryer desulfurization model.	678
Figure 8.43	SO_2 removal as a function of constant-rate drying period from the simplified model prediction.	680

Chapter 9 Low Temperature Dry Scrubbing/LEC Process Support .. 691

Figure 9.1	Generic wet/wet flue gas scrubbing system.	693
Figure 9.2	Typical injection locations for calcium-based sorbents.	694
Figure 9.3	Spray nozzle arrangements for in-duct injection processes.	695
Figure 9.4	Simplified flow schematic of the LEC pilot unit.	706
Figure 9.5	Deactivation of an LEC bed due to surface drying and surface blinding: (Runs are from Prudich et al. [3] with a limestone bed depth of 12 inches and a superficial gas velocity of 1.5 ft/s.)	708
Figure 9.6	Deactivation of an LEC bed due to surface blinding. (Runs from Prudich et al. [3] with a limestone bed depth of 6 inches and a superficial gas velocity of 1.0 ft/s.)	709
Figure 9.7	Simplified schematic of the LEC process.	711
Figure 9.8	Schematic of the moving-bed LEC pilot plant.	711
Figure 9.9	Limestone bed geometry for moving-bed LEC pilot plant.	712
Figure 9.10	LEC SO_2 removal efficiency at high inlet SO_2 concentrations. AASHTO No. 9 limestone, Inlet SO_2: 3500-3900 ppmdv[4].	713
Figure 9.11	Conceptual representation of the sorbent particle in the LEC process.	717
Figure 9.12	Film theory representation of the concentration profiles of sulfur dioxide and llmestone during reaction.	718
Figure 9.13	Flow scheme (a) and differential element (b) for down flow in a packed bed.	723

TABLE OF FIGURES

Figure 9.14　Comparison between the outlet bed temperature predicated by the model and experimental data (Data is from Prudich et al. . . 731
Figure 9.15　Comparison between the SO_2 removal efficiency predicted by the model and experimental data [Data is from Prudich et al. (1988), run 870714A'.] 732
Figure 9.16　Differential element used to develop the model. 733
Figure 9.17　Drying front and outlet SO_2 concentration profiles.............. 735
Figure 9.18　SO_2 contours through the bed. 736
Figure 9.19　Average SO_2 outlet concentration profile. 737
Figure 9.20　Bicarbonate-carbonate equilibria as a function of solultion pH. 740
Figure 9.21　Bisulfite-sulfite distribution as a function of pH. 742
Figure 9.22　Stoichiometric factor, fs, as a function of pH and temperature. 747
Figure 9.23　Experimental setup. 752
Figure 9.24　Basket stirrer. 753
Figure 9.25　Equilibria among calcium and magnesium carbonates [34]. 755
Figure 9.26　Experimental $Ca_2^{++}Mg_2^+$ solubililty for Carey limestone. 759
Figure 9.27　Effect of stirrer speed on observed dissolution rate............... 762
Figure 9.28　Free-drift method. Temperature dependency of Carey limestone d issoution rates... 763
Figure 9.29　Free-drift method. Limestone dissolution rates at 60°C. 764
Figure 9.30　Amount of HCl titrated as a function of time. 765
Figure 9.31　pH-stat method. Limestone dissolution rate at 60°. 766

Chapter 10　Simulation and Optimization of a Granular Limestone Flue Gas Desulfurization Process. 775

Figure 10.1　Schematic representation of the LEC process. Numbered equipment items are described in Table 10.1. 777
Figure 10.2　Schematic representation of the sorbent blinding model. 782
Figure 10.3　Qualitative behavior of precipitate layer thickness. 784
Figure 10.4　Arrangement of small grid elements within large grid element in adaptive technique...................................... 793
Figure 10.5　Regeneration and recycle model diagram. 796

Table of Tables

Chapter 11 Flue Gas Desulfurization for Acid Rain Control 1
- Table 1.1. Comparison of SO_2 Removal Capabilities for Dry SO_2 Control Processes [19]. 9
- Table 1.2 Utility Boilers Controlled by SDA=FGD 35
- Table 1.3. Effect of Sorbent Properties on High Temperature SO_2 Capture Performance [156]. 73
- Table 1.4. Effect of Additives on Lime Reactivity [166] 80

Chapter 12 New Calcium-Based Sorbents for Flue Gas Desulfurization 115
- Table 2.1 Infrared Instrument Settings. 118
- Table 2.2 X-Ray Powder Diffraction Instrument Settings 119
- Table 2.3 Degasser and Surface Area Instrument Parameters and Setting 119
- Table 2.4 Scanning Electron Microscope Settings 120
- Table 2.5 UV-Visible Spectrophotometer Settings 122
- Table 2.6 Sand-Bed Sorber Parameters and Settings 122
- Table 2.7 Names, Formulas, Structural Data, and H_2O Content of Known Calcium Silicates 125
- Table 2.8 Occurrence, Synthesis, and Commercial Sources of Known Calcium Silicates. 130
- Table 2.9 X-ray Powder Pattern Data for C-S-H. 143
- Table 2.10 X-ray Powder Pattern Data for 11-Å Tobermorite, Xonotlite, α-Dicalcium Silicate Hydrate, and Hillebrandite. 143
- Table 2.11 SO_2 Uptake and Ca Utilization of Calcium Silicates and Calcium Hydroxide With SO_2-H_2O-N_2 and SO_2CO_2-H_2O-N_2 Mixtures. 149
- Table 2.12 Synthesis and properties of $Ca(OH)_2$ and fumed SiO_2-$Ca(OH)_2$ products. 159
- Table 2.13 Synthesis and Properties of Natural SiO2 Source-$Ca(OH)_2$ Products. . . 164
- Table 2.14 Hydration of Ca_3SiO_5 and β-Ca_2SiO_4 Under Ordinary Conditionsa . . . 171
- Table 2.15 Synthesis and Properties of Hydrated Ca_3SiO_5 Products.. 173
- Table 2.16 Synthesis and Properties of Hydrated β-Ca_2SiO_4 Products.. 174
- Table 2.17 Synthesis and Properies of Hydrated Cement Prouducts 192
- Table 2.18 Synthesis and Properties of Hydrated Cement-SiO_2 Products.. 194

Chapter 13 Fundamental Studies Concerning Calcium-Based Sorbents 207
- Table 3.1. Effect of Sorbent Properties on High Temperature SO_2 Capture Performance (after Snow et al. [5]) 209
- Table 3.2. Effect of additives on lime reactivity. (Ruiz-Alsop and Rochelle [18]) . 216
- Table 3.3. Treatment Levels—Solubility/Slaking Tests. 228
- Table 3.4. Regressive Coefficients for the Antoine Equation. 235
- Table 3.5. Regression coefficients for Equation (3). 237

TABLE OF TABLES

Table 3.6. Dissolution Times for Ca(OH)$_2$ Particles 238
Table 3.7. Comparison of the Enhanced EER Model with UC Spray Dryer Test Results for the Additive.................249

Chapter 14 Sorbent Transport and Dispersion 255
Table 4.1 Material Properties of Sorbents.................264
Table 4.2. Values of Parameters used in the Hamaker Constant Calculation......267
Table 4.3. Characterization of Particle Classes 318

Chapter 15 Transport Processes Invloved in FGD 343
Table 5.1. Comparison of Ranges of Nondimensional Parameters 355
Table 5.2 Test Facility Parameters and Specifications.................360
Table 5.3 Nominal Properties of Glass Particles.................363
Table 5.4 Test Conditions 376
Table 5.5 Discrimination Verification in the Particle-Laden Flows.............386
Table 5.6 Test Conditions 404
Table 5.7 Procedure for Determination of Flow Reversal 411

Chapter 16 High Temperature Desulfurization of Flue Gas Using Calcium-Based Sorbents............................ 421
Table 6.1. Composition and Initial Structural Properties of Sorbents Investigated .458

Chapter 17 Kinetic Studies on the Medium Temperature Ca(OH)$_2$ Sorbent Injection FGD Process 529
Table 7.1. Duct Injection Processes 533
Table 7.2 Heat Transfer Parameters Obtained from the Literature 544
Table 7.3 Comparison of the Decomposition Temperature Measured by TGA and the Decomposition Temperature Calculated by Equation 7.2 ...552
Table 7.4 Dehydration Reaction of Ca(OH)$_2$ 553
Table 7.5 Comparison of the Dehydration Reaction Model Fits...............558
Table 7.6 Conversion of the Ca(OH)$_2$ Carbonation Reaction from 600°F to 1100°F for Residence Times from 50 ms to 100 ms...............562
Table 7.7 Conversion of the Ca(OH)$_2$ Carbonation Reaction Below 50 ms at 600°F to 1100°F 562
Table 7.8 Conversion of the Ca(OH)$_2$ Carbonation Reaction at Temperatures of 600 to 1200°F and at the residence times of 500 ms and 1000 ms...563
Table 7.9 BET Surface Area from the Carbonation Reaction 565
Table 7.10 Rate Constants for Carbonation Reaction 573
Table 7.11 Sulfation Reaction of Ca(OH)$_2$ at 600-1100°F for Residence Times from 50 ms to 100 ms.................578
Table 7.12 Rate Contents for the Sulfation Reaction 582
Table 7.13 Experimental Results of the Simultaneous Carbonation and Sulfation Reaction of Ca(OH)$_2$.................588

Table 7.14 Conversions in the Simultaneous Carbonation and Sulfation Reaction . 588

Chapter 18 Advances in Spray Drying Desulfurization for High-Sulfur Coals 607

Table 8.1 Comparison between Product Drying and Desulfurization Drying 610
Table 8.2 Comparison of Results for Fisher Reagent Limestone Dissolution Rate 627
Table 8.3 Results of Lime Dissolution Rate Affected by Organic Additives 636
Table 8.4 Solubilities of Calcium Compounds (by replacing ammonium ion with calcium ion from the ammonium compounds) [94] 639
Table 8.5 Results of Lime Dissolution Rate Affected by Inorganic Additives 639
Table 8.6 The Chemical Properties of $NaHCO_3$ (from Church & Dwight Co., Inc.) 645
Table 8.7 Dravo Black River Plant Lime Analysis (Dravo Lime Co.) 646
Table 8.8 Solubilities of Ca^{2+} and Na^+ substances (g per 100g H_2O) [94] 650
Table 8.9 Comparison of the im/Ms Term for Different Substances 652
Table 8.10 Henry's Constants for Certain Dissolved Gases and Liquids 654
Table 8.11 Mineral Analysis of Ashes on a Percent Ignited Basis (Courtesy of Standard Laboratories, Inc., South Charleston, W.V., U.S.A.). ... 658
Table 8.12 Removal Efficiency and Ca Utilization of Heated $MF8/Ca(OH)_2$ Slurries with Different Fly Ash/Quicklime Mass Ratios at 11°C Approach. 662
Table 8.13 SPRAYMOD-M Operating Parameters and Standard Conditions 675

Chapter 19 Low Temperature Dry Scrubbing/LEC Process Support .. 691

Table 9.1 Simulation Conditions. 730
Table 9.2 Simulation Conditions. 734
Table 9.3 Chemical Analysis of Limestones. 754
Table 9.4 Representative Analyses from Manufacturer. 754
Table 9.5 Solubilities of Limestones Reported by Different Investigators 760
Table 9.6 Dissolution Rates Evaluated at pH=5.5, Using the Free-Drift Method. . 764
Table 9.7 Dissolution Rates Obtained for Maxville Limestone at 60°C without Additives 767

Chapter 20 Simulation and Optimization of a Granular Limestone Flue Gas Desulfurization Process 775

Table 10.1 Key to Figure 10.1. 778
Table 10.2 Experimental Results for CSst Determination. 786
Table 10.3 Coal Analysis. 801
Table 10.4 Flue Gas Calculation Assumptions 803
Table 10.5 Flue Gas Flow Rate and Composition. 803
Table 10.6 Optimization Results. 804
Table 10.7 Optimum Values for Search Variables. 806

Preface

The State of Ohio has long encouraged the use of its vast reserves of coal. Support has come primarily through the Ohio Coal Development Office (OCDO), an entity within the Ohio Department of Development. Among other things, OCDO is charged with the development of technologies and processes that can use coal in an economical, environmentally sound manner. Although primarily focused at the demonstration end of the research and development continuum, OCDO is also directed to ". . . ensure that an adequate portion [of its funds] be used for conducting research on fundamental scientific problems related to the utilization of Ohio coal . . . ." [Ohio Revised Code, Sec. 1555.03(B)]

Since its inception in 1984, OCDO has cofunded over 100 projects, the majority of them at the research level. Early on, OCDO noticed that many of the colleges and universities were performing similar but varied types of coal research. As Professor L. Douglas Smoot of Brigham Young University notes in his review of directions in coal research (*Energy and Fuels*, Vol. 7, page 689, 1993), six pertinent areas that merit further attention include: (1) boiler-efficiency, (2) carbon carry over, (3) fouling, (4) SO_x removal, (5) NO_x control, and (6) coal gasification. OCDO, recognizing its charge to hasten the development, installation and use of clean coal technologies, concentrated its research efforts on Item 4, SO_x removal.

OCDO created and funded the Ohio Coal Research Consortium (OCRC) in 1990. Its broad objective was to "improve the efficiency of 'dry' high-sulfur-coal flue gas SO_2 removal processes using calcium-based sorbent injection." The consortium was comprised of four universities—Case Western Reserve University, Ohio University, The Ohio State University, and University of Cincinnati. It was designed to support useful and relevant fundamental research to improve existing processes and to develop new ideas and improved process technologies. The consortium members worked very closely with an industrial and governmental advisory committee representing researchers and experts from U.S. DOE, U.S. EPA, EPRI, utilities, private research institutions and companies.

The consortium was organized into several focus areas, such as dry sorbent processes at high temperature (upper furnace), dry and wet/dry sorbent processes at medium temperature (economizer region), and low temperature processes (in-duct/spray dryer). The technical objectives of the consortium included reaction engineering issues, such as identifying and quantifying fundamental chemical/physical mechanisms of dry and wet/dry SO_2 removal processes, and formulating rate models for these processes. In addition to the above fundamental studies, the consortium identified applied areas of research that needed attention, such as the development and evaluation of chemical additives, better sorbent production, and mechanistic models for enhancement processes. The four universities shared these projects based on their expertise and experience.

Dry desulfurization processes offered significant advantages of low capital and low operating costs when compared to wet desulfurization. It held great economic potential for economically reducing the sulfur emissions from power utilities using high-sulfur coal such as that in the State of Ohio. However, the technology still had not matured sufficiently and was achieving lower-than-desired SO_2 removal efficiencies, and sorbent under-utilization. Hence the consortium served to bring together a specific group of scientists and researchers who were interested in and capable of pursuing research in these areas. The consortium dedicated itself to the common goal of improving the state-of-the-art of dry sorbent processes. Promoting high-sulfur Ohio coals by developing efficient sulfur removal technologies provided the motivation for OCDO to undertake this massive and challenging task.

In the fourth year of the consortium, the scope of work expanded to include hazardous air pollutants and dry processes to control their emissions. The Title III trace toxics were likely to be regulated by the EPA as hazardous emissions from power utilities so the consortium decided to investigate the fundamental characteristics of their formation and control.

The efforts of the consortium members have led to a number of significant scientific breakthroughs, major advancements in our knowledge, and new ideas for process development. Throughout the last five years, a number of fundamental research findings have had a significant and lasting impact in terms of scientific understanding and the development of new and improved processes. A few are worth highlighting:

- The experimental investigation of the upper-furnace sulfur capture obtained time-resolved kinetic data in less than 100 ms time-scales for the first time ever, revealing the true nature of the ultrafast and overlapping phenomena. This was accomplished through the development of a unique, first-of-its-kind entrained flow reactor system.
- Mechanistic investigations of the CaO/SO_2 reaction showed conclusively for the first time the true mechanism of outward ionic diffusion of Ca^{++} and O_2- ions through the product $CaSO_4$ layer.
- Work on the influence of sorbent internal pore structure and size distribution on SO_2 reactivity has led to identification of criteria that can lead to great improvements in reactivity.

- Heavy metals research led to the identification of the true mechanism of selenium interaction with sorbents under the high temperature flue gas conditions.
- The role of chemical additives in SO_2 capture efficiency was elucidated and modeled in the dry scrubbing process.
- Research on the spray dryer for desulfurization resulted in the development of methods and additives that allow these devices to achieve >95% SO_2 removal, along with a discovery that would allow them to reach >99% removal rates of SO_2 in the future. This work is currently undergoing patent review.
- Research into attrition and reaction kinetics of wet/dry gas-sorbent reactions has led to the development of a circulating fluidized bed absorber system that is undergoing demonstration and commercialization.
- Work on the advanced sorbents has resulted in the development of a porous lime-silica sorbent and a mixed portland cement-lime sorbent. These low cost, mixed sorbents are prepared from readily available raw materials and have shown remarkably high SO_2 reactivity. The sorbent and its preparation technique have been patented by researchers. These sorbents are also being developed for selenium and other trace toxics.
- Research into the sorbent dispersion phenomena has led to the identification of optimum particle loading that results in good mixing between particles and cocurrent flue gas flow in ducts. An improved nozzle design, requiring no extra energy, has been suggested for more efficient sorbent injection.
- Research work in the granular limestone scrubbing process resulted in the scale-up and development of a continuously operating pilot plan. This work provided design data for scale-up and operation on a commercial scale.

These major achievements earned the consortium members a number of prestigious awards and honors.

The researchers have contributed to the education and training of graduate students to become skilled researchers and innovators. The four universities granted a total of 19 M.S. and 17 Ph.D. degrees, with 17 more working toward the completion of their degrees. The group has published more than 40 peer-reviewed journal articles and made over 54 presentations at various national and international meetings. The OCDO funds helped generate additional research funds from other government and industrial sources. These efforts have resulted in approximately $3,200,000 in external support from federal agencies and industrial sources between 1990 and 1995.

Another important purpose of the consortium was the training of skilled scientific personnel in the State of Ohio that would continue the development of environmental technologies for the safe and efficient use of the nation's fossil resources.

Funding from OCDO for the entire program was nearly five million dollars, or approximately one million dollars per year. Each year, the projects were reviewed by the Consortium Review Committee, experts from private research institutions, the electric power production sector, the coal production sector, the federal government, and others. Based on these yearly progress reviews, the suite of funded projects evolved. The scope of research was expanded in 1993 to include a number of basic trace metal studies. These are not included in the present document.

The Consortium Review Committee evaluated individual project proposals and recommended broad program direction. The OCRC Steering Committee, comprised of one individual from each school, fine tuned program direction. The OCDO provided program oversight, management, and funding.

Dr. Michael Prudich (Ohio University) served as program manager of the OCRC from 1990 to 1992. His role was to review and edit OCRC publications, to perform other administrative functions, and to arrange meetings. These meetings were often held at Ohio power plants hosting various clean coal technology demonstration projects that the OCRC toured and evaluated. Dr. Prudich was succeeded by Dr. Kendree Sampson (Ohio University) from 1992 through 1996.

Most of the research produced by the OCRC in the 5-year program is presented in this monograph. Each of the chapters is a complete body of work with detailed literature survey, description of the work undertaken, the results obtained, and the significance of the results to dry desulfurization technologies.

Chapter 1, "Flue Gas Desulfurization (FGD) for Acid Rain Control," serves as a general introduction and provides a critical literature review of the state-of-technology of dry, calcium-based sorbents and processes. It contains the original position paper that defined the scope of the consortium.

Chapter 2, "New Calcium-Based Sorbents for Flue Gas Desulfurization." Proceeding from early findings that $Ca(OH)_2$-fly ash sorbents showed a special reactivity related to the silicates in fly ash, a series of "C-S-H" (CaO-SiO_2-H_2O) sorbents was developed and evaluated for SO_2 uptake.

Chapter 3, "Fundamental Studies Concerning Calcium-Based Sorbents." This study responded to the need for advances in sorbent materials, both in terms of the chemistry and the processing methods. This work focused on the role of additives for enhancing SO_2 removal. The lowering of water vapor pressure may be the primary cause for enhancement.

Chapter 4, "Sorbent Transport and Dispersion." This study examines the cohesive nature of powders and its influence on powder dispersion due to shear stress in nozzles to simulate flow structure in a nozzle. A model that simulates flow structure in a nozzle is also discussed in this chapter.

Chapter 5, "Transport Process Involved in FGD," along with Chapter 4, considers the characteristics and optimization of transport and dispersion of sorbents in ducts and the interactions between sorbent physical characteristics and dispersibility. In this program Doppler interferometry/laser Doppler velocimetry are used to develop nonintrusive procedures for examination of flow reversal zones in ducts.

Chapter 6, "High Temperature Desulfurization of Flue Gas Using Calcium-Based Sorbents." High-temperature, fast reaction kinetics are used to examine internal structural properties such as pore structure and surface area. The results are also applied to the development of a model.

Chapter 7, "Kinetic Studies on the Medium Temperature $Ca(OH)_2$ Sorbent Injection FGD Process." Medium temperature (600-1100°F) kinetics are developed for dehydration, sulfation and carbonation processes.

Chapter 8, "Advances in Spray Drying Desulfurization for High-Sulfur Coals." In this work sorbent properties are related to hygroscopicity. Additives are employed which increase dissolution rate and oxidation potential. A model is developed which examines the drying of slurry droplets, mass transfer, and reaction in the droplet.

Chapter 9, "Low Temperature Dry Scrubbing/Limestone Emission Control (LEC) Support," along with Chapter 8, investigates sorbent reactivity in three temperature regions of interest. Chapter 8 reports on low temperature kinetics in spray drying systems and Chapter 9 considers low temperature kinetics in a granular limestone process. Effects of temperature on limestone solubility and dissolution rates show that only the latter is significant.

Chapter 10, "Simulation and Optimization of a Granular Limestone FGD Process," along with Chapters 8 and 9, advances the understanding of particular desulfurization processes. Based on a computer simulation of the Limestone

Emission Control (LEC) system, optimum operating conditions are predicted. LEC appears best suited for high-sulfur coal, small scale applications.

As noted, the consortium has been successful in advancing the dry sorbent processes from their initial, not yet mature state in 1990, to a more evolved and mature technology. The consortium efforts have led to some major improvements that mark the beginning of the next generation of advanced dry sorbent processes. The consortium has also played a significant role in developing dry processes for other pollutants such as trace heavy metals which now represent an entirely new area of application for dry sorbent processes. It also sets the stage for Consortium II efforts which began in September 1996 with the objective of examining advanced dry technologies for acid gas and trace toxics characterization and emission control for Ohio coal. The theme for Consortium II reflects the emphasis on the state-of-the-art processes impacting the utilities' industries and the energy sector in general. Specifically, the advanced low-NO_x pulverized combustors and combined cycle generation systems, such as integrated gasification combined cycle and pressurized fluidized combustors, represent the processes that are of significant interest to the State of Ohio and the U.S. today. These processes will dominate power generation in the future. Consortium II represents the commitment of the State of Ohio to remain at the forefront of research and development in this crucial area of national interest.

The OCRC members wish to gratefully acknowledge the contributions of the Consortium Review Committee, and the funding, support and guidance of the Ohio Coal Development Office. Jacqueline Bird has served as Director of OCDO since 1989.

Kendree J. Sampson,
Program Manager
Ohio University

Acknowledgments

**OHIO COAL DEVELOPMENT OFFICE (OCDO),
OHIO DEPARTMENT OF DEVELOPMENT**

The OCDO supports the research, development and deployment of technologies that can economically use Ohio coal within environmental limits. OCDO was the primary funder of the Ohio Coal Research Consortium (OCRC), providing program oversight, management, and guidance to the project. OCDO staff involved in the OCRC effort were Jacqueline Bird, OCDO Director, Richard Chu, P. E., Howard Johnson, P. E., and Arthur Levy, OCDO consultant.

THE OHIO COAL RESEARCH CONSORTIUM (OCRC)

The consortium was comprised of four universities—Case Western Reserve University, Ohio University, The Ohio State University, and University of Cincinnati. It was designed to support fundamental research to improve the efficiency of flue gas SO_2 removal processes and to develop new ideas and improved process technologies.

Dr. Michael Prudich (Ohio University) served as program manager of the OCRC from 1990 to 1992. Dr. Prudich was succeeded by Dr. Kendree Sampson (Ohio University) from 1992 through 1996, whose strong and active guidance was instrumental in bringing this project to a successful conclusion.

OCRC STEERING COMMITTEE

The OCRC Steering Committee, comprised of one individual from each school, was responsible for program direction. Members were:

Dr. L. S. Fan, Ohio State University
Dr. Jaikrishnam R. Kadambi, Case Western Reserve University
Dr. Ken Sampson, Ohio University
Dr. Timothy C. Keener, University of Cincinnati

CONSORTIUM REVIEW COMMITTEE

The Consortium Review Committee evaluated individual project proposals and recommended broad program direction. Members included:

Manny Babu, Dravo Lime Company
Howard Couch, Ohio Edison

Dr. Brian Gullet, U.S. Environmental Protection Agency
Dr. Joseph Oxley, Battelle Memorial Laboratories/Columbus
Dr. James Porter, Energy and Environmental Engineering
Charles Schmidt, U.S. Department of Energy/Pittsburgh
Dr. Robert Statnick, Consolidation Coal Company
William Downs*, Babcock & Wilcox
Barbara Toole-O'Neil**, Electric Power Research Institute
* Chair, 1990-1992; **Chair, 1992-1996

PRINCIPAL INVESTIGATORS

Case Western Reserve University
Adler, Dr. R. J. (deceased)
Kadambi, Dr. Jaikrishnam R.
Kenney, Dr. Malcolm E.

THE OHIO STATE UNIVERSITY

Fan, Dr. L. S.

OHIO UNIVERSITY

Bayless, Dr. Dave
Prudich, Dr. Michael E.
Sampson, Dr. Kendree

UNIVERSITY OF CINCINNATI

Biswas, Dr. Pratim
Keener, Dr. Timothy C.
Khang, Dr. Soon-Jai
Lin, Dr. Jerry

FOURTH FLOOR DATABASES, INC.

We thank Meredith Angwin, President of Fourth Floor Databases, and Katherine Clay and Laura J. Rinaldi, consultants, for their efforts in the final editing and design of this document.

Jacqueline (Jackie) F. Bird, Director
Ohio Coal Development Office/OCDO
PHN: 614/466-3465; FAX: 614/466-6532
e-mail: jbird@odod.ohio.gov

CHAPTER 1   FLUE GAS DESULFURIZATION FOR ACID RAIN CONTROL

J.R. Kadambi, R.J. Adler,
Case Western Reserve University
M.E. Prudich, Ohio University
L.S. Fan, K. Raghunathan,
The Ohio State University
S.J. Khang, T.C. Keener,
University of Cincinnati

Abstract

This chapter is a reprint of a literature review prepared for the Ohio Coal Development Office by the Ohio Coal Research Consortium in 1990. It represents an assessment of the state of the art in flue gas desulfurization technologies at that time. It also served to motivate the work of the consortium starting in 1991, the results of which are presented in the following chapters.

The discussion begins with a basic review of different technologies which are distinguished first by the amount of moisture in the sorbent feed and product, and second by the temperature range for the sorbent injection. The high temperature range corresponds to furnace injection; the medium temperature range corresponds to economizer injection; and the low temperature range corresponds to any system in place after the air heater. In each case, both process chemistry and contacting technologies are discussed. Low temperature technologies are further divided into duct injection, spray drying, capture in particulate control devices, and other special devices. The chapter concludes with a discussion of the fundamentals of sorbent and sorbent additive chemistry and the fluid mechanics, mass, and heat transfer.

Background

Acid rain continues to be a prominent issue in the United States, Canada, Europe and Japan and will become one of the major environmental problems for many developing nations as their use of fossil fuel grows. The major source of acid rain deposition is still from sulfur dioxide (SO_2), although the role of oxides of nitrogen (NO_x) in acid rain has become an important issue as a result of recent studies that have linked nitrate deposition to vegetation and marine life damages. The reduction of SO_2 emissions is critically important for high-sulfur midwestern coals.

The Ohio Coal Development Office (OCDO) has concentrated its efforts on ameliorating the effects of sulfur in Ohio coal. As a part of its program, the OCDO in late 1988 established a consortium of four Ohio universities to conduct a multiyear fundamental research program. The universities are Case Western Reserve University, Ohio University, The Ohio State University, and the University of Cincinnati.

To focus the work of the consortium, OCDO assigned the subject of dry, calcium-based sorption processes for removing SO_2 from combustion gases produced by high-sulfur Ohio coal. OCDO's goal is to develop a base of fundamental knowledge to complement practical expertise from demonstration projects already under way in the State of Ohio (Limb, Coolside). OCDO wishes to know whether dry, calcium-based sorption processes can ultimately prove to be more advantageous than wet sorption processes. Dry methods may be especially suitable in Ohio for retrofitting existing power plants because of their low space and capital requirements.

With the passage of proposed acid rain legislation in the United States, the mining and use of predominantly medium- to high-sulfur Ohio bituminous coal may be seriously threatened. An article by Streets et al. [1] addressing the regional impact of pending legislation has estimated that Ohio's share of sulfur dioxide reduction for prevention of acid rain could cost between $180 million and $660 million per year, depending on the severity of the control strategy enacted. These enormous costs could have a tremendous impact on both the Ohio coal industry and the rates that utility customers are required to pay. Legislation currently being considered would require a certain fixed percentage removal (>40%) of essentially all large uncontrolled sources east of the Mississippi River. This would include eight sources located in Ohio commercial coal power plants [2] currently burning Ohio coal, with a combined electrical generating capacity of 5890 megawatts.

In addition to the pending dilemma presented by the prospect of acid rain legislation, the interest in coal conversion technologies such as gasification and liquefaction has reached a low point due to the low price of foreign oil and the apparent high costs of these technologies in comparison to conventional combustion methods. As in conventional combustion methods, the cost of sulfur removal for these conversion technologies is extremely high and helps to make these processes uneconomical.

It seems apparent that methods to remove sulfur cheaply and effectively will enhance the prospect of using high-sulfur Ohio coal. Currently, dry injection methods which involve the reaction of sulfur dioxide with a dry solid sorbent appear to offer economic benefits over conventional wet scrubbing methods for some types of coals. While there is an obvious economic benefit for these systems in terms of a lowering of capital cost requirements [3,4], the higher operating costs incurred due to poor sorbent utilization may make them uneconomical for application to high-sulfur coal application [5].

Additionally, low sorbent utilization can significantly increase the waste management problems associated with coal ash, as well as the ability of existing particle control equipment to function efficiently. For instance, a 1000 MW power plant burning 3% sulfur coal will use, on a yearly basis, approximately 2 million tons of coal and generate 120,000 tons of SO_2 (one pound of sulfur produces two pounds of SO_2). The ash generated from such a unit will be approximately 225,000 tons per year assuming an 11% ash coal. Current data indicates that 50% of the SO_2 may be removed with the injection of calcium hydroxide $(Ca(OH)_2)$ at a mole ratio of 2:1 (Ca/S). This represents a sorbent utilization of 25%. This would require the injection of around 277,500 tons per year of calcium hydroxide and result in the formation of 337,500 tons per year of additional scrubber solid waste. The combined scrubber waste and coal ash waste would therefore represent the equivalent of firing coal with an ash content of 28%, or an increase of 17 percentage points. Using the total capacity of Ohio's electric generating capacity which could be affected by acid rain legislation (5,890 MW), the amount of waste generated could be increased by as much as 2 million tons per year. This tremendous increase in waste volume could significantly add to the cost of providing electric power from coal.

With these factors in mind, current technologies of postcombustion dry sorbent and spray drying processes are reviewed while primarily focusing on the utilization of calcium-based sorbents. The technologies are broadly divided according to their injection temperatures and locations: in-furnace, economizer zone, downstream of the air preheater, spray drying and particulate control devices. Sorbent forms and additives are included in a separate section. Since the contacting flow patterns including the multiphase flow are practical scale-up issues, these subjects are also reviewed in a separate section. Finally, conclusions and future research needs are presented.

General Discussion of Current FGD Technology

Overview

Wet lime or limestone FGD systems represent a majority of installed FGD systems for two principal reasons: the technologies are proven and they are currently cost effective. These advantages are countered by problems associated with the large amounts of wet sludge that must be handled and disposed of, and by operational problems such as scaling and plugging.

More recently, dry scrubbing techniques using finely divided lime or limestone particles as sorbents for SO_2 have drawn increasing interest. These dry scrubbing techniques are projected to have lower capital and operating costs than the conventional wet scrubbing techniques. Furthermore, the waste product is a dry powder rather than a wet sludge. This offers potential advantages in the areas of waste handling and disposal.

The dry scrubbing techniques that are currently under consideration can be roughly divided into three categories: in-furnace and over-furnace injection, economizer and convective zone injection, and post-furnace (after the air heater) humidified in-duct injection and spray drying (see figure 1.1).

In-furnace injection and over-furnace injection both take place in or near the combustion zone [6-9]. A direct gas-solid reaction is involved which occurs most favorably in a 1800-2400°F temperature window. The details of the furnace injection technology are reviewed later.

Bortz et al. [10] have investigated dry scrubbing downstream of the combustion zone in the convective zone where economizers are typically located. Much like SO_2 capture in the combustion zone, SO_2 capture in the convective zone involves a direct gas-solid reaction. However, the temperatures in the convective zone are much lower, typically ranging from 800–1200°F.

Dry scrubbing after the air heater takes place at relatively low temperatures (ambient to 350°F). The rate of the gas-solid reaction that drives SO_2 capture in the convective and combustion zones is too slow to be significant at these temperatures. The presence of significant amounts of either liquid-phase or vapor-phase water is required in order to mediate SO_2 capture and produce reasonable capture rates. There is a trend in the current literature to empirically differentiate between systems that require the presence of vapor-phase water only and systems that require direct contact between the sorbent particles and bulk liquid-phase water. For purposes of the following discussion these systems will be called dry/dry and wet/dry, respectively.

Dry/dry sulfur dioxide capture occurs when a humidified flue gas stream is brought into contact with dry sorbent particles. In dry/dry systems, no liquid-phase water is present, but the flue gas stream must be humidified to near saturation. Dry/dry systems have been investigated in the context of both SO_2 capture on fabric filters (in baghouses) and direct humidified in-duct injection [11-15]. The Dravo HALT process is an example of a dry/dry in-duct injection process.

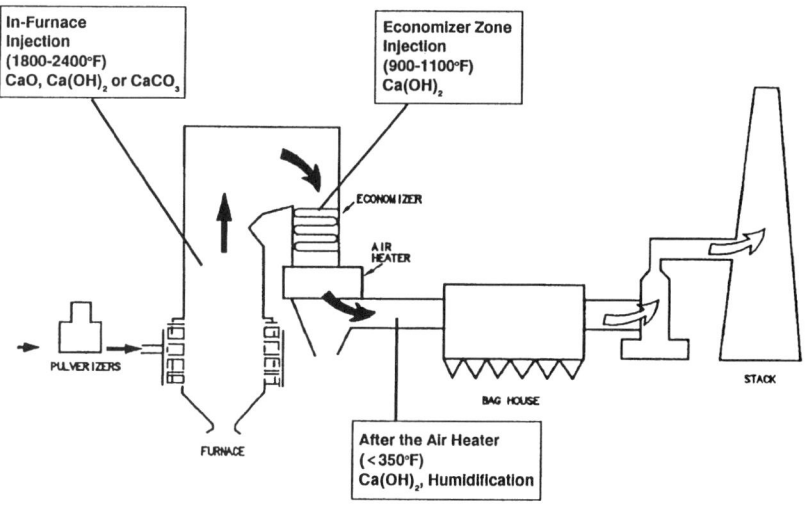

Figure 1.1 Typical injection locations for calcium-based sorbents.

Wet/dry sulfur dioxide capture may also occur when a humidified flue gas stream is brought into contact with sorbent particles. In contrast with dry/dry scrubbing, wet/dry scrubbing requires liquid-phase water to be in direct contact with the sorbent particle. Liquid-phase water is either introduced along with the sorbent in a slurry (in-duct spray drying) or it is injected at a separate location from the solid sorbent as an atomized spray. When the liquid-phase water is injected at a separate location, water droplets are allowed to impact the sorbent particles. Examples of slurry injection include Bechtel's Confined Zone Dispersion process and General Electric's In-Duct Spray Drying process [16]. Consol's Coolside process is an example of a dry scrubbing process that relies on the collision between sorbent particles and atomized water droplets [17].

Dry Scrubbing Technologies for Flue Gas Desulfurization

In addition to humidified in-duct injection and spray drying, the wet/dry sorption process is also being applied to drive ETS' Limestone Emission Control (LEC) process [18]. In the LEC process, water is added to the surface of limestone particles contained in a fixed bed either by direct contact condensation from a humidified flue gas or by the use of an over-bed water spray. Sulfur dioxide containing flue gas is passed through the fixed bed where the SO_2 is captured by the limestone particles. A bed cycle is completed either when the bed has dried out (causing a stoppage of the SO_2-limestone reaction) or when the $CaSO_3/CaSO_4$ layer that forms on the surface of the limestone particles has grown to such an extent that it significantly retards SO_2 capture. The bed is then dried and the reaction product layer is attrited off of the limestone particles and collected for disposal. The regenerated limestone particles are added to a new bed, re-wet, and the SO_2 capture cycle starts again.

Process Research Studies

In addition to the literature available concerning fundamental aspects of the dry scrubbing process, considerable effort has been expended by the electric power industry and others toward development of dry scrubbing processes on a variety of process scales.

Wet/Wet Systems

As a reference point for further discussion, consider the case of wet scrubbers. Conventional wet scrubbing processes have the ability to achieve high sulfur removals (>90%) even when used in conjunction with high-sulfur coals. Their disadvantages lie with their high capital and operating costs. Compared to many dry scrubbing processes, wet scrubbing requires additional steps and equipment such as sorbent slurry makeup and recirculating systems, and settling ponds for sorbent slurry waste disposal. Some dry scrubbing systems eliminate even the scrubbing chamber itself. Existing wet scrubbing technology also suffers due to the tendency of the scrubbers to foul. On the other hand, competitive dry scrubbing processes have poorer SO_2 capture performance than wet scrubbers. This is particularly significant when dry scrubbing processes are used to remove sulfur from high-sulfur coal flue gases. Dry scrubbing processes also tend to use more expensive sorbents than those used in wet scrubbers (which usually use ground and slurried limestone). The potential for processes which are both cost effective and efficient for SO_2 scrubbers can best be understood by reviewing the current technology base for dry scrubbing processes.

Dry/Dry Systems.

Perhaps the simplest (and therefore least expensive) technology under investigation involves the direct injection of dry sorbent into the flue gas. While in contact with the flue gas, calcium-based sorbents will react with SO_2 to form calcium sulfite or sulfate depending on the temperature. The spent sorbent can then be collected in an existing particulate collection device (ESP or baghouse) and discarded or regenerated and recycled. This type of process is classified primarily by the temperature of the flue gas at the point of injection: furnace injection occurs at 1800-2400°F, convective pass injection (at the economizer) occurs at 800-1100°F, and post air heater injection is generally done below 350°F with enough water added to give a 20–50°F approach to saturation [19]. Particular examples of the latter class of dry injection processes are the Dravo HALT process and the Consol Coolside process, both of which use a lime sorbent [16]. Additional scrubbing action will occur at the site of the particulate collection device [20]. ESP performance, which would ordinarily be degraded by the additional sorbent loading, can best be restored by humidifying the flue gas [21,22]. For the case of high sorbent loadings (a high molar ratio of Ca/S) or high-sulfur coals, direct modification of the ESP may be necessary to restore performance [22]. Humidification of the flue gas has a secondary benefit. For the case of sorbent injection after the air heater, it has been shown that the performance of the system improves quickly as the flue gas approaches adiabatic saturation [16].

Several studies of the performance of dry injection systems have been made. Given mediocre performance on low- and medium-sulfur coals (Table 1), it appears that dry sorbent injection can not achieve high sulfur removals from high-sulfur flue gases.

For the case of furnace injection, several pilot-scale studies have shown a maximum SO_2 capture rate at about 2000°F. In these studies, the maximum sulfur removal was 40–50% when the Ca/S molar ratio was 1 and 65–75% when the Ca/S molar ratio was 2 [19-25]. Performance is significantly worse at higher temperatures [26]. The relatively poor performance may be due to the instability of $CaSO_4$ at high temperatures [27]. Other negative factors include pore mouth plugging by $CaSO_4$ and sintering [6,28].

When commercial quality lime was injected into the convective zone of a pilot-scale boiler (at about 1000°F), the maximum sorbent usage was 35% and the maximum sulfur removal was about 50% (at a Ca/S ratio of 2.2 and a feed concentration of 2500 ppm SO_2) [10].

When lime was injected after the air heater on a 2000 ACFM pilot plant (with humidification), the maximum sulfur removal from a 2000 ppm SO_2 feed stream was 70% using a Ca/S ratio of 2 and a recycle ratio of 5:1 [10]. The HALT process has reported sulfur removal efficiencies of 60-70% for a 3.2% sulfur coal using a Ca/S ratio of 2.0 [14].

Wet/Dry Systems

A second class of dry scrubbing processes of widespread current interest is spray drying. In conventional spray drying flue gas desulfurization (FGD), a slaked-lime slurry is atomized in a two-fluid nozzle and sprayed into the spray dryer where it mixes and reacts with the SO_2 in the flue gas. Typical droplet sizes of the atomized slurry range from around 10–200 microns [29]. Within the droplets, individual lime or limestone particles vary in size from about 2–50 microns. When they are exposed to hot flue gas in the spray dryer, the droplets dry out and lose their sulfur capture activity [30]. Typical residence times in the capture zone of back-mixed spray dryers are greater than 10 seconds [29].

Current innovations in this area call for eliminating the spray dryer and spraying the slurry directly into the duct, upstream of the ESP. Examples of the latter type of approach are found in the Bechtel Confined Zone Dispersion process and the General Electric In-Duct Spray Dryer process [16]. Although the in-duct approach eliminates capital, the reduced residence time (less than 3 seconds) leads to reduced sulfur removal efficiencies. Both the Bechtel and General Electric processes have demonstrated only about 50% sulfur dioxide removal to date [16]. Another example of in-duct spray drying is found in the $E\text{-}SO_x$ process. Up to 65% SO_2 removal has been reported on a 1500 ppm SO_2 flue gas in a 30 ACFM pilot plant using the $E\text{-}SO_x$ process [31].

Even though spray drying systems are capable of removing over 90% of the SO_2 from flue gases, they are at present unattractive for high percentage sulfur removals when using high-sulfur coals. The lime/water slurry cannot be concentrated enough to avoid saturating the flue gas when the lime feed rate is increased to accommodate a high-sulfur flue gas [32]. This problem can be counteracted by supplying additional heat to the entering flue gas; however, this adds a considerable economic penalty. A preheater has been used successfully on a pilot scale to help achieve up to 90% SO_2 removal from a 3.2% sulfur coal [33]. There appears to be some confusion in the literature over the application of spray drying with high-sulfur coals: A recent economic study considers a report of 93% SO_2 removal using 3.4% sulfur coal but does not mention the preheater [34]. Of course, another approach to the problem of handling high-sulfur coals would be to combine a spray dryer with another scrubbing technique which does not cause excessive cooling (e.g., furnace injection).

Spray drying also suffers from other negative economic aspects. The capital requirements for a conventional spray dryer are similar to those for a wet scrubbing system with the exception of the settling pond. Moreover, spray drying requires a more expensive sorbent than wet scrubbing (lime costs approximately $50/ton [delivered] while limestone costs approximately $15/ton [delivered]). Economic studies indicate that direct sorbent injection is preferable to spray drying for the cases of 90% sulfur removal from a 1% sulfur coal and 50% sulfur removal from a 1, 2, and 3% sulfur coal [35].

One way to avoid the excessive cooling of spray dryers is to apply a water spray separately from the sorbent injection. An example of this approach can be found in the Tampella Lifac SO_2 removal process [36]. This process combines furnace injection with a downstream water spray. Up to 85% SO_2 removal using a Ca/S ratio of 2 is reported for a 1.7% sulfur coal. An extremely close approach temperature of 5°C was used.

It is apparent from the fact that the efficiency of direct sorbent injection systems is sensitive to flue gas humidity and from the superior performance of spray dryers and wet scrubbers relative to direct sorbent injection, that a film of water on the sorbent surface greatly increases the scrubbing efficiency. It is also apparent from economic studies that a dry product is better than a wet one, that limestone is more attractive than lime, and that sorbent preparation steps such as grinding, slaking, and slurrying, all have a significant negative economic impacts. From these conclusions, it can be deduced that the most attractive process would bring limestone gravel (minimize grinding) directly to a scrubbing device (no limit on the ratio of water to sorbent) and contact it with water and flue gas in such a way as to leave a dry product. This is the approach that is being tested in current work being conducted at Ohio University on ETS' LEC process using pilot-scale equipment [18].

Table 1.1. Comparison of SO_2 Removal Capabilities for Dry SO_2 Control Processes [19].

Process	Alkali	Typical SO_2 Removal. %	Maximum SO_2 Removal. %
Furnace Sorbent Injection	LS	40	50
	L	50	65
Convective Pass Injection	L	50	70
Humidified In-Duct Injection	L	35 (ESP)	50 (ESP)
		50 (FF)	70 (FF)
In-Duct Spray Drying	L	50 (ESP)	60 (ESP)
HYPAS	L	50	85

LS=Limestone; L=Hydrated Lime; ESP=electrostatic precipitator; FF= fabric filter

In-Furnace Injection

In in-furnace injection, naturally occurring limestones, dolomites, or hydrated versions of these materials are injected into the boiler compartment in a temperature window from 1800–2200°F. The point of injection may vary from boiler to boiler depending on the temperature distribution within the boiler but is usually located before the convective heat transfer section. The utilization of sorbents in these systems have generally been observed to be in the range of 25–35% although utilizations as high as 50% have been observed for some of the more exotic hydrates.

One typical example of in-furnace injection technology is the Limestone Injection Multistage Burner (LIMB) process initiated by the U.S. Environmental Protection Agency (EPA) in 1981. This technology has been developed to a full-scale demonstration at Ohio Edison's Edgewater Station located in Lorain, Ohio [37, 38]. SO_2 removals ranging from 55–75% have been achieved at a Ca/S ratio of two using a commercial hydrated lime. This translates to calcium utilizations in the range of 28–38%.

The reaction temperature range for in-furnace injection varies between 1800–2200°F. The reaction time in the furnace is short, typically less than 100 milliseconds. The importance of humidifying and the addition of chemical additives in the downstream duct work has been clearly demonstrated in the literature to further utilize the unconverted sorbents [17,39-41]. Humidification of flue gas, typically to 200°F, ahead of the ESP is needed to capture additional SO_2 and to provide conditioning for the ESP. Recently, a wide variety of sorbents has been proposed and tested in various facilities [42-44] in attempts to identify materials and conditions that can make the dry sorbent injection process commercially viable.

Several chemical processes are important: evolution or liberation of sulfur gases from coal, evolution of active CaO from the raw sorbent, subsequent entrapment of SO_2 by chemical reaction with the sorbent, and sorbent regeneration. In oxidizing atmospheres the ultimate form of sulfur release is SO_2 and SO_3. However, in reducing atmospheres both H_2S and COS may evolve.

The sorbent injected may be $CaCO_3$ or $Ca(OH)_2$ which calcine or decompose to CaO:

$$CaCO_3 \rightarrow CaO + CO_2$$

$$Ca(OH)_2 \rightarrow CaO + H_2O$$

This is the activation step that results in considerable internal surface area and porosity for the CaO, and at typical furnace injection temperatures, it is thought to occur within milliseconds of sorbent injection. In oxidizing atmospheres, the CaO reacts with SO_2 to form calcium sulfate:

$$CaO + SO_2 + 1/2\ O_2 \rightarrow CaSO_4$$

Sulfation causes the active surface area to decrease due to the build up of a product layer. Simultaneously, surface area of the CaO is lost due to thermal sintering. Sintering is the mechanism by which solid grains coalesce when heated at temperatures below their melting point. Fundamental processes involved during sulfur removal are reviewed by Gullett and Kramlich [30]. Thermodynamic considerations favor the formation of the sulfate up to around 1200°C. Below 900°C, the reaction is too slow to be practically effective. The commonly referred to sulfation "window" lies between these two temperatures.

Between the two types of calcium-based sorbents generally used in in-furnace injection, it is well documented that CaO derived from $Ca(OH)_2$ (h-CaO) is more reactive than CaO derived from $CaCO_3$ (c-CaO) [45]. This is attributed in part to the smaller particle size of h-CaO and, more importantly, to the pore structure of the CaO produced. The h-CaO has a slit or plate-like structure while the structure of c-CaO is in the form of cylindrical pores (or spherical grains). The plate-like structure retains its porosity to a greater extent by allowing for particle expansion [46] and results in higher rates of diffusion of the reactant through the product layer [47].

The mechanistic steps during sulfation consist of diffusion of SO_2 to the sorbent-particle interface, diffusion of the gas through the pore structure of the sorbent, diffusion through the product layer, and surface reaction. There is disagreement about the controlling mechanism, which is claimed to be the chemical reaction [48] or ionic diffusion through the product layer [49]. Smaller particle size favors sulfur capture efficiency. Additives can be added to the sorbent to achieve a significant improvement in sulfur removal.

The diffusional control is thought to be governed largely by two resistance terms and their interaction with the sorbent structure. The first of these, *pore diffusional resistance*, dominates for large particles and is weakly dependent on surface area. The second term, *product-layer diffusion*, dominates for small particles and is highly surface-area dependent. Sulfation products may cause a third resistance term, *pore blockage*, which dominates for medium size particles and is strongly dependent on operating temperature. One additional important aspect that should be considered in the development of dry sorbent technology is the competing bi-directional diffusion. For high-temperature desulfurizations, the simultaneous calcination and sulfation reactions cause the opposite diffusion of SO_2 and CO_2, and for low-temperature operations with humidification, the water vapor which is evaporated from the pore internal surface moves out in the opposite direction to SO_2.

Flow behavior within the burner, the location and temperature of sorbent injection and injector configuration also have significant effects on overall performance. Since the time scale for the reaction is quite small, sorbent mixing and temperature history are important parameters.

Fundamentals

Sulfur Evolution

Sulfur appears primarily in two forms in coal: organic sulfur and pyritic sulfur. Small amounts of sulfur may also appear in oxidized coals as sulfates. Evidence shows that the sulfur in coal is evolved in stages, and at different rates and different forms depending on the conditions of reaction [50]. Approximately 50% of the sulfur is evolved in less than 100 ms at an estimated temperature of 1500°C. Smith et al. [51] measured the sulfur species evolved from a Rhenish brown coal; at fuel-rich conditions, H_2S was the major gaseous product and at lean conditions, SO_2 was the major product.

Zghhoul and Grosshandler [52] measured gas species, especially sulfur species, during devolatilization of a low-sulfur subbituminous and a high-sulfur bituminous coal in an opposed-flow diffusion flame (OFDF). The coal was pulverized to 12 mm and the maximum temperature and heating rate in the burner were 1227°C and 20,000°C/s, respectively. They used a stainless steel water-cooled probe, which consists of capillary tubes for cooling water, heated Teflon tubing to prevent condensation, and a cyclone separator and filters to remove char from the gaseous sample.

For the high-sulfur coal, SO_2 is the most dominant sulfur species. H_2S, SO_3, and small amounts of COS and CS_2 were the other species present. Only one-third of the total sulfur in their coal was released in the first 100 ms of combustion. A significant portion of SO_2 was destroyed after just passing through the hottest part of the flame, favoring SO_3 formation. This corresponds to the maximum concentration of SO_3 at 12% of SO_x.

CALCINATION AND SINTERING. Satterfield and Feakes [53] pointed out that the thermal decomposition or calcination of $CaCO_3$ consists of three rate-controlling processes: heat transfer to the surface and then through the CaO product to the reaction surface, mass transfer of CO_2 to the external gas source, and chemical reaction. They concluded that heat transfer is the major controlling factor. This was confirmed by Gallagher and Johnson [54] and Caldwell et al. [55] through experiments using a thermogravimetric analyzer (TGA) at 900–950°C and 550–680°C, respectively. It is interesting to note that even for particles as small as 30 mm, the rate-controlling mechanism was not the chemical reaction.

These claims were disputed by Beruto and Searcy [56] on the basis that when powdered samples were used in a TGA, interparticle diffusion became significant and hence "differential conditions" were not achieved with respect to CO_2. Chemical reaction was demonstrated to be a significant resistance to the overall calcination rate by their work and by the study of Powell and Searcy [57]. They studied calcination under vacuum where diffusional resistances are minimal. Powell and Searcy [57] reported an activation energy of 205action rate of 1.0×10^{-6} mol/cm^2-s at 844°C based on the dimensions of a flat face of the calcite crystal. They also showed that the calcination rate, under chemical reaction control, was proportional to the surface area of the porous particle.

Borgwardt [58] achieved differential conditions with a special design for placing the sample in his reactor. A schematic of his setup is shown in Figure 1.2. About 10 mg of limestone particles were dispersed into a glass wool substrate which was loaded into a sample holder. With this design, particles in the size range 1–90 mm could be studied with velocities of the sweeping gas up to 13 m/s. In this reactor, calcination rates were studied over the range 475–710°C. In a separate entrained flow reactor, studies were made in the temperature range 775–1100°C.

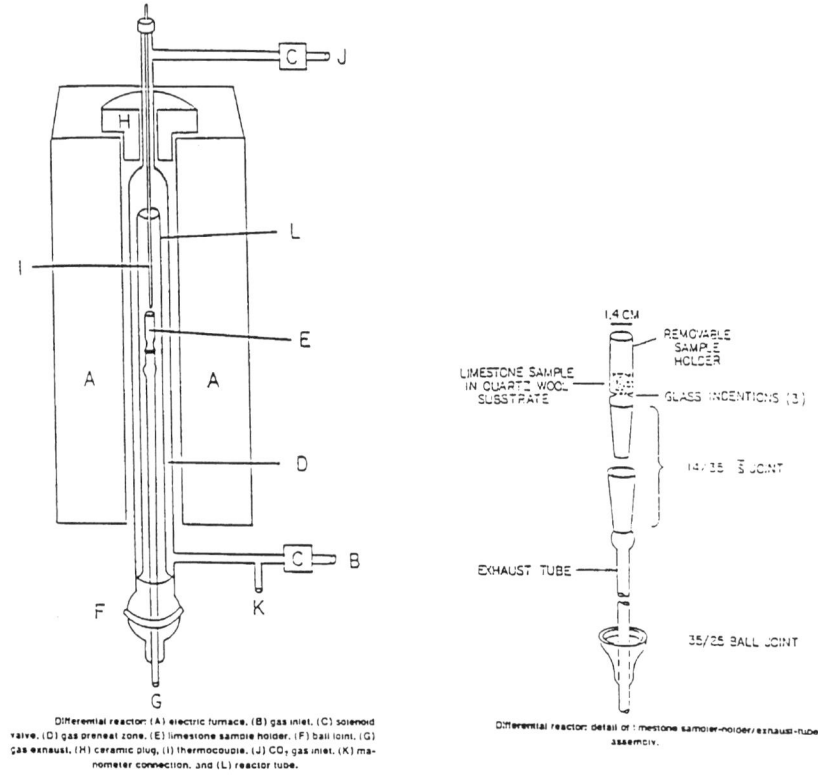

Figure 1.2 Differential reactor setup used by Borgwardt [1985].

He also measured the BET surface area of limestone particles of various sizes and showed that the measured rates were proportional to the BET surface area, indicating that the reaction takes place within the entire pore structure. His calcination model for small particles is

$$ln(1-x) = -k_s S_g t$$

where x is the conversion of $CaCO_3$ to CaO, k_s is the surface reaction rate constant, S_g the BET surface area (cm^2/mol), and t is the reaction time. An important assumption here is that the BET surface area does not change during the course of the reaction.

The maximum attainable surface area of CaO from calcination of limestone was shown to be in the range 50–60 m^2/g of CaO, for particle size range 10–90 mm. Such large surface areas were retained even at high temperatures when small particles were calcined at maximum rate (minimal diffusional resistance). This indicates that the calcination kinetics dominate the sintering kinetics, at least up to the maximum temperature used in his study. Since a larger surface area of the produced CaO implies better sulfur removal, his study shows that the injected limestone should have the minimum particle size and should be well dispersed in the reactor.

When sorbent is injected into a furnace where SO_x is present in a high-temperature environment, although time scales for calcination, sintering and sulfation are different, they may overlap and the degree of overlapping between these three processes depends on the furnace conditions. Therefore, it is important to understand each of the three reactions independently for overall understanding of the mechanisms involved. Borgwardt et al. [43] calcined 1 mm $CaCO_3$ particles at maximum rate and at temperatures 600-950°C, and then under controlled sintering produced CaO particles of varying surface areas. Using this technique, the sulfation process can be isolated and studied in terms of the structure of CaO. At the end of calcination, the BET surface area of CaO particles was 79±6 m^2/g and the corresponding grain radius was 110 Å. The grain radius r_g is related to the BET surface area S_g and density ρ_{CaO}, thus

$$r_g = 3/(S_g \rho_{CaO})$$

They showed that grain size of the CaO particles after calcination can be controlled by stagnant sintering, where CaO is retained in the stagnant reactor for varying time intervals following calcination at a given reactor temperature; sintering with inert sweep gas, and sintering with sweep gas containing CO_2. Their technique now makes it possible to study the sulfation reaction in terms of the pore structure of CaO, without interference from sintering.

Haji-Sulaiman et al. [59] confirmed that CO_2 enhances the rate of sintering. They found that pore volume did not change significantly, but the pore size distribution showed a variation. It appears that under their experimental conditions (850°C) only the first stage of sintering took place. They found that differences in calcination trends are reflected in the sulfation rates of the sorbents.

Silcox et al. [60] developed a mathematical model for the flash calcination of dispersed $CaCO_3$ and $Ca(OH)_2$ particles, which describes the decomposition of the parent material at the reactant-product interface, transport of CO_2 or H_2O through the growing CaO layer, and the sintering of the CaO. Based on the data of Borgwardt et al. [43], they expressed the loss in surface area of the particle due to sintering as

$$dS/dt = -k(S-S_\infty)^2$$

where S is the BET surface area, t is time, k is a rate constant which depends on the CO_2 pressure, and S_∞ is the asymptotic surface area. From their model and data from Slaughter et al. [61], they showed that for a given particle size, an optimum temperature and exposure time exist where the surface area is a maximum and that the smaller the particle size, the shorter the exposure time and the lower the temperature for maximum surface area. Also, hydrate particles produce more reactive CaO than do carbonates because of their small size and rapid rate of calcination.

Sintering was isolated and studied by Borgwardt [62] at 700–1000°C in an inert atmosphere using CaO prepared from ultrapure $CaCO_3$, as well as from limestone and calcium hydroxide. He applied the sintering model of German and Munir [63] which assumes that the grains have an initial spherical shape with multiple points of contact between adjacent grains where necks are formed; the material is transported through the neck, accompanied by growth of the neck. According to the model, the changes in specific surface area can be expressed as

$$(1 - S/S_o)^k = Kt$$

where S and S_o are the instantaneous and initial specific surface areas, and K is a constant. The exponent k takes on values of 1.1, 2.7, 3.3 and 3.5 for plastic flow, lattice diffusion, grain boundary diffusion, and surface diffusion, respectively. The best fit of the model to their data corresponds to k=2.7, in agreement with the hypothesized lattice diffusion mechanism. Pure CaO sinters more slowly because of lack of impurities which tend to catalyze the reaction. The CaO prepared from hydroxide sintered faster in comparison with the carbonate CaO. The reason for this is attributed to the lower porosity (or more points of contact) of the former.

Borgwardt indicated that contacting grains are present in the form of clusters; during the initial stages of sintering, the grains aggregate within a cluster causing the surface area to decrease at constant porosity, and then the clusters aggregate causing a decrease in both surface area and porosity accompanied by particle shrinkage.

Mai and Edgar [64] developed a simple model that accounts for simultaneous calcination and sintering. The calcination rate was assumed to be first order. Second order was assumed for sintering based on previous work. Experiments were conducted in an entrained-flow reactor at 1000 and 1150°C with 12.5 mm $Ca(OH)_2$ particles. The effects of water vapor and CO_2 on the sintering rates and surface areas were studied. They concluded that at the higher temperature, surface area evolution was dominated by calcination rate with a rapid increase in generated surface area. After a shallow maximum, the surface area reached an asymptotic value which depended on temperature and gas concentration. At the lower temperature, the two rates were comparable and a pronounced maximum was observed. Water vapor concentration had the largest effect on the asymptotic surface areas.

It is clear that at this time, sintering is understood only qualitatively, and modeling of this process is only empirical in nature. Such models need to be developed, accompanied by microscopic measurements of the internal structure, in order to achieve a complete description of this mechanism. If an effective means of controlling sintering is developed, it can have a major impact on the desulfurization processes. Sampling and measurement techniques to study the dynamic nature of the sorbent structure and activity at residence times in the order of ms will be extremely useful.

Sulfation. During sulfation, the internal pore structure (not just the exterior surface) of CaO particles takes part in the reaction with SO_2. Hartman and Coughlin [65] studied the sulfation kinetics at 750–1000°C (1400-1800°F). The sizes of particles used in their study varied between 0.5 and 1.25 mm. They observed the progress of the reaction in the particle interior through electron microprobe analysis. They identified an outer shell with partial conversion, and a completely unreacted inner core. Interestingly, even after long time exposure to the reactant gas, a considerable amount of CaO still remained unreacted even near the exterior surface of the particles. They proposed that as the reaction proceeds, the porosity of the outer shell decreases due to the higher molar volume of the product, thus preventing the gas from penetrating to the surface of the unreacted core. They developed a grain model which considers intraparticle diffusion to describe the dynamic changes in the kinetics and pore structure of the sorbent particle.

The sulfation reaction is sensitive to temperature and this observation led to different theories to account for the reaction mechanism. Pigford and Sliger [66] analyzed kinetic data for 96 μm particles at 980°C on the basis of the grain model. From the model fit, they concluded that the diffusion of SO_2 through the product layer controls the overall rate and reported 30 kcal/mol to be the activation energy for this step. However, this value is too high for a process controlled by SO_2 diffusion. James and Hughes [67] measured nearly the same activation energy for 70 μm particles in the temperature range 800–1050°C. However, they concluded that chemical reaction is the controlling step.

Hartman and Trnka [68] noted that the high sensitivity of reaction rate to temperature can be due to diffusion of ions in the solid state instead of diffusion of SO_2 gas. Bhatia and Perlmutter [69] estimated the diffusivity in the product layer at 980°C to be 69×10^{-12} m^2/s which is in the range for ionic diffusion.

Borgwardt and Bruce [70] prepared calcined limestone as described above [43] to obtain CaO particles of different surface areas. Apart from the grain size, they varied the temperature and gas concentration in their study of sulfation kinetics. The grain model for the diffusion through the product layer of unreacted cores,

$$1 - 3(1-X)^{2/3} + 2(1-X) = k_d t$$

where k_d is the diffusion rate constant, gave the best fit to their kinetic data. According to Fick's law of diffusion, k_d is proportional to the square of S_g, which they verified independently through BET measurements. Thus, they confirmed that the fundamental mechanism in the sulfation reaction is diffusion through the product layer. By expressing the diffusivity in an Arrhenius form

$$D_s = D_o \exp(-E/RT)$$

they calculated an activation energy of 37.6 kcal/mol. Since this value is too high for the diffusion of gaseous SO_2, they concluded that the diffusional process is that of ionic migration. They further noted that the dependency of the reaction rate on SO_2 concentration is to the power 0.6, contradicting the commonly assumed first-order dependence. In a later effort, Borgwardt et al. [49] performed experiments with CaO derived from six different sources, and applied the random pore model Bhatia and Perlmutter [71] to the kinetic data. They calculated a value of 33 kcal/mol for the activation energy and 0.118 cm^2/s for D_o. Their results further indicate that the reaction order with respect to SO_2 concentration is 0.64. They suggested that if the transport within the particle is gaseous diffusion, the apparent rate will vary if catalysts of SO_2 oxidation are present. Catalysts had little effect on the rate, from which they confirmed that the diffusion is of an ionic nature.

Milne and Pershing [72] used a phase discrimination probe to sample SO_2 and reported that gas with residence times as low as 30 ms can be sampled with their design. They observed a dramatic sulfur capture within the first 30 ms after injection, and demonstrated that hydrate sorbents are superior to carbonate sorbents, even when derived from the same parent source. They also noted that if the particles are smaller than 5 μm, the calcination rate has little effect on the sulfation rate. At longer exposure times, the hydroxide sorbent was superior, possibly due to the overall CaO structure (grain or plate-like).

Simons and Garman [73] offer an alternate controlling mechanism for the sulfation reaction in terms of the pore-tree model. They contend that complete interconnectivity between pores, as assumed by previous researchers, may be too severe and identified the plugging of the smallest pores to be the rate-controlling mechanism. Simons et al. [48], based on experimental data from their work and from the literature, reported an activation energy of 34 kcal/mol for the intrinsic rate constant and their analysis showed a first order dependency on SO_2 partial pressure.

There is no consensus on the controlling mechanism during sulfation. Thus effective control of the process can be achieved only through trial-and-error procedures and expensive research efforts. Furthermore, in situ measurements are necessary to understand the coupling effects of calcination, sintering and sulfation, and the effects of gas-phase composition and mixing in a burner.

Gullett and Bruce [46] studied the pore structure of carbonate- and hydroxide-based sorbents, prepared under different calcination and sintering conditions, during sulfation at 800°C. They analyzed the nitrogen adsorption data of the sorbents using various theories on adsorption of gases by porous solids [74] and concluded that CaO derived from carbonate can be approximated by a system of cylindrical pores, while CaO from hydroxide exhibits a slit-like or plate-like porous structure. With the progress of sulfation, the pore volume decreases and its distribution shifts to larger pore sizes, implying preferential reaction at the smaller pores. For the hydroxide parent source, the ultimate utilization of CaO was higher than that of the carbonate source. This is a consequence of the plate-like structure of the former which is amenable to greater expansion. They calculated the maximum local conversion of the calcium oxides based on the random pore model [71]:

$$X_m = \varepsilon_o (1-\varepsilon_o)^{-1} (1-Z)^{-1}$$

where Z is the product-to-reactant molar volume ratio. This conversion corresponds to complete pore closure, assuming a constant particle diameter. They measured conversions beyond this maximum which implies that the particle may expand with sulfation and that the intraparticle diffusional resistance is not significant.

Bruce et al. [47] applied the spherical grain model and flat plate model to the sulfation of carbonate- and hydroxide-based CaO respectively, and extracted kinetic parameters by extrapolating the data to zero porosity. They attributed the superior performance of hydrated limes to the smaller particle size and the faster rate of diffusion of the reactant through the growing $CaSO_4$ product layer in the case of the plate-like structure. The difference between the two sorbent performances is exaggerated at longer times and higher temperatures.

Cole et al. [75] studied the reactivity of the sorbents at 1000 and 1200°C, along with the sorbent physical structure. For injection into a furnace without SO_2 at 1000°C, the larger size sorbents had a slightly higher surface area. However, they found no specific correlation between particle size and loss of surface area. They also pointed out that for predictions at elevated temperatures, a model should take sintering into account.

They found the hydrate sorbents to be more reactive on an equal size basis, and no thermal comminution of the hydrate particles was noted within the high-temperature environment. Thus, they concluded that the superiority of hydrates on a common prefiring size basis cannot be explained in terms of their fragmentation into smaller particles upon firing.

Newton et al. [76] studied the mechanisms which limit calcium utilization in a short-time reactor fired with a CO flame, in the temperature range 970–1270°C, and reported that maximum sulfur capture occurs between 1100 and 1200°C. They investigated the relationship between sorbent physical structure and various sulfation parameters through in situ measurements. Calcination of the sorbent in a combustion environment resulted in porosities dramatically less than theoretically possible. The presence of SO_2 was observed to reduce the extent of porosity loss. They indicated that porosity loss and sulfation simultaneously compete for the available CaO sites, and at higher temperatures decomposition of $CaSO_4$ causes the overall sulfur capture and porosity to drop.

Mixing of the sorbent in the reactor also affects the sorbent utilization by delaying the sulfation reaction and by altering the extent of porosity loss by changing the sorbent thermal history. Also, calcines from $CaCO_3$ suffered greater losses in porosity that those from $Ca(OH)_2$ which, along with the larger $CaCO_3$ particle size, accounts for the substantial differences in SO_2 capture between these two sorbents.

The above investigations explain the drop in sulfur capture at higher temperatures in terms of the pore structure of the solid and intraparticle transport mechanisms. Lyngfelt and Leckner [77] analyzed this behavior through thermodynamic considerations. They considered the following reactions of $CaSO_4$:

$$CaSO_4 + CO \rightarrow CaO + SO_2 + CO_2$$

$$CaSO_4 + 4CO/4H_2 \rightarrow CaS + 4CO_2/4H_2O,$$

and constructed an equilibrium diagram between CaS, CaO, CaSO$_4$, SO$_2$, H$_2$S, O$_2$, and H$_2$O, which is shown in figure 3. Under strongly reducing conditions CaS is stable, whereas CaSO$_4$ is stable under oxidizing conditions. Their equilibrium diagram indicated that there is also an intermediate region where CaO is more stable. They verified the reductive decomposition of CaSO$_4$ to CaO in this intermediate region. Thus, the effect of gas-phase composition is important in determining the stability of CaSO$_4$ and hence sulfur capture. This, they suggest, is the reason why sulfur capture in an actual boiler deviates significantly from the laboratory results, where the gas phase properties are well controlled. This study demonstrates that research efforts should consider the variability of the gas phase and thermodynamics.

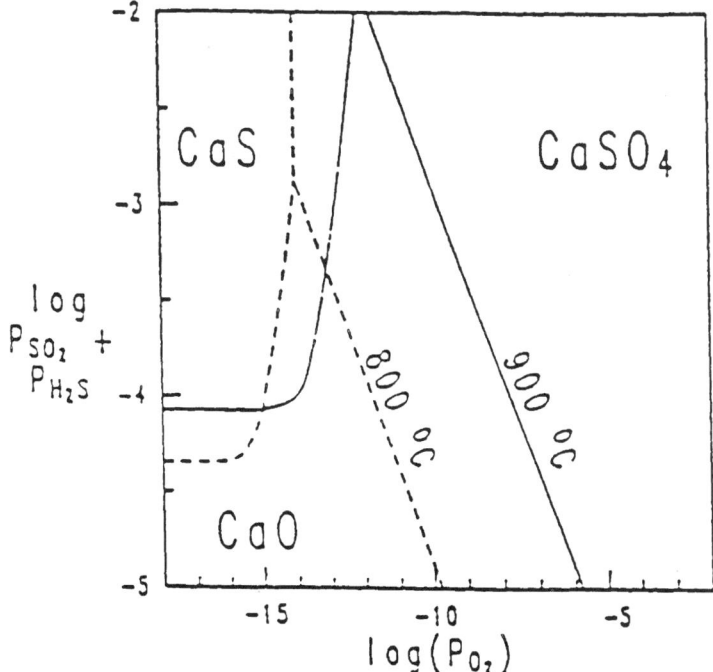

Figure 1.3 Equilibrium diagram for CaO, CaS and CaSO$_4$ with gas species [77].

Modeling of Sorbent Reaction

Modeling of noncatalytic gas-solid reactions has received considerable interest in recent years mainly because of the unresolved questions concerning the removal of sulfur dioxide with dry sorbents. Many models have been developed to describe the solid conversion versus time relationship during the course of the reaction. Each of these models is usually applicable to a particular type of gas-solid reaction. A sharp interface or unreacted-core shrinking model which assumes that the reaction occurs initially at the particle outer surface then moves toward the center of the particle by means of a narrow reaction front is one of the oldest classical models. This model is suitable for a simple reaction of a gas with a nonporous solid that does not undergo any structural change. For a relatively porous solid reactant, a particle-pellet or grain model has been proposed. This model assumes that a solid is composed of many individual nonporous grains which are surrounded by macropores through which the gaseous reactant has to diffuse to reach the grain surface in order for the sharp interface type reaction to take place. Another basic model which is applicable for a very porous solid particle is a volume reaction model. This model assumes that the gaseous reactant diffuses rapidly into the solid particle and the reaction takes place over the entire volume of the particle rather than at a sharp interface.

In order to describe gas-solid reactions in which the solid particle experiences structural changes, modifications have been made to the above basic models. Since the structural change results in a variation of the solid porosity, surface area and the gas effective diffusivity as the reaction proceeds, functions describing how these physical parameters change are usually required for such model modifications [78-82]. The more pictorial approach in dealing with structural type gas-solid reactions is the focus on changes that take place in a pore which represents the behavior of the entire solid particle. This single-pore model considers the change in pore geometry in the model development, in addition to other parameters such as reaction rate constant and gas effective diffusivity [83-86].

The concept of a single pore model has been modified and extended to account for the effect of a pore-size distribution over the solid surface. This leads to the need for a function describing such distributions before the conversion-time behavior of the reaction can be modeled [87,88].

In addition to the gas-solid reaction models discussed above, there have been models developed which attempt to explain the reaction that occurs after a solid has undergone thermal decomposition (calcination) and developed a complex network of pores. The tree-like pore structure model was first developed to describe the heterogeneous reactions that occur in porous coal and char [73,89]. This model assumes a pore structure that is similar in shape to a tree with many branches whose trunk is located at the particle surface. The branches of the tree describe the progression of pores from maximum to minimum radius. The model incorporates expressions to account for the effects of bulk diffusion to the particle exterior surface, continuum diffusion within the large pores, and Knudsen diffusion within the small pores. The model accounts for the effect of product deposits on the walls of the pores and has been used to account for the difference in particle deactivation due to the particle size.

Some questions remain regarding the conversion of calcium-based sorbents which the above models do not answer adequately. These include the observation that calcium hydroxide sorbents achieve higher utilization than limestones. Attempts to correlate these higher utilizations strictly due to particle size effects have not been entirely successful. Recent studies have shown that particle size and surface have a weak effect on sorbent utilization for particles in the 10–20 µm diameter size range [44]. The plate-like pore structure of hydrated lime seems to be responsible for its higher utilization. Additionally, the effect of sintering, which has been shown to be significant at furnace temperatures for these materials, could be responsible for a loss of reactive surface. Fundamental studies which would help to delineate these and other questions will provide direction for new or altered methods of applying these sorbents.

Technology

With the recently established lower emission standards, there is immediate need for process technologies to increase sulfur removal from coal-fired power plants. Such technologies can be broadly divided into two categories [90]. Retrofit technologies are those that add on to the existing equipment and have a small impact on the operation, performance, or capacity of the existing plant. Examples of retrofit technologies include Limestone Injection Multistage Burner (LIMB), slagging combustors, reburning, selective catalytic reduction, etc. Repowering technologies are those that add on to the existing equipment and have major effects on the operation and capacity of the total plant, as well as the final performance of the modified unit. Some of the repowering technologies are gasification combined cycle (GCC) and atmospheric fluidized bed combustion (AFBC). Repowering can have significant implications for the thermal performance of the power plant.

Flue Gas Desulfurization for Acid Rain Control

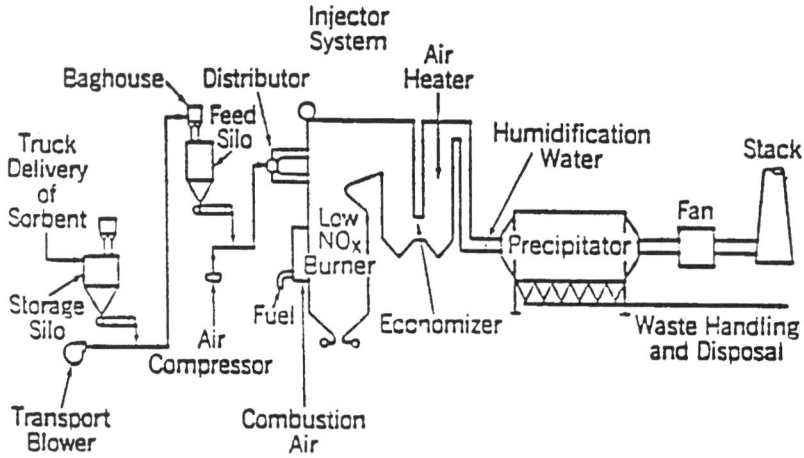

Figure 1.4 A schematic of the LIMB process [91].

Of particular interest to high-temperature, dry sorbent injection is the LIMB process [91]; limestone or lime is injected into the upper furnace of the pulverized coal unit, with expected SO_2 removals of greater than 50%. A schematic of this process is shown in Figure 1.4. At present, DOE and EPA research efforts are underway to increase this efficiency to above 60%.

England et al. [92] tested SO_2 emissions from a 61-MW coal-fired utility boiler with a concentric firing system, and considered design modifications and transport processes around the sorbent injection point. The flow field within the region of the furnace above the burners was a very complex, strongly swirling flow. They indicated that the center of this region contains a large weak circulation zone, and provided useful discussions with respect to the placement of injectors in the burner and their effects. Boiler fouling due to sorbent injection was found to result in increased flue gas temperatures. Because of the complex nature of the flow field, the mixing of the sorbents in the burner becomes an important factor. Vandycke et al. [93] reported a significant difference in capture efficiency with location of sorbent injection level. The reason for this was attributed partly to poor mixing in the upper levels of the boiler. Using cold flow models, they reported that best mixing was obtained with the injectors at the center of each wall and with low transport air velocity.

As opposed to the once-through dry sorbent processes, a sorbent recycle process was tested by Lott et al. [94]. The advantage of recycle is that SO_2 removal and sorbent utilization are improved, while at the same time reducing sorbent feed rates. This is particularly attractive for low-sulfur coals, but can provide process enhancement for high sulfur or retrofit applications. They identified evaluation of recycle effectiveness factors and development of system design and costs as the key areas for further analysis of the process.

Chughtai [95] describes a direct desulfurization process especially for small firing systems. In this process, the sulfated sorbents are reactivated by either flue gas conditioning or fly ash conditioning and recirculation. This process is shown schematically in Figure 1.5. The unreacted CaO present in the flue gas is hydrated with atomized water to form $Ca(OH)_2$. The resultant expansion of the hydroxide and the heat produced during slaking tends to split the solid along the particle boundaries. At these break points new active surface is generated. It is pointed out that application of direct desulfurization with a quench cooler is recommended for boilers equipped with electrostatic precipitators, because flue gas quenching helps upgrade the ESP performance. On the other hand, fly-ash conditioning and recirculation is preferred for boilers with baghouse filters.

One of the areas which has received very little attention is efficient design configurations. Staudinger et al. [96] reported that with a modification of the burner geometry, and all other parameters held constant, the sulfur removal increased from 35 to 63%.

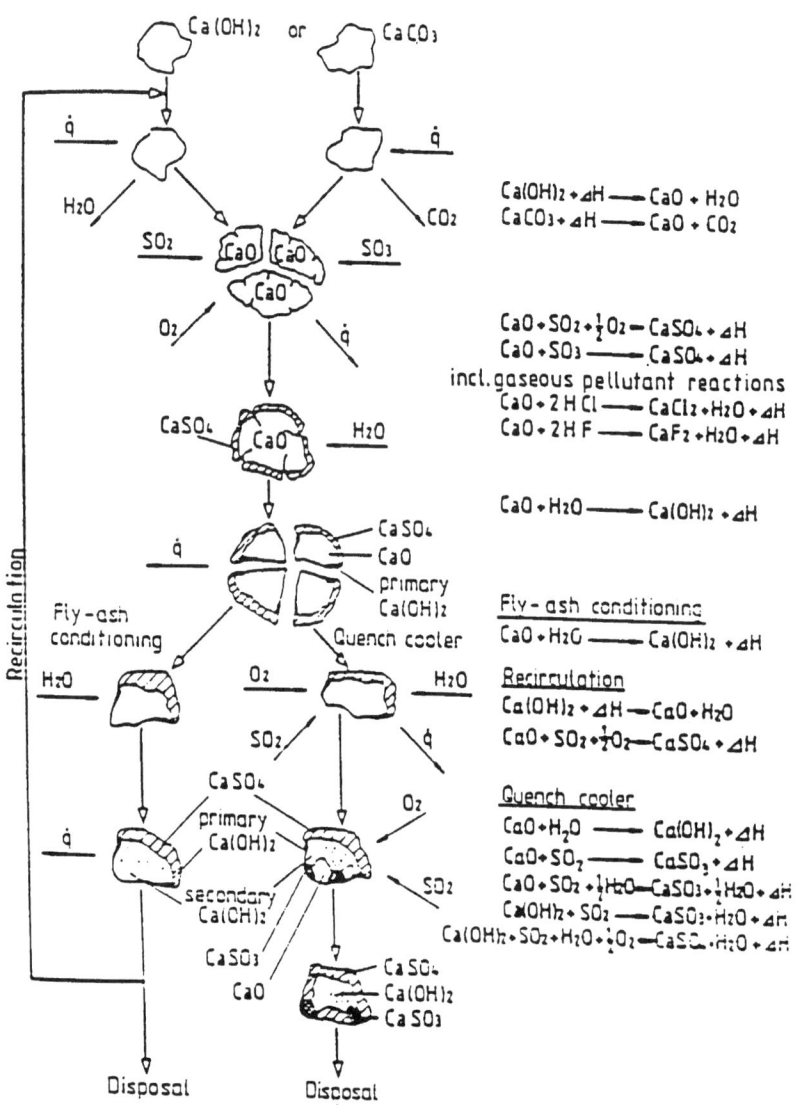

Figure 1.5 Direct desulfurization process proposed by Chughtai (197).

It appears that predictive models are not available for existing technology. Such a model would require mechanistic models for the gas-solid reaction and related phenomena as well as descriptions of three-dimensional flow, heat transfer and mixing. Not only would this be beneficial for achieving better sulfur removal, but also for equipment design to prevent fouling of heat exchanger tubes, etc.

Scope

A review of the literature identifies several fundamental areas for future research in order to achieve improved technology for sulfur emission control. For the case of high-temperature, dry sorbent injection, they are summarized below.

Fundamental mechanisms and controlling processes are not well understood; there is much debate as to which mechanism controls: diffusion through the product layer, pore diffusion, or chemical reaction. Each claim has been supported by choosing an appropriate model and obtaining a good fit to the kinetic data. Only a few studies measure structural data simultaneously. Since the reaction rate and physical structure of the solid are both dynamic in nature, proper understanding of the mechanisms involved can be achieved by systematically measuring both kinetic and structural data, and comparing changes in each data set simultaneously with model expressions.

The experimentally observed sulfation "window" has an upper limit of 1200°C. However, theoretical simulations show that at higher temperatures, dramatically higher surface areas can be obtained for the CaO. There is a general lack of kinetic and structural data above 1100°C, although higher temperatures are encountered under process conditions. At these temperatures, the substantial changes in sorbent utilization and structural evolution take place within a few milliseconds. Experimental techniques for sampling and analyzing the gas phase and the partially reacted sorbents after such low residence times can provide much needed information.

The following observations were made on the reaction kinetics: The reaction rate is roughly proportional to the second power of the initial specific surface area measured by the BET method. The reaction is highly dependent on temperature with an activation energy of about 35 kcal/mol. The product is $CaSO_4$ rather than $CaSO_3$, since $CaSO_3$ is stable only at lower temperatures. Although the sorbent reactivity is governed by both chemical kinetics and diffusional resistance, in practice only the overall reactivity which includes both of these effects can be measured in pilot plant experiments.

While early research has focused mainly on limestone ($CaCO_3$) sorbents, recent research has been focused on hydrated lime ($Ca(OH)_2$) due to the fact that CaO derived from hydrated lime is more reactive than that derived from limestone [97]. It has been reported that CaO utilizations of up to 50% can be achieved within 0.74 second at 1950°F for hydrated lime sorbents of 1-micron diameter with $62~m^2/g$ initial surface area, while calcined limestone of similar physical properties provides only a 38% utilization [98].

Existing experimental data and hence present understanding is based on experiments performed under well-controlled environments using laboratory reactors. However, in a furnace under process conditions, various parameters such as temperature variation, gas-phase species, and composition can have significant effects on the overall process. As a result, considerable disparities exist between laboratory-scale results and pilot-plant results. Therefore, fundamental studies must be complemented by studies under conditions that simulate actual process conditions. In situ analysis of a furnace may offer useful solutions.

Since the time scale in a furnace is rather short, transport processes within the reactor become very important. However, this is an area that has received very little attention. Geometry and location of the sorbent injection nozzle are shown to affect the capture efficiency. With reliable three-dimensional flow, heat transfer and mixing models applicable in a process environment, optimum design configurations and process conditions can be derived. In a furnace, the sorbent particles may attrit, or agglomerate. Since small particle size and good dispersion of these particles are favorable, powder properties of the sorbent, for example mechanical strength, need to be considered.

Economizer Zone Injection, 900-1200°F

Economizer zone sorbent injection is the least understood of the sorbent injection technologies. Hydrated material is injected into a temperature envelope centered around 1000°F. This temperature zone is generally found prior to the economizer section. The reaction is very sensitive to temperature because of the presence of high concentrations of CO_2 which may also react with $Ca(OH)_2$ to form $CaCO_3$. It appears that the ultimate conversion of the sorbent may depend on the degree of porosity which it has attained.

When $Ca(OH)_2$ is injected in the vicinity of the economizer around 1000°F, SO_2 reacts with the dry solids as follows:

$$Ca(OH)_2(solid) + SO_2(gas) \rightarrow CaSO_3(solid) + H_2O(gas)$$

In contrast to in-furnace injection where $Ca(OH)_2$ is calcined to CaO prior to reaction with SO_2, $Ca(OH)_2$ is reacted directly with SO_2 in the economizer since the temperature is too low to support the dehydration reaction. In this temperature range the main product is $CaSO_3$ instead of $CaSO_4$ [15]. The SO_2 capture rate in the economizer section is greater than that observed at 2000°F. The reaction rate falls off sharply below 900°F becoming negligible below 500°F, where no significant SO_2 capture occurs in the absence of water.

It is not clear why SO_2 capture is more rapid in this temperature range and whether or not the reaction is controlled by chemical kinetics or diffusional resistances. The reaction mechanism has not been clearly determined.

Injection of Hydrated Lime Downstream of the Air Preheater

In this reaction/temperature regime (generally less than 350°F), hydrated lime is injected into the flue gas downstream of the boiler air heater. Moisture in the form of steam or liquid water is injected in order to increase the level of the gas relative humidity.

There are two basic types of sorbent injection methods: slurry injection and dry injection with humidification. Due to inherently low gas-solid reaction rates in the intermediate temperature regime below 900°F, the reaction needs to be mediated either by chemical additives or by humidification. Humidification is the most practical effective way of increasing sorbent reactivity. Increased flue gas humidity appears to be a prerequisite for substantial SO_2 removal. The extent of utilization of the sorbent is reported to be a function of the approach to saturation of the flue gas water vapor content and of the type of particulate control device used.

The slurry injection consists of rotary or twin-fluid slurry atomizers located in the duct work. Atomized slurry droplets which include particles of $Ca(OH)_2$ react with SO_2 while they lose water through evaporation into the flue gas stream. The reaction mechanism of slurry injection is similar to that of spray dryer absorption. The main difference is residence time. The residence time for slurry injection is about 1–2 seconds, much shorter than the 10 seconds typical of spray dryer absorber vessel operation.

Fundamental Studies

Dry/Dry Systems

Simply described, a dry/dry system contains sorbent particles that have from zero to about five molecular layers of water physically adsorbed onto their outside surfaces and within their outer pore structures. The water present on the sorbent surface is present due to physical adsorption alone. The amount of water on the surface is less (a somewhat arbitrary distinction) than that which would define a "bulk" water phase (such as the bulk water phase that exists in slurry spray drying or due to collisions between a sorbent particle and an atomized droplet of water).

Klingspor et al. [99], in their seminal paper, reported the first fundamental work aimed at understanding and quantifying the dry/dry capture of SO_2 by limestone particles. Using a single limestone sample, they varied limestone particle size from 3–100 microns, temperature from 104–176°F, SO_2 concentration from 50–4000 ppm, and relative humidity from 0% to 92% in an effort to gain a better understanding of the dry/dry sorption process. Klingspor determined that the amount of water adsorbed onto the limestone surface was directly proportional to the limestone BET (nitrogen sorption) surface area and that the water sorption behavior could be described by a BET Type II sorption isotherm [100].

A simple reaction scheme was proposed:

$$SO_2(g) + CaCO_3(s) \xrightarrow{H_2O,\ O_2} CO_2(g) + \begin{array}{l} CaCO_3(s) \\ CaSO_3 \cdot 2H_2O(s) \\ CaSO_4 \\ CaSO_4 \cdot 1/2H_2O(s) \\ CaSO_4 \cdot 2H_2O(s) \end{array}$$

which can be more simply represented as

$$SO_2(g) + CaCO_3(s) \xrightarrow{H_2O,\ O_2} CO_2(g) + \text{products}$$

Using initial rate data, they found that the rate of reaction varied weakly with the O_2 concentration in their synthetic flue gas and showed no measurable variation with changes in the concentration of CO_2. Most importantly, they found that the rate of SO_2 removal was strongly dependent on the relative humidity. At relative humidities corresponding to less than a single molecular layer of water coverage of the BET surface area of the limestone, the SO_2 capture rate was observed to be virtually zero. The SO_2 removal rate was observed to increase rapidly with increasing relative humidity at values that corresponded to more than a single molecular layer of water coverage. The initial rate of reaction per unit mass of limestone was also found to decrease with increasing limestone particle size. The decrease was directly proportional to the limestone BET surface area as might have been expected since the same limestone was used throughout the testing.

Klingspor et al. [99] arbitrarily chose to represent the fundamental rate of SO_2 removal as a simple nth-order reaction with respect to the SO_2 concentration.

$r = kC^n$

where r is the rate of SO_2 removal, C is the concentration of SO_2 in the gas phase, k is the reaction rate constant, and n is the reaction order. They discovered that this simple reaction rate equation did not adequately describe the dry/dry sorption reaction. Values for both k and n were found to be strong functions of the relative humidity and therefore of the amount of water adsorbed onto the limestone surface. Both k and n were observed to increase rapidly with increasing relative humidity. The value of n increased from 0 to 0.80 and the value of k increased from 0.045 to 118 mol $kg^{-1}h^{-1}(mol\ m^{-3})^{-n}$ as the relative humidity increased from 24% to 92%. It was postulated that this change in the reaction rate equation was due to the transition of the water "phase" from true physically sorbed molecular layers, where the chemistry of the sorbed molecular layers would dominate SO_2 capture behavior, towards a true "bulk" water phase where the chemistry of the molecular layers would not be important and bulk solution chemistry would dominate the SO_2 capture behavior.

Using values of k and n experimentally determined for specific values of relative humidity, Klingspor et al. [99] were able to fit the observed long-term SO_2 capture behavior of 3 to 5 micron particles using a standard unreacted core model.

In more recent work, Jorgensen et al. [101] have concluded that gas-phase diffusion is not the overall rate-controlling step for dry/dry systems. "Ash layer" diffusion through a reacted $CaSO_3/CaSO_4$ layer was assumed to be the rate-determining step. Jorgensen also used the unreacted core model to fit his experimental data. It was found that when using lime as a sorbent, 60–90% of the lime was unreacted after the reaction had terminated.

Wet/Dry Systems

Wet/dry systems are qualitatively and quantitatively different from dry/dry SO_2 sorption systems. This is the type of SO_2 capture mechanism that is expected to dominate the behavior of the Limestone Emission Control (LEC) system currently being developed at Ohio University. A "bulk" water phase dominates the behavior of the wet/dry system, in contrast to the "monolayer" water phase that dominates the dry/dry system. The many potential rate-controlling resistances in the wet/dry system include mass transfer through the gas boundary layer in contact with the water phase, sorption of the SO_2 in the water layer, ionic dissociation of the SO_2 in the water layer, transport of the dissociated SO_2 through the water layer to the reactive interface, dissolution of the sorbent at the sorbent surface, transport of the dissolved sorbent to the reactive interface, diffusion of the dissolved sorbent and/or dissociated SO_2 through the $CaSO_3/CaSO_4$ precipitate layer, and the rate of the intrinsic reaction between the dissolved sorbent and the dissociated SO_2. For in-duct injection, collision/scavenging of the water droplets by the sorbent particles adds yet another important mechanistic step.

Chemical processes that occur in the aqueous phase, ionic dissociation of the SO_2 in the water layer and dissolution of the sorbent at the sorbent, are thought to include

$$SO_2 + H_2O = H^+ + HSO_3^-$$

$$HSO_3^- = H^+ + SO_3^-$$

$$Ca(OH)_2 = Ca(OH)^+ + OH^-$$

$$Ca(OH)^+ = Ca^{++} + OH^-$$

$$Ca^{++} + SO_3^- = CaSO_3$$

$$2H^+ + 2OH^- = 2H_2O$$

All of these ionic chemical reactions appear to be very fast so that none of these steps is thought to be rate limiting. Therefore one or more mass and heat transfer processes would be controlling the overall reaction rate. This topic is elaborated in the following section.

The actual rate-limiting step can be expected to vary depending on such factors as the water/sorbent ratio, the degree of reaction that has already occurred (and therefore the thickness of the $CaSO_3/CaSO_4$ precipitate layer), the temperature, and the process configuration used to affect the sorption process. The process configuration will affect the sorption process via its influence on the relative velocity between the wet sorbent particle and the flue gas stream, the thickness of the water layer, and relative humidity.

Gullett and Kramlich [30] cite unpublished work performed in a single droplet, drop tube reactor which studied the wet/dry sorption reaction "decoupled from evaporation." Results of this study indicated that, for the flue gas velocities studied, the overall reaction rate was gas-film diffusion controlled at gas phase SO_2 concentrations of less than 800 ppm. At concentrations above 800 ppm, it was postulated that the reaction was controlled by the rate of lime dissolution. Doubling the amount of lime present in the droplet did not double the SO_2 removal rate (based on unit weight of lime). As in the case of dry/dry sorption, the performance of wet/dry sorption was found to be a strong function of the relative humidity of the flue gas. However, unlike dry/dry sorption where the humidity in the flue gas supplies the water that is physically sorbed onto the limestone surface, the humidity present in the wet/dry system serves primarily to retard the rate of drying, thereby slowing the rate of disappearance of the "bulk" water phase from the sorbent particle surface. Many other investigators have confirmed the relationship between relative humidity and wet/dry sorption performance [13,16].

Jozewicz and Rochelle [102] attempted to model the wet/dry sorption system by assuming that both the rate of SO_2 removal and H_2O evaporation are controlled by gas-film mass transfer. The reaction of ionized SO_2 with ionized lime in the water film was considered to be instantaneous. Jozewicz also assumed an infinite rate of $Ca(OH)_2$ dissolution. An attempt was made to account for the slurry droplet size distribution. They did not account for accumulation of sulfite/sulfate in the water film. In contradiction to their previous modeling assumptions, Jozewicz stated that the dissolution of $Ca(OH)_2$ may be a rate-controlling step.

Harriot and Kinzey [103] incorporated both gas- and liquid-phase mass-transfer resistances into their modeling of the wet/dry sorption system. Like Jozewicz and Rochelle, they assumed that there was no resistance to the dissolution of $Ca(OH)_2$. They did not account for accumulation of sulfite/sulfate in the water film.

Damle [104] and Maibodi et al. [105] have developed a three resistance model of the wet/dry sorption system. They incorporated both gas- and liquid-phase mass transfer resistances as well as the dissolution resistance of $Ca(OH)_2$. Damle determined that the liquid-phase mass-transfer coefficient was two orders of magnitude greater than the gas-phase mass-transfer coefficient; therefore the liquid-phase resistance was ignored in the model simulations. Maibodi et al. [105] determined that the predominate mass-transfer resistance was that of $Ca(OH)_2$ dissolution.

Karlsson and Klingspor [29] have produced asymptotic models for wet/dry sorption systems dominated by either a gas-phase mass-transfer resistance or by the rate of sorbent (lime or limestone) dissolution. Their models were used to predict the performance of a 0.5-MW spray-dry-scrubbing pilot plant. An anomalous change in the observed reaction rate constant was found to occur with changes in slurry concentration. Furthermore, an observed decrease in the reactivity of lime with increasing extent of reaction was not explained. This decrease was probably due to surface blinding (an additional mass-transfer resistance) caused by the precipitation of $CaSO_3/CaSO_4$. It is expected that this blinding effect will be a strong function of sorbent particle size.

Spray Drying Absorption

Introduction

The process of spray drying involves the transformation of a liquid feed or slurry into a dried solid through spraying the feed into a drying atmosphere. The solid product may take the form of a powder, granule, or an agglomeration, depending on the physical and chemical properties of the feed as well as the spray dryer design and operation. The application of spray drying technology to flue gas desulfurization (FGD) is relatively new and has been called spray drying absorption (SDA).

The history of spray drying is traceable to a U.S. patent issued in 1872. Since that time, the spray drying process has been developed for use in a wide range of industrial applications including the manufacture of powdered soaps and detergents, powdered milk, instant coffee, corn starch, fertilizer production, powdered polymer resins and the production of mineral ores and clays [106]. The production of spray dried ferrites find use in the manufacturing of telephones, televisions and other electronic devices.

The use of spray dryers for environmental control of sulfur dioxide (SO_2) began with pilot-scale studies funded in large part by utility companies. The first full-scale utility application came on line in 1980. To date, seventeen full-scale units treating a total of 6031 megawatts of coal-fired-derived flue gas are in operation in the United States. Table 1.2 is a summary of these units and their various design configurations. As seen from Table 1.2, these systems have been primarily designed for low-sulfur coals (coals with sulfur contents below 2% by weight) and, as such, have resulted in the application of SDA by western utilities. The units in operation have demonstrated the ability to remove 90% SO_2, resulting in capital costs savings averaging $65 per installed kilowatt, Keener et al. [3]. The advantages of SDA over traditional wet scrubbing desulfurization systems (other than the obvious capital costs) include the following:

- The final waste product is a dry powder that does not require the use of expensive sludge handling equipment.
- The flexibility of the feed system allows immediate feed control of sorbent to follow boiler load.
- The materials of construction are typically mild steel.
- The space required is generally smaller than wet FGD systems, making SDA a good choice for retrofit.

Table 1.2. Utility Boilers Controlled by SDA=FGD

Utility/Unit	MW per absorber	Fuel (a)	Re-cycle	Re-heat	Atomizers per absorber
Basin Electric Power					
Antelope Valley 1	88	Lignite (0.7%S)	Yes	Yes	1 rotary
Antelope Valley 2	88	Lignite (0.7%S)	Yes	No	1 rotary
Laramie River 3	143	Western (0.5%S)	Yes	No	16 nozzles
Central Power & Light					
Coleto Creek 2	180	Western (0.4%S)	Yes	Yes	1 rotary
Colorado Ute Electric					
Craig 3	112	Western (0.5%S)	Yes	Yes	12 nozzles
Grand River Dam Authority					
Unit 2	144	Western (1.0%S)	Yes	Yes	3 rotary
Marquette Michigan					
Shiras 3	44	Western 1.5%S	Yes	Yes	1 rotary
Montana-Dakota Utilities					
Coyote 1	110	Lignite (0.9%S)	No	No	3 rotary
Northern States Power					
Riverside 6 & 7 (b)	110	Western (1.2%S)	Yes	No	1 rotary
Sherburne 3	108	Western (1.0%S)	Yes	Yes	1 rotary

Table 1.2. Utility Boilers Controlled by SDA=FGD   cont.

Utility/Unit	MW per absorber	Fuel (a)	Re-cycle	Re-heat	Atomizers per absorber
Pacific Power & Light					
Jim Bridger 2 (c)	100	Western (0.8%S)	Yes	No	10 nozzles
Wyodak 1	110	Western	Yes	Yes	1 rotary
Platt River Power Authority					
Rawhide 1	93	Western (0.3%S)	Yes	Yes	1 rotary
Sierra Power Pacific					
North Valmy 2	92	Western (1.0%S)	Yes	Yes	3 rotary
Sunflower Electric Cooperative					
Holcomb 1	106	Powder River (0.4%S)	Yes	No	1 rotary
Tucson Electric Power					
Springerville 1	123	Subbi-tum-inous (0.7%S)	Yes	Yes	1 rotary
Springerville 2	123	Subbi-tum-inous (0.7%S)	Yes	Yes	1 rotary
United Power Association					
Stanton 10	60	Lignite (0/8%S)	No	No	3 rotary

(a) All labelled "Western" are subbituminous. (b) Originally a demonstration unit; now a commercial installation. (c) Demonstration unit which is no longer in service.

Description of the Spray Drying FGD Process

The SDA-FGD process consists of four operations: reagent preparation, the spray dryer itself, particulate collection, and solids disposal. The reagent is delivered in covered railroad cars and then stored in a suitable container to which it is usually pneumatically conveyed. The reagent is mixed with water (slaked in the case of lime) then classified to prevent any large grit particles from going to the spray dryer where they can cause orifice plugging in the spray nozzles or rotary atomizer.

Hot flue gas is contacted with a fine spray of atomized droplets in the spray drying chamber. The alkaline material in the spray reacts with the SO_2 and the water is evaporated by the flue gas. The reaction product is collected as a dry powder along with the fly ash, typically in a fabric filter, and can then be sent to a suitable disposal site.

Reagent Preparation

Only two reagents have been used commercially. Lime is typically the preferred reagent because of its lower cost, widespread availability, and the fact that its reaction product is not water soluble which simplifies disposal.

Sodium carbonate has seen limited use as a reagent for SDA-FGD, primarily because of reagent cost and the waste disposal difficulties associated with a soluble end product. Hence, reagent preparation is typically a lime slaking operation. One might think that this technology, being well developed, would be the least of any potential problems in the system. This has not proven to be the case because the SDA-FGD application requires a lime slurry with a special set of properties, typically 35-50% solids, and a requirement that very little if any grit be present in order to prevent clogging and abrasion in the atomization process.

There are many different types of lime commercially available, such as dolomitic lime (the cheapest), pulverized lime, granulated lime, and high-calcium pebble lime. Lime comes from limestone which is mined throughout the world and is one of the most widely used chemicals in industry. The lime obtained from the various limestones can vary widely in their physical properties. This makes the choice of a suitable lime type and slaking method critical for optimal reagent preparation.

The reagent is delivered as CaO, "quicklime," and must be converted to a hydroxide slurry in order to maximize SO_2 removal. The slaking process that accomplishes this feat can have a significant impact on the reactivity of the lime slurry and is an important design parameter. The slurry should consist of very small particles in order to maximize the available surface area. Such parameters as quicklime quality, make-up water quality, slaking temperature, water-to-lime ratio, equipment type and operation, and addition of soluble alkaline species all play a role in determining final slurry quality and ultimately in overall SO_2 removal efficiency [108,109].

Ball mill slakers and paste slakers are commonly used. Each method has perceived advantages. Ball mill slakers produce finer particles which may be more reactive. Paste slakers produce a less abrasive slurry and have a lower capital cost and lower power requirement [110]. It has been suggested that the use of a vertical stirred mill for lime slaking has advantages over the more typically used ball mill slaker. Such advantages as finer particle production, simpler operation and maintenance, reduced capital equipment and operating costs, and a decreased net power consumption at higher solids concentrations were noted [111].

For applications utilizing Na_2CO_2, "soda-ash," as the sorbent, reagent preparation is greatly simplified. A solution of the desired concentration is prepared in a suitable mixer. It should be noted, however, that the reagent of choice has overwhelmingly been lime.

The properly prepared slurry or solution is delivered to the spray drying chamber where the desulfurization process begins. Figure 1.6 depicts a typical cylindrically shaped spray dryer with a conical bottom.

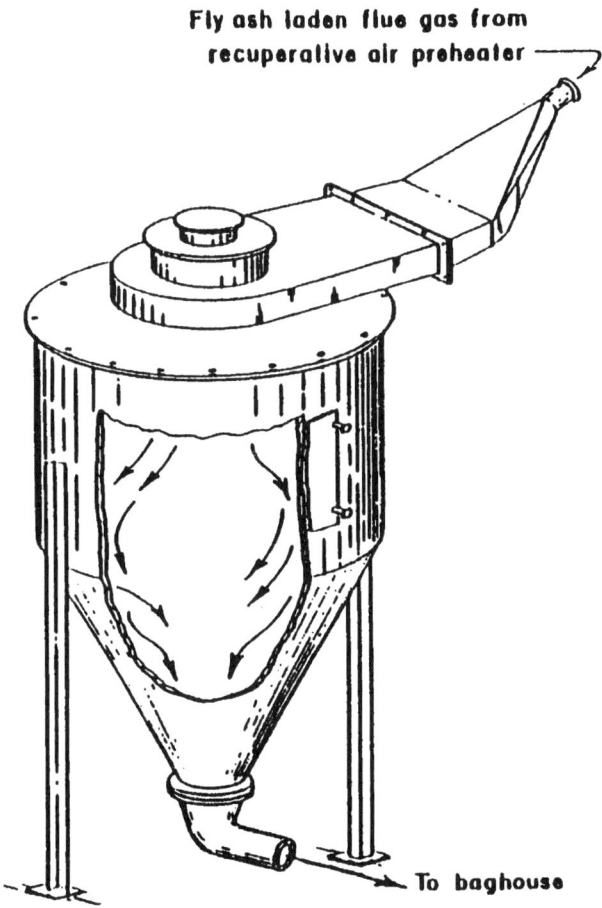

Figure 1.6 Spray dryer absorber [112].

Dryer Configuration

Large commercial SDA-FGD systems may have spray drying chambers of up to 50 feet in diameter. Hot flue gas enters the spray dryer from the combustion air preheater at a temperature typically between 250 and 350°F and is contacted with a finely dispersed spray of reagent slurry droplets. Two processes occur simultaneously in the spray drying chamber during a residence time of about 10 seconds. SO_2 contained in the flue gas is absorbed at the surface of the droplets, allowing the alkaline reagent material to react with the SO_2 forming sulfite and sulfate salts. Evaporation of the water in the droplets occurs due to the high thermal energy of the flue gas with the resultant adiabatic humidification of the flue gas and the formation of a dry powder generally having less than 1% free moisture [110,113]. The dry reaction product is pneumatically conveyed by the flue gas together with the fly ash to a suitable particulate removal system, typically a fabric filter. Usage of an electrostatic precipitator (ESP) for particulates removal has been successfully demonstrated at a location where one was already in place [114]. The disadvantage of using an ESP is that up to 20% of the SO_2 removal can occur in the baghouse on the fabric filters for fly ash which contains a substantial alkalinity value. Some of the dried product may be collected from the bottom of the spray dryer itself in some system designs.

Proper design of the spray drying chamber, gas disperser, gas exit configuration, and atomizer help to ensure successful operation of the system with the goal being formation of a reaction product which is a free-flowing powder. It is desirable to design the spray dryer to effectively contact the flue gas with the sprayed sorbent while having the capability to adjust to varying boiler load. The disperser at the flue gas inlet should create good turbulent mixing of the flue gas and atomized droplets of reagent spray. The relative velocity between the liquid and gas ranges from 400-700 ft/sec with L/G ratio typically 0.002–0.003 gal/cu ft [110].

The atomizer is designed to break up the bulk slurry or solution into a spray of very fine droplets. Typically, one pound of lime slurry can provide a surface area of from 500-1000 ft^2 [111]. It is desirable for the droplets of reagent-containing material to be small to maximize the surface area available for mass transfer. Perhaps most importantly, the atomizer must produce droplets which in conjunction with the other spray dryer design and operating parameters do not deposit upon the spray drying chamber walls.

Atomization can be accomplished by one of two methods: by use of a rotary atomizer or alternatively, through the use of high pressure air in conjunction with a suitable atomizing spray nozzle. Figure 1.7 shows the inlet section for a spray dryer utilizing rotary atomization. The rotary atomizer is typically powered by a 700–800 hp motor (not shown) which is coupled to its shaft. The flue gas directional vanes provide for turbulence and direction of the gas to provide good contact with the atomized slurry.

The atomizer disk is commonly from 12-14 inches in diameter and is spun at speeds of 3000-50,000 rpm, Midkiff [115]. Generally, there is only one rotary atomizer in each spray drying chamber, though three rotary atomizers per spray dryer is another configuration that has been used "because it is questionable whether good atomization can be sustained if spray machines and associated atomizers are built to handle significantly more than 75 gpm of feed" [116]. In rotary atomization, droplet size is a function of the feed slurry characteristics, e.g., solids content, and the rotational speed and diameter of the atomizer disk itself. Droplet size is independent of feed rate for a rotary atomizer and this enables systems so equipped to have good turndown capability. Conversely, a spray nozzle type system will use as many as 10–20 nozzles per absorber. Nozzles operate over a limited range for optimum atomization. Outside of this range, the droplet size can be affected which may require on-off control as boiler load varies. While plugging and abrasion have been problem areas in some spray nozzle equipped systems, proponents of properly designed spray nozzle atomizers cite the advantages of no moving parts and of continued operation while one or more of the spray nozzles is removed for inspection and/or maintenance [117].

The treated gas leaves the spray dryer typically at a temperature 20–50°F above adiabatic saturation. Control of this temperature is one of the most important parameters for successful operation of the SDA-FGD process. If the dewpoint of the existing humidified gas stream is approached too closely, condensation may occur inside the downstream equipment and ductwork, causing problems. The flue gas flow rate entering the spray dryer varies with changing boiler load, hence the exit temperature as well as overall SO_2 removal must be closely monitored and controlled. Measurement of the adiabatic saturation temperature may be complicated by such factors as acid condensation and dust fouling. The harsh nature of this process environment can shorten the life expectancy and reliability of almost any instrument. It has been reported that a wet bulb temperature transmitter performed reasonably well as compared to a solid state in situ probe or an aspirated sample infrared analyzer [114].

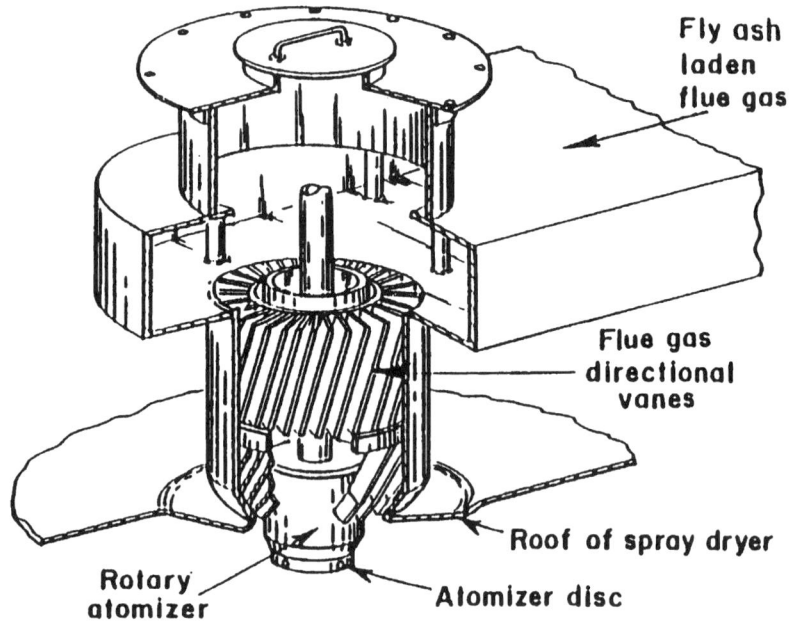

Figure 1.7 Inlet of a spray dryer using rotary atomization [112].

Control of the approach to adiabatic saturation temperature is accomplished by one or a combination of methods:

- by varying the feed rate of the alkaline reagent entering the spray dryer
- through the addition of reheat air as the treated gas leaves the spray dryer (this reheat air is a small portion of the hot, untreated flue gas which is bypassed around the spray dryer from further upstream but its use entails the tradeoff of a decrease in overall SO_2 removal)
- by changing the sorbent stoichiometry of the alkaline reagent feed, thereby changing its concentration or percent solids content before it is delivered to the spray dryer

Bypass gas can be supplied from either of two upstream sources as shown in Figure 1.8. For the case of the hot flue gas immediately downstream of the boiler, significantly less gas can be used to effect temperature control. This hot gas has less of an effect on the effluent SO_2 concentration but a penalty is paid in the decrease of energy available for air preheat. When warm gas from downstream of the combustion air preheater is used, there is no such energy penalty but the amount of this gas that can be used may be limited by overall SO_2 removal requirements.

SO_2 removal increases with increasing sorbent stoichiometry but there are two limiting factors that must be considered. As sorbent stoichiometry is increased the efficiency of sorbent utilization decreases leading to higher reagent and disposal costs. There is an upper limit to the amount of reagent that can be held in suspension in a slurry or dissolved in solution. By recycling a portion of the reaction products collected from the baghouse and/or the bottom of the spray dryer, sorbent utilization can be increased.

Recycle is usually included in SDA-FGD system design because of the lower cost associated with more efficient sorbent utilization.

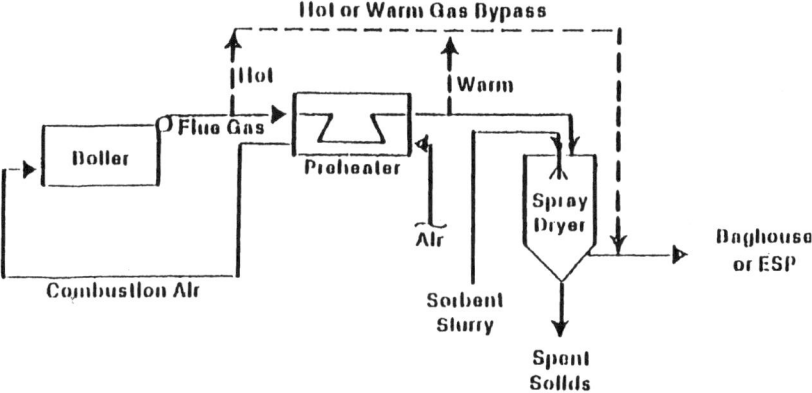

Figure 1.8 Bypass gas options for temperature control [110].

Operating the SDA-FGD system closer to the saturation temperature generally has the effect of greatly increasing the SO_2 removal efficiency and sorbent utilization [110]. The danger here is that in case of a system upset it may not be possible to initiate control quickly enough to avoid condensation and/or deposition of wet material at some point in the system.

Particulate collection SDA-FGD is in effect a two-stage absorption system with a significant portion of the actual SO_2 removal typically taking place in the baghouse on the filters themselves. The amount of actual removal taking place in the baghouse is dependent upon the alkalinity of the fly ash produced in the combustion process. Design parameters include the choice of filter media, the choice of cleaning method, cleaning frequency and cycle, and estimation of gas-to-cloth ratio for sizing the unit. Utilities generally use reverse-air fabric filters while systems for industrial boilers typically have pulse-jet filters [110].

Information about sizing and costing fabric filters has recently been presented [118].

Corrosion in the baghouse has been a problem at some installations due to condensation of water and the subsequent formation of dilute sulfuric acid.

All liquid water from condensation inside the baghouse must be eliminated to control potential corrosion problems. Design recommendations have been made which address this problem [114]:

- elimination of "thru-metal" connections that can conduct cold into the baghouse
- total elimination of possible ambient air infiltration by conducting field leakage tests
- paying close attention to insulation and sealing of adjoining walls when a compartment is taken out of service
- improving insulation around access doors and ports
- maintaining tight quality standards for field installation of insulation

Waste Disposal

Dry spray dryer waste products are typically disposed of in a manner similar to fly ash disposal. The waste products are dampened in rotary unloaders preparatory to their trip to a suitable site to be used as fill material. Tests such as optimum moisture, compressive strength, permeability, and chemical composition are used to determine suitability for land filling.

Establishing a suitable site "can be a very time consuming process" [119]. First, potential sites must be identified, followed by engineering, cost, and environmental feasibility studies. Agreement must be obtained from the site owner and then all of the necessary permits must then be obtained from the appropriate regulatory agencies. Finally the site must be prepared and then opened for operation [119]. Siting of land fills is becoming more and more difficult as the cry of NIMBY (Not in My Back Yard!) rises up across the country. Many organizations are conducting studies to find alternative utilization schemes for FGD waste materials such as pelletization for use as a road bed material.

Process Chemistry

The spray drying process for SO_2 control is comprised of a complicated interaction between the drying droplet and the reaction between SO_2 and $Ca(OH)_2$. The model proposed by Getler et al. [120] attributes the SO_2 uptake into two primary phases: a first phase, called the constant rate drying period, which is of short duration, and a second phase, called the falling rate drying period, where final reaction and drying occurs. A third phase, called the diffusion period develops whenever the moisture level of the drop has been reduced to that required to fill the interparticle voids. Figure 1.9 shows a classical drying curve illustrating these three periods.

The constant rate period has been described by Wentz and Thygeson [121] as that part of the drying process where the evaporation rate is constant and controlled by conditions such as mass velocity, temperature and humidity of the inlet flue gas. During this phase, the atomized droplets of lime-water slurry leave the atomizer or nozzle at high velocities (approximately 450 ft/sec) and decelerate to velocities below 30 ft/sec in less than 0.1 seconds.

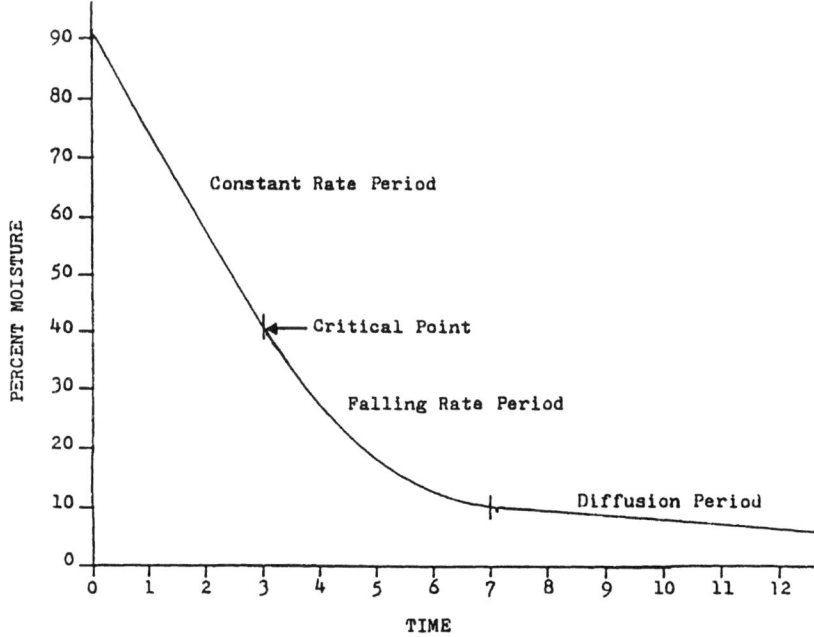

Figure 1.9 Classic drying curve for slurries in spray dryers.

Dry Scrubbing Technologies for Flue Gas Desulfurization

The original atomized lime-water slurry droplet has a mean diameter between 50 and 80 microns and contains a large number (for a 20% slurry each droplet would contain 173,000 particles) of insoluble, discrete alkali particles. The initial moisture evaporation rates are high during this period and culminate in the eventual loss of approximately 75% of the droplet moisture. As the droplet diameter decreases, the distance between discrete particles is reduced until the particles contact one another. At this point, the diffusion paths for the reactants are restricted and diffusion through and around the particles becomes the controlling factor. The time to reach this point in the drying process has been studied by Ranz and Marshall [122]. Their expression for the constant rate drying time is illustrated in Figure 1.9 and given by the expression:

$$t_c = \lambda \rho_o (d_0^2 - d_{cr}^2)/(8k \Delta T_{Gs})$$

where λ is the latent heat of vaporization (kcal/kg), ρ_o is the initial droplet density (kg/m^3), d_0 is the initial droplet diameter (m), d_{cr} is the droplet diameter at the critical point (m), k is the thermal conductivity (kcal/m-hr-°K), and ΔT_{Gs} is the temperature difference between droplet surface and surrounding gas (K).

During the falling rate period, the drying process is controlled by such factors as diffusion of water through and around the solid particles and precipitation of any product "crust" around the particles. The rate of evaporation is greatly reduced during this period due to the formation of the ash layer resistance and the reduced thermal driving force. The time required for the falling rate period may be estimated from the expression Ranz and Marshall [122]:

$$t_F = \lambda d^2_{cr}(W_{cr} - W_2)/(12k \Delta T_{Gs})$$

where W_{cr} is the critical moisture content (kg-H_2O/kg-solids) and W_2 is the final moisture content (kg-H_2O/kg-solids).

The falling rate period is characterized by a reduction in liquid volume to the point where solid particles come into contact. The maximum moisture content of spherical particles arranged in a cubical pattern corresponding to this so called "critical moisture content" is 47.6 percent by volume [123]. Irregularities in the solid may impede the recession of water into the solid matrix during the falling rate period. Pools and channels of water may form on the surface allowing the use of a constant rate assumption during the early stages of the falling rate period, Partridge [123]. As the large volume of water recedes, a coating of reactants may be expected to be deposited on the outer surface of the agglomerated solids. This coating of sulfites and sulfates inhibits further reaction by limiting the accessibility of the unreacted particle interior. This surface enrichment of sulfur has been verified by X-ray analysis of spray dried products [120].

After passing from the spray dryer, the spray-dried solids continue to undergo loss of water until equilibrium with the humid gas is obtained. Reaction between the particle and sulfur dioxide continue but at a substantially reduced rate. This is widely attributed to the physical plugging of the sorbent pores by the reaction products which have a larger molar volume than that of the sorbent.

The reactions involving gaseous SO_2 and a lime-slurry droplet have been postulated as being sequential [119,123] and to consist of the following four steps:

1. ionization of water

$$H_2O\ (l) = H^+(aq) + OH^-(aq)$$

2. two-step ionization of dissolved SO_2

$$SO_2(aq) + H_2O\ (l) = H^+(aq) + HSO^-(aq)$$

and

$$HSO_3^-(aq) = H^+(aq) + SO_3^{-2}\ (aq)$$

3. two-step dissolution and ionization of $Ca(OH)_2$

$$Ca(OH)_2(s) = CaOH^+(aq) + OH^-(aq)$$

and

$$CaOH^+(aq) = Ca^{+2}(aq) + OH^-(aq)$$

4. precipitation of dissolved Ca+2 and SO3-2 ions

$$Ca^{+2}(aq) + SO_3^{-2}(aq) + 1/2\ H_2O(l) = CaSO_3*1/2H_2O(s)$$

The equilibrium constants for the above six reactions are given in Figure 1.10 and Figure 1.11. These constants change relatively little within the temperature range of interest. The disassociation constant of water has the greater slope and changes by over a factor of 100 on going from 25-115°C (77-240°F).

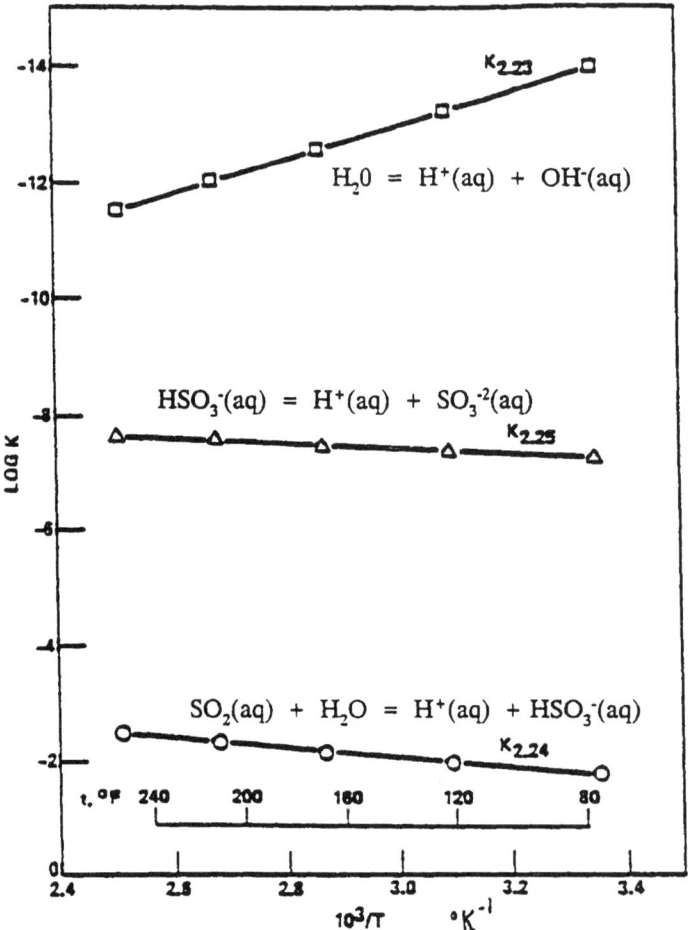

Figure 1.10 Dissociation constants of water and dissolved SO_2.

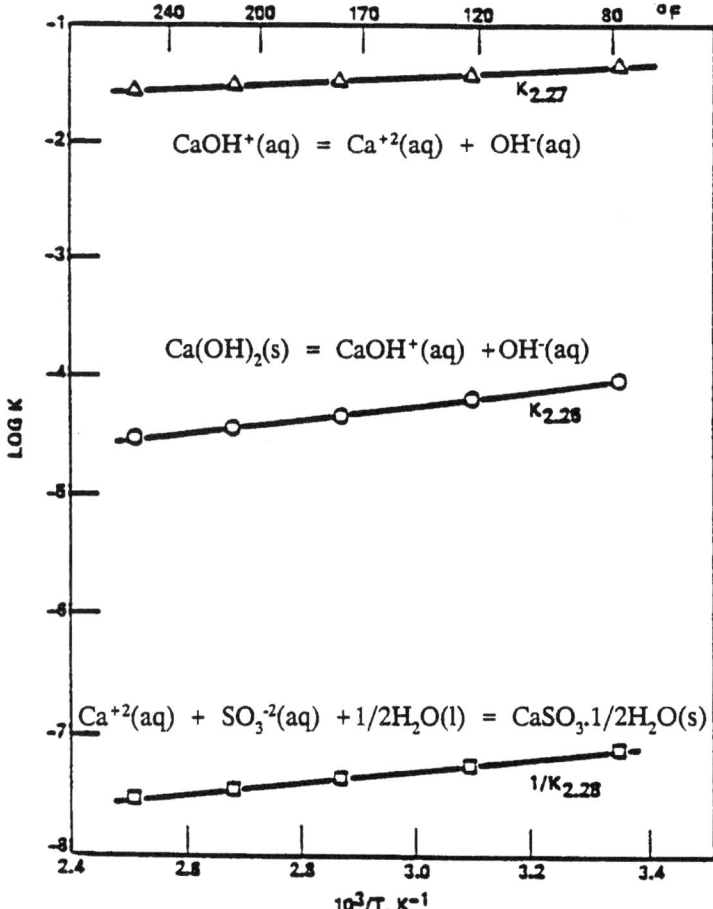

Figure 1.11 Equilibrium constants for the dissociation of $Ca(OH)_2$ and the precipitation of $CaSO_3 \cdot 1/2H_2O$.

The formation of calcium sulfite hemi-hydrate ($CaSO_3 \cdot 1/2H_2O$) is dependent on the local pH value in the drop. In the lime-slurry droplet, the pH will vary from acidic at the gas-liquid interface to alkaline at the liquid-solid interface. Calcium ions are converted to $CaOH^+$ at high pH values according to the precipitation reaction. Also, at low pH values, sulfite ions are depleted by the reverse of the ionization reactions. The ionization of water, the first step in the two-step dissociation of SO_2, and the first step in the two-step dissolution and ionization of

Dry Scrubbing Technologies for Flue Gas Desulfurization 49

$Ca(OH)_2$ are considered to be fast while the second step of the two-step ionization of SO_2 is considered to be slow [120]. The rate-controlling mechanisms may be considered, therefore, to be the diffusion of reactants to the reaction zone and solid dissolution [122].

The most sophisticated modeling effect to date for the spray drying process has been undertaken by Damle [104] and later expanded by Partridge [123]. Partridge's model of the constant rate-drying period is based on film theory and treats the atomized slurry droplet as a sphere of discrete sorbent particles with the fluid phase uniformly distributed around individual sorbent particles. The absorption rate is considered to be a process of absorption of the SO_2 into the slurry drop accompanied by an instantaneous chemical reaction. The rate is governed by the magnitude of four transfer coefficients: the gas-film mass transfer coefficient; the liquid-side SO_2 mass transfer coefficient; the liquid-side $Ca(OH)_2$ mass transfer coefficient; and the solid dissolution rate constant. The overall rate is assumed to be a function of these resistances in series. The model estimates these coefficients from correlations and determines if one resistance is significantly larger than the others. If so, that portion of the transfer process is assumed to control the overall rate.

Particulate Control

The cost-effective control of particulate emissions from coal-fired power plants is a challenging problem that is becoming more complex every day. Changes in regulations for particulate and sulfur dioxide emission control, changes in the state of the art in power plant engineering, and changes in the collection devices themselves, including fabric filters and electrostatic precipitators (ESP), all interact to create a fluid situation. Processes used to remove contaminants like SO_2 often compound particulate-collection problems. This survey is biased towards particulate control methods and devices for coal-fired plants utilizing calcium-based FGD processes. The control technology can be divided into two categories: conventional devices and other mechanical devices [125].

Three types of conventional control devices are used for controlling particulates. These are ESPs, fabric filters (filter bags, baghouses) and wet scrubbers. Historically, ESPs have been the control device of choice. In the past few years, however, fabric filters and wet scrubbers have also been used. The mechanical devices use a combination of centrifugal, inertial and gravitational forces to remove particulates from the flue gas stream.

In designing a control device for the collection of particles from coal-fired boilers, each installation is characterized in terms of fixed, i.e, steady-state, input parameters such as coal and boiler type and the various design parameters for the control device. More recently, the effect of coal variability has also been taken into account. The injection of sorbent for desulfurization of the flue gas further complicates the design process.

Electrostatic Precipitator (ESP)

With the advent of pulverized coal use in power plant boiler systems around 1924, ESPs have become an integral part of large-scale coal combustion. ESPs are useful because pulverized coal-fired boilers generate large volumes of flue gas containing very small dust particles that cannot be removed by other, more conventional devices, such as settling chambers and inertial separators. ESPs offer the following advantages: very high collection efficiency for nearly all sizes of fly ash particles; low power requirements; virtually no pressure drop; minimal maintenance requirements; applicability to very large gas flows; flexibility over a wide range of temperatures, pressures, and gas characteristics; and long working lives.

An ESP consists of a hopper-bottomed box containing rows of plates forming passages through which the flue gas flows. Centrally located in each passage are emitting electrodes energized with high-voltage, negative-polarity, direct current provided by a transformer/rectifier (T/R) set. The voltage applied is high enough to ionize the gas molecules close to the electrodes, resulting in a visible corona. Flow of gas ions from the emitting electrodes across the gas passages to the grounded collecting plates constitutes what is called corona current.

When passing through the flue gas, the charged ions collide with, and attach themselves to, fly ash particles suspended in the gas. The electric field forces the charged particles out of the gas stream toward the grounded plates where they collect in a layer. The plates are periodically cleaned by a rapping system to release the layer as an agglomerated mass into ash hoppers.

The particle velocity toward the plate (also called the drift or migration velocity) results from the electrical and viscous drag forces acting on each particle and can be calculated using the following equation [126]:

$W_p = (q\ E\ C)/(6\ \pi\ V\ r)$

where W_p is the migration velocity, q is the particle charge, E is the electric field strength, C is the Cunningham slip correction factor, V is the gas viscosity, and r is the particle radius. The collection efficiency is a function of the particle migration velocity and any increase in this velocity will result in increased collection efficiency.

Fundamentals and Operating Parameters

ELECTRIC PARAMETERS. The performance of an ESP is dependent upon the operating electrical parameters. The voltage and current characteristics depend upon the geometry of the electrodes, the temperature, pressure and composition of the gas, the particle resistivity, and the mass loading and particle size distribution.

The operating voltage determines the magnitude of the electric fields in the precipitator which affects the particle charge and the electrical force propelling the charged particles to the plate. The particles are charged by two physical mechanisms: field charging and diffusion charging [127-130]. Field charging, produced when ions traveling along electric field lines are intercepted by the particles, is proportional to the electric field strength, and is the primary mechanism for charging particles larger than 1 µm. Diffusion charging occurs when the thermal energy of the ions causes the diffusion of ions to the surface of a particle. Diffusion charging depends upon the ion density and residence time but not the field strength and is the predominant charging mechanism for particles smaller than 0.1 µm. Particles with diameters ranging between 0.1 and 1 µm are charged by both mechanisms.

The velocity of migration is proportional to the electric field strength as well as the particle charge. For larger particles the charge is a function of electric field strength; therefore the collection efficiency for a given particle size is proportional to the square of the electric field strength.

The operating current (flow of ions from the corona wire) charges the particles, providing the driving force for their migration to the collector plate. The electric field strength determines the magnitude of the charge on the particles, while the current density determines the charging rate. It is important that there be sufficient current density to supply the ions to rapidly charge the particles to their maximum levels. The current density also affects the electric field strength. The presence of ions distributed in the inter-electrode space creates an ionic space charge which establishes an electric field that adds to the existing electric field.

It is therefore desirable to operate ESPs at the highest levels of voltage and current possible. Limitations are imposed by electrical breakdown; i.e., sparking of the gas in either the interelectrode space or in the collected dust layer. Breakdown in the gas layer is a function of gas density, and at conditions typical of cold-side ESPs, this breakdown usually occurs at a current density on the order of 50–70 nA/cm2 [131]. Breakdown in the dust layer depends upon both the resistivity and the particle size of the collected dust [132- 134].

PARTICLE RESISTIVITY. Electrical resistivity of collected particles influences the performance of an ESP. It determines the rate at which electrons flow through a dust layer. The optimum resistivity range is 105–1010 ohm-cm. When the resistivity is below 104, the force holding the particles onto the collector plates is reduced and the particles are easily reentrained. In an experiment, Spencer [135] found that at low dust resistivity levels, the corona wind alone was sufficient to reentrain the collected dust.

Particle resistivities greater than the optimum range result in the collected dust layer forming an insulating barrier which prevents the flow of current to the collector plates. The detrimental effects of high-resistivity ash include sparking at reduced operating voltages and the formation of back ionization, which can adversely affect the performance of an ESP [136].

Fly ash resistivity can be divided into three different regions. The regions are described as a function of temperature (figure 13). For temperatures above 300-400°F, the resistivity decreases as the temperature increases. This is known as the volume or bulk resistivity region and is characterized by ionic or electronic conduction. The ash resistivity is predominantly determined by the elemental makeup of the ash particles and is relatively independent of the flue gas characteristics [137-139].

For temperatures below 200-300°F, the resistivity decreases along with decreases in the temperature. This is known as the surface resistivity region. In this region the resistivity is mainly dependent on the interaction between the fly ash and components of the flue gas. Conduction in this region occurs either through adsorbed moisture and chemical films on the particles or is due to an ionic mechanism in which the alkali metal ions serve as the primary charge carriers [140]. Temperature plays an important role as it influences the relative concentration of water vapor and the reactivity between certain gas constituents and the surface of the fly ash. The transition from bulk- to surface-dominated resistivity occurs when the flue gas temperature approaches the water or acid dew point. Most fly ash precipitators operate in the transition region where both fly ash composition and gas stream chemistry are important.

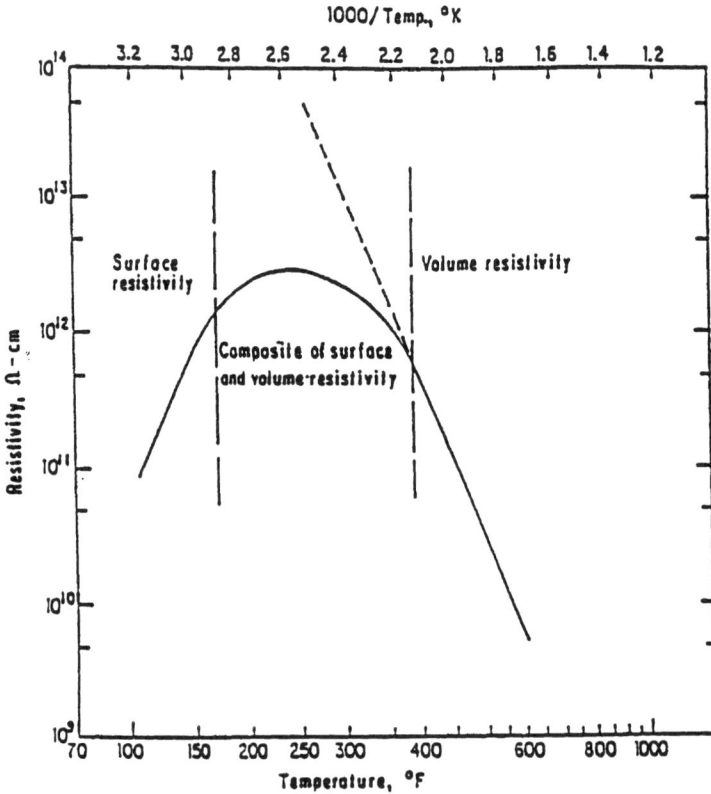

Figure 1.12 Typical temperature-resistivity relationship [126].

An important factor that influences the resistivity of fly ash is the amount of sulfur dioxide in the flue gas. A small proportion (2–3%) of the SO_2 in flue gas gets converted to SO_3. Sulfur trioxide is important because it reacts with water vapor to form sulfuric acid. Even trace quantities of sulfuric acid can have a tremendous effect on the condensation characteristics of flue gas. The water dew point for a flue gas stream with a moisture content of 10%, is approximately 120°F. However, if only 10 ppm of SO_3 is added to the flue gas stream, sulfuric acid will begin to condense at 270°F. This is important in the operation of an ESP because the reaction between the condensing sulfuric acid and the surface of the fly ash greatly reduces the resistivity of the fly ash. There are other chemical components in the ash that affect the resistivity. High concentrations of iron and sodium have been shown to provide acceptable resistivity levels even for low-sulfur coals [136,141].

MEASUREMENT OF RESISTIVITY. The resistivity of fly ash particles can be measured in situ in the field or in the laboratory under simulated conditions. Resistivity measurements are made in the field using a point-plane resistivity device [136]. The resistivity of particles can be obtained in the laboratory using a set-up and procedure defined by the 1981 IEEE Standard 548-1981 *Guidelines for the Laboratory Measurement and Reporting of Fly Ash Resistivity.*

Since SO_3 plays an important role in determining the resistivity of fly ash in many cold-side ESPs, it is necessary to incorporate SO_3 into the test environment when making resistivity measurements. Bickelhaupt [142] developed such a system by replacing most of the stainless steel components in the laboratory instrument with quartz and observed that if a known concentration of SO_3 can be maintained for a sufficient time to equilibrate the ash sample, then it was possible to simulate the effect of SO_3 on particle resistivity.

SPECIFIC COLLECTION AREA (SCA). If the migration velocity is fixed, then the collection efficiency is proportional to the treatment time. This is usually expressed in terms of the total plate area of the ESP divided by the total gas flow rate and is referred to as the specific collection area (SCA). SCA is usually expressed in terms of ft^2 of collector area per 1000 ACFM of gas flow. ESPs for coal-fired boilers usually fall in an SCA range of 150 to 600 ft^2/1000 ACFM. The units of SCA reduce to time/length such that when the migration velocity is multiplied by the SCA the product is a dimensionless parameter that is proportional to collection efficiency.

SPACE CHARGE EFFECTS. The charging of particles in an ESP results in establishing a particulate space charge in the interelectrode spaces. This occurs because the particles are much less mobile than the ions. The distribution of the particulate charge results in an electric field distribution which adds to the fields produced by the electrostatic field and the ionic space charge. The magnitude of the space charge effect is proportional to the number of fine particles in the interelectrode spaces.

The space charge alters the electric field so that the field gradient in the vicinity of the corona glow region is reduced, leading to a decrease in the corona current for a given applied voltage. This effect can be either useful or detrimental depending upon the magnitude of the space charge. For lower concentrations of fine particles, the space charge will often result in an increased operating voltage. The particle space charge has another desirable effect in that it increases the electric field gradient near the collector plate, which produces an increased electrical force for collecting the particles. However, at higher particle concentrations, the quenching or reduction of corona current can be severe, which reduces the efficiency of the ESP. Corona quenching occurs primarily in the inlet sections of an ESP operating on a flue gas with a high particulate loading or a high concentration of small particles.

MASS LOADING. The primary effect of increased mass loading at the inlet of an ESP is to increase the concentration of escaping particles. This is because an ESP works on an efficiency basis. Therefore, with all other factors held constant, anything that increases the inlet concentration will create a corresponding increase in the outlet loading. The mass loading also can influence the electrical characteristics in the inlet fields through space charge effects.

PARTICLE SIZE DISTRIBUTION. The collection efficiency of an ESP is directly proportional to the diameter of the particles. This is because the force driving the particle to the collector plate is proportional to the charge on the particle which is a function of the particle surface area. The drag on the particle is a function only of the diameter, while the charge increases as the square of the diameter, so that the velocity increases with diameter. Particle size also affects the space charge. For a constant mass loading, space charge will be greater for smaller particles.

NON-IDEAL EFFECTS. There are several "non-ideal effects" that can, at times, influence the performance of the ESP. These non-ideal effects include rapping emissions, gas distribution problems, sneakage, and reentrainment.

Operating Conditions

RAPPING. The process of removal of the collected particles from the collection plates and subsequent conveyance of the material from the hoppers is referred to as rapping and is usually accomplished by mechanically jarring the plate to dislodge the collected material. The material breaks away from the plate and falls into the hopper with a portion reentrained into the gas stream. The importance of rapping can be inferred from the fact that field studies indicate that as much as 80% of the emissions from full-scale ESPs occur as a result of rapping [143].

To reduce the rapping emissions it is important to optimize the intensity as well as frequency of the rap [144]. Reduced reentrainment occurs when the dust layer shears from the plate and slides as a sheet down the collector plate. The intensity must be set at the minimum force to cause the dust layer to shear without having it thrown into the gas stream. The frequency of rapping must be adjusted so that a sufficient dust layer thickness is achieved.

Measurements of particle size distribution reported by Spencer [145] showed that the mass median diameter of particles emitted during rapping increased with increased time between raps. Since the larger particles would be easier to collect, this fact alone would account for reduced reentrainment with less frequent rapping. However, with infrequent rapping, the dust layer can build up to a point that it either adversely affects the electrical characteristics or it falls by its own weight into the hopper. This latter case is undesirable because it usually leads to increased reentrainment.

GAS FLOW DISTRIBUTION. Sneakage and poor gas flow distribution adversely affect the collection efficiency. Sneakage occurs when a portion of the gas bypasses the electrified regions of the ESP by flowing through the hopper area and high voltage feed-through areas. This reduces the efficiency because particles in the gas bypassing the energized sections cannot be collected. Additionally, sneakage sometimes is responsible for reentraining material from the hopper. Sneakage is minimized by providing baffles to direct the gas flow away from the hoppers and the high voltage insulator areas.

The ESP inlet must be designed to provide a uniform gas flow. Because of the exponential nature of the collection mechanism in an ESP, uneven gas flow will lead to a reduced collection efficiency because the improved collection in low velocity areas is not sufficient to overcome the reduced collection in high velocity areas. In addition, poor distribution can increase reentrainment and gas sneakage [144]. Acceptable gas flow distribution is provided by proper design and placement of turning vanes and distribution plates at the inlet to the ESP.

REENTRAINMENT. Reentrainment of collected material often occurs from the collector plate where it is remixed with the gas. This can result from scouring by the mean gas flow when the forces holding the material onto the plate are reduced. The collected dust is held onto the plate by both electrical and mechanical forces. The electrical forces are due to the current flowing through the dust to the collector plates. The mechanical forces are due to surface effects on the particles. Since the electrical force is equal to the product of current density and particle resistivity, this force is greatly diminished at low-resistivity conditions. In fact, it is even possible for the electrical forces to reverse and repel the collected particles [135,144]. For these conditions, the mechanical forces must be increased. A greater particle cohesion may increase particle agglomeration, which would decrease reentrainment.

The precipitation process in an ESP is greatly affected by what is happening upstream regarding the quality and type of fuel, combustion process, and the preceding components including the FGD system. Any upstream factors that affect the moisture content of the flue gas, the sulfur content of the coal or the sulfur compounds in the gas, the chemical constituents of the ash, and/or the temperature of the gas will also affect the ESP performance. Mechanical changes to fuel, such as pulverization and other preparation, can often be correlated to ash parameters like resistivity and particle size distribution.

Difficulties in collecting high resistivity fly ash and fine particulates have had profound effects on suppliers and users of ESPs. In some cases, the prospect of difficulties in collecting fly ash has resulted in the specification of very narrow operating limits, or unacceptable increases in ESP power consumption. In extreme cases, this has turned users to fabric filters. However, ESP technology is being improved and upgraded to overcome prior technical limitations and maintain competitiveness with fabric filters.

References [125,126,146] upon which the majority of this section is based, provide significant information regarding ESPs. Particularly, the document by Radian Corp. [126], which is the report on a literature survey undertaken by Radian Corporation for U. S. Department of Energy on Fundamental Investigation of DUCT/ESP Phenomena, provides a lot of information regarding power plants equipped with FGD systems. It lists more than 200 papers in the ESP area: 16 on ESP technology, 44 on the effect of dry FGD processes on ESPs, 17 on ESP upgrades, 3 on wet ESPs, and 28 on modeling ESP performance.

APPLICATION OF ESP TO FGD PROCESSES. One of the areas where ESP development has accelerated recently is the adaption of the process to handling scrubber and non-scrubber SO_2 removal technologies: fluidized-bed boilers, spray-dryer FGD, limestone-injection multi-stage burners (LIMB), furnace sorbent injection, and in-duct sorbent injection. In some cases, especially for the retrofits, the desire is to prevent ESP-performance deterioration. In others, it is to allow the ESP to compete with fabric filters in these services.

ESP Design Considerations

Since dry scrubbing processes represent relatively new technologies, there is not a great deal of long-term data available to determine their impacts on operations and maintenance of ESPs. The full-scale spray dryer systems with ESPs represent about the only experience with continuous operation of an ESP at low approach to saturation temperatures. It is stated [126] that the only problems encountered at the GRDA facility have been corrosion on some of the high voltage insulators. The recommendation is to increase the temperature of the insulator housing. It should be noted that this facility was designed for a low-sulfur coal which at baseline conditions has a very low dew point and high-resistivity ash. Therefore, at the conditions for which the unit was designed, little or no heating of the insulators was necessary. On the other hand, an ESP operating on a high-sulfur coal flue gas will already have an insulator heating system capable of preventing condensation at a dew point of approximately 260°F. Retrofitting duct injection technology at such a site would have more than sufficient insulator heating to prevent condensation at a 125°F dew point.

Sorbents change the flue-gas stream in two primary ways: As a result of the absorption reactions upstream, the dust loading is appreciably higher, and must be reduced at comparable efficiency. There can be orders of magnitude increases or decreases in the particle resistivity. Some research with the spray dryer/ESP combination showed that particulates were easier to collect than the fly-ash-alone case. Particles were found to be larger and the temperature, solids composition, and added moisture combined to significantly lower the resistivity of the ash. This research could ultimately support installation of spray dryers upstream of existing ESPs, many with low SCA values, as an acid-rain-control retrofit strategy.

From the available data on dry sorbent injection programs, three detrimental effects on the performance of the ESP have been identified: increased mass loading, particle space charge quenching in the inlet section, and increased reentrainment due to low resistivity values. Upgrades must be directed specifically at these problems to be beneficial in the duct injection application. Many of the enhancements discussed in references [125,126,146] can be brought to bear on the desulfurization applications. These include increased collector area, wide plate spacing, pulsed energization, intermittent energization, cooled-electrode precharger, and additives for cohesion improvement. There are other operating enhancements being developed, including flue-gas-stream cooling and humidification. However, the overall performance and economics of ESPs compared to fabric filters loom as a large question mark that is still several years from resolution.

Finally, there are methods being developed to make the ESP serve as the desulfurization medium, similar to the fabric filter in sorbent-injection applications. One such process is called E-SOx. Here, the first electrical field in the ESP is removed and replaced with a bank of reagent-spray nozzles similar to those used in spray-dryers. Downstream sections are upgraded with prechargers and improved electrodes so that solids collection efficiency does not suffer.

Extrapolating the E-SOx process to its extreme brings up the wet ESP concept. This technique has been used successfully in Japan [125,126].

Fabric Filters (Bag House)

One of the oldest, simplest, and most efficient methods for removing solid particulate contaminants from gas streams is by filtration through fabric media. The fabric filter is capable of providing high collection efficiencies for particles as small as 0.5 µm and will remove a substantial quantity of particles as small as 0.01 µm [125]. In its simplest form the fabric filter consists of a woven or felted fabric through which dust-laden gases are forced. A combination of factors results in the collection of particles on the fabric fibers. When woven fabrics are used, a dust cake eventually forms; this, in turn, acts predominantly as a sieving mechanism. When felted fabrics are used, this dust cake is minimal or nonexistent. Instead, the primary filtering mechanisms are a combination of inertial forces, impingement, etc., as related to individual particle collection on single fibers. These are essentially the same mechanisms that are applied to particle collection in wet scrubbers, where the collection medium is in the form of liquid droplets rather than solid fibers. At present, fabric filters are contenders for particulate control along with ESPs and are firmly established in the United States for high efficiency particulate control on coal-fired utility power plants. Since 1973, the use of fabric filters has increased from zero to over 24,000-MW equivalent capacity [148].

There are three basic varieties of fabric filters: reverse-air, shake/deflate, and pulse-jet units. They are distinguished by the cleaning mechanism and by the air-to-cloth value (A/C). The A/C is a fundamental fabric-filter descriptor denoting the ratio of the flue-gas flow rate to the amount of fabric, or filtering surface area. Reverse-air units are characterized by low A/C values and by the flow-dirty gas flows from the inside of the bags to the outside. They make use of what is called low-energy cleaning. This means that the fabric filter air flow is periodically reversed, causing a significant bag motion that releases the collected dust.

A shake/deflate unit, which is also a low A/C-type unit, collects dust on the inside of the bags. To clean the bags, the top end is shaken by a drive linkage adjusted for the optimum frequency and amplitude of the shake. This is typically an off-line cleaning technique.

In pulse-jet units, gas flow is from the outside of the bag. A/C ratios are generally double those of reverse-air units, offering a more compact installation in most cases. Cleaning is provided by utilizing a high-pressure burst of air into the open end of the bag. A high-pressure line, maintained by a compressor, runs above the bag openings. At appropriate times, a valve in the line opens to direct a pulse to the open bag. Pulse-jet units can be cleaned on line or off line.

Large, multi-compartment reverse-air and shake/deflate units are designed with suitable valving to isolate one or more compartments while the other compartments service the full rated flow. In reverse-air units, additional duct work, fans, and valves are included to return a portion of the cleaned outlet gas to serve as the reverse air for a compartment taken off line for cleaning.

A substantial number of U. S. utilities [146] use reverse-air units mostly because they are deemed the economic choice for the huge sizes required for central stations (generally larger than 300 MW). However, recent evaluations indicate that shake/deflate units offer economic advantages over reverse-air units and operating experience with several units in utility service has been good.

FLOW PATTERNS. Flow of flue gas into and out of the fabric filter is an important design consideration. The design should minimize pressure drop, maintain appropriate temperature and velocity profiles, and evenly distribute the ash-laden flue gas to the individual compartments and bags. The importance of flow distribution increases as the size of the unit increases.

The velocity through the unit should be minimized without allowing ash to fall out. Inlet and outlet manifolds should be tapered to obtain constant velocity; turning vanes should be used in the outlet manifolds and hopper to ensure ash-flow balance among the compartments and prevent entrainment of ash from the hopper; and internal structural supports in the duct work should be avoided.

Tensioning of bags is another important design parameter, especially for reverse-air units. Bag-suspension mechanisms must be able to apply a predetermined force (or tension) to the bag so it will clean efficiently with a minimum amount of wear. In addition, they must compensate for operating conditions such as increased bag weight from collected ash and variable flows and temperatures brought about by changes upstream to the boiler load.

Design of the hopper and the fly ash evacuation system should ensure that collected dust is not reentrained into the bags and the ash level never gets too high. Mechanical dust collectors upstream of the fabric filter can drastically alter the size distribution and the loading of the ash.

The sulfur content of coal affects the fabric filter operation. It has been conclusively demonstrated that the cohesiveness of fly ash produced from high-sulfur coal-fired plants is greater than from western low-sulfur coal-fired plants [146]. Fabric filters are extremely sensitive to the condensation of acid and maintaining temperature in the unit above the acid dew point is critical in high-sulfur coal applications, especially when operating at reduced loads for significant periods of time. Some form of temperature control is essential. A corollary to this is minimizing any temperature drop in the fabric filter caused by air infiltration or poor insulation.

Characteristics of flue gas and fly ash affect other parameters. One of these is dustcake weight. Over time, bags accumulate a residual dust cake, ash that is never dislodged during the cleaning cycle. This is one reason why pressure drop tends to increase gradually over time and users should interpret pressure-drop guarantees carefully. Research shows that heavier cakes tend to accumulate in high-sulfur-coal units and those that undergo cycling and frequent shutdowns. The residual dustcake weights also affect the bag suspension and tensioning mechanisms.

ENHANCEMENTS. Application of sonic energy to improve bag cleaning and reduction of pressure drop is one of the most recognized and accepted enhancements to fabric filter technology. This technique got its start after early users looked for solutions to the gradual rise in pressure drop over several months of operation. For various reasons, bags became harder and harder to clean back to the original design pressure drop. Sonic horns were then applied and found to work well, reducing pressure drops anywhere between 20% and 50% [146]. Over the last five years, virtually all reverse-air fabric-filter specifications have included sonic horns.

The sound-wave action on the dustcake appears to create a shear force at some boundary layer within the cake. Depth of the boundary layer and the corresponding level of cleaning is, of course, dependent on the dust, flue-gas, and bag material characteristics. Frequency and power are the significant horn parameters affecting bag cleaning. For a given value of sonic pressure, low frequencies are the most effective in terms of dustcake removal and pressure-drop improvement. Sonic frequencies above 250-300 Hz generally are not effective [146].

BAG MATERIAL SELECTION. Operating costs of fabric filter installations consist of two main components: flange-to-flange pressure drop and bag life. These two components are related since proper bag material selection can often relieve pressure-drop problems. Thus, the bag itself can be considered the heart of the fabric filter. As such, pilot studies are often essential before one material can be confidently selected over another.

The selection of cleaning method depends on bag material. Reverse-air units have traditionally been accompanied by woven fiberglass bags with a Teflon B finish because they are an inexpensive choice for the low-energy cleaning application. Over 90% of the bags in U.S.-utility service are of the woven fiberglass variety [146]. Contrary to the conventional expectations, fabrics have been proven to be very sensitive to fly ash and flue gas characteristics. Filter fabric usually does not last the life of the unit. Therefore, the ultimate choice must include both relative first-time cost and the anticipated replacement cost.

The filter bag acts as a surface on which the dustcake is formed. Materials for bag construction are chosen for temperature and chemical resistance, mechanical stability, ability to collect the particulates as a cake, and ability to release the cake during the cleaning cycle. The bag performance is rated in terms of its permeability, cleanability, and durability. References [148,149] provide the fluid dynamic aspects of fabric filter design. Further details of fabric filters (baghouses) regarding performance and construction are available in selected references [147-150].

The impact of fabric filters on dry calcium sorbent-based FGD is of great interest to us. It has been shown that fabric filters can provide a key advantage in desulfurization [146]. The unused sorbent gets collected on the bag and absorbs more SO_2 while it is still on the bag. The sorbent can also be collected and later recycled mechanically. In spray-dryer FGD systems, SO_2 removal capability may increase by 25% above that accomplished in the spray dryer [146]. Even with the higher particulate loading, the fabric filter is little affected by having the spray dryer upstream, unless a process upset sends wet, sticky particles to the unit. Products of the SO_2 absorption reaction form a cementatious substance which when sticky can "blind" the bags. Collection of finer particles in the fiber could also cause pressure-drop problems. It is imperative to maintain temperatures above the water and acid dew points, especially during startups and shutdowns. Maintaining the dust layer on the bags by carefully controlling the cleaning cycle is another objective during operation.

One potential problem area is corrosion of the clean side of the baghouse compartments. Some installations have had to implement new design concepts to protect against corrosion. Some of these include flue gas reheat between the spray dryer and the fabric filter, wall-temperature monitoring, increasing insulation, reducing of air in-leakage, thermal gaskets on hopper flanges, eliminating air cavities in casings by using wall stiffeners, and isolating of individual casings with guillotine dampers.

In many sorbent-injection processes, the fabric filter is intended to serve as the primary desulfurizing medium. Extensive tests with injecting sodium compounds has shown that sorbent injection has little impact on the operation of the fabric filter. A change to the cleaning cycle may be required to compensate for variations in pressure drop or the increased particulate loading. One area that deserves further research for desulfurization applications is choosing the fabric finish that assists in providing the most efficient SO_2 removal in terms of the dustcake build up and cake-release properties.

A comparison of the performance of fabric filters (baghouse) and ESPs for calcium-based dry sorbent FGD will be made later in this section.

Wet Scrubber

Wet scrubbers are used on coal-fired boilers because of their inherent ability to effectively remove both particulate and gaseous (i.e., sulfur dioxide and nitrogen oxides) pollutants. The primary particulate collection mechanism involved in the conventional wet scrubbing operation may include some or all of the following: inertial impaction, direct interception, diffusion (Brownian movement), and condensation.

Inertial impaction occurs when the droplet placed in the path of a particle-laden gas stream causes the gas to diverge and flow around it. Larger particles, however, tend to continue in a straight path because of their inertia. They may impinge on the obstacle and be collected. Since the trajectories of particle centers can be calculated, it is possible to determine theoretically the probability of collision.

Deposition by direct interception occurs when the particles, moving along streamlines of fluid, approach the droplet within a distance equal to the radius of the particle. As previously stated, the trajectory of particle centers can be calculated; however, even though the center may bypass the target object, a collision might occur, since the particle has finite size. A collision occurs due to direct interception if the dust particle's center misses the target object by some dimension less than the particle's radius.

The diffusion mechanism results from the fact that very small (submicron) particles suspended in a gas stream have an individual oscillatory motion, known as Brownian movement. In this case, particle and target collide as a result of relative motion within a limited space. In all diffusional processes, the rate of transfer is proportional to the surface area available for diffusion. Thus, small liquid droplets with high surface-to-volume ratios lead to high collection efficiencies.

The effects of condensation may also come into play. Condensation occurs if the gas is rapidly cooled below its dew point. When moisture is condensed out of the gas stream, fogging occurs, and the dust particles can serve as condensation nuclei. The dust particles can become larger as a result of the condensed liquid, and the probability of removal by impaction and diffusiophoresis is increased.

Four types of wet scrubbers are used for particulate control in the electric utility industry. These are moving bed scrubbers, flooded fixed-bed scrubbers, preformed spray scrubbers, and venturi scrubbers. Descriptions of these devices are provided in the ANL Report [125].

Wet scrubbers are used in wet FGD processes; e.g., in the conventional wet limestone scrubbing process [151]. In this process, limestone slurry contacts the SO_2-laden flue gas in a countercurrent spray tower absorber. The waste slurry is oxidized in separate tanks to convert the sulfite to sulfate, and the waste slurry is pumped to a wet stacking operation for ultimate waste disposal. The wet limestone scrubbing system is installed downstream of an ESP which removes sufficient entrained particulate matter to meet new source performance standards requirements.

Mechanical Particulate Control Devices

A combination of centrifugal, inertial, and gravitational forces are used in mechanical particulate control devices to remove fly ash from the flue gas stream. Although different kinds of these devices are available, only the multiple cyclone is of interest.

The multiple cyclone consists of many long, small-diameter cyclones operating in parallel and having a common gas inlet and outlet. By increasing body length and decreasing the diameter on a cyclone, one can gain two advantages: greater centrifugal force is exerted on the particulates and the dust particles are retained longer. Both of these help to improve collection efficiency. Operation, however, is more expensive, because pressure drop, and therefore fan power requirements, increase as the diameter decreases.

The diameter of cyclones used in mechanical particulate control devices ranges from about 6-24 inches. Sizes as small as 2 inches have been tried, but plugging problems make these sizes impractical [146].

Manufacturers of these devices claim efficiencies of up to 95-96% for their mechanical collection systems (usually two multiple cyclones arranged in a series), such high efficiencies are generally achieved only under ideal circumstances. Changes in gas volume, temperature, and density caused by load swings and other operating conditions; particulate loading and dust characteristics caused by variations in fuel characteristics (sizing, ash content, etc.); and other variables influencing dust collection, can adversely affect performance.

The best outlet dust loading one can expect under operating conditions typical of most process plants is approximately 0.3 lb/10^6 Btu heat input. One reason for this is that cyclones are ineffective in collecting particles below 5 microns in size. This poses another regulatory problem—that of meeting opacity standards.

Mechanical collectors, in spite of these shortcomings, have a future in industrial and institutional steam plants. For example, in stoker-fired power plants they can be used to protect the principal collector against possible fire damage, particularly if it is a fabric filter, by eliminating glowing char upstream. Or they can be used as a scalping device ahead of wet scrubbers or precipitators, to remove large particles inexpensively.

Mechanical collection devices offer several advantages over some of the alternative dust collection systems for industrial boilers. The most important advantage is the simplicity of design, which makes cyclones inherently reliable. Erosion of internals, which had been a problem in many old units, has been slowed by providing a more even distribution of flue gas flow throughout the collector and by using white cast iron with a minimum Brinell hardness of 450 or NiHard for critical parts [146]. These materials last about four years in abrasive service. Fabricated steel internals are suitable for nonabrasive applications such as in the collection of fly ash from oil-fired boilers. Other advantages include compactness and the need for minimal maintenance.

Manufacturers have refined their equipment designs to reduce power requirements while maximizing collection efficiency over a wider load range and reducing reentrainment of fine particulates. Use of so-called recovery vanes at the entrance to the clean-gas outlet tube, for example, helps eliminate the cyclonic flow pattern after it separates fly ash from the flue gas, thereby reducing the overall pressure drop through the collector.

One European design has an adjustable vane right at the inlet to the primary collection tube, which is moved inward as gas flow decreases, outward as it increases on load change. (Adjustments can be either automatic, using a differential-pressure controller, or manual.) A benefit of the adjustable vane is that the pressure drop is held constant over the load range. This means that as boiler output and gas flow drop, a greater centrifugal force is exerted on the dust particles, improving collection efficiency at partial loads. In fact, when the system is operated properly, its efficiency is highest at about 60% of the full-load rating, the level at which many industrial boilers operate. Performance is about the same at 30% load as it is at 100%.

A secondary fan downstream of a small cyclone pulls out the fly ash continuously from the hopper. Clean gas from the secondary collector is returned to the flue gas inlet of the primary unit, while ash is discharged to a silo. Use of the secondary collector ensures that ash is always pulled from each of the primary collector's cyclones, minimizing its reentrainment.

Other manufacturers have applied a variation of this design to improve the efficiency of their systems and to reduce opacity. What they do is draw off some 10-20% of the total flue gas flow through the dust hopper, to capture light ash particles, which tend to float upward and escape through the primary collector outlet tube. (Fly ash collected at the bottom of the hopper is removed separately.) This slipstream is pulled through a high-efficiency scrubber or fabric filter, where 98-99% of the fines it contains are removed. The only regular attention normally required by cyclones is appropriate discharge of fly ash into collection hoppers.

Comparison of SO_2 Removal in Fabric Filter (Baghouse) and ESPs

Of the three conventional devices—ESPs, fabric filters (baghouses) and wet scrubbers—we will consider ESP and fabric filters only, since wet scrubbers are generally associated with wet FGD processes and the emphasis here is placed on dry sorbent processes which also include in-duct processes.

In low temperature dry scrubbing processes, most of the sulfur dioxide removal occurs while the sorbent is in slurry form. A significant amount of removal of SO_2 also occurs in the particulate control device. The removal of SO_2 depends upon the type and operating conditions of the particulate control device. For a commercial FGD spray drying process, the SO_2 removed in the particulate control device can be as high as 20% of SO_2 removal for the system [127,146]. The SO_2 removal in the particulate control device is also very important for FGD systems involving duct injection. However, there are only a few studies that separate SO_2 removal in the duct and in the device.

The removal of SO_2 in the particulate control device is a function of many variables. One is the approach to the saturation temperature of the flue gas because the reactivity of sorbent with SO_2 increases with decreasing approach temperatures; however, the effect on SO_2 removal depends upon the type of control device [147,153]. Figure 1.13 shows that SO_2 removal is a stronger function of the approach temperature for a fabric filter (baghouse) than for an ESP, and that level of SO_2 removal is greater in a baghouse than in an ESP. The reagent ratio also affects the SO_2 removal [146,152]. Figure 1.15 shows that reagent ratio has a greater impact for the baghouse and that the level of SO_2 removal is again greater for the baghouse than the ESP. This phenomena is explained by Damle [104] for baghouses and ESPs. The solids on the bag filter are in intimate contact with the flue gas, so the concentration of SO_2 at the particle surface is equal to the bulk gas concentration. Therefore the gas phase mass transfer resistance to SO_2 is not significant. SO_2 removal in the bag filter should depend only on the unreacted sorbent concentration, air-to-cloth ratio, filter cake thickness and the reaction rate constant for the sorbent and SO_2. The SO_2 removal in the baghouse is very sensitive to the approach to the saturation temperature because the reaction rate constant increases exponentially with the relative humidity in the flue gas [99].

For ESPs, the situation is different, as the sorbent is concentrated on the ESP plates and does not come into good contact with the flue gas. The bulk mass transfer for SO_2 is therefore a significant resistance to SO_2 removal in an ESP. Damle [104] states that SO_2 removal in an ESP depends upon the mass-transfer coefficient of SO_2 in the flue gas, the specific collection area (SCA) of the ESP, the concentration of unreacted sorbent, and the thickness of dustcake. A simple mathematical analysis based upon diffusion of SO_2 through a stagnant film was developed by Yoon and Stouffer [153]. The analysis estimated that maximum SO_2 removal in the ESP was about 20%. The SO_2 removal in an ESP does not strongly depend upon the approach to saturation temperature because the gas-phase mass transfer coefficient is not a strong function of this parameter. The removal in an ESP increases with increasing reagent ratio because the reagent ratio affects the concentration of unreacted sorbent and thickness of the dustcake.

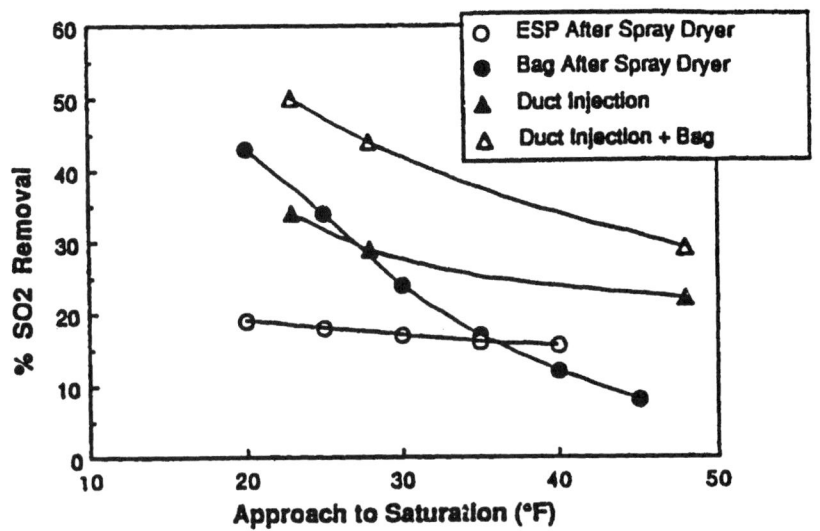

Figure 1.13 Comparison of ESP with bag filter for SO_2 removal [126].

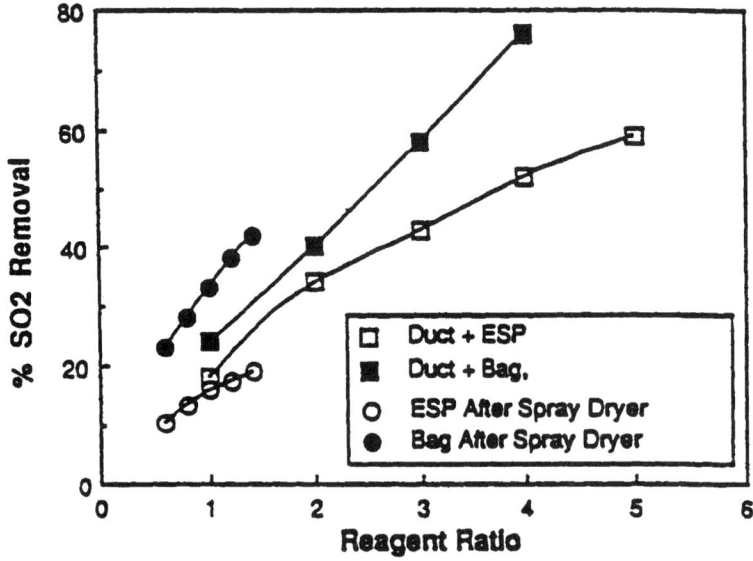

Figure 1.14 Comparison of ESP with bag filter for SO_2 removal [126].

These discussions can lead one to conclude that bag filters are preferred over ESPs for the dry scrubbing process. However, one has to be careful since the minimum operating approach to saturation temperature is lower for an ESP than for a bag filter. The bag filter can clog up (i.e., get blinded) if the collected solids become sticky (if the humidity is too high). Therefore, a duct injection system that utilizes an ESP can operate at a lower approach to saturation temperature than a system using bag filters. Thus, it may result in a higher SO_2 removal in the duct work upstream of ESP, and this increased removal may offset the lower removal in the particulate control device.

Scope

The survey of the existing particulate control systems for coal-fired boilers indicates that only two devices, a fabric filter (baghouse) and an electrostatic precipitator (ESP), are used in conjunction with dry sorbent-based FGD systems. Both systems, as discussed, have certain advantages and disadvantages. Both ESPs and fabric filters are equally sensitive to flue gas and fly ash characteristics, although in different ways. Addition of sorbent in FGD systems also results in changing particulate characteristics. Products of SO_2 absorption reactions, especially calcium sulfate, become cementatious when combined with moisture.

This makes blinding of fabric bags more of a concern. Finer particles tend to work their way into interstices of the fabric causing pressure drop problems. It is imperative to maintain temperatures above the water and acid dew points to ensure that no wet sticky particles blind the bag and acid and attack the bag material. On the other hand, unused reagent collected on the bag enhances sorbent utilization and improves SO_2 removal. Difficulties in collecting high-resistivity fly ash and fine particulates have had profound effects on ESP performance and have resulted in the specification of very large units or unacceptable increases in power consumption. ESP manufacturers have improved ESP technology and enhanced ESP capabilities. Some of the enhancements in ESPs are pushing ESPs in a direction that might make them cost competitive with fabric filters for FGD systems. There are methods being developed (E-SOx) [126,146] to make the ESP serve as a desulfurizing medium. Neither particle removal technique is very conducive to easy sorbent recycling. In both cases, the partially spent sorbent has to be removed from the hopper and then fed back into the FGD system. There might be a need to crush the sorbent into fine particles. One desirable aspect for a new device would be the capability to recycle the partially spent sorbent without the need to remove the spent sorbent from a hopper and recrushing it into finer particles. The new device, of course, has to be free from the types of problems associated with ESPs and fabric filters.

Sorbent Forms and Additives

Calcium-Based Sorbents

For each type of dry scrubbing described in this chapter, increased calcium utilization would lead to reduced raw sorbent requirements and to reduced mass loadings on the particulate-removing equipment. Even more significantly, increases in calcium utilization would extend the practical range of application of dry scrubbing processes to higher sulfur coals. Calcium utilization can be increased by recycling unreacted sorbent and/or by increasing the sorbent conversion per individual process pass.

This section of our critical literature review deals with methods currently under investigation for improving the reaction rate/calcium utilization of the sorbent through alterations in the composition and/or particle morphology of the sorbent itself. This snapshot review of the current state of the art suggests research areas in which meaningful contributions can be made.

The study of calcium-based sorbents for sulfur dioxide capture is complicated by two factors: little is known about the chemical mechanisms by which the "standard" sorbent preparation and enhancement techniques work. A sorbent preparation technique that produces a calcium-based sorbent that enjoys enhanced calcium utilization in one temperature regime may not produce a sorbent that enjoys enhanced calcium utilization in the other reaction zones. Again, an in-depth understanding of the mechanism of sorbent enhancement is necessary if a systematic approach to sorbent development is to be used.

The sorbent modification studies described below relate to both SO_2 capture in the 1600-2400°F range and the <350°F range in a roughly 80:20 ratio. For the purposes of this section, studies performed in the temperature range 1600-2400°F will be called high-temperature studies and studies performed in the temperature range <350°F will be called low-temperature studies. No additive studies are offered to describe the application of promoted sorbents to the problem of SO_2 capture in the 800-1100°F reaction range.

Chemical Form

HIGH TEMPERATURE STUDIES. Cole et al. [154], using an isothermal reactor at 2000°F, Ca/S ratio of 2, and 2000 ppm SO_2, showed that calcitic hydrates were more reactive than calcitic carbonates. They attributed this observation to a combination of three reasons:

1. Dehydration occurs much more rapidly than CO_2 evolution; therefore, the time available for sulfation after the calcination process is significantly greater for hydrates than for carbonates.
2. Hydrates begin with a much higher initial surface area.
3. The dehydration process generally produces a reduction in the mean particle size of the sorbent, thereby reducing the internal diffusion resistance to SO_2 diffusion.

Gullett and Bruce [155] have found that the reactivity of $CaCO_3$ and $Ca(OH)_2$ is dependent on particle size, sorbent type, surface area, and pore volume and shape. They state that $Ca(OH)_2$-derived sorbents react to a greater extent than do $CaCO_3$-derived sorbents of the same particle size due in part to differences in grain shapes and to the fact that the flat plate structure of the activated $Ca(OH)_2$ grains allows greater product layer diffusion than the spherical grains found in activated $CaCO_3$. They also suggest that ultimate sorbent reactivity is limited by the degree of developed porosity.

Calcium/Magnesium Ratio

HIGH TEMPERATURE STUDIES. Snow et al. [156], injecting sorbents into a flue gas stream produced by the combustion of Pittsburgh No. 8 coal (2210°F injection temperature, 468°F/s quench rate, 1.3 s average residence time at reaction temperature, a Ca/S mole ratio of 2, 1900 ppm SO_2), found that sorbent SO_2

capture reactivity varied in the order of dolomitic hydrates highest, then calcitic hydrates, then carbonates (based on equal molar Ca/S ratios). However, for a given sorbent type there was no clear correlation between physical properties (BET surface area, mass mean particle size, elemental stoichiometry) and SO_2 capture performance. A summary of their data is given in Table 3.

Table 1.3. Effect of Sorbent Properties on High Temperature SO_2 Capture Performance [156].

Sorbent	Mass Mean Particle Dia (um)	BET Surface Area (m^2/g)	Percent by Weight		SO_2 Removal (%)
			Ca (%)	MG (%)	
Vicron Limetone	11.0	1.01	39.01	0.49	38
Mercer CAH	7.10	17.8	50.5	0.04	59
Kemikal CAH	3.88	18.0	49.0	1.0	62
Marblehead CAH	7.79	14.3	50.3	0.49	69
Detroit Lime CAH	8.33	14.9	50.2	0.51	60
Black River CAH	5.53	13.3	49.7	1.6	61
Kemidol CAH	19.71	20.6	29.9	18.5	73
Ivory Finish DPH	13.72	18.4	28.6	18.1	75

CAH = Calcitic Atmospheric Hydrate
DAH = Dolomitic Atmospheric Hydrate
DPH = Dolomitic Pressure Hydrate
Percent SO_2 removal at Ca/S=2; 2210°F injection temperature.

Teixeira et al. [157], testing sorbents for SO_2 removal from flue gases generated by burning low-sulfur Western coal (2000°F injection temperature; 0.33 s average residence time; 500 ppm SO_2), also found that SO_2 capture reactivity varied in the order dolomitic hydrates highest, then calcitic hydrates, then carbonates.

This observation was further confirmed by Beittel et al. [6], Overmoe et al. [158], and Bortz and Flament [159].

Gooch et al. [160] have observed that the increased utilization of pressure-hydrated dolomitic lime is often only sufficient to compensate for the decreased calcium content. This balance results in no net reduction of the mass of sorbent required to accomplish a given SO_2 reduction.

Morphology

HIGH TEMPERATURE STUDIES. Gooch et al. [160] tested 11 commercial calcitic hydrates under furnace injection conditions. No obvious correlation of properties of the raw sorbent with SO_2 removal was identified. However, when testing four specially prepared sorbents with widely varying BET surface areas, particle sizes, and pore structures, it was found that a two-fold increase in raw sorbent surface area resulted in a 20% relative increase in calcium utilization. The increased surface area also served to decrease the dependence of sulfur dioxide capture efficiency on injection temperature.

McCarthy et al. [161] also agree that reactivity in furnace injection does not correlate with hydrate surface area. They did show, however, that SO_2 capture reactivity did correlate with the surface area of the calcined hydrate, with an increase in reactivity being associated with a higher calcined surface area (Figure 1.15).

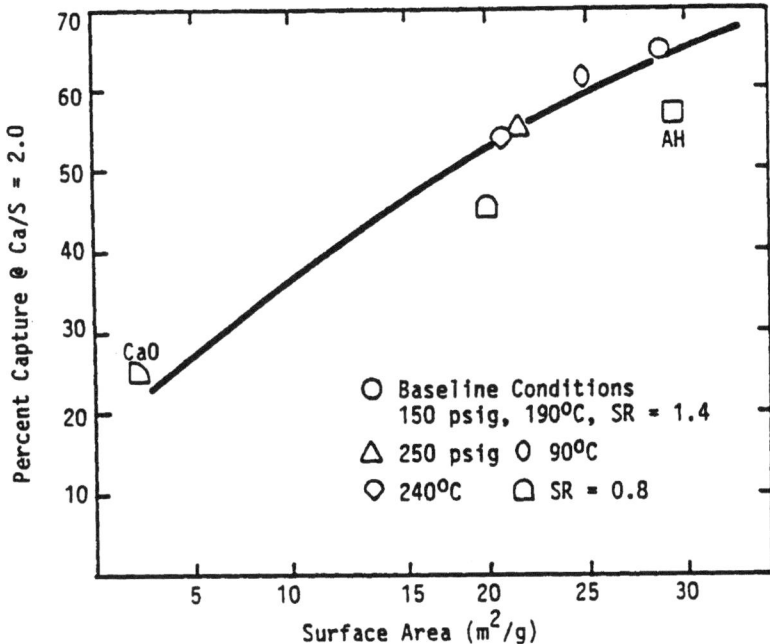

Figure 1.15 SO$_2$ capture reactivity versus the surface area of the calcined hydrate [161].

LOW TEMPERATURE STUDIES. Yoon et al. [17] have prepared high BET surface area calcitic hydrates using atmospheric hydration under N$_2$ followed by filtration and vacuum drying. Using these hydrates Yoon was able to show a strong relationship between sorbent utilization in dry/dry capture (150°F, 60% relative humidity, 20°F approach to saturation, 1000 ppm SO$_2$) and raw sorbent BET surface area (Figure 1.16). Using hydrates whose BET surface areas varied between 10 and 50 m^2/g, he was able to effect calcium utilizations from between 12% and 45%.

Borgwardt and Bruce [162] confirmed the work of Yoon et al. [17] by preparing a series of high surface area hydrates and showing the strong effect of hydrate BET surface area on dry/dry humidified SO$_2$ capture (Figure 1.16).

Dry Scrubbing Technologies for Flue Gas Desulfurization

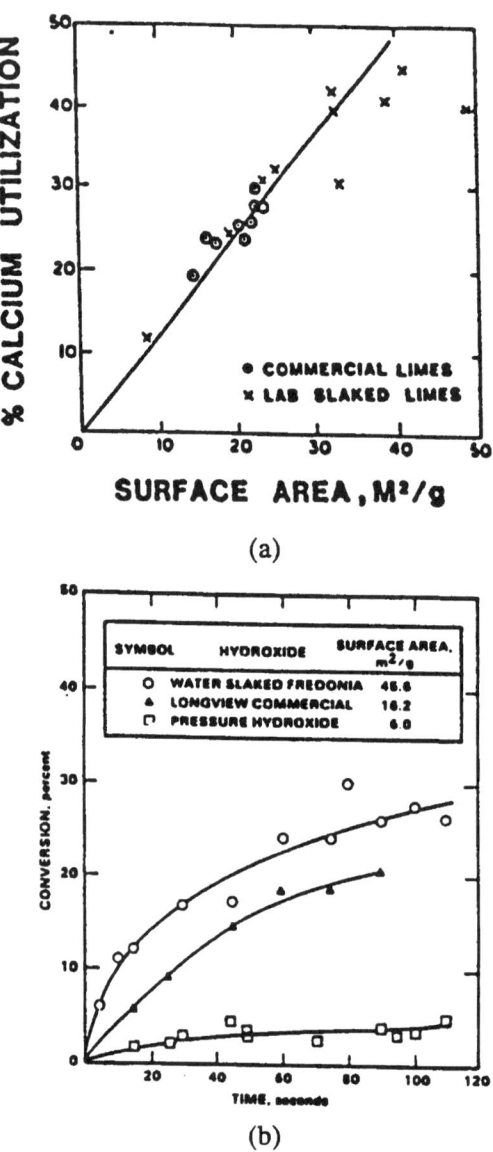

Figure 1.16 Effect of sorbent surface area on low temperature SO_2 capture performance: (a) 150°F, 60% relative humidity, 1000 ppm SO_2, 60 minutes, Yoon et al. [17]. (b) 150°F, 20 vol % H_2O, 1500 ppm SO_2, Borgwardt and Bruce [162].

Preparation by Aqueous Hydration

HIGH TEMPERATURE STUDIES. Kirchgessner et al. [163], using calcines and hydrates prepared from Eldorado limestone, found a slight decrease in the SO_2 capture reactivity of the product hydrate with increasing particle size of the parent calcine. As the mean calcine particle size increased from 0.5 mm to 9.5 mm the percent Ca utilization of the product hydrate dropped from 14.5% to 13.0%. The observed effect was attributed to the larger particle surface area present for the smaller calcine.

McCarthy et al. [161] reported that pressure hydrates generated under well-controlled conditions are more reactive than commercially produced atmospheric hydrates. They reported that important hydration process variables include size and composition of the parent quicklime, the hydration temperature and pressure, the rate of water addition to the hydration reactor, and the pressure progression during the discharge of the hydrate from the hydration reactor. However, Gooch et al. [160] found that pressure-hydrated calcitic lime produced and evaluated in laboratory-scale apparatus could not be distinguished from its companion atmospheric hydrate on the basis of either particle morphology, chemical composition, or SO_2 capture ability.

LOW TEMPERATURE STUDIES. Borgwardt and Bruce [162] prepared calcitic hydrates using steam hydration. The hydrate sorbents prepared in this manner show inferior BET surface areas (9-11 m^2/g) when compared to hydrates produced through the use of a liquid water phase. The SO_2 capture performance of the steam hydrated samples was substantially less than that of liquid water hydrated samples. Borgwardt suggested that this observation might explain the sometimes contradictory results obtained during pressure hydration. The performance of the pressure hydration process might in part be controlled by a delicate balance between the positive effects of decompressive drying and the negative effects of steam exposure.

The Use of Additives to Enhance Reactivity

The use of additives to enhance the $Ca(OH)_2$ reactivity with SO_2 have been explored by a number of investigations. The type of additives investigated to date may be classified as deliquescent inorganic salts and related inorganic salts, sodium-containing basic compounds, organic compounds, or oxidation catalysts [124].

The majority of the compounds tested fall into the first category above, although there is some cross classification among the compounds evaluated to date. For example, NaOH is capable of reacting directly with SO_2 as a basic solution but it is also a deliquescent compound.

Deliquescents, Buffers, and Sodium Additives

HIGH TEMPERATURE STUDIES. Sodium compounds have also been tried, with varying success, in the in-furnace (1600-2400°F) regime. Generally these additives are added to the water phase during the hydration process. Teixeira et al. [157] found that sodium carbonate, sodium bicarbonate, and sodium hydroxide all enhance the sulfur-removing capability of dolomitic hydrates when removing sulfur from SO_2-doped natural gas flue gases. However, the benefit of the sodium compounds disappeared when western coal-derived fly ash was present in the flue gas stream. No ash analyses were offered in their paper.

Weber et al. [164] also remarked on the capability of NaOH to enhance the behavior of a pressure-hydrated calcitic hydrate for in-furnace SO_2 capture when using doped natural gas flue gas. However, they also saw this advantage disappear when they used a flue gas generated by burning a Beulah lignite. The high calcium and sodium content of the ash from lignite (15.5wt% CaO, 4.0 wt% Na_2O) probably contributed significantly to this effect.

Snow et al. [156] investigated $NaHCO_3$ as an additive for capture of SO_2 at 2210°F from a flue gas generated by burning Pittsburgh coal. At a Ca/S ratio of 2, the addition of the $NaHCO_3$ additive increased the capture efficiency of a calcitic atmospheric hydrate from 63-69% to 83% and for atmospheric and pressure dolomitic hydrates from 73-75% to 88%. Ashes from eastern bituminous coals are not expected to contain significant amounts of calcium or sodium.

Muzio et al. [165] have investigated the effects of nine additives added to the hydration water on SO_2 capture. The nine additives studied include NaOH, NaCl, Na_2CO_3, Li_2SO_4, $LiNO_3$, K_2CO_3, Cs_2SO_4, $Fe(NO_3)_3$, and $FeCl_3$. For experiments at 2100°F, a Ca/ ratio of 2, and a metal promoter/CaO weight ratio of 0.03 the use of the promoters resulted in the following increases in SO_2 capture when compared to a nonpromoted, base hydrate: The highest was Cs(33%), followed by K(29%), Na(20%), and Li(0%) and Fe(0%). In all cases where improvement was observed, the improvement was greater than that which would have been predicted if all of the promoting material was transformed to its sulfate form. Muzio also showed that the form of sodium added had essentially no effect on the sulfur capture performance of the promoted sorbent. This was particularly interesting in light of the wide variation in BET surface areas of the raw hydrated sorbents (Na_2CO_3, 16 m^2/g; NaOH, 7.8 m^2/g; NaCl, 4.5 m^2/g). Muzio et al. [165] speculated that for the promoted hydrates, alkali crystals may block the pores at room temperature but, as a result of melting and vaporization in the combustion zone, the pore structure may reopen at reaction conditions. Physical mixtures of CaO and Na_2CO_3 were found to be less effective than addition of the same amount of Na_2CO_3 to the hydration water for the same CaO sorbent (38% SO_2 capture versus 45% SO_2 capture). The presence of fly ash in the flue gas served to totally eliminate any promotion effect of the sodium metal additives.

LOW TEMPERATURE STUDIES. Ruiz-Alsop and Rochelle [166] tested the effectiveness of 18 additives (two buffers, three organic deliquescents, and thirteen inorganic deliquescents) towards improving calcium utilization in dry/dry SO_2 capture. Their experiments were performed in a sand bed reactor at 54-74% relative humidity. Reaction conditions were set to simulate the conditions found in bag filters during flue gas spray drying. It has been postulated that deliquescent salts should increase the efficiency of sulfur capture in dry/dry systems by enhancing the moisture content in the sorbent solids. This study found that the two buffers and the three organic deliquescents caused a degradation in SO_2 capture efficiency while the inorganic deliquescents caused a increase in the SO_2 capture efficiency, in some cases almost doubling the efficiency (Table 4 on page 80). The most effective inorganic deliquescents were LiCl, KCl, NaBr, and Na_2NO_3. Ruiz-Alsop and Rochelle [166] found a poor correlation between the relative deliquescence of the inorganic salts and their ability to enhance SO_2 capture. They speculated that some of the salts also acted to favorably modify the sulfite reaction product layer formed on the surface of the sorbent particles.

Table 1.4. Effect of Additives on Lime Reactivity [166]

Additive	Lime Conversion at 60 minutes	
	74% RH 64.4°C	54% RH 66°C
No Additive	22.4	11.8
Buffers		
5 wt% Glycolic Acid	11.3	----
1 wt% Adipic Acid	20.3	----
Organic Deliquescents		
5 wt% Monoethanolamine	19.6	----
5 wt% Ethylene Glycol	20.3	----
5 wt% TEG	20.5	----
Inorganic Deliquescents		
5 mole% Na_2SO_4	28.3	----
5 mole% Na_2SO_3	29.8	16.1
5 mole% $CaCl_2$(*)	34.6	16.4
10 mole% NaCl	38.5	27.0
10 mole% NaOH	38.8	17.3
5 mole% $Ca(NO_3)_2$(*)	39.4	12.3
10 mole% $NaNO_2$	40.0	-----
10 mole% $NaNO_3$	41.5	27.2
$BaCl_2*2H_2O$	---	19.4
$Na_2S_2O_3$	----	21.6
KCl	----	37.3
$NaBr*2H_2O$	----	42.0
LiCl	----	43.9
100% SO_2 Removal	48.2	48.2

(*) Solid phase are $CaCl_2*Ca(OH)_2*H_2O$ and $CaN_2O_7*2H_2O$ respectively.

Huang et al. [167] have reported that the addition of NaOH in the lime slurry used for an industrial-sized spray dryer caused the partial oxidation of NO to

NO_2. This is a significant development in that NO_2 is more reactive with a variety of gaseous compounds and could conceivably be removed as a particle [168].

Yoon et al. [17] used additives to promote the activity of samples of hydrated lime in conjunction with tests on the Coolside process (a process operating in the <350°F regime). The action of both "cosorptive" and "non-cosorptive" additives were evaluated. Examples of cosorptive additives include NaOH, Na_2CO_3, and possibly NaCl, Na_2SO_3, and Na_2NO_3. Examples of non-cosorptive additives include $CaCl_2$, KCl, $FeCl_3$, and $MgCl_2$. The additives were either added to the hydrated lime in an aqueous solution or were added to the lime during the hydration process. Both sets of compounds were found to be highly effective in enhancing the capture behavior of the hydrated lime, even when the cosorptive properties of the sodium salts were subtracted out. Treating -325 mesh hydrated lime particles with NaCl using a promotion mole ratios of 0.05 to 0.2 Na+/Ca++ in a laboratory reactor (operated in the dry/dry sorption mode) resulted in relative calcium utilization increases of 80-114% over those achieved using unpromoted hydrated lime samples. It was speculated that the additives might act to enhance SO_2 capture in any one of three ways: changing the sorbent particle's physical properties (particularly the surface area of the hydrate), enhancing the basicity of the sorbent, and increasing or retaining moisture at the sorbent surface. No evidence was offered to support any of these three proposed mechanisms.

Organic acids and buffers have been studied as wet-FGD additives and have been found to be effective in increasing the overall rate of SO_2 capture in CaO or $CaCO_3$ slurries. This wet FGD slurry work should shed some light on behavior that might be expected for wet/dry SO_2 capture systems.

Jarvis et al. [169] evaluated the performance of several organic acid additives (adipic acid, maleic acid, formic acid, glutamic acid, succinic acid) in a bench-scale wet FGD system. The use of these additives increased the removal of SO_2 by 15%. Other investigators have confirmed the effectiveness of organic acid and buffer additives for wet FGD systems. These include Chang and Brna [170] (adipic acid, citric acid, sodium formate), Wang and Burbank [171] (adipic acid), Rochelle et al. [172] (sulfopropionic acid, sulfosuccinic acid, acetic acid, adipic acid, hydroxypropionic acid, aluminum sulfate). Works by Chan and Rochelle [173] and Rochelle and King [174] have provided models for the mass transfer enhancement actions of organic acids, alkali additives, and buffers in wet FGD systems.

Alcohol and Sucrose Hydration

HIGH TEMPERATURE APPLICATIONS. Gooch et al. [160] evaluated alcohol (methanol, ethanol, isopropanol), acetone, and sucrose hydration techniques. It was observed that while the alcohols are all removed from the final product by evaporation (and therefore can be recovered and reused) the sucrose remains in the final product. For hydration tests conducted at 60-70°C, an atmospheric hydrate with a BET surface area of 22 m^2/g was produced starting with a parent CaO with a surface area of 2.2 m^2/g. Hydration with aqueous acetone and methanol solutions under "optimum" conditions produced hydrates with surface areas of 50 m^2/g and 80 m^2/g, respectively. Hydrates produced using an aqueous mixture of sucrose and methanol had a BET surface area of 85 m^2/g.

Gooch et al. [160] presented the speculation that alcohols act to improve sorbent surface area by abstracting the heat of hydration from the sorbent surfaces by evaporation and that sucrose acts by increasing the solubility of CaO in water (0.1 wt% in water and 9.8 wt% in 35% sucrose/65% water at 25°C). No experimental evidence was reported to support either of these proposed mechanisms.

Silicate Additives

LOW TEMPERATURE STUDIES. The use of product recycle (with included fly ash) has been shown to improve spray dryer performance in pilot-plant tests [107].

Also, bench-scale studies by Jozewicz and Rochelle [175], using a packed-bed reactor, have shown substantial improvements in SO_2 uptake for $Ca(OH)_2$ slurried with several different fly ashes. They speculate that the fly ash reacts with $Ca(OH)_2$ to produce calcium silicates with more reactive surface area than the original $Ca(OH)_2$. Reagent-grade Al_2O_3, Fe_2O_3, and H_2SiO_3 were also found to enhance calcium utilization for dry/dry SO_2 scrubbing systems.

A potential problem in the use of fly ash as a silica source is the apparent increased solids loading to the atomizer that must be used in order to achieve increased SO_2 removal. The quality of sorbent material entrained in the gas stream is directly proportional to the amount of recycle, and thus represents an increased load to the particle collection device. A direct injection technology utilizing silica-enhanced $Ca(OH)_2$ is being developed and is known as the ADVACATE process [124].

The impact of coal chloride concentration on SO_2 removal in a spray dryer has been reported by Brown et al. [176]. In tests conducted at the TVA 10-MW spray dryer/ESP test facility a SO_2 removal level of 85% was achieved over an extended test period for a 4.0% sulfur, 0.05% chloride coal with a reagent ratio of 1.6 for operation at an 18°F approach to adiabatic saturation and an inlet gas temperature of 320°F. For similar operating conditions, a SO_2 removal level of 93% was achieved when the coal was changed to one containing 4.0% sulfur and 0.25% chloride. On high chloride coal (i.e., greater than 0.2% chloride), 90% removal appears to be achievable at a reagent ratio between 1.4 and 1.5. The role of chloride in promoting greater SO_2 reactivity is postulated to be caused by the formation of higher concentrations of HCl in the flue gas which subsequently react with the slurry in the spray dryer. Calcium chloride, a classical deliquescent salt, has been reported by others [12] to greatly improve SO_2 uptake in calcium-based sorbents.

Miscellaneous Additives - High Temperature Applications

EFFECTS OF SINTERING AND PORE STRUCTURE. Borgwardt et al. [49] pointed out that diffusion through the product layer in a solid, when it occurs through solid-state mechanisms, increases with the concentration of lattice defects. These can exist as point defects, which involve individual atoms, or as extended defects, which involve lines or planes of disorder in the crystal structure [177]. At a given temperature, the concentration of point defects in the product layer may depend on the concentration of foreign ions. Thus, higher diffusivities can be expected in impure materials, when a solid-state mechanism prevails. From their results, they concluded that the higher reactivity of impure CaO is due to defects inherent with the crystal structure of the limestone derivatives. Impurities in the form of aliovalent ions are known to generate defects in the crystal structure [178,179].

Accordingly, Borgwardt et al. [49] studied the effects of alkali metal ions doped on the surface of well-annealed CaO. They added sulfates of lithium, sodium or potassium to pre-calcined and pre-sintered CaO by grinding in a mortar, and observed a significant increase in sorbent utilization during sulfation. However, doping $CaCO_3$ or $Ca(OH)_2$ with Na+ prior to calcination and sintering was not successful. This, they explain, is because these ions enhance the diffusion and hence the overall rate during sintering as well, which in turn causes a decrease in the surface area of the sorbent available for sulfation.

According to Haji-Sulaiman et al. [59], higher impurity content in the sorbent increases the extent of calcination. Thus, sorbents with a more open pore structure are obtained, resulting in improved sulfation efficiency by preventing early pore blockage. Shadman and Dombek [180] view the role of additives solely as structure modifiers. They prepared flakes of modified sorbents by mixing the additives with a slurry of hydrated lime and spreading the mixture in a thin layer followed by drying. The additives used were bauxite, silica and kaolin. For all three additives, both the reaction rate and the maximum achievable conversion increased significantly. Among these additives, when their concentration and particle size were the same, they observed little difference in their performance. Hence they concluded that the effect of the additives is purely physical in nature, specifically an increase in macroporosity.

Following SEM micrograph analysis, they assumed that the sorbent flakes consist of spherical grains of lime mixed with inert additives. This leads to a bimodal pore size distribution where micropores account for pores in the particles and macropores represent the void space between particles. They developed a diffusion-controlled model with micropore and macropore diffusivities as the adjustable parameters. Model simulation supported their experimental finding that initial macroporosity is a critical factor in determining the sulfation performance of the modified sorbent.

ORGANIC SURFACTANTS. Kirchgessner and Lorrain [181] modified the sorbent $Ca(OH)_2$ with calcium lignosulfonate, an additive added with the water of hydration. They observed that the utilization of the modified sorbent increases with increasing additive content, reaches a maximum, and then decreases, as shown in Figure 1.17.

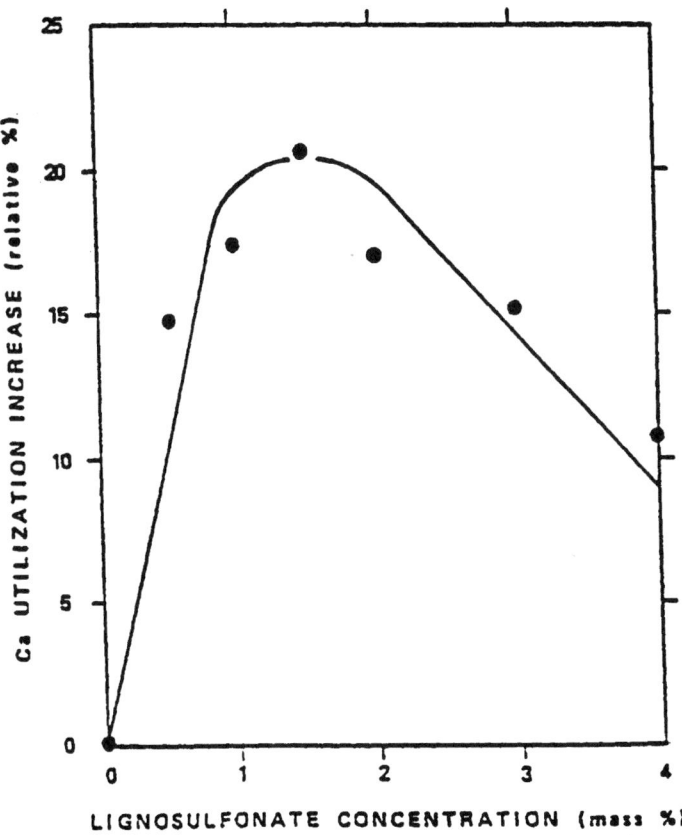

Figure 1.17 Effect of additive concentration on sorbent utilization Kirchgessner and Lorrain [181].

The maximum occurred at an additive content of 1.5 dry weight percent, where the sorbent utilization was 20% higher than the unmodified hydroxide. Through size analysis of the modified particles, they showed that the superior performance is due to particle size reduction achieved primarily through de-agglomeration and secondarily through crystal size reduction. They suggested that above the optimal level of 1.5%, the large lignosulfonate molecule may block access of the SO_2 molecule to the reactive CaO sites, causing the relative utilization to decrease.

Subsequently, Kirchgessner and Jozewicz [182] performed extensive studies of the changes in pore structure during sintering of CaO produced from $Ca(OH)_2$ modified with the 1% calcium lignosulfate. They first noted that modified $Ca(OH)_2$ calcined more quickly and retained more of its original surface area and porosity than the conventional $Ca(OH)_2$. With increasing time and temperature the difference between the surface area and pore structure between modified and original sorbents due to sintering became more pronounced. Their pore size measurements confirm that sintering involves pore filling, with the smallest pores filled first. This is reflected in the dramatic difference in pore volume with time and temperature for pore sizes less than 50 Å. They noted that the drastic changes in surface area and pore size are complete within 1.5 seconds at 700°C and 0.8 second at 1000°C.

Since sintering of CaO is known to be catalyzed by H_2O, differences in the rates of water loss for modified and unmodified $Ca(OH)_2$ may explain the differences in the rates of surface area loss and in pore structure. Their measurements show that water loss from modified hydroxide is indeed greater than the loss from unmodified hydroxide. However, the difference in water loss between the two types took place before 0.6 second, whereas the difference in surface areas do not become pronounced until after 0.6 second. Therefore, water loss does not fully explain the difference between the structure of the two sorbents.

One of the mechanisms of sintering is the mobility of grain boundary. Because of the large size of the hydrated lignosulfate molecule, it is believed to be located at the grain boundaries and the surfaces of $Ca(OH)_2$ rather than within the crystal structure. In this structure, it reduces grain mobility, thus reducing the sintering effects. Thus, smaller particle size, increased extent of calcination, and lower sintering all contribute to the better performance of the hydrate sorbent modified with lignosulfate.

METAL OXIDES. Slaughter et al. [183] investigated the addition of chromium (Cr_2O_3), sodium ($NaHCO_3$), and iron (Fe_2O_3) compounds to the sorbent for injection at 2150°F and 2600°F. Unlike the previously cited studies, the promoters were injected in powder form along with the calcine instead of being added to the hydration water during the preparation of the hydrate. Both the chromium compound (18% SO_2 capture without and 39% SO_2 capture with at 2600°F) and the sodium compound (12% SO_2 capture without and 42% SO_2 capture with at 2600°F) were effective in promoting SO_2 capture by the calcine when added at a 15:1 calcium:promoter metal atomic ratio. The iron compound was not effective in promoting sulfur capture. Slaughter indicated that both chromium and sodium react with the calcium sorbent creating large cracks or pores, particle fragmentation and the formation of a liquid phase, all of which serve to increase SO_2 accessibility to the CaO sites. The presence of mineral matter ash in the flue gas stream acted to reduce the effectiveness of both promoters. The effectiveness of the sodium promoter showed a strong dependence on mineral ash concentration while the effectiveness of the chromium promoter showed only a weak dependence on mineral ash concentration.

This is an important area of research. Alternate additives should be considered, chosen on the basis of solid state chemistry. Systematic characterization of reactivity of similar sorbents through crystal structure analysis may explain the difference between them and lead to a better choice of sorbents.

Contacting Flow Patterns and Multiphase Flows

Contacting Flow Patterns and Some Novel Techniques

The various dry processes involved in FGD discussed in earlier sections include dry/dry systems, wet/dry systems, in-furnace injection, economizer-zone injection, in-duct-hydrated-line injection downstream of air preheater and spray-drying absorption. It is noteworthy that these processes have cocurrent contacting patterns; i.e., the sorbent particles and the flue gases flow cocurrently. Theoretically, the best contacting pattern is countercurrent flow of the sorbent particles and the gases. This brings the new unused sorbent particles into contact with the cleanest flue gas and progressively allows the most consumed sorbent particles to contact the flue gas with highest SO_2 concentration. The process that allows full countercurrent contacting is the wet FGD process [151]. The Lurgi circulating fluidized bed (CFB) dry FGD process [184] has some aspects of cocurrent as well as countercurrent flows.

An interesting question is whether it is possible to develop a countercurrent version of some of the cocurrent dry/dry and wet/dry processes. Such contacting may substantially enhance the sorbent utilization efficiencies. The contacting geometry used in dry/dry processes is similar to a cyclone. The flue gas from the boiler enters the top of the spray dryer and is swirled downward in a concentric ring surrounding the rotary atomizer. Droplets from the atomizer are sprayed into the hot flue gas and both droplets and flue gas pass cocurrently downwards through the spray dryer. In the Lurgi CFB process, hydrated lime is injected into the absorber where it is intimately mixed with the recycle material in the fluidized bed. The high superficial velocity in the absorber gradually moves the sorbent up through the bed, out of the top of the absorber and into the initial stage of the ESP. Water is also injected into the absorber to enhance the reaction with SO_2 and humidify and cool the flue gas. Most of the particulate matter removed in the ESP is recycled into the CFB absorber while some is removed from the system into a waste silo. A comparison of various FGD processes in operation worldwide is provided by Hollinden, Burnett and Torstrick [151].

RECYCLING. Relatively lower sorbent reactivity and high loading of solids are some of the factors that limit the application of the dry sorbent processes. One possible technique of increasing the SO_2 removal and sorbent utilization efficiency substantially is the recycling of the sorbent material. Lott et al. [94] have developed a dry sorbent recycle process model for examining SO_2 removal as a function of recycle ratio. The model also takes into account factors like recycle sorbent material, furnace sorbent/ash particulate loading, and sorbent/ash separation. They have also conducted laboratory tests of a continuous recycle system to examine the SO_2 removal levels, recycle ratios, and the recycle effectiveness factors for several sorbents.

The recycle concept is not a new concept and has been used to improve reagent utilization and increase product yield in process industries. The recycling of sorbent for SO_2 removal seems to be a promising method for improving sorbent utilization leading to reduction in the sorbent feed rates. Important areas that need to be addressed include design process trade offs between fresh sorbent and recycle sorbent feeds, process development for continuous recycle including designing of continuous recycling hardware, development of ash/sorbent separation techniques, system design, and costs.

Novel Techniques with FGD Applications

Some of the novel techniques that have applications in the area of flue gas desulfurization will be discussed in this section. These techniques include rotating packed beds, circulating fluidized bed absorbers, the limestone emission control (LEC) system, a novel impeller fluidizer, and other devices (acoustic and 3S methods).

FLUE GAS DESULFURIZATION BY ROTATING PACKED BED. Mass transfer in processes such as vapor-liquid contacting can be substantially intensified through the use of high gravitational fields (g-force). Increasing g has two effects—gas velocity can be increased and higher surface area packings can be used; as a result, high volumetric masstransfer coefficients can be achieved. This enhanced mass transfer enables the physical size of an absorber to be greatly reduced. Other benefits include much smaller equipment and supporting structure, reduced solvent inventory, shorter startup and shutdown times, faster response with improved process control, and reduced pressure drops.

In the late 1970s, the New Science Group of Imperial Chemical Industries (ICI) developed HIGEE based on the above concept for absorption and distillation applications [185,186]. The high gravitational fields are achieved using rotating packed beds which are one to two orders of magnitude smaller in volume than conventional packed columns. The device consists of a rotating packed bed in the shape of a toroidal disk. The liquid is sprayed on the inner radius of the packed bed, accelerated and distributed through the packed bed, and then thrown out by the centrifugal force. The gas is introduced from the outer radius of the rotating bed and, due to imposed pressure gradient, flows inward, countercurrent to the liquid. The HIGEE packed bed was operated at mean gravitational fields of 200-500 g and the reported volumetric mass transfer coefficients were up to 100 times larger than those observed in conventional packed columns. The height of a transfer unit (HTU) is much reduced, from about 2-5 ft in conventional columns (operating at 1 g) to 1/4-2 inches in rotating packed beds (operating at 300 g and higher). Glitch Inc. (Dallas, Texas), which has acquired worldwide marketing rights, has already installed a unit for the U.S. Coast Guard at a facility near Traverse City, Michigan to airstrip aviation fuel from contaminated ground water. Field tests have been performed for an undisclosed natural gas producer in the U.S. [187].

Although the rotating packed bed technology appears very promising for vapor-liquid contacting, there is only limited information available in the literature. ICI and Glitch, in spite of their extensive tests, reported very little information about their results. No information is available on the operating characteristics of the rotating packed beds such as residence time distributions, pressure drops, and power consumptions. There are some reports about mass transfer of the rotating packed beds. Vivian et al. [188] studied the effect of gravitational fields on gas absorption in a packed column by mounting a cylindrical packed bed on the horizontal arm of a large centrifuge. The gravitational force was varied from 1-6.4 times the gravity force. They reported the liquid phase volumetric mass transfer coefficient varies with the gravitational force to the 0.41-0.48 power. Dudukovic et al. [189] reported theoretical and experimental work on rotating packed beds. They studied the effects of rotational speed on gas-liquid interfacial area and liquid-side mass transfer coefficients; their results on mass transfer coefficients are in agreement with those reported by Vivian et al. [188].

The rotating packed bed has been applied to FGD at Case Western Reserve University [190,191]. The rotating bed packing is a one piece torous-shaped unit (18 inches O.D., 10 inches I.D., 1-3/4 inches thick), made of a foam metal which is dimensionally stable in the high gravitational fields. The packing has a specific surface area of 450 ft^2/ft^3. The rotating bed can operate from 600-2000 rpm, which corresponds to 100-1000 g. Numerous tests and experiments have been performed to evaluate the operating characteristics of the rotating packed beds. These include measurements of power consumption, residence time and pressure drop. The major power consumption in the rotating bed is the acceleration of the liquid to the angular velocity of the rotor; measured power consumptions for liquid acceleration are less than 1.5 W per g/s of liquid. An empirical correlation for the total power consumption in rotating beds has been developed. The mean liquid residence time depends primarily on the rotational speed and liquid flow rate; typical values range from 0.5-2 seconds. The pressure drop across the rotating bed depends primarily on the rotational speed and gas flow rate, and depends only slightly on the liquid flow rate; measured values range from 100-1500 Pa (0.4 to 6 in-H2O). For the SO_2 removal studied here, the pressure drop is typically 100 Pa per transfer unit.

Measurements of overall mass transfer coefficients have been initiated for SO_2 removal by lime/limestone slurries. Preliminary results indicate that substantial increases in the overall volumetric mass transfer coefficients can be achieved using the same operating conditions employed in conventional packed towers. An SO_2-air mixture containing 5000 ppm SO_2 was scrubbed with a 2 wt% lime slurry in the rotating bed at a rotational speed of 700 rpm (100g). The gas and liquid flow rates were 50 acfm and 3 gal/min, respectively (the liquid-to-gas ratio was 60 gal/1,000 acf). The experiments were performed at room temperature and 1 atmosphere. Under these conditions, the calculated overall volumetric mass-transfer coefficient, Kga, was 230 lb-mole/ft^3-hr. The removal efficiency based on measured SO_2 concentrations at the gas inlet (5,000 ppm) and outlet (35 ppm) was 99.3%. Typical Kga values are 5-30 lb-mole/ft^3-hr.

Circulating Fluidized Bed Absorber

A circulating fluidizer-bed absorber (CFBA) has been built at the University of Cincinnati [192] and is being evaluated for control of SO_2. This device can be retrofitted in an older plant or incorporated with a new design. It involves contacting a pulverized dry sorbent such as sodium bicarbonate, limestone, lime or calcium hydroxide with flue gases in a fluidized-bed reactor operated at high velocities. A solid/gas separator, located downstream of the reactor, concentrates unused sorbent for recycle and subsequent reuse. A schematic diagram of the experimental setup is shown in Figure 1.19. This contacting method offers ease of retrofitting to existing plants, isothermal operation, applicability of a variety of sorbents and high sorbent utilization. The test results provided by Keener, Jiang and Hao [192] show over 90% SO_2 removal with sodium bicarbonate sorbents with sorbent utilization of over 90% for reactor temperatures of 254-370°F. The sodium bicarbonate particles undergo a change in particle size due to attrition within the reactor, which allows for higher sorbent utilization.

Additionally, the use of CaO has been found to provide high removal efficiencies when contacted with the gas at high relative humidities. Efficiencies of over 60% using a coarse-sized calcium oxide injected at stoichiometric conditions were obtained when water was injected into the CFBA to bring the approach to saturation temperature within 60°F. A model based on gas phase mass transfer limitations has indicated that the critical reactor parameters include the reactor void fraction as a function of reactor height and the sorbent particle relative velocity. Also, sorbent attrition is seen as being fundamental to achieving high sorbent utilization as the attrited core provides such unreacted sorbent.

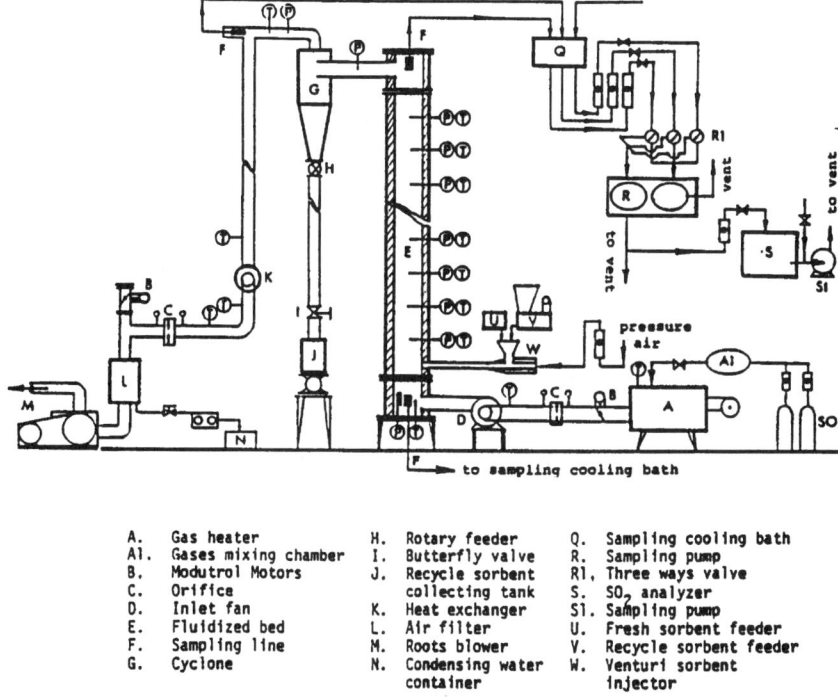

A.	Gas heater	H.	Rotary feeder	Q.	Sampling cooling bath
A1.	Gases mixing chamber	I.	Butterfly valve	R.	Sampling pump
B.	Modutrol Motors	J.	Recycle sorbent collecting tank	R1.	Three ways valve
C.	Orifice			S.	SO_2 analyzer
D.	Inlet fan	K.	Heat exchanger	S1.	Sampling pump
E.	Fluidized bed	L.	Air filter	U.	Fresh sorbent feeder
F.	Sampling line	M.	Roots blower	V.	Recycle sorbent feeder
G.	Cyclone	N.	Condensing water container	W.	Venturi sorbent injector

Figure 1.18 Scheme of experimental set-up for circulating fluidized bed absorber [192].

Limestone Emission Control (LEC) System

The Limestone Emission Control (LEC) process is a FGD system that utilizes a bed of granular-sized limestone to remove SO_2 from flue gases. Furlong, Wright and Prudich [194] describe the results of a pilot demonstration plant which was installed and tested on a slip stream of Ohio University's coal-fired boiler.

The slip stream was withdrawn from the outlet of the steam plant ESP by an induced draft fan and the hot flue gases were led into the spray chamber. Water spray at the top of the chamber cooled and humidified the gas. Steam injection ports provided in the chamber also allowed the humidification of the gases without lowering the gas temperature. The humidified gas then entered the LEC reactor. The LEC reactor assembly consisted of an inner basket of limestone contained in an outer insulated reactor housing. The outer housing was equipped with a multiple head water sprayer arrangement which allowed the introduction of water onto the limestone bed. After leaving the reactor bed, the gases were sampled for SO_2 concentration, temperature and humidity. Details of the setup are provided by Prudich et al. [18]. Over 100 experimental runs were made with the SO_2 concentration in the flue gas ranging from 500-3500 ppm. Results indicate the LEC system is capable of 99% SO_2 removal [194]. An average of greater than 90% removal was achieved over extended operation during which the limestone bed was regenerated several times. Favorable economic comparison of the LEC with wet scrubbing and spray drying technologies are also presented in the paper. These results indicate that the LEC system offers significant advantages over current technology.

Novel Impeller Fluidizer

The impeller fluidizer is a fluid mechanical method of trapping, concentrating, and processing fine particles within a fluid particle mixture. The device was developed at Case Western Reserve University [195]. The particles are trapped by the combined action of centrifugal force and convection, both produced by an impeller-driven swirling flow (Figure 1.19). Depending on the impeller rotation rate, the particles either disperse or concentrate in a locally fluidized state (Figure 1.20). At low rotation rates, the particles suspend and mix uniformly as expected, but at higher rotation rates, the particles concentrate in the upper part of the cylinder. The fluid at the impeller end of the cylinder becomes completely free of particles. The trapped particles form a fluidized bed of toroidal geometry.

The phenomenon is robust; it occurs with pitched as well as flat-blade impellers, and is little affected by reasonable flows through feed and exit ports. The direction of impeller rotation and orientation in the earth's gravitational field are unimportant. A wide range of geometries, speeds, particles, and fluid-solid concentrations are workable. The cause of the particle trapping is a combination of strong flow swirling around the cylinder axis, and weaker secondary flow. The swirling flow produces a centrifugal force field which causes particles denser than the fluid to move radially outward toward the curved wall (Figure 1.19). The secondary flow moves radially outward near the impeller, upward along the walls to the top of the cylinder, radially inward at the top, and then downward along the cylinder axis back to the impeller. The upward portion of the flow along the wall convects particles to the top of the cylinder. There, the radially inward portion of the flow drags particles inward until a radial position is reached where the steeply increasing centrifugal force balances the drag force. At that point the fluid disengages from the particles and continues its inward, and then downward, recirculation. Hence, the particles are trapped dynamically by the interaction of the centrifugal force field and the recirculating flow drag.

The trapped particles resemble a fluidized bed which rotates about the cylinder axis and is confined to a portion of the vessel without using mechanical constraints such as screens or filters. As the rate of impeller rotation increases, the trapped zone of particles becomes smaller, but does not become a tightly packed bed. There is substantial flow through the bed under all conditions. Some systematic investigations of the phenomenon with liquid/solid mixtures show that small particles can be trapped and held in the form of an expanded bed in the presence of substantial through flows [196]. Qualitative exploratory experiments on solids and gases indicate good prospects for processing fine solids in gases. Calculations indicate particles as small as 0.2-5 microns can be trapped. Centrifugation depends strongly on the density difference between the particles and the fluid, so particle separation in gas-solid systems promises to be easier than in liquid-solid systems. The lower viscosity of gases compared to liquids is also favorable.

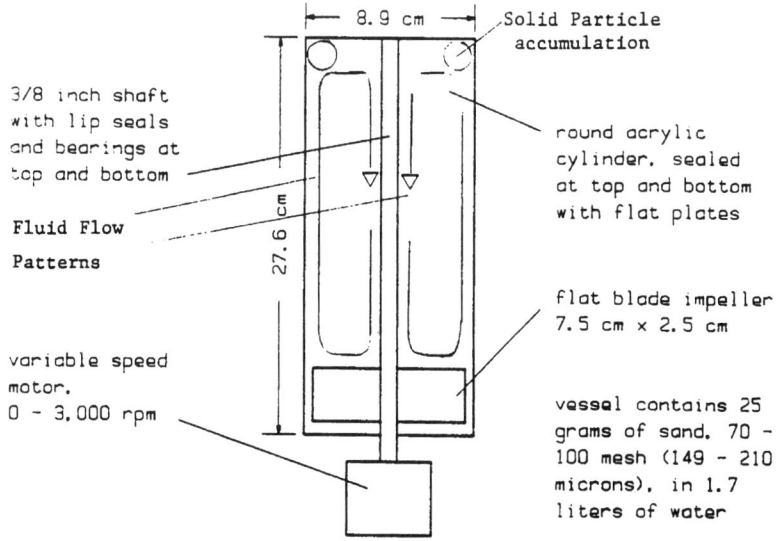

Figure 1.19 Single impeller experiment.

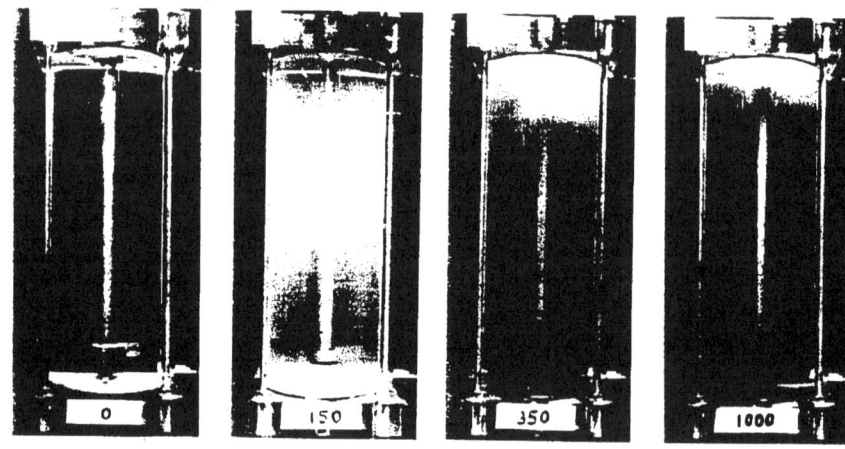

Figure 1.20 Effect of rotation rate, single impeller. Impeller speeds are marked in rpm.

Of practical importance is the pressure gradient induced in the radial direction by the spinning flow. This pressure is sufficient to transport gas and/or gas-particle mixtures through contactors without external pumps; the contactor combines the functions of contacting, filtering, and pumping. The impeller fluidizer can be useful in the FGD processes in the following areas: recycling of absorbent particles, ash removal, and sorbent removal; providing an efficient limestone regeneration device for the LEC system.

RECYCLING OF ABSORBENT PARTICLES AND ASH REMOVAL. From the earlier discussions, recycling of calcium-based sorbents appears to be important for achieving high sorbent utilization. The recycling process is made difficult by the presence of fly ash, and there is a need to develop continuous sorbent recycling equipment. The impeller fluidizer characteristics may be useful for separating the fly ash prior to the introduction of the sorbent, and may also be useful for recycling sorbent particles. Figure 23 illustrates a possible configuration for recycling sorbent particles. The contactor separates, concentrates, and recycles particles to the entrance of the duct admitting flue gas from the air preheaters.

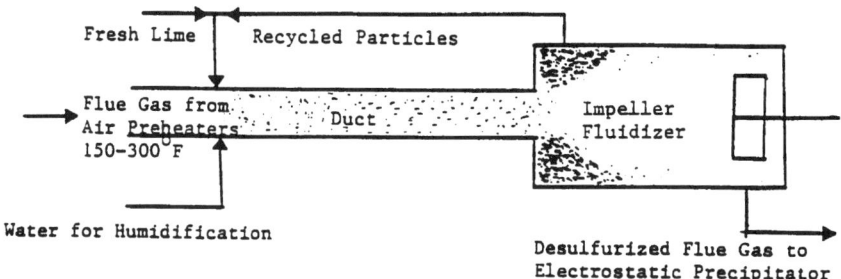

Figure 1.21 Cocurrent recycling scheme utilizing the impeller fluidizer.

LIMESTONE RECOVERY IN THE LIMESTONE EMISSION CONTROL (LEC) PROCESS. The impeller fluidizer may be useful for wet grinding of used limestone particles in the LEC process. The impeller fluidizer is a closed cylindrical vessel with an impeller located at one end spinning around the axis of the cylinder. The fluid flow field produced by the impeller causes the limestone particles to form a dense, rotating, toroidal, fluidized bed confined to the end of the cylinder opposite the impeller. The sulfite-sulfate coating is selectively removed by self comminution of the particles grinding against each other. The recovered limestone particles are pumped in slurry form to form a new LEC bed, and a fine slurry of sulfite and sulfate particles is pumped to an auxiliary sedimentation process. All of the functions (grinding, separating the sulfite-sulfate slurry, and pumping) are performed by the impeller fluidizer. An initial test of the fluidizer in this application showed convenient, controlled wet grinding.

Acoustic and Other Techniques

In addition to the techniques discussed in the preceding section, work has been reported in the area of the applicability of high intensity acoustic technology in FGD processes [197]. Studies of acoustic agglomeration of fly ash and acoustic enhancement of particulate removal devices like ESPs, baghouses, etc. are being conducted. Acoustic enhancement of SO_2 absorption by calcined limestone particles has been reported.

The 3S process, simultaneous (acid gas and NO_x) sodium-based sorption, is in the early stages of development by Lindbauer and Mair [198]. They have used sodium bicarbonate and other additives; e.g., activated bauxite. The laboratory tests have included fixed bed and fluidized bed experiments.

Fundamental Fluid Mechanics, Mass and Heat Transfer Consideration in FGD Processes Involving Multiphase Flows

The introduction of sorbents into flue gases for FGD involves issues related to complicated multiphase flows. These flows include solid-gas flows (dry sorbent injection), solid-liquid-gas flows (sorbent slurry injection into gas), solid-liquid flows (flow of wet sorbent slurry), and liquid-gas flows (spraying of water to humidify and cool flue gas). Multiphase flow by itself is not well understood and the understanding of the effect of these flows on sorbent SO_2 absorption is even less well understood. Research needs related to multiphase flows exist in the area of fluid mechanics and mass and heat transfer. We will briefly discuss these areas of research.

Transport Properties of Dry Sorbent and Sorbent Slurries

Issues related to the flow of dry sorbent; i.e., powder, (solid-air) and wet sorbent slurries (solid-liquid) have to be addressed. These problems are related to material handling aspects which include clogging or plugging of pipe lines, pumping power requirements, and ensuring that the solid-liquid and solid-gas slurries stay in the homogenous flow regime for different solid loading or concentrations as required for the FGD processes. By homogeneous flow, we mean that fluid and solid move as a uniform mixture. Liquid-solid slurry aspects are also to be considered in sorbent waste handling and stabilization.

WET AND DRY SORBENT INJECTION. Some of the difficulties associated with wet and dry sorbent injection are described below. Flow jet mixing problems impact the mass transfer aspects. Research to understand the contribution of mixing on overall mass transfer; i.e., SO_2 absorption, is needed. Nozzle design for sorbent injection has to be optimized to provide proper mixing of the sorbent jet with flue gases, resulting in maximum sorbent utilization. Study of the particle size distribution and agglomeration problems in dry sorbent injection is needed. Ideally small particles are desired with enough interparticle spacing to ameliorate agglomeration. Study of wet slurry spray (atomization) injection is needed. Smaller diameter droplets are desired in drying. The mixing of flue gas and the sorbent spray jet impacts the mass transfer and SO_2 absorption. Essentially, contacting between sorbent spray and the flue gas has to be optimized. Dry sorbent/water spray (droplet) impingement is also important. Optimization of the contacting between the droplet and dry sorbent to enhance SO_2 absorption and humidification of the flue gases is needed. Finally, 3D jet mixing models and 3D reaction models are needed.

Essentially, research areas of interest include fundamental studies of wet and dry slurry injection to understand the fluid mechanics, mass and heat transfer aspects involved and their effects on SO_2 absorption by the sorbent. Development of such an understanding will lead to better sorbent utilization. Understanding of the transport properties of the dry and wet slurries is of importance from the material handling aspects.

Conclusions

In-Furnace Injection

The optimum sulfation temperature is around 2400°F. The substantial changes in sorbent utilization and structural evolution take place within only a few milliseconds. The desulfurization rate is roughly proportional to the second power of the initial BET surface area. The reaction is highly dependent on temperature with an activation energy of about 35 kcal/mol. The limestone calcination rate is roughly proportional to the BET surface area. Thermal sintering significantly reduces the surface area resulting in a lower desulfurization rate. The reaction product is $CaSO_4$. Although pore plugging and pore diffusion are suspected as the rate-limiting steps, fundamental mechanisms and controlling processes are not well understood. CaO derived from $Ca(OH)_2$ is more reactive than CaO derived from $CaCO_3$. Downstream in-duct humidification enhances the sorbent utilization. The current sorbent utilization is about 25% (a 50% removal of SO_2 with a Ca/S ratio of 2). Gas-solid mixing and transport processes within the furnace are very important. The single pore model and the grain model have been used to describe the sulfation reaction. Structural changes and thermal sintering have not been successfully incorporated into these models. The majority of single-particle models are not practical for commercial and pilot-plant applications.

Economizer Zone Injection

The optimum reaction temperature is about 1000°F. The reaction rate is comparable or even higher than that at 2400°F. Hydrated lime is the main form of sorbent. The main reaction product is $CaSO_3$. Fundamental mechanisms and controlling processes are not well understood, although the apparent improvement in reaction rates has been explained by chemical reaction steps controlling in this temperature range.

Injection of Hydrated Lime Downstream of the Air Preheater

Sorbent utilization is a function of the approach to saturation and of the type of particulate control device used. The sorbent reaction mechanism in the duct is similar to that in a spray dryer. Liquid water injection enhances sorbent reactivity to a greater extent than does steam injection. Water may be injected either before, during or after sorbent injection. The rate of SO_2 removal is seen as an exponential function of the degree of relative humidity. The fact that water, rather than steam, enhances reactivity indicates that relative humidity, rather than absolute humidity, is important. However, a sufficient amount of liquid water is required (i.e., direct impaction of drops with particles) which indicates that the importance of relative humidity is to retard evaporation of the bulk water phase from the particles' surface. A "bulk" water phase dominates the behavior of the wet/dry (e.g., the LEC) system. Wet/dry removal may be gas-film mass transfer controlled for low SO_2 concentrations and controlled by lime dissolution for higher SO_2 concentrations. Modelling efforts on the wet/dry system have focused on gas-film mass transfer control and $Ca(OH)_2$ dissolution.

Spray Drying Absorption

Reagent preparation is very important in the production of slurry with a maximum of surface area. Proper design of the dryer is important to achieve the required degree of product dryness. The degree of adiabatic saturation of the entering flue gas is an important consideration in terms of operation and degree of SO_2 removal. The extent of additional SO_2 removal by means of increased sorbent concentration in the slurry is limited due to product accumulation effects in the slurry drop. Removal of 90% of the SO_2 from a high-sulfur coal flue gas by in-duct injection requires a substantial amount of product recycle in order to raise the effective sorbent stoichiometric ratio and increase utilization. Baghouses have been shown to substantially increase the overall degree of SO_2 removal when applied to the fly ash from some Western coals. Corrosion of the particulate collection device has been a problem for some full-scale installations. Descriptions of the mechanisms involved in the reaction of the slurry drop with SO_2 have centered around the constant rate drying period of the drop. The influence of water evaporation on the degree of SO_2 uptake is not well understood. Recycle effects have not been considered in modelling efforts to date. The use of additives to enhance spray dryer performance on high sulfur coal appears promising. Nitrogen oxides have been reported to be removed under certain conditions in spray dryer systems.

Particulate Control

Only fabric filters or ESPs are used in conjunction with dry sorbent FGD systems. ESP efficiencies are directly proportional to inlet mass loading. ESP performance can be dominated by non-ideal effects such as rapping emissions, gas distribution problems, and reentrainment. Duct injection technologies produce two conditions that may affect ESP performance: an increase in inlet dust loading and an order of magnitude increase or decrease in particle resistivity. Spray dryers may improve ESP collection efficiency. Methods to offset a decline in

ESP performance due to duct injection include increased collector area, pulsed energization, intermittent energization and additives for cohesion improvement. The E-SO_x program is designed to use existing ESPs for FGD service. Fabric filters offer the possibility of increased SO_2 removal through reaction across the filter cake. Testing indicates that duct injection technologies have little impact on fabric filter performance. The degree of SO_2 removal in a fabric filter is dependent on the approach to saturation temperature. The degree of SO_2 removal in an ESP is dependent on the sorbent mass ratio.

Future Research Needs

The preceding literature review points out two major points that hinder the adoption of dry scrubbing technologies: poor sorbent utilization and low SO_2 removals. Anything that can be done to overcome these limitations has the potential to improve the economic feasibility of dry scrubbing processes.

Greater in-depth knowledge can reveal the factors that control process performance, and lead to systematic, substantial enhancements of FGD processes. Basic research is needed in three broad areas: reaction mechanisms and sorbent material, transport effects and modeling, and process development.

Reaction Mechanisms and Sorbent Material

In-depth knowledge of the inherent kinetics and controlling resistances in the dry scrubbing processes is needed in all areas of dry SO_2 capture such as in- and over-furnace injection, convective pass injection, and low temperature (wet/dry, dry/dry) injection.

Sorbent preparation techniques used today have generally been adapted from those traditionally used by the lime industry. Traditional "dry" hydration and slaking techniques have been optimized for a product intended for the building industry. These preparation techniques should be examined with an eye to optimization of the properties important to the SO_2 capture process.

Little attention has been paid to the selection of limestones for the SO_2 capture process. Current demonstrations rely on highly reactive limes which are often transported large distances to the demonstration site. These large transportation costs would not be acceptable for routine use. Criteria must be established for limestone selection so that locally available limestones can be evaluated and used.

There is a need to elucidate the chemical mechanism(s) that allow successful sorbent additives to be effective. An understanding of the activating mechanism(s) will enable the systematic selection of new additive materials as well as the optimization of sorbent/additive blends. There is also a need for basic research to study reactivity of new sorbents, such as calcium silicate sorbents formed from lime and fly ash.

Transport Effects and Modeling

FGD processes involve complicated multiphase flows (solid-gas, liquid-gas, solid-liquid-gas and solid-liquid). Multiphase flows by themselves are not well understood and their impact on FGD processes are even less well understood. A basic knowledge of fluid mechanics, and heat and mass transfer rate process effects is needed in order to understand and achieve substantial improvements in many key FGD processes and obtain reliable scale-up criteria. Examples of key processes include dry sorbent injection and dispersion (gas-solid flow), sorbent humidification by water injection (gas-solid-liquid flow), and material handling aspects of slurries and fly ash.

Opportunity exists to find methods of improving the efficiency of contacting operations, for example by contacting countercurrent rather than cocurrent. New and/or improved contacting devices need to be developed.

The ability to obtain results from small-scale experiments that can be used to predict large-scale performance is a problem that is common to many areas of engineering endeavor. All the fundamental research areas obviously impact on this problem. Reliable small-scale testing is needed in order to minimize the expense of proving out process and sorbent improvements.

Process Development

An understanding of how the process configuration interacts with the process chemistry to influence SO_2 capture performance is needed. As an example, Consol's Coolside, Keener's circulating fluidized bed, ETS' LEC, and high centrifugal field absorption can all be operated in the "same" temperature range/mode. Their uniqueness as SO_2 capture systems relies on the interaction between the inherent process chemistry and their process configuration/flow patterns. An understanding of this interaction can be crucial to defining a "best" system for a specific application.

If high sorbent utilizations can be achieved, the recycle and separation of unreacted sorbent materials is not a problem. In the event that low one-pass utilizations are inevitable, unreacted sorbent separation and recycle can serve to give high overall utilizations and reduce the volume of solid waste.

Innovative separation technologies such as membrane separation and electrochemical processes may lead to new process concepts applicable to FGD process.

References

1. Streets, D. G. et al. "Selected Strategies to Reduce Acidic Deposition in the U.S.," *Environmental Science and Technology*, 10, 474A-485A (1983)
2. NUS Co, *Commercial Coal Power Plants*, 5th Edition, NUS Corp., Gaithersburg, MD, (1983)
3. Keener, T. C. and S. U. Keener, "Current Status of Flue Gas Desulfurization in the United States," *Proceedings, ASCE Environmental Engineering Division National Conference*, July 1986
4. Ireland, P. A. et al. "Site Specific Evaluation of Six Sorbent Injection Processes," *FGD and Dry SO_2 Control Symposium*, St. Louis, MO, October 25-28, 1988
5. Offen, G. R. et al. "Assessment of Dry Sorbent Emission Control Technologies - Part II. Applications," *J. of Air Pollution Control Association*, 37 (8), 968 (1987)
6. Beittel, R., J. P. Gooch, E. B. Dismukes and L. J. Muzio, "Studies of Sorbent Calcination and SO_2-Sorbent Reactions in a Pilot-Scale Furnace," paper 16, *Proceedings: First Joint Symposium on Dry SO_2 and Simultaneous SO_2/NO_x Control Technologies*, 1, EPRI CS-4178, June 1985
7. Fink, C. E., N. S. Harding, B. J. Koch, D. C. McCoy, R. M. Statnick and T. J. Hassell, "Demonstration of Boiler Limestone Injection in an Industrial Boiler," paper 18, *Proceedings: First Joint Symposium on Dry SO_2 and Simultaneous SO_2/NO_x Control Technologies*, 1, EPRI CS-4178, June 1985
8. Farzan, H., L. Rodgers, G. Maringo and G. R. Offen, "Application of Upper Furnace Sorbent Injection for SO_2 Control in Coal-Fired Cyclone-Equipped Boilers," paper 20, *Proceedings: First Joint Symposium on Dry SO_2 and Simultaneous SO_2/NO_x Control Technologies*, 1, EPRI CS-4966, December 1986
9. Greene, S. B., S. L. Chen, D. W. Pershing, M. P. Heap and W. R. Seeker, "Bench Scale Process Evaluation of In-Furnace NO_x and SO_x Reduction by Reburning and Sorbent Injection," paper 20, *Proceedings: First Joint Symposium on Dry SO_2 and Simultaneous SO_2/NO_x Control Technologies*, Vol. 1, EPRI CS-4178, June 1985
10. Bortz, S. J., V. P. Roman, R. J. Yang, and G. R. Offen, "Dry Hydroxide Injection at Economizer Temperatures for Improved SO_2 Control," paper 21, *Proceedings: First Joint Symposium on Dry SO_2 and Simultaneous SO_2/NO_x Control Technologies*, 2, EPRI CS-4966, December 1986
11. Seeker, W. R., S. L. Chen, J. C. Kramlich, S. B. Greene and B. J. Overmoe, "Fundamental Studies of Low-Temperature Sulfur Capture by Dry Calcitic Sorbent Injection," paper 32, *Proceedings: First Joint Symposium on Dry SO_2 and Simultaneous SO_2/NO_x Control Technologies*, 2, EPRI CS-4966, December 1986
12. Karlsson, H. T., J. Klingsper, M. Linne and I. Bjerle, "Activated Wet-Dry Scrubbing of SO_2," *J. of Air Pollution Control Association*, 33 (1), 23 (1983)
13. Yoon, H., M. R. Stouffer and F. P. Burke, *Coolside Desulfurization Reactions and Mechanisms*, Am. Chem. Soc. Div. Fuel Chem. Prepr., 32 (4), 484 (1987)
14. Babu, M., R. C. Forsythe, C. V. Runyon, E. Evans and J. L. Thompson, "Results of 1.0 MM Btu/Hour Testing of HALT (Hydrate Addition at Low Temperature) for SO_2 Control," paper 34, *Proceedings: 1986 Joint Symposium on Dry SO_2 and Simultaneous SO_2/NO_x Control Technologies*, 2, EPRI CS-4966, December 1986

15. Blythe, G., R. Smith, M. McElroy, R. Rhudy, V. Bland and C. Martin, "EPRI Pilot Testing of SO_2 Removal by Calcium Injection Upstream of a Particulate Control Device,", paper 35, *Proceedings: 1986 Joint Symposium on Dry SO_2 and Simultaneous SO_2/NO_x Control Technologies*, 2, EPRI CS-4966, December 1986
16. Statnick, R. ., F. P. Burke, B. J. Koch, D. C. McCoy and H. Yoon, "Status of Flue Gas Sorbent Injection Technologies," *Proceedings: Fourth Pittsburgh Coal Conference*, p250 (1987)
17. Yoon, H., J.A. Withum, W. A. Rosenhoover and F. P. Burke, "Sorbent Improvement and Computer Modeling Studies for Coolside Desulfurization," paper 33, *Proceedings: 1986 Joint Symposium on Dry SO_2 and Simultaneous SO_2/NO_x Control Technologies*, 2, EPRI CS-4966, December 1986
18. Prudich, M. E., K. W. Appell, M. J. Visneski, J.D. McKenna, D. A. Furlong, J. C. Mycock, J. F. Szalay and J. E. Wright, *Small Pilot Plant Demonstration of ETS' Limestone Emission Control System - Volumes 1 and 2*, OCDO Grant No. CDO/R-86-24 (1988)
19. McElroy, M. W., "Overview of Dry Sorbent Injection Technology for Retrofit SO_2 Control," paper 84, *Fourth Pittsburgh Coal Conference*, Pittsburgh, PA (1987)
20. Pohl, F. G., M. W. McElroy and R. G. Rhudy, "Pilot Evaluation of Combined SO_2 and Particulate Removal on a Fabric Filter," paper 27, *Proceedings: First Joint Symposium on Dry SO_2 and Simultaneous SO_2/NO_x Control Technologies*, 1, EPRI CS-4178, June 1985
21. DuBard, J. L., J. P. Gooch, R. Beittel, S. L. Rakes and G. R. Offen, "Particle Properties Related to ESP Performance with Sorbent Injection and Gas Conditioning," paper 36, *Proceedings: 1986 Joint Symposium on Dry SO_2 and Simultaneous SO_2/NO_x Control Technologies*, 2, EPRI CS-4966, December 1986
22. Helfritch, D. J., P. L. Feldman, B. Weinstein and M. W. McElroy, "Electrostatic Precipitator Upgrades for Furnace Sorbent Injection," paper 37, *Proceedings: 1986 Joint Symposium on Dry SO_2 and Simultaneous SO_2/NO_x Control Technologies*, 2, EPRI CS-4966, December 1986
23. Beittel, R., S. J. Bortz, G. R. Offen and D. C. Drehmel, "Effects of Injection Temperature and Quench Rate on Sorbent Utilization," paper 19, *Proceedings: 1986 Joint Symposium on Dry SO_2 and Simultaneous SO_2/NO_x Control Technologies*, 1, EPRI CS-4966, December 1986
24. Tokuda, K., M. Sakai, T. Sengoku, N. Murakami, M. W. McElroy and K. Mouri, "Evaluation of SO_2 Removal by Furnace Limestone Injection with Tangentially Fired Low-NO_x Burner," paper 12, *Proceedings: First Joint Symposium on Dry SO_2 and Simultaneous SO_2/NO_x Control Technologies*, 1, EPRI CS-4178, June 1985
25. Mozes, M. S., R. Mangal and R. Thampi, "Pilot Scale Studies of Limestone Injection Process," paper 50, *Proceedings: 1986 Joint Symposium on Dry SO_2 and Simultaneous SO_2/NO_x Control Technologies*, 2, EPRI CS-4966, December 1986
26. Brice, H., G. Chelu, G. Flament, R. Manhaval and M. Vandycke, "Reduction of SO_2 Emissions from a Coal Fired Power Station by Direct Injection of Calcium Sorbents in Furnace," paper 40, *Proceedings: First Joint Symposium on Dry SO_2 and Simultaneous SO_2/NO_x Control Technologies*, 2, EPRI CS-4178, June 1985
27. Reid, W. T., "Basic Factors in the Capture of Sulfur Dioxide by Limestone and Dolomite," *J. Engineering for Power*, p11, January 1970
28. Yu, H. C. and S. V. Sotirchos, "A Generalized Pore Model for Gas-Solid Reactions Exhibiting Pore Closure," *AIChE J.*, 33 (3), 382 (1987)
29. Karlsson, H. T. and J. Klingspor, "Tentative Modelling of *Spray-drying Scrubbing of SO_2*," *Chem. Eng. Technol.*, 10 (2), 104 (1987)

30. Gullett, B. K. and K. C. Kramlich, "Fundamental Process Involved in SO_2 Capture by Calcium-Based Adsorbents," *Proceedings: Fourth Annual Pittsburgh Coal Conference*, 219, 1987
31. Sparks, L. E., N. Plaks, G. H. Ramsey and R. E. Valentine, "Investigation of Combined Particulate and SO_2 Using E-SO_x," page 8-15, *Proceedings: Ninth EPA/EPRI Symposium on Flue Gas Desulfurization*, Vol. 2, EPRI CS-4390-V2, January, 1986
32. Sawyers, L. E., P. V. Smith and T. B. Hurst, "Flue Gas Desulfurization by Combined Furnace Limestone Injection and Dry Scrubbing," paper 26, *Proceedings: First Joint Symposium on Dry SO_2 and Simultaneous SO_2/NO_x Control Technologies*, 1, EPRI CS-4178, June 1985
33. Robards, R. F., R. W. Aldred, T. A. Burnett, L. R. Humphries and M. J. Widico, "High-Sulfur Spray Dryer Evaluations," 9-43, *Proceedings: Ninth EPA/EPRI Symposium Flue Gas Desulfurization*, Vol. 2, EPRI CS-4390-V2, January, 1986
34. Keeth, R. J., M. J. Krajewski and P. A. Ireland, *Economic Evaluation of FGD Systems - Volume 5: The NOXSO and SOXAL Sodium-Based Processes and Four Additional Calcium-Based Processes*, EPRI CS-3342-V5, 4-1 (1986)
35. Bobman, M. H., G. F. Weber and T. P. Dorchak, "Comparative Costs of Flue Gas Desulfurization: Advantages of Furnace Injection of Pressure Hydrated Lime Over Dry Scrubbing," paper 27, *Proceedings: 1986 Joint Symposium on Dry SO_2 and Simultaneous SO_2/NO_x Control Technologies*, 2, EPRI CS-4966, December 1986
36. Kenakkala, T., M. Suokos and J. Hautanen, "The Tampella LIFAC SO_2 Removal Process," paper 45, *Proceedings: 1986 Joint Symposium on Dry SO_2 and Simultaneous SO_2/NO_x Control Technologies*, 2, EPRI CS-4966, December 1986
37. Nolan, P. S. et al. "Operation of the LIMB/Humidifier Demonstration at Edgewater," *First Combined FGD and Dry SO_2 Control Symposium*, October 25-28, 1988b
38. Nolan, P. S. and R. V. Hendriks, *Initial Test Results of the Limestone Injection Multistage Burner (LIMB) Demonstration Project*," 81st Annual Meeting of the Air Pollution Control Association, Dallas, TX, June 20-24, 1988
39. Yoon, H. et al. "Pilot Process Variable Study of Coolside Desulfurization," *Environmental Progress*, 7(2), 104 (1988)
40. Muzio, L. J. and G. R. Offen, "Assessment of Dry Sorbent Emission Control Technologies," *J. of Air Pollution Control Association*, 37 (5), 642 (1987)
41. Rakes, S. L. et al. "Performance of Sorbents With and Without Additives Injected into a Small Innovative Furnace," *Proceedings: First Joint Symposium on Dry SO_2 and Simultaneous SO_2/NO_x Control Technologies*, Vol. I., NTIS PB85-232353, EPA 600/9-85-020a, July 1985
42. Roman, V. P. et al. "Flow Reactor Study of Calcination and Sulfation," *Proceedings: First Joint Symposium on Dry SO_2 and Simultaneous SO_2/NO_x Control Technologies*, 1, NTIS PB85-232353, EPA 600/9-85-020a, July 1985
43. Borgwardt, R. H., N. F. Roache and K. R. Bruce, "Method of Variation of Grain Size of Gas-Solid Reactions Involving CaO," *Industrial & Engineering Chemistry*, 25 (1), 165 (1986)
44. Cole, J. A. et al. *The Influence of Sorbent Physical Properties Upon Reactivity With Sulfur Dioxide - Task Final Report*, Report for US EPA-AEERL, Contract No. 68-02-3995, by Energy and Environmental Research Corporation, Irvine, CA (1987)
45. Silcox, G. D. et al. *Status and Evaluation of Calcitic SO_2 Captures: Analysis of Facilities Performance*, Vol 1, EPA 600/7-87-014, June 1987
46. Gullett, B. K. and K. R. Bruce, "Pore Distribution Changes of Calcium-Based Sorbents Reacting with Sulfur Dioxide," *AIChE J.*, 33(10), 1719 (1987)

47. Bruce, K. R., B. K. Gullett and L. O. Beach, "Comparative SO_2 Reactivity of CaO Derived from $CaCO_3$ and $Ca(OH)_2$," *AIChE J.*, 35(1), 37 (1989).
48. Simons G. A., A. R. Garman and A. A. Boni, "The Kinetic Rate of SO_2 Sorption by CaO," *AIChE J.* 33 (2), 211, (1987).
49. Borgwardt, R. H., K. R. Bruce and J. Blake, "An Investigation of Product-Layer Diffusivity for CaO Sulfation," *Industrial & Engineering Chemistry*, 26 (10), 1993 (1987).
50. Pohl, J. H. and R. H. Essenhigh, "Sulfur/sorbent Chemistry in Flame," presented at the Fourth Annual Pittsburgh Coal Conference, Pittsburgh, PA (1987).
51. Smith, M. Y., W. H. Beck and K. Hein, "Equilibrium Calculations of Fireside Products Formed During the Combustion of Rhineland Brown Coals with Special Emphasis on Fouling Constituents," *Combust. Sci. Tech.*, 42, 115 (1985).
52. Zghoul, A. M. and W. L. Grosshandler, "Evolution of Coal Sulfur in a Diffusion Flame," *Proceedings: Fourth Annual Pittsburgh Coal Conference*, 119 (1987).
53. Satterfield, C. N. and F. Feakes, "Kinetics of the Thermal Decomposition of Calcium Carbonate," *AIChE J.*, 5, 115 (1959).
54. Gallagher, P. K. and D. W. Johnson, "Kinetics of the Thermal Decomposition of $CaCO_3$ in CO_2 and Some Observations on the Kinetic Compensation Effect," *Thermochimica Acta*, 14, 255 (1976).
55. Caldwell, K. M., P. K. Gallagher and D. W. Johnson, "Effect of Thermal Transport Mechanisms on the Thermal Decomposition of $CaCO_3$," *Thermochimica Acta*, 8, 15 (1977).
56. Beruto, D. and A. W. Searcy, "Use of Langmuir Method for Kinetic Studies of Decomposition Reactions," *J. Chem. Soc.* Faraday Trans., 7(2), 145 (1974).
57. Powell, E. K. and D. W. Searcy, "The Rate and Activation Enthalpy of Decomposition of $CaCO_3$," *Metallurgical Trans.*, 11b, 427 (1980).
58. Borgwardt, R. H., "Calcination Kinetics and Surface Area of Dispersed Limestone Particles," *AIChE J.*, 31 (1), 103 (1985).
59. Haji-Sulaiman, Scaroni, A. W. and S. Yavuzkurt, "Sorbent Performance During Fluidized Bed Combustion," *Proceedings: Fourth Annual Pittsburgh Coal Conference*, 129 (1987).
60. Silcox, G. D., J. C. Kramlich and D. W. Pershing, "A Mathematical Model for the Flash Calcination of Dispersed $CaCO_3$ and $Ca(OH)_2$ Particles," *Ind. Eng. Chem. Res.*, 28, 155 (1989).
61. Slaughter, D. M., P. M. Lemieux, D. W. Pershing and D. Kirchgessner, "Chemical and Physical Characteristics of Calcium Oxide for Enhanced SO_2 Reactivity," *AIChE J.*, in press, (1989).
62. Borgwardt, R. H., "Sintering of Nascent Calcium Oxide," *Chem. Eng. Sci.*, 44 (1), 53 (1989).
63. German, R. M. and Z. A. Munir, "Surface Area Reduction during Isothermal Sintering," *J. Am. Ceram. Soc.*, 59, 379 (1976).
64. Mai, M. C. and T. F. Edgar, "Surface Area Evolution of Calcium Hydroxide During Calcination and Sintering," *AIChE J.*, 35 (1), 30 (1989).
65. Hartman, M. and R. W. Coughlin, "Reaction of Sulfur Dioxide with Limestone and the Grain Model," *AIChE J.*, 32 (3), 490 (1976).
66. Pigford, R. L. and G. Sliger, "Rate of Diffusion-Controlled Reaction Between a Gas and a Porous Solid Sphere," *Ind. Eng. Chem. Proc. Des. Dev.*, 12, 85 (1973).
67. James, N. J. and R. Hughes, "Rate of SO_2 Absorption in Calcined Limestones and Dolomites," Univ. of Salford, *Second International Conference on Control of Gaseous Sulfur and Nitrogen Compound Emissions*, Salford, England (1976).

68. Hartman, M. and O. Trnka, "Influence of Temperature on the Reactivity of Limestone Particles with Sulfur Dioxide," *Chem. Eng. Sci.*, 35, 1189 (1980)
69. Bhatia, S. K. and D. D. Perlmutter, "The Effect of Pore Structure on Fluid-Solid Reactions: Application to the SO_2-Lime Reaction," *AIChE J.*, 27, 226 (1981)
70. Borgwardt, R. H. and K. R. Bruce, "Effects of Specific Surface Area on the Reactivity of CaO with SO_2," *AIChE J.*, 32, 239 (1986)
71. Bhatia S. K. and D. D. Perlmutter, "A Random Pore Model for Fluid-Solid Reactions: II. Diffusion and Transport Effects," *AIChE J.*, 27 (2), 247 (1981)
72. Milne, C. R. and D. W. Pershing, "Time Resolved Sulfation Rate Measurements for Sized Sorbents," *Proceedings: Fourth Annual Pittsburgh Coal Conference*, 232 (1987)
73. Simons G. A. and A. R. Garman, "Small Pore Closure and the Deactivation of the Limestone Sulfation Reaction," *AIChE J.*, 32 (9), 1491 (1986)
74. Gregg, S. J. and K. S. W. Sing, *Adsorption, Surface Area and Porosity*, 2nd ed., Academic Press (1982)
75. Cole, J. A., J. C. Kramlich, W. R. Seeker and G. D. Silcox, "Activation and Reactivity of Calcareous Sorbents Towards Sulfur Dioxide, "*Environmental Science and Technology*, 11, 239 (1985)
76. Newton, G. H., D. K. Moyeda, G. Kindt, J. M. McCarthy, S. L. Chen, J. A. Cole and J. C. Kramlich, "Fundamental Studies of Dry Injection of Calcium-Based Sorbents for SO_2 Control in Utility Boilers," Project Summary prepared for EPA, March 1989
77. Lyngfelt, A. and B. Leckner, "SO_2 Capture in Fluidised-Bed Boilers: Re-emission of SO_2 Due to Reduction of $CaSO_4$," *Chem. Eng. Sci.*, 44 (2), 207 (1989)
78. Szekely, J., J. W. Evans and H. Y. Sohn, *Gas-Solid Reaction*, Academic Press, Inc. (1976)
79. Ramachandran, P. A. and L. K. Doraiswamy, "Modeling of Noncatalytic Gas-Solid Reactions," *AIChE J.*, 28 (6), 881 (1982)
80. Hartman, M. and R. W. Coughlin, "Reaction of Sulfur Dioxide with Limestone and the Influence of Pore Structure," *Ind. Eng. Chem. Process Des. Dev.*, 13 (3), 248 (1974)
81. Linder B. and D. Simonson, "Comparison of Structural Models for Gas-Solid Reactions in Porous Solids Undergoing Structural Changes," *Chem. Eng. Sci.*, 36 (9), 1519 (1981)
82. Garza-Garza, O. and M. P. Dudukovic, "Some Observations on Gas-Solid Noncatalytic Reactions with Structural Changes," *Chem. Eng. Sci.*, 36, 1257 (1981)
83. Szekely, J. and J. W. Evans, "A Structural Model for Gas-Solid Reactions with a Moving Boundary-II : The Effect of Grain Size, Porosity and Temperature on the Reaction of Porous Pellets," *Chem. Eng. Sci.*, 26, 1901 (1971)
84. Lee D. C. and C. Georgakis, "A Single, Particle-Size Model for Sulfur Retention in Fluidized Bed Coal Combustors," *AIChE J.*, 27 (3), 472 (1981)
85. Ramachandran P. A. and J. M. Smith, "A Single-Pore Model for Gas-Solid Noncatalytic Reactions," *AIChE J.*, 23 (3), 353 (1977)
86. Chrostowski J. and W. C. Georgakis, "Pore Plugging Model for Gas-Solid Reactions," *ACS Symp. Ser.*, 65, 225 (1978)
87. Yortsos, Y. C. and K. Shankar, "Asymptotic Analysis of Pore Closure Reactions," *Ind. Eng. Chem. Fundam.*, 23 (1), 132 (1984)
88. Bhatia S. K. and D. D. Perlmutter, "A Random Pore Model for Fluid-Solid Reactions: I. Isothermal Kinetic Control," *AIChE J.*, 26 (3), 379 (1980)
89. Christman P. G. and T. F. Edgar, "Distributed Pore-Size Model for Sulfation of Limestone," *AIChE J.*, 29 (3), 388 (1983)

90. Gupta, K. K., D. G. Sloat, A. W. Wendorf and J. Yasin, "Retrofit or Repowering for SO_x and NO_x Reductions for Coal-Fired Power Plants," *Proceedings: Fourth Annual Pittsburgh Coal Conference*, 893 (1987)

91. Stern, R.D., "EPA's Program for Evaluation and Demonstration of Low-Cost Retrofit LIMB Technology," *Proceedings: Fourth Annual Pittsburgh Coal Conference*, 232 (1987)

92. England, G. C., B. A. Folsom, R. Payne, T. M. Sommer, M. W. McElroy, P. J. Chappell and I. A. Huffman, "Prototype Evaluation of Sorbent Injection with Humidification," Presented at the First Combined FGD and Dry SO_2 Control Symposium, St. Louis, MO (1988)

93. Vandycke, M., B. Auric and J. L. Merry, "4 Years of Operating Experience with Direct In-Furnace SO_2 Reduction at the 600 MWe Unit of Provence Power Station," Presented at the First Combined FGD and Dry SO_2 Control Symposium, St.Louis, MO (1988)

94. Lott, T. A., D. P. Teixeira, L. J. Muzio and M. E. Teague, "Dry Sorbent Recycle Processes," Presented at the First Combined FGD and Dry SO_2 Control Symposium, St.Louis, MO (1988). EPRI Report GS-6307, April 1989

95. Chughtai, M. Y., "Desulfurization Results for Three Different Firing Concepts," Presented at the First Combined FGD and Dry SO_2 Control Symposium, St. Louis, MO (1988)

96. Staudinger, G., P. Melcher and K. Eckersdorfer, "Austrian Experience with Furnace Limestone Injection," Presented at the First Combined FGD and Dry SO_2 Control Symposium, St.Louis, MO (1988)

97. Cole, J. A., J. C. Kramlich, W. R. Seeker, G. D. Silcox, G. H. Newton, D. J. Harrison and D. W. Pershing, "Fundamental Studies of Sorbent Reactivity in Isothermal Reactors," paper 16, *Proceedings: 1986 Joint Symposium on Dry SO_2 and Simultaneous SO_2/NO_x Control Technologies*, 1, EPRI CS-4966, December 1986

98. Gullett, B. K., "Porosity, Surface Area, and Particle Size Effects of CaO Reacting With SO_2 at 1100°C," *Reactivity of Solids*, 6, 263 (1988)

99. Klingspor, J., H. T. Karlsson and I. Bjerle, "A Kinetic Study of the Dry SO_2-Limestone Reaction at Low Temperature," *Chem. Eng. Commun.*, 22, 88 (1983)

100. Brunauer, S., P. H. Emmett and E. Teller, *J. Am. Chem. Soc.*, 60, 309 (1938)

101. Jorgensen, C., J. C. S. Chang and T. G. Brna, "Evaluation of Sorbents and Additives for Dry SO_2 Removal," *Environmental Progress*, 6 (2), 26 (1987)

102. Jozewicz, W. and G. T. Rochelle, "Modeling of SO_2 Removal by Spray Dryers," *Proceedings: First Pittsburgh Coal Conference*, 663 (1984)

103. Harriot, P. and M. Kinzey, "Modeling the Gas and Liquid Phase Resistances in the Dry Scrubbing Process for SO_2 Removal," *Proceedings: Third Pittsburgh Coal Conference*, 220 (1986)

104. Damle, A. S., *Modelling of SO_2 Removal in Spray-Dryer Flue-Gas Desulfurization System*, EPA-600/7-85-038, December (1985)

105. Maibodi, M. M., T. L. Pearson, R. M. Counce, and W. T. Davis, "Simulation of Spray Dryer Absorber for Removal of SO_2 from Flue Gases," pp 7/19-7/23, *Proceedings: Tenth Symposium on Flue Gas Desulfurization*, 1, EPRI CS-5167, May 1987

106. Masters, K., *Spray Drying Handbook*, John Wiley and Sons (1979)

107. Blythe, G. M., J. M. Burke, M. E. Kelly, L. A. Rohlack and R. G. Rhudy, "EPRI Spray Drying Pilot Plant Status and Results," *Proceedings: Symposium on Flue Gas Desulfurization*, 2,EPA-600/9-83-0206 (NTIS PB84-110576) (1983)

108. Reinauer, T. V., J. P. Monat and M. Mutsakis, "Reducing Plant Pollution Exposure: Dry FGD on an Industrial Boiler," *Chemical Engineering Progress*, 79 (3), March 1983
109. Beals, J., L. Cannell and J. Hengel, "How FGD Reagent Quality Affects System Performance," *Power*, 128 (3) (1984)
110. Kelly, M. E., J. D. Kilgroe and T. G. Brna, "Current Status of Dry SO_2 Control Systems," *Proceedings: Symposium on Flue Gas Desulfurization*, 2, EPRI, Palo Alto, CA., March 1983
111. Dharmarajan, N. N., "Stirred Mill Proves Its Worth for Lime-Slaking Duties," *Power*, 129 (10), October 1985
112. Yeh, J. T., R. J. Demski, D. F. Gyorke and J. I. Joubert, "Experimental Evaluation of Spray Dryer Flue Gas Desulfurization for Use With Eastern U.S. Coals," *Proceedings: Symposium on Flue Gas Desulfurization*, 2, EPRI, Palo Alto, CA., March 1983
113. Stearns Catalytic Corp., *Economic Evaluation of Dry-Injection FGD Technology*, EPRI, Palo Alto, CA., January 1986
114. Kaplan, S. M., Y. J. Chen, R. C. Hyde, C. A. Sannes, Jr. and M. F. Skinner, "Dry Scrubbing at Northern States Power Company Riverside Generating Plant," *Proceedings: Symposium on Flue Gas Desulfurization*, 2, EPRI, Palo Alto, CA., March 1983
115. Midkiff, L. A., "Spray-Dryer System Scrubs SO_2," *Power*, 123 (1), January 1979
116. Lewis, M. F. and D. C. Gehri, "Atomization - The Key to Dry Scrubbing at the Coyote Station," *Proceedings: Symposium on Flue Gas Desulfurization*, 2, EPRI, Palo Alto, CA., March 1983
117. Maurin, P. G., "Controlling SO_2 Emissions: A Look at Wet and Dry Flue Gas Desulfurization Systems," *Plant Engineering*, 39, May 9, 1985
118. Turner, J. H., A. S. Viner, J. D. McKenna, R. E. Jenkins and W. M. Vatavuk, "Sizing and Costing of Fabric Filters: Parts I & II," *J. of Air Pollution Control Association*, 37 (6,7), June and July 1987
119. Donnelly, J. R., R. P. Ellis and W. C. Webster, "Dry Flue Gas Desulfurization End-Product Disposal Riverside Demonstration Facility Experience," *Proceedings: Symposium on Flue Gas Desulfurization*, 2, EPRI, Palo Alto, CA, March 1983
120. Getler, J. L., H. L. Shelton and D. A. Furlong, "Modeling the Spray Absorption Process for SO_2 Removal," *J. of Air Pollution Control Association*, 29 (12), 1270 (1979)
121. Wentz, T.H. and J.R. Thygeson, *Handbook of Separation Techniques for Chemical Engineers*, Chapter 4.10, McGraw-Hill, NY (1979)
122. Ranz, W. F. and W. R. Marshall, "Evaporation from Drops," *Chem. Eng. Prog.*, 48 (3), 141 and 48 (4), 173 (1952)
123. Partridge, G. P., *A Mechanistic Spray Dryer Mathematical Model Based On Film Theory To Predict Sulfur Dioxide Absorption and Reaction by a Calcium Hydroxide Slurry in the Constant Rate Drying Period*, Doctoral Dissertation, University of Tennessee, Knoxville, December 1987
124. Gooch, J. P., E. B. Dismukes, R. S. Dahlin and M. G. Faulkner, "Scaleup Tests and Supporting Research For the Development of Duct Injection Technology," *Draft Topical Report No. 1 - Literature Review*, U.S. DOE, Pittsburgh, PA, March 31, 1989
125. ANL Report, *Control Technology for Fine-Particulate Emissions*, Report prepared by Chemical Engineering Department, Manhattan College, Environmental Control-Coal Utilization Program, Argonne National Laboratory Report No. ANL/ECT-5, Argonne, IL, October 1978
126. Radian Corp., *Fundamental Investigation of DUCT/ESP Phenomena, Literature Review*, Report prepared by Radian Corporation, Austin, Texas, for the U. S. Department of Energy, April 1989

127. Liu, B. and Y. H. Yeh, "On the Theory of Charging and Aerosol Particles in an Electric Field," *J. of Applied Physics*, 39 (3), February 1968
128. Smith, W. B. and J. R. McDonald, "Development of a Theory for the Charging of Particles by Unhappier Ions," *J. of Aerosol Science*, 7 (1976)
129. Liu, B. Y. H. and A. Kapodia, "Combined Field and Diffusion Charging of Aerosol Particles in the Continuum Regime," *J. Aerosol Science*, 9, 227 (1978)
130. Mizuno, A., "Review of Particle Charging Research," *International Conference on Electrostatic Precipitation*, Monterey, CA, October 14-16, 1981
131. Faulkner, M. G. and J. L. DuBard, *A Mathematical Model of Electrostatic Precipitation*, (Revision 3): EPA-600/7-84-069 a, b, c, Vol. I, *Modeling and Programming*, NTIS PB84-212-679, Vol. II, *User's Manual*, NTIS PB84-212-687, *FORTRAN Source Code Tape*, NTIS PB84-232-990, August 6, 1984
132. Moslehi, G. B. and S. A. Self, "Electromechanics of Particulate Layers," *Proceedings: EPA/EPRI Fourth Symposium on the Transfer and Utilization of Particulate Control Technology*, EPA-600/9-84-025b, pp. 306-321 November 1984
133. Moslehi, G. B. and S.A. Self, "Electrical Breakdown of Particulate Layers," *Proceedings: EPA/EPRI Fourth Symposium on the Transfer and Utilization of Particulate Control Technology*, EPA-600/9-025b, 288-305, November 1984
134. Young, R. P., J. L. DuBard and L. E. Sparks, "The Onset of Electrical Breakdown in Dust Layers," *Proceedings: EPA/EPRI Fifth Symposium on the Transfer and Utilization of Particulate Control Technology*, EPRI CS-4404, 2, p. 25-1, February 1986
135. Spencer, H. W. III, *Electrostatic Precipitators: Relationship between Resistivity, Particle Size, and Sparkover*, EPA-600/2-76-144, May 1976
136. White, H. J., "Resistivity Problems in Electrostatic Precipitation," *J. Air Pollution Control Association*, 24 (4), April 1974
137. Bickelhaupt, R. E., "Electrical Volume Conduction in Fly Ash," *JAPCA*, 24, 251 (1974)
138. Bickelhaupt, R. E., *Influence of Fly Ash Compositional Factors on Electrical Volume Resistivity*, EPA-650/2-74-074, July 1974
139. Bickelhaupt, R. E., "Volume Resistivity - Fly Ash Composition Relationships," *Environmental Science & Technology*, 9(4), 336 (1975)
140. Bickelhaupt, R. E., "Surface Resistivity and the Chemical Composition of Fly Ash," *J. of the Air Pollution Control Association*, 25 (2), 148 (1975)
141. Dunson, J. F., Jr., "Effects of Ash Chemistry on Electrostatic Precipitator Performance," Paper 81-17.3, Presented at the 74th Annual Meeting of the Air Pollution Control Association, Philadelphia, PA, June 21-26, 1981
142. Bickelhaupt, R. E., *Measure of Fly Ash Resistivity Using Simulated Flue Gas Environments*, NTIS, PB-278758, Southern Research Institute, Birmingham, AL, March 1978
143. Gooch, J. P. and G. H. Marchant, *Electrostatic Precipitator Rapping Reentrainment and Computer Model Studies*, EPRI Report No. FP-792.3, EPRI Contract RP 413-1, June 1978
144. Sproull, W. T., "Minimizing Rapping Loss in Precipitators at 2000 Megawatt Coal Fired Power Station," *J. of the Air Pollution Control Association*, 22 (3), 181 (1972)
145. Spencer, H. W. III, *Rapping Reentrainment in a Nearly Full Scale Pilot Electrostatic Precipitator*, EPA-600/2-76-140:76 (1976)
146. Makansi, J., "Particulate Control: Optimizing Precipitators and Fabric Filters for Today's Power Plants, Special Report," *Power*, December 1986

147. Piulle, W., "1985 Update - Operating History and Current Status of Fabric Filters in the Utility Industry," *Proceedings of the Third Conference on Fabric Filter Technology for Coal Fired Power Plants*, Scottsdale, AZ, November 1985

148. Carr, P. C., "Pulse-Jet Fabric Filters for Particulate Control in the U. S. Electric Utility Industry," *International Coal Conference*, University of Pittsburgh, Pittsburgh, PA, September 1988

149. EPRI, *Fluid Dynamic Design Guidelines for Utility Fabric Filter Systems*, EPRI Report CS-3811, Electric Power Research Institute, Palo Alto, CA, October 1984

150. Bibb, R., "Baghouses: Facts and Fiction," *Kentucky Industrial Coal Conference*, University of Kentucky, Lexington, KY, April 1985

151. Hollinden, G. A., T. A. Burnett and R. L. Torstrick, "A Worldwide Review of Advanced SO_2 Control Technologies," *International Coal Conference*, University of Pittsburgh, Pittsburgh, PA, September 1988

152. Babu, M., J. College, R. Forsythe, R. Herbert, D. Kanary, D. Kerivan and K. Lee, *5-MW Toronto HALT Pilot Plant Testing - Final Test Results*, DOE Contract DE-AC22-85PC81012, December 1988

153. Yoon, H., M. R. Stouffer, W. A. Rosenhoover and R. M. Statnick, "Laboratory and Field Development of Coolside SO_2 Abatement Technology," *Second Annual Pittsburgh Coal Conference*, Pittsburgh, PA, September 1985

154. Cole, J. A., J. C. Kramlich, W. R. Seeker, G. D. Silcox, G. H. Newton, D. J. Harrison and D. W. Pershing, "Fundamental Studies of Sorbent Reactivity in Isothermal Reactors," *Proceedings: Symposium on Dry SO_2 and Simultaneous SO_2/NO_x Control Technology*, NTIS PB87-120456, EPA-600/9-86/029a (1986)

155. Gullett, B. K. and K. R. Bruce, "Effect of CaO Sorbent Physical Parameters Upon Sulfation," paper 5A-7, *EPA/EPRI First Combined FGD and Dry SO_2 Control Symposium*, St. Louis, MO (1988)

156. Snow, G. C., J. M. Lorrain and S. L. Rakes, "Pilot Scale Furnace Evaluation of Hydrated Sorbents for SO_2 Capture," paper 6, *Proceedings: 1986 Joint Symposium on Dry SO_2 and Simultaneous SO_2/NO_x Control Technologies*, 1, EPRI CS-4966, December 1986

157. Teixeira, D. P., T. A. Lott and L. J. Muzio, "Dry Sorbent SO_2 Control for New Power Plants Burning Low-Sulfur Western Coals," paper 7, *Proceedings: 1986 Joint Symposium on Dry SO_2 and Simultaneous SO_2/NO_x Control Technologies*, 1, EPRI CS-4966, December 1986

158. Overmoe, B. J., S. L. Chen, L. Ho, W. R. Seeker, M. P. Heap and D. W. Pershing, "Boiler Simulator Studies on Sorbent Utilization for SO_2 Control," paper 15, *Proceedings: First Joint Symposium on Dry SO_2 and Simultaneous SO_2/NO_x Control Technologies*, 1, EPRI CS-4178, June 1985

159. Bortz, S. J. and P. Flament, "Recent IFRF Fundamental and Pilot Scale Studies on the Direct Sorbent Injection Process," paper 17, *Proceedings: First Joint Symposium on Dry SO_2 and Simultaneous SO_2/NO_x Control Technologies*, 1, EPRI CS-4178, June 1985

160. Gooch, J. P., E. B. Dismukes, R. Beittel, J. L. Thompson and S. L. Rakes, "Sorbent Development and Production Studies," paper 11, *Proceedings: 1986 Joint Symposium on Dry SO_2 and Simultaneous SO_2/NO_x Control Technologies*, Vol. 1, EPRI CS-4966, December 1986

161. McCarthy, J. M., S. L. Chen, J. C. Kramlich, W. R. Seeker and D. W. Pershing, "Reactivity of Atmospheric and Pressure Hydrated Sorbents for SO_2 Control," paper 10, *Proceedings: 1986 Joint Symposium on Dry SO_2 and Simultaneous SO_2/NO_x Control Technologies*, 1, EPRI CS-4966, December 1986

162. Borgwardt, R. H. and K. R. Bruce, "EPA Study of Hydroxide Reactivity in a Differential Reactor," paper 15, *Proceedings: 1986 Joint Symposium on Dry SO_2 and Simultaneous SO_2/NO_x Control Technologies*, 1, EPRI CS-4966, December 1986

163. Kirchgessner, D. A., B. K. Gullett and J. M. Lorrain, "Physical Parameters Governing the Reactivity of $Ca(OH)_2$ with SO_2," paper 8, *Proceedings: 1986 Joint Symposium on Dry SO_2 and Simultaneous SO_2/NO_x Control Technologies*, 1, EPRI CS-4966, December 1986

164. Weber, G. F., M. E. Cowlings and M. H. Bobman, "Enhanced Utilization of Furnace Injected Calcium-Based Sorbents," paper 9, *Proceedings: 1986 Joint Symposium on Dry SO_2 and Simultaneous SO_2/NO_x Control Technologies*, 1, EPRI CS-4966, December 1986

165. Muzio, L. J., A. A. Boni, G. R. Offen and R. Beittel, "The Effectiveness of Additives for Enhancing SO_2 Removal with Calcium Based Sorbents," paper 13, *Proceedings: 1986 Joint Symposium on Dry SO_2 and Simultaneous SO_2/NO_x Control Technologies*, 1, EPRI CS-4966, December 1986

166. Ruiz-Alsop, R. N. and G. T. Rochelle, "Effect of Deliquescent Salt Additives on the Reaction of Sulfur Dioxide with Dry $Ca(OH)_2$," Amer. Chem. Soc. Div. Fuel Chem. Prepr., 30 (2), 88 (1985)

167. Huang, H. S., P. S. Farber, C. D. Livengood, J. T. Yeh, J. M. Markussen and C. J. Drummond, *Simultaneous NO_x and SO_2 Removal in a Spray Dryer System*, AIChE 1987 Spring National Meeting, Houston, TX., March 29-April 2, 1987

168. Keener, T. C., *Simultaneous Removal of Sulfur Oxides and Particulate Matter From Stoker-Fired Boiler Flue Gas With a Pilot Plant Fabric Filter Collector*, Master's Thesis, University of Tennessee, Knoxville, August 1977

169. Jarvis, J. B., R. W. Farmer and D. A. Stewart, "Description and Mechanism of Limestone FGD Operating Problems Due to Aluminum/Fluoride Chemistry," pp 7/79-7/83, *Proceedings: Tenth Symposium on Flue Gas Desulfurization*, 1, EPRI CS-5167, May 1987

170. Chang, J. C. S. and T. G. Brna, "Enhancement of Wet Limestone Flue Gas Desulfurization by Organic Acid/Salt Additives," paper 6b, *Tenth Symposium on Flue Gas Desulfurization*, Atlanta, GA, November 18-21, 1986

171. Wang, S. C. and D. A. Burbank, "Adipic Acid-Enhanced Lime/Limestone Test Results at the EPA Alkali Scrubbing Test Facility," *Flue Gas Desulfurization*, J.L. Hudson and G.T. Rochelle, eds., Amer. Chem. Soc. Symposium, Ser 188, pp267-306, Washington, (1982)

172. Rochelle, G. T., W. T. Weems, R. J. Smith and M. W. Hsiang, "Buffer Additives for Lime/Limestone Slurry Scrubbing," *Flue Gas Desulfurization*, J.L. Hudson and G.T. Rochelle, eds., Amer. Chem. Soc. Symposium Ser 188, pp243-265, Washington, (1982)

173. Chan, P. K. and G. T. Rochelle, "Limestone Dissolution: Effects of pH, CO_2, and Buffers Modeled by Mass Transfer," *Flue Gas Desulfurization*, J.L. Hudson and G.T. Rochelle, eds., Amer. Chem. Soc. Symposium 188, Washington, 75-97 (1982)

174. Rochelle, G. T. and C. J. King, "The Effect of Additives on Mass Transfer in $CaCO_3$ and CaO Slurry Scrubbing of SO_2 from Waste Gases," *Ind. Eng. Chem. Fundamentals*, 16 (1), 67 (1977)

175. Jozewicz, W. and G. T. Rochelle, "Fly Ash Recycle in Dry Scrubbing," *Environmental Progress*, 5 (4), 218 (1986)

176. Brown, C. A., G. M. Blythe, L. R. Humphries, R. F. Robends, R. A. Runzan and R. G. Rhudy, "Results From the TVA 10-MW Spray Dryer/ESP Evaluation," *EPA/EPRI First Combined FGD and Dry SO_2 Control Symposium*, St. Louis, MO., October 25-28, 1988

177. West, A. R., *Solid State Chemistry and Its Applications*, Wiley, NY (1984)
178. Berniere, F. and C. R. A. Catlow, "Mass Transport in Solids," *Plenum*, NY (1983)
179. Bardakci, T., *Thermochim. Acta*, 76, 287 (1984)
180. Shadman, F. and P. E. Dombek, "Enhancement of SO_2 Sorption on Lime by Structure Modifiers," *Can. J. Chem. Eng.*, 66, 930 (1988)
181. Kirchgessner, D. A. and J. M. Lorrain, "Lignosulfonate-Modified Calcium Hydroxide for Sulfur Dioxide Control," *Ind. Eng. Chem. Res.*, 26 (11), 2397 (1987)
182. Kirchgessner, D. A. and W. Jozewicz, "Structural Changes in Surfactant-Modified Sorbents During Furnace Injection," *AIChE J.*, 35 (3), 500 (1989)
183. Slaughter, D. M., S. L. Chen and W. R. Seeker, "Enhanced Sulfur Capture by Promoted Calcium-Based Sorbents," paper 12, *Proceedings: 1986 Joint Symposium on Dry SO_2 and Simultaneous SO_2/NO_x Control Technologies*, 1, EPRI CS-4966, December 1986
184. Sauer, H., J. D. Riley and G. Haug, "Operating Experience with a Dry FGD Plant Using a Circulating Fluid Bed (CFB) at Lignite-Fired Power Station of PREAG in Borken, F.D.R.," *Proceedings: First Combined Flue Gas Desulfurization and Dry SO_2 Control Symposium*, St. Louis, MO, October 1988, EPRI Report GS-6307, April 1989
185. Ramshaw, C. and R. H. Mallinson, *Mass Transfer Process*, U.S. Patent #4,283,255, assigned to Imperial Chemical Industries, London, England (1981)
186. Wen, J. W., *Centrifugal Gas-Liquid Contact Apparatus*, U.S. Patent #4,382,900, assigned to Imperial Chemical Industries, London, England (1983)
187. Basta, N., "Facelift for Distillation," *Chem. Eng.*, 14-16, March 2, 1987
188. Vivian, J. E., P. L. T. Brian and V. J. Krudonis, "The Influences of Gravitational Force on Gas Absorption in a Packed Column," *AIChE J.*, 1088 (1965)
189. Munjal, S., M. P. Dudukovic and P. Ramachandran, "Mass Transfer in a Rotating Packed Bed with Countercurrent Gas-Liquid Flow," presented at the Annual AIChE Meeting in Chicago, November 10-15, 1987
190. Gardner, N. C. and M. Keyvani, "Separation Processes in Rotating Packed Beds," presented at American Chemical Society Meeting in Los Angeles, CA, September 25-30, 1988
191. Gardner, N.C., M. Keyvani, "Residence Time, Liquid Holdup, and Power Requirement in Rotating Packed Beds," presented at AIChE Annual Meeting, Session 21, in Washington, DC, November 27-December 2, 1988
192. Keener, T. C. et al. "Study of the Reaction of SO_2 with Na_2HCO_3," *J. Air Pollution Control Assoc.*, 33 (4), 651 (1984)
193. Keener, T. and X. Jiang, and J. Hao, "Test Results on the Use of a Circulating Fluidized Bed Absorber (CFBA) for Control of SO_2," *Proceedings: First Combined Flue Gas Desulfurization and Dry SO_2 Control Symposium*, St. Louis, MO, October 1988. EPRI Report GS-6307, April 1989
194. Furlong, D., J. Wright and M. E. Prudich, "A Pilot Demonstration of the Limestone Emission Control (LEC) System," *Proceedings: First Combined Flue Gas Delsulfurization and Dry SO_2 Control Symposium*, St. Louis, MO, October 1988. EPRI Report GS-6307, April 1989
195. Adler, R. J., M. M. Balasko, R. V. Edwards and S. B. Adler, "Initial Studies of Particle Trapping in Impeller Fluidizers," *AIChE Symposium Series*, 83 (255), 112 (1987)
196. Adler, R. J., M. A. Patrick and J. Wineland, *Methods and Apparatus for Treating a Mixture of Particles and Fluids*, U.S. Patent Application No. 843, 055, March 1986

197. Reethof, G., "Applicability of High Intensity Acoustic Technology to Several Emission Control Technologies," *Proceedings of the Emissions Control for Small-Scale Combustors Workshop*, U.S. Department of Energy, PETC, Pittsburgh, PA, November 1987
198. Lindbauer, R. and F. Mail, "The 3-S Process: Simultaneous (Acid Gas and NO_x) Sodium Based Sorption of Flue Gases," *Proceedings: First Combined Flue Gas Desulfurization and Dry SO_2 Control Symposium*, St. Louis, MO, October, 1988. EPRI Report GS-6307, April 1989

CHAPTER 2
NEW CALCIUM-BASED SORBENTS FOR FLUE GAS DESULFURIZATION

Ray-Kuang Chiang, Gwanghoon Kwag and Malcolm E. Kenney
Department of Chemistry
Case Western Reserve University
Cleveland, OH

Abstract

Three sorbents for SO_2 in flue gas have been investigated. One is the mixture made by agitating a high water-to-solids slurry of $Ca(OH)_2$ and a reactive SiO_2 such as diatomite or a reactive SiO_2 source such as perlite or pumice. This mixture is composed largely of a porous form of the calcium silicate hydrate known as C-S-H. This C-S-H is the main active component in it. A second is the mixture prepared by vigorously agitating a high water-to-solids slurry of $Ca(OH)_2$ and an amount of fumed SiO_2 which is sufficient to react with only part of the $Ca(OH)_2$. This mixture is made up of $Ca(OH)_2$ particles imbedded in porous C-S-H. Both the C-S-H and the $Ca(OH)_2$ contribute to the effectiveness of this sorbent. It is presumed that similar sorbents can be prepared from mixtures in which diatomite, perlite or pumice are used in place of the fumed SiO_2. The third sorbent is prepared by vigorously agitating a high water-to-solids slurry of type I (ordinary) portland cement. This sorbent is composed largely of a mixture of porous C-S-H, $Ca(OH)_2$ and the aluminate phase known as AFm. All three of these species contribute to its effectiveness. Each of these sorbents is effective and of practical interest. Each is simple to prepare. The first and third can be made from readily available, low-cost reactants, and the second probably also can be made from such reagents. The cement sorbent is unique and appears to be of the most interest.

Introduction

Legislation by Congress, particularly the Clean Air Act Amendments of 1990, has created a national need to develop and install commercially cost-effective technologies that will reduce SO_2 emissions from fossil-fuel-fired utility boilers. Among the technologies that show promise for this purpose are those based on in-duct injection of sorbents or sorbent slurries. However, while previous work has shown that in-duct injection sorbent techologies are practical, fully suitable sorbents and sorbent slurries have not yet been reported. To a substantial extent, the properties needed in such sorbents are governed by the properties of SO_2 and CO_2, and thus it is appropriate to consider the properties of these two compounds first.

Properties of SO_2

Sulfur dioxide is a colorless gas which neither burns nor supports combustion. Its melting point is -75.5°C and its boiling point is -10.1°C. It is thermally stable, dissociation becoming significant only above 2000°C. Sulfur dioxide is readily soluble in H_2O, 3.9 L dissolving in 100 g of H_2O at 25°C. Its solubility in H_2O increases with increasing partial pressure and decreases with increasing temperature. Its oxidation to SO_3 with O_2 is very favorable thermodynamically but very slow in the absence of a suitable catalyst [4-7].

Solutions of SO_2 in H_2O are often referred to as sulfurous acid, H_2SO_3. However, H_2SO_3, if it is present, is present in only infinitesimal amounts. The first acid dissociation constant for "H_2SO_3" is properly given as

$K_1 = [H^+][HSO_3^-] / [total\ dissolved\ SO_2] - [HSO_3^-] - [SO_3^{2-}]$

K_1 has a value of $1.3 \cdot 10^{-2}$. "Sulfurous acid" is thus a relatively strong acid [4].

Properties of CO_2

Carbon dioxide is also a colorless gas which neither burns nor supports combustion. Its melting point is -56.5°C (5.2 atm) and its sublimation temperature is -78.5°C. It also is thermally stable, dissociation only becoming significant above 2000°C. Carbon dioxide is soluble in H_2O but much less so than SO_2, 0.36 L dissolving in 100 g of H_2O at 20°C. Its solubility in H_2O increases with increasing partial pressure and decreases with increasing temperature. Its reduction to CO is relatively difficult to accomplish [4,5,7].

Solutions of CO_2 in water are often referred to as "carbonic acid," H_2CO_3. However, while H_2CO_3 is present in these solutions, a considerable amount of loosely hydrated CO_2 is also present. The first acid dissociation constant for "H_2CO_3" is generally given as

$$K_1 = [H^+][HCO_3^-]/[H_2CO_3].$$

K_1 has a value of $4.16 \cdot 10^{-7}$. "Carbonic acid" is thus a very weak acid. (When the true activity of H_2CO_3 is taken into account, K_1 is $2 \cdot 10^{-4}$ and thus more nearly in agreement with expectations based on structure and bonding considerations [4].)

Properties Required in Sorbents

While the susceptibility of SO_2 to oxidation could be used as a basis for sorbents for it, generally its ability to neutralize bases is used. Thus SO_2 sorbents are commonly bases.

A base suitable for use as a sorbent must meet a number of requirements. It must react rapidly with SO_2 in the presence of H_2O vapor or liquid H_2O and it must have a high capacity for SO_2. At the same time it must react slowly with CO_2 in the presence of H_2O vapor or liquid H_2O. This is of considerable importance because the concentration of CO_2 is much higher than that of SO_2 in flue gas (typically flue gas contains up to about 19% CO_2 as against up to about 4000 ppm or 0.4% SO_2 [8]).

In addition the base must be low cost. This requirement eliminates most bases. Further, it must react with flue gas to give an environmentally acceptable product. Thus, if the spent sorbent cannot be sold or recycled and, as a consequence, must be disposed of in a landfill, it must not contain appreciable concentrations of toxic ions. Also it must form a landfill that is physically stable.

Beyond all this, the sorbent should have good handling characteristics. That is, it should flow readily in handling systems and it should not form wall deposits in ducts.

Previously Considered Sorbents

Among the previously considered sorbents are Na_2CO_3, $NaHCO_3$ and Na_2SO_3 [9]. These sodium-based sorbents are of considerable interest for a variety of reasons. However, the spent sorbents they yield contain easily leachable Na^+ ions and they thus cannot be disposed of satisfactorily in landfills.

Other sorbents of interest are $CaCO_3$, CaO and $Ca(OH)_2$ [10]. The first of these is abundant naturally and the other two can be made from it easily. The spent sorbents yielded by all three do not contain easily leachable ions. However, these sorbents do not react with SO_2 sufficiently rapidly under acceptable conditions. Another calcium-based sorbent that has been considered is one made from an aqueous slurry of fly ash and $Ca(OH)_2$ (11,12). It is still in an experimental stage.

If more reactive calcium-based sorbents could be found, fully practical sorbents could result. With this in mind, this work on calcium-based sorbents was undertaken.

Techniques and Apparatus

Infrared Spectroscopy

A Perkin-Elmer Model 16 PC Fourier Transform Infrared Spectrophotometer (Perkin-Elmer, Norwalk, CT) interfaced (GPIB-PCII) to a DECpc 433dx LP 486 PC (Digital Equipment, Westminster, MA) and a Perkin-Elmer IR 598 Spectrophotometer were used to record the infrared spectra. KBr plates were used to support those samples which were ground in mineral oil. The spectra were calibrated with the 1602.0 cm-1 band of polystyrene. Typical instrument settings used are given in Table 2.1.

Table 2.1 Infrared Instrument Settings

Perkin-Elmer 16 PC FT-IR	
range	4000-400cm^{-1}
resolution	4 cm-^{1}
apodization	weak
background scans	2
number of scans	8
Perkin-Elmer 598	
range	4000-200 cm^{-1}
scan time	12 min
slit width	medium

X-Ray Powder Diffractometry

The X-ray powder diffraction patterns were recorded on a Philips APD 3520 Diffractometer (Philips Electronic Instruments, Mahwah, NJ) equipped with a Philips XRG 3100 Generator (Cu target) and a Philips graphite monochromator. The instrument was calibrated with the 3.342 Å reflection of a quartz plate. Typical instrument settings used are given in Table 2.2.

Table 2.2 X-Ray Powder Diffraction Instrument Settings

generator power	40 KV, 20 mA
scan mode	step scan
2θ scan rate	1°/min
2θ step size	0.05
2θ range	4-60

BET Nitrogen Adsorption Surface Area Determinations

Surface areas were determined with a Micromeritics FlowPrep 060 Degasser (Micromeritics Instrument, Norcross, GA) and a Gemini II 2370 Surface Area Analyzer (Micromeritics) interfaced (RS 232) to a Gateway 2000 386/25 PC (Gateway 2000, North Sioux City, SD) (He free space adsorbate, N_2 surface area adsorbate). Typical instrument settings are given in Table 2.3.

Table 2.3 Degasser and Surface Area Instrument Parameters and Setting

outgasing		
	drying temperature	105°C
	drying time	2 h
area determination		
	sampling interval	equilibrium
	temperature	-195.8°C
	relative pressure	0.05-0.3
	number of points	5

Scanning Electron Microscopy

Morphological studies were carried out with JEOL 35CF and JEOL 100 SX Scanning Electron Microscopes (Jeol USA, Peabody, MA). The JEOL 100 SX microscope was equipped with a Tracor Northern EDAX Quantitative Element Dispersive Spectrum accessory (TN 5500 X-ray Analyzer, Tracor Northern, Middleton, WI). The samples were coated with C and Au or with C and Pd. Typical instrument settings are given in Table 2.4.

Table 2.4 Scanning Electron Microscope Settings

Scanning electron microscope	
accelerating voltage	25 kV
probe current	$1 \cdot 10^{-7}$ A
mode	SEI
magnification	2,000-20,000 X
energy dispersive spectra	
accelerating voltage	5 kV
probe current	$1 \cdot 10^{-9}$ A

SO_2 Uptake Measurements

A custom-built sorption system patterned after a system described by Jozewicz [11] was used to measure the SO_2 uptake of the sorbents (Figure 2.1). The sorption system was designed to simulate typical baghouse conditions in a power plant. The feed gases were N_2 (Dry Grade, 99.99%, Linde, Somerset, NJ) and a mixture of N_2 and SO_2 or a mixture of SO_2, CO_2 and N_2 (calibrated, Matheson, Montgomeryville, PA). The N_2 was directed by stainless steel tubing to mass flowmeter 1 (258C, MKS Instruments Inc., Andover, MA), and the SO_2-containing gas was directed by stainless steel tubing to mass flowmeter 2 (258C, MKS Instruments). These mass flowmeters were controlled by a mass flowmeter controller (247C, MKS Instruments). The N_2 from mass flowmeter 1 was directed to a stainless steel H_2O-evaporation chamber filled with stainless steel packing. Simultaneously, H_2O was introduced into the chamber with a peristaltic pump (Ismatec, Omega, Stamford, CT). The chamber was heated with a heating mantle and its temperature was controlled with a thermocouple (type J, Omega) and a PID temperature controller (Shimaden, Tokyo, Japan). A pressure relief valve

(Nupro, Swagelock, Willoughby, OH) was connected to the chamber (to relieve any inadvertent pressure rises).

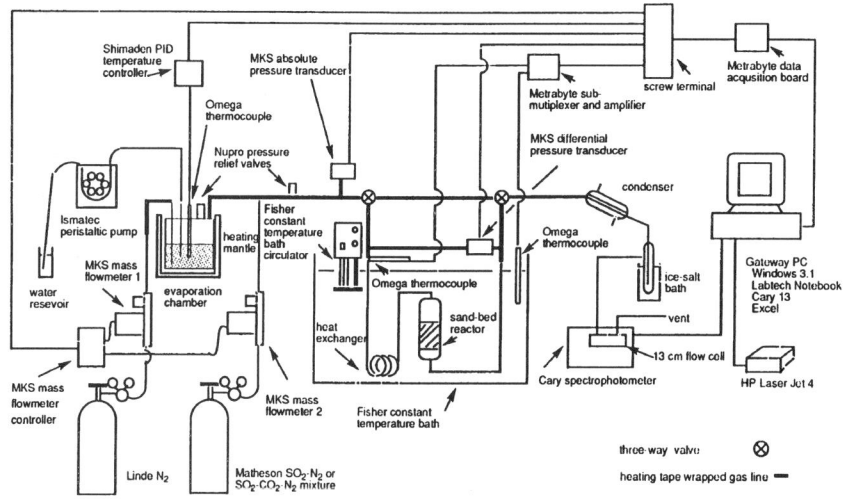

Figure 2.1 Custom-built sand-bed SO_2 sorption system.

The humidified N_2 gas produced in the chamber was directed to a tee. The SO_2-containing gas from mass flowmeter 2 was directed to this same tee and was mixed with the humidified N_2 at this point. Just beyond the tee, a second relief valve (Nupro) was connected to the system. Further on, an absolute pressure transducer (220A, MKS Instruments) was connected to it. Beyond this transducer, a three-way valve (Nupro) was connected. Here, the gas was directed either to the reactor bypass or to the reactor. If it was directed to the reactor, its temperature was monitored with a thermocouple (type J, Omega) and then it was passed through a stainless steel heat-exchanger coil and through the reactor. The reactor was a sand-bed type reactor and was made of glass and Teflon. Both the coil and reactor were immersed in a heated constant-temperature water bath (Fisher Scientific, Pittsburgh, PA). The temperature of the bath was monitored with a thermocouple (type J, Omega). From the reactor, the gas was directed to another three-way valve (Nupro). The gas from the bypass was also directed to this valve. A differential pressure transducer (220D, MKS Intruments) was connected to the system both before the heat exchanger coil and after the reactor. The gas from the three-way valve (Nupro) was then directed to a glass condenser. From the far side of the evaporation chamber to the condenser, the exposed tubing was heated with heating tape. The gas from the condenser was directed by Tygon tubing to a trap immersed in an ice-salt mixture held in a dewar. The resulting dehumidified gas was directed to a flow-through gas cell (fused SiO_2, 13 cm, Spectrocell Corp., Oreland, PA) mounted in a UV-Visible spectrophotometer (Varian Cary 1, Varian Associates, Palo Alto, CA) interfaced (IEEE 488) to a Gateway 2000 386/25 PC (Gateway 2000). Finally the gas was directed to a vent. Typical instrument settings used with the spectrophotometer are given in Table 2.5

Table 2.5 UV-Visible Spectrophotometer Settings

wavelength	286 nm
photometric mode	absorbance
abscissa range	0-70 min
spectral band width	2.0 nm
signal average time	3 sec
lamp select	UV

The data from the mass flowmeter controller, the evaporation chamber thermocouple, the absolute pressure transducer, the gas line thermocouple, the differential pressure transducer, and the water bath thermocouple were fed to an acquisition board (Metrabyte DAS 16, Kiethley, Taunton, MA) and then to the PC. The data from the spectrophotometer also were fed to the PC. All these data were processed with Labtech Notebook for Windows (Labtech, Wilmintong, MA) and Cary 13 (Varian) software. The processed data were displayed on the PC monitor and, if desired, printed with the printer (HP Laser Jet 4, Hewlett-Packard, Boise, ID) and saved in a file. Typical parameters and instrument readouts are given in Table 2.6

Table 2.6 Sand-Bed Sorber Parameters and Settings

sample	0.50 g
sand	20 g
N_2 flow rate	800 mL/min
SO_2 mixture flow rate	200 mL/min
SO_2 concentration	2000 ppm
pump flow rate	6.5 mL/h

evaporation chamber temperature	90±5°C
relative humidity	60%
absolute pressure	750± 8 torr
gas line temperature	65±5°C
water bath temperature	60±1°C
differential pressure	10± 1 torr
absorbance data interval	3 sec
run duration	1 h

The relationship between absorbance and SO_2 concentration was established with a calibration curve. This curve was constructed with a group of three gas mixtures of known SO_2 concentration (calibrated, Matheson). From the UV absorbance data collected and the calibration curve, the SO_2 uptake value of sorbent was determined with the equation

$$SO_2 \text{ uptake (mmol/g)} = (\sum_{i=1}^{1200} (C_i - C_0)rt - \sum_{i=1}^{1200} (C_i - C_b)rt/w)$$

where C_i is the SO_2 concentration of the incoming gas in mmol/g, C_b is the SO_2 concentration of the gas after being passed through the sand alone in mmol/L, Co is the SO_2 concentration of the gas after being passed through the sorbent in mmol/L, r is the rate of flow of the incoming gas in L/sec, t is the time between absorbance measurements in sec, and w is the sample weight in g.

Milling

A ball mill (U.S. Stoneware, Mahwah, NJ) was used to carry out some sorbent synthesis work. A typical rate of rotation used was 120 rpm.

Shaking

A wrist shaker (Burrell, Pittsburg, PA) also was used to carry out some sorbent synthesis work. A typical rate of shaking used was 2-3 Hz.

Survey of Calcium Silicates

Introduction

The results of studies done earlier on $Ca(OH)_2$-fly ash (Advocate-type) sorbents suggested that the active components of these sorbents include one or more calcium silicates [12]. With this as a lead, work on calcium silicate sorbents was undertaken. Work designed to yield a relatively complete compilation of the names, the formulas, and some of the important properties of the known calcium silicates; and work designed to show the potential of members of a selected group of these silicates was carried out. The work on the compilation is described in this section and the work on the potential of the selected silicates is described in the next section.

Tabulation of Known Calcium Silicates

A relatively complete tabulation of the names, the formulas, and some of the important features of the known silicates in which Ca is the only metal cation is given in Table 2.7. A parallel tabulation of the occurrences, syntheses, and commercial sources of the silicates listed in Table 2.7 is given in Table 2.8. As noted, in some cases the name given refers to a compound whose exact composition is unknown, in other cases to a phase and not a compound, and in at least one case, that of C-S-H, to a family of closely related phases. Because of this, the formulas are idealized or are approximated in a number of instances.

Further Description of C-S-H

The family of phases called C-S-H is of particular interest and merits further discussion. This leads to the need for more detail on the structures of 11-Å tobermorite, 14-Å tobermorite and jennite.

11-Å TOBERMORITE, $Ca_5(Si_6O_{16}(OH)_2) \cdot 4H_2O$. 11-Å Tobermorite, as indicated in Table 2.7, is a phase and has a layer structure. The silicate layers of this phase have a core of Ca and O. The Ca of the core is seven coordinated and the formula of the core is CaO_2. All the oxygen atoms of the core are shared with Si-O chains (Figure 2.2). On both sides of the core are end-to-end oligomeric silicate chains. These chains have the formula $(Si_3O_9H)_n$ (Figure 2.3). They share oxygens with the core and form ribs on it, thus giving a ribbed layer. Layers of this type and those composed of Ca ions and H_2O molecules stack one upon another to give the phase (Figure 2.4). The repeat spacing of the layer stack is 11.3 Å, hence the name of this silicate [74].

Table 2.7 Names, Formulas, Structural Data, and H$_2$O Content of Known Calcium Silicates

	Formulas			CA Coord	Silicate Ion	H$_2$O (%)
	Structural	Oxide	Cement			
Anhydrous						
high pressure wollastonite	CaSiO$_3$	CaO·SiO$_2$	CS	6, 7	ring	
kilchoanite	Ca$_6$(SiO$_4$)(Si$_3$O$_{10}$)	6CaO·4SiO$_2$	C$_3$S$_2$	6, 8	ortho & tri	
parawollastonite	β-CaSiO$_3$	CaO·SiO$_2$	CS	6, 8	chain	
pseudowollastonite	αCaSiO$_3$	CaO·SiO$_2$	CS	8	ring	
rankinite	Ca$_3$Si$_2$O$_7$	3CaO·2SiO$_2$	C$_3$S$_2$	7	di	
wollastonite	β-CaSiO$_3$	CaO·SiO$_2$	CS	6, 7	chain	
γ-Ca$_2$SiO$_4$,	Ca$_2$SiO$_4$	2CaO·SiO$_2$	γ-C$_2$S	6		
β-Ca$_2$SiO$_4$	Ca$_2$SiO$_4$	2CaO·SiO$_2$	β-C$_2$S	8, 10	ortho	
α-, α'-Ca$_2$SiO$_4$	Ca$_2$SiO$_4$	2CaO·SiO$_2$	α-, α'-C$_2$S	8, 10	ortho	
T$_1$-, T$_2$-, T$_3$-Ca$_3$SiO$_5$	Ca$_3$O(SiO$_4$)	3CaO·SiO$_2$	T-C$_3$S	6, 7		
M$_1$-,M$_2$-,M$_3$-Ca$_3$SiO$_5$	Ca$_3$O(SiO$_4$)	3CaO·SiO$_2$	M-C$_3$S	6, 7		

Dry Scrubbing Technologies for Flue Gas Desulfurization

New Calcium-Based Sorbents for Flue Gas Desulfurization

Table 2.7 Names, Formulas, Structural Data, and H_2O Content of Known Calcium Silicates (cont.)

	Formulas			CA Coord	Silicate Ion	H_2O (%)
	Structural	Oxide	Cement			
R-Ca_3SiO_5	$Ca_3O(SiO_4)$	$3CaO \cdot SiO_2$	R-C_3S	7	ortho	
Hydrated, crystalline						
afwillite	$Ca_3(SiO_3OH)_2 2H_2O$	$3CaO \cdot 2SiO_2 \cdot 3H_2O$	$C_3S_2H_3$	7	ortho	15.8
calciochondrodite[a]	$Ca_5(SiO_4)_2(OH)_2$	$5CaO \cdot 2SiO_2 \cdot H_2O$	C_5S_2H	6^b	ortho	4.3
dellaite	$Ca_6(Si_2O_7)(SiO_4)(OH)_2$	$6CaO \cdot 3SiO_2 \cdot H_2O$	C_6S_3H	6,7	ortho & di	7.9
α-dicalcium silicate hydrate	$Ca_2(SiO_3OH)OH$	$2CaO \cdot SiO_2 \cdot H_2O$	C_2SH	6,7	ortho	9.5
foshagite	$Ca_4(Si_3O_9)(OH)_2$	$4CaO \cdot 3SiO_2 \cdot H_2O$	C_4S_3H	6	chain	4.3
gyrolite[a]	$Ca_{16}(Si_8O_{20})(Si_8O_{20})_2 (OH)_8 \cdot 14H_2)^{c,d}$	$2CaO \cdot 3SiO_2 \cdot 2H_2O$	$C_2S_3H_2$	$6,7^d$	sheet	12.2
hilebrandite[a]	$Ca_2(SiO_3)(OH)_2$	$2CaO \cdot SiO_2 \cdot H_2O$	C_2SH		chain	9.5
jaffeite (tricalcium silicate hydrate)	$Ca_6(Si_2O_7)(OH)_6$	$3CaO \cdot SiO_2 \cdot 3H_2O$	C_3SH_3		di	10.6
jennite[a]	$Ca_9(Si_6O_{16}(OH)_2)(OH)_8 \cdot 6H_2O$	$9CaO \cdot 6SiO_2 \cdot 11H_2O$	$C_9S_5H_{11}$		chain	18.6

Table 2.7 Names, Formulas, Structural Data, and H_2O Content of Known Calcium Silicates (cont.)

	Formulas			CA Coord	Silicate Ion	H_2O (%)
	Structural	Oxide	Cement			
kilchoanite[a]	$Ca_6(SiO_4)(Si_3O_{10}) \cdot H_2O$[e]	$6CaO \cdot 4SiO_2 \cdot H_2O$	C_6S_4H	6,8e	ortho & tri	4.4
killalite[a]	$Ca_3Si_2O_7 \cdot 0.5H_2O$	$3CaO \cdot 3SiO_2 \cdot 0.5H_2O$	$C_3S_3H_{0.5}$			3.0
K-phase	$Ca_7(Si_{16}O_{38})(OH)_2$	$7CaO \cdot 6SiO_2 \cdot H_2O$	C_7S_6H	6	sheet	1.3
nekoite	$Ca_3(Si_6O_{15}) \cdot 7H_2O$	$3CaO \cdot 6SiO_2 \cdot 7H_2O$	$C_3S_6H_7$	6	sheet	19.3
okenite	$Ca_{10}(Si_6O_{16})(Si_6O_{15}) \cdot 18H_2O$	$5CaO \cdot 6SiO_2 \cdot 9H_2O$	$C_5S_6H_9$	6	ladder & sheet	16.5
rosenhahnite	$Ca_3(Si_3O_8(OH)_2)$	$3CaO \cdot 3SiO_2 \cdot H_2O$	C_3S_3H	6,7	tri	4.9
suolunite	$Ca_2(Si_2O_5(OH)_2) \cdot 4H_2$	$CaO \cdot SiO_2 \cdot 3H_2O$	CSH_3	8	di	13.4
11-Å tobermorite[c]	$Ca_5(Si_6O_{16}(OH)_2) \cdot 4H_2O$	$5CaO \cdot 6SiO_2 \cdot 5H_2O$	$C_5S_6H_5$	7	chain	12.3
14-Å tobermorite[a,c]	$Ca_5(Si_6O_{16}(OH)_2) \cdot 8H_2O$[f]	$5CaO \cdot 6SiO_2 \cdot 9H_2O$	$C_5S_6H_9$	7[f]	chain	20.2
trabzonite[a]	$Ca_4Sl_3O_{10} \cdot 2H_2O$	$4CaO \cdot 3SiO_2 \cdot 2H_2O$	$C_4S_3H_2$			8.2
truscottite[a]	$Ca_{14}(Si_8O_{20})(Si_{16}O_{38})(OH)_8 \cdot 2H_2O$	$7CaO \cdot 12SiO_2 \cdot H_2O$	$C_7S_{12}H$		sheet	4.6

Table 2.7 Names, Formulas, Structural Data, and H_2O Content of Known Calcium Silicates (cont.)

	Formulas			CA Coord	Silicate Ion	H_2O (%)
	Structural	Oxide	Cement			
xonotlite	$Ca_6(Si_6O_{17})(OH)_2$	$6CaO \cdot 6SiO_2 \cdot H_2O$	C_6S_6H	6,7	ladder	2.5
Z-phase[a]	$Ca_8(Si_8O_{20})_2 \cdot 16H_2O$	$CaO \cdot 2SiO_2 \cdot H_2O$	C_2SH		sheet	17.0
Hydrated, amorphous						
C-S-H	$CaO \cdot SiO_2 \cdot H_2O$	$CaO \cdot SiO_2 \cdot H_2O$	C-S-H		chain	~30.0
Additional Anion						
bultfonteinite	$Ca_4Si_2O_{10}F_2H_6$			7	ortho	13.4
cuspidine	$Ca_4Si_2O_7(OH,F)$				di	
fukalite	$Ca_4Si_2O_6(OH,F)(CO_3)$				di	
hydroxyellestadite	$Ca_{10}(SiO_4)_3(SO_4)_3(OH,Cl,F)_2$			6	ortho	
rustmite	$Ca_{10}(Si_2O_7)_2(SiO_4)Cl_2(OH)_2$				di & ortho	
octacalcium chlorosilicate	$Ca_8Si_4O_{12}Cl_8$				ring	
scawtite	$Ca_7(Si_6O_{18})(CO_3)2H_2O$	$7CaO \cdot 6SiO_2 \cdot CO_2 \cdot 2H_2O$	$C_7S_6\overline{C}H_2$	7,8	ring	4.3
spurrite	$Ca_5(SiO_4)_2(CO_3)$	$5CaO \cdot 2SiO_2 \cdot CO_2$	$C_5S_2\overline{C}$	7,8	ortho	

Table 2.7 Names, Formulas, Structural Data, and H_2O Content of Known Calcium Silicates (cont.)

	Formulas			CA Coord	Silicate Ion	H_2O (%)
	Structural	Oxide	Cement			
tilleyite	$Ca_5(Si_2O_7)(CO_3)_2$	$5CaO \cdot 2SiO_2 \cdot 2CO_2$	$C_5S_2\overline{C}_2$	6,7	di	
thaumasite	$[Ca_3Si(OH)_6 \cdot 12H_2O](SO_4)(CO_3)$	$3CaO \cdot SiO_2 \cdot SO_3 \cdot CO_2 \cdot 15H_2O$	$C_3S\overline{S}\overline{C}H_{15}$	8	Si $(OH)_6^{2-}$	
zeophyllite	$Ca_{13}Si_{10}O_{28-6}F_{7-4}(OH)_{2-6}H_2O$			8,9	sheet	8.2

[a] Structure not known. [b] Based on structure of chondrodite. [c] Formula approximate. [d] Based on structure of natural mineral, $NaCa_{16}Si_{23}A10_{60}(OH)_8 \cdot 14H_2O$. [e] Based on data for anhydrous kilchoanite. [f] Based on data for 11-Å tobermorite

New Calcium-Based Sorbents for Flue Gas Desulfurization

Table 2.8 Occurrence, Synthesis, and Commercial Sources of Known Calcium Silicates

		Synthesis					
	Occurrence	Reactants	Conditions			Commercial sources	Leading References
			Temp (°C)	Time (Day)	Pressure (kbar)		
Anhydrous							
high pressure wollastonite		$CaSiO_3$ glass	1300		65		13
kilchoanite		CaO-SiO_2gel, H_2O	715	1-3			14
parawollastonite		CaO, SiO_2	1100				15,16
pseudowollastonite		$CaSiO_3$	1400				17
rankinite	Antrim, Ireland	CaO, SiO_2	1400				18
wollastonite	Willsboro, NY	CaO, SiO_2	1100			Ward's[a], NYCO[b]	19, 20
γ-Ca_2SiO_4, β-Ca_2SiO_4, α–, α^1-Ca_2SiO_4		CaO, SiO_2	1450	0.3-1		CTL(β,γ)[c]	21,22

Table 2.8 Occurrence, Synthesis, and Commercial Sources of Known Calcium Silicates (cont.)

		Synthesis					
				Conditions			
	Occurrence	Reactants	Temp (°C)	Time (Day)	Pressure (kbar)	Commercial sources	Leading References
T_1-,T_2-,T_3-Ca_3SiO_5		CaO, SiO_2	1500	0.5			23,24
M_1-, M_2-, M_3-Ca_3SiO_5						CTL	
R-Ca_3SiO_5							
Hydrated, crystalline							
afwillite	Crestmore, CA	CaO, SiO_2, H_2O	125-175	1-4			25
calciochondrodite		CaO-SiO_2gel, H_2O	600-700	1			26,27
dellaite		CaO-SiO_2gel, H_2O	800	1			28,29
α–dicalcium silicate hydrate		CaO, SiO_2, H_2O	90-200	2			30,31
foshagite	Crestmore, CA	CaO, SiO_2, H_2O	300-350	5			32
gyrolite	Poona, India	CaO, SiO_2, H_2O	150-400	2		Ward's	33,35
hillebrandite	Durango, Mexico	CaO, SiO_2, H_2O	170-250	1			36,37

Table 2.8 Occurrence, Synthesis, and Commercial Sources of Known Calcium Silicates (cont.)

		Synthesis					
	Occurrence	Reactants	Conditions			Commercial sources	Leading References
			Temp (°C)	Time (Day)	Pressure (kbar)		
jaffeite (tricalcium silicate hydrate)	Namibia	Ca_3SiO_5, H_2O	200-450	2			
jennite	Crestmore, CA	CaO, fumed SiO_2, H_2O	80	40			40, 41
kilchoanite[d]	Kilchoan, Scotland	CaO, SiO_2, H_2O	<715				42
killalite	Killala Bay, Ireland						43
K-phase		CaO, SiO_2H_2O	400	0.2			44, 45
nekoite	Riverside, CA						46
okenite	Bombay, India					Ward's	47
rosenhahnite	Mendocino, CA						48
suolunite	Inner Mongolia						49
11-Å tobermorite	Crestmore, CA	CaO, Quartz, H_2O	175	1			50-52

Table 2.8 Occurrence, Synthesis, and Commercial Sources of Known Calcium Silicates (cont.)

	Occurrence	Synthesis					Leading References
		Reactants	Conditions			Commercial sources	
			Temp (°C)	Time (Day)	Pressure (kbar)		
14-Å tobermorite	Crestmore, CA	CaO, fumed SiO$_2$, H$_2$O	60	300			53, 54
trabzonite	Trabzon, Turkey						55
truscottite	Yellow Stone, WY						56
xonotlite	Mendocino, CA	CaO, SiO$_2$, H$_2$O	180-350	2-3			57, 58
Z-phase		CaO, SiO$_2$, H$_2$O	400	0.2			59
Hydrated, amorphous							
C-S-H		CaO, SiO$_2$, H$_2$O	20-140	6		CTL	60, 61
Additional Anion							
bultfonteinite	Kimberly, South Africa						62
cuspidine	Crestmore, CA						63
fukalite	Fuka, Japan						64

Table 2.8 Occurrence, Synthesis, and Commercial Sources of Known Calcium Silicates (cont.)

		Synthesis					
	Occurrence	Reactants	Conditions			Commercial sources	Leading References
			Temp (°C)	Time (Day)	Pressure (kbar)		
hydroxyellestadite	Crestmore, CA						65
octacalcium chlorosilicate	Crestmore, CA	$CaCl_2$, CaO, SiO_2	770	2			66, 67
rustmite	Kilchoan, Scotland						68
scawtite	Crestmore, CA						69
spurrite	Crestmore, CA	SiO_2, $CaCO_3$, CaF_2	800	1			70
thaumasite	Crestmore, CA						71
tilleyite	Crestmore, CA	CaO, $H_4C_2O_4 \cdot 2H_2O$, H_2O	650-700	3-7	1		72
zeophyllite	Monte Somma, Italy						73

[a] Ward's, Ward's Natural Science Establishment, Rochester, NY. [b] NYCO, NYCO Minerals, Willsboro, NY.
[c] CTL, Construction Technology Laboratories, Skokie, IL. [d] Kilchoanite (synthesized at low temperature) containing water.

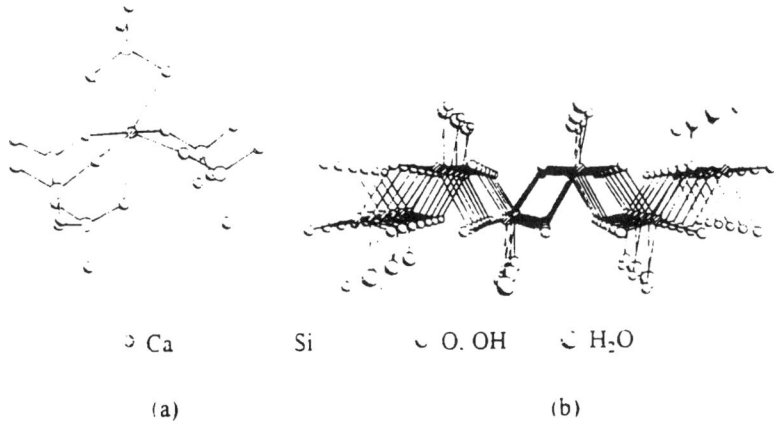

Figure 2.2 (a) Seven-coordination of Ca in CaO_2 core of calcium silicate layer in 11-Å tobermorite. (b) CaO_2 core of this layer (Hamid's data [74]).

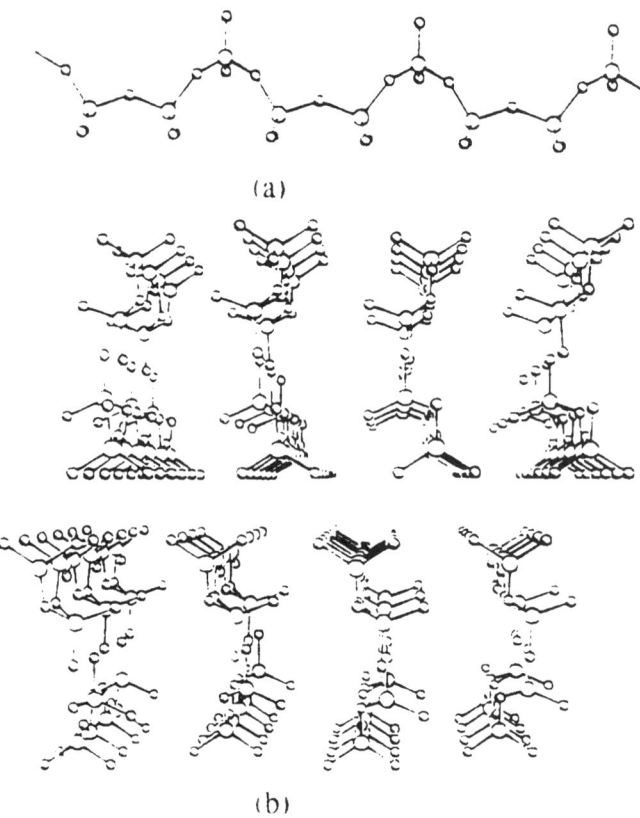

Figure 2.3 (a) Part of an oligomeric (Si3O9H)n chain of calcium silicate layer in 11-Å tobermorite. (b) (Si3O9H)n chains oriented as in layer (Hamid's data [74]).

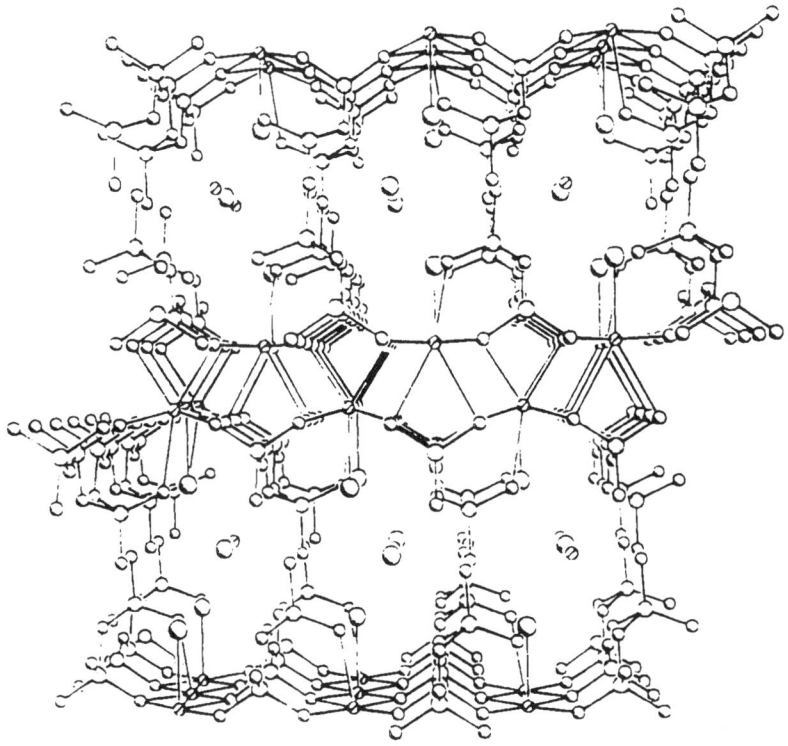

Figure 2.4 Structure of 11-Å tobermorite (Hamid's data [74]).

14-Å TOBERMORITE, $Ca_5(Si_6O_{16}(OH)_2)\cdot 8H_2O$. As also indicated on Table 2.7, 14-Å tobermorite is a phase and has a layer structure. The full structure of this compound is unknown, but it is clear that it is similar to that of 11-Å tobermorite with the major difference being that its Ca-H_2O layer has more H_2O molecules. Not surprisingly, 14-Å tobermorite can be converted to 11-Å tobermorite by very mild treatment; e.g., by heating it at 55°C. The spacing between the silicate layers in this compound is 14.0 Å [53].

JENNITE, $Ca_9(Si_6O_{16}(OH)_2)(OH)_8\cdot 6H_2O$. Jennite, like the two tobermorites, is a phase and, like them, has a layer structure. The full structure of this compound also is not known. However, it is believed that it generally resembles that of 14-Å tobermorite with major differences being that it has alternating rows of OH groups and oligomeric silicate chains on its CaO_2 core, that its oligomeric silicate chains are quite short with n being perhaps 1 and 2, and that its CaO_2 core is corrugated [61].

C-S-H, $CaO \cdot SiO_2 \cdot H_2O$. The family of calcium silicates phases called C-S-H is poorly understood in spite of the very large amount of effort that has been devoted to the study of it (because of its very great technological importance). Two relatively crystalline members of this family have been described. One of these is called C-S-H (I) [75]. The Ca-to-Si ratio in this phase ranges from about 0.8-1.5. Its degree of crystallinity decreases as its Ca-to-Si ratio increases. Its structure is not well understood but is believed to be like that of 14-Å tobermolite except that it has shorter oligomers in its oligomeric silicate chains, it has more Ca ions in its Ca-H_2O layers, and it has little more than two-dimensional order [76a].

The second relatively crystalline C-S-H that has been described is called C-S-H (II) [75]. Its Ca-to-Si ratio ranges somewhat below 2. The structure of this silicate is also not well understood but is thought to be similar to that of jennite except that it has shorter oligomers in its oligomeric silicate chains and it has less than three-dimensional order [76a].

The C-S-H in the product formed when portland cement is hydrated and hardened, C-S-H gel, is believed to have elements of the structure of both 14-Å tobermorite and jennite [76a].

Nanostructure of C-S-H Gel. It has been proposed that C-S-H gel (the C-S-H of cement) is composed of colloidal particles made up of two or three C-S-H repeat layers stacked in an orderly fashion; i.e., stacked like the repeat layers in a clay particle. It has further been proposed that the surfaces of the particles are covered with H_2O molecules (Figure 2.5a) [77].

Alternatively, it has been proposed that C-S-H gel is composed of extended irregular sheets and that the open parts of the sheets are covered with H_2O molecules (Figure 2.5b) [78].

Drying of C-S-H. The water content of C-S-H gel depends on the way it has been dried. C-S-H gel which is free of so-called evaporable water can be prepared by subjecting it to so-called D-drying; i.e., by drying it at -70°C and 0.5 mtorr or at 105°C and 760 torr in the absence of CO_2. D-dried C-S-H gel typically contains ca. 15% H_2O [79, 80]. Because of the high surface tension of H_2O, the drying of C-S-H gel leads to compression of its structure [81]. Because D-drying compresses the structure of the C-S-H gel, it decreases its surface area. It likewise decreases its pore size [81].

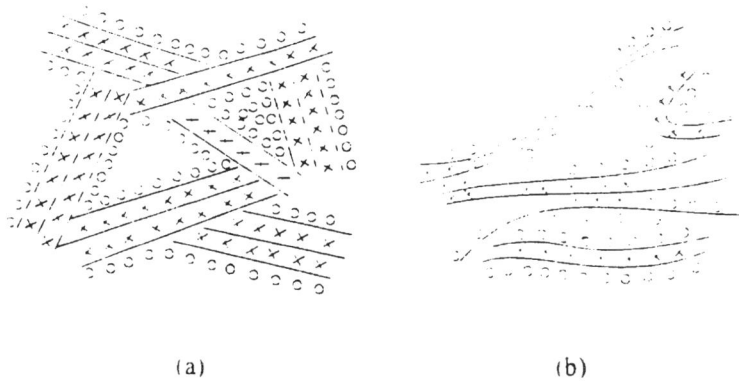

Figure 2.5 (a) Colloidal-particle model of C-S-H, (b) extended-sheet model of C-S-H.

SURFACE AREA AND PORE SIZE OF C-S-H. The surface area of D-dried C-S-H gel in cement paste as measured by BET N_2 adsorption ranges from 10-100 m^2/g and as measured by BET H_2O adsorption ranges from 250-450 m^2/g [81]. It has been suggested that the N_2 adsorption range is low and reflects incomplete N_2 adsorption, and that the H_2O adsorption range is correct [78]. Alternatively, it has been suggested that the H_2O adsorption is high and reflects multiple H_2O layer formation, and that the N_2 adsorption range is correct [77]. It is not known which hypothesis is right.

Conclusions

A number of the calcium silicates have property sets that make them of interest as SO_2 sorbents or of interest as sources of data that could allow insights to be gained into the SO_2 sorption characteristics of such silicates. Some of these silicates are readily enough available to make it practical to consider using them for these purposes.

Pure Calcium Silicate Sorbents

Introduction

As just indicated, a number of calcium silicates appear to have the potential to be good SO_2 sorbents and are sufficiently accessible through purchase or synthesis to make it practical to carry out studies on them. This section describes an investigation of some of these silicates.

APPROACH OF WORK. In the work done, three calcium silicates were purchased and five were synthesized. The SO_2 uptake and Ca utilization values of all eight silicates were studied with a simulated flue gas composed of SO_2, H_2O and N_2. In addition, the SO_2 uptake and Ca utilization values of four of the eight were studied with a simulated flue gas composed of SO_2, CO_2, H_2O and N_2. The conditions used in both studies were chosen so as to simulate those in a typical baghouse.

Experimental

CHEMICALS. β-Ca_2SiO_4 (stabilized with B_2O_3) and γ-Ca_2SiO_4 (stabilized with Al_2O_3 and MgO) were purchased from Construction Technology Laboratories. $CaSiO_3$ (wollastonite) was purchased from Ward's Natural Science Establishment. These three calcium silicates were ground (- 270 mesh) before use. $Ca(OH)_2$ was purchased from Fisher Scientific. It was representative of ordinary $Ca(OH)_2$. CaO, fumed SiO_2 (99.9 %, 350 m^2/g), and quartz (-325 mesh) were purchased from Alfa/Johnson Matthey (Ward Hill, MA). Quartz (99.5%, <10µ) was obtained as a gift from U.S. Silica (Berkeley Springs, WV).

SYNTHESIS OF LOW CALCIUM-CONTENT SILICATES. C-S-H, CaO-SiO_2-H_2O. In work patterned on that of Taylor [82], a mixture of fumed SiO_2 (3.00 g, 49.9 mmol), $Ca(OH)_2$ (3.70 g, 49.9 mmol) and H_2O (80 mL, 4.44 mol) in a capped polypropylene bottle was stirred while being warmed (70°C) for 7 days. The resulting mixture was filtered, dried (ca. 60°C, ca. 60 torr) for 20 h and weighed (6.83 g). IR data (Nujol mull, cm^{-1}) for the product are: 3418 (s br, OH), 1650 (w br, OH), 969 (s, SiO), 458 (w, SiO). XRD (d(Å) (I/I_0)) for it are: 13.6 (11), 5.36 (7), 3.04 (100), 2.80 (43), 2.11 (14), 1.82 (37), 1.67 (14).

11-Å TOBERMORITE, $Ca_5Si_6O_{16}(OH)_2$•$4H_2O$. With the work of Roy [83] as a guide, a mixture of quartz (<10µ, 0.621 g, 10.3 mmole), $Ca(OH)_2$ (0.599 g, 8.08 mmole), and H_2O (12 mL, 0.666 mole) contained in a bomb (Teflon-lined, 23 mL, Parr 4749, Parr Instrument Co., Moline, IL) was heated in a furnace (180°C) for 20 h. The resultant was filtered (Whatman #2), dried in a dessicator (Drierite, ca. 20°C, ca. 60 torr) for 48 h, and weighed (1.22 g, 97%). IR data (Nujol mull, cm^{-1}) for the product are: 3480 (s br, OH), 1630 (w br, OH), 1215 (w br), 1150 (w), 1060 (sh), 1020 (sh), 973 (s br, SiO), 935 (sh), 900 (s), 690 (w br), 530 (sh), 480 (sh), 450 (s, SiO). XRD data (d(Å) (I/I_0)) for it are: 11.3 (23), 5.67 (3), 5.48 (18), 3.78 (5), 3.64 (10), 3.53 (15), 3.31 (16), 3.08 (100), 2.98 (63), 2.82 (53), 2.74 (9), 2.53 (6), 2.42 (13), 2.23 (14), 2.26 (18), 2.14 (20), 2.07 (18), 2.00 (22), 1.84 (58), 1.67 (37), 1.62 (17).

XONOTLITE, $Ca_6(Si_6O_{17})(OH)_2$. With the work of Taylor [82] as a basis, a mixture of quartz (-325 mesh, 0.601 g, 10.0 mmole), CaO (0.561 g, 10.0 mmole) and H_2O (12 mL, 0.667 mole) contained in a bomb (Teflon-lined, 23 mL, Parr 4749) was heated in a furnace (220°C) for 48 h. The resultant was filtered (Whatman #2), dried (ca. 60°C, ca. 60 torr) for 10 h, and weighed (1.29 g, 100%). IR data (Nujol mull, cm^{-1}) for the product are: 3612 (s br, OH), 1200 (s br), 1075 (w br), 1000 (sh), 972 (s), 930 (sh), 670 (w), 620 (w br), 535 (s br), 450 (s). XRD data (d(Å) (I/I_o)) for it are: 7.00 (10), 4.26 (16), 3.65 (27), 3.50 (5), 3.25 (31), 3.09 (100), 2.83 (48), 2.70 (35), 2.63 (6), 2.51 (26), 2.34 (14), 2.26 (10), 2.04 (33), 1.95 (46), 1.84 (40), 1.75 (12), 1.71 (19).

SYNTHESIS OF HIGH CALCIUM-CONTENT SILICATES, α-DICALCIUM SILICATE HYDRATE, α-$Ca_2(SiO_3OH)OH$. With work of Taylor [84] as a guide, a mixture of ß-Ca_2SiO_4 (5.00 g, 29.0 mmole) and H_2O (50 mL, 2.78 mole) was stirred (stirring bar) while being heated (oil bath, 100°C) without open access to the atmosphere for 4 days (during this time the mixture changed from greenish gray to white). The resultant was filtered (Whatman #2), dried (ca. 60°C, ca. 60 torr, 12 h), and weighed (5.50 g, 100%). IR data (Nujol mull, cm-1) for the product are: 3550 (s, OH), 2450 (m br, OH), 1720 (w br, OH), 1282 (s, OH), 1152 (w br), 982 (s, SiO), 946 (s), 854 (s), 767 (m), 754 (s), 715 (m), 520 (m), 498 (s, SiO), 472 (m). XRD data (d(Å) (I/I_o)) for it are: 5.32 (6), 4.22 (26), 4.14 (4), 3.91 (16), 3.86 (5), 3.54 (14), 3.30 (28), 3.27 (100), 2.88 (28), 2.82 (25), 2.71 (16), 2.69 (16), 2.66 (23), 2.61 (30), 2.57 (21), 2.53 (29), 2.50 (4), 2.47 (5), 2.42 (58), 2.34 (6), 2.32 (10), 2.30 (12), 2.29 (4), 2.24 (14), 2.19 (8), 2.16 (8), 2.11 (11), 2.08 (15), 2.06 (22), 2.03 (11), 1.96 (15), 1.93 (18), 1.91 (4), 1.89 (4), 1.79 (35), 1.74 (12), 1.72 (10), 1.71 (14), 1.65 (26).

HILLEBRANDITE, $Ca_2(SiO_3)(OH)_2$. In work patterned after that of Mitsuda [85], a mixture of quartz (<10 m, 0.300 g, 4.99 mmole), $Ca(OH)_2$ (0.740 g, 10 mmole) and H_2O (12 mL, 0.667 mole) contained in a bomb (Teflon-lined, 23 mL, Parr 4749) was heated in a furnace (250°C) for 20 h. The resulting mixture was filtered (Whatman #2), dried (ca. 60°C, ca. 60 torr) for 10 h, and weighed (0.89 g, 94%). IR data (Nujol mull, cm^{-1}) for the product are: 3630 (s, OH), 3595 (s, OH), 1072 (s br), 1030 (s br), 960 (s), 920 (s), 650 (w br), 550 (w br), 450 (s). XRD data (d(Å) (I/I_o)) for it are: 8.21 (9), 6.72 (6), 5.79 (4), 4.82 (21), 4.71 (23), 4.06 (15), 3.52 (18), 3.02 (28), 2.92 (100), 2.82 (31), 2.76 (33), 2.67 (15), 2.63 (18), 2.45 (15), 2.37 (27), 2.26 (21), 2.23 (30), 2.10 (13), 2.06 (18), 1.87 (30), 1.81 (25), 1.75 (18), 1.74 (25), 1.72 (12).

SORPTION STUDIES. For the studies in which the uptake of SO_2 in the absence of CO_2 was examined, the usual conditions and the usual simulated flue gas were used. For those in which the uptake of SO_2 in the presence of CO_2 was examined, the usual conditions and a simulated flue gas composed of 2000 ppm SO_2, 11.4% CO_2, and the balance N_2, humidified to 60% at 60 °C, were used.

Results and Discussion

DIFFRACTION DATA ON SILICATES PREPARED. The X-ray powder diffraction data for the silicates made together with literature data for these silicates are shown in Tables 9 and 10 and figures 6 and 7. These data verify that C-S-H, 11-Å tobermorite, xonotlite, a-dicalcium silicate hydrate, and hillebrandite were made. They further suggest that the C-S-H made was probably close to being C-S-H (I). In view of the amounts of reactants used, it appears that the Ca-to-Si ratio of the C-S-H was near 1. Because of the ambiguity in its identification, this C-S-H is referred to below as simply C-S-H.

SO_2 UPTAKE OF KNOWN CALCIUM SILICATES. SO_2. *Sorption Studies*. The results of the SO_2 sorption studies on the eight known calcium silicates and $Ca(OH)_2$ are summarized in Table 11. As is seen, the Ca utilization values for the silicates ranged from poor to very good.

Various factors could have contributed to these results. One is the Ca-to-Si ratio of the silicates. However, as is seen from Table 11, there is no clear correlation between the Ca-to-Si ratios and the Ca utilization values (compare, for example, the values for wollastonite and C-S-H).

Another factor that could have contributed to the results is the H_2O content of the silicates (counting, as mineralogists do, two hydroxyl groups as one H_2O molecule and one oxide ion). Again, however, there is no clear correlation (for example, compare the values for a-dicalcium silicate hydrate and 11-Å tobermorite). A further factor that could have been of importance is the degree of crystallinity of the silicates. Here too there is not a good correlation (for example, compare the values for wollastonite and xonotlite).

Still another factor that could have been of significance is the surface area of the silicates. In this case there appears to be a good correlation.

Table 2.9 X-ray Powder Pattern Data for C-S-H

Experimental		C-S-H (I)[a]		C-S-H (I)[b]	
d(Å)	I/I_0	d(Å)	I/I_0	d(Å)	I/I_0
13.6	11				
		12.5	100	12.5	vs
5.36	7	5.3	10	5.3	vvw
3.04	100	3.07	100	3.04	vs
2.80	43	2.80	80	2.80	s
		2.40	30	2.40	w,d
					w,d
2.11	14	2.10	30	2.10	s
1.82	37	1.83	80	1.82	mw
1.67	14	1.67	50	1.67	vw

[a]Reference 76b.   [b]Reference 86.

Table 2.10 X-ray Powder Pattern Data for 11-Å Tobermorite, Xonotlite, α-Dicalcium Silicate Hydrate, and Hillebrandite

Experimental		Literature		Experimental		Literature	
d(Å)	I/I_0	d(Å)	I/I_0	d(Å)	I/I_0	d(Å)	I/I_0
11-Å Tobermorite[a]							
11.3	23	11.3	80	2.14	20	2.15	16
5.67	3	5.67	4	2.07	18	2.08	10
5.48	18	5.48	25	2.00	22	2.00	20
3.78	5	3.78	6	1.84	58	1.84	40
3.64	10	3.64	8	1.67	37	1.67	20

Dry Scrubbing Technologies for Flue Gas Desulfurization

Table 2.10 X-ray Powder Pattern Data for 11-Å Tobermorite, Xonotlite, α-Dicalcium Silicate Hydrate, and Hillebrandite (cont.)

Experimental		Literature		Experimental		Literature	
d(Å)	I/I_0	d(Å)	I/I_0	d(Å)	I/I_0	d(Å)	I/I_0
3.53	15	3.53	20	1.62	17	1.63	10
3.31	16	3.31	18				
3.08	100	3.08	100				
2.98	63	2.98	65				
2.82	53	2.82	40				
xonotlite[b]							
2.74	9	2.74	10	7.00	10	6.98	36
2.53	6	2.53	12	4.26	16	4.24	24
2.42	13	2.43	10	3.65	27	3.63	35
2.29	14	2.29	8	3.50	6	3.50	18
2.26	18	2.26	14	3.25	31	3.24	50
3.09	100	3.08	100	2.69	16	2.69	25
2.94	6	2.82	50	2.66	23	2.66	30
2.83	48	2.70	36	2.61	30	2.61	45
2.70	35	2.63	14	2.57	21	2.57	35
2.63	6			2.53	29	2.53	40
2.51	26	2.51	22	2.50	4	2.50	9
2.34	14	2.34	16	2.47	5	2.47	8
2.26	10	2.25	12	2.42	58	2.42	60
2.04	33	2.04	22	2.34	6	2.34	11
1.95	46	1.95	30	2.32	10	2.32	17
1.84	40	1.84	20	2.30	12	2.30	16
1.75	12	1.75	8	2.29	4	2.29	7
1.71	19	1.71	16	2.24	14	2.24	20
				2.19	8	2.19	11
				2.16	8	2.16	16

Table 2.10 X-ray Powder Pattern Data for 11-Å Tobermorite, Xonotlite, α-Dicalcium Silicate Hydrate, and Hillebrandite (cont.)

Experimental		Literature		Experimental		Literature	
d(Å)	I/I_0	d(Å)	I/I_0	d(Å)	I/I_0	d(Å)	I/I_0
α-dicalcium silicate hydrate[c]							
5.32	6	5.32	13	2.11	11	2.11	18
		4.64	4	2.08	15	2.08	19
		4.60	3	2.06	22	2.06	30
4.22	26	4.22	50	2.03	11	2.03	12
4.14	4	4.14	6	1.99	6	1.99	9
3.91	16	3.91	30	1.96	15	1.96	18
3.86	5	3.86	7	1.93	18	1.93	20
3.54	14	3.54	35	1.91	4	1.91	4
		3.48	4	1.89	4	1.89	8
3.30	28	3.30	45	1.87	4	1.87	4
3.27	100	3.27	100	1.84	8	1.84	8
		3.12	4	1.79	35	1.79	35
2.88	28	2.88	50	1.74	12	1.74	14
2.82	25	2.82	45	1.72	10	1.72	10
2.71	16	2.71	30	1.71	14	1.71	15
				1.65	26	1.65	30
hillebrandite[d]							
8.21	9	8.2	50	2.37	27	2.37	80
6.72	6	6.7	20	2.26	21	2.26	70
5.79	4	5.8	40	2.23	30	2.23	70
4.82	21	4.76	90	2.10	13	2.10	10
4.71	23			2.06	18	2.06	70
4.06	15	4.06	50			1.96	70

Table 2.10 X-ray Powder Pattern Data for 11-Å Tobermorite, Xonotlite, α-Dicalcium Silicate Hydrate, and Hillebrandite (cont.)

Experimental		Literature		Experimental		Literature	
d(Å)	I/I_0	d(Å)	I/I_0	d(Å)	I/I_0	d(Å)	I/I_0
3.52	18	3.52	50			1.93	60
		3.33	90	1.87	30	1.87	70
3.02	28	3.02	80			1.85	60
2.92	100	2.92	100	1.81	25	1.81	80
2.82	31	2.82	80	1.75	18	1.75	60
2.76	33	2.76	80	1.74	25	1.74	50
2.67	15	2.67	40	1.72	12	1.72	10
2.63	18	2.63	50				
2.45	15	2.45	40				

[a]Reference 87. [b]Reference 88. [c]Reference 89. [d]Reference 90.

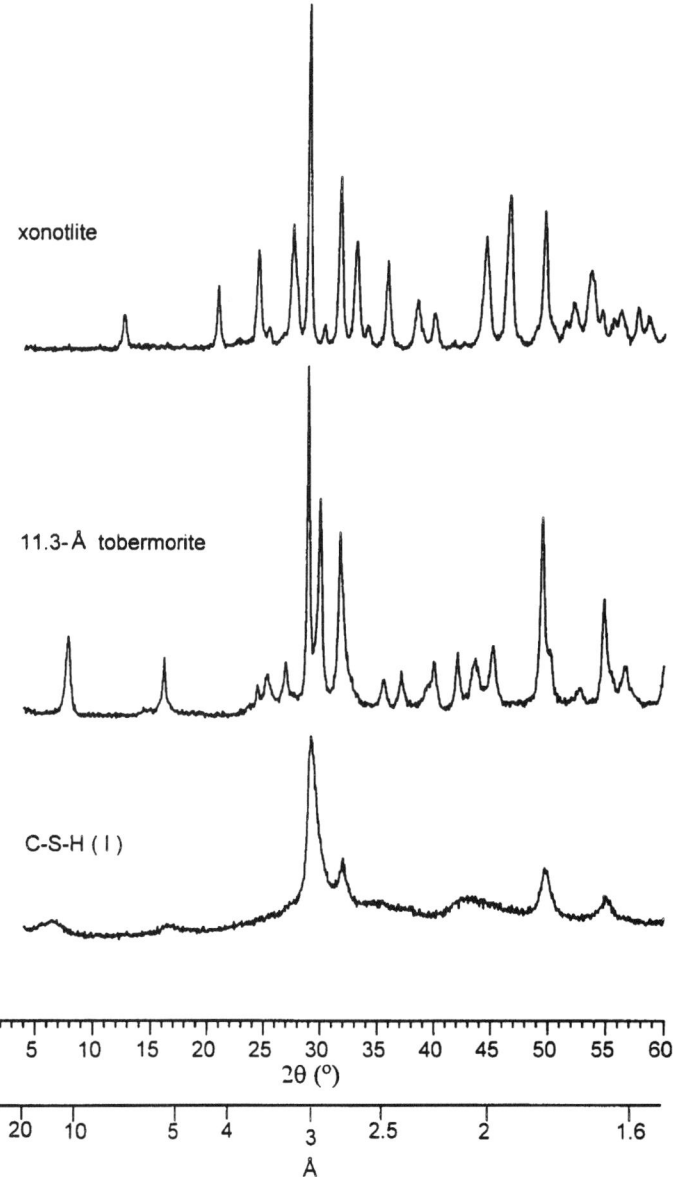

Figure 2.6 X-ray powder patterns of xonotlite, 11Å-tobermorite and C-S-H.

New Calcium-Based Sorbents for Flue Gas Desulfurization

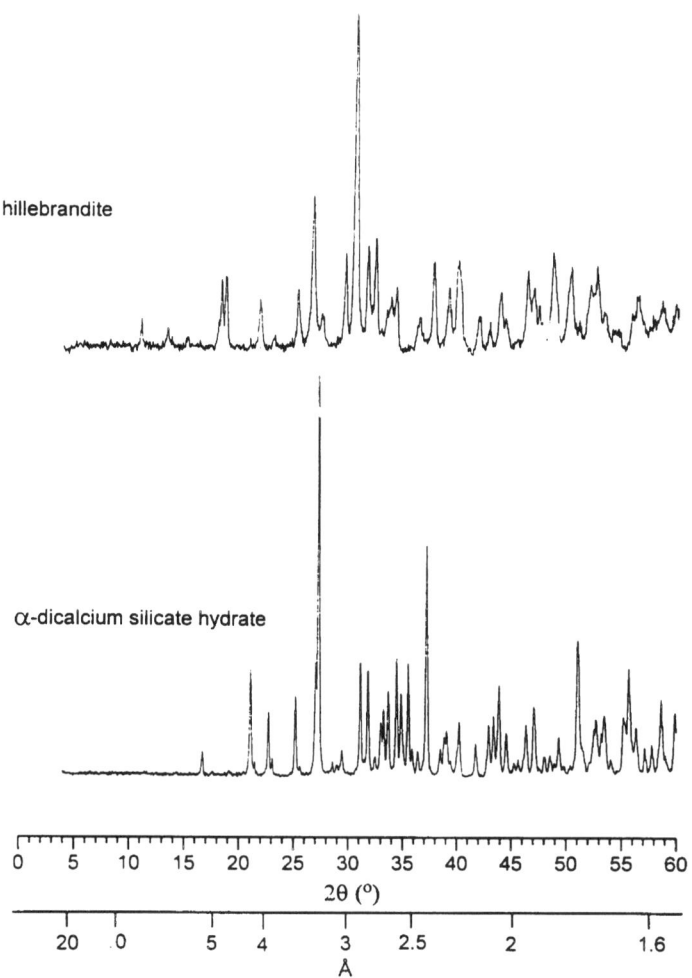

Figure 2.7 X-ray powder patterns of hillebrandite and a-dicalcium silicate hydrate.

Table 2.11 SO_2 Uptake and Ca Utilization of Calcium Silicates and Calcium Hydroxide With SO_2-H_2O-N_2 and SO_2CO_2-H_2O-N_2 Mixtures

Name	Formula	Source	CA/Si Mole Ratio	H_2O (%)	Degree Cryst	Particulars			SO_2 Uptake (mmol/g)		CA Util (%)	
						Size	Packing	Type	SO_2	SO_2-CO_2	SO_2	SO_2-CO_2
β-dicalcium silicate	β–Ca_2SiO_4	CTL	2:1	0	high	coarse	loose	rounded[a] grains	0		0	
γ-diacalcium silicate	γ–Ca_2SiO_4	CTL	2:1	0	high	coarse	loose	rounded[a] grains	0		0	
wollastonite	β–$CaSiO_3$	Ward's	1:1	0	high	coarse	tight	fibers[a]	0.2		2	
α-dicalcium silicate hydrate	α–$Ca_2(SiO_3OH)OH$	synth	2:1	10	high	fine	tight	tablets[b]	2.2	1.9	21	18
hillebrandite	$Ca_2(SiO_3)(OH)_2$	synth	2:1	10	high	fine	loose	fibers[b]	3.4		32	
xonotlite	$Ca_6Si_6O_{17}(OH)_2$	synth	1:1	3	high	fine	loose	fibers[b]	6.3	4.3	75	52
C-S-H	CaO-SiO_2-H_2O	synth	1:1	15	very low	fine	loose	foils fibers[b]	6.0	5.4	82	74

New Calcium-Based Sorbents for Flue Gas Desulfurization

Table 2.11 SO_2 Uptake and Ca Utilization of Calcium Silicates and Calcium Hydroxide With SO_2-H_2O-N_2 and SO_2CO_2-H_2O-N_2 Mixtures

Name	Formula	Source	CA/Si Mole Ratio	H_2O (%)	Degree Cryst	Particulars			SO_2 Uptake (mmol/g)		CA Util (%)	
						Size	Packing	Type	SO_2	SO_2,CO_2	SO_2	SO_2,CO_2
11-Å tobermorite	$Ca_5Si_6O_{16}(OH)_2 \cdot 4H_2O$	synth	0.8:1	12	low	fine	loose	platelets[b]	5.9	4.6	86	67
calcium hydroxide	$Ca(OH)_2$	Smith Lime	1:0	24	high	coarse	tight	grains[b]	2.6	1.9	19	14

[a] By visual examination. [b] By scanning electron microscope examination.

Thus, those silicates that are composed of particles of a size, type, and packing that can be expected to lead to a high surface area have high Ca utilization values, and those that are of a size, type, and packing that can be expected to lead to a low surface area have low Ca utilization values (e.g., compare wollastonite with its coarse, tightly packed, short fibers with xonotlite with its fine, loosely packed, short fibers). This result is not surprising because with heterogeneous reactions a high surface area is known to lead to fast reactions.

For those silicates with high Ca utilization values, the uptake values can be expected to be relatively low if the Ca-to-Si ratios are low. No examples of this are provided by the data accumulated.

SO_2-CO_2 SORPTION STUDIES. The results of the SO_2-CO_2 sorption studies are also summarized in Table 2.11. The Ca utilization values show that the presence of CO_2 in the simulated flue gas reduces the SO_2 uptake only moderately (ca. 10-30%). This can be ascribed to the low acidity of carbonic acid ($pK_a = 6.37$) relative to that of sulfurous acid ($pK_a = 1.81$).

Conclusions

The high Ca utilization value of C-S-H relative to that of ordinary $Ca(OH)_2$ and the comparative ease with which C-S-H can be synthesized suggest that C-S-H based sorbents have considerable potential. The presence of high concentrations of CO_2 in the SO_2 gas stream does not greatly affect the uptake of SO_2 by calcium silicates.

SiO_2-Modified $Ca(OH)_2$ Sorbents

Introduction

Calcium hydroxide has a high capacity for SO_2, contains an environmentally acceptable cation, and is relatively inexpensive (ca. $60/ton in Cleveland in 1994) [91]. Because of these and other features of it, considerable effort has been devoted in the past to finding ways of preparing practical $Ca(OH)_2$ sorbents and practical $Ca(OH)_2$-based sorbents. In this work, additional efforts on finding ways of preparing such sorbents were carried out. The results of these efforts are described in this section.

SYNTHESIS OF CA(OH). Calcium hydroxide is generally prepared from $CaCO_3$ by a two-step synthesis. In the first step, the $CaCO_3$ is calcined:

$CaCO_3 \rightarrow CaO + CO_2$.

If the calcination temperature is below about 1000°C, the CaO has a relatively low density, while if the calcination temperature is above about 1000°C, it has a higher density [92]. In the second step, the CaO is hydrated with water or steam:

$CaO + H_2O \rightarrow Ca(OH)_2$.

The low density CaO hydrates much more rapidly than the high density CaO [93a]. The mechanism of the hydration of the low density form is not understood. That of the high density form is believed to entail the adsorption of H_2O on the CaO, the formation of $CaO \cdot 2H_2O$, and finally the formation of $Ca(OH)_2$ [94]. Typically about 3-4 moles of water per mole of CaO are used when CaO is hydrated [95]. The excess water is used to compensate for the loss of H_2O as steam during the hydration (the hydration of CaO is quite exothermic, DH = 64.8 KJ/mol).

The surface area of the $Ca(OH)_2$ produced when CaO is hydrated with a low water-to-solids ratio or with steam is in the range of ca. 13-22 m^2/g [96]. This is insufficient for this $Ca(OH)_2$ to be useful as an in-duct flue gas sorbent.

Calcium hydroxide produced when the CaO is hydrated with a high water-to-solids ratio is initially colloidal. However, this colloidal $Ca(OH)_2$ quickly flocculates and agglomerates. This is attributable to the high density of the OH groups on the surface of the particles and to the tendency of such groups to hydrogen bond. The agglomerated $Ca(OH)_2$ has a surface area that is relatively low [97]. Again the surface area of this $Ca(OH)_2$ is not sufficient for it to be useful as an in-duct flue gas sorbent.

STRUCTURE AND MORPHOLOGY OF $Ca(OH)_2$. Calcium hydroxide crystallizes in the brucite structure. It thus has a layer structure (Figure 2.8) [98].

When prepared by mixing aqueous solutions of $CaCl_2$ and NaOH, $Ca(OH)_2$ occurs as hexagonal prisms [99]. In contrast, $Ca(OH)_2$ prepared by repeatedly heating and cooling suspensions of irregular $Ca(OH)_2$ particles occurs as hexagonal platelets [100].

The presence of foreign species during the precipitation of $Ca(OH)_2$ can affect its morphology. Thus, $Ca(OH)_2$ prepared by mixing aqueous solutions of $CaCl_2$ and NaOH with ethanol occurs as hexagonal plates [101]. The plate morphology of this $Ca(OH)_2$ is attributable to the adsorption of ethanol molecules on its crystal faces which are parallel to its OH planes (Figure 8). Calcium hydroxide formed during the hydration of cement also occurs as hexagonal plates [102]. This can ascribed to the adsorption of silicic acids on its OH faces.

MODIFIED $Ca(OH)_2$ AND $Ca(OH)_2$-BASED SORBENTS. A number of different approaches have been tried in an effort to get useful $Ca(OH)_2$ or $Ca(OH)_2$-based sorbents. Thus, to provide H_2O to aid the SO_2-$Ca(OH)_2$ reaction, mixtures of calcium hydroxide and deliquescent salts have been prepared [103]. To modify its morphology and increase its surface area, $Ca(OH)_2$ has been treated with lignosulfonate [104]. $Ca(OH)_2$ has likewise been prepared in the presence of ethanol to accomplish the same objective [105]. It has also been prepared in the presence of kaolinite in an attempt to favorably alter its pore structure [106]. In addition, it has been milled in an attempt to increase its surface area [107]. None of these approaches has been carried beyond the pilot plant stage.

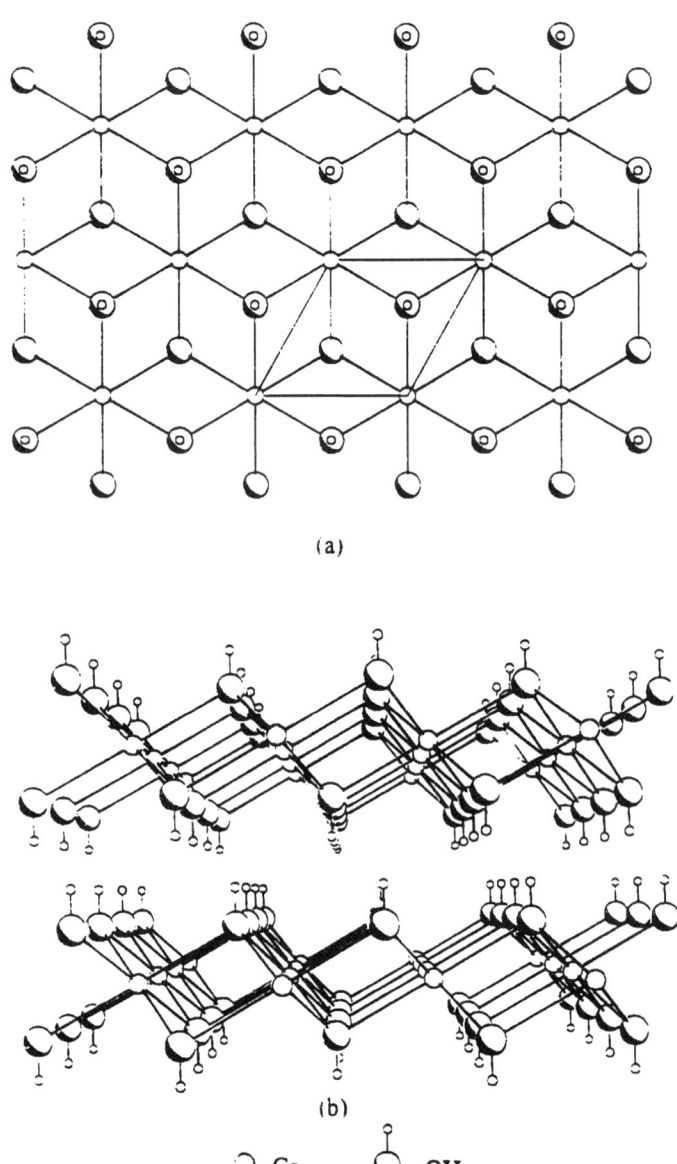

Figure 2.8 Structure of Ca(OH)$_2$: (a) a single layer projected along <001>; (b) edge view of two layers (Petch's data [98]).

APPROACH OF WORK. In this work, products derived by treating $Ca(OH)_2$ with fumed SiO_2 were investigated. In an extension of this approach, sorbents derived by fully reacting $Ca(OH)_2$ with SiO_2 or an SiO_2 source were also investigated. The sorbents were analyzed by infrared spectroscopy, X-ray powder diffractometry, scanning electron microscopy, BET surface area measurement, and SO_2 uptake.

Experimental

CHEMICALS. $CaCO_3$ and $Ca(OH)_2$ were purchased from Fisher, and fumed silica (99.9% SiO_2, 350 m^2/g) and anatase (TiO_2) were obtained from Alfa/Johnson Matthey (Ward Hill, MA). Uncalcined diatomite was obtained as a gift from Celite (Lompoc, CA), and pumice (Tamez Mountains, NM) and tripoli were purchased from Ward's Natural Science Establishment. Perlite was obtained as a gift from Harborlite (Vicksburg, MI), and rottenstone was obtained as a gift from Keystone Filler and Manufacturing (Muncy, PA). MgO-stabilized ZrO_2 beads were purchased from US Stoneware (Mahwah, NJ).

SYNTHESIS OF SORBENTS. *CaO. Shaken Reaction Without Beads.* A mixture of CaO (3.00 g) and H_2O (30 mL) in a capped polyethylene bottle (80 mL) was shaken with a shaker (ca. 2-3 Hz) at room temperature for 5 min. The resulting slurry was filtered (Whatman #2) and the solid was washed with water (5 mL), vacuum dried (ca. 60°C, ca. 60 torr) for 12 h, and weighed (3.87 g). IR data (Nujol mull, cm-1) for the product are: 3644 (s, OH). XRD data (d(Å) (I/Io)) for it are: 4.90 (64, CH), 3.11 (38, CH), 2.63 (100, CH), 1.93 (34, CH), 1.80 (32, CH), 1.69 (18, CH). The sorbent was a moderate density, white powder.

FUMED SiO_2, SHAKEN REACTION WITHOUT BEADS. In a representative reaction (Table 2.12, run 2), a mixture of CaO (produced by calcining $CaCO_3$ at 960°C for 6 h, 2.24 g) and H_2O (25 mL) in a capped polyethylene bottle (80 mL) was shaken with a shaker (ca. 2-3 Hz) at room temperature for 5 min. Fumed SiO_2 (0.48 g) was added to the mixture and it was shaken (ca. 2-3 Hz) at room temperature for an additional 6 h. The resulting slurry was filtered (Whatman #2) and the solid was washed with ethanol (ca. 5 mL) and ether (ca. 10 mL), vacuum dried (ca. 60°C, ca. 60 torr) for 9 h, and weighed (3.28 g). IR data (Nujol mull, cm-1) for the product are: 3645 (s, OH), 3620 (s br, OH), 1650 (w br, OH), 972 (s br, SiO). XRD data (d(Å) (I/Io)) for it are: 4.90 (63, CH), 3.04 (12, CSH), 3.11 (40, CH), 2.63 (100, CH), 1.93 (33, CH), 1.80 (32, CH), 1.69 (19, CH). The sorbent was a low density, white powder.

NATURAL SiO_2 SOURCE, STIRRED REACTIONS WITHOUT BEADS. In a typical reaction (Table 2.13, run 8), a mixture of $Ca(OH)_2$ (3.00 g), diatomite (2.16 g), and H_2O (50 mL) was stirred (magnetic stirring bar, ca. 150 rpm) without open access to the atmosphere while being heated (oil bath, 100°C) for 5 h and filtered (Whatman #2). The solid was washed with water (ca. 5 mL), vacuum dried (ca. 60°C, ca. 60 torr) for ca. 24 h, weighed (5.92 g), and crushed. IR data (Nujol mull, cm-1) for the product: 3418 (s br, OH), 1650 (w br, OH), 969 (s, SiO), 458 (w, SiO). XRD data (d(Å) (I/Io)) for it are: 3.04 (100, CSH), 2.80 (27, CSH), 1.82 (26, CSH). The sorbent was a low density, off-white powder.

NATURAL SiO_2 SOURCE, STIRRED REACTIONS WITH BEADS. In a representative reaction (Table 2.13, run 10), a mixture of $Ca(OH)_2$ (3.00 g), pumice (2.16 g), H_2O (50 mL), and ZrO_2 beads (MgO-stabilized, 1.4-1.6 mm, ca. 15 g) was stirred (magnetic stirring bar) without open access to the atmosphere while being heated (oil bath, 100°C) for 10 h. The resulting mixture was filtered (60-mesh bronze screen) and the beads were washed with H_2O (ca. 5 mL). The filtrate and washings were combined and filtered (Whatman #2), and the solid was washed with water (ca. 5 mL), vacuum dried (ca. 60°C, ca. 60 torr) for ca. 24 h, weighed (5.64 g) and crushed. IR data (Nujol mull, cm-1) for the product are: 3414 (s br, OH), 1650 (w br, OH), 971 (s, SiO), 498 (s, SiO). XRD data (d(Å) (I/Io)) for it are: 3.04 (100, CSH), 2.80 (26, CSH), 1.82 (27, CSH). The sorbent was a low density, off-white powder.

ANALYSIS OF FUMED SiO_2, SHAKEN REACTION SORBENTS. The relative intensities of the 2.63 Å and 3.52 Å powder pattern reflections of $Ca(OH)_2$ and anatase in a 1:1 weight-weight mixture of these compounds were obtained. The relative intensities of the same reflections in mixtures containing known weights of the sorbents and anatase also were obtained. These data were then used to determine analytical values pertaining to the sorbents. First, the relative intensity factor for anatase and $Ca(OH)_2$ in the sorbent-anatase mixtures, f_{rel}, was calculated as

$$f_{rel} = \frac{I'_{an}}{I'_{CH}} \cdot \frac{w'_{CH}}{w'_{an}}$$

where I'_{an} and I'_{CH} are the intensities of the 2.63 Å anatase and 3.52 Å $Ca(OH)_2$ reflections of the 1:1 anatase-$Ca(OH)_2$ mixture and W'_{CH} and W'_{an} are the weights of anatase and $Ca(OH)_2$ in it. Then the weights of $Ca(OH)_2$ in the sorbent-anatase mixtures, W_{CH}, were calculated as

$$W_{CH} = W_{an} \cdot f_{rel} \cdot \frac{I_{CH}}{I_{an}}$$

where w_{an} is the weight of anatase in a mixture and I_{CH} and I_{an} are the intensities of the 2.63 Å anatase and 3.52 Å $Ca(OH)_2$ reflections of the mixture. Next the weight fractions of $Ca(OH)_2$ and C-S-H in the sorbents in these mixtures, f_{CH} and f_{CSH}, were calculated as

$$f_{CH} = \frac{W_{CH}}{W_{tot}}$$

and

$$f_{CSH} = 1 - f_{CH}$$

where w_{tot} is the total weight of the sorbent in a mixture. Finally, the Ca-to-Si ratio of the C-S-H in the mixtures, r, was calculated with the formulas

$$Ca_{CH} = \frac{w_{CH}}{74.09}$$

and

$$r = [CA_{tot}(1 - Ca_{CH})]/Si_{tot}$$

where CaCH is the number of moles of Ca in the $Ca(OH)_2$ in the sorbent in a mixture, and Ca_{tot} and S_{tot} are the number of moles of Ca and Si in the sorbent in a mixture.

Because of the difficulties inherent in this type of analytical methodology, the calculated values obtained probably are subject to considerable systematic error (perhaps 30%).

Results And Discussion

FUMED SiO_2-$Ca(OH)_2$ STUDIES. *Effect of Si-to-Ca Reactant Ratio on Composition and Nonsorption Properties of Products.* In the fumed SiO_2-$Ca(OH)_2$ work, studies of the effect of the Si-to-Ca reactant ratio on the composition and nonsorption related properties of the fumed SiO_2-$Ca(OH)_2$ product were carried out. Results of six runs made in these studies are summarized in Table 2.12. The values of the composition of the products are based, as indicated in the previous section, on diffraction data.

With an Si-to-Ca reactant ratio of 0.2, a 6-hour reaction time, an ca. 20°C reaction temperature, and mild agitation, complete utilization of the SiO_2 was obtained. The product was composed of $Ca(OH)_2$ and C-S-H (Figures 2.9 and

2.10, run 2, and Table 2.12, run 2). The relative rapidity of the reaction is attributable to the very high surface area of the fumed SiO_2 and to the tendency of fumed SiO_2 to adsorb on $Ca(OH)_2$ [108]. The fact the product was composed of C-S-H is as expected since $Ca(OH)_2$ and SiO_2 can react to yield C-S-H. Given the conditions under which the reaction was carried out, it appears that the SiO_2 reacted with the surface of the $Ca(OH)_2$ particles and that this occurred in such a way as to leave them rough. It further appears that the C-S-H produced was deposited on the surfaces of the particles and that this led to their envelopment by C-S-H and their isolation from one another, (Figure 2.11.)

Table 2.12 Synthesis and properties of Ca(OH)$_2$ and fumed SiO$_2$-Ca(OH)$_2$ products.

SiO$_2$ Source	Si/CA Mole Ratio	Conditions			Composition[a]			CSH CA/SI mole ratio	BET Area (m^2/g)	SO$_2$ Uptake (mmol/g)	CA Util (%)
		Time (h)	Temp (°C)	Media Agit	CH (%)	CSH (%)					
1.	0.0	6	~20	shaking	100	0		24	5.2	39
2. fumed SiO$_2$	0.2	6	~20	shaking					7.4	61
3. fumed SiO$_2$	0.4	6	~20	shaking	22	78	1.8		7.9	74
4. fumed SiO$_2$	0.5	6	~20	shaking	15	85	1.6		8.1	77
5. fumed SiO$_2$	0.6	6	~20	shaking	10	90	1.4		8.0	84
6. fumed SiO$_2$	0.8	6	~20	shaking	4	96	1.2		7.3	85
7. fumed SiO$_2$	1.0	6	~20	shaking	0	100	1.0		5.9	86

[a]CH, Ca(OH)$_2$; CSH, CaO-SiO$_2$-H$_2$O

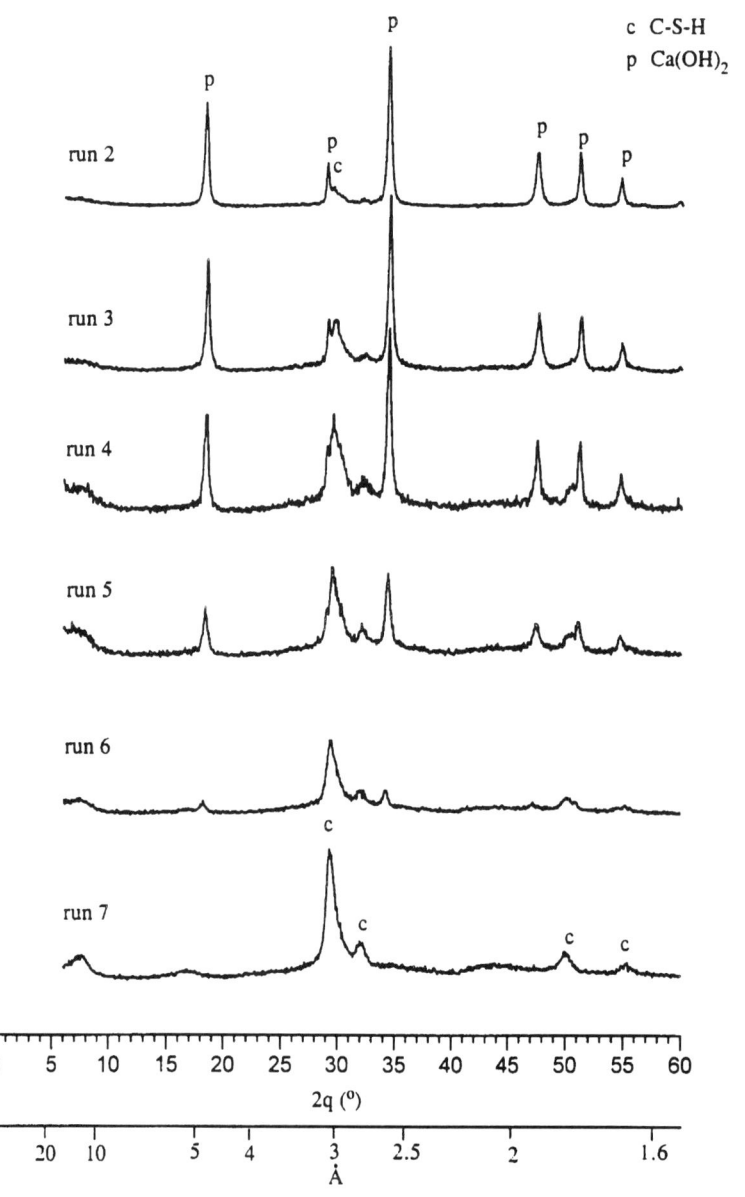

Figure 2.9 X-ray powder patterns of products of runs 2 through 7, Table 2.12.

New Calcium-Based Sorbents for Flue Gas Desulfurization

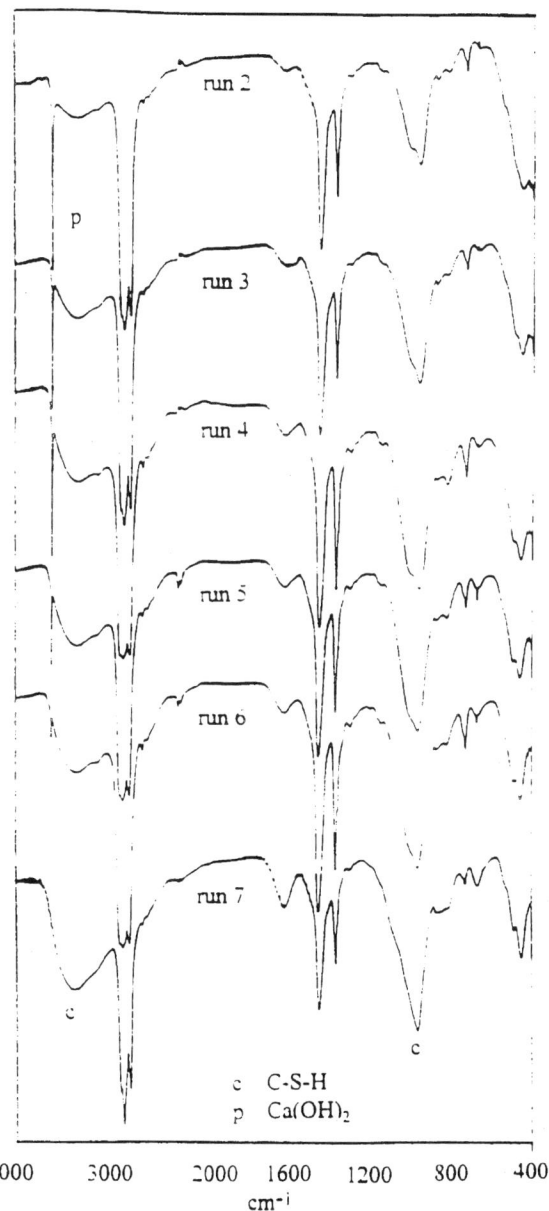

Figure 2.10 Infrared spectra of products of runs 2 through 7. Table 2.12.

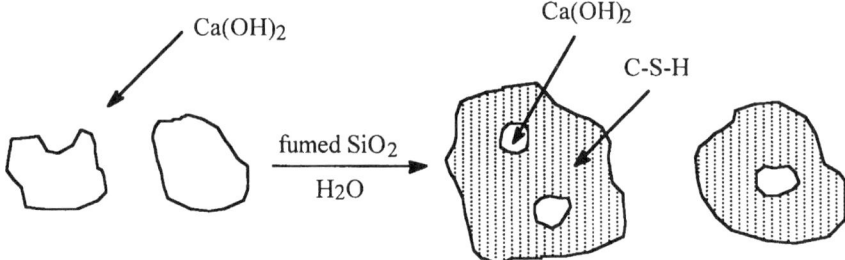

Figure 2.11 Schematic representation of the Ca(OH)$_2$-fumed SiO$_2$ reaction.

With an Si-to-Ca reactant ratio of 0.4-1.0 and the same reaction conditions, again full utilization of the SiO$_2$ was obtained. The products were composed of C-S-H and some to no Ca(OH)$_2$. The proportion of C-S-H increased with increasing Si-to-Ca reactant ratio, and the Ca-to-Si ratio in the C-S-H formed decreased with increasing Si-to-Ca reactant ratio. These results are as expected [109]. The proportion of C-S-H in the products was relatively high when the Si-to-Ca reactant ratio was low. This is attributable to the formation of C-S-H low in Si. In this set of products, it also appears that the Ca(OH)$_2$ occurred, when it was present, as small, rough particles enveloped in C-S-H and isolated from each other.

REFERENCE Ca(OH)$_2$. An investigation of Ca(OH)$_2$ prepared in such a way as to make it acceptable as a reference sorbent was also done. This Ca(OH)$_2$ had a relatively low surface area, 24 m^2/g (Table 12, run 1). This is as expected in view of the way it was prepared [110].

EFFECT OF Si-TO-Ca REACTANT RATIO ON SO$_2$ SORPTION PROPERTIES OF PRODUCTS. Studies of the effect of the Si-to-Ca reactant ratio on the SO$_2$ sorption properties of the fumed SiO$_2$-Ca(OH)$_2$ products were also carried out. The results are summarized in Table 12. They show that all the products took up SO$_2$ very well. They also show that the higher the Si-to-Ca reactant ratio up to a value of ca. 0.5, the larger the amount of SO$_2$ that was taken up per gram, and that the higher the ratio above ca. 0.5, the smaller the amount that was taken up per gram. This can be ascribed to the combined effects of the ratio on the size of Ca(OH)$_2$ particles and on the amount of Ca in the C-S-H. The sorption results also show that the Ca utilization of all the products was high, and that it increased with increasing Si-to-Ca reactant ratio. The high Ca utilization of the low Si-to-Ca reactant ratio products is attributable to the high proportion of C-S-H in the products. The high Ca utilization of the high Si-to-Ca reactant ratio products is as expected.

NATURAL SiO_2 OR SiO_2-SOURCE SORBENTS. *Effect of SiO_2 or SiO_2-Source on Composition and Nonsorption Properties of Sorbents.* In the natural SiO_2 or SiO_2-source studies, the effect of the source of the SiO_2 or SiO_2-source reactant on the composition and nonsorption-related properties of the SiO_2 or SiO_2-source sorbents was investigated. The results of six runs made in these studies are summarized in Table 13.

The values listed in this table for the estimated composition of the products made are estimates based on careful inspection of the X-ray and infrared data. They should not be taken as values derived from quatitative composition determinations.

With diatomite as the SiO_2 source, a 1:1 Si-to-Ca reactant ratio, a 5-hour reaction time, a 100°C reaction temperature, and mild agitation, complete reaction was obtained. The only species identified in the product was C-S-H. Its surface area was very high, 245 m^2/g (Figure 12, run 8, and Table 13, run 8). The relative rapidity of the reaction is attributable to the porosity and high surface area of the diatomite [111]. The high surface area of the product is ascribed to these same properties since their existence permits the C-S-H to grow relatively unhindered physically.

Table 2.13 Synthesis and Properties of Natural SiO$_2$ Source-Ca(OH)$_2$ Products.

SiO$_2$ Source	Si/CA Mole Ratio	Conditions				Estimated Composition			BET Area (m^2/g)	SO$_2$ Uptake (mmol/g)	CA Util (%)
		Time (h)	Temp (°C)	Media	Agit	CH (%)	CSH (%)	C$_2$SH (%)			
8 diatomite (Celite)	1:1	5	100		stirring		100		245	5.8	85
9 diatomite (Celite)	1:2	5	100		stirring	40	60		170	7.9	76
10 pumice	1:1	10	100	ZrO$_2$[b]	stirring		100		85	6.0	83
11 perlite	1:1	12	100	ZrO$_2$[b]	stirring		100		95	6.1	84
12 rottenstone	1:1	12	100	ZrO$_2$[b]	stirring	60	20	20	22		
13 tripoli	1:1	18	100	ZrO$_2$[b]	stirring	60	20	20	24		

[a] CH, Ca(OH)$_2$; CSH, CaO-SiO$_2$-H$_2$O; C$_2$SH, α–Ca$_2$(SiO$_3$OH)(OH). [b] 1.4-1.6-mm Mg-stabilized ZrO$_2$ beads

With diatomite as the SiO_2 source, a 1:2 Si-to-Ca reactant ratio, and the same conditions as before, complete reaction of the diatomite was again obtained. However, the product contained both C-S-H and unreacted $Ca(OH)_2$. The surface area of the product was again very high, 170 m^2/g (Figure 2.12, run 9, and Table 2.13, run 9). The presence of $Ca(OH)_2$ in the product is attributable to the low Si-to-Ca ratio.

The use of pumice as the SiO_2 source, a 1:1 Si-to-Ca reactant ratio, a reaction time of 10 hours, a reaction temperature of 100 °C, and very vigorous agitation gave a product in which the only species identified was C-S-H. As expected, the surface area of the product was high, 85 m^2/g (Figure 2.12, run 10, and Table 2.13, run 10). The fate of the cations in the pumice is unknown. Some may have been incorporated with the C-S-H and some may have been carried away in the wash water. The lower surface area of the product is ascribed to the less porous nature of pumice.

The use of perlite as the SiO_2 source, a 1:1 Si-to-Ca reactant ratio, and the same reaction conditions also gave a product in which the only species was C-S-H. Again the surface area of the product was high, 95 m^2/g (Figure 2.12, run 11, and Table 2.13, run 11). The lower surface area of the product compared to that of the diatomite product is attributable to the nonporous nature of perlite.

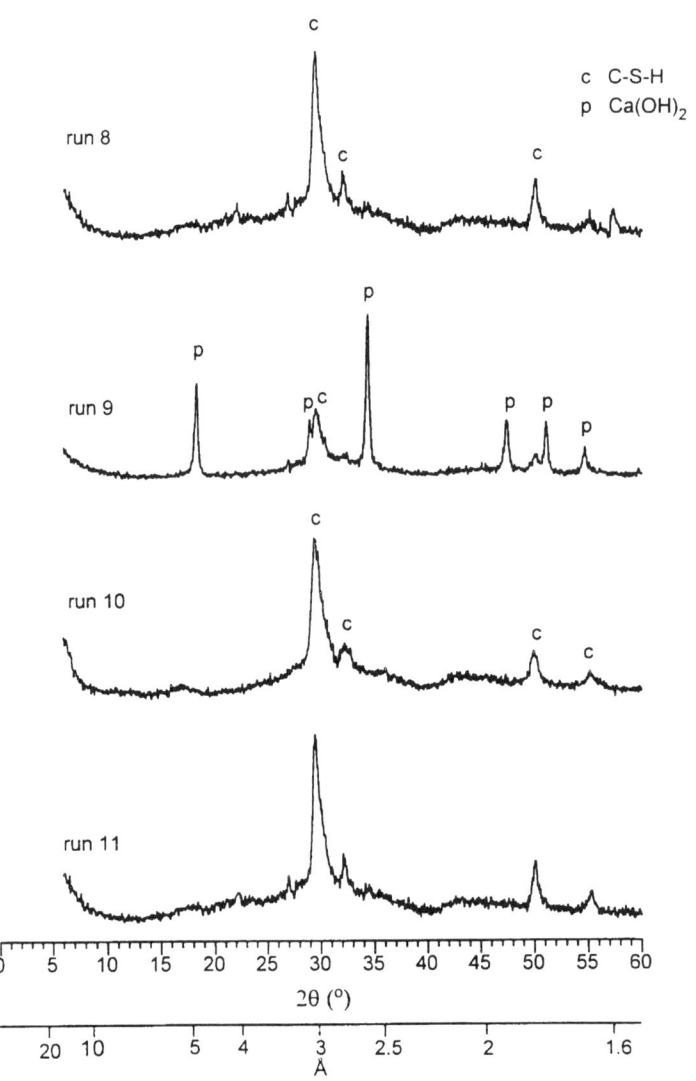

Figure 2.12 X-Ray power patterns of products of runs 8 through 11, Table 2.13

With rottenstone, a 1:1 Si-to-Ca reactant ratio, and the same reaction conditions and with tripoli, a 1:1 ratio, and similar reaction conditions, complete reaction of the silicate was not obtained. Here $Ca(OH)_2$, C-S-H, a-$Ca_2(SiO_3OH)(OH)$ and unreacted silicate were identified in the products. The surface area of the products was low, 22 and 24 m^2/g (Figure 2.13, runs 12 and 13, and Table 2.13, runs 12 and 13). The slowness of the reaction is attributed to the inertness of the silicates. The presence of a-Ca$_2$(SiO$_3$OH)(OH) in the products is attributed to the combination of the high reaction temperatures and the extended reaction times used.

EFFECT OF SiO$_2$ OR SiO$_2$-SOURCE ON SO$_2$ SORPTION PROPERTIES OF THE PRODUCTS. Studies of the effect of the source of the SiO$_2$ on the SO$_2$ uptake and Ca utilization of the products were also carried out. The results are summarized in Table 2.13. The uptake and Ca utilization values of the diatomite, pumice and perlite products show that good sorbents can be made from various inexpensive SiO$_2$ sources.

Conclusions

The results from the studies with the fumed SiO$_2$ and Ca(OH)$_2$ show that the minimum Si-to-Ca reactant ratio required for a good SiO$_2$-modified sorbent is low. This is attributable to the envelopment of the Ca(OH)$_2$ particles of the sorbent in a porous, reactive coating of C-S-H. The results from the studies with natural SiO$_2$ and natural SiO$_2$ sources show that good sorbents rich in C-S-H can be made from diatomite, pumice and perlite. This and the outcome of the fumed SiO$_2$-Ca(OH)$_2$ studies indicate that it should be possible to prepare good SiO$_2$-modified sorbents that have a low Si-to-Ca reactant ratio and are practical and inexpensive. This is important because, in some areas, Ca(OH)$_2$ is more available than is reactive SiO$_2$ [112].

Hydrated Ca$_3$SiO$_5$ AND β-Ca$_2$SiO$_4$ Sorbents

Introduction

It is apparent from the work described above on pure silicates that the silicate phase C-S-H has the potential to be a useful SO$_2$ sorbent. However, while several ways of synthesizing C-S-H have been described, ways of preparing it inexpensively in a form that is suitable for use as an SO$_2$ sorbent have not. The known ways of preparing C-S-H can be grouped into three approaches [113]. One involves the reaction of SiO$_2$ and Ca(OH)$_2$:

$$Ca(OH)_2 + SiO_2 + xH_2O \rightarrow CaO\cdot SiO_2\cdot xH_2O$$

A second involves the reaction of a calcium salt and a soluble silicate; e.g., CaCl$_2$ and Na$_2$SiO$_3$:

$$CaCl_2 + Na_2SiO_3 + xH_2O \rightarrow CaO\text{-}SiO_2\text{-}xH_2O + 2NaCl$$

The third involves the reaction of a calcium silicate and water; e.g., Ca_3SiO_5 and H_2O or $\beta\text{-}Ca_2SiO_4$ and H_2O:

$$Ca_3SiO_5 + xH_2O \rightarrow 1.7CaO\text{-}SiO_2\text{-}(x\text{-}1.3)H_2O + 1.3Ca(OH)_2$$

or

$$\beta\text{-}Ca_2SiO_4 + xH_2O \rightarrow 1.7CaO\text{-}SiO_2\text{-}(x\text{-}0.3) + 0.3Ca(OH)_2$$

The most attractive of these three approaches for making C-S-H for sorbent use is the third because much of ordinary (type I) portland cement is made up of two hydrolyzable calcium silicates. These two silicates are alite and belite. Alite is a phase with a composition similar to that of Ca_3SiO_5 and belite is a phase with composition similar to that of $\beta\text{-}Ca_2SiO_4$.

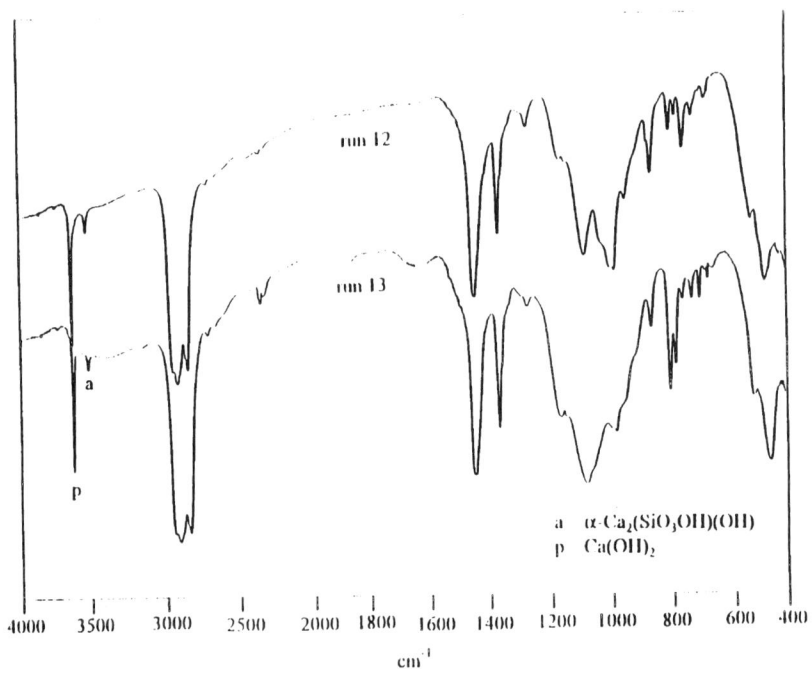

Figure 2.13 Infrared spectras of products of runs 12 and 13, Table 2.14.

However, while the use of cement for the production of a sorbent rich in C-S-H is attractive, there are difficulties. With the low water-to-solids ratio usually used in the hydrolysis of cement (i.e., that used in the formation of concrete), the product is a high-density, low-permeability, unreactive material. Further, the rate at which this material is formed is slow, essentially complete reaction taking years.

This section describes fundamental work done in an effort to find ways of rapidly preparing high surface area sorbents that are rich in C-S-H from cement.

Ca_3SiO_5 AND β-Ca_2SiO_4. Ca_3SiO_5 AND β-Ca_2SiO_4 AS MODELS FOR CEMENT AND BELITE. Besides alite and belite, type I portland cement contains significant amounts of other components. Typically it contains 50-70% alite (ca. Ca_3SiO_5), 15-30% belite (ca. Ca_2SiO_4), 5-10% aluminate phase (ca. $Ca_3Al_2O_6$), 5-15 % ferrite phase (ca. Ca_2AlFeO_5), and 3-4% gypsum ($CaSO_4.2H_2O$) [76c].

Stabilized monoclinic Ca_3SiO_5 serves as a fairly good and relative simple model for cement itself and for its alite component, and has frequently been used for this purpose. (Pure monoclinic Ca_3SiO_5 is not suitable for use as a model because it is not stable at room temperature.) Generally the stabilizing ions used in stabilized monoclinic Ca_3SiO_5 are Mg^{2+} and Al^{3+} or Mg^{2+} and Fe^{3+} [114].

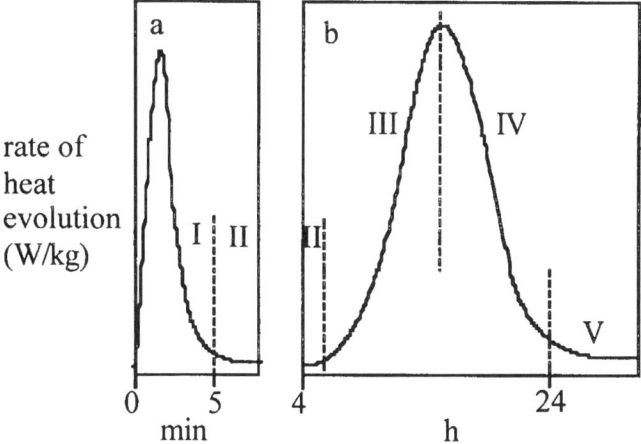

Figure 2.14 Calorimetric curve for hydration of Ca_3SiO_5 at a 1:2 water-to-solids ratio during (a) early part of hydration reaction and (b) intermediate part of hydration reaction (Jennings [120]).

Likewise, stabilized β-Ca_2SiO_4 serves as a fairly good and relatively simple model for the belite phase in cement and has been used for this purpose. (As with

Ca_3SiO_5, the pure compound again cannot be used as a model because it is not stable at room temperature.) Generally the stabilizing ions used in stabilized β-Ca_2SiO_4 are B^{3+} or Al^{3+}, Mg^{2+} and Fe^{3+} [115].

HYDRATION OF Ca_3SiO_5. The hydration of stabilized Ca_3SiO_5 has been reviewed by Taylor et al. [76a, 116, 117]. At room temperature and a low water-to-solids ratio, the product is mainly, as already indicated, C-S-H and $Ca(OH)_2$. The C-S-H has a calcium-to-silicon ratio of ca. 1.7 and its silicate ions are mainly $Si_2O_7^{6-}$ ions [118,119].

At room temperature and a low water-to-solids ratio, the rate of heat evolution curve for Ca_3SiO_5 shows two maxima. With this curve as a basis, the hydration process is generally divided into five stages (Figure 2.14). During the first stage, a discontinuous layer is formed on the Ca_3SiO_5 [121,122] while during the second more C-S-H is added to this layer. In these two stages, it has been suggested that hydroxylation of the surface of the Ca_3SiO_5, congruent dissolution of this hydroxlated surface, and precipitation of metastable C-S-H on it occur [123]. Alternatively, it has been suggested that incongruent dissolution of the surface of the Ca_3SiO_5 and the formation of metastable C-S-H on this surface occur [120,124]. In either case, it is postulated that the buildup of the layer of metastable C-S-H accounts for the slowing of the rate of heat evolution in the latter part of stage I. The observed decrease in the length of steps I and II with an increase in temperature is attributed to an increase in the rate of diffusion through the layer of ions which are required by and produced by the hydration [125].

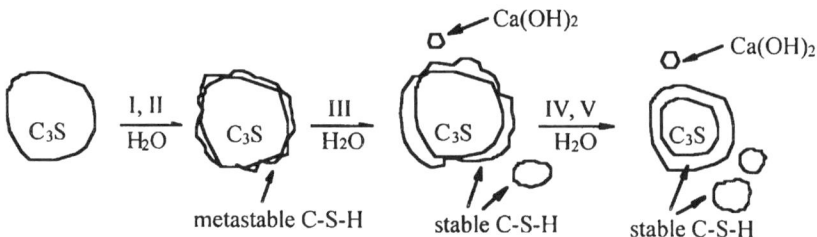

Figure 2.15 Schematic representation of Ca_3SiO_5 hydration mechanism.

In stages III and IV, it has been suggested the metasable C-S-H layer is converted to a stable C-S-H and that this results in disruption of the layer, or alternatively that osmotic pressure disrupts the layer and that this leads to conversion of the metastable C-S-H to stable C-S-H [126,127]. In either case, it is thought that a layer of stable C-S-H is formed on the surface of the Ca_3SiO_5, and that this layer accounts for the slowing of the rate in stage IV. Finally in stage V, it is thought that a diffusion-controlled buildup of C-S-H on the Ca_3SiO_5 occurs (Figure 2.15).

Full hydration of Ca_3SiO_5 is very slow (Table 2.14). This is consistent with the mechanism postulated for stage V.

Kantrol et al have shown that the hydration of Ca_3SiO_5 at room temperature and a 9:1 water-to-solids ratio is accelerated by ball milling [128]. However, under these conditions the product is not a mixture of C-S-H and $Ca(OH)_2$, but rather is a mixture of $Ca_3(SiO_3OH)_2 \cdot 2H_2O$ (afwillite) and $Ca(OH)_2$. It has been suggested that a dispersion of the orthosilicate ions into the solution caused by the milling leads to the formation of the afwillite [129].

Others have shown that the hydration of Ca_3SiO_5 at room temperature and a 1:1 to 2:1 water-to-solids ratio is accelerated by reactive SiO_2. In this case, the composition of the product is changed with the product becoming richer in C-S-H [130]. It has been suggested that the reaction is accelerated because the SiO_2 provides nucleation sites for C-S-H and because it reacts with the $Ca(OH)_2$ formed in the hydration [131]. Also of interest is the fact that C-S-H grows epitaxially on the surface of the calcium germanate hydrate $1.6CaO \cdot GeO_2 \cdot nH_2O$ [131].

Table 2.14 Hydration of Ca_3SiO_5 and β-Ca_2SiO_4 Under Ordinary Conditions[a]

			Conditions				Completion
Reactants	Wt Ratio	Time (d)	Temp (°C)	Media	Agit		(%)
1 Ca_3SiO_5, H_2O	ca 2:1	28	ca.20	none	none		70
2 Ca_2SiO_4, H_2O	ca.2:1	28	ca.20	none	none		30

[a]Reference 93 b

HYDRATION OF β-Ca_2SiO_4. Studies of the hydration of β-Ca_2SiO_4 at a low water-to-solids ratio and room temperature indicate that the mechanism of hydration of β-Ca_2SiO_4 is similar to that of Ca_3SiO_5 (although at least one report suggests that no passivating layer is formed in the early stages [132]). The effect of impurity ions on the hydration of β-Ca_2SiO_4 is not yet clear [133].

The role of hydration of β-Ca_2SiO_4 is slower than that of Ca_3SiO_5 (Table 2.14). This has been attributed to the presence of an oxygen unconnected to the silicon in Ca_3SiO_5 but not in β-Ca_2SiO_4 [134]. Support for this conclusion is found in the fact that $Ca_3Si_2O_7$ and $CaSiO_3$, which lack such oxygens, are very resistant to hydration. The fact that b-Ca_2SiO_4 has some reactivity has been ascribed to the way in which its CaOx polyhedron are linked [135].

High-surface-area β-Ca_2SiO_4 prepared by the dehydration of the hillebrandite has been shown to hydrate more rapidly than ordinary β-Ca_2SiO_4 [85]. This is as expected on the basis of the hydration mechanism postulated for β-Ca_2SiO_4.

APPROACH OF WORK. In the work done, the hydration of Ca_3SiO_5 and β-Ca_2SiO_4 was investigated. The objective was to find conditions that would yield products rich in C-S-H and reactive to SO_2. The products were analyzed by infrared spectroscopy, powder X-ray diffractometry and SO_2 uptake measurements.

Experimental

CHEMICALS. Monoclinic Ca_3SiO_5 (stabilized with Al2O3 and MgO) and β-Ca_2SiO_4 (stabilized with B2O3) were obtained from Construction Technology Laboratories. They were ground to pass 270 mesh before use. Fumed SiO_2 (99.9% SiO_2) was obtained from Alfa/Johnson Matthey. Quartz (99.5% SiO_2, <10m) was obtained as a gift from U.S. Silica (Berkley Springs, WV), and $CaSO_4 \cdot 2H_2O$ was obtained from Aldrich Chemical (Milwaukee, WI).

SYNTHESIS OF SORBENTS. *Stirred Reactions Without Beads.* In a typical reaction (Table 2.15, run 1), a slurry of Ca_3SiO_5 (3.00 g) and water (30 mL) was stirred (magnetic stirring bar, ca. 150 rpm) without open access to air while being heated (oil bath, 100°C) for 72 h, and then filtered (Whatman #2). The solid was vacuum dried (ca. 60°C, ca. 60 torr) for ca. 12 h, weighed (3.39 g) and crushed. IR data (Nujol mull, cm-1) for the product are: 3644 (s, OH), 3538 (s, OH), 2466 (m br, OH) 1720 (w br, OH), 1282 (s, OH), 1152 (w br), 982 (s, SiO), 946 (s), 767 (m), 754 (s), 710 (m), 520 (m), 498 (s, SiO), 472 (m). XRD data (d(Å) (I/Io)) for it are: 5.32 (11, C_2SH), 4.90 (100, CH), 4.22 (29, C_2SH), 3.91 (16, C_2SH), 3.54 (24, C_2SH), 3.30 (29, C_2SH), 3.27 (100, C_2SH), 3.11 (20, CH), 2.88 (29, C_2SH), 2.82(27, C_2SH), 2.71 (18, C_2SH), 2.69 (15, C_2SH), 2.66 (30, C_2SH), 2.63 (95, CH), 2.61 (34, C_2SH), 2.57 (18, C_2SH), 2.53 (27, C_2SH), 2.42 (55, C_2SH), 2.34 (12, C_2SH), 2.32 (12, C_2SH), 2.30 (14, C_2SH), 2.29 (15, C_2SH), 2.24 (15, C_2SH), 2.16 (11, C_2SH), 2.11 (11, C_2SH), 2.08 (15, C_2SH), 2.06 (22, C_2SH), 2.03 (11, C_2SH), 1.96 (15, C_2SH), 1.93 (69, CH), 1.80 (39, CH), 1.79 (35, C_2SH), 1.69 (28, CH).

Table 2.15 Synthesis and properties of hydrated Ca_3SiO_5 products.

Reactants[a]	Wt Ratio	Conditions				Estimated Composition[b]					SO_2 Uptake (mmol/g)
		Time (h)	Temp (°C)	Media	Agit	CSH	CH	$C_3S_2H_3$	C_2SH	C_3S	
1 Ca_3SiO_5, H_2O	1:10	72	100		stirring		40		60		2.4
2 Ca_3SiO_5, H_2O	1:10	12	~20	ZrO_2[c]	shaking	40	20	40			5.9
3 Ca_3SiO_5, H_2O	1:10	6	60	ZrO_2[c]	shaking	40	20	40			
4 Ca_3SiO_5, H_2O	1:10	3	100	ZrO_2[d]	stirring	60	40				
5 Ca_3SiO_5, $CaSO_4 \cdot 2H_2O$, H_2O	1:0.5:10	6	60	ZrO_2[c]	shaking	60	40				5.3
6 Ca_3SiO_5, SiO_2[e], H_2O	1:0.26:13[f]	72	100		stirring	40	20		40		
7 Ca_3SiO_5, SiO_2[g], H_2O	1:0.26:13[f]	72	100		stirring	100					
8 Ca_3SiO_5, SiO_2[g], H_2O	1:0.26:13[f]	72	~20	ZrO_2[c]	shaking	100					6.8

[a]Ca_3SiO_5<270mesh. [b]CSH,CaO-SiO_2-H_2O;Ch,Ca(OH)$_2$;$C_3S_2H_3$,$Ca_3(SiO_3OH)_2 \cdot 2H_2O$, afwillite; C_2SH,a-$Ca_2(SiO_3OH)(OH)$. [c]2-mm diameter ZrO_2 (Mg-stabilized) beads. [d]0.6-0.8 mm diameter ZrO_2 (Mg-stabilized) beads. [e]Quartz (<10m). [f]Ca:Si, 1.5:1. [g]Fumed SiO_2 (350m²/g).

Dry Scrubbing Technologies for Flue Gas Desulfurization

Table 2.16 Synthesis and properties of hydrated β-Ca_2SiO_4 products.

Reactants[a]	Wt Ratio	Conditions				Estimated Composition[b]					SO_2 Uptake (mmol/g)
		Time (h)	Temp (°C)	Media	Agit	CSH	CH	$C_3S_2H_3$	C_2SH	βC_2S	
1. Ca_2SiO_4, H_2O	1:10	72	100		stirring				100		2.2
2. Ca_2SiO_4, H_2O	1:10	72	~20	ZrO_2^C	shaking	50	20			30	6.4
3. Ca_2SiO_4, H_2O	1:10	24	80	ZrO_2^C	shaking	30	20		50		
4. Ca_2SiO_4, SiO_2^d, H_2O	1:0.35:14[e]	72	100		stirring	100					
5. Ca_2SiO_4, SiO_2^d, H_2O	1:0.35:14[e]	72	~20	ZrO_2^C	shaking	100					6.6

[a]β-Ca_2SiO_4<270 mesh. [b]CSH,CaO-SiO_2-H_2O;CH,Ca(OH)$_2$; $C_3S_2H_3$, $Ca_3(SiO_3OH)_2 \cdot 2H_2O$, afwillite; C_2SH, $Ca_2(SiO3OH)(OH)$; β-C2S, Ca_2SiO_4. [c]2-mm diameter ZrO_2(MG-stabilized) beads. [d]Fumed SiO_2 (350m^2/g). [e]Ca/Si, 1:1.

SHAKEN REACTIONS WITH BEADS. In a typical reaction (Table 2.15, run 3), a mixture of Ca_3SiO_5 (3.00 g), water (30 mL), and ZrO_2 beads (2 mm, ca. 30 g) in a capped Teflon bottle (125 mL) was shaken (2-3 Hz) while being heated (heating tape, ca. 60°C) for 6 h. The resulting mixture was filtered (60 mesh bronze screen), and the beads were washed with water (ca. 10 mL). The washings and filtrate were combined and filtered (Whatman #2), and the solid was vacuum dried (100°C, 60 torr) for ca. 12 h, weighed (5.13 g), and crushed. IR data (Nujol mull, cm-1) for the product are: 3644 (s, OH), 3346 (s br, OH), 1654 (w br, OH), 962 (s br, SiO), 910 (s br), 814 (w), 472 (s br, SiO). XRD data (d(Å) (I/Io)) for it are: 6.52 (46, $C_3S_2H_3$), 4.90 (100, CH), 4.14 (13, $C_3S_2H_3$), 3.88 (13, $C_3S_2H_3$), 3.27 (36, $C_3S_2H_3$), 3.17 (64, $C_3S_2H_3$), 3.11 (7, CH), 2.83 (100, $C_3S_2H_3$), 2.73 (84, $C_3S_2H_3$), 2.63 (34, CH), 2.35 (26, $C_3S_2H_3$), 2.31 (20, $C_3S_2H_3$), 2.15 (45, $C_3S_2H_3$), 1.98 (22, $C_3S_2H_3$), 1.95 (52, $C_3S_2H_3$), 1.93 (26, CH), 1.80 (12, CH), 1.77 (39, $C_3S_2H_3$), 1.69 (9, CH).

STIRRED REACTIONS WITH BEADS. In a typical reaction (Table 15, run 4), a mixture of Ca_3SiO_5 (3.00 g), water (30 mL), and ZrO_2 beads (0.6-0.8, ca. 150 g) was stirred (mechanical paddle stirrer, ca. 500 rpm) without open access to air while being heated (oil bath, 100°C) for 3h. The resulting mixture was filtered (60-mesh bronze screen) and the beads were washed with water (ca. 10 mL). The filtrate and washings were combined and filtered (Whatman #2), and the solid was vacuum dried (ca. 60°C, ca. 60 torr) for ca. 12h, weighed (4.96 g), and crushed. IR data (Nujol mull, cm-1) for the product are: 3644 (s, OH), 3418 (s br, OH), 1648 (w br, OH), 972 (s br, SiO), 452 (m br, SiO). XRD data (d(Å) (I/Io)) for it are: 4.90 (90, CH), 3.11 (30, CH), 3.04 (100, CSH), 2.80 (56, CSH), 2.63 (100, CH), 1.93 (38, CH), 1.82 (38, CSH), 1.80 (28, CH), 1.69 (18,CH)

Results And Discussion

RAPID HYDRATION OF Ca_3SiO_5 AND β-Ca_2SiO_4. *Rapid Hydration of Ca_3SiO_5.* Various approaches were explored in an effort to find ways to hydrate Ca_3SiO_5 rapidly at a high water-to-solids ratio to products rich in C-S-H. One was based on the use of compounds in the hydration mixture known to accelerate the hydration of Ca_3SiO_5 (and the hydration of portland cement) at low water-to-solids ratios. The compounds used were $CaCl_2$, $CaSO_4 \cdot 2H_2O$, and $Ca(OH)_2$. The use of these compounds did not lead to a significant acceleration of the hydration reaction and their use was not followed further.

Other approaches were based on the use of various combinations of high-shear agitation, elevated temperatures, and the addition of SiO_2. The results of eight runs based on the use of these approaches are summarized in Table 2.15.

The values listed in this table and in the related table, Table 2.16, for the estimated composition of the products are estimates based on careful inspection of the X-ray and infrared data. They should not be taken as values derived from quantitative composition determinations.

The use of a three-day reaction time and a 100°C reaction temperature gave a complete reaction (Table 2.15, run 1). However, the product was $Ca(OH)_2$ and a-$Ca_2(SiO_3OH)(OH)$, a poor sorbent and thus an undesired compound (Figure 2.16). This silicate is probably the stable silicate product of the hydration of Ca_3SiO_5 at ca. 100 °C. Thus, while C-S-H could have been formed, if so, it was subsequently consumed because of the length of the reaction time. This suggests that if the hydration of Ca_3SiO_5 is to be carried out at ca. 100°C, it must be done quickly.

The use of a 12-hour reaction time, an ca. 20°C reaction temperature and high-shear agitation also gave a complete reaction (run 2). The observed acceleration of the reaction is attributed to the continuous removal of the passivating coating assumed to be continuously formed on the Ca_3SiO_5 during the reaction. Again however, the product was not the desired one, in this case being composed of C-S-H, $Ca(OH)_2$, and $Ca_3(SiO_3OH)_2.2H_2O$, another poor sorbent and thus another unwanted compound. Whether $Ca_3(SiO_3OH)_2.2H_2O$ is the stable silicate under these conditions is not clear, and it is thus difficult to account for its formation. Perhaps the Ca_3SiO_5 reacted to give a high SO_{44}- ion concentration under these low-temperature conditions and this led to the precipitation of the $Ca_3(SiO_3OH)_2.2H_2O$.

With the use of a shorter reaction time, a 60°C reaction temperature and high-shear agitation, the product was again C-S-H, $Ca(OH)_2$, and $Ca_3(SiO_3OH)_2.2H_2O$ (run 3). Probably the SiO_{44}- ion concentration again became high and led to the precipitation of $Ca_3(SiO_3OH)_2.2H_2O$. The X-ray powder patterns of products formed under the same conditions, except for the use of shorter reaction times, are consistent with this proposal (Figure 2.17).

Interestingly, the use of a still shorter reaction time, a 100°C reaction temperature, and high-shear agitation gave a product composed of C-S-H and $Ca(OH)_2$ (run 4 and Figure 2.16). At this elevated temperature, it may be that the concentration of the SiO_{44}- ions was low because of silicate ion condensation reactions. If so, this low concentration could have favored the formation of C-S-H since C-S-H can accommodate oligomeric silicate ions, and it would have disfavored the immediate formation of $Ca_3(SiO_3OH)_2.2H_2O$ and a-$Ca_2(SiO3OH)(OH)$ since these silicates contain SiO_{44}- ions. Consistent with this, it has been suggested that the chain lengths of the silicate ions in C-S-H formed from Ca_3SiO_5 increase with temperature [136]. Probably the average chain length of the silicate ions of the C-S-H of run 4 was relatively long. It could be that a similar run with a 6-hour, instead of 3-hour, reaction time would have given α-$Ca_2(SiO_3OH)(OH)$ as a product.

Conditions the same as those of run 3, except for the addition of a small amount of $CaSO_4.2H_2O$ to the reaction mixture, led to a product composed of C-S-H and $Ca(OH)_2$ (run 5). Perhaps the formation of $Ca_3(SiO_3OH)_2.2H_2O$ was inhibited by the deposition of SO_4^{2-} ions in SiO_4^{4-} sites on the surface of the $Ca_3(SiO_3OH)_2.2H_2O$ nuclei.

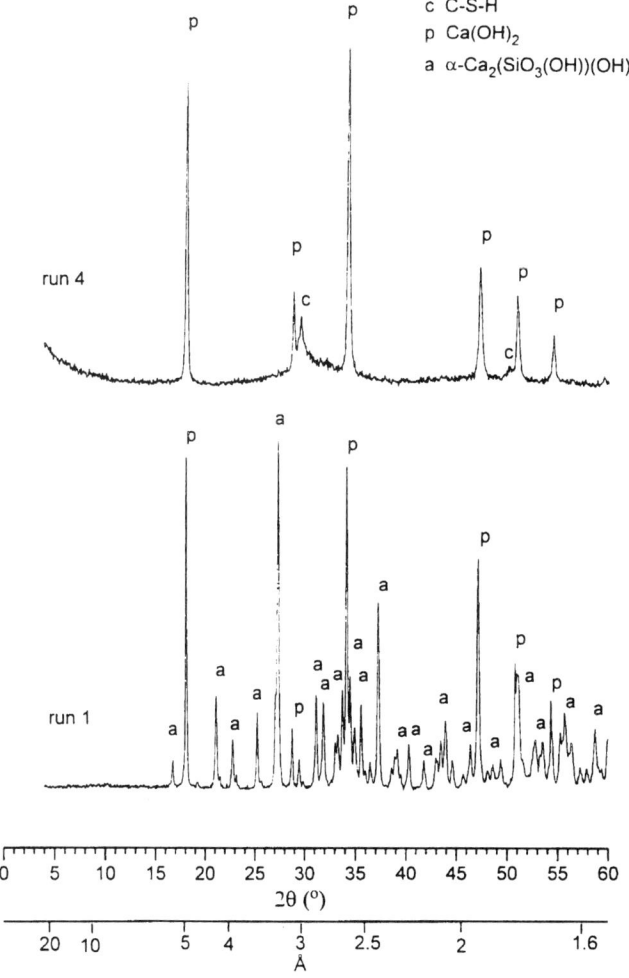

Figure 2.16 X-ray powder patterns of products of runs 1 and 4, Table 2.15.

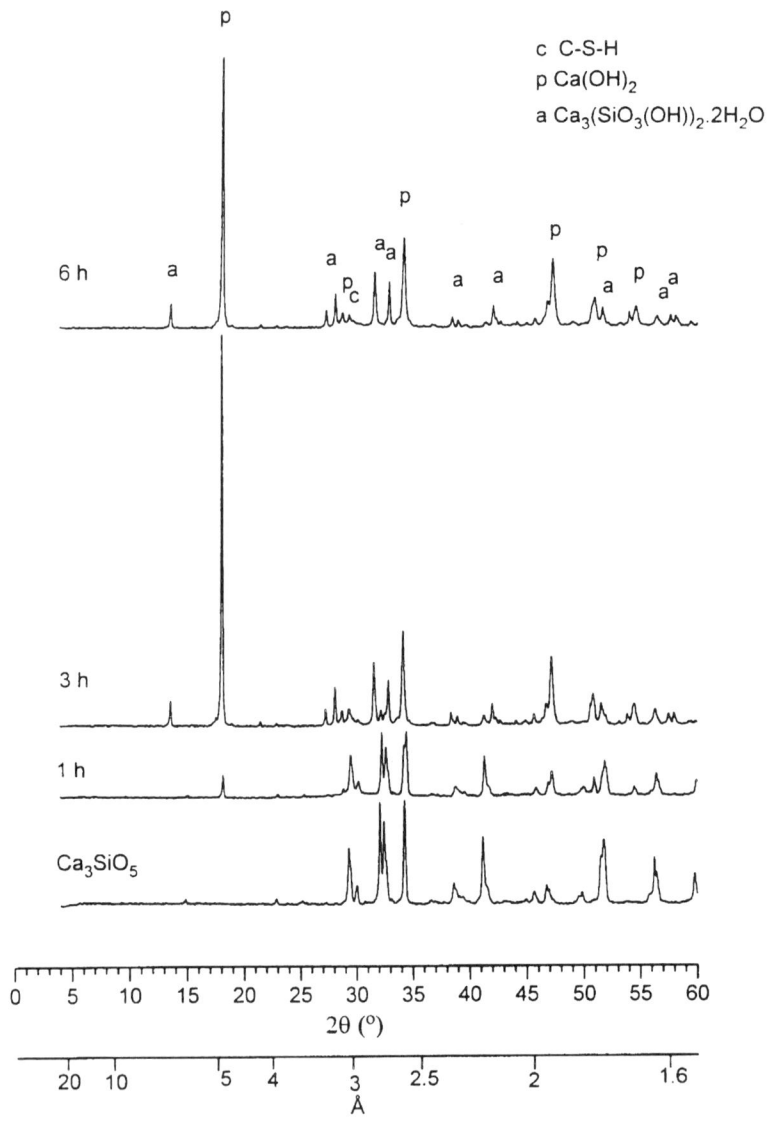

Figure 2.17 X-ray powder patterns of intermediate and final products of run 3, Table 2.15.

New Calcium-Based Sorbents for Flue Gas Desulfurization

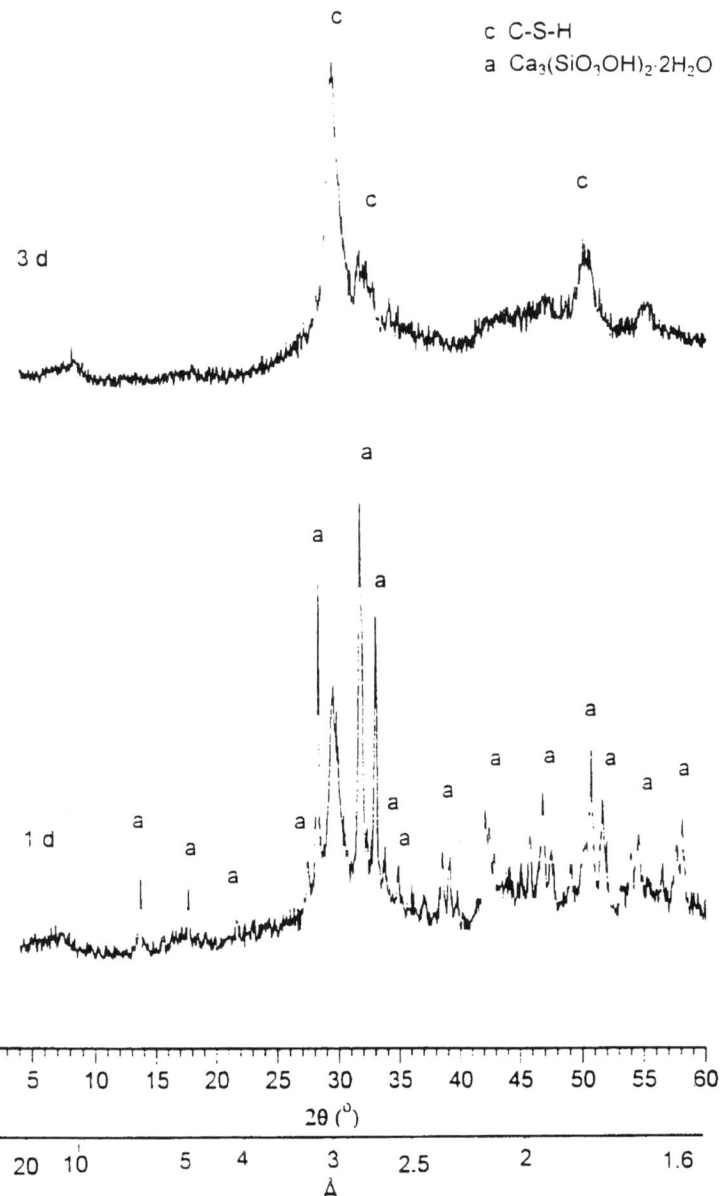

Figure 2.18 X-ray powder patterns of intermediate and final products of run 8, Table 2.15.

The use of conditions like those employed in run 1, except for the addition of 10 μ quartz to the reaction mixture, led to a product composed of C-S-H, $Ca(OH)_2$, and α-$Ca_2(SiO_3OH)(OH)$ (run 6). Probably the finely divided silica reacted with the $Ca(OH)_2$ formed by the hydration of Ca_3SiO_5 and gave C-S-H. It may be that it also provided nucleation sites for the precipitation of the C-S-H.

In contrast, the use of conditions the same as those of run 6, except for the substitution of fumed SiO_2 for the powdered quartz, led to a product composed of only C-S-H (run 7). Again, probably the silica reacted with the $Ca(OH)_2$ and gave C-S-H, and again it may have provided nucleation sites for C-S-H precipitation. Since fumed SiO_2 is more reactive than powdered quartz, it is not surprising that more C-S-H is formed in run 7 than in run 6.

Finally, a three-day reaction time, a reaction temperature of ca. 20°C, high-shear agitation, and the addition of fumed SiO_2 led to a product composed of only C-S-H (run 8). Similar runs with shorter reaction times gave products containing $Ca_3(SiO_3OH)_2.2H_2O$ (Figure 2.18). This suggests that $Ca_3(SiO_3OH)_2.2H_2O$ is not the stable silicate product when active SiO_2 is present.

RAPID HYDRATION OF β-Ca_2SiO_4. Various approaches also were explored in an effort to find ways to rapidly hydrate β-Ca_2SiO_4 at a high water-to-solids ratio to products rich in C-S-H. These approaches were, as before, based on the use of various combinations of high-shear agitation, elevated temperatures and the addition of SiO_2 to the hydration mixture. The results of five runs utilizing these approaches are summarized in Table 16.

The use of a three-day reaction time and a 100°C reaction temperature gave a complete reaction (Table 2.16, run 1). The product, however, consisted entirely of α-$Ca_2(SiO_3OH)(OH)$ and thus had an undesired composition. This silicate probably is the stable hydration product at a high water-to-solids ratio not only for Ca_3SiO_5 at 100°C but also for β-Ca_2SiO_4 at this temperature. Its apparent stability and the three-day reaction time likely account for its presence in the product.

The use of a three-day reaction time, an ca. 20°C reaction temperature, and high-shear agitation gave a product containing C-S-H, $Ca(OH)_2$ and some unreacted b-Ca_2SiO_4 (run 2). This shows that the hydration of b-Ca_2SiO_4 under these water-to-solids, temperature, and agitation conditions is, as expected, slower than the hydration of Ca_3SiO_5 under the same conditions (compare to Table 2.14, run 2). It seems likely that this can be attributed to the greater difficulty of hydroxylating the ß-Ca_2SiO_4 surface.

With a one-day reaction time, an 80°C reaction temperature, and high-shear agitation, the product obtained was C-S-H, $Ca(OH)_2$, and $\alpha\text{-}Ca_2(SiO_3OH)(OH)$ (run 3). Probably as in the case of run 1, the presence of $\alpha\text{-}Ca_2(SiO_3OH)(OH)$ in the product reflects the apparent stability of this silicate at elevated temperatures.

This conclusion is supported by the fact that a run with a 100°C reaction temperature and moderate shear agitation gave a product that consisted of C-S-H, $Ca(OH)_2$ and $\beta\text{-}Ca_2SiO_4$ when a short reaction time was used, and $\alpha\text{-}Ca_2(SiO_3OH)(OH)$ when a three-day reaction time was used.

The use of conditions like those used in run 1 except for the addition of fumed SiO_2 led to a product made up of only C-S-H (run 4). This suggests that $\alpha\text{-}Ca_2(SiO_3OH)(OH)$ is not the stable product in the presence of active SiO_2.

With conditions like those used in run 2 except for the addition of fumed SiO_2, the product was again C-S-H only (run 5). This shows that active SiO_2 accelerates the reaction. Perhaps this is attributable to the provision of nucleation sites for C-S-H by the SiO_2.

SO_2 **UPTAKE OF PRODUCTS.** The SO_2 uptake values of the products of Table 15, runs 1, 2, 5 and 8 and Table 16 runs 1, 2, and 5 are as expected. Thus, those products that contain $\alpha\text{-}Ca_2(SiO_3OH)(OH)$ and $Ca_3(SiO_3OH)_2 \cdot 2H_2O$ are not as good as the products that contain only C-S-H (taking the amount of Ca in the products into account).

Conclusions

It appears that the hydration of Ca_3SiO_5 and $\beta\text{-}Ca_2SiO_4$ to C-S-H and $Ca(OH)_2$ is slowed by the formation of passivating coatins on these silicates, that it is hindered by the stability of $\alpha\text{-}Ca_2(SiO_3OH)(OH)$ at high temperatures, and that it is hindered by the ability of $Ca_3(SiO_3OH)_2 \cdot 2H_2O$ to form at low temperatures. It further appears that the hydration of Ca_3SiO_5 and $\beta\text{-}Ca_2SiO_4$ to C-S-H only is promoted by the presence of active SiO_2, hindered by the formation of a passivating coating on these silicates, and hindered by the stability of $\alpha\text{-}Ca_2(SiO_3OH)(OH)$ at high temperature.

On the basis of this, it is concluded that Ca_3SiO_5 and $\beta\text{-}Ca_2SiO_4$ could be very rapidly hydrated to C-S-H and $Ca(OH)_2$ by the use of very high shear conditions (such as those provided by commercial high-shear mills) and high temperatures. It is further concluded that $Ca_3SiO_5\text{-}SiO_2$ and $b\text{-}Ca_2SiO_4\text{-}SiO_2$ mixtures could be hydrated very rapidly to C-S-H alone with the same type of conditions.

Hydtrated Portland Cement Sorbents

Introduction

As already mentioned, portland cement is of interest in this study because it has the potential to yield good SO_2 sorbents. This section deals with experimental work done to uncover this potential.

Portland cement is an exceedingly important material for modern society. It is produced cheaply and on a very large scale worldwide. In Cleveland in 1994 the price of cement was about $65 per ton in bulk quantities [137]. World production of cement in 1990 was about 1.3 billion tons (about 0.3 ton for each person in the world) [138].

Cement is produced in a two-step process. In the first step, a mixture of carefully chosen common rocks is heated. Typically the rocks are a limestone and a clay or a shale and the reaction temperature is about 1450°C. Then in the second step, the product of the first step, clinker, is ground with $CaSO_4 \cdot 2H_2O$ to obtain cement [139].

Cement is classified into types based on its composition. Ordinary (type I) cement typically contains, as already stated, 50-70% alite (ca. Ca_3SiO_5), 15-30% belite (ca. Ca_2SiO_4), 5-10% aluminate phase (ca. $Ca_3Al_2O_6$), 5-15% ferrite phase (ca. Ca_2AlFeO_5), and 3-4% gypsum ($CaSO_4 \cdot 2H_2O$) [76c].

An enormous literature on the preparation and properties of cement exists. Much of it has been summarized recently [76,140].

HYDRATION OF THE MAJOR COMPONENTS OF CEMENT. All the main components in portland cement except gypsum are anhydrous. When brought into contact with water most of the anhydrous components are attacked and form hydrated compounds or phases. Because of the complexity of cement, it is useful to discuss the hydration of its major components separately.

ALITE. The hydration of the alite phase of cement under the low water-to-solids ratio normally used (0.5:1) is relatively slow, essentially complete hydration requiring about a year. This hydration gives C-S-H and $Ca(OH)_2$ [141].

$$Ca_3SiO_5 + xH_2O \rightarrow 1.7CaO \cdot 1.0SiO_2 \cdot (x - 1.3)H_2O + 1.3Ca(OH)_2$$

$$(C_3S + H \rightarrow 1.7C \cdot 1.0S \cdot (x - 1.3)H + 1.3\ CH)$$

(To make this section more informative, equations in cement chemist's notation are given in parenthesis in addition to the equations in standard chemist's notation.)

Belite. The hydration of the belite phase of cement under the same conditions is still slower, essentially complete hydration requiring more than four years. It also gives a mixture of C-S-H and Ca(OH)$_2$, but the proportion of C-S-H in this mixture is greater [142].

$\beta\text{-}Ca_2SiO_4 + xH_2O \rightarrow 1.7CaO\text{-}1.0SiO_2\text{-}(x - 0.3)H_2O + 0.3Ca(OH)_2$

$(\beta\text{-}C_2S + H \rightarrow 1.7C\text{-}1.0S\text{-}(x - 0.3)H + 0.3\ CH)$

The C-S-H formed by the hydration of the alite and belite in cement apparently has a structure which is similar to but less ordered than the structure of the C-S-H formed by the hydration of Ca_3SiO_5 and $\beta\text{-}Ca_2SiO_4$. It does not contain appreciable amounts of Al^{3+} and Fe^{3+}, but it is intimately mixed with the iron-aluminum phase AFm (see below). Its admixture with AFm probably is a cause of its low degree of order [76d]. Taylor [76d] has suggested that cement C-S-H be called cement gel in order to recognize its unique features.

ALUMINATE PHASE. As already indicated, the composition of the aluminate phase of cement is similar to that of $Ca_3Al_2O_6$. Typically ca. 3% of the Ca_2+ in this phase is replaced interstitially and isomorphously with 2 Na^+, ca. 11% of the Al^{3+} is replaced isomorphously by Fe^{3+}, and ca. 9% of the Al^{3+} is replaced isomorphously by Si^{4+} [76e]. Not surprisingly, the hydration of the aluminate phase of cement is similar to that of $Ca_3Al_2O_6$. For clarity, it is helpful to consider the hydration of $Ca_3Al_2O_6$ and mixtures of it with $CaSO_4.2H_2O$ and $Ca(OH)_2$ before considering the hydration of the aluminate phase itself.

HYDRATION OF $Ca_3Al_2O_6$. The hydration of $Ca_3Al_2O_6$ is much more rapid than that of Ca_3SiO_5. Within 30 minutes it gives a mixture of $[Ca_2Al(OH)_6](OH).6H_2O$ and $[Ca_2Al(OH)_6][Al(OH)_4].3H_2O$ (Figure 2.19).

New Calcium-Based Sorbents for Flue Gas Desulfurization

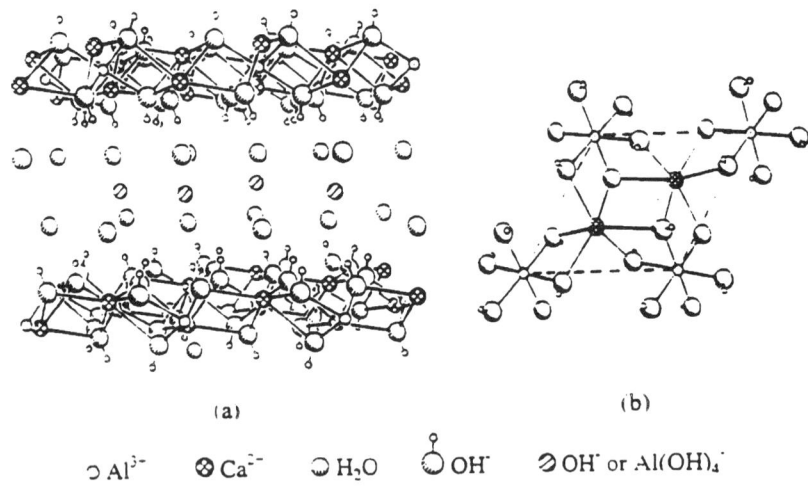

Figure 2.19 (a) Schematic structure of $[Ca_2Al(OH)_6](OH) \cdot 6H_2O$ and $[Ca_2Al(OH)_6][Al(OH)_4] \cdot 3H_2O$ showing $[Ca_2Al(OH)_6]^+$ layers and OH^- or $Al(OH)_4^-$ ions between layers. (b) Structure of $[Ca_2Al(OH)_6]^+$ layers (Taylor's data [143]).

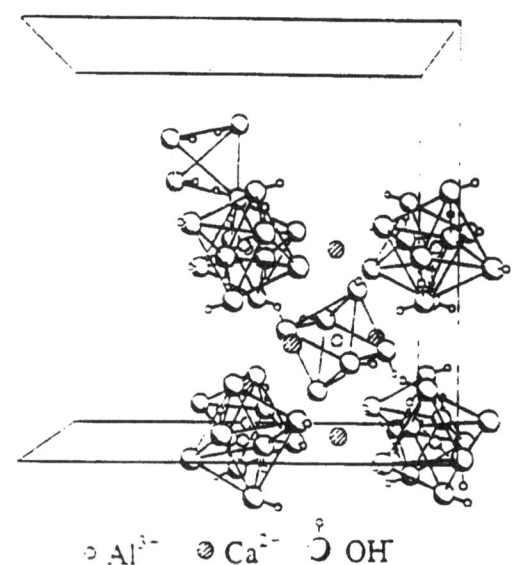

Figure 2.20 Portion of unit cell of $Ca_3Al_2(O_4H_4)3$, hydro-garnet, showing $Al(OH)_6$ and $(OH)_4$ units in structure (Foreman's data [144]).

$2Ca_3Al_2O_6 + 27H_2O \rightarrow [Ca_2Al(OH)_6](OH).6H_2O + [Ca_2Al(OH)_6][Al(OH)_4].3H_2O$

$(2\ C_3A + 27\ H \rightarrow C_4AH_{19} + C_2AH_8)$.

These compounds then interact to give $Ca_3Al_2(O_4H_4)_3$, hydrogarnet (Figure 2.20), within 24 to 48 hours [116].

$2[Ca_2Al(OH)_6](OH).6H_2O + Ca_2Al(OH)_6][Al(OH)_4].3H_2O \rightarrow 2Ca_3Al_2(O_4H_4)_3$

$(C_4AH_{19} + C_2AH_8 \rightarrow 2\ C_3AH_6)$.

HYDRATION OF $Ca_3Al_2O_6$-$CaSO_4$ and $Ca_3Al_2O_6$-$Ca(OH)_2$ MIXTURES. The hydration of a 1:3 mixture of $Ca_3Al_2O_6$ and $CaSO_4.2H_2O$ is rapid. In about 10-12 minutes it gives $[Ca_3Al(OH)_6.12H_2O]2(SO_4)3.2H_2O$, trisulfoaluminate hydrate, or the Al^{3+} and SO_4^{2-} end-member of the ettringite family [116] (Figure 2.21).

$Ca_3Al_2O_6 + 3CaSO_4.2H_2O + 26H_2O \rightarrow [Ca_3Al(OH)_{6.12}H_2O]2(SO_4)3.2H_2O$

$(C_3A + 3\ CH_2 + 26 \rightarrow H\ C_6A3H_{32})$.

The hydration of a mixture richer in $Ca_3Al_2O_6$ gives trisulfoaluminate hydrate as an intermediate and then this reacts with more $Ca_3Al_2O_6$ to give $[Ca_2Al(OH)_6]2(SO_4).6H_2O$, monosulfoaluminate hydrate [116] (Figure 22).

$2Ca_3Al_2O_6 + [Ca_3Al(OH)_{6.12}H_2O]2(SO_4)3.2H_2O\ +$
$4H_2O \rightarrow [Ca_2Al(OH)_6]2(SO_4).6\ H_2O$

$(2C_3A + C_6A_3H_{32} + 4H \rightarrow 3C_4AH_{12})$

Thus a 1:1 ratio of $Ca_3Al_2O_6$ and $CaSO_4.2H_2O$ ultimately gives monosulfoaluminate hydrate [116].

$Ca_3Al_2O_6 + CaSO_4.2H_2O + 10H_2O \rightarrow [Ca_2Al(OH)_6]2(SO_4).6H_2O$

$(C_3A + CH_2 + 10H \rightarrow C_4AH_{12})$

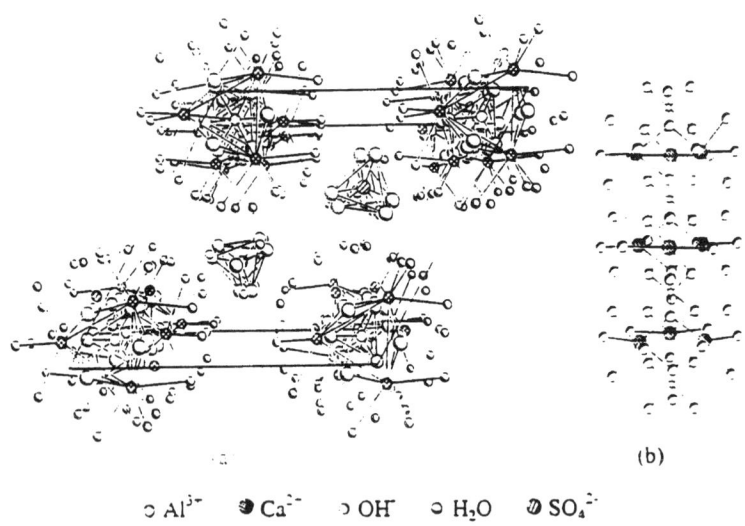

$\circ$ Al^{3+} ● Ca^{2+} $\circ$ OH$^-$ $\circ$ H$_2$O   ⊘ SO$_4^{2-}$

Figure 2.21(a) Structure of [Ca$_3$Al(OH)$_6$·12H$_2$O]2(SO$_4$)3·2H$_2$O, trisulfoaluminate hydrate, projected on the <001> showing the columns and channels present. (b) Structure of [Ca$_3$Al(OH)$_6$·12H$_2$O]3+ columns (Taylor's data [145]).

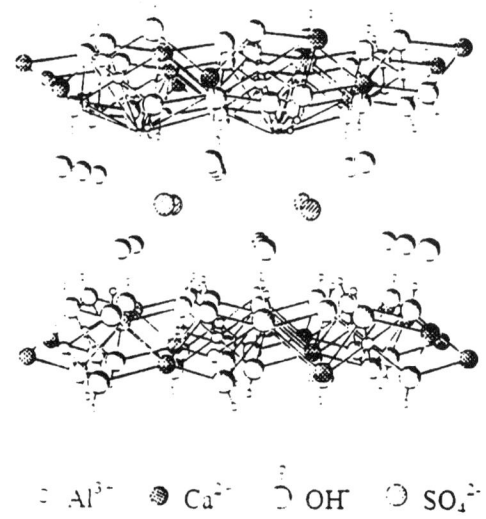

$\circ$ Al^{3+} ● Ca^{2+} $\circ$ OH$^-$ $\circ$ SO$_4^{2-}$

Figure 2.22 Structure of [Ca$_2$Al(OH)$_6$]SO$_4$·6H$_2$O, monosulfo-aluminate hydrate, showing layers of [Ca$_2$Al(OH)$_6$·]+ and H$_2$O and SO$_4{}^{2-}$ ions between layers (Allmann's data [146]).

The SO_4^{2-} ions in this phase can be exchanged for other ions such as OH^-. Not surprisingly in view of this, the hydration of a 1:1 mixture of $Ca_3Al_2O_6$ and $Ca(OH)_2$ gives the hydroxy analogue of monosulfoaluminate hydrate, $[Ca_2Al(OH)_6](OH).3H_2O$ [116].

$$Ca_3Al_2O_6 + Ca(OH)_2 + 12H_2O \rightarrow 2[Ca_2Al(OH)_6](OH).3H_2O$$

$(C_3A + CH + 12H \rightarrow C_4AH_{13})$.

HYDRATION OF ALUMINATE PHASE IN CEMENT. The hydration of the aluminate phase in cement involves not only this phase and water but also the ferrite phase, $CaSO_4.2H_2O$, and CO_2 (from the atmosphere). The main product of the hydration is $[Ca_2(Al, Fe)(OH)_6]2(SO_4, CO_3, 2OH).6H_2O$, or AFm. This phase has a structure like that of monosulfoaluminate hydrate with Fe^{3+} isomorphously replacing Al^{3+}, and CO_3^{2-} and OH^- isomorphously replacing SO_4^{2-}. As is to be expected, the anions of this phase can undergo ion exchange [76f].

The hydration of the mixture of the aluminate phase, ferrite phase, and $CaSO_4.2H_2O$ in cement together with CO_3^{2-} also can lead to $Ca_3(Al,Fe)(OH)_{6.12}H_2O]2(SO_4,CO_3,2OH)_{3.2}H_2O$ or AFt. This has a structure similar to trisulfoaluminate hydrate with Fe_3+ ion isomorphously replacing $Al3+$ and CO_3^{2-} and OH^- isomorphously replacing SO_4^{2-} [76d].

In cement, this phase loses its water at about 130°C and becomes amorphous. It is postulated that this dehydrated synthetic AFt still has a column-like structure but with open channels. Interestingly, this dehydrated synthetic AFt shows a high Ca utilization with SO_2 and NO_x [147].

FERRITE PHASE. The ferrite phase in cement is actually a solid solution with a composition between $Ca_2Fe_2O_5$ and $Ca_6Al_4Fe_2O_{15}$. The formula often used to represent this phase, Ca_2AlFeO_5, corresponds to an average composition [76g]. Again it is helpful to consider the hydration of the pure phase and mixtures of it with $CaSO_4.2H_2O$ before considering the hydration of it in cement.

HYDRATION OF Ca_2AlFeO_5. Although its rate is slower, the hydration of $Ca_4Al_2Fe_2O_{10}$ is similar to the hydration of the aluminate phase. First $[Ca_2(Al,Fe)(OH)_6](OH).6H_2O$ and $[Ca_2(Al,Fe)(OH)_6][(Al,Fe)(OH)_4] \cdot 3H_2O$ are formed and then $Ca_3(Al,Fe)2(O_4H_4)_3$ is formed [148]. Possible simplified equations are

$$4Ca_2AlFeO_5 + 38H_2O \rightarrow 4[Ca_2(Al_{0.5}Fe_{0.5})(OH)_6](OH).6H_2O +$$

$2[Ca_2(Al_{0.5}Fe_{0.5})(OH)_6][(Al_{0.5}Fe_{0.5})(OH)_4] \cdot 3H_2O + (Al_{0.5}Fe_{0.5})(OH)_3$

$(2C_4AF + 38H \rightarrow C4(A_{0.5}F_{0.5})H_{19} + 2C_2(A_{0.5}F_{0.5})H_8 + (A_{0.5}F_{0.5})H_3)$

and

$2[Ca_2(Al_{0.5}Fe_{0.5})(OH)_6](OH) \cdot 6H_2O +$

$[Ca_2Al_{0.5}Fe_{0.5})(OH)_6][(Al_{0.5}Fe_{0.5})(OH)_4] \cdot 3H_2O \rightarrow 2Ca_3(Al_{0.5}Fe_{0.5})2(O_4H_4)_3$

$(C_4(A_{0.5}F_{0.5})H_{19} + C_2(A_{0.5}F_{0.5})H_8 \rightarrow 2C_3(A_{0.5}F_{0.5})H_6)$.

The structures of $[Ca_2(Al,Fe)(OH)_6](OH) \cdot 6H_2O$ and $[Ca_2(Al,Fe)(OH)_6][(Al,Fe)(OH)_4] \cdot 3H_2O$ are like those of $[Ca_2Al(OH)_6](OH) \cdot 6H_2O$ and $[Ca_2Al(OH)_6][Al(OH)_4] \cdot 3H_2O$ and the structure of $Ca_3(Al,Fe)2(O_4H_4)_3$ is like that of $Ca_3Al_2(O_4H_4)_3$.

HYDRATION OF $Ca_4Al_2Fe_2O_{10}$-$CaSO_4 \cdot 2H_2O$ MIXTURES. The hydration of 1:1 mixture $Ca_4Al_2Fe_2O_{10}$ and $Ca(SO_4)_2 \cdot 2H_2O$ is similar to that of a 1:1 mixture of $Ca_3Al_2O_6$ and $Ca(SO_4)_2 \cdot 2H_2O$. First it gives AFt and then it gives AFm, both apparently are mixed with some $(Al,Fe)(OH)_3$ gel which may contain Ca [149]. Possible simplified equations for this are

$3(CaAlFeO_5)2 + 12CaSO_4 \cdot 2H_2O + 106H_2O \rightarrow$

$4[Ca_3(Al_{0.5}Fe_{0.5})(OH)_6 \cdot 12H_2O]2(SO_4)_3 \cdot 2H_2O + 4(Al_{0.5}Fe_{0.5})(OH)_3$

$(3C_4AF + 12C\bar{S}H_2 + 106H \rightarrow 4C_6(A_{0.5}F_{0.5})\bar{S}3 \cdot H_{31} + 2(A_{0.5}F_{0.5})H_3)$

and

$3(CaAlFeO5)2 + 2[Ca_3(Al_{0.5}Fe_{0.5})(OH)_6 \cdot 12H_2O]2(SO_4)_3 \cdot 2H_2O \rightarrow$

$6[Ca_2(Al_{0.5}Fe_{0.5})(OH)_6]2(SO_4) \cdot 6H_2O + 4(Al_{0.5}Fe_{0.5})(OH)3$

$(3C4AF + 2C6(A_{0.5}F_{0.5})\bar{S}_3 \cdot H_{31} + 16H \rightarrow 6C_4(A_{0.5}F_{0.5})\bar{S}H_{12} + A_{0.5}F_{0.5})H_3)$.

Probably the $(Al,Fe)(OH)_3$ gel forms because of the amount of Fe that can be accommodated in AFt is less than that in the anhydrous ferrite and the amount of Fe that can be accommodated in AFm is less than that in AFt. When the ferrite is hydrated under the above conditions except for the additional presence of $Ca(OH)_2$, it has been reported that the AFt has a greater ability to accommodate iron in its structure [150]. A possible simplified equation for this modified hydration is

$$2Ca_2AlFeO_5 + 2Ca(OH)_2 + 6CaSO_4.2H_2O + 50H_2O \rightarrow$$
$$[Ca_3Al_{0.5}Fe_{0.5}(OH)_6 \cdot 12\,H_2O]2(SO_4)\,3.2H_2O$$

$$(C_4AF + 2CH + 6C\bar{S}H_2 + 50H \rightarrow 2C_6(A_{0.5}F_{0.5})\bar{S}_3H_{32}).$$

HYDRATION OF FERRITE PHASE IN CEMENT. The hydration of the ferrite phase in cement is not yet understood. Probably this phase together with $CaSO_4.2H_2O$ and $Ca(OH)_2$ in the hydrating mixture react to give AFm. The iron from this phase is accommodated largely in the AFm. Apparently no $(Al,Fe)(OH)_3$ gel is formed during the hydration [151].

COMPONENTS IN HYDRATED CEMENT. In sum, the hydration of cement gives a product containing ca. 60% C-S-H, ca. 20% $Ca(OH)_2$ and ca. 20% of a mixture of AFm and AFt. The C-S-H is amorphous, the $Ca(OH)_2$ is crystalline, and the AFm and AFt can be either crystalline or amorphous.

Relatively little work has been done on the hydration of cement at a high (10:1) water-to-solids ratios. However, it appears that the same types of reactions take place. The products obtained are not massive but rather are powders.

APPROACH OF WORK. The approach used in this work is similar to that used in the work on the hydration of the cement models ß-Ca_2SiO_4 and Ca_3SiO_5, that is, hydration of cement under various conditions and then analysis of the products by infrared spectroscopy, powder X-ray diffractometry, BET surface area measurements, and SO_2 uptake measurements.

Experimental

CHEMICALS. Type I portland cement (SiO_2 20.73%, Al_2O_3 5.22%, Fe_2O_3 3.30%, CaO 63.71%, MgO 1.33%, SO_3 2.63%, Na_2O 0.15%, K_2O 0.83%, TiO_2 0.27%, P_2O_5 0.15%, MN_2O_3 0.13%, SrO 0.15%, LOI 1.78%) was obtained through a retail store from Medusa Cement (Cleveland, OH). Byproduct SiO_2 fume (Microsilica EMS 960) was obtained as a gift from Elkem Materials (Pittsburgh, PA) and uncalcined diatomite was obtained as a gift from Celite (Lompoc, CA). Pumice (Tamez Mountains, NM) was purchased from Ward's Natural Science Establishment and perlite was obtained as a gift from Harborlite (Vicksburg, MI). $Ca_3Al_2O_6$ was purchased from Construction Technology Laboratory and $CaSO_3 \cdot 0.5H_2O$ was purchased from Pfaltz and Bauer (Waterbury, CT) and $CaSO_4.2H_2O$ was purchased from Aldrich Chemical. MgO-stabilized ZrO_2 beads were purchased from US Stoneware.

SYNTHESIS OF SORBENTS. Unagitated Reactions. In a typical reaction (Table 17, run 1), a mixture of cement (type I, 10.0 g) and water (100 mL) in a capped polyethylene bottle (250 mL) was shaken by hand for several minutes and then held without agitation at room temperature for 30 days. (During this time the color of the slurry changed from greenish-gray to off-white). The resulting mixture was filtered (Whatman #2), and the solid was dried (ca. 60°C, ca. 60 torr) and weighed (10.1 g). IR data (Nujol mull) (cm-1) for the product are: 3644 (s br, OH), 1650 (w br, OH) and 974 (s br, SiO). XRD data (d(Å) (I/Io)) for it are: 7.92 (9, AFm), 4.90 (100, CH), 3.11 (10, CH), 3.04 (25, CSH), 2.79-2.73 (31, b-C2S), 2.63 (60, CH), 1.93 (55, CH), 1.80 (50, CH) and 1.69 (15, CH).

STIRRED REACTIONS WITHOUT BEADS. In a typical reaction (Table 2.17, run 3), a mixture of cement (type I, 10.0 g) and water (100 mL) was stirred (magnetic stirring bar, ca. 150 rpm) while being heated (oil bath, 100°C) without access to the atmosphere for 72 h and then filtered (Whatman #2). (During this time the color of the slurry changed from greenish-gray to off-white). The solid was vacuum dried (ca. 60°C, ca. 60 torr) for ca. 24 h, weighed (10.2 g), and crushed. IR data (Nujol mull, cm-1) for the product are: 3644 (s, OH), 3538 (s, OH), 2466 (m br, OH) 1720 (w br, OH), 1282 (s, OH), 1152 (w br), 982 (s, SiO), 946 (s), 767 (m), 754 (s), 710 (m), 520 (m), 498 (s, SiO) and 472 (m). XRD data (d(Å) (I/Io)) for it are: 5.32 (4, C_2SH), 4.90 (100, CH), 4.22 (5, C_2SH), 3.91 (4, C_2SH), 3.54 (5, C_2SH), 3.30 (4, C_2SH), 3.27 (7, C_2SH), 3.11 (7, CH), 2.88 (6, C_2SH), 2.82 (6, C_2SH), 2.71 (3, C_2SH), 2.69 (3, C_2SH), 2.66 (7, C_2SH), 2.63 (8, CH), 2.61 (4, C_2SH), 2.57 (3, C_2SH), 2.53 (4, C_2SH), 2.42 (12, C_2SH), 2.34 (3, C_2SH), 2.32 (4, C_2SH), 2.30 (3, C_2SH), 2.29 (3, C_2SH), 2.24 (4, C_2SH), 2.16 (3, C_2SH), 2.11 (4, C_2SH), 2.08 (3, C_2SH), 2.06 (4, C_2SH), 2.03 (4, C_2SH), 1.96 (3, C_2SH), 1.93 (6, CH), 1.80 (7, CH), 1.79 (5, C_2SH) and 1.69 (5, CH).

SHAKEN REACTIONS WITH BEADS. In a typical reaction (Table 2.17, run 5), a mixture of portland cement (type I, 5.00 g), water (50 mL) and ZrO_2 beads (MgO-stabilized, 2 mm, ca. 50 g) in a capped polyethylene bottle (125 mL) was shaken with a shaker (ca. 2-3 Hz) at room temperature for 3 days. (During this time the color of the cement-water slurry changed from greenish-gray to off-white). The resulting mixture was filtered (60 mesh bronze screen), and the beads were washed with water (ca. 10 mL). The washings and filtrate were combined and refiltered (Whatman #2), and the solid was vacuum dried (ca. 60°C, ca. 60 torr) for ca. 24 h, weighed (5.79 g) and crushed. IR data (Nujol mull) (cm-1) for the product are: 3644 (s br, OH), 1650 (w br, OH) and 974 (s br, SiO). XRD data (d(Å) (I/Io)) for it are: 7.92 (10, AFm), 4.90 (96, CH), 3.11 (45, CH), 3.04 (25, CSH), 2.63 (100, CH), 1.93 (55, CH), 1.80 (55, CH) and 1.69 (21, CH).

STIRRED REACTIONS WITH BEADS. In a typical reaction (Table 17, run 10), a mixture of cement (10.0 g), water (100 mL), and ZrO_2 beads (MgO-stabilized, 0.6-0.8 mm, ca. 600 g) was stirred (mechanical paddle stirrer, ca. 500 rpm) while being heated (oil bath, 100°C) for 3 h. The resulting mixture was filtered (60 mesh bronze screen) and the beads were washed with water (ca. 10 mL). The filtrate and washings were combined and filtered (Whatman #2), and the solid was vacuum dried (ca. 60°C, ca. 60 torr) for ca. 24 h, weighed (10.8 g) and crushed. IR data (Nujol mull, cm-1) for the product are: 3644 (s, OH), 3418 (s br, OH), 1648 (w br, OH), 972 (s br, SiO) and 452 (m br, SiO). XRD data (d(Å) (I/Io)) for it are: 4.90 (90, CH), 3.11 (30, CH), 3.04 (15, CSH), 2.63 (100, CH), 1.93 (36, CH), 1.82 (10, CSH), 1.80 (28, CH) and 1.69 (18, CH).

Results and Discussion

RAPID HYDRATION OF CEMENT AND CEMENT-SiO_2 SOURCE MIXTURES. *Rapid Hydration of Cement.* In parallel with the work done on the hydration of Ca_3SiO_5 and β-Ca_2SiO_4, various approaches were explored in an effort to find ways to hydrate cement rapidly at high water-to-solids ratios to products rich in C-S-H, AFm and, in some cases, also $Ca(OH)_2$. These approaches were based on the use of a number of different combinations of reaction time, reaction temperature, and degree of agitation. The results of ten runs based on these approaches are summarized in Table 2.17. (The runs in this work were carried out so as to yield information on the uptake of SO_2 by the sorbent and not on the percent capture of SO_2 in the gas stream. Capture work can be done in the future.)

The values given in Table 2.17 and in the related table, Table 2.18, for the composition of the products are estimates based on careful inspection of the X-ray and infrared data. They should not be taken as values derived from quantitative composition determinations

New Calcium-Based Sorbents for Flue Gas Desulfurization

Table 2.17 Synthesis and properties of hydrated cement products

Reactants	Wt Ratio	Time (h)	Temp (°C)	Media[b]	Agit	CSH (%)	CH (%)	AFm[c]	C_2SH (%)	cem (%)	BET Area (m^2/g)	SO_2 Uptake (mmol/g)	CA Util (%)
1. cem, H_2O	1:10	720	~20			45	20	25[d]		10		3.9	35
2. cem, H_2O	1:10	360	~20			35	10	25[e,f]		30			
3. cem, H_2O	1:10	72	100		stirring		20	20[d]	50	10		4.2	38
4. cem, H_2O	1:10	24	100		stirring	15	15	20[d]	30	20			
5. cem, H_2O	1:10	72	~20	ZrO^2	shaking	45	20	25[d]		10	64	6.5	67
6. cem, H_2O	1:10	24	~20	ZrO^2	shaking	30	15	25[d]		30			
7. cem, H_2O	1:10	24	~20	ZrO^2	stirring	45	20	25[d]		10	62		
8. cem, H_2O	1:10	5	~20	ZrO^2	stirring	30	15	25[e,f]		30	26		
9. cem, H_2O	1:10	5	80	ZrO^2	stirring	45	20	25[d]		10	63	6.4	63
10. cem, H_2O	1:10	3	100	ZrO^2	stirring	45	20	25[d]		10	85	6.4	63

[a]CSH, CaO-SiO_2-H_2O:CH, Ca(OH)$_2$; AFm, [Ca$_2$(Al,Fe)(OH)$_6$]$_2$ (SO$_4$,CO$_3$,2OH)·6H$_2$O or a related amorphous phase; C_2SH, α-Ca$_2$(SiO$_3$OH)(OH); cem, mainly unreacted belite; [b]2-mm diameter MG-stabilized ZrO$_2$ beads; [c]Amount infered from cement composition and literature data; [d]Relatively amorphous; [e]Probably [Ca$_2$Al(OH)$_6$](OH)·3H$_2$O, C$_4$AH$_{12}$. [f]Relatively ordered.

The values given in Table 2.17 and in the related table, Table 2.18, for the composition of the products are estimates based on careful inspection of the X-ray and infrared data. They should not be taken as values derived from quantitative composition determinations.

The use of a 30-day reaction time, an ca. 20°C reaction temperature and very little agitation gave a relatively complete reaction with the hydrated components of the product being mainly C-S-H, Ca(OH)$_2$ and AFm. (AFm is used to mean [Ca$_2$(Al,Fe)(OH)$_6$]2(SO$_4$,CO$_3$,2OH).6H$_2$O or a related amorphous phase, here and below.) The 7.92 and 3.99 Å AFm lines and the 3.04 and 1.82 Å C-S-H lines of the product were weak and its 4.90, 3.11, 1.93, 1.80, and 1.69 Å Ca(OH)$_2$ lines were split. The 3650 cm-1 Ca(OH)$_2$ hydroxy band, the 3450 cm-1 C-S-H hydroxy band, and the 970 cm-1 C-S-H Si-O band were all distinct. The habit of the Ca(OH)$_2$ crystals as shòwn by electron micrographs was tabular with typical face dimensions being ca. 60 m. Not surprisingly, the main component in the unreacted cement was belite (Table 2.17, run 1).

In view of known cement chemistry, the composition of the hydrated components of the product is as expected. The weakness of the AFm lines indicates that the AFm was relatively amorphous.

Probably the AFm was intermixed intimately with C-S-H and effects arising from this intermixing led to the relatively amorphous nature of the AFm. The splitting of the Ca(OH)$_2$ lines indicates that the Ca(OH)$_2$ had a defective structure (e.g., extra ions between its layers or layer stacking faults). The relatively large size of the Ca(OH)$_2$ crystals is attributable to the long reaction time and the absence of vigorous agitation during the reaction. In general, the results from this run are consistent with those from similar runs described in the literature [152].

Table 2.18 Synthesis and Properties of Hydrated Cement-SiO$_2$ Products.

Reactants	Wt Ratio	Conditions				Estimated Composition[a]				BET Area	SO$_2$ Uptake	CA Util
		Time (h)	Temp (°C)	Media	Agit	CSH (%)	CH (%)	AFm[b,c]	cem (%)	(m^2/g)	(mmol/g)	(%)
cem, SiO$_2$[d], H$_2$O	1:0.3:13	6	80	ZrO$_2$[e]	stirring	70		15[f]	15	102	7.4	86
cem, diatomite, H$_2$O	1:0.3:13	6	80	ZrO$_2$[g]	stirring	70		15[f]	15	211	6.4	79
cem, diatomite, H$_2$O	1:0.3:13	6	100	ZrO$_2$[e]	stirring	70		15[f]	15	215	6.3	77
cem, pumice, H$_2$O	1:0.3:13	8	80	ZrO$_2$[g]	stirring	70		15[f]	15	154	6.5	73
cem, perlite, H$_2$O	1:0.3:13	7	80	ZrO$_2$[g]	stirring	55	15	15[f]	15	174	6.1	71

[a]CSH, CaO-SiO$_2$-H$_2$O;CH, Ca(OH)$_2$; AFm, [Ca$_2$(Al, Fe)(OH)$_6$]$_2$(SO$_4$, CO$_3$, OH)·6H$_2$O or a related amorphous phase; cem, mainly unreacted belite; [b]Existence inferred mainly from cement composition and literature data; [c]Amount inferred from cement composition and literature data; [d]Byproduct SiO$_2$ fume; [e]2-mm diameter Mg-stabilized ZrO$_2$ beads; [f]Relatively amorphous. [g]0.6-0.8 mm diameter Mg-stabilized ZrO$_2$ beads.

With the same reaction conditions except for the use of a 15-day reaction time, the reaction was incomplete. The AFm powder pattern lines of the product were stronger and the $Ca(OH)_2$ lines were unsplit (Table 2.17, run 2). The less amorphous nature of the AFm is attributable to a reduced intermixing of it with C-S-H. Based on the position and sharpness of 7.92 Å AFm line, it appears this AFm may be $[Ca_2Al(OH)_6](OH).3H_2O$. The reason the $Ca(OH)_2$ from this run was less defective than that from run 1 is not known.

The use of a 3-day reaction time, a 100°C reaction temperature and vigorous agitation gave a relatively complete reaction. In addition to C-S-H, $Ca(OH)_2$ and AFm, the main hydration components in the product also included α-$Ca_2(SiO_3OH)(OH)$. The AFm was relatively amorphous. The habit of this silicate, as shown by electron microscopy, was rod-like with typical axial dimensions being ca. 3 m (run 3). In view of the results obtained from the studies on the hydration of Ca_3SiO_5 at 100°C (Table 15, run 1), and the apparent stability of α-$Ca_2(SiO_3OH)(OH)$ at 100°C, the presence of this calcium silicate in the product of this run is not surprising.

When a shorter reaction time, 1 day, was used, the product contained less, but still considerable α-$Ca_2(SiO_3OH)(OH)\alpha$-$Ca_2(SiO_3OH)(OH)$ (run 4). This result emphasizes the relative stability of a-$Ca_2(SiO_3OH)(OH)$ at 100°C.

With a 3-day reaction time, an ca. 20°C reaction temperature, and very vigorous agitation, a relatively complete reaction was obtained. As with runs 1 and 2, the hydrated components of the product were mainly C-S-H, $Ca(OH)_2$, and AFm. The AFm was relatively amorphous, and the product had an appreciable surface area (run 5).

The relative completeness of the reaction in spite of the shortness of the reaction time is attributable to the continuous removal of the passivating film formed on the cement particles by the high-shear agitation used. The substantial surface area of the product is not surprising because C-S-H and AFm can have high surface areas [153]. When a shorter reaction time, 1 day, was used, the reaction was less complete (run 6). This is unsurprising given the slowness of the hydration of cement.

The use of a 1-day reaction time, an ca. 20 °C reaction temperature, and very vigorous agitation (this time provided by stirred beads) gave a relatively complete reaction with the hydrated components of the product being mainly C-S-H, $Ca(OH)_2$, and AFm. The AFm in this product, like that in the product of run 1, was relatively amorphous, and the surface area of the product was appreciable

(Table 2.17, run 7). The greater completeness of the reaction compared to that of run 6 is attributable to the more efficient agitation used.

With the same reaction conditions except for the use of a 5-hour reaction time, the reaction was relatively incomplete, the product had a low surface area, and the AFm was less amorphous (Table 2.17, run 8). Again the AFm may have been $[Ca_2Al(OH)_6](OH) \cdot 6H_2O$.

The use of the same conditions except for an 80°C reaction temperature gave a relatively complete reaction with the hydrated components of the product being mainly C-S-H, $Ca(OH)_2$ and AFm. The AFm in the product was relatively amorphous and had a substantial surface area (Table 2.17, run 9). The lack of α-$Ca_2(SiO_3OH)(OH)$ in this product is significant and shows that if the hydration time is sufficiently short, this silicate is not favored at elevated temperatures.

Finally, the use of the same reaction conditions except for a 3-hour reaction time and a 100°C reaction temperature gave a like product (Table 2.17, run 10). This result is noteworthy because it shows that cement can be hydrated rapidly to a desirable mixture if sufficiently vigorous agitation and an elevated temperature are used.

RAPID HYDRATION OF CEMENT-SiO_2 SOURCE MIXTURES. In an extension of the work on the hydration of cement, similar work was carried out on the hydration of cement-SiO_2 source mixtures.

The use of a 6-hour reaction time, an 80°C reaction temperature and very vigorous agitation with a 1:0.3 cement-SiO_2 fume mixture gave a relatively complete reaction with the hydrated components of the product being mainly C-S-H and AFm. The AFm in the product was very amorphous and, in contrast to the cement products, the surface area of the product was quite high. As with the cement products, the unreacted cement in this product was mainly belite (Table 2.18, run 1).

The slow rate of the reaction compared to the rates of similar cement reactions (e.g., Table 2.17, run 9) can be ascribed to the relative inertness of SiO_2.

The absence of $Ca(OH)_2$ in the product is attributable to the presence of sufficient SiO_2 in the reaction mixture to convert the $Ca(OH)_2$ formed by the hydration of the cement to C-S-H. The high surface area of the product is consistent with the fact it was composed largely of C-S-H and highly amorphous AFm. The presence of AFm in the product was inferred on the basis of the composition of cement and literature data indicating that only a small amount of the Si in C-S-H that is high in Ca_2+, such as that found in hydrated cement product, can be replaced by Fe and Al [76d].

Under the same reaction conditions, a cement-diatomite mixture gave a relatively complete reaction with the main hydrated components of the product being the same. This product had a very high surface area (Table 2.18, run 2). Perhaps its very high surface area is attributable to the filigree structure of the diatom skeletons in diatomite since this could allow the C-S-H to grow with fewer restrictions than usual.

When the same reaction conditions were used except for a 100°C reaction temperature, the product was very similar (run 3). Here the use of a higher temperature did not appear to significantly shorten the required reaction time.

The use of reaction conditions like those used in runs 1 and 2 with a cement-pumice mixture gave a product like that of run 3 except that it had a somewhat lower surface area (run 4). No evidence for byproducts arising from the metal ions in the pumice was observed.

When similar reaction conditions were used with a cement-perlite mixture, a product, having as its main hydration components C-S-H, $Ca(OH)_2$ and AFm, was obtained (again no evidence for byproducts was observed). The surface area of this product was quite high (run 5). The presence of $Ca(OH)_2$ in the product is attributable to the low reactivity of perlite.

SO_2 UPTAKE OF PRODUCTS. *Hydrated Cement Products.* The SO_2 uptake and utilization values of the hydrated cement products are generally as expected, Table 2.17. The low value of the product of run 1, Table 2.17, is consistent with the large size of its $Ca(OH)_2$ crystals and hence the unavailability of its $Ca(OH)_2$. The low value of the product of run 3, Table 2.17, is consistent with the presence of considerable α-$Ca_2(SiO_3OH)(OH)$ in it. For the products of runs 5, 9 and 10, Table 2.17, the high values fit with the composition and high surface areas of these products.

HYDRATED CEMENT-SiO_2 SOURCE PRODUCTS. The SO_2 uptake and Ca utilization values of the hydrated cement-SiO_2 source products are also generally as expected, Table 18. Their high values fit with their compositions and their high surface areas.

Conclusions

The results from the cement hydration work show that excellent sorbents composed mainly of C-S-H, $Ca(OH)_2$, and AFm can be made from cement. They further show that sorbents of this type can be made with short reaction times. Thus such sorbents are attractive for practical use.

The results from the cement-SiO_2 source hydration studies show that very good sorbents composed mainly of C-S-H and AFm can be made from cement and readily available and cheap SiO_2 sources. In addition, they show that sorbents of this type can be made with moderate reaction times. Thus sorbents of this type also are attractive for practical use.

Summary

The mixture made by agitating a high water-to-solids slurry of $Ca(OH)_2$ and a reactive SiO_2 such as diatomite or a reactive SiO_2 source such as perlite or pumice is a good SO_2 sorbent. This mixture is largely composed of porous C-S-H, and this silicate is the main active component in it.

The mixture prepared by vigorously agitating a high water-to-solids slurry of $Ca(OH)_2$ and an amount of fumed SiO_2 which is sufficient to react with only part of the $Ca(OH)_2$ is also a good SO_2 sorbent. This mixture is composed of $Ca(OH)_2$ particles embedded in porous C-S-H. Both the C-S-H and the $Ca(OH)_2$ contribute to the effectiveness of this sorbent. It is presumed that a sorbent largely composed of a $Ca(OH)_2$ embedded in porous C-S-H can be prepared from high water-to-solids slurries of $Ca(OH)_2$ and diatomite, perlite, or pumice as well.

Another good sorbent is one prepared by vigorously agitating a high water-to-solids slurry of type I (ordinary) portland cement. This sorbent is composed largely of a mixture of porous C-S-H, $Ca(OH)_2$ and AFm. All three of these species contribute to its effectiveness.

Each of these three sorbents, the $Ca(OH)_2$-SiO_2 sorbent, the SiO_2-deficient $Ca(OH)_2$-SiO_2 sorbent, and the cement sorbent is of practical interest. Each is simple to prepare. Two can be made from readily available, low cost reactants and the third probably can be made from such reactants also. The cement sorbent is unique. It appears to have the most practical potential.

The reactions used in making these three sorbents are heterogeneous and complex. Much more work is needed to provide a good understanding of these reactions.

Porous C-S-H is a good sorbent partly because it can take up SO_2 in the presence of CO_2. It is able to do this because it is only moderately basic.

List of Symbols and Abbreviations

A	Al_2O_3
AFm	a phase in hydrated cement which can be approximated as $Ca_2(Al,Fe)(OH)_6]_2(SO_4,CO_3,2OH) \cdot 6H_2O$
AFt	a phase in hydrated cement which can be approximated as $[Ca_3(Al,Fe)(OH)_6 \cdot 12H_2O]_2(SO_4,CO_3,2OH)_3 \cdot 2H_2O$
BET	Brunauer-Emmett-Teller
C	CaO
$\bar{C}$	CO_2
C-S-H	the phase $CaO\text{-}SiO_2\text{-}H_2O$
EDS	energy dispersive spectra
F	Fe_2O_3
H	H_2O
S	SiO_2
$\bar{S}$	SO_3
SEI	secondary electron image
SEM	scanning electron microscopy

Notes and References

1. Much of this chapter is taken directly from a report [2a] and a thesis [2b]. It also draws on a patent [3].
2. (a) Kenney, M. E., R.-K. Chiang and K. L. Fillgrove, *New High Capacity Calcium Based Sorbents—Calcium Silicate Sorbents,* Final Report, OCDO OCRC/93-1.8, 1994; (b) Chiang, R.-K., Ph.D. Thesis, Case Western Reserve University, November 1994
3. Kenney, M. E. and R.-K. Chiang, U. S. Patent 5 403 808 (1995)
4. Cotton, F. A. and G. Wilkinson, *Advanced Inorganic Chemistry,* 5th ed., Wiley Interscience, New York, (a) pp244-246, (b) pp 519-521 (1988)
5. Greenwood, N. N. and A. Earnshaw, *Chemistry of the Elements,* Pergamon, Oxford, (a) p301, (b) p 824, (1990)
6. Muller, H., In *Ullmann's Encyclopedia of Industrial Chemistry,* 5th ed., Elvers, B. Ed., VCH, Weinheim, Germany, Vol. A25, p569 (1994)
7. *The Merck Index,* 11th ed., Budavar, S., Ed., Merck, Rahway, NJ, (a) p274, (b) p1417 (1989)
8. Dawson, C. W., In *Institute of Chemical Engineering Symposium, No. 123,* Kyte, W. S., Ed., Institute of Chemical Engineering, Warwickshire, U. K., p25, (1991)
9. (a) Malone, P. G. and J. H. May, In *Lime for Environmental Uses,* Gutschick K., Ed., American Society for Testing and Materials, Philadelphia, PA, p 42, (1987); (b) Smith, C. L., In *Lime for Environmental Uses,* Gutschick K., Ed., American Society for Testing and Materials, Philadelphia, PA, p52, (1987)
10. Makansi, J., "Clean Air Act Amendments: The Engineering Response," *J. Power,* 11 (1991)
11. Jozewicz, W., J. C. S. Chang, C. B. Sedman and T. G. Brna, "Silica-Enhanced Sorbents for Dry Injection Removal of SO_2 from Flue Gas," *J. Air Pollution Control Association,* 38, 1027 (1988)
12. Jozewicz, W., J. C. S. Chang, C. B. Sedman and T. G. Brna, "Characterization of Advanced Sorbents for Dry SO_2 Control," *Reactivity of Solids,* 6, 243 (1988)
13. Trojer, F. J., "The Crystal Structure of a High-Pressure Polymorph of $CaSiO_3$," *Z. Kristallogr.,* 130, 185 (1969)
14. Heller, K.-F., "Hydrated Calcium Silicates. Part IV. Hydrothermal Reactions: Lime:Silica Ratios 2:1 and 3:1," *J. Chem. Soc.,* 2835 (1952)
15. Hesse, K.-F., "Refinement of the Crystal Structure of Wollastonite-2M(parawollastonite)", *Z. Kristallogr.,* 168, 93 (1984)
16. Trojer, F. J., "The Crystal Structure of Parawollastonite," *Z. Kristallogr.,* 127, 291 (1968)
17. Yamanaka, T. and H. Mori, "The Structure and Polytypes of α-$CaSiO_3$(Pseudowollastonite)," *Acta Cryst.,* B37, 1010 (1981)
18. Saburi, S., I. Kusachi, C. Henmi, A. Kawahara, K. Henmi and I. Kawada, "Refinement of the Structure of Rankinite," *Mineral. J.,* 8, 240 (1976)
19. Peacor, D. R. and C. T. Prewitt, "Comparison of the Crystal Structures of Bustamite and Wollastonite," *Am. Mineral.,* 48, 588 (1963)
20. Angel, R., "Structural Variation in Wollastonite and Bustamite," *Mineral. Mag.,* 49, 37 (1985)
21. Ghosh, S. N., P. B. Rao, A. K. Paul and K. Raina, "The Chemistry of Dicalcium Silicate Mineral," *J. Mater. Sci.,* 14, 1554 (1979)
22. Barbier, J. and B. G. Hyde, "The Structures of the Polymorphs of Dicalcium Silicate, Ca_2SiO_4," *Acta Cryst.,* B41, 383 (1985)
23. Jeffery, J. W., "The Crystal Structure of Tricalcium Silicate," *Acta Cryst.,* 5, 26 (1952)

24. Maki, I. and K. Kato, *Phase Identification of Alite in Portland Cement Clinker,* Cem. Concr. Res., 12, 93 (1981)
25. Malik, K. M. A. and J. W. Jeffery, "A Re-investigation of the Structure of Afwillite," *Acta Cryst.*, B32, 475 (1976)
26. Buckle, E. R. and H. F. W. Taylor, "A Calcium Analogue of Chondrodite," *Am. Mineral.*, 43, 818 (1958)
27. Gibbs, G. V., P. H. Ribbe and C. P. Anderson, "The Crystal Structures of the Humite Minerals. II. Chondrodite," *Am. Mineral.*, 55, 1182 (1970)
28. Ganiev, R. M., V. V. Ilyukhin and A. N. V. Belov, "Crystal Structure of Cement Phase Y = $Ca_6[Si_2][SiO_4](OH)_2$," *Sov. Phys. Dokl.*, 15, 85 (1970)
29. Safronov, A. N., N. N. Nevskii, V. V. Ilyukhin and N. V. Belov, "Refinement of the Crystal Structure of the Cementite Phase Y-C_6S_2H," *Sov. Phys. Dokl.*, 26(2), 129 (1981)
30. Heller, L., "The Structure of Dicalcium Silicate α-Hydrate," *Acta Cryst.*, 5, 724 (1952)
31. Yano, T., K. Urabe, H. Ikawa, T. Teraushi, N. Ishizawa and S. Udagawa, "Structure of Dicalcium Silicate Hydrate," *Acta Cryst.*, C49, 1555 (1993)
32. Gard, J. A. and H. F. W. Taylor, "The Crystal Structure of Foshagite," *Acta Cryst.*, 13, 785 (1960)
33. Merlino, S., "Gyrolite: Its Crystal Structure and Crystal Chemistry," *Mineral. Mag.*, 52, 377 (1988)
34. Stevula, L. and J. Petrovic, "Formation of an Intermediate C-S-H Phase During the Hydrothermal Synthesis of Gyrolite," *J. Cem. Concr. Res.*, 13, 684 (1983)
35. Miyake, M., M. Iwaya and T. Suzuki, "Aluminum-Substituted Gyrolite as Cation Exchanger," *J. Am. Ceram. Soc.*, 73(11), 3524 (1990)
36. Heller, L., "X-Ray Investigation of Hillebrandite," *Mineral. Mag.*, 30, 150 (1953)
37. Ishida, H., K. Sasaki and T. Mitsuda, "Highly Reactive β-Dicalcium Silicate: I. Hydration Behavior at Room Temperature," *J. Am. Ceram. Soc.*, 75(2), 353 (1992)
38. Sarp, H. and D. R. Peacor, "Jaffeite, a New Hydrated Calcium Silicate from the Kombat Mine, Namibia," *Am. Mineral.*, 74, 1203 (1989)
39. Kazak, V. F., V. V. Blinov and N. V. Belov, Dokl. Akad. Nauk SSSR, 219(2), 340 (1974)
40. Gard, J. A., H. F. W. Taylor, G. Cliff and G. Lorimer, "A Reexamination of Jennite," *Am. Mineral.*, 62, 365 (1977)
41. Hara, H. and N. Inoue, "Formation of Jennite from Fumed Silica," *Cem. Concr. Res.*, 10, 677 (1980)
42. Roy, D. M., "Studies in the System $CaO-Al_2O_3-H_2O$ IV; Phase Equilibria in the High-Lime Portion of The System $CaO-SiO_2-H_2O$," *Am. Mineral.*, 43, 1009 (1958)
43. Nawaz, R., "Killalaite, a New Mineral from Co. Sligo, Ireland," *Mineral. Mag.*, 39, 544 (1974)
44. Gard, J. A., K. Luke and H. F. W. Taylor, "Crystal Structure of K-Phase, $Ca_7Si_{16}O_{40}H_2$," *Sov. Phys. Crystallogr.*, 26(6), 691 (1981)
45. Gard, J. A., K. Luke and H. F. W. Taylor, "$Ca_7Si_{16}O_{40}H_2$, a New Calcium Silicate Hydrate Phase of the Truscottite Group," *Cem. Concr. Res.*, 11, 659 (1981)
46. Alberti, A. and E. Galli, "The Structure of Nekoite, $Ca_3Si_6O_{15} \cdot 7H_2O$, a New Type of Sheet Silicate," *Am. Mineral.*, 65, 1270 (1980)
47. Merlino, S., "Okenite, $Ca_{10}Si_{18}O_{46} \cdot 18H_2O$: the First Example of a Chain and Sheet Silicate," *Am. Mineral.*, 68, 614 (1983)

48. Wan, C. and S. Ghose, "Rosenhahnite, $Ca_3Si_3O_8(OH)_2$: Crystal Structure and the Stereochemical Configuration of the Hydroxylated Trisilicate Group, $[Si_3O_8(OH)_2]$," *Am. Mineral.*, 62, 503 (1977)
49. Lin, Z., "The Crystal Structure of Suolunite," *Scientia Geologica Sinica* (Chinese), 2(5), 117 (1974)
50. Labhasetwar, N. K., O. P. Shrivastava and Y. Y. Medikov, "Mossbauer Study on Iron-Exchanged Calcium Silicate Hydrate: $Ca_{5-x}Fe_xSi_6O_{18}H_2 \cdot nH_2O$," *J. Solid St. Chem.*, 93, 82 (1991)
51. Mitsuda, T. and H. F. W. Taylor, "Influence of Aluminum on the Conversion of Calcium Silicate Hydrate Cels into 11Å Tobermorite at 90oC and 120oC," *Cem. Concr. Res.*, 5, 203 (1975)
52. Okada, Y., M. Shimoda, T. Mitsda and H. Toraya, "Synthesis of Tobermorite: NMR Spectroscopy and Analytical Electron Microscopy," *Onoda Kenkyu Hokoku*, 42(2), 199 (1990)
53. Hara, N., C. F. Chan and T. Mistuda, "Formation of 14 Å Tobermorite," *Cem. Concr. Res.*, 8, 113 (1978)
54. Suzuki, S. and E. Sinn, "1.4nm Tobermorite-like Calcium Silicate Hydrate Prepared at Room Temperature from $Si(OH)_4$ and $CaCl_2$ Solutions," *J. Mater. Sci. Let.*, 12, 542 (1993)
55. Sarp, H. and G. Burri, "Trabzonite $Ca_4Si_3O10 \cdot H_2O$, A New Mineral," *Bull. Geol. Soc.Turkey*, 30, 57 (1987)
56. Luke, K. and H. F. W. Taylor, "Equilibria and Non-Equilibria in the Formation of Xonotlite and Truscottite," *Cem. Concr. Res.*, 14, 657 (1984)
57. Taylor, H. F. W., "The Dehydration of Xonotlite," *Acta Cryst.*, 9, 1002 (1956)
58. Kudoh, Y. and Y. Takeuchi, "Polytypism of Xonotlite: (I) Structure of an A1 Polytype," *Mineral J.* (Japan), 9(6), 349 (1979)
59. Gard, J. A., T. Mitsuda and H. F. W. Taylor, "Some Observations on Assarsson's Z-Phase and its Structural Relations to Gyrolite, Truscottite, and Reyerite," *Am. Mineral.*, 40, 325 (1975)
60. Taylor, H. F. W., "Proposed Structure for Calcium Silicate Hydrate Gel," *Am. Ceram. Soc.*, 69(6), 464 (1986)
61. Taylor, H. F. W., "Tobermorite, Jennite, and Cement Gel," *Z. Kristallgr.*, 202, 41 (1992)
62. McIver, E. J., "The Structure of Bultfonteinite, $Ca_4Si_2O_{10}F_2H_6$," *Acta Cryst.*, 16, 551 (1963)
63. Saburi, D., A. Kawahara, C. Henmi and K. Kihara, "The Refinement of the Crystal Structure of Cuspidine," *Mineral. J.* (Japan), 8(5), 286 (1977)
64. Henmi, C., I. Kusachi, A. Kawahara and K. Henmi, "Fukalite, a New Calcium Carbonate Silicate Hydrate Mineral," *Mineral. J.* (Japan), 8(7), 374 (1977)
65. Sudarsanan, K., "Structure of Hydroxyellestadite," *Acta Cryst.*, B36, 1636 (1980)
66. Golovastikov, N. J. and V. F. Kazak, "The Crystal Structure of Calcium Chlorosillicate Ca_2SiO_3Cl," *Sov. Phys. Crystallogr.* (Engl. Transl.), 22, 549 (1977)
67. Chukhlantsev, V. G., "Calcium Chlorosilicates. The $CaCO_3$-SiO_2-$CaCl_2$ System," *Dokl. Phys. Chem.* (Engl. Transl.), 246, 530 (1979)
68. Howie, R. A. and V. V. Iltukhin, "Crystal Structure of Rustumite," *Nature*, 269, 231 (1977)
69. Pluth, J. J. and J. V. Smith, "The Crystal Structure of Scawtite, $Ca_7(Si_6O_{18})(CO_3) \cdot H_2O$," *Acta Cryst.*, B29, 73 (1973)
70. Colville, A. A. and P. A. Colville, "The Crystal Structure of Spurrite, $Ca_5(SiO_4)CO_3$. II. Description of Structure," *Am. Mineral.*, 62, 1003 (1977)
71. Edge, R. A. and H. F. W. Taylor, "Crystal Structure of Thaumasite, $[Ca_3Si(OH)_6 \cdot 12H_2O](SO_4)(CO_3)$," *Acta Cryst.*, B27, 594 (1971)

72. Louisnathan, S. J. and J. V. Smith, "The Crystal Structure of Tilleyite," *Z. Kristallgr.*, 132, 288 (1970)
73. Sudarsanan, K., "The Crystal Structure of Zeophyllite," *Acta Cryst.*, B36, 1636 (1980)
74. Hamid, S. A., "The Crystal Structure of the 11 Å Natural Tobermorite $Ca_{2.25}[Si_3O_{7.5}(OH)_{1.5}] \cdot 1H_2O$," *Z. Kristallgr.*, 154, 189 (1981)
75. Gard, J. A., J. W. Howison and H. F. W. Taylor, "Synthetic Compounds Related to Tobermorite: an Electron-Microscope, X-Ray, and Dehydration Study," *Mag. Concr. Res.*, 11(33), 151 (1959)
76. Taylor, H. F. W., *Cement Chemistry*, Academic Press, London, (a) pp 123-166, (b) p 146, (c) pp 1-3, (d) pp 199-242, (e)pp 23-25, (f) p 168, (g) p 28, (h) p 278, (i) p 287 (1990)
77. Brunauer, S., I. Older and M. Yudenfreund, "The New Model of Hardened Portland Cement Paste," *Highway Research Record*, 328, 89 (1970)
78. Feldman, R. F. and R. J. Sereda, "A New Model for Hydrated Portland Cement and Its Practical Implications," *Engineering J.*, (Canada), 53, 53 (1970)
79. Copeland, L. E. and J. C. Hayes, "Determination of Non-Evaporable Water in Hardened Portland-Cement Paste," *ASTM Bull.*, 194, 70 (1953)
80. Odler, I. and H. Dorr, "Early Hydration of Tricalcium Silicate I. Kinetics of the Hydration process and the Stoichiometry of the Hydration Products," *Cem. Concr. Res.*, 9, 239 (1979)
81. Mindess, S. and J. F. Young, *Concrete*, Prentice-Hall, Inc., Englewood, NJ, 1981, Chapter 4
82. Heller, L. and H. F. W. Taylor, "Hydrated Calcium Silicates. Part II. Hydrothermal Reactions: Lime:Silica Ratio 1:1," *J. Chem. Soc.*, 2397 (1951)
83. Komarneni, S. and D. M. Roy, "New Tobermorite Cation Exchangers," *J. Mat. Sci.*, 20, 2930 (1985)
84. Heller, L. and H. F. W. Taylor, "Hydrated Calcium Silicates. Part V. The Water Content of Calcium Silicate Hydrate (I)," *J. Chem. Soc.*, 1018 (1952)
85. Sasaki, K., H. Ishida, Y. Okada and T. Mitsda, "Highly Reactive β-Dicalcium Silicate: V. Influence of Specific Surface Area on Hydration," *J. Am. Ceram. Soc.*, 76(4), 870 (1993)
86. JCPDS file 34-2
87. JCPDS file 19-1364
88. JCPDS file 23-125
89. JCPDS file 29-373
90. JCPDS file 42-538
91. *Chemical Marketing Reporter*, August 22, 1994
92. McClellan, G. H. and L. J. Eades, In *The Reaction Parameters of Lime*, American Society for Testing and Materials, Philadelphia, PA, pp 209-227 (1969)
93. Lea, F. M., *The Chemistry of Cement and Concrete*, Chemical Publishing Company, NY, (a) p29, (b) pp179-181, (c) pp454-489 (1971)
94. Backman, A., "Physikalisch-Chemische Probleme Bei Der Industriellen Kalkhydratherstellung," *Zement-Kalk-Gips*, 9, 262 (1956)
95. Boynton, R. S., In *Kirk Othmer Enclopedia of Chemical Technology*, Vol 14, John Wiley, NY, p343 (1977)
96. Stouffer, M. R., *Pilot Support Test for Edgewater Coolside Demostration: Part 1- Sorbent Evaluation*, DOE Cooperative Agreement, No. DE-FC22-87PC79798, February 1989
97. Zander, H. V., "Kornform und Korngrosse Trocken und nass Geloschter Kalkhydrate," *Zement-Kalk-Gips*, 11, 41 (1958)

98. Petch, H. E., "The Hydrogen Positions in Portlandite, $Ca(OH)_2$, as Indicated by the Electron Distribution," *Acta Cryst.*, 14, 950 (1961)
99. Tadros, M. E., J. Skalny and R. Kalyoncu, "Kinetic of Calcium Hydroxide Crystal Growth from Solution," *J. Colloid Interface Sci.*, 55(1), 20 (1976)
100. Yasue, T., Y. Kojima and Y. Arai, "Synthesis of Hexagonal Plate-like Crystal of Calcium Hydroxide by Hydration of Calcium Oxide," *Gypsum and Lime*, 206, 3
101. Arai, Y. and T. Yasue, "Controls of Size, Shape, Modification and Composition on Precipitated Particles," *Shikizai Kyokaishi*, 64(1), 2 (1991)
102. Zurz, A. and I. Odler, "XRD Studies of Portlandite Present in Hydrated Portland Cement Paste," *Adv. Cem. Res.*, 1, 27 (1987)
103. Ruiz-Alsop, R. N., *Effect of Relative Humidity and Additives on the Reaction of Sulfur Dioxide with Calcium Hydroxide,* Ph.D. Thesis, U. of Texas, Austin, 1986
104. Barker, A. P., N. H. Brett and J. H. Sharp, "Influence of Organic Additives on the Morphology and X-Ray Diffraction Line Profiles," *J. Mater. Sci.*, 22, 3253 (1987)
105. Schwarzkopf, F., H. P. Hennecke and A. Roeder, In *Innovations and Uses of Lime,* Walker, D. D., T. B. Hardy, D. C. Hoffman, D. D. Stanley, Eds., American Society for Testing and Materials, Philadelphia, PA, pp96-111 (1992)
106. Shadman, F. and P. E. Dombek, "Enhancement of SO_2 Sorption on Lime by Structure Modifiers," *Can. J. Chem. Eng.*, 66, 930 (1988)
107. Dharmarajan, N. N., "Stirred Mill Proves its Worth for FGD Lime-Slaking Duties," *Power*, 61, October 1985
108. Wu, Z.-Q. and J. F. Young, "Formation of Calcium Hydroxide from Aqueous Suspensions of Tricalcium Silicate," *J. Am. Ceram. Soc.*, 67(1), 48 (1984)
109. Atkinson, A., J. A. Hearne and C. F. Knights, "Aqueous Chemistry and Thermodynamic Modeling of $CaO-SiO_2-H_2O$ Gels," *J. Chem. Soc. Dalton Trans*, 2371 (1989)
110. Bestek, H., W. Ewald, H.-P. Hennecke, A. Roeder and F. Schmitz, Patent 4, 636,379 (1987)
111. Remmers, T. E. and H. Kranich, *Pigment Handbook,* Vol.1, John Wiley, NY, pp125-135 (1988)
112. Kondo, R. and S. Ohsawa, "Reactivities of Various Silicates with Calcium Hydroxide and Water," *J. Am. Ceram. Soc.*, 62, 447 (1979)
113. Taylor, H. F. W., "Hydrated Calcium Silicates. Part I. Compound Formation at Ordinary Temperatures," *J. Chem. Soc.*, 3682 (1950)
114. Abdul-Maula, S. and I. Odler, In *The Chemistry and Chemically-Related Properties of Cement,* Glasser, F. P. Ed., The British Ceram. Soc., London, pp83-91 (1984)
115. Chan, C.-J., W. M. Kriven and J. F. Young, "Analytical Electron Microscopic Studies of Doped Dicalcium Silicates," *J. Am. Ceram. Soc.*, 71(9), 713 (1988)
116. Jawed, I., J. Skalny and J. F. Young, In *Structure and Performance of Cement,* Barnes, P., Ed., Applied Science Publishers, Chapter 6 (1983)
117. Garter, E. M. and J. M. Gaidis, In *Materials Science of Concrete,* Skalny, J. P., Ed., Am. Ceram. Soc., Westerville, OH, pp95-125 (1989)
118. Jennings, H. M. and L. J. Parro, "Microstructure analysis of Hydrated Alite Paste," *J. Mater. Sci.*, 21, 4053 (1986)
119. Gard, J. A. and H. F. W. Taylor, "Calcium Silicate Hydrate (II) ("C-S-H(II)")," *Cem. Concr. Res.*, 6, 667 (1976)
120. Jennings, H. M., B. J. Dalgleish and P. L. Pratt, "Morphological Development of Hydrating Tricalcium Silicate as Examined by Electron Microscopy Techniques," *J. Am. Ceram. Soc.*, 64, 567 (1981)

121. Fuji, K. and W. Kondo, "Kinetics of the Hydration of Tricalcium Silicate," *J. Am. Ceram. Soc.*, 57(11), 492 (1974)

122. Menetrier, D., I. Jawed and T. S. Sun, "ESCA and SEM Studies on Early C_3S Hydration," *Cem. Concr. Res.*, 9(4), 473 (1979)

123. Damidot, D. and A. Nonat, In *Hydration and Setting of Cement,* Nonat, A., J. C. Mutin, Eds., E &FN Spon, London, pp23-24 (1992)

124. Young, J. F., H. S. Tong and R. L. Berger, "Composition of Solutions in Contact with Hydrating Tricalcium Silicate Pastes," *J. Am Ceram. Soc.*, 60, 193 (1977)

125. Boyer, J. P. and R. L. Berger, "Influence of Temperature Increases During the Induction period of C_3S Hydration on the Microstructure and Strength of C_3S Mortars," *J. Am. Ceram. Soc.*, 63, 675 (1980)

126. Kondo, R. and M. Daimon, "Early Hydration of Tricalcium Silicate: A Solid Reaction with Induction and Acceleration Periods," *J. Am. Ceram. Soc.*, 52, 502 (1969)

127. Double, D. D., *New Developments in Understanding the Chemistry of Cement Hydration,* Phil. Trans. R. Soc. Lond., A310, 53 (1983)

128. Kantro, D. L., S. Brunauer and W. C. H., "The Ball-Mill Hydration of Tricalcium Silicate at Room Temperature," *J. Colloid Sci.*, 14, 363 (1959)

129. Rodger, S. A., G. W. Groves, N. J. Clayden and C. M. Dobson, "Hydration of Tricalcium Silicate Followed by ^{29}Si NMR with Cross-Polarization," *J. Am. Ceram. Soc.*, 71(2), 91 (1988)

130. Wu, Z.-Q. Y., J. F., "The Hydration of Tricalcium Silicate in the Presence of Colloidal Silica," *J. Am. Mater. Sci.*, 19, 3477 (1984)

131. Sample, D. and P. W. Brown, "Hydration of Tricalcium Germanate-Tricalcium Silicate Solid Solutions," *J. Am. Ceram. Soc.*, 75(11), 1070 (1992)

132. Menetrier, D., D. K. McNamara, I. Jawed and J. Skalny, "Early Hydration of β-C_2S: Surface Morphology," *Cem. Concr. Res.*, 10, 107 (1980)

133. Pritts, I. M. and K. E. Daugherty, "The Effect of Stabilizing Agents on the Hydration Rate of β-C_2S," *Cem. Concr. Res.*, 6, 783 (1976)

134. Regourd, M., In *Structure and Performance of Cement,* Barnes, P., Ed., Applied Science Publishers, Chapter 3 (1983)

135. Jost, K. H. Z., B., "Relations between the Crystal Structures of Calcium Silicates and Their Reactivity Against Water," *Cem. Concr. Res.*, 14, 177 (1984)

136. Hirljac, J., Z.-Q. Wu and J. F. Young, "Silicate Polymerization During the Hydration of Alite," *Cem. Concr. Res.*, 13, 877 (1983)

137. Personal Communication, Edward Kraus, Medusa Cement, Cleveland, OH (1994)

138. *Mineral Year Book,* U.S. Department of the Interior, Bureau of Mines (1990)

139. Kosmatka, S. P., W., *Design and Control of Cement Mixtures, 13th ed.,* Portland Cement Association, Skokie, IL, Chapter 2 (1990)

140. *Material Science of Concrete, Vol. 1-3,* Skanly, J., Ed., Am. Ceram. Soc., Westerville, OH, (1990-1992)

141. Gard, J. A., K. Mohan and H. F. W. Taylor, "Analytical Electron Microscopy of Cement Pastes: I, Tricalcium Silicate Pastes," *J. Am. Ceram. Soc.*, 63, 336 (1980)

142. Fujii, K. and W. Kondo, "Rate and Mechanism of Hydration of β-Dicalcium Silicate," *J. Am. Ceram. Soc.*, 62, 161 (1979)

143. Ahmed, S. J. and H. F. W. Taylor, "Crystal Structure of the Lamellar Calcium Aluminate Hydrates," *Nature*, 215, 623 (1967)

144. Foreman, W., "Neutron and X-Ray Diffraction Study of $Ca_3Al_2(O_4D_4)_3$ a Garnetoid," *J. Chem. Phys.*, 48, 3037 (1968)
145. Moore, A. E. and H. F. W. Taylor, "Crystal Structure of Ettringite," *Acta Cryst.*, B26, 386 (1970)
146. Allmann, R., "Calcium Aluminum Hydrate$[Ca_2Al(OH)_6]_2SO_4 \cdot 6H_2O$," *Neues Jahrb. Mineral.*, Montsh, 136 (1977)
147. Hattori, H. and H. Kumagai, *Mechanistic Study of Desulfurization by Absorbent Prepared from Coal Fly Ash*, p367
148. Fortune, J. M. and C. J. M. D., "Hydration Products of Calcium Aluminoferrite," *Cem. Concr. Res.*, 13, 696 (1983)
149. Brown, P. W., "Early Hydration of Tetracalcium Aluminoferrite in Gypsum and Lime-Gypsum Solutions," *J. Am. Ceram. Soc.*, 70(7), 493 (1987)
150. Liang, T. Y., N., "Hydration Products of Calcium Aluminoferrite in the Presence of Gypsum," *Cem. Concr. Res.*, 24, 150 (1994)
151. Harchand, K. S., R. Kumar and K. Chandra, "Mossbauer and X-Ray Investigations of Some Portland Cements," *Cem. Concr. Res.*, 14, 170 (1984)
152. Copeland, L. E. and E. G. Schulz, "Electron Optical Investigation of the Hydration Products of Calcium Silicates and Portland Cement," *J. PCA Research and Development Laboratories*, 4(1), 2 (1962)
153. Poeppelmeier, K. P., C. K. Chiang and D. O. Kipp, "Synthesis of High-Surface-Area α-$LiAlO_2$," *Inorg. Chem.*, 27, 4524 (1988)

CHAPTER 3 *FUNDAMENTAL STUDIES CONCERNING CALCIUM-BASED SORBENTS*

D. Mandal, R. Venkataramakrishnan, K.J. Sampson, M.E. Prudich
Department of Chemical Engineering
Ohio University
Athens, OH

Abstract

Sorbent preparation techniques used today have generally been adapted from techniques traditionally used by the lime industry. Traditional "dry" hydration and slaking processes have been optimized to produce materials intended for use in the building industry. These preparation techniques should be examined with an eye to optimization of properties important to the SO_2 capture process.

The study of calcium-based sorbents for sulfur dioxide capture is complicated by two factors: (1) little is known about the chemical mechanisms by which the "standard" sorbent preparation and enhancement techniques work, and (2) a sorbent preparation technique that produces a calcium-based sorbent that enjoys enhanced calcium utilization in one regime of operation [flame zone (>2400°F), in-furnace (1600-2400°F), economizer (800-1100°F), after air preheater (<350°F)] may not produce a sorbent that enjoys enhanced calcium utilization in the other reaction zones. Again, an in-depth understanding of the mechanism of sorbent enhancement is necessary if a systematic approach to sorbent development is to be used.

The long-term goals of the experimental program that resulted in this report were (1) defining the effects of slaking conditions on the properties of calcium-based sorbents, (2) determining how the parent limestone properties and preparation techniques interact to define the SO_2 capture properties of calcium-based sorbents, and (3) elucidating the mechanism(s) relating to the activity of various dry sorbent additives.

This chapter documents (1) the collection, production, and characterization of a series of limestone/hydrated lime/quicklime samples representing a range of Ohio limestone products, (2) an investigation of the importance of lime solubility enhancement in the evolution of surface area during the slaking process, and (3) a model/paper study of the effects of chemical additives on the performance in spray drying and in-duct injection processes.

Literature

Background

The sorbent modification studies reviewed below relate to both SO_2 capture in the 1600-2400°F range and the <350°F range in a roughly 80:20 ratio. For the purposes of the review section, studies performed in the temperature range 1600-2400°F will be called high-temperature studies and studies performed in the temperature range <350°F will be called low-temperature studies. No additive studies are offered to describe the application of promoted sorbents to the problem of SO_2 capture in the 800-1100°F reaction range.

Sorbent Preparation

CHEMICAL FORM/HIGH-TEMPERATURE STUDIES. Cole et al. [1], using an isothermal reactor at 2000°F, Ca/S=2, and 2000 ppm SO_2, showed that calcitic hydrates were more reactive than calcitic carbonates. They attributed this observation to a combination of three reasons: (1) Dehydration occurs much more rapidly than CO_2 evolution; therefore, the time available for sulfation after the calcination process is significantly greater for hydrates than for carbonates; (2) hydrates begin with a much higher initial surface area; and (3) the dehydration process generally produces a reduction in the mean particle size of the sorbent, thereby reducing the internal diffusion resistance to SO_2 diffusion.

Between the two types of calcium-based sorbents generally used in in-furnace injection, it is well documented that CaO derived from $Ca(OH)_2$ (h-CaO) is more reactive than CaO derived from $CaCO_3$ (c-CaO) [2]. This is attributed in part to the smaller particle size of h-CaO, and more importantly, to the pore structure of the CaO produced. The h-CaO has a slit or plate-like structure while the structure of c-CaO is in the form of cylindrical pores (or spherical grains). The plate-like structure retains its porosity to a greater extent by allowing for particle expansion [3], and results in higher rates of diffusion of the reactant through the product layer [4].

CALCIUM/MAGNESIUM RATIO/HIGH TEMPERATURE STUDIES. Snow et al. [5], injecting sorbents into a flue gas stream produced by the combustion of Pittsburgh No. 8 coal (injection temperature = 2210°F; quench rate = 468°F/s; average residence time at reaction temperature = 1.3 s; Ca/S mole ratio = 2; 1900 ppm SO_2) found that sorbent SO_2 capture reactivity varied in the order dolomitic hydrates > calcitic hydrates > carbonates (based on equal molar Ca/S ratios). However, for a given sorbent type there was no clear correlation between physical properties (BET surface area, mass mean particle size, elemental stoichiometry) and SO_2 capture performance. A summary of their data is given in Table 3.1.

Teixeira et al. [6], testing sorbents for SO_2 removal from flue gases generated by burning low-sulfur Western coals (injection temperature = 2000°F; average residence time = 0.33 s; 500 ppm SO_2) also found that SO_2 capture reactivity varied in the order dolomitic hydrates > calcitic hydrates > carbonates. This observation was further confirmed by Beittel et al. [7], Overmoe et al. [8], and Bortz and Flament [9].

Table 3.1. Effect of Sorbent Properties on High Temperature SO_2 Capture Performance (after Snow et al. [5])

| Sorbent | Mass Mean Particle (Diam. μm) | BET Surface Area (m^2/g) | Percent by Weight | | SO_2 Removal (%) |
			Ca (%)	Mg (%)	
Vicron Limestone	11.0	1.01	39.01	0.49	38
Mercer CAH	7.10	17.8	50.5	0.4	59
Kemikal CAH	3.88	18.0	49.0	1.0	62
Marblehead CAH	7.79	14.3	50.3	0.49	69
Detroit Lime CAH	8.33	14.9	50.2	0.51	60
Black River CAH	5.53	13.3	49.7	1.6	61
Kemidol DAH	19.71	20.6	29.9	18.5	73
Ivory Finish DPH	13.72	18.4	28.6	18.1	75

CAH----Calcitic Atmospheric Hydrate
DAH----Dolomitic Atmospheric Hydrate
DPH----Dolomitic Pressure Hydrate
Percent SO_2 removal at Ca/S=2; 2210°F injection temperature.

Gooch et al. [10] observed that the increased utilization of pressure-hydrated dolomitic lime is often only sufficient to compensate for the decreased calcium content. This balance results in no net reduction of the mass of sorbent required to accomplish a given SO_2 reduction.

MORPHOLOGY/HIGH TEMPERATURE STUDIES. Gooch et al. [10] tested 11 commercial calcitic hydrates under furnace injection conditions. No obvious correlation of properties of the raw sorbent with SO_2 removal was identified. However, when testing four specially-prepared sorbents with widely varying BET surface areas, particle sizes, and pore structures, it was found that a two-fold increase in raw sorbent surface area resulted in a 20% relative increase in calcium utilization. The increased surface area also served to decrease the dependence of sulfur dioxide capture efficiency on injection temperature.

McCarthy et al. [11] also agree that reactivity in furnace injection does not correlate with hydrate surface area. They did show, however, that SO_2 capture reactivity did correlate with the surface area of the calcined hydrate, with an increase in reactivity being associated with a higher calcined surface area (Figure 3.1).

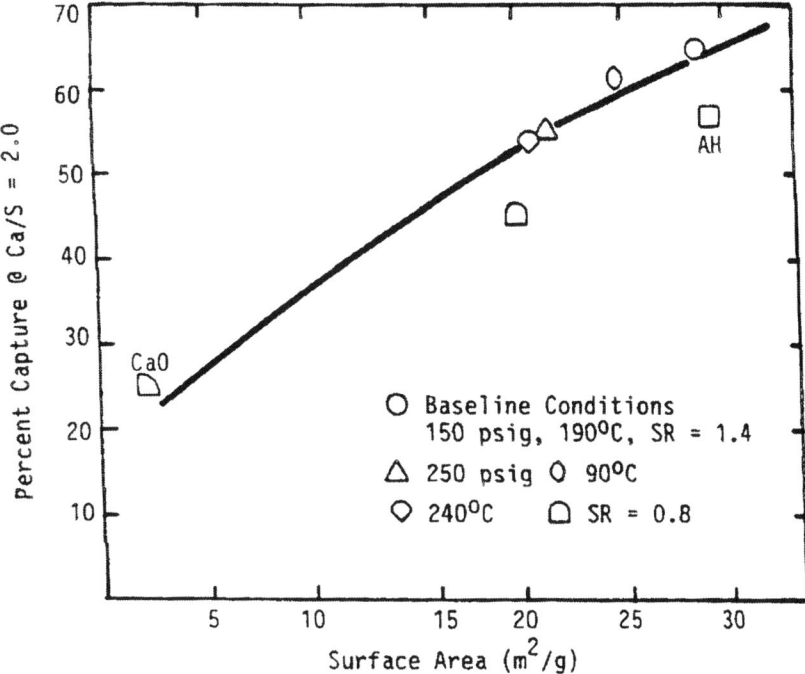

Figure 3.1 SO_2 capture reactivity versus surface area of the calcined hydrate [11].

MORPHOLOGY/LOW-TEMPERATURE STUDIES. Yoon et al. [12] have prepared high BET surface area calcitic hydrates using atmospheric hydration under N_2 followed by filtration and vacuum drying. Using these hydrates Yoon was able to show a strong relationship between sorbent utilization in dry/dry capture (150°F, 60% relative humidity, 20°F approach to saturation, 1000 ppm SO_2) and raw sorbent BET surface area (Figure 3.2a). Using hydrates whose BET surface areas varied between 10 and 50 m^2/g he was able to effect calcium utilizations between 12% and 45%.

Borgwardt and Bruce [13] confirmed the work of Yoon et al. [12] by preparing a series of high surface area hydrates and showing the strong effect of hydrate BET surface area on dry/dry humidified SO_2 capture (Figure 3.2b).

PREPARATION BY AQUEOUS HYDRATION/HIGH TEMPERATURE STUDIES.
Kirchgessner et al. [14], using calcines and hydrates prepared from Eldorado limestone, found a slight decrease in the SO_2 capture reactivity of the product hydrate with increasing particle size of the parent calcine. As the mean calcine particle size increased from 0.5 mm to 9.5 mm the percent Ca utilization of the product hydrate dropped from 14.5% to 13.0%. The observed effect was attributed to the larger particle surface area present for the smaller-sized calcine.

McCarthy et al. [11] report that pressure hydrates generated under well-controlled conditions are more reactive than commercially produced atmospheric hydrates. They reported that important hydration process variables include size and composition of the parent quicklime, the hydration temperature and pressure, the rate of water addition to the hydration reactor, and the pressure progression during the discharge of the hydrate from the hydration reactor. However, Gooch et al. [10] found that pressure-hydrated calcitic lime produced and evaluated in a laboratory-scale apparatus could not be distinguished from its companion atmospheric hydrate on the basis of either particle morphology, chemical composition, or SO_2 capture ability.

PREPARATION BY AQUEOUS HYDRATION/LOW TEMPERATURE STUDIES.
Borgwardt and Bruce [13] prepared calcitic hydrates using steam hydration. The hydrate sorbents prepared in this manner show inferior BET surface areas (9-11 m^2/g) when compared to hydrates produced through the use of a liquid water phase. The SO_2 capture performance of the steam hydrated samples was substantially less than that of liquid water hydrated samples. Borgwardt suggested that this observation might explain the sometimes contradictory results obtained during pressure hydration. The performance of the pressure hydration process might in part be controlled by a delicate balance between the positive effects of decompressive drying and the negative effects of steam exposure.

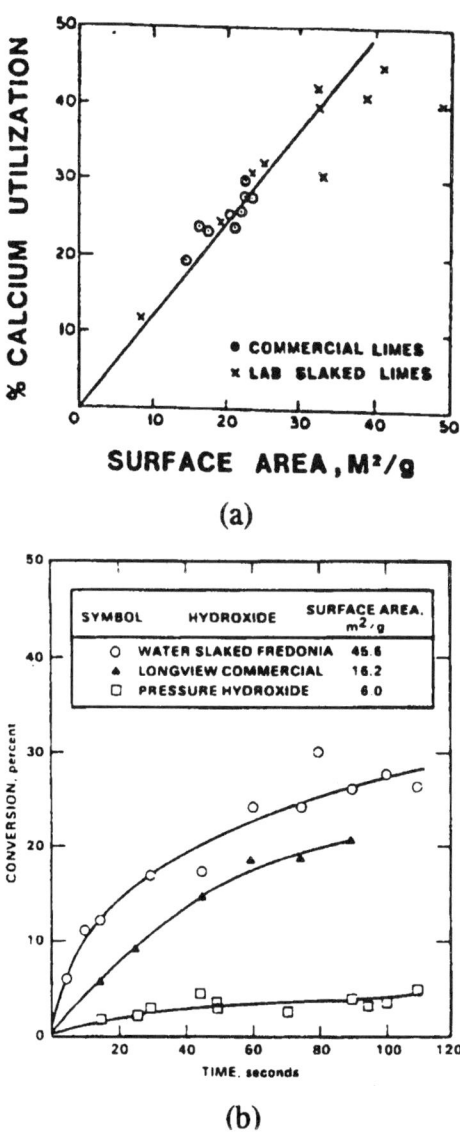

Figure 3.2 Effect of sorbent surface area on low temperature SO_2 capture performance: (1) 150°F, 60% relative humidity, 1000 ppm SO_2, 60 minutes [12]; (b) 150°F, 20 vol% H_2O, 1500 ppm SO_2 [13].

THE USE OF ADDITIVES TO ENHANCE REACTIVITY. The use of additives to enhance the $Ca(OH)_2$ reactivity with SO_2 have been explored by a number of investigators. The type of additives investigated to date may be classified as [15] (1) deliquescent inorganic salts and related inorganic salts, (2) sodium–containing basic compounds, (3) organic compounds, and (4) oxidation catalysts.

The majority of the compounds tested fall into the first category above, although there is some cross-classification among the compounds evaluated to date. For example, NaOH is capable of reacting directly with SO_2 as a basic solution but it is also a deliquescent compound.

DELIQUESCENTS, BUFFERS, AND SODIUM ADDITIVES/HIGH TEMPERATURE STUDIES. Sodium compounds have also been tried, with varying success, in the in-furnace (1600-2400°F) regime. Generally these additives are added to the water phase during the hydration process. Teixeira et al. [6] found that sodium carbonate, sodium bicarbonate, and sodium hydroxide all enhance the sulfur-removing capability of dolomitic hydrates when removing sulfur from SO_2-doped natural gas flue gases. However, the benefit of the sodium compounds disappeared when western coal-derived fly ash was present in the flue gas stream. No ash analyses were offered in their paper.

Weber et al. [16] also remarked on the capability of NaOH to enhance the behavior of a pressure-hydrated calcitic hydrate for in-furnace SO_2 capture when using doped natural gas flue gas. However, they also saw this advantage disappear when they used a flue gas generated by burning a Beulah lignite. The high calcium and sodium content of the ash from lignite (15.5wt% CaO, 4.0 wt% Na_2O) probably contributed significantly to this effect.

Snow et al. [5] investigated $NaHCO_3$ as an additive for capture of SO_2 at 2210°F from a flue gas generated by burning Pittsburgh No. 8 coal. At a Ca/s ratio of 2, the addition of the $NaHCO_3$ additive increased the capture efficiency of a calcitic atmospheric hydrate from 63-69% to 83% and for atmospheric and pressure dolomitic hydrates from 73-75% to 88%. Ashes from eastern bituminous coals are not expected to contain significant amounts of calcium or sodium.

Muzio et al. [17] have investigated the effects of nine additives added to the hydration water on SO_2 capture. The nine additives studied include: NaOH, NaCl, Na_2CO_3, Li_2SO_4, $LiNO_3$, K_2CO_3, Cs_2SO_4, $Fe(NO_3)_3$, and $FeCl_3$. For experiments at 2100°F, Ca/S=2, and a metal promoter/CaO weight ratio of 0.03 the use of the promoters resulted in the following increases in SO_2 capture when compared to a nonpromoted, base hydrate: Cs(33%) > K(29%) > Na(20%) > Li(0%), Fe(0%). In all cases where improvement was observed, the improvement was greater than that which would have been predicted if all of the promoting material was transformed to its sulfate form. Muzio also showed that the form of sodium added had essentially no effect on the sulfur capture performance of the promoted sorbent. This was particularly interesting in light of the wide variation in BET surface areas of the raw hydrated sorbents (Na_2CO_3, 16 m^2/g; NaOH, 7.8 m^2/g; NaCl, 4.5 m^2/g). Muzio et al. [17] speculated that for the promoted hydrates, alkali crystals may block the pores at room temperature but, as a result of melting and vaporization in the combustion zone, the pore structure may reopen at reaction conditions. Physical mixtures of CaO and Na_2CO_3 were found to be less effective than addition of the same amount of Na_2CO_3 to the hydration water for the same CaO sorbent (38% SO_2 capture versus 45% SO_2 capture). The presence of fly ash in the flue gas served to totally eliminate any promotion effect of the sodium metal additives.

DELIQUESCENTS, BUFFERS, AND SODIUM ADDITIVES/LOW TEMPERATURE STUDIES. Ruiz-Alsop and Rochelle [18] tested the effectiveness of 18 additives (two buffers, three organic deliquescents, and thirteen inorganic deliquescents) towards improving calcium utilization in dry/dry SO_2 capture. Their experiments were performed in a sand bed reactor at 54-74% relative humidity. Reaction conditions were set to simulate the conditions found in bag filters during flue gas spray drying. It has been postulated that deliquescent salts should increase the efficiency of sulfur capture in dry/dry systems by enhancing the moisture content in the sorbent solids. This study found that the two buffers and the three organic deliquescents caused a degradation in SO_2 capture efficiency while the inorganic deliquescents caused a increase in the SO_2 capture efficiency, in some cases almost doubling the efficiency (Table 2). The most effective inorganic deliquescents were LiCl, KCl, NaBr, and Na_2NO_3. Ruiz-Alsop and Rochelle [18] found a poor correlation between the relative deliquescence of the inorganic salts and their ability to enhance SO_2 capture. They speculated that some of the salts also acted to favorably modify the sulfite reaction product layer formed on the surface of the sorbent particles.

Huang et al. [19] have reported that the addition of NaOH in the lime slurry used for an industrial-sized spray dryer caused the partial oxidation of NO to NO_2. This is a significant development in that NO_2 is more reactive with a variety of gaseous compounds and could conceivably be removed as a particle [20].

Yoon et al. [12] used additives to promote the activity of samples of hydrated lime in conjunction with tests on the Coolside process (a process operating in the <350°F regime). The action of both "cosorptive" and "noncosorptive" additives were evaluated. Examples of cosorptive additives include NaOH, Na_2CO_3, and possibly NaCl, Na_2SO_3, and Na_2NO_3. Examples of noncosorptive additives include $CaCl_2$, KCl, $FeCl_3$, and $MgCl_2$. The additives were either added to the

Fundamental Studies Concerning Calcium-Based Sorbents

hydrated lime in an aqueous solution or were added to the lime during the hydration process. Both sets of compounds were found to be highly effective in enhancing the capture behavior of the hydrated lime, even when the cosorptive properties of the sodium salts were subtracted out. Treating -325 mesh hydrated lime particles with NaCl using promotion mole ratios of 0.05 to 0.2 Na^+/Ca^{++} in a laboratory reactor (operated in the dry/dry sorption mode) resulted in relative calcium utilization increases of 80% to 114% over those achieved using unprompted hydrated lime samples. It was speculated that the additives might act to enhance SO_2 capture in any one of three ways: (1) changing the sorbent particle's physical properties, particularly the surface area of the hydrate, (2) enhancing the basicity of the sorbent, and (3) increasing or retaining moisture at the sorbent surface. No evidence was offered to support any of these three proposed mechanisms.

Organic acids and buffers have been studied as wet FGD additives and have been found to be effective in increasing the overall rate of SO_2 capture in CaO or $CaCO_3$ slurries. This wet FGD slurry work should shed some light on behavior that might be expected for wet/dry SO_2 capture systems.

Table 3.2. Effect of additives on lime reactivity. (Ruiz-Alsop and Rochelle [18])

Additive	Lime Conversion at 60 minutes	
	74% RH 64.4°C	54% RH 66°C
No additive	22.4	11.8
Buffers		
5 wt% Glycolic Acid	11.3	---
1 wt% Adipic Acid	20.3	---
Organic Deliquescents		
5 wt% Moneothanolamine	19.6	---
5 wt% Ethylene Glycol	20.3	---
5 wt% TEG	20.5	
Inorganic Deliquescents		
5 mole% Na_2So_4	28.3	---
5 mole% Na_2SO_3	29.8	16.1
5 mole% $CaCl_2$ (*)	34.6	16.4
10 mole% NaCl	38.5	27.0

Table 3.2. Effect of additives on lime reactivity. (Ruiz-Alsop and Rochelle [18])

Additive	Lime Conversion at 60 minutes	
	74% RH 64.4°C	54% RH 66°C
10 mole% NaOH	38.8	17.3
5 mole% Ca(NO$_3$)$_2$(*)	39.4	12.3
10 mole% NaNO$_2$	40.0	---
10 mole% NaNO$_3$	41.5	27.2
BaCl$_2$*2H$_2$O	---	19.4
Na$_2$S$_2$O$_3$	---	21.6
KCl	---	37.3
NaBr*2H$_2$O	---	42.0
LiCl	---	43.9
100% So$_2$ Removal	48.2	48.2

(*) Solid phases are CaCl$_2$*Ca(OH)$_2$*H2O and CaN$_2$O$_7$*2H$_2$O, respectively.
RH Relative Humidity

Jarvis et al. [21] evaluated the performance of several organic acid additives (adipic acid, maleic acid, formic acid, glutamic acid, succinic acid) in a bench-scale wet FGD system. The use of these additives increased the removal of SO$_2$ by 15%. Other investigators have confirmed the effectiveness of organic acid and buffer additives for wet FGD systems (Chang and Brna [22]—adipic acid, citric acid, sodium formate; Wang and Burbank [23]—adipic acid; Rochelle et al. [24]—sulfopropionic acid, sulfosuccinic acid, acetic acid, adipic acid, hydroxypropionic acid, aluminum sulfate). Works by Chan and Rochelle [22] and Rochelle and King [25] have provided models for the mass transfer enhancement actions of organic acids, alkali additives, and buffers in wet FGD systems.

ALCOHOL AND SUCROSE HYDRATION/HIGH TEMPERATURE APPLICATIONS.
Gooch et al. [10] evaluated alcohol (methanol, ethanol, isopropanol), acetone, and sucrose hydration techniques. It was observed that while the alcohols are all removed from the final product by evaporation (and therefore can be recovered and reused), the sucrose remains in the final product. For hydration tests conducted at 60-70°C, an atmospheric hydrate with a BET surface area of 22 m^2/g was produced starting with a parent CaO with a surface area of 2.2 m^2/g. Hydration with aqueous acetone and methanol solutions under "optimum" conditions

produced hydrates with surface areas of 50 m^2/g and 80 m^2/g, respectively. Hydrates produced using an aqueous mixture of sucrose and methanol had a BET surface area of 85 m^2/g.

Gooch et al. [10] presented the speculation that alcohols act to improve sorbent surface area by abstracting the heat of hydration from the sorbent surfaces by evaporation and that sucrose acts by increasing the solubility of CaO in water (0.1 wt% in water and 9.8 wt% in 35% sucrose/65% water at 25°C). No experimental evidence was reported to support either of these proposed mechanisms.

SILICATE ADDITIVES—LOW TEMPERATURE STUDIES. The use of product recycle (with included fly ash) has been shown to improve spray dryer performance in pilot plant tests [26]. Also, bench scale studies with a packed bed reactor have shown substantial improvements in SO_2 uptake for $Ca(OH)_2$ slurried with several different fly ashes. [27]. They speculate that the fly ash reacts with $Ca(OH)_2$ to produce calcium silicates with more reactive surface area than the original $Ca(OH)_2$. Reagent-grade Al_2O_3, Fe_2O_3, and H_2SiO_3 were also found to enhance calcium utilization for dry/dry SO_2 scrubbing systems.

A potential problem in the use of fly ash as a silica source is an increase in the solids loading to the atomizer which must be used in order to achieve increased SO_2 removal. The quality of sorbent material entrained in the gas stream is directly proportional to the amount of recycle, and thus, represents an increased load to the particle collection device. A direct injection technology utilizing silica-enhanced $Ca(OH)_2$ is being developed and is known as the ADVACATE process [15].

The impact of coal chloride concentration on SO_2 removal in a spray dryer has been reported by Brown et al. [28]. In tests conducted at the TVA 10-MW spray dryer/ESP test facility, an SO_2 removal level of 85% was achieved over an extended test period for a 4.0% sulfur, 0.05% chloride coal with a reagent ratio of 1.6 for operation at an 18°F approach to adiabatic saturation and an inlet gas temperature of 320°F. For similar operating conditions, an SO_2 removal level of 93% was achieved when the coal was changed to one containing 4.0% sulfur and 0.25% chloride. On high chloride coal (i.e., chloride > 0.2%), 90% removal appears to be achievable at a reagent ratio between 1.4 and 1.5. The role of chloride in promoting greater SO_2 reactivity is postulated to be caused by the formation of higher concentrations of HCl in the flue gas which subsequently react with the slurry in the spray dryer. Calcium chloride, a classical deliquescent salt, has been reported by others [29] to greatly improve SO_2 uptake in calcium-based sorbents.

MISCELLANEOUS ADDITIVES—HIGH TEMPERATURE APPLICATIONS/ EFFECTS OF SINTERING AND PORE STRUCTURE. Borgwardt et al. [30] pointed out that diffusion through the product layer in a solid, when it occurs through solid-state mechanisms, increases with the concentration of lattice defects. These can exist as point defects, which involve individual atoms, or as extended defects, which involve lines or planes of disorder in the crystal structure [31]. At a given temperature, the concentration of point defects in the product layer may depend on the concentration of foreign ions. Thus, higher diffusivities can be expected in impure materials, when a solid-state mechanism prevails. From their results, they concluded that the higher reactivity of impure CaO is due to defects inherent with the crystal structure of the limestone derivatives. Impurities in the form of aliovalent ions are known to generate defects in the crystal structure [32, 33].

Accordingly, Borgwardt et al. [30] studied the effects of alkali metal ions doped on the surface of well-annealed CaO. They added sulfates of lithium, sodium or potassium to pre-calcined and pre-sintered CaO by grinding in a mortar, and observed a significant increase in sorbent utilization during sulfation. However, doping $CaCO_3$ or $Ca(OH)_2$ with Na^+ prior to calcination and sintering was not successful. This, they explain, is because these ions enhance the diffusion and hence the overall rate during sintering as well, which in turn causes a decrease in the surface area of the sorbent available for sulfation.

According to Haji-Sulaiman et al. [34], higher impurity content in the sorbent increases the extent of calcination. Thus, sorbents with a more open pore structure are obtained resulting in improved sulfation efficiency by preventing early pore blockage. Shadman and Dombek [35] view the role of additives solely as structure modifiers. They prepared flakes of modified sorbents by mixing the additives with a slurry of hydrated lime and spreading the mixture in a thin layer followed by drying. The additives used were bauxite, silica and kaolin. For all three additives, both the reaction rate and the maximum achievable conversion increased significantly. Between these additives, when their concentration and particle size were the same, they observed little difference in their performance. Hence they concluded that the effect of the additives is purely physical in nature, specifically an increase in macroporosity.

Following SEM micrograph analysis, they assumed that the sorbent flakes consist of spherical grains of lime mixed with inert additives. This leads to a bimodal pore size distribution where micropores account for pores in the particles and macropores represent the void space between particles. They developed a diffusion-controlled model with micropore and macropore diffusivities as the adjustable parameters. Model simulation supported their experimental finding that initial macroporosity is a critical factor in determining the sulfation performance of the modified sorbent.

MISCELLANEOUS ADDITIVES—HIGH TEMPERATURE APPLICATIONS/ ORGANIC SURFACTANTS. Kirchgessner and Lorrain [36] modified the sorbent $Ca(OH)_2$ with calcium lignosulfonate, an additive added with the water of hydration. They observed that the utilization of the modified sorbent increases with increasing additive content, reaches a maximum, and then decreases. The maximum occurred at an additive content of 1.5 dry weight percent, where the sorbent utilization was 20% higher than the unmodified hydroxide. Through size analysis of the modified particles, they showed that the superior performance is due to particle size reduction achieved primarily through de-agglomeration and secondarily through crystal size reduction. They suggested that above the optimal level of 1.5%, the large lignosulfonate molecule may block access of the SO_2 molecule to the reactive CaO sites, causing the relative utilization to decrease.

Subsequently, Kirchgessner and Jozewicz [37] performed extensive studies of the changes in pore structure during sintering of CaO produced from $Ca(OH)_2$ modified with the 1% calcium lignosulfate. They first noted that modified $Ca(OH)_2$ calcined more quickly, and retained more of its original surface area and porosity than the conventional $Ca(OH)_2$. With increasing time and temperature the difference between the surface area and pore structure between modified and original sorbents due to sintering became more pronounced. Their pore size measurements confirm that sintering involves pore filling, with the smallest pores filled first. This is reflected in the dramatic variation in pore volume with time and temperature for pore sizes less than 50 Å. They noted that the drastic changes in surface area and pore size are complete within 1.5 seconds at 700°C and 0.8 second at 1000°C.

Since sintering of CaO is known to be catalyzed by H_2O, differences in the rates of water loss for modified and unmodified $Ca(OH)_2$ may explain the differences in the rates of surface area loss and in pore structure. Their measurements show that water loss from modified hydroxide is indeed greater than the loss from unmodified hydroxide. However, the difference in water loss between the two types took place before 0.6 second, whereas the difference in surface areas do not become pronounced until after 0.6 second. Therefore, water loss does not fully explain the difference between the structure of the two sorbents.

One of the mechanisms of sintering is the mobility of the grain boundary. Because of the large size of the hydrated lignosulfate molecule, it is believed to be located at the grain boundaries and the surfaces of $Ca(OH)_2$ rather than within the crystal structure. In this structure, it reduces grain mobility, thus reducing the sintering effects. Thus, smaller particle size, increased extent of calcination, and lower sin-

Fundamental Studies Concerning Calcium-Based Sorbents

tering all contribute to the better performance of the hydrate sorbent modified with lignosulfate.

MISCELLANEOUS ADDITIVES—HIGH TEMPERATURE APPLICATIONS/METAL OXIDES. Slaughter et al. [38] investigated the addition of chromium (Cr_2O_3), sodium ($NaHCO_3$), and iron (Fe_2O_3) compounds to the sorbent for injection at 2150°F and 2600°F. Unlike the previously cited studies, the promoters were injected in powder form along with the calcine instead of being added to the hydration water during the preparation of the hydrate. Both the chromium compound (18% SO_2 capture without, 39% SO_2 capture with at 2600°F) and the sodium compound (12% SO_2 capture without, 42% SO_2 capture with at 2600°F) were effective in promoting SO_2 capture by the calcine when added at a 15:1 calcium: promoter metal atomic ratio. The iron compound was not effective in promoting sulfur capture. Slaughter [38] indicated that both chromium and sodium react with the calcium sorbent creating large cracks or pores, particle fragmentation and the formation of a liquid phase, all of which serve to increase SO_2 accessibility to the CaO sites. The presence of mineral matter ash in the flue gas stream acted to reduce the effectiveness of both promoters. The effectiveness of the sodium promoter showed a strong dependence on mineral ash concentration while the effectiveness of the chromium promoter showed only a weak dependence on mineral ash concentration.

This is an important area of research. Alternate additives should be considered, chosen on the basis of solid state chemistry. Systematic characterization of reactivity of similar sorbents through crystal structure analysis might explain the difference between them and lead to a better choice of sorbents.

Limestone/Lime Characterization Study

Limestone/Lime Samples

Several limestone and limestone products were selected for characterization and evaluation as a part of this project. The limestones and limes selected are as follows: (1) Maxville limestone. Approximately 80:20 weight basis $CaCO_3/MgCO_3$ ratio. This limestone was used in the fixed-bed LEC work. This limestone was also used in the moving-bed LEC work. This is an Ohio limestone. (2) Vanport limestone. Approximately 97:03 weight basis $CaCO_3/MgCO_3$ ratio. This limestone was used in the fixed-bed LEC work. This is an Ohio limestone (3) "Mid-Ohio" limestone, hydrate, quicklime. This quicklime is being used as the in-duct injection work taking place at the Ohio Power Company Muskingum River Station. This is an Ohio limestone. (4) Bucyrus limestone, hydrate. Approximately 80:20 weight basis $CaCO_3/MgCO_3$ ratio. This limestone has been used in several FBC and CFBC projects. Bucyrus limestone has been used in TRW's slagging coal combustor. This is an Ohio limestone. (5) Care limestone, hydrate, quicklime. Approximately 55:45 weight basis $CaCO_3/MgCO_3$ ratio. This limestone was used in the fixed-bed LEC work. This is an Ohio limestone. (6) Mississippi limestone, hydrate, quicklime: This limestone and its lime products have been used in many national FGD tests (including LIMB and Coolside). This is not an Ohio limestone.

All collected limestone and lime samples were characterized using the following test: (a) BET Surface Area. Raw BET surface areas were determined for all samples using a Quantisorb Jr. BET sorption apparatus. A BET surface area versus average particle size curve was determined for each limestone sample in order to identify the individual contributions of its internal and external surface areas. (b) Total Pore Volume and Pore Size Distribution. These analyses were performed using a mercury porosimter (Micrometries AutoPore II 9220). Intrusion pressure of up to 60,000 psia were used in determining the pore size distribution. (c) Chemical Composition. Limestone samples were subjected to elemental analysis in order to confirm their chemical composition.

Calcine Preparation

A bench-scale calcination reactor (figure 3.4) was designed and constructed for use in this study. The heart of the calcination reactor was a 1700W Lindberg furnace. This furnace had the ability to reach and maintain temperatures as high as 1200°C. In each calcination test, approximately 250 grams of powdered limestone (either- 70M or -80+100M) were calcined at 950°C for six hours. A steady flow of air (25cc/s) was used as a sweep gas during the calcination runs.

Fundamental Studies Concerning Calcium-Based Sorbents

1. Air Cylinder
2. Air Inlet
3. Calcium Carbonate
4. Cylindrical Vessel
5. Vent
6. Temperature Recorder and Controller
7. Furnace
8. Wire-Screen
9.,10. Thermocouple
11. Temperature Recorder

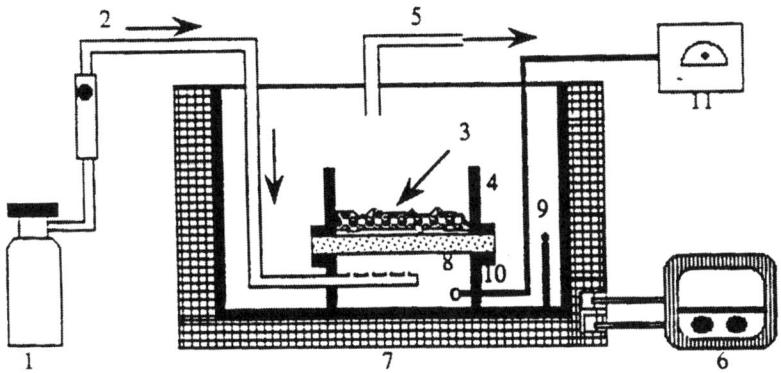

Figure 3.3 Schematic of the bench-scale calcination reactor.

Hydrate Preparation

A bench-scale slaking reactor (Figure 3.4) was constructed for use in this model. This reactor system was primarily comprised of a 1000 ml jacketed reactor flask, a K-Tron solids feeder, and a Brookfield EX-100 constant temperature circulator. This slaking reactor could be operated in any one of three modes: (1) batch in both water and lime; (2) batch in water but with a continuous lime flow to the reactor; and (3) continuous in both water flow and lime flow. Additionally, the slaking reactor could be operated in both constant temperature and variable temperature modes.

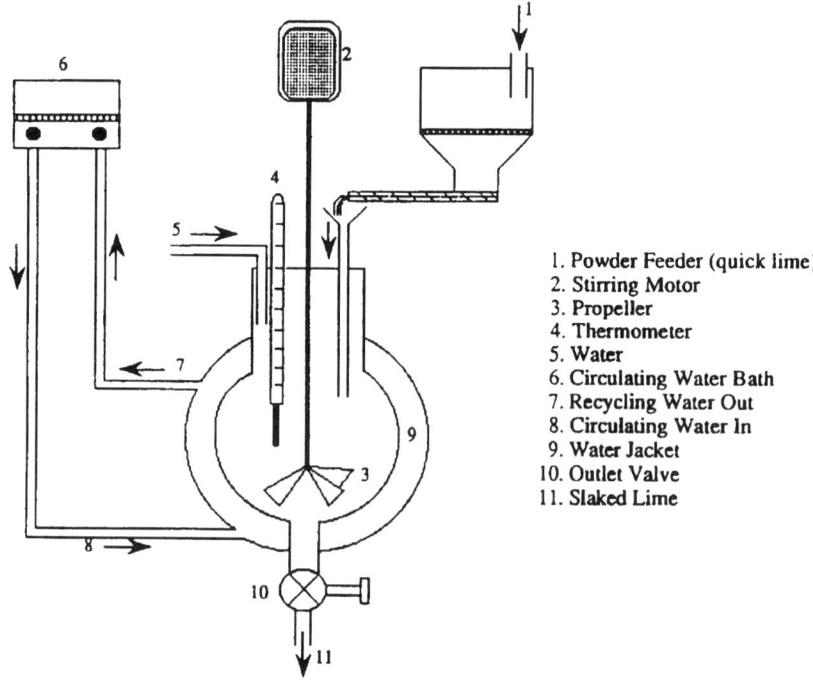

Figure 3.4 Schematic of the bench-scale slaking reactor.

For this study, hydrates were prepared from calcines in a semi-batch mode using the slaking reactor. The entire charge of water (500 ml) was added to the reactor vessel at the beginning of each hydrate run. Calcine was then added to the slaking water at a constant rate of 4 grams per minute for a period of 20 minutes. Calcine/water contacting was continued for a period of 10 minutes after the calcine addition had ceased. This procedure resulted in a final 20:1 H_2O: calcine mole ratio in the reactor. The slaking reactor was held at a temperature of 70-75°C for the duration of the slaking process.

Sorbent Characterization

Duplicate limestone characterizations (BET and mercury porosimeter) of the limestones collected for study were made. After the limestones were size separated, the
-4+5 mesh, -70 mesh and -80+100 mesh size fractions were characterized with

the -80+100 mesh size fractions being selected for calcination and subsequent hydration testing. The complete results of these tests can be found in Reference 39.

Figure 3.6 shows a comparison of the BET surface areas measured for the six different feed limestones while figure 3.7 shows a comparison of the BET surface area of the calcines produced from the selected -80+100 mesh limestones. Only small changes in BET surface area were observed after calcination. If any trend is observed, it is that the BET surface area of the calcitic and magnesia limestones decreased after calcination whereas the BET surface area of the dolomitic limestones (Carey and Bucyrus) increase after calcination.

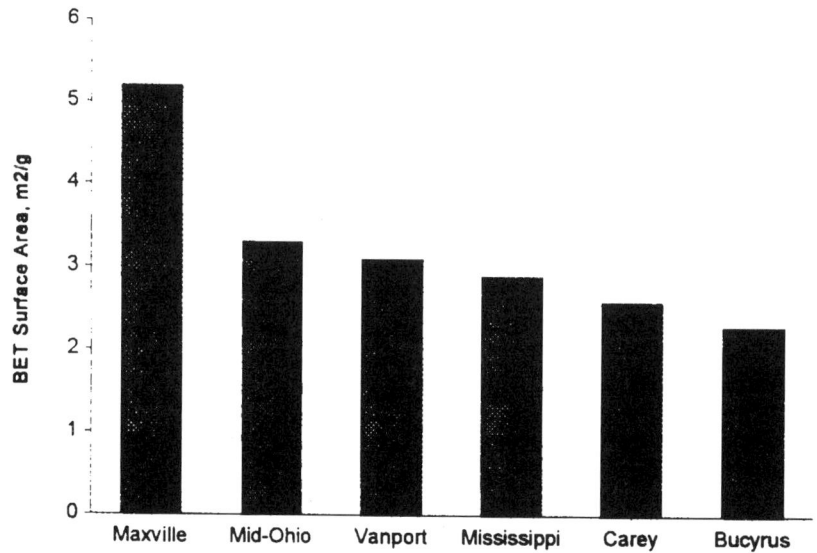

Figure 3.5 Comparison of the BET surface areas of -80+100M limestone samples.

Dry Scrubbing Technologies for Flue Gas Desulfurization

Fundamental Studies Concerning Calcium-Based Sorbents

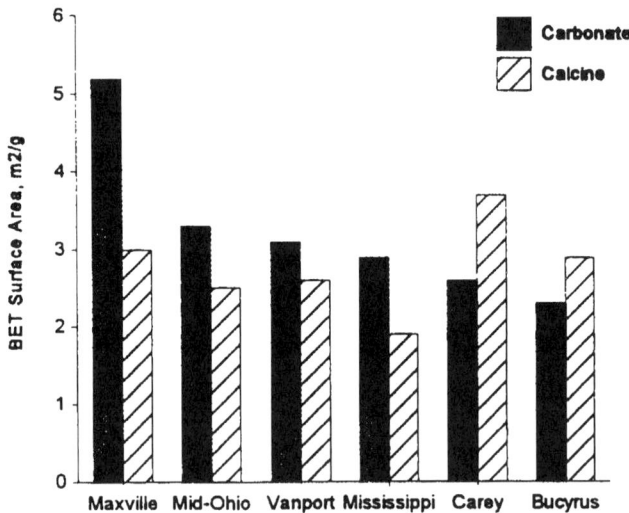

Figure 3.6 Comparison of the BET Surface area -80z+100 limestone samples and their calcines.

Figure 3.8 shows a comparison of the BET surface area of the hydrates produced from the calcines made from the -80+100 mesh parent limestone. In all cases, the surface area of the hydrated products were significantly larger (by a factor of 5 to 10) than those of the parent limestones. No relationship of hydrate surface area to the surface area of the parent limestone is evident from the data collected.

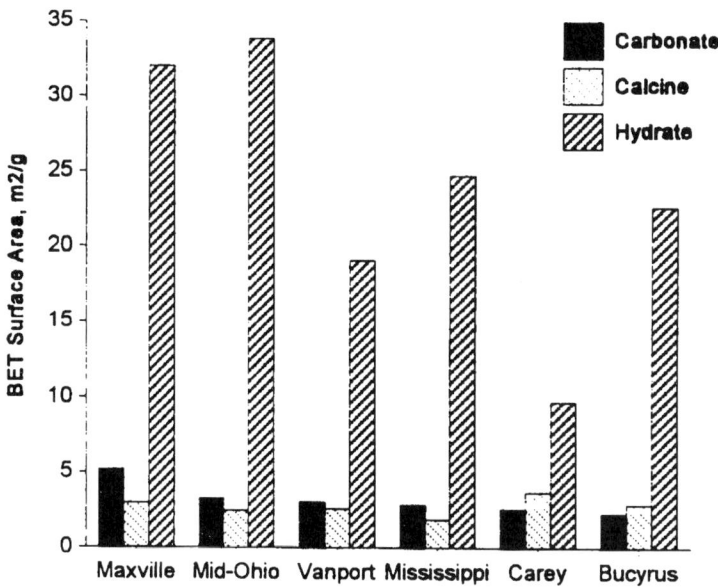

Figure 3.7 Comparison of BET surface area of -80+100M limestone samples, their calcines, and their hydrates.

Fundamental Studies Concerning Calcium-Based Sorbents

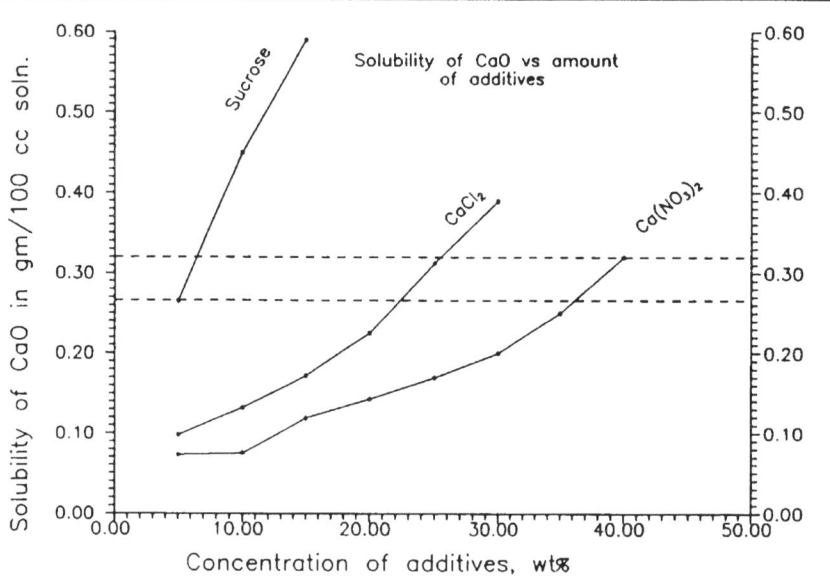

Figure 3.8 CaO solubility versus additive concentration of 70°C.

Table 3.3. Treatment Levels—Solubility/Slaking Tests

Additive	
Treatment Level "A"	**Treatment Level "B"**
4.9g/100cc solution	6.7g/100cc solution
22.4g/100cc solution	25.6g/100cc solution
36.3g/100cc solution	40.0g/100cc solution
Treatment Level "A" corresponds to roughly a 270% increase in CaO/ $Ca(OH)_2$ solubility in the slaking water solution.	Treatment Level "B" corresponds to roughly a 340% increase in CaO/ $Ca(OH)_2$ solubility in the slaking water solution.

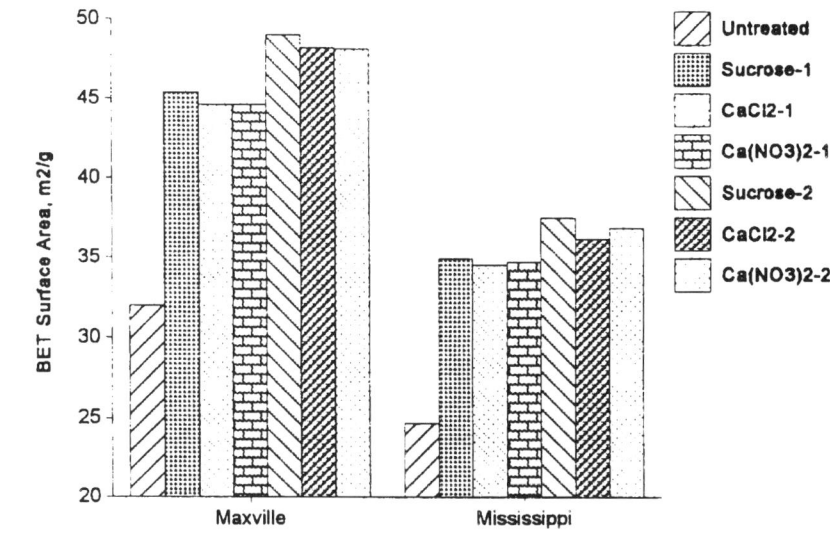

Figure 3.9 Effect of solubility additives on hydrate BET surface area.

Effect of Chemical Additives on Duct Injection/Spray Drying Performance

Introduction/General Description

The primary objective of this project task was the identification of the factor or factors that contribute to the SO_2 capture enhancement and synergism due to additive addition in low temperature (<350°F) spray drying processes. The sorbent injection strategies included are: (1) injection of slurry consisting of the sorbent $Ca(OH)_2$ directly into the duct and (2) injection of dry sorbent followed by injection of water spray (Coolside process).

Additives are generally added to calcium-based sorbents in order to enhance SO_2 capture reactivity. The reasons for this increased SO_2 capture performance can be attributed to one of, or a combination of, the following chemical effects: (1) Water Vapor Pressure Depression: The addition of a nonvolatile solute (additive) lowers the vapor pressure of water and, consequently, the evaporation rate. This increases the reaction time between the wetted sorbent and SO_2 and thus enhances the SO_2

capture. (2) pH Effects: The additive (base) would change the pH of the aqueous phase and contribute to enhanced SO_2 (acid) capture. (3) $Ca(OH)_2$ and SO_2 Solubility Enhancements. This list is not meant to be exhaustive.

The in-duct SO_2 capture model formulated by the Energy and Environmental Research Corporation (EER) [40] was selected as the basic vehicle for determining the additive effects in spray drying processes as each effect can be evaluated separately (although this does not imply that the effects are independent). Amongst the various models available for duct injection spray drying processes, the EER model was chosen because it is a comprehensive model which accounts for the following important processes occurring concurrently with evaporation: (1) diffusion of SO_2 from the bulk gas phase to the droplet surface, (2) absorption of SO_2 at the droplet surface, (3) dissolution of SO_2 to form H_2SO_3 and ionization of H_2SO_3 to HSO_3^- and $SO_3^=$, (4) diffusion of these liquid-phase sulfur species inward, (5) dissolution of the $Ca(OH)_2$ particle at the sorbent surface, and (6) diffusion of $Ca(OH)_2$ from the sorbent surface to the bulk liquid phase.

The model also considers product recycle and the reaction of sorbent (wet agglomerate) with SO_2 after evaporation has ceased. A sub-model which considers the scavenging of sorbent particles is included for the Coolside type process. The EER model does not, however, have the ability to deal with the effects of varying concentrations and types of additives in its present form. One of the major tasks of this project was altering the EER model to better accommodate the additive effects.

The low temperature, dry capture model of Jozewicz and Rochelle [41] assumed that the rate of SO_2 capture is controlled by external SO_2 mass transfer. Karlsson and Klingspor [42] developed two models; one assumed that external SO_2 mass transfer is controlling (high slurry concentrations) and the other assumed that lime dissolution rate is rate-limiting. The model provided by Damle and Sparks [43] assumed that liquid-phase mass transfer is fast, and that the slurry droplet could be viewed as a well-mixed reactor. Kinzey and Harriott [44], on the contrary, assumed that sorbent dissolution is fast and considered both external and liquid phase mass transfer of sulfur species. The diffusion of calcium species away from the droplet center has not been considered.

The EER slurry droplet model (which is of prime interest to us) handles both the slurry injection strategy and the scavenging injection strategy. The model assumes that sorbent particles are initially uniformly distributed throughout the droplet and that they do not circulate within the droplet. As water evaporates from the droplet, the droplet surface recedes until it encounters particles which were initially near the droplet surface. As evaporation continues, the surface particles are pushed inward and an accumulation of sorbent particles at the droplet surface occurs (figure 11). A particle concentration gradient is created between the particles crowded together at the surface and those particles within the droplet which are still at the initial particle concentration.

Due to this physical model of sorbent particle concentration, the instantaneous liquid-phase reaction between the sulfur and calcium species can occur at two types of reaction fronts within the slurry droplet. As the liquid-phase sulfur species diffuse into the droplet they pass individual sorbent particles which are simultaneously diffusing calcium outward. According to film theory, calcium that has dissolved at the surface of a sorbent particle will diffuse outward through a film whose thickness can be calculated. A calcium/sulfur reaction front will therefore be present at the outer edge of this film. The sulfur species in the surrounding bulk liquid phase will also diffuse inward. The distance that calcium must diffuse outward will therefore be less than that calculated by film theory and, since the rate of sulfur diffusion inward is dependent on the bulk liquid-phase concentration of the sulfur species, the distance of the reaction front from the surface of the sorbent particle will vary depending on the local concentration of the liquid-phase sulfur species. The local concentration will, of course, be a function of the spacial location within the slurry droplet.

A total of nine different equations governing droplet deceleration, droplet temperature, droplet evaporation, gas temperature, particle scavenging, particle diffusion within the droplet, liquid-phase SO_2 diffusion of calcium, and sorbent utilization must be solved simultaneously by the EER model. The major assumptions of the EER model listed by its authors [45] are as follows: (1) Liquid-phase ionic reaction between dissolved calcium and sulfur species is instantaneous. (2) Water and sorbent particles do not circulate within the droplet. (3) Calcium sulfite is considered insoluble and precipitates as free-standing crystals. (4) The heats of reaction and sorbent dissolution are small and can be ignored. (5) Thermal gradients within the droplet can be ignored. (6) The droplets and sorbent particles are assumed to be spherical.

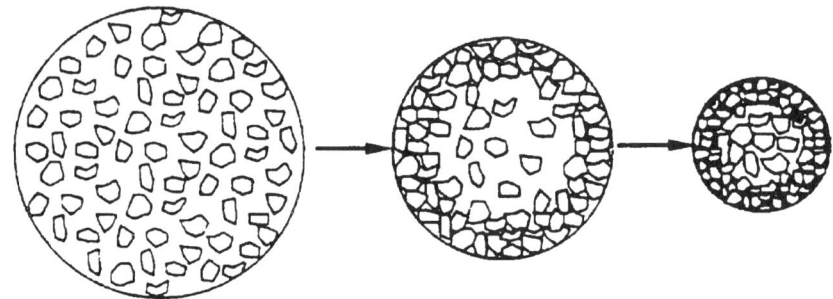

Figure 3.10 Illustration of the accumulation of sorbent and product at the droplet surface as evaporation shrinks the surface [40].

The EER model which considers simultaneous humidification and SO_2 capture processes is a one-dimensional model. However, recent modeling efforts by Oberjohn et al. [46] have indicated that a one-dimensional model is adequate for predicting SO_2 capture and droplet evaporation. The original EER model has set the lime dissolution rate constant to be rather high, implying that the resistance due to lime dissolution is negligible. However, as part of modifying the code (model) to include the effects of additives, a reasonable value has been set for the dissolution rate constant. The modified EER code (including the correlations listed below), as well as example input and output files, can be found in Reference 47.

The additives that have been tested for enhancement in sulfur-dioxide capture in spray drying processes can be classified broadly into the following categories: (1) cosorptive additives, namely NaOH, Na_2SO_3, Na_2CO_3, $NaHCO_3$, NaCl, and $NaNO_3$ (sodium additives) and (2) noncosorptive additives, namely $CaCl_2$, $FeCl_3$, KCl, and $MgCl_2$ (chloride additives). Classification of additives can be also be done in terms of basicity, chemical nature (organic or inorganic), deliquescence, etc.

Sodium additives have been considered as one of the most promising additives in duct injection processes by many researchers. Yoon et al. [12], while conducting laboratory-scale studies on the Coolside process (duct injection strategy with the lime added prior to humidification) observed that there was a relative increase of 30-114% in saturated calcium utilization, when the additive was added to lime during hydration, and a 30-90% increase in saturated calcium utilization, when the additive was added to lime after hydration. NaCl was the most promising additive followed by $NaNO_3$, Na_2CO_3, Na_2SO_3, and NaOH. NaOH and Na_2CO_3 acted as co-sorbents and the other sodium additives were converted to NaOH by reacting with hydrated lime during the process of promotion. The Na^+/Ca^{++} molar ratio was between 0.05-0.2 for all the additive tests carried out in the above study. In the bench scale studies carried out by Chu and Rochelle [48] there was a relative enhancement of 40% in SO_2 removal as compared to the base case, when 0.08M NaOH was added to the system (the temperature in the system was 66°C). Although Na_2SO_3 contains twice as much sodium as NaOH contains, the results obtained with it were just about the same. Probably, some SO_2 removal can be attributed to NaOH itself. The sorbent evaluation studies conducted by Stouffer et al. [49], as part of the pilot plant demonstration in support of the Coolside process, clearly showed the important role played by NaOH in the enhancement of SO_2 removal. Aqueous NaOH was added to the humidification water at Na^+/Ca^{++} mole ratios up to 0.2. For tests conducted with Mississippi hydrated lime with a Ca/S ratio of 2, SO_2 inlet concentration of 1620 ppm, gas inlet temperature of 300°F, and an approach to saturation of 25°F, a 90% removal of SO_2 was obtained with a Na/Ca molar ratio of 0.2. The SO_2 removal without any NaOH, for the same conditions, was 60%. The sorbent utilization studies conducted using a mini-pilot spray dryer by Keener et al. [50] also illustrated the beneficial effect of sodium additives in SO_2 removal. The additive that effected the maximum increase in SO_2 removal over the base case was NaOH (16% increase), followed by $NaHCO_3$ (12%), NaCl (11%), and $CaCl_2$ (4%), for an additive concentration of 300 mg/l-slurry. The tests were performed with an approach to saturation of 28°F, Ca/S ratio of 1.0, SO_2 inlet concentration of 2500 ppm and gas inlet temperature of 300°F.

Yoon et al. [12] found that chloride additives ($CaCl_2$, KCl, $FeCl_3$) were as effective as the sodium additives. Calcium chloride, when added during the hydration of lime (0.1 mole/mole $Ca(OH)_2$), increased the saturated calcium utilization by around 100%. The second best result was obtained with KCl (77% increase), followed by $FeCl_3$ (45%), and $MgCl_2$ (14%). Organic additives (glycerin, adipic acid, sugar) tested in the above study showed no significant positive effect. The effectiveness of using a hygroscopic salt like $CaCl_2$ as an additive was highlighted by Brown et al. [51]. In the pilot-scale tests, SO_2 removal increased from 40% (base case) to 72% when recycle was used and $CaCl_2$ was added (3.4%) to the humidification water. The base conditions were a Ca/S ratio of 2.0, 30°F approach to adiabatic saturation, no recycle, and lime injection upstream of humidification. With recycle alone (i.e., additive not added) the SO_2 removal was 54%. However when $CaCl_2$ alone was added to water (without any recycle), there was no significant increase in SO_2 removal performance.

Fundamental Studies Concerning Calcium-Based Sorbents

In light of the above literature results, four typical additives were chosen for this study. Among the sodium additives, NaOH (cosorbent) is a strong base, Na_2CO_3 (has twice the amount of sodium compared to NaOH) is a weak base, and NaCl is a neutral salt. $CaCl_2$ is highly hygroscopic and is a chlorine additive.

Chemical Effects of Additives

VAPOR PRESSURE DEPRESSION. The addition of a non-volatile solute (additive) lowers the water vapor pressure and, consequently, the evaporation rate. This increases the drying time of the droplet and thus enhances the SO_2 capture. The dissolved solids (additives) cause the evaporation to stop before the droplet dries to completion. As a result, moisture is retained in the pores of the solid agglomerate. This core of moisture is called the equilibrium moisture content of the solid. The core diameter or the moisture content is solely determined by the concentration of the dissolved salts, the temperature, and the humidity of the gas. Increasing the equilibrium moisture content increases the capture of sulfur dioxide during the post-evaporation stage of the process. According to Kinzey [52], in typical spray drying processes, though the equilibrium moisture content represents less than 1% of the original water in the droplet, about 50% of the solid in the agglomerate remains wet.

For this purpose, vapor pressure lowering data was obtained from the literature [53] for the selected additives ($CaCl_2$, NaCl, Na_2CO_3, NaOH) and the data was fit to the following (Antoine) equation:

Equation 3.1

$$P_w = a*\exp\left(A - \frac{B}{T+C}\right)$$

where P_w is the vapor pressure of water, a is an activity coefficient, T is the temperature, and A, B, and C are Antoine constants. a, the activity coefficient, is represented by a polynomial of the form

Equation 3.2

$$a = a0 + a1*X_m + a2*X_m^2 + a3*X_m^3 + a4*X_m^4$$

where a0, a1, a2, a3, and a4 are fitting constants for the polynomial and X_m is the weight fraction of additive in solution. The values of the above constants for the four additives mentioned are given in Table 3.4.

Table 3.4. Regressive Coefficients for the Antoine Equation

Additive	a_0	a_1	a_2	a_3	a_4	A	B	C
$CaCl_2$	0.9899	-8.67E-3	-3.68E-2	4.624E-3	-1.67E-4	18.3036	3816.44	46.13
NaCl	0.9899	-3.71E-2	1.416E-3	-1.06E-3	2.50E-5	18.5208	3935.38	41.37
Na_2CO_3	0.9899	-3.65E-2	-5.74E-3	3.09E-3	-1.05E-3	17.9918	3623.98	53.38
NaOH	0.9899	-2.33E-2	-3.11E-3	1.66E-4	-2.0E-6	18.3028	3722.23	53.44

Fundamental Studies Concerning Calcium-Based Sorbents

There is an enhancement of around 3–7 percentage points in SO_2 capture due to the vapor pressure depression effect alone. Several sets of simulation results are represented graphically in Figure 3.12.

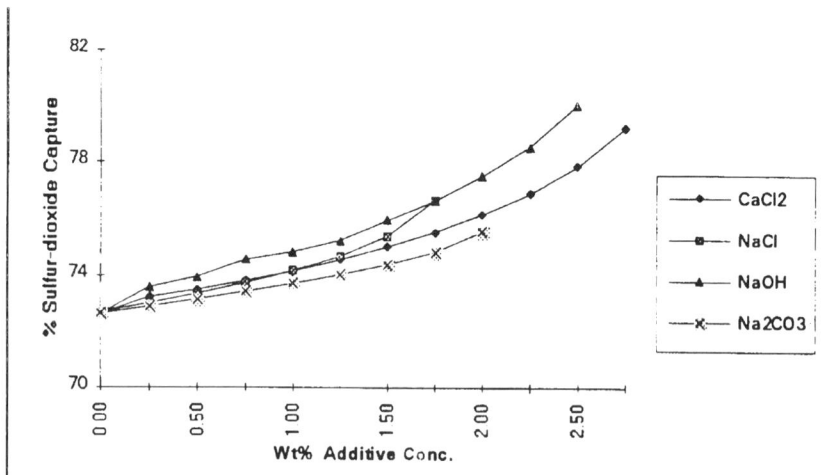

Figure 3.11 SO_2 capture efficiency versus additive concentration (vapor pressure depression effect).

$CA(OH)_2$ SOLUBILITY ENHANCEMENT. With the addition of the additive, the solubility of lime is found to increase for most of the cases studied here (the exception being NaOH). An increase in the solubility of lime enhances the diffusion of the calcium species, and thus increases calcium utilization and SO_2 capture. For the case of NaOH, the solubility of lime decreases due to the common-ion effect. The solubility data for ternary mixtures of $Ca(OH)_2$, water, and additive (e.g., $CaCl_2$) have been obtained and as the solubility of $Ca(OH)_2$ was found to be linear function of the concentration of the additive (for an additive concentration of within 10 wt%, our area of focus), the data was fit to the following equation.

Equation 3.3

$$CLime = \frac{A}{T} + BX + C$$

where CLime is the concentration of $Ca(OH)_2$, X is the weight percent of the additive, and T is the temperature. A, B, and C are fitting constants. The values of the fitting constants for three of the additives studied are given in Table 3.5.

Table 3.5. Regression coefficients for Equation (3)

Additive	A	B	C
$CaCl_2$	16550	1.13556	35.62
*NaOH	16550	-17.12	35.62
NaCl	16550	0.45105	35.62

The value for the dissolution rate at the sorbent surface was set arbitrarily high in the original EER model. This model deficiency needs to be corrected prior to the evaluation of the effect of $Ca(OH)_2$ solubility enhancement/reduction on the sulfur dioxide capture. The importance of the lime dissolution rate is dealt with in the next section.

DISSOLUTION RATE OF LIME. The EER model assumes that there is no resistance to the dissolution of $Ca(OH)_2$; in other words, the dissolution of lime does not affect the rate of sulfur dioxide capture. Hence, the dissolution rate constant for lime has been set unrealistically high in the original EER model. The concentration of lime at the sorbent particle surface (C_L^*) is related to the equilibrium concentration of lime ($C_{L,e}$) by the following equation (as given in the EER model):

Equation 3.4

$$C_L^* = \frac{C_{L,e}}{1 + \frac{D_L}{\sigma \delta K_d}}$$

where D_L is the diffusivity of lime in m²/s, δ is the film thickness in meters (when diffusion of liquid-phase sulfur species to the particle is considered), s is the roughness factor, and K_d is the dissolution rate constant of lime. The original model sets K_d equal to 1.0 m³/m²xs (i.e. $C_L^* = C_{L,e}$). Typical values of D_L and s are 1.67E-9 m²/s (at a temperature of 310K) and 20, respectively. The value of d is on the order of 10^{-7}-10^{-8}m (a typical droplet diameter is 50 μm).

The dissolution times for an isolated $Ca(OH)_2$ particle, as given by Kinzey [52] are tabulated in Table 3.6.

Table 3.6. Dissolution Times for Ca(OH)$_2$ Particles

Particle Size, (µm)	1	2	3	4	5
Dissolution Time, (sec)	0.08	0.33	0.75	1.33	2.08

Although the residence time of a droplet in a duct is typically 2.5–5 seconds, its drying time is usually much less. For droplets around 50µm in size, the drying time is generally between 0.3 to 1.2 seconds. Thus, the dissolution rate becomes important for larger sorbent particles (4-5µm).

A sensitivity study was performed on the dissolution rate constant of lime to give an insight into its role in the SO$_2$ capture process. The upper and lower limits for the dissolution rate constant were fixed at 1.0 and 10^{-4} m^3/m^{2}xs, respectively. If it is assumed that C_L^* is 80% of $C_{L,e}$ in Equation 3.4, then a typical value of K_d (dissolution rate constant) would be 3×10^{-4} m^3/m^{2}xs. Furthermore, typical values of the lime dissolution rate constant obtained by Ritchie et al. [54] by using rotating discs prepared from calcium hydroxide were 10^{-5} to 10^{-6} m^3/m^{2}xs. However, for powdered samples, the dissolution rate constant is expected to have a two-order-of-magnitude higher value. It can be seen from figure 13 that there is a steep increase in sulfur dioxide removal until a K_d value of around 10^{-2} m^3/m^{2}xs is reached. There is not a considerable increase in SO$_2$ removal after this point. The dissolution rate is expected to play a major role in the Coolside process where the sorbent is injected upstream of the water spray, as the sorbent does not have a large time to mix with water before the droplet dries. In typical spray drying processes, the dissolution rate constant would be on the order of 10^{-2} to 10^{-4} m^3/m^{2}xs. The presence of CO$_2$ in the flue gas might reduce the dissolution rate of lime as CO$_2$ may react with the sorbent particles to form a layer of relatively insoluble calcium carbonate. The value for the dissolution rate constant has been set at 3×10^{-4} m^3/m^{2}xs (the point at which the slope changes in Figure 3.12) for all of the model simulations, unless otherwise mentioned.

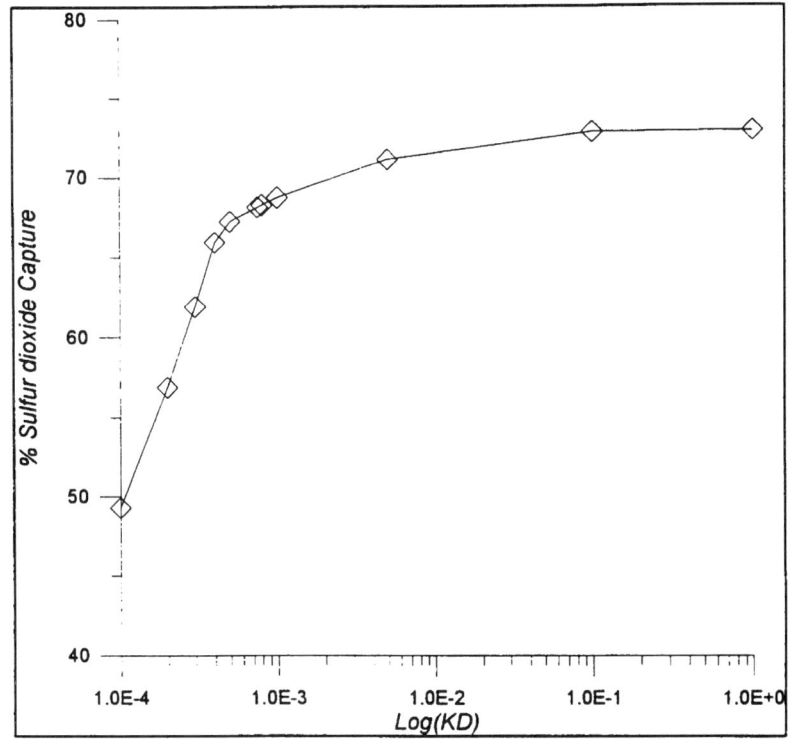

Figure 3.12 Percentage sulfur dioxide capture versus dissolution rate constant for lime (slurry injection case).

With the addition of additives, the dissolution rate constant would increase or decrease depending on the type of additive used. For the case of NaOH, there is a considerable decrease in the dissolution rate. Data were taken from the literature [54] for the reduction in the dissolution rate constant with the addition of NaOH and was regressed to obtain the following equation:

Equation 3.5

$$K_{d(NaOH)} = K_d \left(\frac{x^{-0.925}}{917.37} \right)$$

where K_d is the dissolution rate constant of lime in $m^3/m^2 xs$ and x is the weight percent of NaOH in the water. Equation 3.5 is good only for x values between

Fundamental Studies Concerning Calcium-Based Sorbents

0.08 and 5%. Figure 14 shows the reduction in the lime dissolution rate constant as a function of NaOH concentration.

The effect of changes in the equilibrium solubility of lime on SO_2 removal efficiency at various dissolution rate constants is considered next. It can be seen from Figure 3.14 that with an increase in solubility of lime, the enhancement in sulfur dioxide removal is much greater for lower lime dissolution rates. Thus, for baseline conditions, the sulfur-dioxide removal would be around 60% (as compared to 72% with K_d being equal to 1.0 m³/m²xs).

The common process parameters in all the above simulations were as follows: inlet SO_2 concentration = 1500 ppm; no sorbent recycle; Ca/S ratio = 2.0; approach to saturation = 23°F; mean droplet diameter = 50 μm; sorbent particle diameter = 4 μm; and residence time = 2.5 seconds. It should be noted that the value of the dissolution rate constant in Figure 3.11 (vapor pressure depression effect) is 1.0 m³/m²xs.

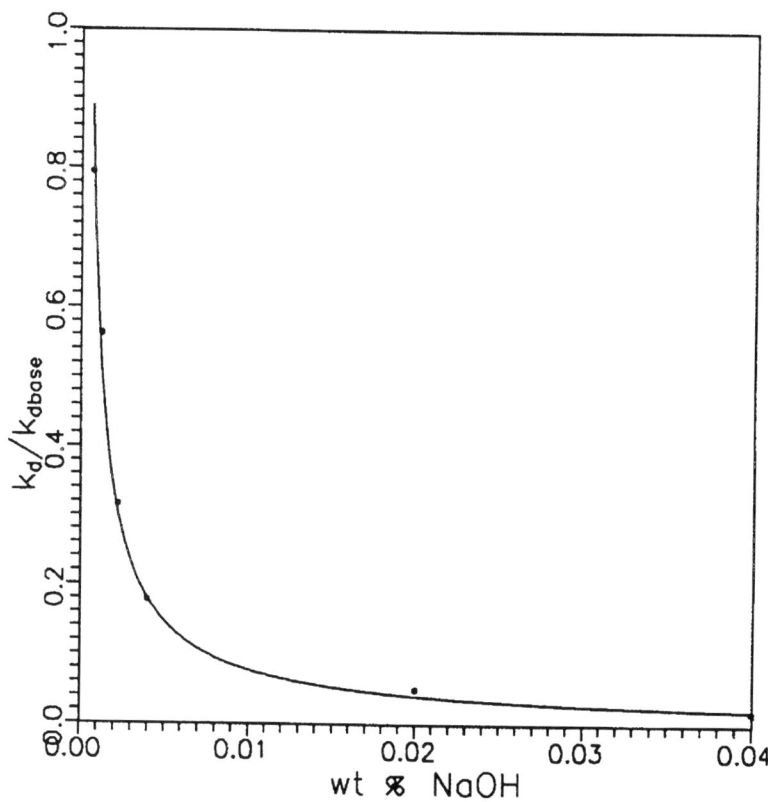

Figure 3.13 Reduction in dissolution rate constant as a function of NaOH concentration.

pH EFFECTS. The pH of the droplet decreases as SO_2 diffuses to sites near the $Ca(OH)_2$ particles. This enhances the lime dissolution rate. On the other hand, the addition of NaOH into the system increases the pH and thus decreases the dissolution rate of lime. This has been dealt with in the previous section. However, the increase in pH of the droplet has some beneficial effects.

Dry Scrubbing Technologies for Flue Gas Desulfurization

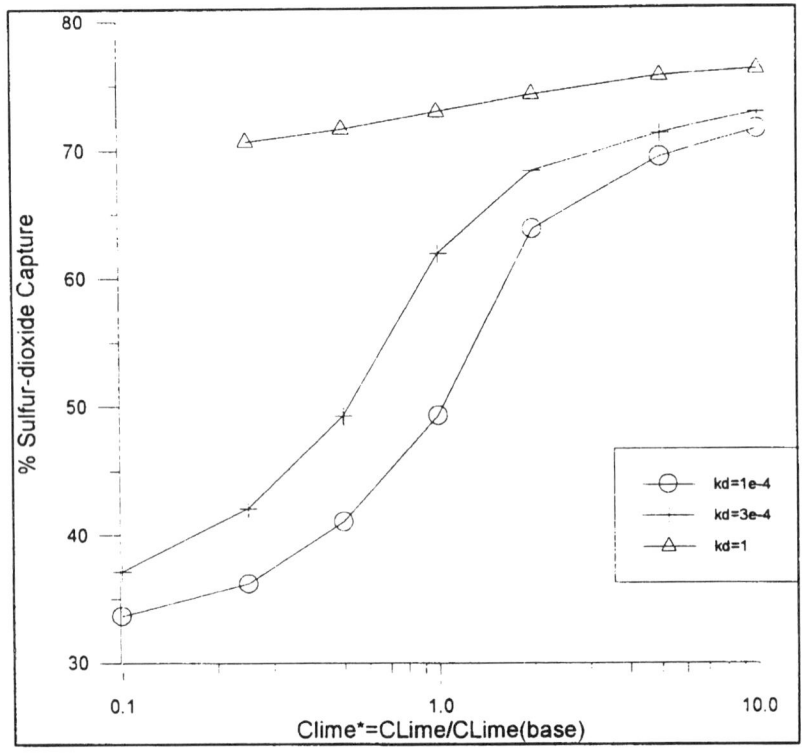

Figure 3.14 Percentage sulfur dioxide capture versus equilibrium solubility of lime at various lime dissolution rates constants.

SO$_2$ Absorption. With the addition of NaOH to the slurry, there is an enhancement in the capture of sulfur dioxide. Literature [56] values were obtained for the enhancement factor (ø) for SO$_2$ absorption as a function of concentration of NaOH. The sulfur absorption rate is given by

Equation 3.6

$$N_{SO_2} = -\phi\left(D_1\frac{dC_1}{dr} + D_2\frac{dC_2}{dr}\right)$$

where D_1 and D_2 are the diffusivities of H_2SO_3 and HSO_3^-, respectively, in m^2/s, and C_1 and C_2 are the concentration of H_2SO_3 and HSO_3^-, respectively, in moles/m^3.

A sensitivity study was done to determine whether the SO_2 absorption was rate controlling in the overall process. For this purpose, the mass transfer enhancement factor was varied over the maximum expected range and the predicted percentage SO_2 capture was determined for each case. It can be seen from figure 16 that there is not much of an enhancement in the net SO_2 removal rate due to enhanced absorption of SO_2 into the droplet, in turn due to NaOH addition.

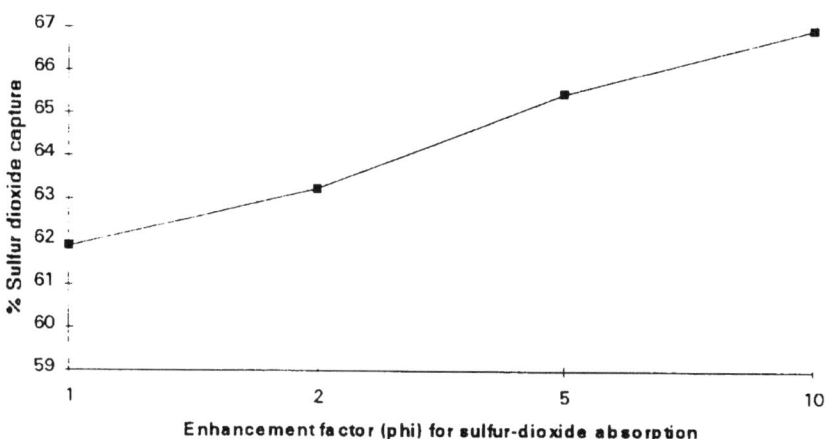

Figure 3.15 Percentage SO_2 capture versus the SO_2 absorption enhancement factor.

SOLUBILITY OF SO_2. The effect of the solubility of SO_2 on SO_2 capture was studied by carrying out a sensitivity study on He, the Henry's law constant for sulfur dioxide. The value of the Henry's law constant (Rabe and Harris, 1963) for the base case is given by

Equation 3.7
$$He = \alpha * \exp\left(2.4717 - \frac{2851.1}{T_d}\right)$$

where T_d is the droplet temperature (K). For different values of α, He and the corresponding SO_2 capture are reported in Figure 3.16. It can be seen that with increasing values of He, though there is a decrease in SO_2 capture as expected, this decrease is not very significant. This indicates that the solubility of SO_2 is not a limiting factor in SO_2 removal. In this particular simulation, the dissolution rate constant was maintained at 3×10^{-4} $m^3/m^2 xs$.

Fundamental Studies Concerning Calcium-Based Sorbents

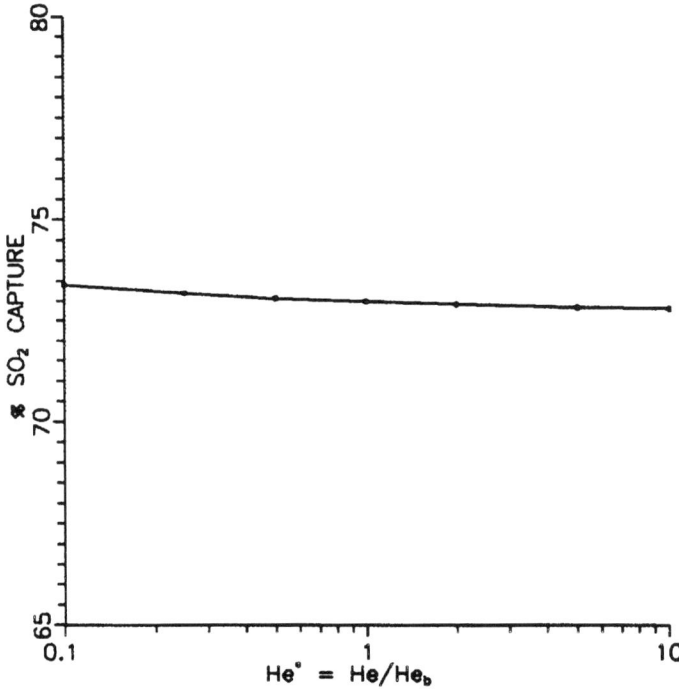

Figure 3.16 Percentage SO_2 capture efficiency versus Henry's law constant.

Thus, the liquid-phase diffusional resistance is not rate limiting. It may become significant towards the end of the droplet lifetime when the sorbent particles near the surface are consumed and the sulfur species have to diffuse inward through the product layer to react with $Ca(OH)_2$.

ALKALINITY OF THE WET AGGLOMERATE. The most important effect of pH would be to increase the alkalinity of the wet agglomerate. As water evaporates, the mass fraction of the additive (NaOH) increases (though some of it may react with SO_2), thereby affecting the wet agglomerate alkalinity. Thus, the dried solids reactivity towards SO_2 will increase for the case where NaOH is the additive. The increase in alkalinity in the agglomerate due to NaOH addition is much more than that in the liquid phase. SO_2 capture is drastically reduced once the evaporation has stopped; hence, any factor that would increase the reactivity of the wet solids would have a significant effect in SO_2 removal efficiency.

Once the evaporation has stopped, the moisture retained in the pores of the solid agglomerate aids in the capture of sulfur dioxide. The chemical reaction rate constant (K_r) for the lime-SO_2 reaction in the wet agglomerate is known to be dependent on the relative humidity of the gas. Experiments performed by Damle and Sparks [43] showed that the value of K_r under conditions of 50% relative humidity is thrice that obtained under completely dry conditions. There are no similar experimentally proven results obtained so far for the dependence of K_r on the pH of the wet agglomerate. Nevertheless, the resistance to SO_2 absorption is bound to decrease as a result of increased alkalinity of the agglomerate. This would be reflected in an increase in the magnitude of the reaction constant. A sensitivity study has been performed on the effect of a change in the reaction constant (K_r); i.e., the alkalinity of the wet agglomerate, on the sulfur dioxide removal efficiency. The results are depicted in Figure 3.17.

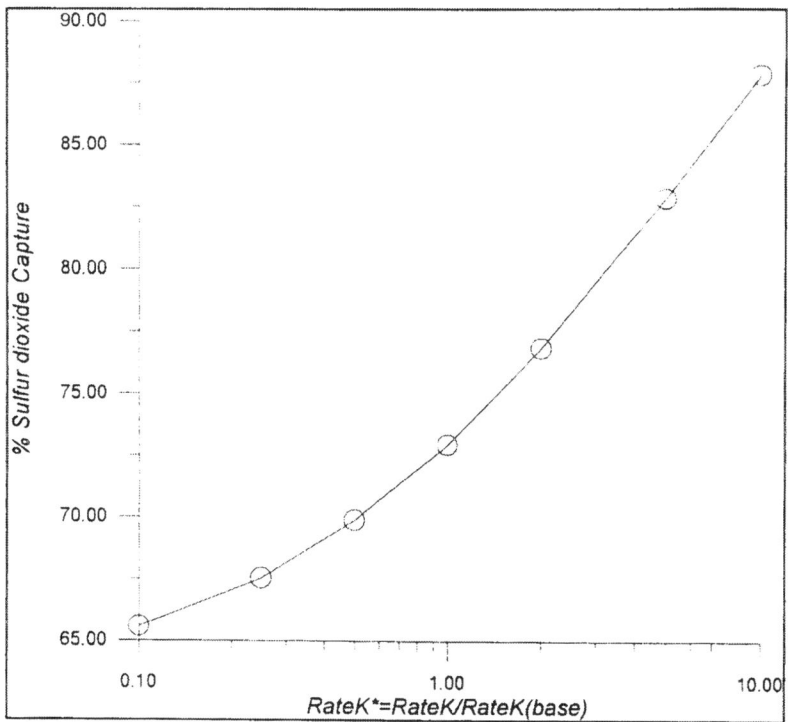

Figure 3.17 Sulfur dioxide capture efficiency as a function of solid-phase alkalinity.

SUMMARY OF SYNERGISTIC EFFECTS. The additive effects described in the preceding sections were evaluated independently. The combined results of the chemical effects of the additives listed for both the injection strategies are presented here. The percentage sulfur dioxide removal for the case of calcium chloride and sodium chloride is given in Figure 3.18. The baseline conditions are as before. The enhancement in sulfur dioxide capture due to these additives is primarily due to an increase in droplet lifetime, which in turn is due to reduced evaporation. For 2 wt% $CaCl_2$, the droplet lifetime increases from 0.45 to 0.6 seconds. The 0.15 second increase, though small, represents a significant percentage of the total residence time in the duct (2.5 seconds) calculated for this simulation. Also, the equilibrium water content of the wet agglomerate increases. When compared with the vapor pressure lowering effect, the effect on SO_2 capture due to enhanced lime solubility is not very significant. As the pH of the slurry is affected only marginally by addition of $CaCl_2$ or NaCl, the effect on the lime dissolution rate is also negligible. The pH of saturated $Ca(OH)_2$ decreases from 12.47 to 12.43 when 3 wt% NaCl is added to the slurry [50].

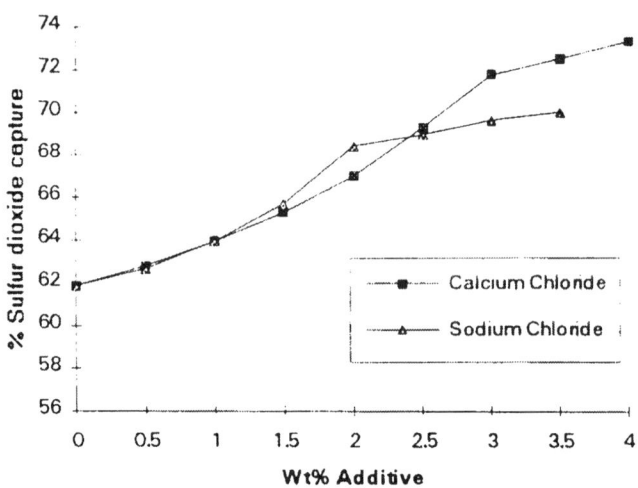

Figure 3.18 Percentage SO_2 capture efficiency versus additive concentration.

Figure 3.19 indicates that the performance of $CaCl_2$ as an additive is slightly better than NaCl in slurry injection case. For a 3.5 wt% additive concentration (based on the initial mass of water), the increase in SO_2 removal is 10 and 7 percentage points for $CaCl_2$ and NaCl, respectively. For both cases, there is a steep increase in SO_2 removal initially (until around 3 and 2 wt% for $CaCl_2$ and NaCl, respectively). After this level of additive concentration, the net increase in sulfur dioxide removal over the base case is almost constant.

The effect of a change in the solubility of lime in water due to Na_2CO_3 addition could not be evaluated as data could not be obtained for this ternary system. However, the net SO_2 removal is expected to increase, with the vapor pressure lowering effect alone contributing to around 4-5 percentage points for a typical concentration of 3 wt%.

For the case of NaOH, there is a significant decrease in the solubility of lime in water and in the lime dissolution rate, which tends to counter the increase in droplet lifetime and enhancement in the liquid phase flux of SO_2 inward during the evaporation and post-evaporation stages (wet agglomerate). The enhancement of the SO_2 mass-transfer coefficient in the droplet due to the increased alkalinity has a lesser impact on the SO_2 removal rate than the decrease in dissolution rate of lime due to the same reason. However, in the post-evaporation stage, when the wet agglomerate, consisting of unreacted sorbent particles and reaction product ($CaSO_3 \cdot H_2O$) with an equilibrium amount of moisture, is formed, the alkalinity provided by the NaOH is expected to result in significant removal of SO_2. The data for this phenomena could not be obtained from the literature. Also, the effective stoichiometric ratio increased marginally, considering the fact that NaOH can be treated as co-sorbent. The net increase in SO_2 removal without the effect due to increased alkalinity of the wet agglomerate is 6.5 percentage points for an NaOH concentration of 3 wt%.

For the sorbent injection followed by water spray case, the baseline SO_2 capture is 25 percentage points. This is an underprediction of SO_2 removal considering that the reported SO_2 removal is 40 percentage points in the pilot-scale studies conducted at Meredosia station as part of the demonstration of the Coolside process. However, the droplet size distribution was not reported (the average droplet diameter in these simulations is 50 μm). The importance of the droplet size is illustrated in Figure 3.19.

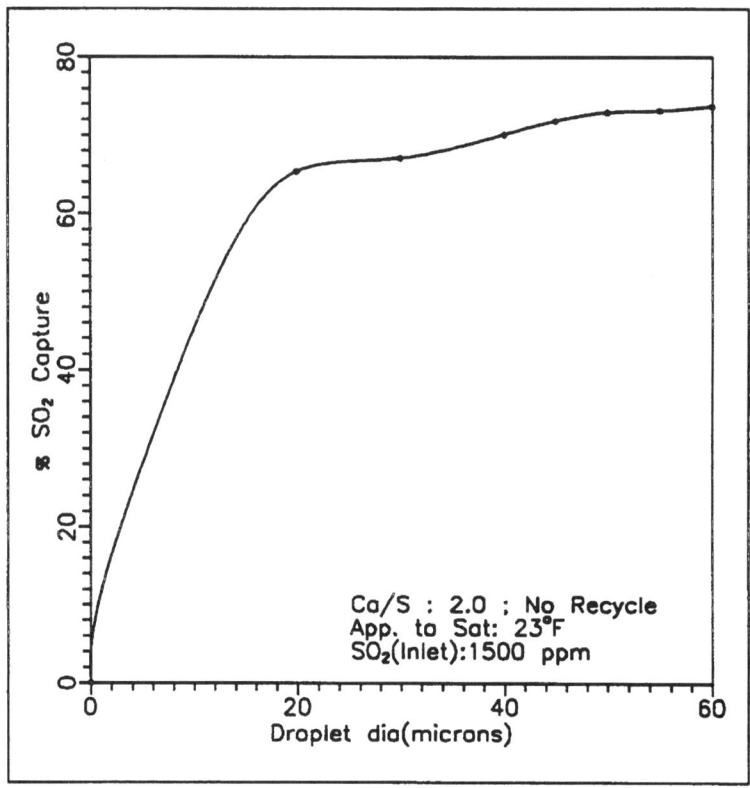

Figure 3.19 Percentage SO_2 capture efficiency versus droplet size.

It can be seen from Figure 3.19 that below a droplet size of around 20 μm, there is a steep increase in SO_2 capture as droplet size increases. This increase is due to an increase in the lifetime of the droplet. Above 20 μm, as the droplet size increases, the decreasing rate of external mass transfer almost counters the effect of increasing lifetime and the net increase in SO_2 capture is not significant. Also, the droplet size distribution plays an important role as a larger sized droplet will take longer time to evaporate in the presence of smaller sized droplets than it would otherwise take.

There may also be an underprediction of the number of sorbent particles scavenged by the decelerating droplet in the duct, which may lead to underprediction of the SO_2 capture. The simulations run for the various additive cases listed predict only marginal improvement in SO_2 removal. The probable reasons for this observation should be investigated in future work.

Model Comparisons

Comparisons between model predictions for in-duct slurry injection and results obtained experimentally at the University of Cincinnati (UC spray dryer tests) are presented in Table 3.7.

Table 3.7. Comparison of the Enhanced EER Model with UC Spray Dryer Test Results for the Additive

Wt % additive	$CaCl_2$ (additive)	
	Model	Exptl.
0.0 (Base case)	38.31	44.2
1.0	40.48	--
2.0	43.93	--
3.0	46.51	48.8

The process parameters for the comparisons provided are as follows: Ca/S ratio = 1.0, SO_2 concentration (inlet) = 2200 ppm, no recycle, and approach to saturation = 26°F. Droplet size was not reported as a part of the experimental results (droplet size in the model predictions = 50 μm). The residence time in the experiment was approximately 10 seconds. For the simulations, the residence time was taken as 2.5 seconds. The difference in the experimental and model results could be due to the different residence times used. In fact, when the residence time was increased, the difference in the model and experimental values did come down. However, the percentage SO_2 removal was predicted using a 2.5 second residence time, as this is a typical value in in-duct injection processes.

It was intended at the initiation of this project that a more extensive comparison of the EER model with UC results of spray drying with additives be performed. As it turned out, additional UC experimental results were not available to us for model comparison and validation.

Conclusions

The summary of the chemical effects of additives for in-duct injection (slurry) spray drying processes are given as follows: (1) The lowering of the water vapor pressure with the addition of additives (dissolved solids) is the primary reason for the enhancement in sulfur dioxide removal. The droplet lifetime is increased and so is the equilibrium moisture content of the wet agglomerate. (2) The enhancement in lime solubility due to the addition of certain additives like $CaCl_2$ and NaCl also increases the SO_2 capture, although this effect is not as significant as the first effect. (3) With the addition of NaOH, the decrease in solubility of lime in water due to the common-ion effect and the decreased dissolution rate of lime due to an increase in the pH of the slurry is countered by the increased absorption of SO_2; however, the liquid-phase diffusion of SO_2 is not rate controlling throughout most periods of time except when the sorbent particles near the droplet surface are consumed. This happens only towards the end of the droplet lifetime. (4) The alkalinity of the wet agglomerate is increased with the addition of NaOH, leading to increased SO_2 removal in the post-evaporation stage.

Recommendations

Experiments to evaluate parameters like the dissolution rate of lime, reactivity of wet solids, equilibrium moisture content in the wet agglomerate, etc. need to be made for typical spray drying processes with and without various additives. The modified EER model can then be evaluated/validated and additional modifications could then be added as needed.

Nomenclature

A	Constant in Equation 3.1. different in Equation 3.3
A'	Constant in Equation 3.3.
a	Activity coefficient.
a0	Constant in Equation 3.2.
a1	Constant in Equation 3.2.
a2	Constant in Equation 3.2.
a3	Constant in Equation 3.2.
a4	Constant in Equation 3.2.
B	Constant in Equation 3.1.
B'	Constant in Equation 3.3.
C	Constant in Equation 3.1.
C'	Constant in Equation 3.3.
C_L^*	Concentration of lime at the particle surface.
$C_{L,e}$	Equilibrium concentration of lime.
Clime	Concentration of $Ca(OH)_2$.
C_1	Concentration of H_2SO_3.
C_2	Concentration of HSO_3^-.
D_L	Diffusivity of lime.
D_1	Diffusivity of H_2SO_3.
D_2	Diffusivity of HSO_3^-.
He	Henry's Law constant.
K_d	Dissolution rate constant for lime.
$K_{d(NaOH)}$	Dissolution rate constant for lime in the presence of NaOH.
N_{SO2}	Flux of SO_2.
P_w	Vapor pressure of water.
r	Distance.
T	Temperature.
T_d	Droplet temperature.
X	Weight percent of additive.
X_m	Weight fraction of additive in solution.

Fundamental Studies Concerning Calcium-Based Sorbents

x	Weight percent NaOH in solution.
α	Fitting constant in Equation 3.7.
δ	Film thickness.
σ	Roughness factor.
ϕ	Enhancement factor.

References

1. Cole, J. A., J. C. Kramlich, W. R. Seeker, G. D. Silcox, G. H. Newton, D. J. Harrison and D. W. Pershing, "Fundamental Studies of Sorbent Reactivity in Isothermal Reactors," paper 16, *Proceedings: 1986 Joint Symposium on Dry SO_2 and Simultaneous SO_2/NO_x Control Technologies*, 1, EPRI CS-4966 (1986)
2. Silcox, G. D., *Status and Evaluation of Calcitic SO_2 Capture: Analysis of Facilities Performance*, EPA 600/7-87-014 (1987)
3. Gullett, B. K. and K. R. Bruce, "Pore Distribution Changes of Calcium-Based Sorbents Reacting with Sulfur Dioxide," *AIChE J.*, 33(10), 1719 (1987)
4. Bruce, K. R., B. K. Gullett and L. O. Beach, "Comparative SO_2 Reactivity of CaO Derived from $CaCO_3$ and $Ca(OH)_2$," *AIChE J.*, 35(1), 37 (1989)
5. Snow, G. C., J. M. Lorrain and S. L. Rakes, "Pilot Scale Furnace Evaluation of Hydrated Sorbents for SO_2 Capture," paper 6, *Proceedings: 1986 Joint Symposium on Dry SO_2 and Simultaneous SO_2/NO_x Control Technologies*, 1, EPRI CS-4966 (1986)
6. Teixeira, D. P., T. A. Lott and L. J. Muzio, "Dry Sorbent SO_2 Control for New Power Plants Burning Low Sulfur Western Coals," paper 7, *Proceedings: 1986 Joint Symposium on Dry SO_2 and Simultaneous SO_2/NO_x Control Technologies*, 1, EPRI CS-4966 (1986)
7. Beittel, R., J. P. Gooch, E. B. Dismukes and L. J. Muzio, "Studies of Sorbent Calcination and SO_2-Sorbent Reactions in a Pilot-Scale Furnace," paper 16, *Proceedings: First Joint Symposium on Dry SO_2 and Simultaneous SO_2/NO_x Control Technologies*, 1, EPRI CS-4178 (1985)
8. Overmoe, B. J., S. L. Chen, L. Ho, W. R. Seeker, M. P. Heap and D. W. Pershing, "Boiler Simulator Studies on Sorbent Utilization for SO_2 Control," paper 15, *Proceedings: First Joint Symposium on Dry SO_2 and Simultaneous SO_2/NO_x Control Technologies*, 1, EPRI CS-4178 (1985)
9. Bortz, S. J. and P. Flament, "Recent IFRF Fundamental and Pilot Scale Studies on the Direct Sorbent Injection Process," paper 17, *Proceedings: First Joint Symposium on Dry SO_2 and Simultaneous SO_2/NO_x Control Technologies*, 1, EPRI CS-4178 (1985)
10. Gooch, J. P., E. B. Dismukes, R. Beittel, J. L. Thompson and S. L. Rakes, "Sorbent Development and Production Studies," paper 11, *Proceedings: 1986 Joint Symposium on Dry SO_2 and Simultaneous SO_2/NO_x Control Technologies*, 1, EPRI CS-4966 (1986)
11. McCarthy, J. M., S. L. Chen, J. C. Kramlich, W. R. Seeker and D. W. Pershing, "Reactivity of Atmospheric and Pressure Hydrated Sorbents for SO_2 Control," paper 10, *Proceedings: 1986 Joint Symposium on Dry SO_2 and Simultaneous SO_2/NO_x Control Technologies*, 1, EPRI CS-4966 (1986)
12. Yoon, H., J. A. Withum, W. A. Rosenhoover and F. P. Burke, "Sorbent Improvement and Computer Modeling Studies for Coolside Desulfurization," paper 33, *Proceedings: 1986 Joint Symposium on Dry SO_2 and Simultaneous SO_2/NO_x Control Technologies*, 1, EPRI CS-4966 (1986)
13. Borgwardt, R. H. and K. R. Bruce, "Effects of Specific Surface Area on the Reactivity of CaO with SO_2," *AIChE J.*, 32, 239 (1986)
14. Kirchgessner, D. A., B. K. Gullett and J. M. Lorrain, "Physical Parameters Governing the Reactivity of $Ca(OH)_2$ with SO_2," paper 8, *Proceedings: 1986 Joint Symposium on Dry SO_2 and Simultaneous SO_2/NO_x Control Technologies*, 1, EPRI CS-4966 (1986)
15. Gooch, J. P., E. B. Dismukes, R. S. Dahlin and M. G. Faulkner, "Scaleup Tests and Supporting Research for the Development of Duct Injection Technology," *Draft Topical Report No. 1 - Literature Review*, U.S.D.O.E. (1989)

CHAPTER 4 **SORBENT TRANSPORT AND DISPERSION**

L.-S. Fan, E. Abou-Zeida,
S.-C. Liang, X. Luo
The Department of Chemical Engineering
The Ohio State University
Columbus, OH

Abstract

Three topics are covered in this chapter: powder characterization, powder dispersion, and modeling of powder attrition and dispersion. In the powder characterization section, powder surface properties, the results of powder transport under the influence of attraction forces, the effects of additives and temperature, and the powder mechanical properties are presented. Four sorbents are characterized—calcite, dolomite, hydrated lime and dolomitic hydrate. Calcite and dolomite have comparable median particle sizes and dolomite has a wider size distribution than the calcite. The dolomite hydrates are composed of fine particles. All the sorbents have similar morphology. It is found that the van der Waals force is dominant, compared to the electrostatic and gravitational forces, in powder agglomeration. Powder dispersion experiments show that the sorbents with the lower average van der Waals forces have greater dispersibility. Temperature has an effect on the dispersion of powders. The properties of hydrate can be modified by additives, such as lignosulfanate. SEM results show that the agglomerate size of the modified hydrate is smaller than that of the pure hydrate and the modified hydrates are less cohesive. Experimental results indicate more agglomeration and/or loss of fines at high temperatures for hydrates.

The second section compares three types of nozzles in terms of the local solid concentration and velocity at the entrance region. A particle image velocimetry (PIV) technique is developed to quantify these parameters. PIV is used to analyze the recorded agglomerate images. The uniqueness of this PIV system is its capability to measure the instantaneous characteristics of particle motions; e.g., velocity, solids concentration, and size distribution. It is found that powder dispersion is mainly due to the shear stress generated by high turbulence intensity inside the nozzles. From the experimental study on the agglomerate size distribution and the numerical simulation on the flow structures in different nozzles, the expansion nozzle with two booster jets has the optimum performance.

Attrition of sorbents can take place during handling, transport, and injection of sorbent powders, and attrition may be due to thermal, chemical, static mechanical and kinetic stresses. A stochastic model for sorbent attrition is developed. Variations of particle size distribution along with the mean and variance of the distribution during the attrition process are also presented. To simulate the powder dispersion in the nozzle, an integral model is developed. In this model, particle-particle interaction force and the hydrodynamic stress on the powder agglomerates are taken into account. The interparticle force is mainly van der Waals force. The flow structure in the nozzle is solved via computational fluid dynamics. The agglomerate size distribution is simulated by this model. Simulation results show that this model can reasonably fit the experimental data.

Introduction

The focus of this chapter is on sorbent injection technologies using calcium-based sorbents for high-sulfur coal flue gas desulfurization (FGD). The goal is to provide research findings on transport and dispersion of sorbent powder and increasing sorbent utilization in a cost-effective fashion. This chapter aims to cover the fundamental aspects of powder technology relevant to the fine sorbent powders and to provide a means of improving sorbent performance through superior dispersion and reduced dispersed particle size.

Handling of fine powders can be very challenging because of the cohesive nature of the powders. The sizes of the dispersed particles in the furnace are often larger than the primary particle size because of agglomeration; the resulting effect is a reduction in the overall sulfur capture efficiency. At high temperatures, sintering becomes very pronounced and can also result in agglomeration. Powders with a tendency to agglomerate are likely to adhere to the walls of the transport systems, which can impede the flow. Typical attraction forces responsible for powder agglomeration are the van der Waals, electrostatic, and capillary forces. These forces are especially significant for fine powders such as calcium-based sorbents with sizes less than 50 mm. Therefore, a fundamental understanding of these attraction forces is essential from both reactivity and flow behavior viewpoints.

In the LIMB process, sorbent powders are stored in a feed silo before being injected into furnace. Interparticle cohesion forces lead to the formation of agglomerates. Proper dispersion of these agglomerates in the feed jets is essential to provide a well-mixed gas-powder mixture for efficient SO_2 removal.

Three topics will be covered in this chapter: powder characterization, powder dispersion, and modeling of powder attrition and dispersion. In the powder characterization section, powder surface properties, the results of powder transport under the influence of attraction forces, the effects of additives and temperature and the powder mechanical properties are presented. The second section compares three types of nozzles in terms of the local solid concentration and velocity at the entrance region. A Particle Image Velocimetry (PIV) technique is developed to quantify these parameters. In the last section, a stochastic model is described which characterizes the attrition behavior of the particle during transport and a population balance model is developed to account for the powder dispersion behavior.

Powder Characterization

General Theory of Interparticle Forces

There are three possible interparticle forces acting between powder particles: van der Waals or molecular forces, electrostatic forces, and capillary forces. These forces are a function of the material and the environment. A brief description of the general theory of these three forces is presented below.

VAN DER WAALS FORCES. The van der Waals force is electromagnetic in origin. It is an attraction force between particles due to the dynamic polarization of the interacting atoms. The first description of these forces [1] was based on the microscopic case of two single interacting atoms.

One important conclusion in London's development [1] was that the interaction between several molecules is an additive superposition of single forces between pairs. Hamaker [2], using the assumption of the additive nature of these forces, developed an expression for the interaction between macroscopic bodies. For a sphere and flat plate geometry, the van der Waals force is

Equation 4.1

$$F_{VDW} = AR/6d^2$$

where A is the Hamaker constant equal to $\pi^2 q^2 B$, d is the distance of separation of the two atoms and R is the sphere radius. For two interacting spheres of radii R_1 and R_2, the force is

Equation 4.2

$$F_{VDW} = (AR_1R_2)/(12d^2(R_1+R_2))$$

For two flat surfaces in contact, the van der Waals force is given as a force per unit area of contact or contact pressure. The force of contact will be determined by the area of the smaller contact surface. This pressure is given by

Equation 4.3

$$P_{VDW} = A/6\pi d^3$$

For materials in contact, such as in 'flocs or agglomerates, z is commonly approximated as 4Å. Therefore, comparing F_{VDW} at contact for similar geometries, say a sphere of known radius on a flat plate, the only unknown in Equation 4.1 is the Hamaker constant, a function of the material. For similar contacting geometries, comparisons between Hamaker's constants alone should indicate the relative magnitude of F_{VDW} between different materials.

Hamaker's constant, A, is determined from the macroscopic development (modern theory) by Lifshitz [3], based on dielectric spectra of interacting materials. This approach provides the most accurate method to determine A. The modern theory equations require the complete dielectric spectra of the material but high accuracy can be obtained using only the refractive indices in the visible region and the static dielectric constant. This approach is described in previous studies [3-6] and is used in this study.

The modern theory equation for Hamaker constant for materials 1 and 2 separated by a vacuum or gas is

Equation 4.4

$$A = \frac{3kT}{2} \sum_{n=0}^{\infty} \sum_{s-1}^{\infty} \frac{(\Delta_1 \Delta_2)^s}{s^3}$$

where

Equation 4.5

$$\Delta_j = \frac{\varepsilon_j(i\xi_n) - 1}{\varepsilon_j(i\xi_n) + 1}$$

and

Equation 4.6

$$\xi_n = n\left[\frac{2\pi kT}{\eta}\right]$$

For identical solid materials interacting in a vacuum or gas, Equation 4.4 reduces to

Equation 4.7

$$A = = \frac{3kT}{2} \sum_{n=0}^{\infty} \sum_{s-1}^{\infty} \frac{(\Delta^{2s})}{s^3}$$

In these equations, k is the Boltzman's constant, T is temperature, η is Planck's constant, ξ is frequency, and ε is dielectric constant at complex frequency $i\xi$. The only terms that are unknown in these equations are C_{UV} and ω_{UV}. The data required to calculate these terms are the three indices of refraction, η, in the visible frequency range. The visible range must be used because the Cauchy plot is linear in this region. The Cauchy plot is defined by plotting [$\eta 2(\omega) - 1$] versus [$\eta^2(\omega) - 1$] ω^2 where $\eta(\omega)$ is the index of refraction at frequency ω. From the resulting straight line, the slope is $1/\omega^2_{UV}$ and the intercept is C_{UV}.

For this work, the indices of refraction were determined experimentally by McCrone Research Institute in Chicago for the four sorbent materials investigated.

Once C_{UV} and ω_{UV} have been determined from the refractive index data, ε ($i\xi$) can be determined from the Equation 4.8 as Equation 4.6

Sorbent Transport and Dispersion

Equation 4.8

$$\varepsilon(i) = 1 + \frac{C_{UV}}{1 + \left(\frac{\xi}{\omega_{UV}}\right)^2}$$

If the value of the Hamaker constant is desired for a solid made up of a mixture of components, ε ($i\xi$) can be determined using the Clausius-Mosetti representation [6]

Equation 4.9

$$\frac{\varepsilon(i\xi) - 1}{\varepsilon(i\xi) + 2} = \Phi\left[\frac{\varepsilon_1(i\xi) - 1}{\varepsilon_1(i\xi) + 2}\right] + (1 - \Phi)\left[\frac{\varepsilon_2(i\xi) - 1}{\varepsilon_2(i\xi) + 2}\right]$$

ELECTROSTATIC FORCES. Handling and transport of fine dielectric powders such as calcium-based sorbents involves extensive contact and friction between the powder and the surfaces of the handling and the transport apparatuses. As a result, the powder becomes electrically charged. This type of electric charging is known as triboelectrification. In addition to the explosion hazard problem, the presence of charges in gas-solid systems can significantly alter the dynamic behavior of the powder and the hydrodynamics of the system. In addition, electrostatic forces can give rise to powder adhesion and agglomeration.

The attraction force between a point charge, q, and its image charge, -q, in a dielectric medium such as the air can be obtained by Coulomb's law.

Equation 4.10

$$F = \frac{qq^1}{16\pi\varepsilon_0 d^2}$$

where q and q^1 are the amounts of charges on the particles and ε_0 and ε_r represent the dielectric constants of vacuum and medium, respectively [7]. If a sphere is in contact with a plane surface, d is two times larger than the sphere diameter. In the absence of external fields, the attraction force in a vacuum between a charged particle and an adjacent uncharged particle with a fully developed image charge is given by:

Equation 4.11

$$F = \frac{q^2}{16\pi\varepsilon_0 d^2}$$

Since the above equation is only valid under the assumption of a point charge on the particle and a plane metal surface of infinite extent, the image charge on a particle is smaller than q and the attraction force is reduced to [8]

Equation 4.12

$$F^1 = \frac{q^2\left(1 - \frac{d}{\sqrt{R^2 + d^2}}\right)}{16\pi\varepsilon_0 R^2}$$

In practice, it is generally agreed that van der Waals forces are the dominating forces at particle-particle contact and so are responsible for preserving agglomerates once they form. Electrostatic forces, on the other hand, play an important role in bringing particles together since they act over a longer range than van der Waals forces.

CAPILLARY FORCES. Capillary forces can act at high humidities (> 65%) to bind particles together. This force is due to the capillary condensation of liquid in the void between two particles in close contact. For two smooth, spherical particles of radius R and a fluid of surface tension γ, the capillary force is given by

Equation 4.13

$$F_c = 2\pi\gamma R$$

If capillary forces are present, they will act along with the van der Waals forces, increasing attraction. However, at the high humidities required for capillary forces, electrostatic forces will not be a factor.

The different adhesion forces described will contribute in varying magnitudes to the behavior of powder transport. Quantification of the contribution is difficult due to the complex nature of these forces. In general, the van der Waals forces dominate over electrostatic for all particle diameters shown. Note that all forces dominate over particle weight except at very large particle sizes.

Sorbent Transport and Dispersion

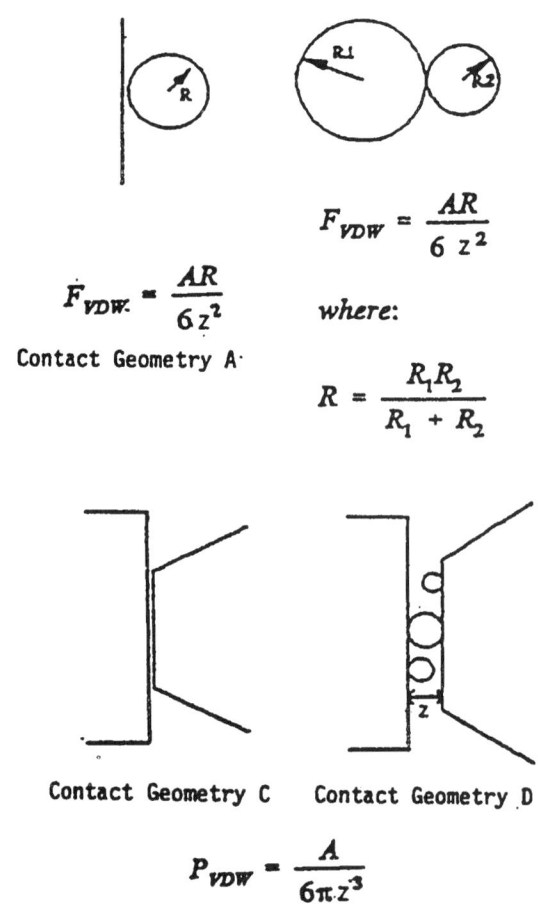

Figure 4.1 Sorbent contact geometries.

MODELING OF SORBENT CONTACT GEOMETRIES. There are four types (A, B, C, and D) of modeling of sorbent contact geometries. Type A models small, approximately spherical sorbent particles ($\approx \leq 1\mu m$) attached to flat surfaces of larger particles (see Figure 4.1) and accounts for approximately 6% of the total mass. The van der Waals force for this contact geometry can be modeled by Equation 4.1. In this model, the sorbent types are all carbonates since they are in their original crystalline form (e.g., calcite and dolomite). Type B is for sphere-sphere contact geometry (Figure 4.1). Spheres may be the same or different radius. The van der Waals force for this contact geometry can be modeled by Equation 4.2 where R is the reduced radius for different sized spheres. Sorbent types in this model are small (≤ 1μm) particle-particle contact for carbonates and all hydrates (including lignolime) for the entire size range. Surfaces of even the large hydrate particles have rounded protrusions and are best represented by a sphere-sphere contact geometry. Type C describes two flat regions in contact (see Figure 4.1). Van der Waals force for this contact geometry can be modeled by Equation 4.3 multiplied by the contacting surface area. Type D is for two flat regions separated by small particles (see Figure 4.1). Van der Waals force for this contact geometry can be modeled by Equation 4.3 as in Type C except that the separation distance will equal the largest diameter of the separating particles. Sorbent types in both model C and D are all carbonates.

Results of Powder Characterization

PHYSICAL PROPERTIES AND PARTICLE MORPHOLOGY. Table 4.1 lists the chemical compositions, the Hamaker constants and the dielectric constants, of four commonly used sorbents: calcite, dolomite, hydrated lime, and dolomitic hydrate. The static dielectric constants of the sorbent materials are used to give an indication of the magnitudes of the van der Waals forces and the electrostatic forces at the air breakdown field strength. Values of the Hamaker constant are determined from optical data; i.e., refractive index as a function of frequency in the visible range, for each sorbent. Since the average particle size is small for the dolomitic hydrate, Equation 4.9 is used for the two components: hydrated lime and periclase. The values for hydrated lime and dolomitic hydrate, which are not available in the literature are obtained by taking capacitance measurements on pressed powder pellets.

Scanning electron microscope (SEM) micrographs of the four sorbent materials are shown in Figure 4.2. The calcite and dolomite have more angular morphologies than the hydrates. The fine particles of the carbonates (D ≤ 1 μm) have a tendency to cohere to the flat surfaces of the larger particles as shown in figures 4.2(a) and 4.2(b). The hydrates, shown in Figures 4.2(c) and 4.2(d), have more rounded morphologies due to their processing from limestone. The individual particles are made up of smaller particles fused together during the hydration process. The results are single particles with rounded surface protrusion or nodules.

Table 4.1 Material Properties of Sorbents

	Calcite	Dolomite	Hydrated Lime	Dolomitic Hydrate
Chemical Composition WT%				
$CaCO_3$	96.4	54.6-55.7	0.35-0.75	-
$MgCO_3$	3.1	44.6-45.8	--	-
$Ca(OH)_2$	-	-	97.0-98.0	63.10
SiO_2	-	0.2-0.3	0.35-0.48	0.80
Al_2O_3	0.33	0.06-0.1	0.15-0.25	0.20
Fe_2O_3	0.07	0.06-0.08	0.05-0.055	0.20
S	1.0	0.01-0.014	0.08-0.098	0.10
MgO	-	-	0.40-0.45	33.50
P (q/cm^3)	2.71	2.83	2.23	2.60
ε_r	8.20	7.34	13.82	18.64
λ (x10^{-20}) J	9.86	7.34	6.39	7.65

Sorbent Transport and Dispersion

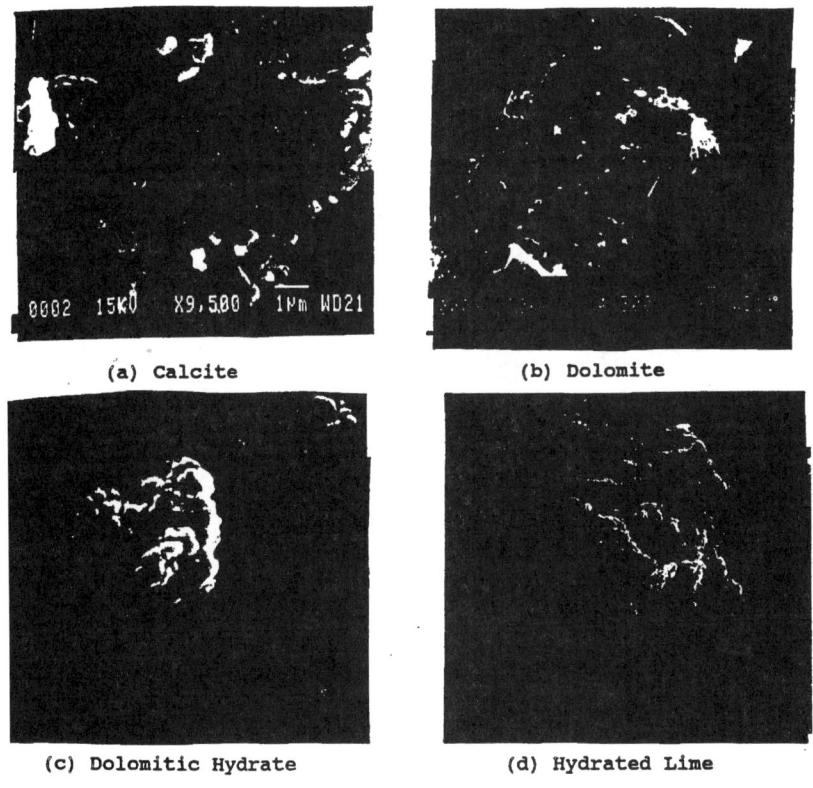

Figure 4.2 SEM micrographs of four sorbent powders.

The particle size distributions for the sorbent powders determined from the Sedigraph are shown in Figure 4.3. The results show that calcite and dolomite have comparable median particle sizes and that dolomite has a wider size distribution than the calcite. Comparing the Sedigraph results for the two hydrates shows that the dolomitic hydrate is composed of finer particles. Though of similar particle morphologies, the difference in densities is greater between these two sorbents than for the two carbonates. The effect of a greater density is to increase the aerodynamic diameter reported for a given size.

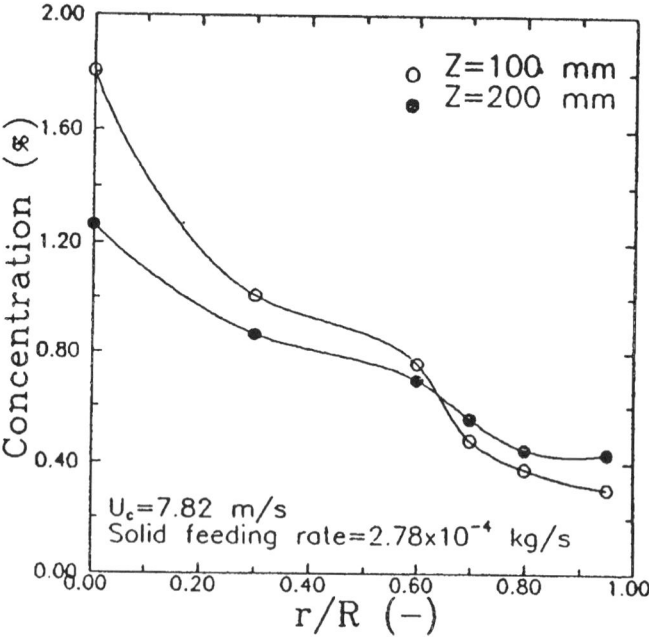

Figure 4.3 Sedigraph analyses of four sorbents.

CALCULATION OF HAMAKER CONSTANTS AND INTERPARTICLE FORCES. The static dielectric constant of each sorbent material is listed in Table 4.2. Table 4.2. lists the values obtained from the Cauchy plots of the parameters needed for the Hamaker constant determinations. Average values are listed for the uniaxial materials.

The Hamaker constant is calculated for each sorbent material using the values of the parameters in Table 4.2 combined with Equations 4.4, 4.5, 4.6 and 4.8. For the dolomitic hydrate, Equation 4.12 is solved using refractive indices data for each component, MgO and $Ca(OH)_2$, and the results are used in Equation 4.8 to generate $\varepsilon_1(i\xi)$ and $\varepsilon_2(i\xi)$. Knowing the volume fraction ϕ, Equation 4.9 is solved for an average $\varepsilon(i\xi)$ at each value of ξ.

Table 4.2. Values of Parameters used in the Hamaker Constant Calculation

Material	$\eta^2 (O)$	C_{UV}	$\omega_{UV} (\times 10^{16})$
Calcite			
ordinary ray	2.686	1.686	1.668
extraordinary ray	2.196	1.183	2.171
average	2.523	1.518	1.836
Dolomite			
ordinary ray	1.650	1.722	1.359
extraordinary ray	1.484	1.203	1.604
average	1.595	1.549	1.441
Hydrated Lime			
ordinary ray	2.332	1.332	1.577
extraordinary ray	2.346	1.345	1.461
average	2.337	1.336	1.538
Periclase	2.953	1.953	1.760

The calculated values of the Hamaker constant for the four sorbent materials are listed in Table 4.1. Calcite has the highest and hydrated lime has the lowest Hamaker constant among the four sorbent materials. Figure 4.4 shows theoretical values for three forces for hydrated time. The results show that the van der Waals forces dominate over both the electrostatic forces and the gravitational force. Figure 4.5 shows how the particle size distribution can result in greater van der Waals forces for a sorbent. This result indicates the sorbent material (Hamaker constant) and average sorbent contact geometry are both important effects to be considered in powder agglomeration.

POWDER AGGLOMERATION DURING TRANSPORT. Due to the small particle sizes in sorbent powders, the primary particles may exist in agglomerate form, reducing the powder's effectiveness for sulfur removal. Significant agglomeration can occur during powder transport and a relationship between the sorbent agglomeration tendency during transport and powder properties is desired. Powder properties of interest are particle morphology and relative magnitudes of interparticle forces between powders. This section investigates that relationship through experiments involving pneumatic transport of sorbent powders.

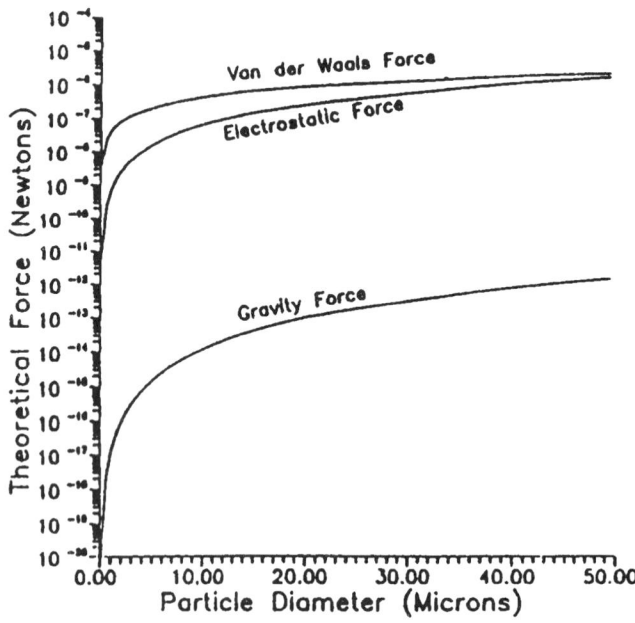

Figure 4.4 Theoretical force versus particle diameter for hydrated lime.

Sorbent Transport and Dispersion

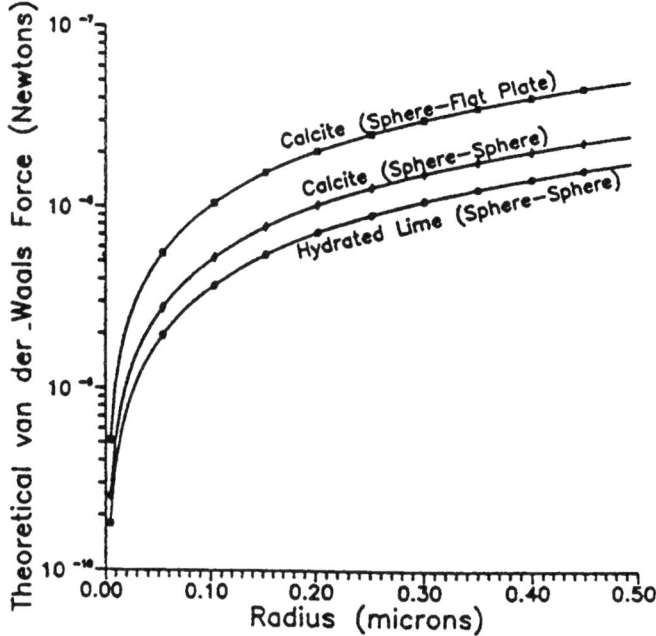

Figure 4.5 Comparison between calcite (sphere-flat plate) calcite (sphere-sphere) and hydrated lime (sphere-sphere) van der Waals forces.

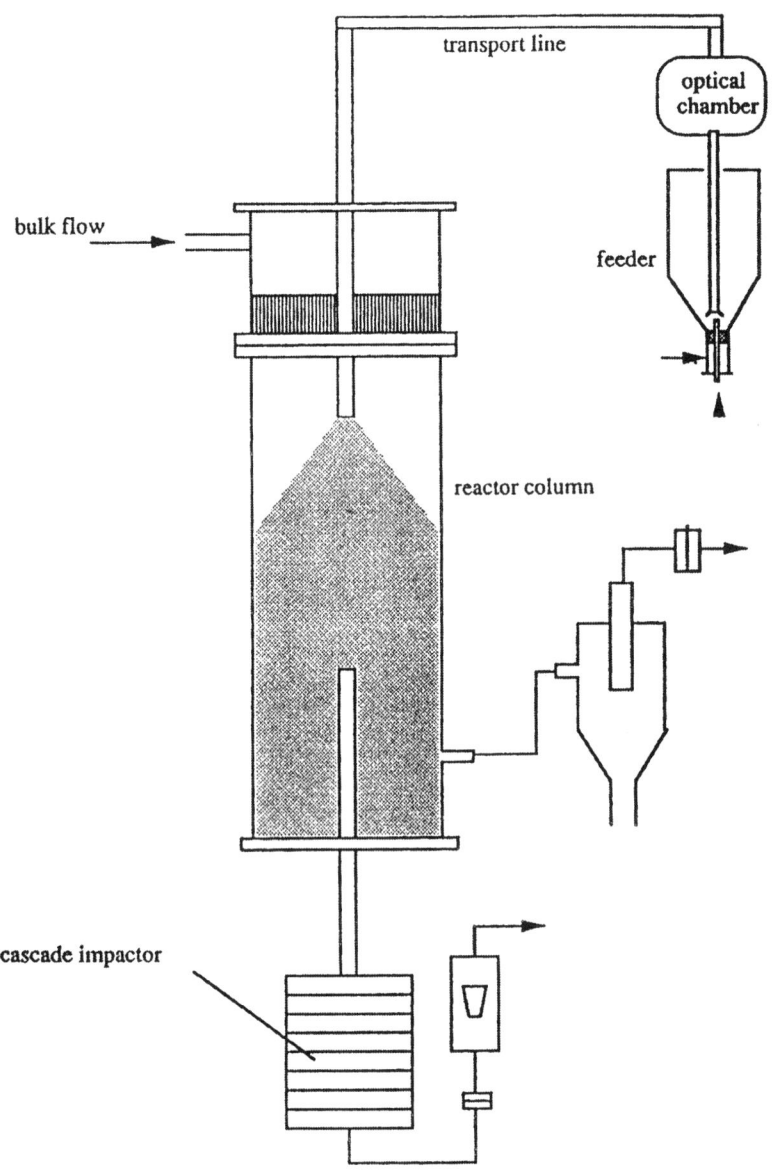

Figure 4.6 Experimental apparatus used in transport experiments.

Sorbent Transport and Dispersion

The experimental apparatus used in the transport experiments is shown in Figure 4.6. The powder is dispersed out of the feeder system (described below) into the transport tube. The gas used in the transport experiments is high-purity nitrogen from a cylinder. This gas is used to assure that moisture into the system is minimal and does not affect agglomeration tendencies.

As the powder leaves the transport tube, it is diluted with dry air in the column. This is required so that the powders can be sampled in the particle size analyzer isokinetically. Isokinetic sampling assures that the movement of particles into the sampling system does not affect the resulting size distribution. The particle size analyzer used was an Anderson Mark II cascade impactor.

The feeder system consists of a plexiglass column with a cylindrical top and a conical bottom. Inserted into the bottom region is a 1.2 cm ID tube fitted with a porous disc. A capillary tube located at the center allows jet flow into the feeder. Gas is fed into the annular region surrounding the capillary tube and flows up through the porous plate. The flow of gas through the porous plate loosens or locally fluidizes the powder in the immediate vicinity. Pulsing of the annular flow with a solenoid has been found to provide for better particle loosening, especially for the sticky hydrated powders. The off-take tube is attached to a vibrator to help with particle loosening.

Based on the low mass of powder fed per pulse ($\approx$1.5 mg/pulse, 30 pulses/minute) and the high transport velocities, the mode of transport is assumed to be dilute solid-gas pneumatic transport.

The pre- and post-transport size distributions for four sorbent materials are shown in Figure 4.7. The distributions presented are the average of three runs with approximately the same mass of solid transported per run. Also, since the powder flow is not continuous in this system, a solids flow rate is not reported. Particle deposition on the tube wall was observed for all sorbent materials. The transport tube was cleaned after every run to assure the same probability of adhesion between runs.

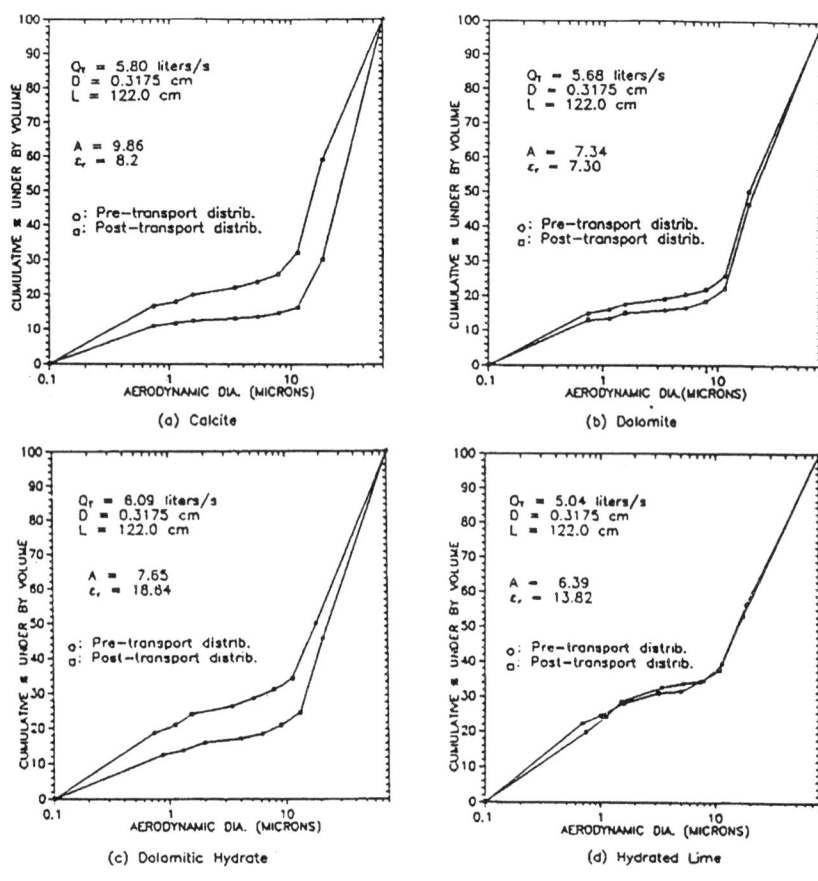

Figure 4.7 Comparison of pre- and post- transport volume distributions.

Comparing the calcite and dolomite results in Figure 4.7, the calcite shows a greater difference between size distributions for pre- and post-transport. Like calcite, dolomitic hydrate also shows a significant loss of fine particles with transport. The hydrated lime shows essentially no loss of fines above 1μm, and a small loss of fines below 1μm. Comparing the hydrates, the dolomitic hydrate has a much greater coarsening of size with transport than the hydrated lime.

One other factor that should be mentioned is the effect of inertia on the dispersibility of the sorbents. Greater particle inertia may also play a role in the slightly

Sorbent Transport and Dispersion

higher dispersibility of the hydrated lime compared to the dolomitic hydrate since inertia varies with the cube of the particle radius. For the carbonates, the effect of inertia should be slight when comparing dispersion results since the densities and particle sizes are not significantly different.

Comparing all sorbent materials, it appears the sorbents with the lower average van der Waals forces have greater dispersibility. In addition to the relative magnitudes of the Hamaker constants, differences in contact geometries are expected to have a significant effect on the relative van der Waals forces. The presence of nodules on the hydrates, as seen in the SEM, results in a reduction of van der Waals forces and an increase in the dispersion. A greater percentage of fine particles for the dolomite, as determined by the Sedigraph, may be a factor in its increased dispersion. The influence of electrostatic forces on dispersion did not seem significant.

As immediately obvious from the results, the van der Waals forces alone cannot explain the observed differences between pre- and post-transport behavior. The dolomitic hydrate shows as much coarsening with transport as the calcite, though van der Waals forces are lower as indicated by the Hamaker coefficient and contact geometry. Differences in median diameters are no longer a factor with dispersion since all powders show a median size between 10 and 20 μm prior to transport. When density effects are taken into account (since the results show aerodynamic diameters) even smaller differences in median diameters are observed. This unexpected high level of agglomeration/adhesion observed with the dolomitic hydrate may be due to an increased role of electrostatic forces during the transport process. However, considering the higher efficiency of transport of the hydrated lime compared with the calcite, though having higher electrostatic forces, may indicate that van der Waals forces may also play a significant role in transport processes.

The van der Waals forces for transport now include not only the particle-particle cohesive interactions but also particle-wall adhesive interactions. A particle more tightly held to the tube wall with greater van der Waals forces will have a smaller probability of being reintroduced into the flow stream. The relative effect of the Hamaker constants and contact geometries on the van der Waals forces will remain unchanged for the case of adherence to the tube wall.

The importance of van der Waals forces to transport processes is also reflected in the comparison in the results between the calcite and dolomite. Electrostatic forces are assumed not to be a factor in the observed results based on the similarities in the pre-transport average median diameters and static dielectric constants for the carbonates. Random effects are expected to average out. Therefore, the observed difference appears to be due to the difference in the van der Waals forces between the carbonates.

HIGH TEMPERATURE EFFECTS. The high-temperature reactor consists of two concentric alumina tubes (5.08 cm and 7.62 cm OD respectively) which are housed within a 3-zone, 1500°C, Lindberg vertical furnace. Furnace heating is controlled for each zone separately by a control console. The reactant gas enters from the bottom into the annular region between the two alumina tubes. While traveling upwards through a heated length of 91.4 cm (36 inches), it is heated up to the desired reaction temperature, and at the top makes a 120° turn to enter into the inner tube as two hot jets. Pulses of sorbent particles are introduced into the reactor by the injection probe, whose tip is located just above the incoming jets. Thus the sorbent stream from the injection probe is impacted by the hot reactant gas jets.

Figure 4.8 compares the size distributions of the black river hydrate exiting the furnace at room temperature and high temperature (825°C). The results indicate even more agglomeration and/or loss of fines at high temperatures. This result may be due to the effect of calcination of the hydrate on the van der Waals forces. As noted, the Hamaker constant for lime is double in magnitude the value for hydrated lime. With higher van der Waals forces, the probability of agglomeration after contact between particles would increase. Also, the van der Waals forces acting between the wall material and particle would increase with increasing particle Hamaker constant. The increase in these forces may explain the observed increase in particle size with transport.

Figure 4.9 shows the results for the 1.3% Lignosulfonated BRH at low (30°C) and high temperature conditions (839°C). The effect of temperature on the size distribution is not as significant as with the unmodified hydrate. The lignosulfonate molecules adsorbed on the surface of the solid particles appear to act to keep the particles from agglomerating at both high and low temperature. The deagglomerating effect of the lignosulfonate molecules overshadows the increase in van der Waals forces at high temperatures (above the calcination temperature).

LIGNOSULFONATE-MODIFIED HYDRATED LIME POWDERS. Studies have shown that calcium hydrates precipitated in the presence of lignosulfonate salts have 15-20% more reactivity to sulfation compared to unmodified hydrates [9,10]. Results show an optimum powder reactivity at approximately 1.5 wt% calcium lignosulfonate. The increased reactivity is thought to be due to a decrease in the primary particle size combined with a decrease in the tendency to agglomerate.

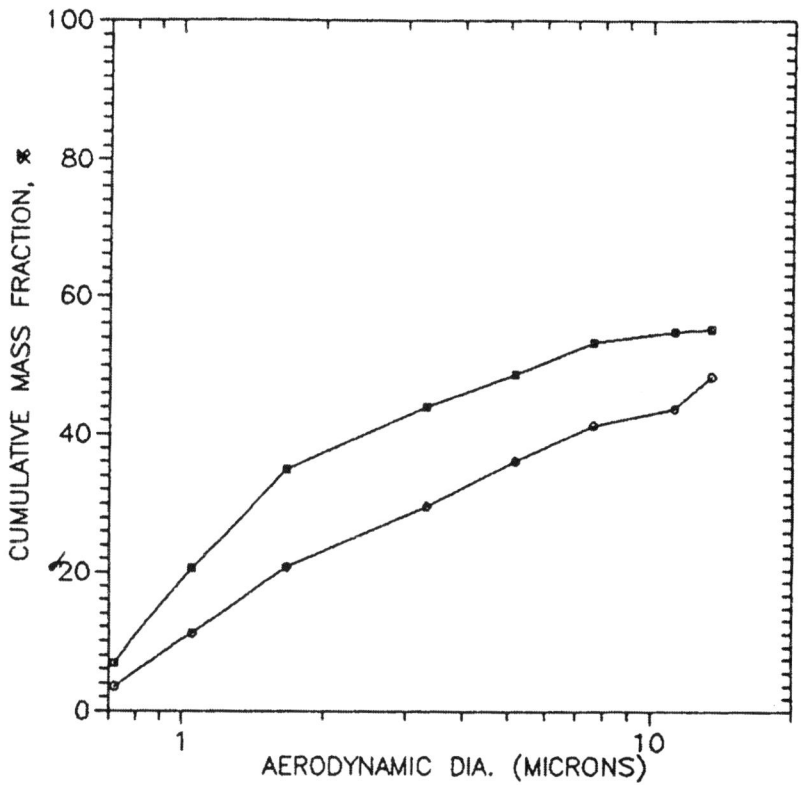

Figure 4.8 Comparison of Black River Hydrate size distributions from the furnace at 21°C (~) and 825°C (O).

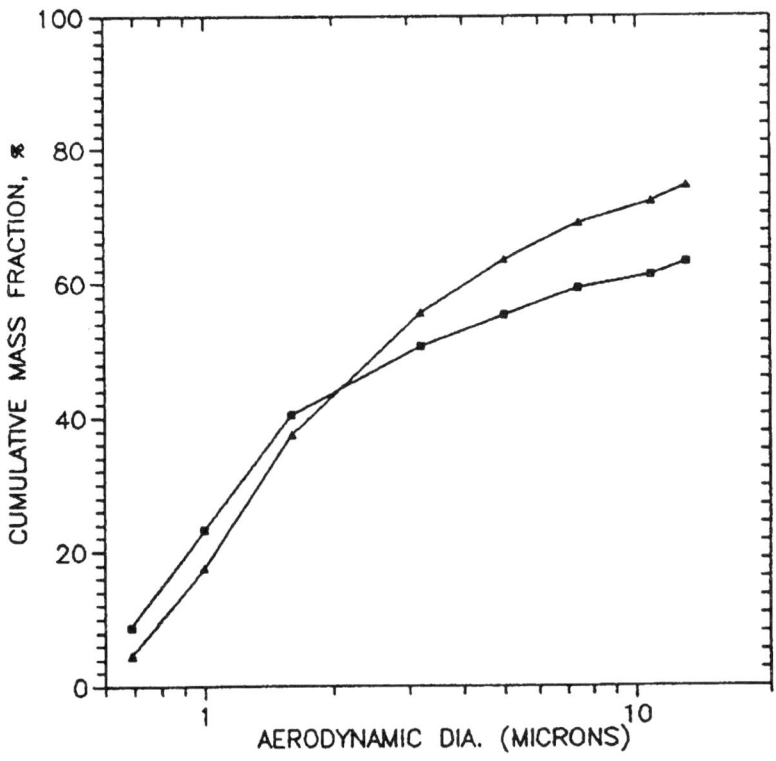

Figure 4.9 Comparison of 1.3% lignosulfonated hydrate size distributions from the furnace at 30°C (■) and 839°C (~).

Lignosulfonate salts are ionic dispersing agents, having ionic groups distributed over the entire surfactant molecule and hydrophobic groups containing polarizable structures such as aromatic rings and ether linkages [11]. The dispersing action of lignosulfonate can be due to a number of factors. The molecules adsorbed to the solid surface may cause the formation of electrical barriers to aggregation due to the repulsive force felt between the ionic charges of similar sign on adsorbed molecules.

Calcium hydrates are made in the laboratory using 2 grams of reagent-grade (high purity) CaO and 20 ml distilled water per sample. This volume of water gives 30 times the theoretical amount needed for complete hydration. In making the hydrates, the CaO powder is placed in a beaker and distilled water in the temperature range 70-80°C is added. The beaker is placed on a hot plate to boil off the excess water. After most of the excess water has evaporated, the beaker is placed in a vacuum oven at 50°C for 24 hours. Finally, the dried cake is lightly ground in a mortar and pestle.

The calcium lignosulfonate used in these studies is made by Pfaltz and Bauer, Inc. and contains 80% calcium lignosulfonate, 9.2% sugars, and 1% insolubles. For the modified hydrates, the necessary mass of calcium lignosulfonate to equal the final product mass percent is dissolved in the water of hydration prior to adding it to the CaO. The mass percentages of calcium lignosulfonate in the final product investigated are 0.0%, 0.38%, 0.75%, 1.13%, 1.5%, 3.7% and 7.1%. A portion of the powder samples up to 1.5 wt% are filtered after the drying process with distilled water at 70-80°C and re-dried.

Differences in the morphology of the dried powders are examined under a SEM. The powders are sprinkled onto double-sided conductive tape and pressed down to ensure good electrical contact. The samples are placed on aluminum stubs and gold coated. Surface areas of the resulting hydrates are measured using the nitrogen adsorption/desorption BET method.

The reactivities of the hydrates are tested at room temperature using the carbonation reaction. To conduct the experiments, the powdered hydrates are placed in petri dishes and set in an evacuated chamber at room temperature. Carbon dioxide from a gas cylinder is first bubbled through distilled water before entering the chamber. The pressure in the chamber is adjusted to one atmosphere. The hydrates are exposed to the CO_2 and moisture for 12 hours. The extent of carbonation of the hydrates is measured using thermogravimetric analysis (TGA) with a Perkin-Elmer TGA 7 instrument.

CaO powders have a NaCl crystal structure and are cubic in shape under the SEM. Originally, the average CaO particle size of the powder used in these experiments is about 7x7 μm. After hydration, micrographs of the pure $Ca(OH)_2$ product; e.g., as shown in figure 10, indicate a much finer crystal size compared to its precursor and a loss of particle distinctness due to agglomeration of the fine crystals. For a given amount of mass, a high nucleation rate will result in a large number of small crystals while a low nucleation rate will generate fewer crystals larger in size. By comparing the sizes between pure CaO and $Ca(OH)_2$, it is reasonable to conclude that the hydration reaction must involve a large degree of nucleation and low growth of the $Ca(OH)_2$ crystals.

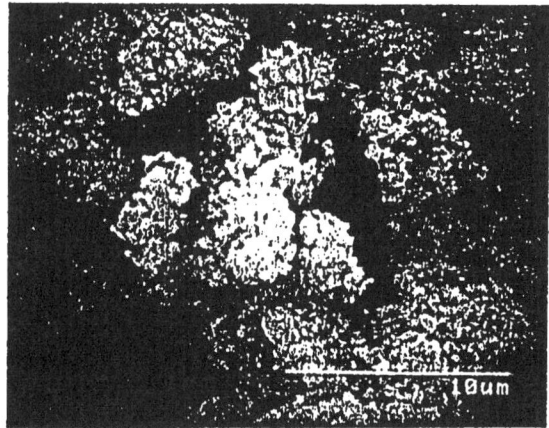

Figure 4.10 SEM micrograph of pure, as-hydrated Ca(OH)$_2$ sample. The hydrate is composed of numerous ill-defined solid crystals.

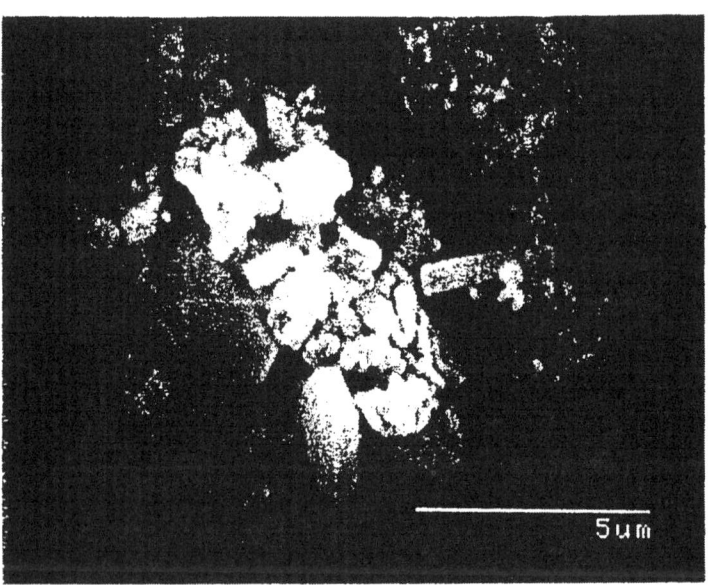

Figure 4.11 SEM micrograph of 0.38 wt% lignosulfonate-modified sample. Large, hexagonal single crystals are clearly observed.

Figure 4.11 shows the formation of hexagonal $Ca(OH)_2$ plate-like single crystals in the 0.38 wt% lignosulfonate-modified hydrate. These single hexagonal crystals are observed in all modified hydrate samples. In addition to the normal nucleation and growth process as in the pure $Ca(OH)_2$ sample, the lignosulfonate-modified samples seem to have a second mechanism which generates the hexagonal single crystals. It is well known that the presence of additives in a precipitating solution will alter the crystal morphology or the habit of crystal growth [12, 13]. As reported by Smith and Alexander [14], adsorbed impurities can lower the transport from solution to the solid by (1) reducing the area of exposed crystal, (2) generating electric fields which repel ionic species, or (3) affecting the growth steps on the solid crystal surfaces. In addition, polymeric molecules have relatively long chains which can adsorb onto a solid surface at more than one site, increasing adsorption stability.

The following explanation of the observed SEM results for the lignosulfonate-modified hydrates is proposed. In the early stages of hydration, the water content is high, the degree of supersaturation is low and, therefore, the nucleation rate is low. Moreover, the presence of lignosulfonate molecules in the solution may interfere with the nucleation process. The relatively large plate-like hexagonal single crystals seen under the SEM in all modified samples indicate preferential crystal growth in the a-axes directions of the developing single crystals and, at the same time, suggest both a retarded growth in the c-axes directions and an overall low nucleation rate. The plate-like structure of the hexagonal single crystals indicates that the basal planes of the developing crystallites may have been "poisoned." In other words, the lignosulfonate molecules preferentially adsorb to the basal planes of the developing single crystals thereby retarding growth in the c-axes directions. As water evaporates, the degree of super-saturation increases significantly which results in the formation of the tiny $Ca(OH)_2$ crystallites. The final product is therefore a mixture of large well-defined plate-like hexagonal single crystals and many small crystals.

SEM observations also indicate an increase in the number of hexagonal single crystals with an increase in the wt% lignosulfonate. This effect is probably the result of a larger amount of lignosulfonate available for developing crystallites at higher lignosulfonate weight percentages.

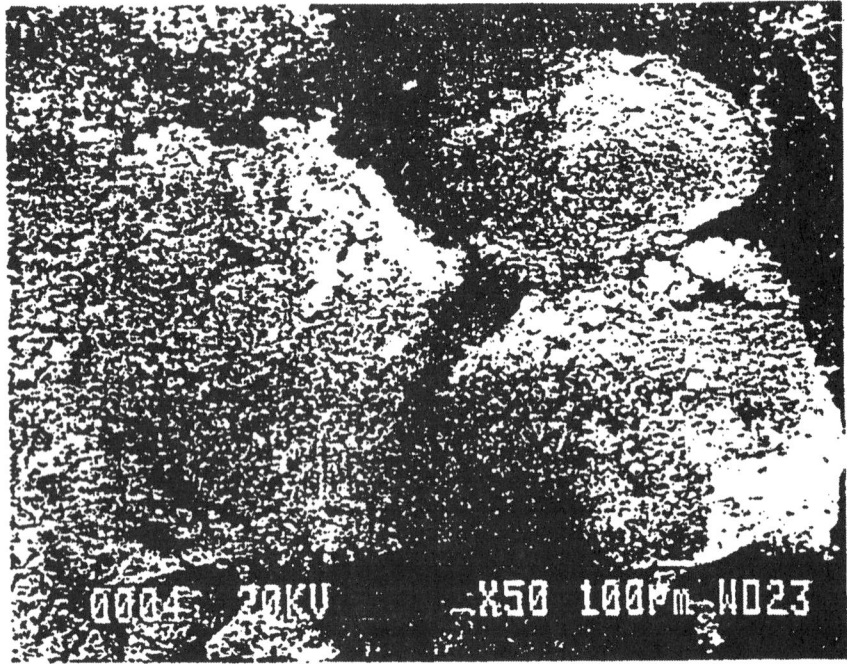

Figure 4.12 SEM micrograph of pure hydrate. The ill-defined, tiny crystals appear to form large agglomerates after drying.

Figure 4.12 is a low magnification micrograph of the pure $Ca(OH)_2$ powder showing an example of the severe agglomeration noted in this study. When modified by lignosulfonate, agglomeration is still observed at all percentages but to a lesser extent. Figure 4.13 shows a typical agglomerate of a modified hydrate. Note that the agglomerate size is smaller compared with the pure hydrate and that the pressing action of sample preparation broke the agglomerate apart. This breakage indicates that the modified hydrates are less cohesive.

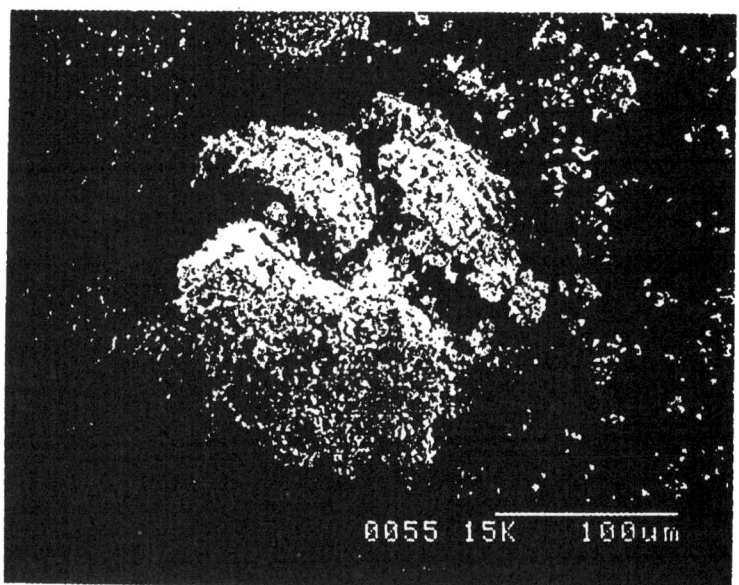

Figure 4.13 SEM micrograph of 1.5 wt% lignosulfonate-modified sample. This breakage of agglomerates is not observed in the pure, unmodified hydrate sample.

When the lignosulfonate-modified hydrate samples are washed with boiling distilled water and re-dried, large, cohesive agglomerates are observed as in the pure hydrate sample. This behavior indicates that the adsorption is reversible at these conditions and that the presence of lignosulfonate reduces agglomeration in the dried solid as well as changes the crystallization process in solution.

Results of the surface area analyses and carbonation experiments are plotted in Figure 4.14 as a function of wt% lignosulfonate. The surface area steadily increases with increasing wt% lignosulfonate for unfiltered hydrates up to 1.5%. The surface area decreases for wt% lignosulfonate greater than 1.5%. There appears to be a critical concentration below which the surface area increases with% lignosulfonate. The increase in surface area is thought to be due to the presence of lignosulfonate molecules in the solution. The large molecules sterically hinder the close approach of other crystallites in the solution that would otherwise agglomerate as seen in the pure hydrate sample. Another factor that may play a role is the electrostatic repulsion between lignosulfonate molecules adsorbed on the separate crystals acting to prevent agglomeration. The decrease in surface area at the higher lignosulfonate concentrations is probably due to the

Dry Scrubbing Technologies for Flue Gas Desulfurization

Sorbent Transport and Dispersion

combination of a lower nucleation rate caused by the increasing interference of the lignosulfonate molecules and the formation of a greater number of hexagonal single crystals.

The filtered modified hydrates show a decrease in surface area from the unfiltered samples. This observation supports the proposal that the samples re-agglomerate to some extent after filtering and re-drying. Interestingly, the 0.38 wt% sample shows an even smaller surface area than the pure (0%) hydrate after filtering. This result indicates even more severe agglomeration occurred in this sample than the pure sample.

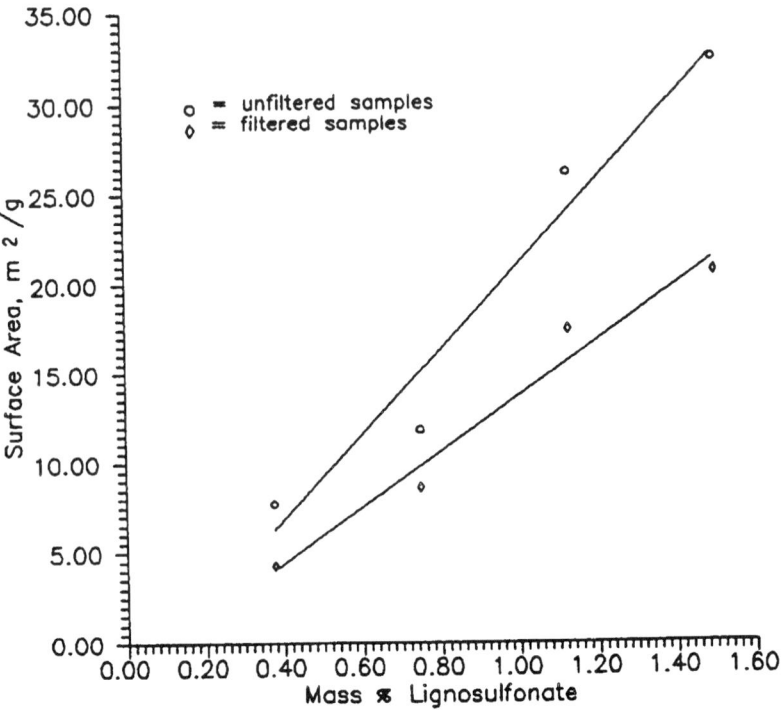

Figure 4.14 Surface area versus mass% lignosulfonate added to final product hydrate.

Figure 4.15 shows both the changes in surface area and conversion as a function of the% lignosulfonate in the product hydrate. This graph indicates the direct relationship between these two factors.

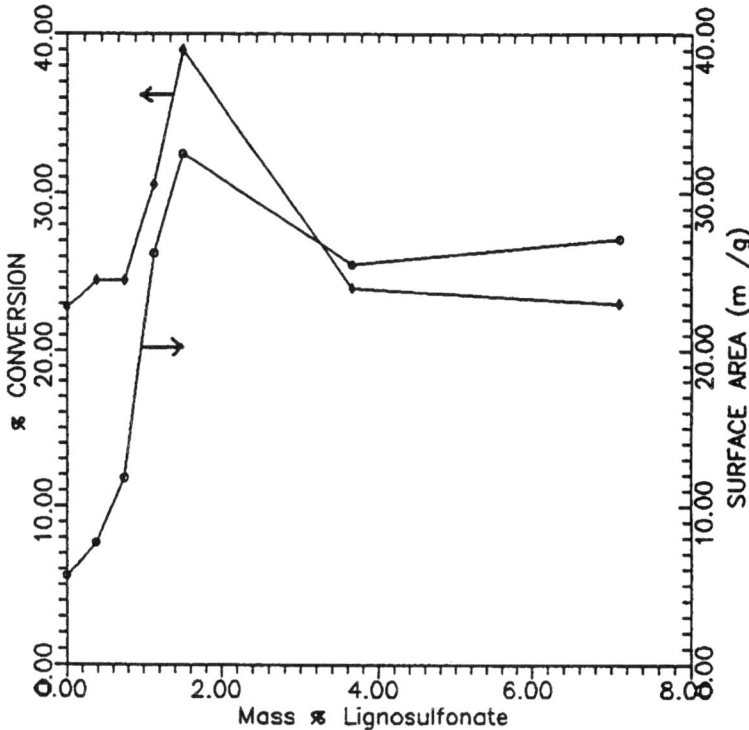

Figure 4.15 % conversion and surface area versus mass % lignosulfonate for the unfiltered hydrates.

CHARGING BEHAVIOR. This section is aimed at examining the effect of transport conditions and sorbent materials on the charge accumulation on calcium-based sorbent powder during transport.

Figure 4.16 shows the schematic of the experimental setup, which can be divided into two parts: the transport system and the charge measurement apparatus. For the transport system, the sorbent powders are transported by a Plexiglas feeder with powder flow coming from the side of the feeder. The tube for powder injection is placed horizontally and powder is carried into this tube from a small hole at the bottom. The powder flow rate can be adjusted by changing the size of the hole. Fitted at the bottom of the feeder is a tube for providing an air jet to loosen the powder. A solenoid valve hooked to a pulse generator/timer is connected to the air jet line to provide intermittent flow in order to prevent the formation of the channel flow in the vicinity of the small hole.

Prior to conducting experiments, the powder is dried in a vacuum oven to eliminate the moisture factor. Moisture content is measured to assure the drying process was completed.

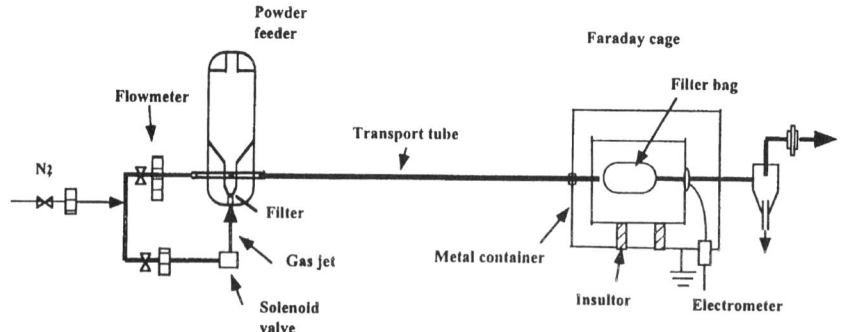

Figure 4.16 Schematic of the experimental setup.

The transport line was designed such that tubes made of different materials can be installed and tested. The transport tube is of 0.635 cm OD and 100 cm length. Three kinds of tube materials were tested: brass, carbon steel, and copper. Carbon steel was tested because it is one of the most commonly used materials in high temperature processes. High-purity nitrogen was used as a carrier gas to eliminate the humidity effect while studying the electrostatic properties of different sorbents. The gas flow controlled by needle valves ranges from 25 to 45 SCFH.

The charge measurement apparatus works as a Faraday cage. It used to measure the net charges carried on powders. The Faraday cage is connected to the transport tube through a short copper tube. The Faraday cage essentially consists of two conductive containers. The dimensions of the inner and outer containers are 9 cm diameter by 30 cm long and 20 cm diameter by 40 cm long, respectively. The inner container is connected to Keithley electrometer by a copper wire and is isolated from the outer container by highly insulating materials, (PTFE). The distance between the inner and outer container is long enough to ensure there is no decay of charge. It is important to obtain the charge/mass ratio; therefore, a filter bag is installed inside the inner container to retain the particles and allow the gas to escape. Since the Sedigraph analysis for powder size distribution obtained in previous studies showed that approximately 95% of the sorbent powder is larger than 1 γm, a filter with porosity of 1 γm is used. The filter bag can be removed from the Faraday cage and weighed after particles are collected.

A thorough examination of various effects on the charging tendency and the charge polarity has been carried out for the black river hydrate. Effects investigated include powder mass flow rate, gas flow rate, and tube materials.

Because of the cohesive nature of the powder and the configuration of the feeder, it is difficult to deliver powder at a steady flow rate; therefore, the data presented are values averaged over several sets of experimental results. Error bars are provided to indicate the deviations. Figure 4.17 shows the variation of the charge-to-mass ratio with the powder mass flow rate for a brass tube at a gas flow velocity of 16.8 m/s. The top and bottom sets of data are experimental values without and with the transport tube, respectively; the one in the middle represents the calculated values of charges attributed to the transport tube. The hydrated lime is found to be charged negatively and the charge-to-mass ratio is in the range of -10^{-5} to -10^{-4} C/kg. The absolute values of charge-to-mass for hydrated lime obtained in this study seem to be slightly lower than that for micron sized particles in air reported by Soo [15], which were about 10^{-4} C/kg. Such a discrepancy can possibly be attributed to the lower Reynolds number and the number of contacts between particles and the wall in this study.

Sorbent Transport and Dispersion

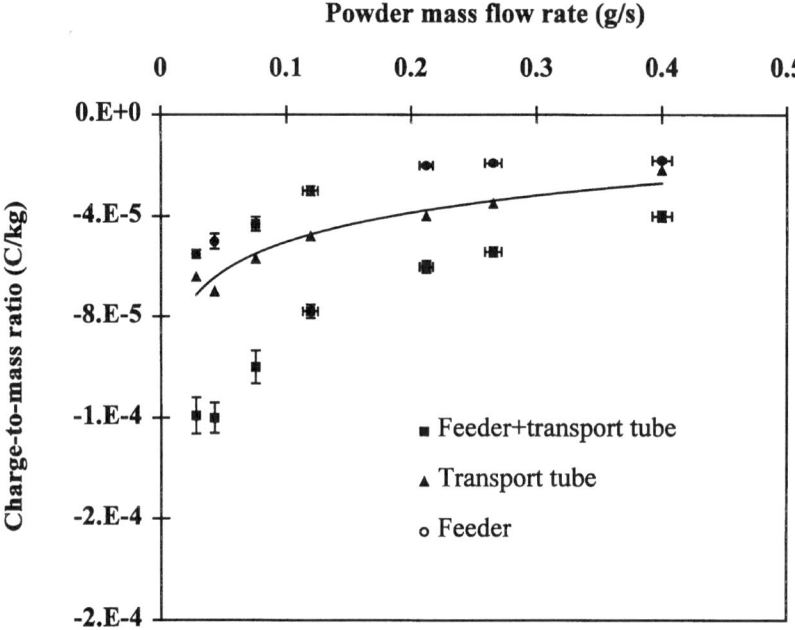

Figure 4.17 Variation of charge-to-mass ratio with powder mass flow rate for hydrated lime.

There has been evidence showing that the charging of polymers is closely associated with their chemical nature [41,42]. For example, basic materials are more likely to be charged positively and polymers with OH side-groups also have the tendency to lose electrons. This functionality relationship, however, cannot be used to explain the negative polarity of the basic hydrated lime. It is also observed that the charge-to-mass ratio decreases with increasing mass flow rate, and it decreases at a faster rate at low powder mass flow rates. Kittaka [43] reported similar charging behavior; i.e., an increase in the initial rate of charge with decreasing solids loading. Nifuku et al. [44] and Gajewski [45] also reported comparable results for coal dust and polystyrene particles, respectively. This tendency may be explained for dense gas-solid flows because of a decreasing collision frequency between individual particles and the wall surface at higher powder mass flow rates. As the powder mass flow rate increases, the particle-particle collisions become more significant, deflecting particles from their paths to collide with the wall. This explanation, however, can not be applied to dilute suspensions such as the system of the present study because of the negligible particle-particle interactions. One possible explanation is that an increase in powder mass flow rate (or solids loading) results in a reduction in turbulent intensity, which in turn reduces the collision frequency of particles with the wall [46].

Assuming that the particles travel at the same velocity as the gas, the solids loading can be obtained from

Equation 4.14

$$\phi = \frac{w}{(\bar{\mu}\pi d^2)/4}$$

where ϕ is the solids loading, w is the powder mass flow rate, $\bar{\mu}$ is the mean gas velocity, and d is the pipe diameter. It is noted that, in the theoretical formulation of the electrostatic charging during transport, the effect of the solids loading has always been neglected. Both Cole et al. [47] and Musuda et al. [48] assumed that the collision frequency per unit area of the pipe is proportional to the solids loading. Musuda et al. [48] proposed the following equation:

Equation 4.15

$$I_m = \frac{\gamma \bar{\mu}^a w}{\bar{d}_p}$$

where I_m is the current generated along a certain length of pipe, γ and α are constants, and $\bar{d}_p$ is the mean particle size. They reported that the value of α is around 1.9. The charge-to-mass ratio, q/m_p, can be obtained from Equation 4.15 as

Equation 4.16

$$\frac{q}{m_p} = \frac{(\int I_m dt)/(w\Delta t/m_p)}{m_p} = \gamma \bar{\mu}^a / \bar{d}_p$$

It can be seen that the solids loading was not accounted for in their prediction for charge-to-mass ratio.

Cheng and Soo [49] indicated that the charge acquired by a particle in turbulent flow is independent of the powder mass flow rate and they attributed the observed decreasing charge-to-mass ratio with increasing powder mass flow rate to changes in wall deposition. The equation for predicting charge accumulation takes the form of $q_{12} \alpha d_p^2 v^{0.6}$ where q_{12} is the charge transferred per particle and v is the particle turbulent intensity. Their theoretical model is also based on the assumption that the total charge is proportional to the square of the particle diameter.

Figure 18 shows the charge-to-mass ratios of the hydrated lime at four different gas velocities. The symbols represent the experimental data. As expected, the charge-to-mass ratio increases with increasing gas velocity. It is found that the absolute values of the charge-to-mass ratio is more dependent on the change of the gas velocity than the powder mass flow rate. Using a linear regression, the charge-to-mass ratio can be expressed as a function of powder mass flow rate and gas velocity as follows:

Equation 4.17

$$\frac{q}{m_p} = -1.4 \times 10^{-6} w^{-0.29} \bar{u}^{0.98}$$

Equation 4.17 has a correlation coefficient, r^2, value of 0.92. The solid lines in Figure 4.18 represent fitted curves. It should be noted that the curves on all the other figures except Figure 4.18 are used to indicate only the trend of the data. It is found that q/m_p is proportional to the 0.98 power of the gas velocity, which is slightly higher than that proposed in Cheng and Soo [49] assuming that the turbulent intensity is proportional to the average gas velocity. Nifuku et al. [44] reported that a maximum charge is reached at a certain gas velocity and beyond that the charge accumulation decreases with further increasing gas velocity. Kittaka et al. [43] proposed a similar form of correlation in which the initial charge-to-mass ratio is proportional to the -1.07 power mass flow rate.

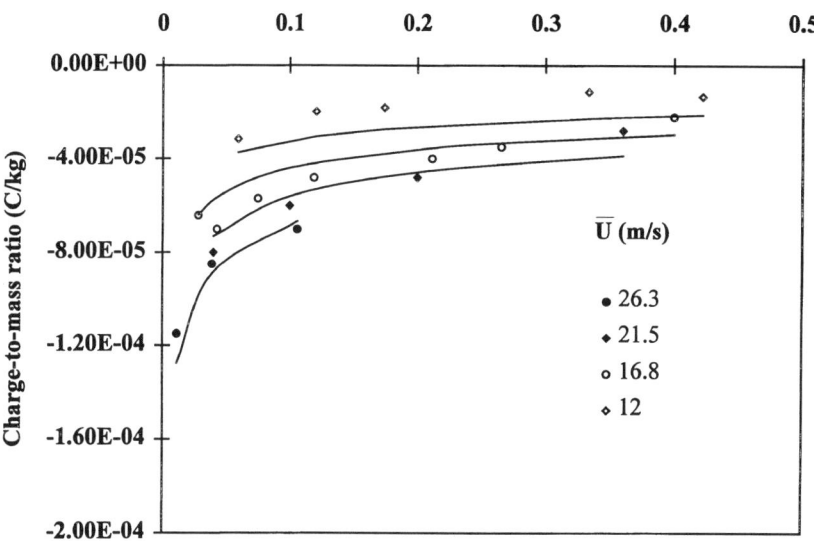

Figure 4.18 Effect of gas velocity on the charge-to-mass ratio for hydrated lime.

For gas-solid pipe flows, the maximum charges a particle can acquire are in general governed by two criteria which are based on Gauss's law: neither the electric field generated from a particle surface nor that from the pipe wall can exceed the breakdown field strength of the air. Whichever criterion is met first determines the maximum value of the charges on powders. The former gives a maximum charge-to-mass ratio of

Equation 4.18

$$\left(\frac{q}{m_p}\right)_{max} = \frac{3\varepsilon_0 E_B}{r_p \rho_p}$$

where ε_0 (=8.854 • 10^{-12} F/m) is the permittivity of the free space, E_B (=3 • 10^6 V/m) represents the breakdown field strength of the air, r_p is the particle radius, and ρ_p is the particle density. The latter sets a limit on the charge-to-mass ratio of

Equation 4.19

$$\left(\frac{q}{m_p}\right)_{max} = \frac{2\varepsilon_0 E_B}{\Phi R}$$

where R is the pipe radius. As the particle size decreases, this upper bound on the charge-to-mass ratio becomes lower than that based on the maximum field strength on a single particle surface. It seems that although the charge-to-mass ratios of the hydrated lime powder are lower for high solids loadings, the corresponding highest sustainable amount of charges on a powder particle is also reduced, which indicates an increase in the possibility of discharging. To characterize the effect of powder mass flow rate on discharging, one can examine the electric field on the pipe surface, E, given as

Equation 4.20

$$E = \frac{R(q/m_p)\Phi}{2\varepsilon_0}$$

Substituting Equation 4.17 into the above equation gives

Equation 4.21

$$E = \frac{(-5.02 \times 10^{-7})R^{0.42}\bar{u}^{0.69}\Phi^{0.71}}{\varepsilon_0}$$

Figure 4.19 shows the comparison of the electric field strengths at the wall surface with the theoretical maximum values for various powder mass flow rates at a gas velocity of 16.8 m/s. It can be seen that as the powder mass flow rate increases, the absolute values of the field strength increases; however, they are two orders of magnitude lower than the breakdown field strength of the air. It is noted that the sharp points on the hydrated lime usually possess higher charge densities, which are more likely to reach the breakdown field strength of the air even though the field strength based on the average amount of charge on a particle is not high enough.

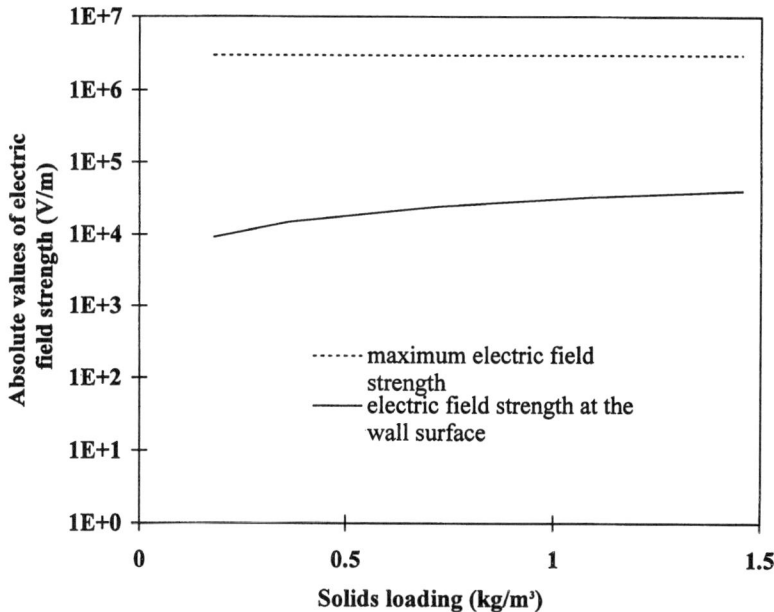

Figure 4.19 Electric field at the wall surface as a function of powder mass flow rate.

Sorbent Transport and Dispersion

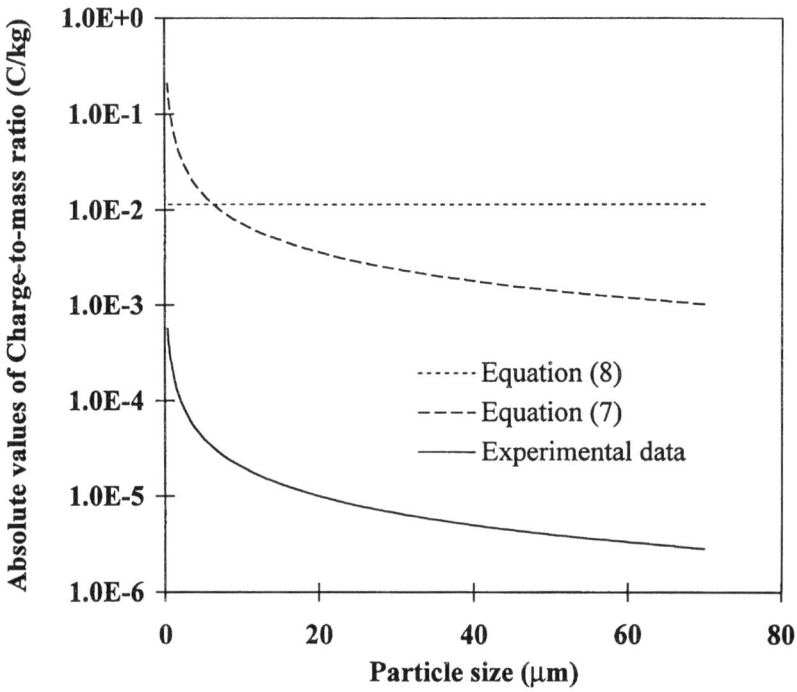

Figure 4.20 Comparison of experimental charge-to-mass ratio with maximum values based on the breakdown field strength of the air.

For a cloud of particles with various sizes, the average charge density might not be sufficient for gauging the possibility of discharging. It is therefore of interest to evaluate the charge density on individual particles. The total charges on powders per unit time are

Equation 4.22

$$\left(\frac{q}{m_p}\right)_{avg} w = \sum n_1 q_1 = \sum \left(\frac{w \alpha_{di}}{\rho_p \pi d_{pi}^3 / 6}\right) q_1$$

where n_i is the number of charges of particles of d_{pi} diameter, q_i is the amount of charges of particles of d_{pi} diameter, and α_{di} is the weight percentage of particles of d_{pi} diameter. Assuming that the charges on an individual particle are proportional to the square of the particle diameter ($q_i = \beta d_{pi}^2$), the above equation can be reduced to

Equation 4.23

$$\left(\frac{q}{m_p}\right)avg = \frac{\Sigma \alpha_{di} \beta d_{pi}^{-1}}{\rho_p \pi / 6}$$

With a further assumption that there is no change in the particle size during transport, the values of α_{di} are obtained from the Sedigraph results and the value of β is found to be $-2.3 \cdot 10^{-7}$ units for an average charge-to-mass ratio of $-8 \cdot 10^{-5}$ C/kg. The charge-to-mass ratios are found to vary from $-2.85 \cdot 10^{-6}$ to $-5.7 \cdot 10^{-4}$ C/kg for particles of 70 to 0.35 μm. As shown in Figure 4.20, the absolute values of the experimental charge-to-mass ratio for various particle sizes are also 2-3 orders of magnitude lower than the ones theoretical maximum values on the breakdown field strength of the air.

The charge measurements for all four transport tube materials were carried out at a gas velocity of 16.8 cm/s. As shown in Figure 4.21, copper, brass, and aluminum tubes charge the hydrated lime powder negatively, but the powder seems to lose or acquire only a small amount of charges during the contact with the stainless steel tube. The copper tube charges the hydrated lime the most followed by the brass, aluminum, and stainless steel tubes. The work function theory states that materials with higher work functions tend to acquire electrons, which leads to negative charges on the materials. Given the negative polarity of the powder, it can be concluded that the hydrated lime has a higher work function than all four metals, and the ascending order of the work function would be copper, brass, aluminum, stainless steel, and hydrated lime.

Figure 4.22 shows the UPS spectrum of Cu. The work function of the tube metals are calculated from the following equation:

Equation 4.24

$$\varphi = h\nu - W$$

where φ is the work function of the metal, $h\nu$ is the energy of the UV photon (=21.2 ev) and W is the width of the spectrum. The work functions of copper and brass are found to be about 4.3 ev and for aluminum and stainless steel to be about 4.35 ev and 4.25 ev, respectively. The order of work functions determined from UPS does not seem to agree with that predicted based on the charge-to-mass ratios. Data from the CRC handbook indicate that aluminum has a work

Sorbent Transport and Dispersion

function value lower than that of copper. One possible reason for this disagreement is that the work function theory cannot adequately describe the charging tendency of hydrated lime. It is also noted that the difference between the experimentally determined work functions for the metals is very small and is likely to be within the experimental error, thus making it difficult to determine the order of work functions of these metals. Further investigations using metals with significant differences in their work function can help solve this problem.

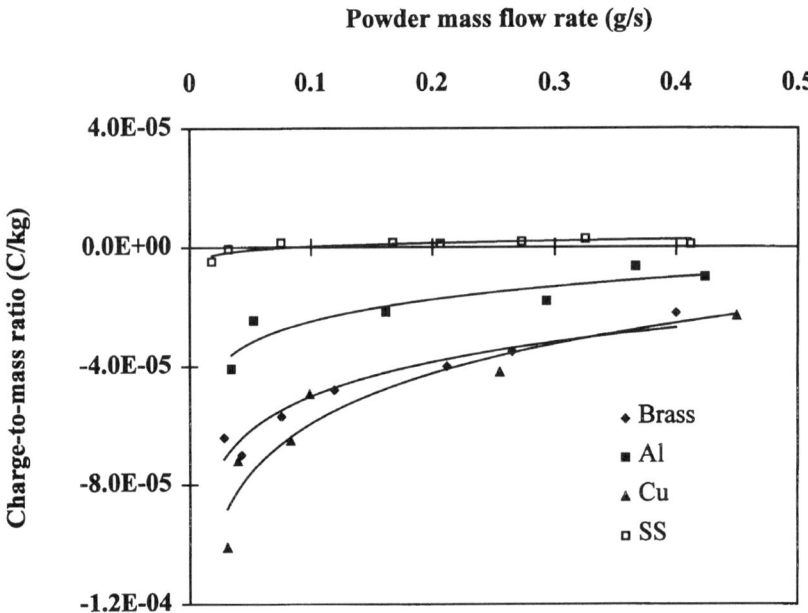

Figure 4.21 Effect of four tube materials on the charge-to-mass ratio for hydrated lime.

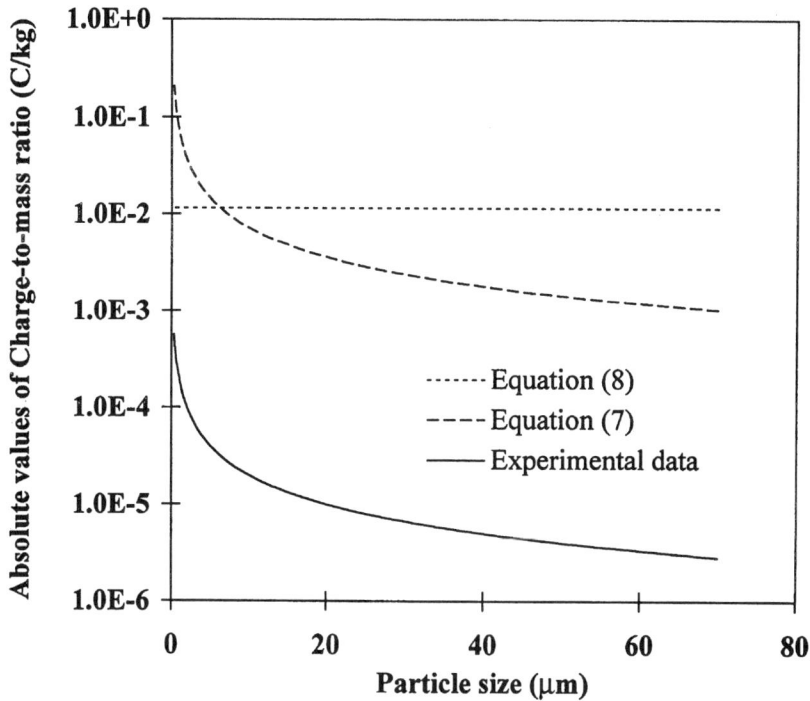

Figure 4.22 UPS spectrum of Cu at 25°C.

Figure 4.23 shows the charging curves at three low gas velocities of 4.3, 3.1 and 2.1 m/s in which the gas passes a fixed amount of powder which is placed in one end of the transport tube. Positive charges are observed for lower gas velocities. This appears to contradict the previous results. To help understand this charging behavior, similar experiments were carried out using a Plexiglas tube in order to visualize the movement of the particles. It was observed that at a gas velocity of 4.3 m/s the particles travel essentially as a huge particle through the tube. In this case, the primary charging source is the contact between the powder and the transport tube. Other possible charging sources are the contact among particles and the entrainment of the smaller particles in the gas stream from the powder layer deposited on the wall surface. For example, at a gas velocity of 2.1 m/s a number of powder dunes are observed on the bottom of the tube and it appears that most of the particles flowing into the tube are the ones which are entrained in the flow. For the former mechanism, the contact among particles generates

Sorbent Transport and Dispersion

equal but opposite charges, so no change in the total charges is observed. For the latter mechanism, only small particles are swirled up into the gas stream and the detected charges are attributed to these small particles. The sudden decrease and gradual increase in charge with time at velocities of 4.3 and 3.1 m/s seems to indicate that the polarity of the small particles is opposite to the predominant polarity of the particles. Based on the above results, it is suggested that the contact of friction between hydrated lime and the tube walls generates only negative charges. Although a number of smaller particles carry positive charges through either mechanism discussed above, most particles carry negative charges and therefore the net charges are negative.

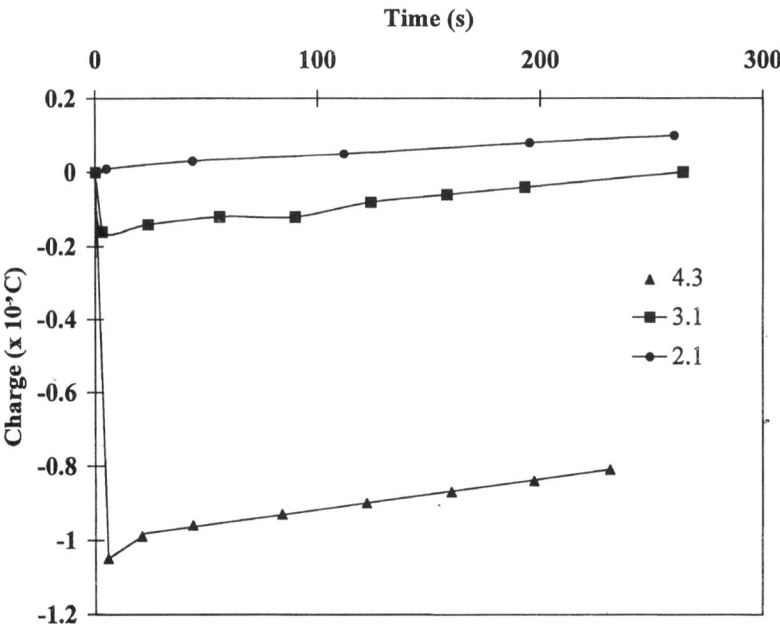

Figure 4.23 Variation of charge accumulation with time at three low gas velocities.

Mechanical Properties

Flow behavior of powders and bulk solid materials becomes a major parameter in designing plants which involve handling and transport of these particulate materials. Early designs of bulk handling and transport equipments such as bins, hoppers, feeders, and standpipes were based on structural and space economy. Unfortunately, following this design technique led to generations of equipment with poor performance. Arching, segregation, flooding, and inconsistent flow rate are examples of the problems that may disrupt the handling and transport processes when the flow properties of bulk solids are ignored. There are several techniques available in the literature for overall evaluation of the degree of flowability of bulk solids [19-22]. In these techniques, powder properties like the angle of repose, angle of internal friction, compressibility, dispersibility, and cohesion are evaluated and related to the flow behavior of these powders in handling and transport equipment. The flow properties of a particulate material are not constant parameters but they may vary widely depending on the conditions under which they are handled. Consolidating pressure and moisture content strongly affect these properties. A listing of these properties which may be found in standard handbooks for a given material is always misleading. The best way to calculate the flow properties of a particulate material is to test this material under the conditions that are specific for its transport and handling.

TESTING EQUIPMENT AND MEASURED PROPERTIES. The following powder testing instruments are used for the measurements of sorbents flow properties: (1) Ajax cohesion tester for the measurements of the cohesive strength of the sorbents; (2) Jenike translatory shear tester for the determination of the angle of internal friction based on measurements of the yield locus of the sorbents; (3) Hosokawa powder characterization tester for the measurements of angles of repose, spatula, fall, and difference, loose and packed bulk densities, compressibility and Hausner ratio, dispersibility and cohesion number, and degrees of flowability and floodability; and (4) moisture analyzer for the determination of the moisture content in the sorbents.

Tests performed on five types of calcium-based sorbents are included in the next section to show the testing procedures and the methodology of flow properties assessment. The examined sorbents are Delta calcium carbonate DCC and four modified hydrates with four mass percentages of lignosulfonate: BRH (0.0%), LH-109 (0.5%), LH-108 (1.0%), and CH-3-L (1.3%).

TESTING RESULTS. The loose (aerated) bulk density is the bulk density of the powder when it is as loose as it can be. The packed bulk density is the bulk density of the powder when it is packed. Results of the tests performed indicate that the bulk density of the carbonate, DCC, is much higher than that of the hydrates (more than double) which suggests that a higher gas or air velocity is required for transportation of the DCC.

Compressibility, C, of a powder is defined as

Equation 4.25

$$C = 100\,(P - A)\,/\,P\ (\%)$$

where P and A are the packed and aerated bulk densities of the powder, respectively. As the compressibility of a powder increases, its flowability decreases. Powders with compressibility more than 20% will not be "free-flowing." Powders with compressibility 40-50% are very hard to flow and have the tendency to block the flow and form bridges. The Hausner ratio is defined as the ratio between the packed and aerated bulk densities of the powder (i.e., P/A). Studies on the fluidization behavior of fine powders [23, 24] classify powders according to this ratio as follows: (1) Powders having a Hausner ratio < 1.25 are considered as group A powders (i.e., free-flowing easy-to-fluidize powders); (2) Powders having a Hausner ratio > 1.4 are considered as group C powders (i.e., cohesive difficult-to-fluidize powders); (3) Powders having the ratio in the range 1.25 to 1.4 may have some properties of both groups A and C.

Results shown in Figure 4.24 show that all powders tested are of group C (Hausner ratio > 1.4) and are very compressible (47.5 - 52.7%). Accordingly, a high degree of flow blocking must be expected. However, increasing the percentage of lignosulfonate slightly decreases the compressibility and Hausner ratio for the modified hydrates.

The angle of repose is the angle between the horizontal and the free surface of a powder poured freely on a horizontal surface from a given height. The angle of spatula is a rupture angle that is similar to the angle of internal friction of a powder. Powders with low angles of repose and spatula are more flowable. Powders with an angle of spatula less than 40 degrees are "free-flowing" powders. The angle of fall is the new angle of repose of a pile of powder which has been disturbed by a falling object. The angle of difference is the difference between the angle of repose and the angle of fall. A powder with low angle of fall is more likely to have an unsteady floatable flow. On the other hand, the greater the angle of difference of a powder is, the greater its potential for flooding or fluidization will be.

All tested powders show a low angle of fall and a high angle of difference. Hence, it is expected for these powders to have unstable and floatable flow. Also, increasing the percentage of lignosulfonate in the modified hydrates, increases the potential for flooding behavior, since it appears that the angle of fall for the modified hydrates decreases with increasing lignosulfonate percentage. Also, the angle of difference increases with increasing lignosulfonate percentage in the modified hydrate. This indicates that increasing the lignosulfonate percentage in the modified hydrate increases its tendency to be fluidized, yet uncontrolled flooding flow must be expected in this case. Results also show that both the angle of repose and the angle of spatula are considerably higher for all tested powders. Accordingly, poor flowability is expected for these powders.

Dispersibility of a powder is defined as the ratio of the weight of the dispersed portion of a sample of powder to the total weight of the sample when it is allowed to fall freely from a given height. Dispersibility is a measure for the potential of a powder to flood or be fluidized. The higher the dispersibility of a powder is, the more floatable it will be. The cohesion number is defined in this work as a percentage number rather than its conventional definition as the shear strength of a material under zero stress. To measure the dispersibility, a sample of 10 gm of the powder is permitted to fall freely through a glass cylinder from a height of 20 inches. The undispersed powder is collected in a glass of diameter 4 inches. Dispersibility of the powder is then calculated as

Equation 4.26

Dispersibility=10x(10-wt. of undispersed powder)(%)

For cohesion number measurement, three mesh screens (#60, #100, #200) are fitted on top of each other and mounted on the vibrating unit in the tester. These screens are chosen based on the average bulk density of tested powder. A sample of 10 gm of the powder is placed on the #60 mesh screen. The unit is then set to vibrate for a period of time, T, given by

Equation 4.27

$T = 20 + [(1.6 - W)/0.016]$

where T is the vibrating time and W is the working bulk density of the powder (W=(P-A)C+A). After the vibration is completed, the net weight of remaining powder on each screen is measured and the cohesion number is calculated as

Equation 4.28

Cohesion number=$10x(W_{60}+0.6W_{100}+0.2W_{200})$(%)

where W_{60}, W_{100}, and W_{200} are the net weight of the remaining powder on the 60, 100, and 200 mesh screens, respectively. From Figure 4.25, it can be concluded that the tested modified lignosulfonated hydrates are less cohesive than the BFH and the DCC. Also, the carbonate, DCC, has a very low dispersibility compared to the hydrate and modified hydrates. Results confirm that increasing the lignosulfonate percentage improves the dispersibility of the modified hydrates.

Sorbent Transport and Dispersion

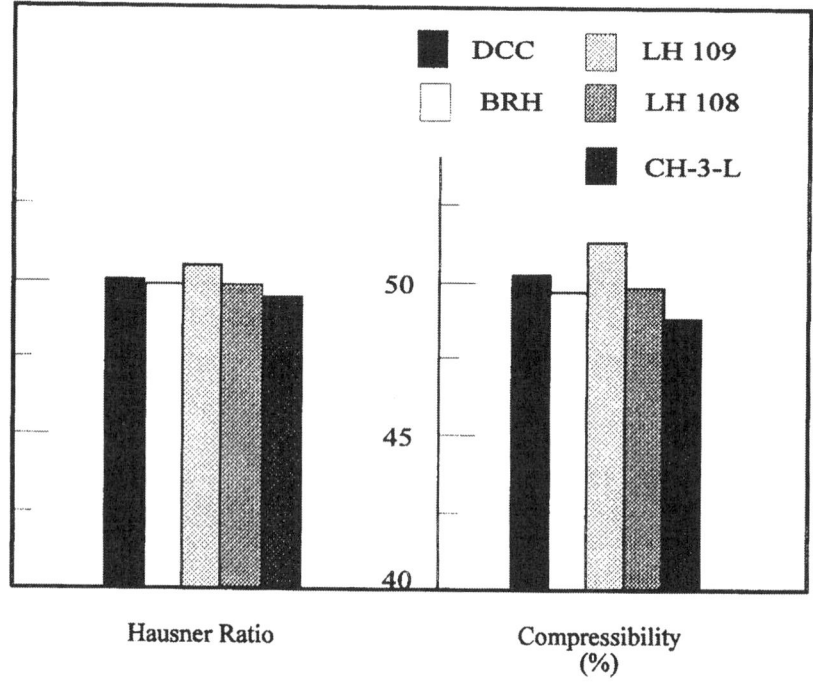

Figure 4.24 Hausner ratio and compressibility.

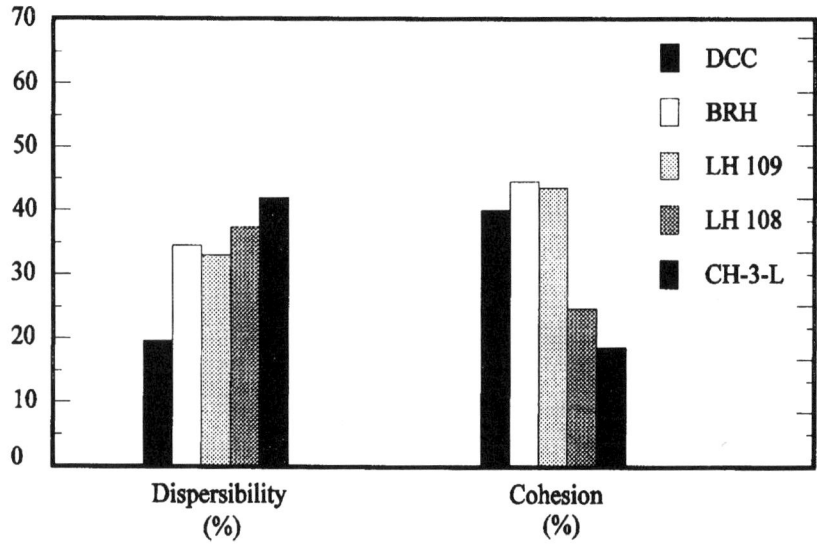

Figure 4.25 Dispersibility and cohesion number.

The flow properties may be classified into two categories: 1) properties which are measures of flowability (angle of repose, angle of spatula, compressibility, and cohesion number) and 2) properties which are measures of floodability (angle of fall, angle of difference, flowability, and dispersibility). According to Carr's evaluation [15], each of these properties is given an index number based on its value. Summation of these indices categorizes the corresponding degree of flowability or floodability of the powder over the range of 0 to 100. For example, a "free-flowing" material will have a flowability index (summation of corresponding flowability properties indices) of 90-100, while a material with "very bad" flowability will have a flowability index of 0-19.

All the tested powders have a "bad" degree of flowability (flowability index of 20-39). Also, all the tested powders have a "fairly high" degree of floodability (floodability index of 60-79), except for the CH-3-L which has a "very high" degree of floodability (floodability index of 80-100).

The angle of internal friction and cohesive strength defines the relation between the normal and shear stresses in powders at failure and flowing conditions. The Jenike shear tester is used to determine the combined shear and normal stresses required to cause failure along a plane within the sorbent mass. The relation between normal and shear stresses along the failure plan is given by

Equation 4.29

$$\tau = c + \sigma \tan \phi$$

where c is the cohesion, ϕ is the angle of internal friction, and σ and τ are the normal and shear stresses, respectively. The curve represented by this equation is called the yield locus or the Mohr-Coulomb failure envelope for the material.

The cohesive strength of a powder is the shearing stress in the powder when it is sheared under zero normal stresses. The capacity of powders to gain internal strength allows them to develop stable arches and resist gravity flow in the transport lines and handling facilities. The Ajax cohesion tester is used for the cohesive strength measurements.

For the tested modified lignosulfonated hydrates, sorbents with a higher lignosulfonate percentage have lower angles of friction. This agrees with the conclusion that increasing the lignosulfonate percentage enhances the flow properties of the modified hydrates. However, all tested sorbents have high angles of internal friction and gain more cohesive strength when compacted. Accordingly, dense phase transport is not recommended for these sorbents.

EFFECT OF MOISTURE. Tests were conducted on the Hosokawa tester to study the effect of moisture content on the flow properties of the sorbents. Moisture content is defined as the weight ratio of the water contained by a sample of the powder to the original weight of the sample. Ambient humidity during storage and transport affects the moisture content of sorbents [24]. Changes in the moisture content of powders affect most of their mechanical properties. In this work, the effect of moisture on the flow properties of sorbents was examined within the range of 0–6% moisture content.

The presence of moisture in the tested sorbents changes their flow properties. Cohesion and dispersibility are among the properties that are greatly affected by moisture. This is because of the water bonds which increase the amount of cohesive forces in the sorbent and cause the particles to agglomerate. The degrees of flowability and floodability of sorbents decrease with an increase in moisture content.

Powder Dispersion

Introduction

Powder interparticle cohesion forces lead to the formation of agglomerates. The problem in dispersing powders in a gas stream lies in the re-dispersion of these agglomerates. The mechanisms of powder deagglomeration in a gas stream are uncertain. Thomas [26] investigated the turbulent disruption of flocs in a liquid suspension and indicated that the pressure difference on opposite sides of the floc is the main mechanism leading to floc rupture under turbulent conditions. Assuming that an aggregate is spherical in shape, Bagster and Tomi [27] analyzed the stresses within a spherical particle in simple flow fields, specifically the stress arising from a simple uniform flow and an induced shear field. They found that the maximum stresses in a simple shear flow field were $\sigma\gamma_{lmax}=5\mu\gamma$ for tension and $\tau\gamma_{lmax}=8.5\mu\gamma$ for shear. However, their analysis was based on a viscous liquid. A gas-fine particle mixture is usually dispersed by forcing the mixture through a nozzle or wound tube. Masuda et al. [28] investigated the deagglomeration of metallic silicon and calcium carbonate in a mixer type dispersor, a fluidized bed, and a capillary pipe. Their results showed that in such a flow, agglomerates were disrupted mainly by acceleration or deceleration of powders. Kousaka et al. [29] explored the disruption mechanisms of aerosol agglomerates by analyzing the disruption forces. They claimed that impaction of powder on obstacles and acceleration or deceleration of the powder were effective ways for powder dispersion in an air stream. It appears that the key to powder dispersion, from a hydrodynamic viewpoint, is to organize the flow structure inside a powder disperser so that strong turbulence can be generated.

The purpose of this investigation is to first develop a non-intrusive, in situ Particle Image Velocimetry (PIV) technique for the measurement of local velocity, concentration, and size distribution of the powder agglomerates. Following this, the effect of different entrance geometries on powder dispersion is studied experimentally by this technique and theoretically by numerical simulation of the turbulent flow field inside the entrance. Finally, an optimum design is selected to test the effects of powder additives and the influence of operating conditions on the powder dispersion.

Experimental

POWDER DISPERSION SYSTEM. Figure 4.26 shows the schematic of the powder dispersion system. It consists of a powder feeder, a transport column, a powder separation sub-system, and a powder recirculation device. The entrance of the transport column is replaceable for the examination of entrance geometric effects on powder dispersion. The vertical transport column has two sections;

namely, the bottom section and the upper section. The bottom section is a transparent cylindrical column of 6 inches diameter. The entrance to this column is 2 inches in diameter and about 12 inches in length. The sorbent powder is fed into the system by a powder feeder through the entrance and the dispersion of the powder inside the entrance can be observed through a viewing window made of PYREX glass. The powder entrained by the gas stream is separated by a two-stage filter from the gas and then collected in the powder hopper. To determine the solids flow rate in the column, a calibration of the powder feeding rate is conducted against the revolution rate of the screw in the feeder for a given powder by a weighing method. The sorbent powders are stored in the feeder at least 24 hours before the experiments, to eliminate the electrostatic charge generated during the handling of the powder.

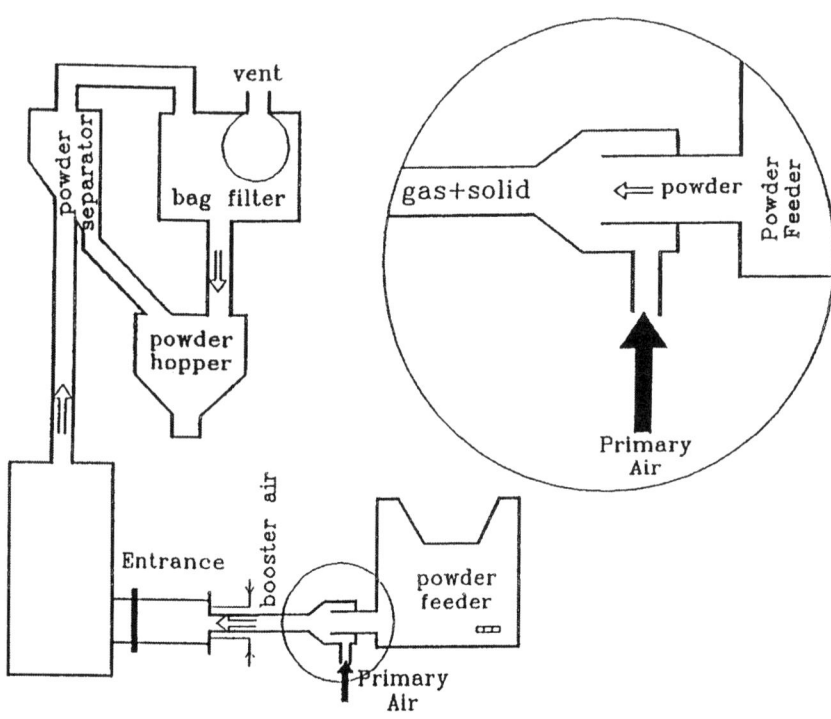

Figure 4.26 Schematic diagram of powder dispersion system.

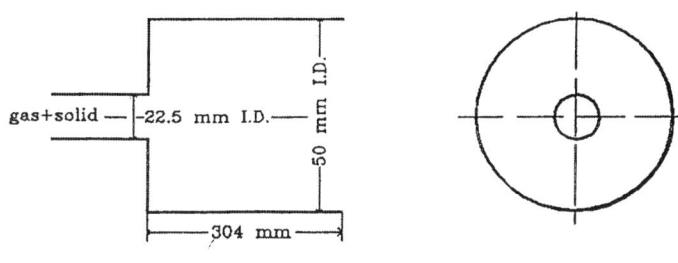

Expansion Nozzle

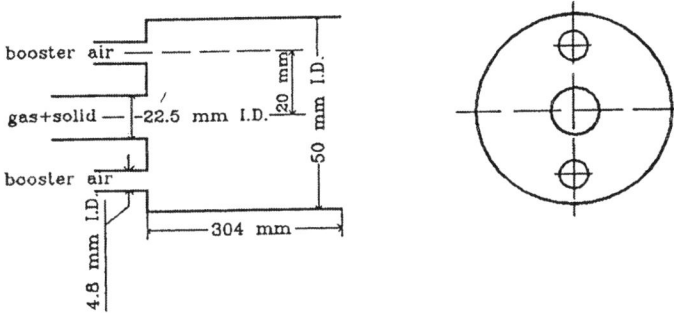

Expansion Nozzle with a Two-Jet Booster Air

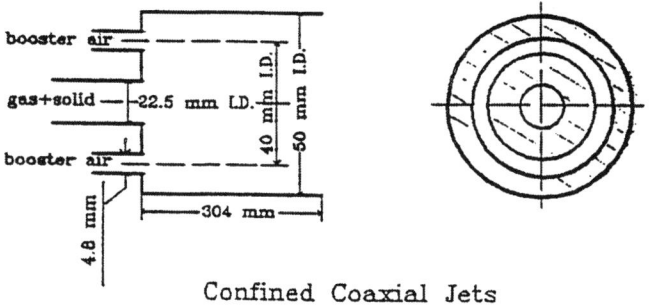

Confined Coaxial Jets

Figure 4.27 Detailed configuration of different powder injection nozzles.

Dry Scrubbing Technologies for Flue Gas Desulfurization

Sorbent Transport and Dispersion

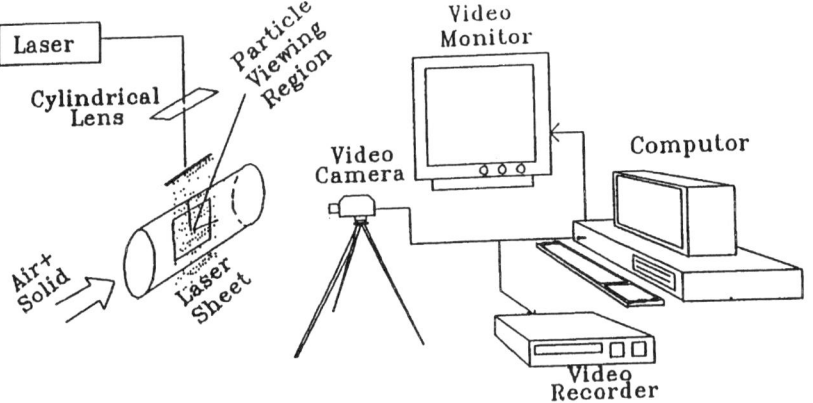

Figure 4.28 Schematic diagram of the powder dispersion visualization system.

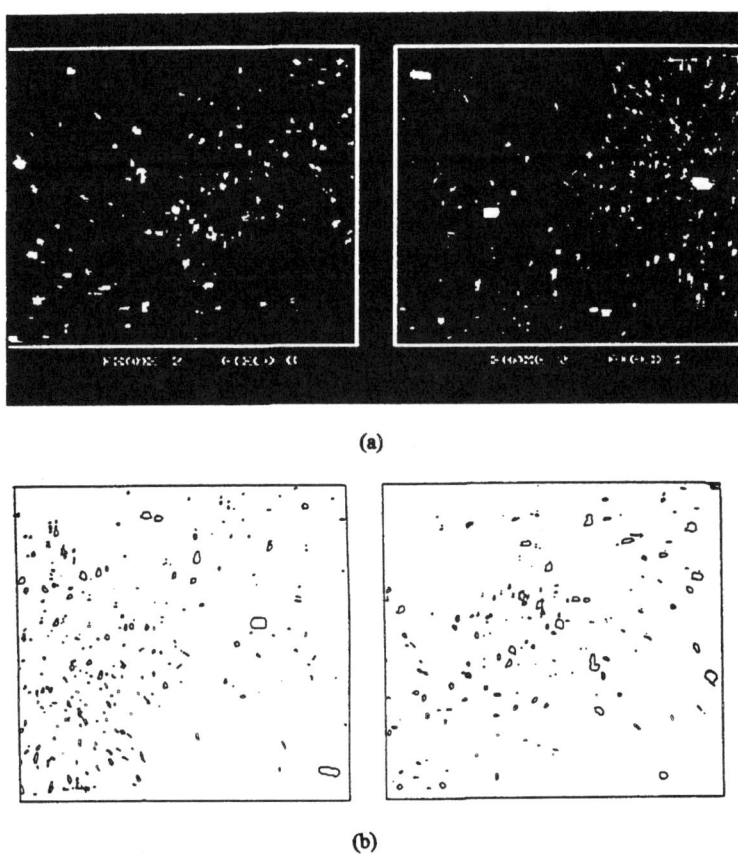

Figure 4.29 Samples of the original images and the processed images: (a) original images, (b) processed images.

In the LIMB process, the sorbent powder injector usually includes a primary jet and an annular coflowing jet of booster air. In this section, three types of transparent entrances, namely an expansion nozzle, an expansion nozzle with a two-jet booster air, and a confined coaxial jet nozzle, are constructed and tested. The detailed configurations and dimensions of various entrances used in the present study are shown in Figure 4.27. Different flow structures in these nozzles can be achieved by adjusting the flow rates of the primary jet and booster air independently.

Sorbent Transport and Dispersion

VISUALIZATION OF POWDER DISPERSION. Figure 4.28 shows the schematic diagram of a powder visualization system for the measurements of the local velocity, concentration, and size distribution of the powder agglomerates.

The flow field inside the nozzle is illuminated by a laser sheet. The laser sheet is generated by a Level 3500, 4 watt Argon Ion laser system and Flow View optical accessories, including a 10-foot long fiber optics cable, adapters and a cylindrical lens. The thickness of the laser sheet is about 2 mm. The continuous laser sheet is used in combination with a CCD camera which controls the exposure time by an adjustable shutter. The laser sheet is shifted to focus on various planes along the radial direction and the pictures of powder motion at various locations are recorded. Powder tracks or streaks are obtained from which powder velocities are calculated by dividing the length of the track by the time duration of exposure. For particles with high velocities, a strobe light of 3 µs exposure time is used instead of a laser sheet, because of the limitation of the shutter speed of the CCD camera. The frequency of the strobe light must be adjusted to match the camera shutter speed so that multiple exposures on one image are avoided. When the strobe light is used, the CCD camera is always set to the lowest shutter speed due to the low light intensity of the strobe. A TV lens is coupled with the CCD camera to picture the whole flow field inside the entrance. For a 10 µm particle moving at a velocity of 2 m/s, the exposure time should be shorter than 5 µs. The resolution of this system can be improved by means of the magnification of the image. Figure 28 also shows the camera system used for the powder particles/agglomerates size and shape measurement. A camera lens with a 5.5 cm focal length is coupled with an extension tube and a bellows is mounted between the extension tube and the CCD camera. This system can provide a magnification of 30X to 150X with a focal length of about 30 mm. Higher magnifications can be obtained by using longer extension tubes between the bellows and CCD camera. This pre-magnification provides a significant increase in the resolution for the powder sizing. However, the view area becomes small, in the range of 10 by 10 to 20 by 20 mm^2.

The camera used in this study is a high resolution 800 by 490 pixel Pulnix TM-745 frame transfer CCD array equipped with a synchronous variable electronic shutter ranging from 1/60 second to 1/8000 second. A frame grabber board (Data Translation DT-2861) with a resolution of 512 by 480 by 8 bits is used to digitize the RS-70 analog voltage output from the CCD array. The DT-2861 board provides 246 gray levels and a total of 4 megabytes of storage buffer, yielding a maximum storage ability of up to 16 frames on the board. The maximum frame grab speed is 1/300 second and each frame consists of two interlaced video fields with a time interval of 1/60 second. A subroutine library DT-IRIS is used to support all operations on the frame-grabber and to manipulate the image.

DEVELOPMENT OF PIV PROGRAM FOR IMAGE ANALYSIS. A PIV software package was developed to analyze the recorded images. The uniqueness of this PIV system is its capability to measure the instantaneous characteristics of particle motions; i.e., velocity, solids concentration, and size distribution. The PIV interrogation software can identify images, determine solid concentration, calculate solid sizes, locate centroids, and compute displacement between image pairs. The instantaneous velocities are determined from the calculations of displacements. Additionally, the statistical flow information such as mean and fluctuation velocities and concentrations at specified locations can be obtained. The time interval between two consecutive frames is 1/30 second and each frame consists of two (even and odd) fields with a time interval of 1/60 second. The PIV interrogation technique utilizes sequentially recorded frames, instead of one frame, as a processing unit.

Each field sent from the frame grabber is first preprocessed to remove background and electronic noise to improve the image quality. The noise removal is achieved by using gray level thresholding (both high and low pass filtering). The image quality can be further improved by generating a bilevel image. The process of identification of the particle images is then initiated. A scanning subroutine is executed first to find a pixel with a saturated gray value. When such a pixel is located, an image boundary search is performed in a clockwise manner to find the upper right, lower right, lower left and upper left corners of the particle image. This search is completed when the initial registered saturated pixel is encountered. An integration along the x and y axes is performed inside the identified boundary of the particle image. The centroid is then calculated from the projection. The mean radius of the particle image is also computed. The scanning and boundary searching subroutines are then resumed until the whole field is processed completely. From the mean radius of each particle image and the precalibrated scale factor, the real size of the particle can be obtained. After the identification of phases, the local number concentrations of dispersed phases are then calculated by counting the centroids of the specified area, those sizes can be specified for each field. Figure 4.29 shows the comparison of the original and processed images. The intensity of particle image is not important in the particle identification process as long as the particle image is distinguishable from the background. In fact, all the particle images have the same saturated intensity level after the process of filtering and generating bilevel images.

With the assumption of linear and constant flow during the short recording intervals, the velocities of the particle images are obtained by computing the displacements of the centroids found from different fields. The starting field and the number of fields used for determining velocity vectors can be specified. The "instantaneous" means velocities and fluctuations of defined locations are available with a 16-frame real-time handling capability. A displacement scanning subroutine was developed to find the instantaneous particle displacement vectors. With the information on displacements and the time interval of the adjacent field (1/60 sec.), the instanteous velocities of a plane are obtained. The method used to calculate the turbulent intensity is a time/volume ensemble technique. In this technique, the local mean and fluctuation velocities are obtained based on a small defined area instead of a point usually used in single-phase flow measurements.

A bench test was conducted with a single particle settling in a two-dimensional column to verify the validity of the present PIV system. Different sizes of glass beads ranging from 50 µm to 5 mm were used. A ruler was attached to the column wall to allow the moving distance of the settling particle during the field interval (1/60 sec.) to be accurately recorded by video camera. The comparison of the particle settling velocity obtained from the measured distance and the elapsed time and from the PIV system shows a deviation within ± 1 %. With a field of view of 7 by 7 cm^2, the particle size is found to be overestimated by the PIV for particle sizes below 1.5 mm and the degree of overestimation increases with a decrease in actual particle size. The overestimation of particle size is due to light diffraction effects and the resolution of the CCD camera. The overestimation of 50 µm particles can be as large as two to three times the actual particle size. However, for particle sizes larger than 2 mm, it is found that the measured size is almost identical to the actual size. In the above, the scale factor is about 146 mm/pixel. Therefore, if the particle images can be magnified and recorded over ten pixel numbers, the overestimation of size can be minimized.

IMAGE COLLECTION AND PROCESSING. Using the visualization technique described above, the powder motion under various operating conditions for all four entrance types can be recorded on video tape. Images are clarified with an image enhancement and filtering technique prior to the analysis. Under the perfect condition (maximum contrast) the background should be black and the powder should be bright. In practice, three possible states of focus can be obtained: (1) particles are perfectly focused with sharp edge; (2) particles are non-perfectly focused with fuzzy boundaries; (3) particles are out of focus. The boundary of an agglomerate is determined by defining a curve connecting all the pixels on the perimeter having the same gray level higher than a given threshold value. This boundary changes as the threshold gray level varies. The threshold level is determined by plotting the covered area by the boundary versus a cut-off gray level. The threshold level is the cut-off gray level corresponding to a sharp decrease of this plot.

Simulation of the Gas Flow Field in the Entrances

To understand the mechanism by which the flow structure affects the powder dispersion, a computer simulation of the gas flow inside the entrance is conducted. In this simulation, a k-ε turbulent model is employed for the calculation of turbulent intensity. The finite-difference equations of this model are set up with a semi-integral approach originally suggested by Patankar [30]. The discretized equations are solved using a tri-diagonal matrix algorithm (TDMA).

Results and Discussion

In the simple expansion nozzle the central gas-powder stream diverges slightly due to the large initial momentum of the air stream and the powder agglomerates. In this case, most agglomerates are concentrated in the central jet zone as shown in figure 4.30.

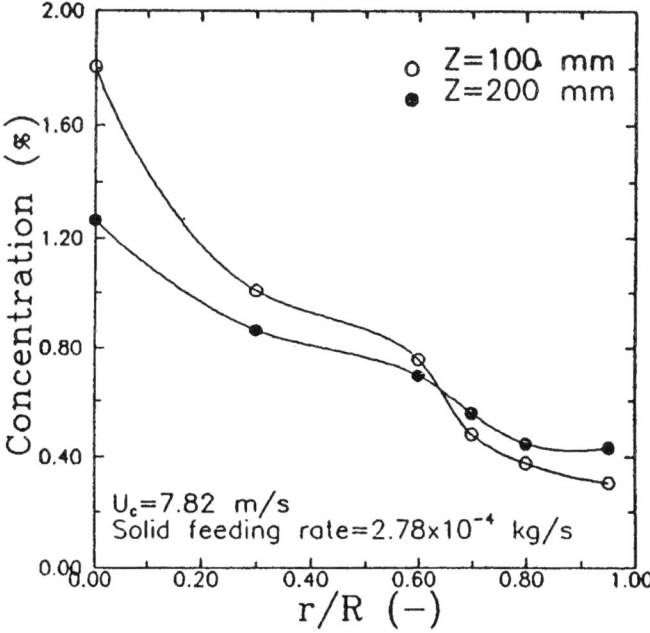

Figure 4.30 Radial distribution of solid concentration at two axial locations in the simple expansion nozzle.

Dry Scrubbing Technologies for Flue Gas Desulfurization

Near the inlet region in this nozzle, most of the powders are in the central zone. In the downstream region, the powder diffusion is enhanced by the turbulence generated during the central jet development and the agglomerate concentration becomes more uniform in the radial direction. The size distribution of the powder agglomerates is also studied and the typical results are shown in Figure 4.31(a). Agglomerates with large sizes and a wide distribution are observed in this type of nozzle. Powder dispersion can be enhanced significantly by adding two booster air jets to the simple expansion nozzle in the vertical direction. The typical agglomerate size distribution obtained in the expansion nozzle with two booster air jets is shown in Figure 4.31(a). The comparison of the mean agglomerate size distributions and powder concentration profiles at different axial positions are shown in Figures 4.32 and 4.33 for this type of nozzle. In the confined coaxial jet nozzle, powder agglomerates are mostly dispersed in the free shear layer located between the primary central jet and the booster air jet. The typical size distribution for a confined coaxial jet type entrance is shown in Figure 4.31(c). From the comparison of the agglomerate size distribution and the mean size in these three nozzles, it is clear that the expansion nozzle with two booster air jets provides the optimum powder dispersion.

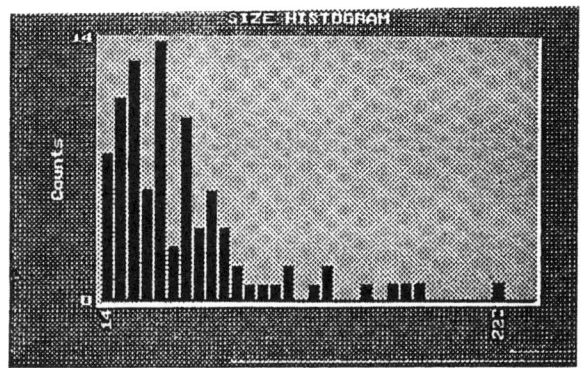
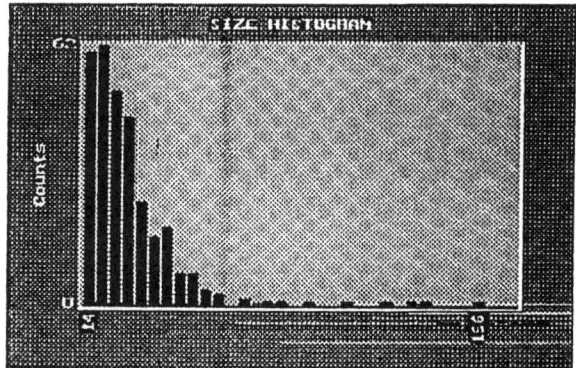
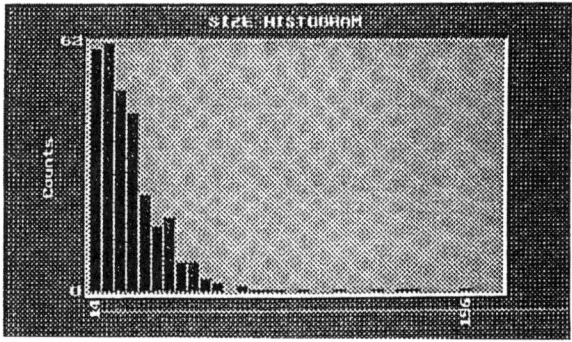

Figure 4.31 Comparison of agglomerate size distribution in different nozzles (U_c = 6.1 m/s, U_b = 22.2 m/s, solid feeding rate = 2.78 x 10^{-4} kg/s): (a) Simple expansion nozzle (average agglomerate size = 68 µ, solid concentration = 0.908%), (b) expansion nozzle with two booster jets (average agglomerate size = 40 µ, solid concentration = 1.11%, c) Confined coaxial jets nozzle (average agglomerate size = 62 µ, solid concentration = 1.93%).

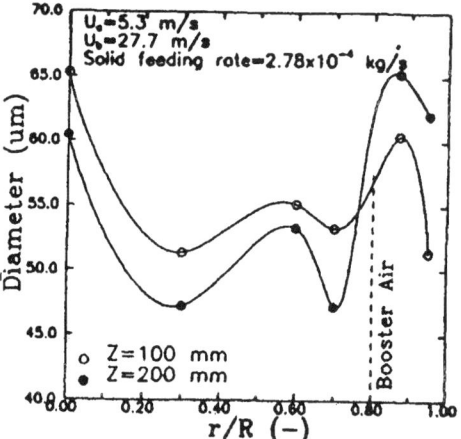

Figure 4.32 Radial profiles of the agglomerate size at two axial locations in the expansion nozzle with two booster jets.

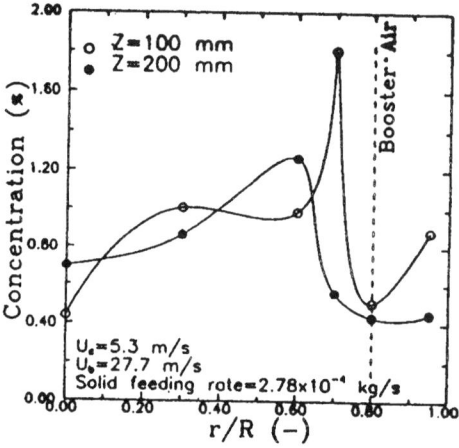

Figure 4.33 Radial profiles of solid concentration at two axial positions in the expansion nozzle with two booster jets.

It is believed that powder dispersion is mainly due to the shear stress generated by high turbulence intensity. This can be explained by an analysis of the flow structures inside these three types of nozzles. The typical flow structure inside the simple nozzle is shown in Figure 4.34(a), which is obtained by numerical simulation. The sudden increase in the flow area creates a reverse flow region.

Some powder agglomerates are trapped into this dead zone and deposit on the bottom wall. Thus, powder accumulation can be observed in this zone. For the confined coaxial jet nozzle, Figure 4.34(b), the recirculation zone depends on the velocity difference between the primary central jet and the annular booster air jet. A large velocity difference is required to create a large region with high turbulence intensity. However, this is not practical for the LIMB process. In the expansion nozzle with two booster air jets, the flow pattern is significantly different from the above two (Figure 4.34(c)). A very high booster air velocity creates low pressure zones around the entrance, leading to strong recirculations and hence high shear stresses in the recirculation zones. In addition, the dead zone observed in the simple expansion nozzle is suppressed due to a three-dimensional vortex stretch effect and consequently high momentum exchange rates between the booster air jets and the surrounding fluid. As can be seen in this figure, the booster air can create a flow in the tangential direction, which enhances the powder and air mixing and turbulence intensity in addition to the strong recirculation in the axial direction. Compared with the flow in the coaxial entrance type, Figure 4.34(b), the flow pattern shown in Figure 4.34(c) is characterized by a three-dimensional flow structure instead of axi-symmetric flow.

From the experimental study of the agglomerate size distribution and the numerical simulation of the flow structures in different nozzles, it is clear that the expansion nozzle with two booster jets has the optimum performance. Therefore, further experimental study was conducted to investigate the effects of powder additives and the influence of operating conditions on powder dispersion in this type of nozzle. Figure 4.35 shows the effects of the total air flow rate on powder dispersion in the expansion nozzle with two booster jets for two different powders; i.e., pure Black River hydrate (BRH) and 1.3% Lignosulfonated Black River hydrate. Each point shown in this figure is the averaged result of over 20 instantaneous images at a fixed condition. It can be seen that the mean agglomerate size decreases with the total air flow rate. Comparing these two types of sorbent, it appears that the dispersion of the 1.3% Lignosulfonated Black River hydrate is better than that of BRH.

Sorbent Transport and Dispersion

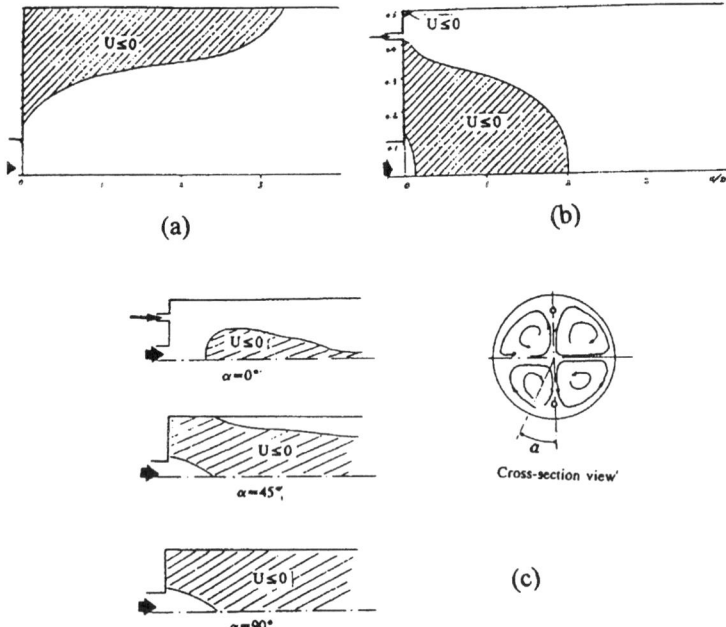

Figure 4.34 Flow patterns in different nozzels: (a) the simple expansion nozzle, (b) the confined coaxial jets nozzle, (c) the expansion nozzle with two booster jets.

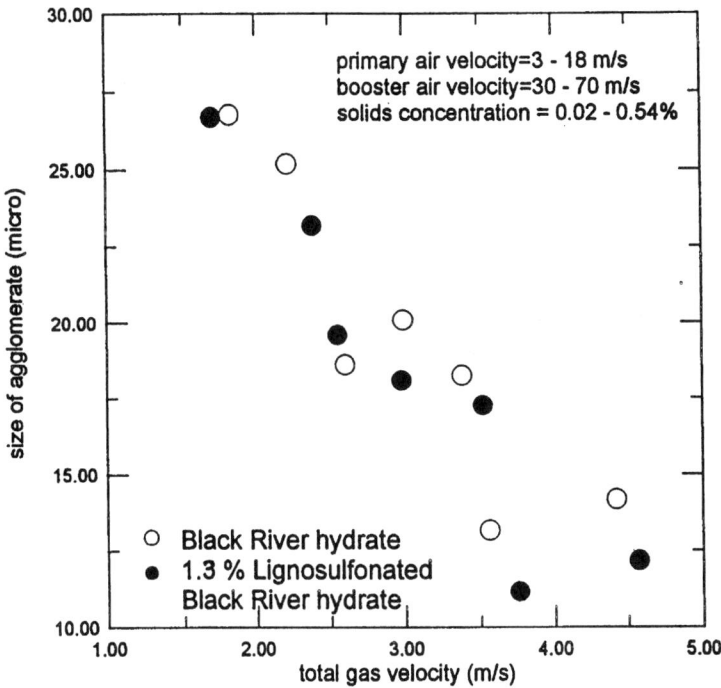

Figure 4.35 Effects of total air flow rate on powder dispersion.

Modeling

A Stochastic Model for Attrition of Sorbent Particles

INTRODUCTION. Attrition can take place during handling, transport, and injection of sorbent powders. And attrition may be due to thermal, chemical, static mechanical, and kinetic stresses [31]. Thermal stresses result from rapid temperature changes in a particle causing uneven expansion and intra-particle stress. Chemical attrition is due to gas-solid chemical reactions accompanied by large volume changes. Static mechanical stresses are due to the effect of external loads applied to the particle surface. Kinetic stresses are associated with impacts due to the velocity of the gas-solid flow. All these mechanisms cause a reduction in particle size and an increase in the number of particles. The theories in the literature describing the attrition rate of solid particles are primarily based on the population balance approach [32]. Generally, these theories use deterministic models in their applications. Since the attrition process behaves randomly in

Sorbent Transport and Dispersion

nature, a model based on a stochastic approach should be used to simulate the attrition process. In this work, a stochastic model for sorbent attrition is developed. Variations in the particle size distribution along with the mean and variance of the distribution during the attrition process are also presented.

MODEL FORMULATION. Suppose sorbent particles in the system are divided into k classes based on their diameters as shown in Table 4.3.

Table 4.3. Characterization of Particle Classes

Class	Average diameter (μm)	Average weight at time t (g)
1	D_1	$W_1(t)$
2	D_2	$W_2(t)$
.	.	.
.	.	.
.	.	.
k	D_k	$W_k(t)$

First, we consider the attrition of a sorbent particle which is in class k at time t=0. This particle will be attrited into a particle of class k-1 and a large number of particles of class 1, say, X_{k1}. Similarly, a particle of class i will be attrited into a particle of class i-1, where i = 2, 3, ..., k-1 and produce X_{i1} particles in class 1. The state of a particle can be labeled as follows:

Equation 4.30

$$S_k = \{ 0, 0, 0, \ldots, 1_k \},$$

$$S_{k-1} = \{ X_{k1}, 0, 0, \ldots, 1_{k-1}, 0 \},$$

$$S_2 = \{ (X_{k-1} + X_{(k-1)1} + \ldots + X_{31}), 1_2, 0, 0, \ldots, 0 \}, \text{ and}$$

$$S_1 = \{ (X_{k1} + X_{(k-1)1} + \ldots + X_{21}), 0, 0, \ldots, 0 \}.$$

Each set in Equation 4.30 contains k elements. Since the total mass of the system remains unchanged during the course of particle attrition, the following equation holds for particle disintegration from class k to class k-1.

Equation 4.31

$$X_{k1}D_1^3 + D_{k-1}^3 = D_k^3$$

Thus, X_{k1} can be expressed by

Equation 4.32

$$X_{k1} = (D_k^3 - D_{k-1}^3)/D_1^3$$

For particle disintegration from class i to class i-1, we have

Equation 4.33

$$X_{i1} = (D_i^3 - D_{i-1}^3)/D_1^3$$

$i=3, 4, ..., k$

and

$$X_{21} = D_2^3/D_1^3$$

The total number of particles of class 1 generated from particle disintegratiaon from class k to class 1 becomes

Equation 4.34

$$X^*_{k,1} = \sum_{i=0}^{k-2} X_{(k-i)1}$$

In the development of a stochastic model for the attrition process, it is assumed that the probability of transition $S_i \to S_{i-1}$ in the time interval (t, t+h) is $\lambda_i h + 0(h)$, where λ_i is a constant and $0(h)$ is a function of h such that $0(h)/h \to 0$ as h $\to 0$. It is also assumed that the probability of transition $S_i \to S_{i-j}$, j> 1 in the time interval (t, t+h) is $0(h)$.

Let $P_{ki}(t)$ be the probability that a sorbent particle is in state S_i (or class i) at time t given that it was in state S_k (or class k) at time t=0); i= 1, 2, 3, ... Taking a probability balance over state k, the continuous time, discrete state stochastic process yields

Sorbent Transport and Dispersion

Equation 4.35

$$P_{kk}(t+h) = P_{kk}(t)[1 - \lambda_k h + O(h)]$$

When h approaches zero, the equation above becomes

Equation 4.36

$$\frac{d}{dt}[P_{kk}(t)] = \lambda_k P_{kk}(t)$$

The initial condition for Equation 4.36 is

Equation 4.37

$$P_{kk}(0) = 1.$$

Equations 4.36 and 4.37 yield

Equation 4.38

$$P_{kk}(t) = e^{-\lambda_k t}$$

Likewise, the probability balance over state k-1 gives rise to

Equation 4.39

$$P_{k(k-1)}(t+h) = P_{k(k-1)}(t)[1 - \lambda_{k-1}h + O(h)] + P_{kk}(t)\lambda_k h + O(h)$$

When h approaches zero, Equation 4.39 becomes

Equation 4.40

$$\frac{d}{dt}[P_{k(k-1)}(t)] = \lambda_{k-1} P_{k(k-1)}(t) + \lambda_k P_{kk}(t)$$

Substituting Equation 4.38 into Equation 4.40, we have

Equation 4.41

$$\frac{d}{dt}[P_{k(k-1)}(t)] + \lambda_{k-1} P_{k(k-1)}(t) = \lambda_k e^{-\lambda_k t}$$

The initial condition for Equation 4.41 is

Equation 4.42

$$P_{k(k-1)}(0) = 0$$

Thus, Equations 4.41 and 4.42 give rise to the solution for $P_{k(k-1)}(t)$ as

Equation 4.43

$$P_{k(k-1)}(t) = (-1)\lambda_k \left(\frac{e^{-\lambda_k t}}{\lambda_k - \lambda_{k-1}} + \frac{e^{-\lambda_{k-1} t}}{\lambda_{k-1} - \lambda_k} \right)$$

The expression for $P_{k(k-2)}(t)$ and the general expression for $P_{k(h-1)}(t)$, i=1,2,..,k-2, can also be obtained in a similar manner as follows:

Equation 4.44

$$P_{k(k-2)}(t) = \lambda_k \lambda_{k-1} \left(\frac{e^{-\lambda_k t}}{(\lambda_k - \lambda_{k-1})(\lambda_k - \lambda_{k-2})} + \frac{e^{-\lambda_{k-1} t}}{(\lambda_{k-1} - \lambda_k)(\lambda_{k-1} - \lambda_{k-2})} + \frac{e^{-\lambda_{k-2} t}}{(\lambda_{k-2} - \lambda_{k-1})(\lambda_{k-21} - \lambda_k)} \right)$$

and

Equation 4.45

$$P_{k(k-1)}(t) = (-1)^i \left[\prod_{j=0}^{i-1} \lambda_{k-j} \sum_{j=0}^{i} \frac{e^{-\lambda_{k-j} t}}{\prod_{\substack{m=0 \\ m \neq j}}^{i} (\lambda_{k-j} - \lambda_{k-m})} \right]$$

i=1,2,....,k-2

It is reasonably postulated that the larger the sorbent particle size, the greater the attrition rate for the particle. Thus, the following inequality is valid:

Sorbent Transport and Dispersion

Equation 4.46

$$\lambda_k > \lambda_{k-1} > \lambda_{k-2}... > \lambda_2$$

Assuming that the intensity of attrition for a class i particle, λ_i, is linearly proportional to the volume of the same class particle and the proportionality constants for all class particles are identical, we have

Equation 4.47

$$\lambda_{i-1} = (D_{i-1}/D_i)^3 \lambda_i$$

i=2, 3...,k

and

Equation 4.48

$$\lambda_1 = 0.$$

Thus, λ_k is the only undetermined or adjustable parameter in the model.

Next, we consider the attrition of multiple particles in the system. Suppose at time t=0, there are $N_i(0)$, i=2,3,...,k, particles in class i in the system. We will assume that the particles behave independently during the attrition process. Let N_i, i=2,3,...,k, denote the number of particles in class i at time t>0. The $N_i(0)$ particles which were in class i at time t=0 are distributed into classes i, (i-1),...,2, according to a multinomial distribution

Equation 4.49

$$\left[\frac{N_i(0)!}{\prod_{j=2}^{i} N_{ij}(t)! \left[N_i(0) - \sum_{j=2}^{i} N_{ij}(t) \right]!} \right] \prod_{j=2}^{i} [P_{ij}t]^{N_{ij}(t)} \left[1 - \sum_{j=2}^{i} P_{ij}(t) \right]^{N_i(0) - \sum_{j=2}^{i} N_{ij}(t)}$$

where $N_{ij}(t)$ denotes the number of particles in class j at time t>0 resulting from $N_i(0)$ in class i at time t=0. Note that throughout this derivation, capital $N_{ij}(t)$ and $N_i(t)$ denote random quantities and the corresponding lower case $n_{ij}(t)$ or $n_i(t)$ denote their expected values. $P_{ki}(t)$, i=2,3,...,k, in Equation 4.48 can be expressed by Equation 4.38 and Equation 4.45.

Given the initial number of particles in class k and number of particles in class i, i=2,3,...,k, at time t, the number of particles in class 1 at time t can readily be obtained from the mass balance of the sorbent paricles. Thus, we have

Equation 4.50

$$N_{kl}(t) = [N(0) - N_{kk}(t)]$$

$$X_{kl} + \lceil N_k(0) - N_{kk}(t) - N_{k(k-1)}(t) \rceil X_{k-1} 1 + \lceil N_k(0) - N_{kk}(t) - ... - N_{k2}(t) \rceil X_{21}$$

The expected value for $N_{kl}(t)$ becomes

Equation 4.51

$$n_{kl}(t) = E[N_{kl}(t)] = [N_k(0) - n_{kk}(t)]X_{kl} + \lceil N_k(0) - n_{kk}(t) - n_{k(k-1)}t \rceil X_{(k-1)i} +$$

$$+[N_k(o) - n_{kk}(t) - ... - n_{k2}(t)] X_{2i}$$

The expected number and variance of particles in class i at time t are (from binomial consideration)

Equation 4.52

$$n_{kj}(t) = E[N_{kj}(t)] = N_k(0)P_{kj}(t)$$

$j=2,3,...,k$

and

Equation 4.53

$$v[N_{kj}(t)] = N_k(0)P_{kj}(t)[1 - P_{kj}(t)]$$

$j=2,3,...,k$

The expected total weight of the particles in class j at time t becomes

Sorbent Transport and Dispersion

Equation 4.54

$$w_{kj}(t) = E[W_{kj}(t)] = n_{kj}(t)(\pi/6)D_j^3 \rho_p$$

$j=1,2,...,k$

where n_{k1} is given by Equation 4.51 and ρ_p is the density of the sorbent particle. Let $N_k(t)$ denote the number of particles in class j. Clearly, we have

Equation 4.55

$$N_j(t) = \sum_{i=j}^{k} N_{ij}(t)$$

It can be seen that the joint probability distribution of $N_{ij}(t)$, i=2,3,...,k; j=2,3,...,i, at time t given $N_i(0)$, i=2,3,...,k, is the product of the multinomial distribution given by Equation 4.49. The conditional distribution of $N_j(t)$ is the sum of the random variables with multinomial distribution given by Equation 4.49.

The expected number of particles in class j and the variance at time t are (from binomial consideration)

Equation 4.56

$$n_j(t) = E[N_j(t)] = \sum_{i=j}^{k} N_i(0) P_{ij}(t)$$

$j=2,3,...,k$

and

Equation 4.57

$$v[N_j(t)] = \sum_{i=j}^{k} N_i(0) P_{ij}(t)[1 - P_{ij}(t)]$$

$j=2,3,...,k$

Given the number of particles in class i, i=2,3,...,k, at the initial state and time t, the expected number of particles in class 1 at time t can readily be obtained from the mass balance of the sorbent particles. Thus we have

Equation 4.58

$$= \frac{1}{D_1^3}([N_k(0) - n_k(t)]D_k^3 + [N_{k-1}(0) - n_{k-1}(t)]D_{k-1}^3 + [N_2(0) - n_2 t]D_2^3))$$

The expected weight of the particles in class j at time t becomes

Equation 4.59

$$(w_j(t) = E[W_j(t)] = n_j(t)(\pi/6))D_j^3 \rho_p$$

j=1,2,...k.

When time t approaches infinity, the steady state weight-particle size distribution is reached. Under this condition, Equation 4.38 and Equation 4.45 reduce to

Equation 4.60

$$P_{kk}(\infty) = 0$$

and

Equation 4.61

$$P_{k(k-i)}(\infty) = 0$$

i=1,2,...,k-2

Thus, we have

Equation 4.62

$$P_{k1}(\infty) = 1$$

Clearly, the steady state weight-particle size distribution in the system follows a delta function. When l_k varies with time t in the following manner

Sorbent Transport and Dispersion

Equation 4.63

$$\lambda_k = \begin{cases} \text{Constant } (>0), & t < t^* \\ 0, & \text{for } t \geq t^* \end{cases}$$

Equation 4.38 and Equation 4.45, for $t \geq t^*$, become

Equation 4.64

$$P_{kk}(t) = 1$$

and

Equation 4.65

$$P_{k(k-1)}(t) = 0$$

$i = 1, 2, \ldots k,$

The steady state weight-particle size distribution can thus be represented by ω_i $(t^*)/\omega$, $i = 1,2,3,\ldots,k$. Here, w is the total weight of the sorbent particles in the system. The time t^* defined here is equivalent to that required to reach steady state for the sorbent particle size distribution. This time, t^*, and accordingly λ_k, depend on the mechanical and physical properties of the sorbent particles as well as the internal temperature gradients within these particles.

Thus, a stochastic model is developed to describe the performance of sorbent particles undergoing attrition. In this model, the sorbent particles are classified into i classes, $i=1,2,3,\ldots,k$. For each class, D_i, $N_i(0)$, and $w_i(0)$ can be known from mechanical and optical analyses. The parameter λ_k depends on the thermal, physical, and mechanical properties of the sorbent particles, and is treated as an adjustable parameter in the model. Then, λ_i, $i=1,2,\ldots,k-1$, can be evaluated from Equation 4.47. The probability that the particle is in class i at time t that was in class k at time $t=0$ can be obtained from Equation 4.38 and Equation 4.45.

Finally, the expected total weight of the sorbent particles in class i and the variance of weight-particle size distribution with time can be calculated by using Equation 4.56 through Equation 4.59.

MODEL SIMULATION RESULTS. A computer program has been written to perform the mathematical computations described in the previous section. The program determines the weight fraction of the particles in each class at any given time. In the program calculations, the sorbent is classified into five different size groups, but it can be extended for any number of classes.

Five size classes with particle diameters of 50, 250, 400, 700 and 1000 μm are considered, and the initial number of particles in these classes are taken to be 20, 50, 70, 100 and 150, respectively. The change in weight distribution due to attrition is shown in figure 4.36 for t = 0.25, 0.5, and 3.0 seconds. The effect of parameter λ_5 is also shown for each time. As expected, the weight distribution shifts to a lower size with increasing time. An increase in the value of λ_5 leads to substantially faster attrition of particles.

Sorbent Transport and Dispersion

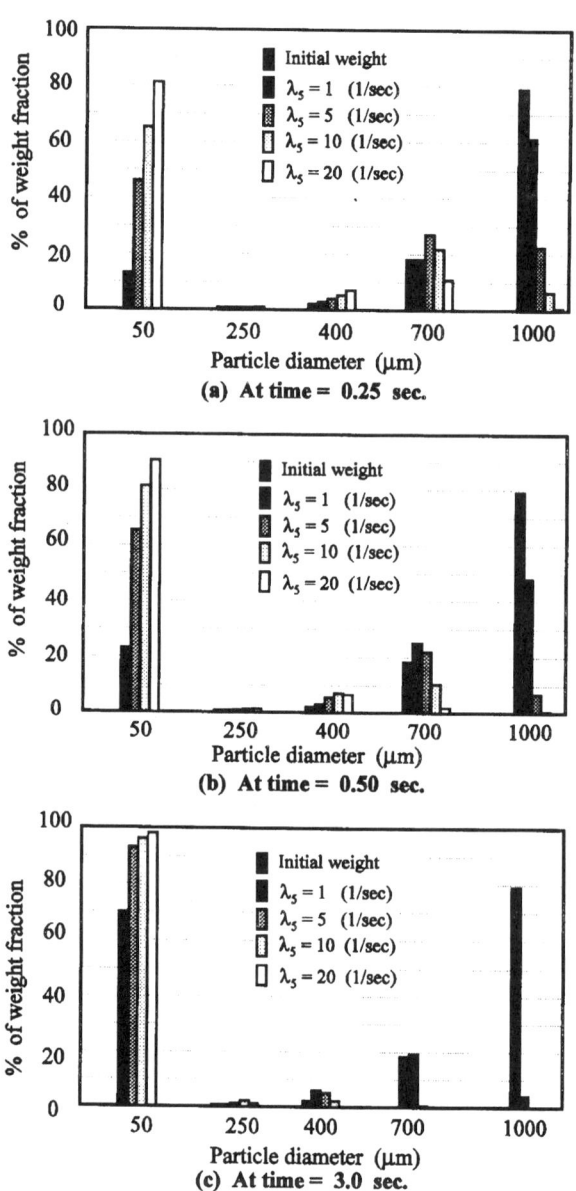

Figure 4.36 Size distribution of attrited particles for different values of

parameters λ_5.

The parameter λ_k depends on various stresses on the particle. Thermal stresses result from rapid temperature changes of particles causing thermal attrition, which is a result of uneven expansion and intra-particle stresses. Chemical attrition results when gas-solid reaction occurs with a large molecular volume change or structural difference. Static mechanical stresses are due to external loads on the particle, while kinetic stresses are related to the transport conditions. The parameter λ_k is expected to depend on these attrition mechanisms, and its dependence needs to be determined experimentally.

Integral Model for Powder Dispersion

INTRODUCTION. In the Powder Dispersion section experimental studies have been described on powder dispersion in powder injection nozzles and one optimum nozzle design (an expansion nozzle with two-jet booster jets) has been selected. For a better understanding and, furthermore, for predicting the performance of the nozzle or agglomerate size distribution at the exit of the nozzle, previous studies on powder characterization and powder handling have been incorporated into a mathematical model.

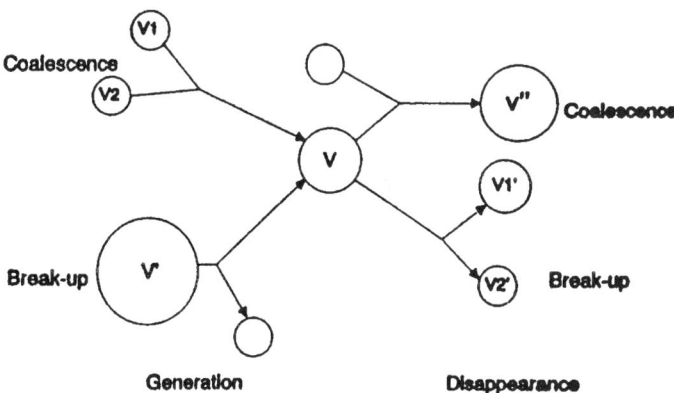

Figure 4.37 Dynamics of agglomerate size variation.

Sorbent Transport and Dispersion

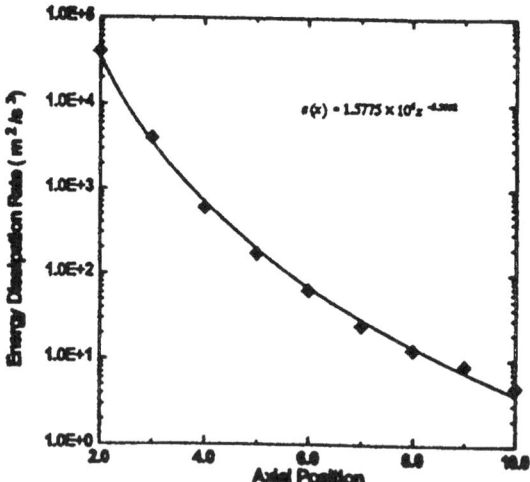

Figure 4.38 Axial variation of cross-sectionally averaged turbulent energy dissipation rate.

General population-balance models have been extensively applied in liquid-liquid dispersion systems to simulate the drop population variation in terms of drop size, age, concentration, etc. [33, 34, 35, 36]. This model can also be applied to gas-solid systems [37]. However, little has been reported on modeling the agglomerate size distribution variation in the dispersion of cohesive particles in turbulent gas-solid flows; i.e., powder injectors.

Although the effect of particles on the fluid turbulence may be important even under low particle concentration conditions, existing models for describing this influence are still incomplete [38]. In the literature, the mean flow field in turbulent gas-solid systems is assumed to be not significantly affected by particles in dilute systems and can be determined by the standard k-e model [38, 39].

The agglomerate size distribution is dictated by two simultaneous processes: coalescence and break-up of agglomerates. Both processes are controlled by hydrodynamic forces and interparticle forces in the system, as shown in Figure 4.37. Therefore, to simulate the agglomerate size distribution variation in a powder injector, both hydrodynamic and interparticle forces should be taken into account.

In this section, a population balance model is developed to simulate the agglomerate size distribution variation in the optimum powder injector; i.e., an expansion nozzle with two booster air jets.

MODEL DESCRIPTION. The flow field inside the nozzle is calculated from the standard k-ε turbulent model, as discussed above. Simulation results of the flow field show that the turbulence intensity is very high in the nozzle (10^2 - 10^5 m²/s³) and drops exponentially along the nozzle, as shown in Figure 4.38. It was found that under the experimental conditions in the Powder Dispersion section, most agglomerates are smaller than the energy dissipation eddy (Kolmogorov eddy) in the turbulent flow, and therefore, viscous shear stress is dominant.

The population balance model is expressed by a one-dimensional mass balance. The mass balance for agglomerates of volume v over a differential axial location range from x to dx, under steady state conditions, is

Equation 4.66

$$G_s \frac{\partial M_{(v,x)}}{\partial x} = M_{B(v,x)} - M_{D(v,x)}$$

where G_s, $M_{(v,x)}$, $M_{B(v,x)}$, and $M_{D(v,x)}$ are the solid mass flow rate, the probability density distribution function (based on mass fraction) of agglomerates of volume v at axial location x, and the mass rate of birth and death rate of agglomerates, respectively. This equation can also be expressed in terms of the number distribution of the agglomerate size.

Equation 4.67

$$G_s \frac{\partial (v n_{(v,x)}/\alpha)}{\partial x} = M_{B(v,x)} - M_{D(v,x)}$$

The physical meaning of Equation 4.67 is shown in Figure 4.37.

The coalescence of two agglomerates is due to their cohesion upon collision. Only binary collisions are considered, since the probability of collisions among three or more particles is negligible. Similar to the kinetic theory of gas, the collision frequency of agglomerates depends on the number density of the agglomerates $n_{(v,x)}$, averaged particle velocity fluctuations $_pV'$, and the size of the collision agglomerates v_1, v_2. Thus, the collision frequency between two agglomerates can be written as

Sorbent Transport and Dispersion

Equation 4.68

$$F_{C(v_1,v_2,x)} = k_1 n_{(v_1,x)} n_{(v_2,x)} V_p^1 (v_1 + v_2)$$

However, only a fraction of the collisions lead to adhesion or coalescence. The controlling factors on the probability of adhesion include interparticle forces (only van der Waals forces are considered) represented by the Hamaker constant of the sorbent powder (A), the size of the particles (v_1 and v_2), and averaged solid velocity fluctuations V_p'.

Equation 4.69

$$P_{A(v_1,v_2,x)} = k_2 A / (V_p^1 v_1 v_2)$$

From the above equations, the mass rate of coalescence of two agglomerates of size v_1 and v_2 ($v=v_1+v_2$) is

Equation 4.70

$$M_{c(v_1,v_2,x)} = F_{c(v_1,v_2,x)} P_{A(v_1,v_2,x)} (v_1 + v_2) \rho_A$$

where ρ_A is the density of agglomerates. More specifically, the generation rate of agglomerates of size v by coalescence is

Equation 4.71

$$\dot{M}_{B,C(v,x)} = \int_{v^1} F_{C(v^1, v-v^1, x)} P_{A(v^1, v-v^1, x)} [v^1 + (v-v^1)] \rho_A dv^1$$

$$= K_1 \rho_A A v_2 \int_{v_{min}}^{v - v_{min}} \frac{n_{(v-v^1, x)} n_{(v^1, x)}}{v^1 (v - v^1)} dv^1$$

The disappearance rate of agglomerates of size v by coalescence is

Equation 4.72

$$\dot{M}_{D,C(v,x)} = \int_{v^1} F_{C(v,v^1,x)} P_{A(v,v^1,x)} (v+v^1) \rho_A dv^1$$

$$= K_1 A \frac{\rho_A}{v} n_{(v,x)} \int_{v_{min}}^{v_{max}} \frac{n_{(v^1,x)}(v+v^1)^2}{v^1} dv^1$$

The breakup of agglomerates can be attributed to different reasons; e.g., hydrodynamic forces, impacting forces, thermal stress, and mechanical stresses. However, in this nozzle, only hydrodynamic forces are important to the breakup of agglomerates.

According to the theory of Hinze [40], the generalized Weber number plays a dominant role in the deformation and breakup of a globule. The Weber number represents the ratio of hydrodynamic forces to the surface tension of a globule. However, what is important in agglomerate breakup is interparticle forces rather than surface tension. Therefore, a similar number is defined as

Equation 4.73

$$\Omega = \frac{\text{hydrodynamic forces}}{\text{interparticle forces}}$$

In this model, the viscous shear stress is dominant, as the simulation of the flow field shows

Equation 4.74

$$\tau = k\sqrt{\rho_f \mu \varepsilon_{1(x)}}$$

The van der Waals force is represented by the Hamaker constant of the sorbent powder (A), since the packing of primary particles in an agglomerate is random. It is further assumed that the breakup frequency is proportional to Ω.

Equation 4.75

$$F_{Br(v,x)} = k_3 \frac{v}{A} \sqrt{\rho_f \mu \varepsilon_{t(x)}} n_{(v,x)}$$

Hence, the mass rate of breakup of agglomerates of volume v is

Sorbent Transport and Dispersion

Equation 4.76

$$\dot{M}_{Br(v,x)} = F_{Br(v,x)}^{v} \rho_A = -K_2 \frac{\rho_A}{A} \sqrt{\rho_f \mu \varepsilon_{1(x)}} v^2 n_{(v,x)}$$

and the disappearance rate of agglomerates of volume v by breakup is in the same form.

Equation 4.77

$$\dot{M}_{D,B(v,x)} = \left(-K_2 \frac{\rho_A}{A} \sqrt{\rho_f \mu \varepsilon_{1(x)}} v^2 n_{(v,x)} \right)$$

But, the generation rate of agglomerates of volume v by breakup is different.

Equation 4.78

$$\dot{M}_{B,B(v,x)} = \int_{v^1} F_{Br(v^1,x)}^{v^1} \rho_A B_{vv^1(x)} dv^1$$

where $B_{vv=}$ is the size (v) distribution (based on mass fraction) of the fragments of a larger agglomerate (v'). Since no experimental data for this term is available, it is assumed that

Equation 4.79

$$B_{vv^1(x)} = \frac{1.17 v^{0.17}}{v^{1.17} - v_{min}^{1.17}}$$

Hence, the generation rate of agglomerates of volume v by breakup of larger agglomerates is:

Equation 4.80

$$\dot{M}_{B,B(v,x)} = K_2 \frac{\rho_A}{A} \sqrt{\rho_f \mu \varepsilon_{t(x)}} \int_{v+v_{min}}^{v_{max}} \frac{1.17 v^{1/2} n_{(v^1,x)} v^{0.17}}{v^{1.17} - v_{min}^{1.17}} dv^1$$

Substituting Equation 4.71, 4.72, 4.77, and 4.80 into Equation 4.67 yields the general form of the governing equation of this model

Equation 4.81

$$\left(G_s \frac{\partial \left[\frac{vn_{(v,x)}}{\alpha}\right]}{\partial x} = K_1 \rho_A A v^2 \int_{v_{min}}^{v-v_{min}} \frac{n_{(v-v^1,x)} n_{(v^1,x)}}{v^1(v-v^1)} dv^1 + K_2 \frac{\rho_A}{A} \sqrt{\rho_f \mu \varepsilon_{t(x)}} \right.$$

$$\int_{v+v_{min}}^{v_{max}} \frac{1.17 v^{/2} n_{(v^1,x)} v^{0.17}}{v^{/1.17} - v_{min}^{1.17}} dv^1$$

$$\left. -K_1 A \frac{\rho_A}{v} n_{(v,x)} \int_{v_{min}}^{v_{max}} \frac{n_{(v^1,x)} (v+v^1)^2}{v^1} dv^1 - K_2 \frac{\rho_A}{A} \sqrt{\rho_f \mu \varepsilon_{t(x)}} v^2 n_{(v,x)} \right)$$

However, in this study, two factors contribute to simplifying this general equation: (1) The solid volume fraction can be considered as a constant; (2) The solid volume fraction and number density are so low that the coalescence of agglomerates is negligible and the breakup of agglomerates is dominant in determining the agglomerate size distribution variation. The simplified governing equation is thus obtained.

Equation 4.82

$$\frac{\partial n_{(v,x)}}{\partial x} = \left(\frac{1}{G_s} K \frac{\rho_A \alpha}{A} \sqrt{\rho_f \mu}\right) \frac{1}{v} \sqrt{\varepsilon_{t(x)}} \left[\int_{v+v_{min}}^{v_{max}} \frac{v^2 n_1(v^1,x)}{v^1 - v_{min}} dv^1 - v^2 n_{(v,x)}\right]$$

$$c \frac{\sqrt{\varepsilon_{t(x)}}}{v} \left[\int_{v^1+v_{min}}^{v_{max}} \frac{v^2 n(v^1,x)}{v^1 - v_{min}} dv^1 - v^2 n_{(v,x)}\right]$$

Simulation results. The simplified equation is solved by a finite difference method. A typical result is given in figure 4.39. Figure 4.39 shows the comparison between the experimental data of the size distribution of agglomerates at the exit of the nozzle and the simulation results. From this figure, it can be seen that this model reasonably fits the experimental data.

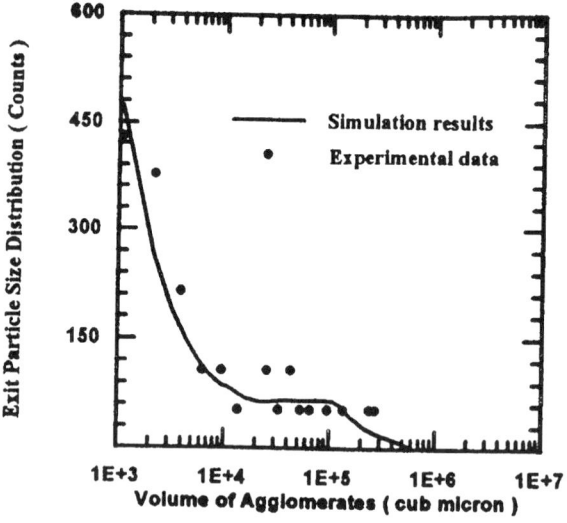

Figure 4.39 Comparison of the simulation results with experimental data.

Nomenclature

A	Aerated bulk density, Hamaker constant.
c	Cohesion.
C	Compressibility. Constant in Equation 4.82
D_i	Diameter of a particle in class i.
d	Pipe diameter.
d_p	Particle diameter.
d_{pi}	Diameter of class i particle.
E	Electric field on the pipe surface.
E_B	Breakdown field strength of the air.
F_c	Collision frequency.
G_s	Solid mass flow rate.
I_m	Current generated along a certain length of a pipe.
K_1, K_2	Constants in Equation 4.81.
k	Turbulent kinetic energy.

$M_{(v,x)}$	Probability density distribution function (based on mass fraction) of agglomerates of volume v at axial location x.
$M_{B(v,x)}$	Mass rate of birth.
$M_{D(v,x)}$	Death rate of agglomerates.
$N_i(t)$	Random variable representing the number of particles in class i at time t.
$N_{ij}(t)$	Random variable representing the number of sorbent particles in class j at time t that were in class i at time t=0.
$n_i(t)$	Expected number of particles in class i at time t.
$n_{ij}(t)$	Expected number of particles in class j at time t that were in class i at time t=0.
$n_{(v,x)}$	Number density of agglomerates of volume v at axial location x per unit volume.
P	Packed bulk density.
$P_{ki}(t)$	Probability that the particle is in class i at time **t** that were in class k at time t=0.
q/m_p	Charge-to-mass ratio.
q_i	Amount of charges on particles of d_{pi} in diameter.
q_{12}	Charge transferred per particle.
R	Pipe radius.
$\tilde{a}$	Mean gas velocity.
V_p'	Averaged solid velocity fluctuation.
W	Working bulk density.
W_i	Random variable representing the weight of the particles in class i.
w	Powder mass flow rate.
w_i	Expected weight of the particles in class i.
X_{i1}	Number of particles in class i which are generated from the disintegration of a particle from class 1 to class i-1.
α	Solid volume fraction.
λ_k	Intensity of particle disintegration from class k to k-1.
ρ_A	Density of agglomerates.
ρ_f	Density of the fluid phase.
ρ_p	Density of the sorbent particles.

Sorbent Transport and Dispersion

τ	Shear stress.
τ_s	Viscous shear stress.
σ	Normal stress.
ϕ	Angle of internal friction.
$E, \varepsilon_{t(x)}$	Dissipation rate of turbulent kinetic energy.
μ	Viscosity.

References

1. London, F., Z. *Physik*, 63, 245 (1930)
2. Hamaker, H. C., "The London-van der Waals Attraction Between Spherical Particles," *Physica IV*, No. 10, 1058 (1937)
3. Lifshitz, E. M., "The Theory of Molecular Attractive Force Between Solids," *J. Phys. Chem. Solids*, 32, 1657 (1971)
4. Prieve, D. C. and W. B. Russel, "Simplified Predictions of Hamaker Constants from Lifshitz Theory," *J. Colloid Interface Sci.*, 125 (1), 1 (1988)
5. Parsegian, V. A. and G. H. Weiss, "Spectroscopic Parameters for Computation of van der Waals Forces," *J. Colloid and Interface Sci.*, 81, 285 (1981)
6. Hough, D. B. and R. W. Lee, "The Calculation of Hamaker Constants from Lifshitz Theory with Applications to Wetting Phenomena," *Adv. Colloid Interface Sci.*, 14, 3 (1980)
7. Cross, J., *EctrostaticsÑPrinciples, Problems and Applications,* Adam Hilger, Bristol, England (1980)
8. Bailey, A. G., "Electrostatic Phenomena During Powder Handling," *Powder Technology*, 37, 71 (1984)
9. Jozewicz, W. and D. A., Kirchgessner, "Activation and Reactivity of Novel Calcium-Based Sorbents for Dry SO_2 Control in Boilers," *Powder Technology.* 58, 221 (1989)
10. Kirchgessner, D. A. and J. M. Lorrain, "Lignosulfonate-Modified Calcium Hydroxide for Sulfur Dioxide Control," *Ind. Eng. Chem. Res.*, 26, 2397 (1987)
11. Rosen, M. J., *Surfactants and Interfacial Phenomena,* John Wiley and Sons, Inc., NY (1978)
12. Tadros, M. E. and I. Meyes, "Linear Growth Rates of Calcium Sulfate Dihydrate Crystals in the Presence of Additives," *J. of Colloid and Interf. Sci.*, 72, 245 (1964)
13. Sarig, S., "Crystal Habit Modification by Water Soluble Polymers," *J. of Cryst. Growth*, 24, 338 (1974)
14. Smith, B. R. and A. E. Alexander, "The Effect of Additives on the Process of Crystallization: II. Further Studies on Calcium Sulfate (I)," *J. of Colloid and Interf. Sci.*, 34 (1), 81 (1970)
15. Soo, S. L., "Dynamics of Charged Suspension in International Reviews," *Aerosol Physics and Chemistry*, edited by Hidy, G. M. and Brock, J., 2, 61 (1971)
16. Lowell, J., A. C. Rose-Innes, "Contact Electrification," *Advances in Physics*, 29 (6), 947 (1980)
17. Medley, J. A., *Nature*, 171, 1077 (1953)
18. Shaw, P. E., Proc. R. Soc., 94, 16 (1917)
19. Carr, R. L., "Evaluating Low Properties of Solids," *Chemical Engineering*, 163, January 18, 1965
20. Jozewicz, W. and B. K.Gullett, "The Effect of Storage Conditions on Handling and SO_2 Reactivity of $Ca(OH)_2$-Based Sorbents," *ZKG International*, 5, 242 (1991)
21. Jenike, A. W., *Storage and Flow of Solids*, University of Utah, Engineering Experiment Station, Salt Lake City, Bulletin No. 123 (1964)
22. Johanson, J. R. and A. W. Jenike, *Stress and Velocity Fields in Gravity Flow of Bulk Solids*, University of Utah, Engineering Experiment Station, Salt Lake City, Bulletin No.1. 116 (1962)
23. Geldart, D., N. Harnby and A. C. Wong, "Fluidization of Cohesive Powders," *Powder Technology*, 37, 25 (1984)

Sorbent Transport and Dispersion

24. Rastogi, S., S. V. Dhodapkar, F. Cabrejos, J. Baker, M.Weintraub and G. E. Klinzing, "Survey of Characterization Techniques of Dry Ultrafine Coals and Their Relationships to Transport, Handling and Storage," *Powder Technology*, 74, 47 (1993)
25. EFCE, *Standard Shear Testing Technique for Particulate Solids Using the Jenike Shear Cell*, The Institution of Chemical Engineers, England (1989)
26. Thomas, D. G., "Turbulent Disruption of Flocs in Small Particle size Suspension" *AIChE J.*, 10, 517 (1964)
27. Bagster, D. F. and D. Tomi, "The Stresses within a Sphere in Simple Flow Fields," *Chem. Eng. Sci.*, 29, 1733-1741 (1974)
28. Masuda, H., S. Fushiro and K. Inoya, "Experimental study on the dispersion of fine particles in air," *J. Assoc. Powder Tech. Japan*, 14, 3-10 (1977)
29. Kousaka, Y., K. Okuyama, A. Shimizu and T. Yosia, "Mechanism of dispersion of aggregate particles in air," *J. Chem. Eng. Japan*, 12, 152-158 (1979)
30. Patakankar, S. V.,*Numerical Heat Transfer and Fluid Flow*, McGraw-Hill Book Company, NY (1980)
31. Chraibi, M. A. and G. Flamant, *Powder Technology*, 59, 97 (1989)
32. Wei, J., W. Lee and R. F. Krambech, *Chem. Eng. Sci.* 32, 1211 (1977)
33. Valentas, K. J., O. Bilous and N. R. Amundson, "Analysis of Breakage in Dispersed Phase Systems," *Ind. and Eng. Chem. Fund.*, 5, 271 (1966)
34. Coulaloglou, C. A. and L. L. Tavlarides, "Description of Interaction Processes in Agitated Liquid-Liquid Dispersions," *Chem. Eng. Sci.*, 32, 1289 (1977)
35. Ramkrishna, D., "The Status of Population Balances," *Rev. Chem. Eng.*, 3, 49 (1985)
36. Tsouris, C. and L. L. Tavlarides, "Breakage and Coalescence Models for Drops in Turbulent Dispersions," *AIChE J.*, 40, 395 (1994)
37. Yang, Y., H. Arastoopour and M. H. Hariri, "Agglomeration of Polyolefin Particles in a Fluidized Bed With a Central Jet, Part II - Theory," *Powder Technology*, 74, 239 (1993)
38.
39. 38.Sommerfeld, Martin, "Particle Dispersion in Turbulent Flow: The Effect of Particle Size Distribution," *Part. Part. Syst. Charact.*, 7, 209 (1990)
40. Picart, A., A. Berlemont and G. Gouesbet, "Modelling and Predicting Turbulence Fields and The Dispersion of Discrete Particles Transported by Turbulent Flows," *Int. J. Multiphase Flow*, 12, 2, 237 (1986)
41. Hinze, J. O., "Fundamentals of the Hydrodynamic Mechanism of Splitting in Dispersion Processes," *AIChE J.*, 1, 289 (1955)
42. Shaw, P. E., "Experiments on Tribo-Electricity I-The Tribo-Electric Series," *Proc. R. Soc.*, 94,16 (1917)
43. Lowell, J. and A. C. Rose-Innes, "Contact Electrification," *Adv. in Physics*, 29, 947 (1980)
44. Kittaka, B., Masui and Y. Murata, "A Method for Measuring the Charging Tendency of Powder in Pneumatic Conveyance through Metal Pipes," *J. of Electrostatics*, 6, 181 (1979)
45. Nifuku, M., T. Ishikawa and T. Sasaki, "Static Electrification Phenomena in Pneumatic Transportation of Coal," *J. of Electrostatics*, 23, 45 (1989)
46. Gajewski, A., "Measuring the Charging Tendency of Polystyrene Particles in Pneumatic Conveyance," *J. of Electrostatics*, 23, 55 (1989)
47. Soo, S. L. *Fluid Dynamics of Multiphase Systems*, Science Press, Beijing, China (1990)
48. Cole, B. N., M. R. Baum and F. R. Mobbs, "An Investigation of Electrostatic Charging Effects in High-Speed Gas-Solids Pipe Flows," *Proc. Inst. Mech. Engrs.*, 184, Pt 3C, 77 (1970)

49. Masuda, H., T. Komatsu and K. Iinoys, "The Static Electrification of Particles in Gas-Solids Pipe," *AIChE J.*, 22, 558 (1976)
50. Cheng, L. and S. L. Soo, "Charging of Dust Particles by Impact," *J. Appl. Phys.*, 41, 585 (1970)

CHAPTER 5 **_TRANSPORT PROCESSES INVOLVED IN FGD_**

J.R. Kadambi, P., Chinnapalaniandi
C.U. Yurteri, V.P. Kadaba, M.A. Assar
Mechanical and Aerospace Engineering
Case Western Reserve University
Cleveland, Ohio

Abstract

This chapter describes the investigations conducted to improve the understanding of the transport processes related to dry flue gas desulfurization (FGD). It is important to improve the understanding of the various processes involved in FGD; e.g, dry sorbent injection and dispersion, sorbent and flue gas mixing, sorbent humidification by water injection, so that the sorbent utilization can be enhanced. The objectives of this study were to experimentally obtain a basic understanding of turbulent flow structure of the mixing zone and its influence on particle dispersion; the effect of particle loading on turbulent properties and mixing; the effect of jet entrainment, water spray sorbent cocurrent flow interactions, sorbent wetting and mixing; and to investigate the flow field where certain ratios of jet velocity to flue gas velocity result in regions of negative flow.

A sorbent injection facility which can simulate the conditions encountered in COOLSIDE was designed and built. Non-intrusive laser-based diagnostic tools, PDA/LDA, were used for characterization of flow field. Sorbent injection schemes were investigated. Spray-cocurrent flow interaction and particle-laden jet cocurrent flow interactions were investigated and a criterion was developed for predicting the flow reversal. Reversal results in deposition of water droplets/ sorbent particles on the duct wall and occurs when the spray/particle-laden jet has entrained all the cocurrent flowing stream. This criterion will allow the designer to avoid flow conditions which result in flow reversal and deposition of droplet and sorbent particles on the duct wall and floor.

Tests were conducted with specially designed swirl nozzles and the simple nozzle for injection of particle-laden jet in the cocurrent duct flow. The test data indicated better mixing characteristics for the swirl nozzles. The results indicated that the mixing is enhanced and the turbulence intensities in the radial direction show an increase of 10-30%. The swirl nozzle has four blades mounted on its outer diameter which swirl the flow. No additional energy is required. Further optimization and improvement in nozzle design is recommended which may be very useful in enhancing sorbent utilization in dry FGD methods.

Introduction

Issues Regarding Transfer Processes for FGD Processes

Complicated multiphase flows (solid-gas, gas-liquid-solid and liquid-gas) are involved in many FGD processes. Multiphase flows by themselves are not well understood and their impact on FGD processes are even less understood. Drummond et al. [1], in their discussion of duct injection technologies for SO_2 control, have concluded that the transport and chemical processes within the duct, fluid mechanics of the system, nozzle design and operation, and humidification system design are areas of concern that require research to develop a better understanding of the process. Improvements in the basic knowledge of fluid mechanics, heat transfer and mass transfer rate effects are needed [2,3] to understand and achieve substantial improvements in many key areas of FGD processes and to obtain scale-up criteria. Dry sorbent injection and dispersion; sorbent humidification by water injection; sorbent slurry injection; and material handling aspects of sorbent slurry, sorbent powder, and ash are some examples. In the utilization of Ohio coal, emphasis has been placed on increasing the efficiency of dry, high sulfur flue gas scrubbing processes using calcium-based sorbents (e.g., LIMB and COOLSIDE). This entails the need to improve the understanding of the mixing processes which play a crucial role in enhancing sorbent utilization.

The reaction between SO_2 and sorbent particles involves mixing resulting from gross flow patterns and turbulence, mass transfer due to diffusion and dispersion and, finally, chemical reaction. Since the sorbent particles are injected into the flue gas, a prerequisite for fast reaction is the rapid mixing of the sorbent particles from the injection nozzle with the flue gas stream. Free jet entrainment can be a very important mechanism of rapid mixing. Clearly, mixing is determined by flow patterns and turbulent flow processes which are influenced by the characteristics of the injection jet and interactions with the surrounding stream.

Additionally, the wetting of sorbent particles with water spray seems to improve the sorbent utilization. A review of literature, presented in the next section, indicates that the experimental data available in the area of in-duct FGD process is sparse and does not cover the range of interest for dry FGD processes. Many of the fundamental questions involved in solid-gas and solid-liquid-gas flows have not been resolved. In addition, the mixing of the sorbent particles in FGD processes can possess different features because of different particle size, jet Reynolds number, particle loading, and agglomeration properties.

Particle-laden flows are inherently more complex than single-phase flows. This follows from the fact that the particles are distinct from the fluid, the continuous phase element. First, unlike the fluid element, the particles have finite size and shape and cannot deform under strain. Viscous forces result when particle surface velocity is different than the velocity of the surrounding fluid. Second, there is usually a substantial density difference between solids and the continuous gas phase. The particles have much more inertia than the fluid elements moving at similar velocities and are therefore unable to follow every scale of fluctuation of the fluid flow. In the case of a mismatch between the fluid and particle velocities, the particles exchange energy and momentum with the fluid through viscous drag. Faster moving particles may increase local fluid velocities by dragging along the fluid and vice versa. Particles may dampen or amplify the fluid fluctuation levels in different scales of eddies as a result of this interaction. A sufficiently large particle mass loading can result in the modification of the overall turbulence levels. For high particle loading, interaction among particles and the interaction of particle wakes can further complicate the flow.

The inertia of a particle in fluid flow can be assessed quantitatively by calculating a particle time constant. Assuming Stokesian flow and that the particle density is substantially larger than fluid density, the time, t_p, taken for the particle to accelerate from rest to 63% of the free stream velocity is given by

$$t_p = \rho_p d_p^2 / 18 \mu$$

where ρ_p is the particle density, d_p is the particle diameter, and μ is the fluid viscosity.

A fluid time scale, t_f, based upon a given eddy is given by

$$t_f = L/U$$

where L is the eddy length scale and U is the eddy velocity scale.

Based upon the relationship between t_p and t_f, the particle and fluid time scales, we can divide the possibilities of particle responses to turbulent flow into three regimes. In the first regime, if $t_p \gg t_f$, a particle will not respond to fluctuations in the flow. For example, paths of heavy large steel balls dropping through air will not be affected by turbulent eddies. In the second regime, if $t_p \ll t_f$, a particle will completely follow the flow, including any fluctuations. The seed particles used in laser doppler anemometry (LDA) measurements to obtain flow velocities are assumed to behave in such a manner. The third possible regime occurs when both t_p and t_f are of the same order of magnitude. In such a case, particles respond partially to the fluctuations in the flow. Here, the particles are not able to follow the fluid elements exactly, but their paths are altered by fluid fluctuations. The flow in this regime is most poorly understood and is very difficult to model. This regime may encompass the in-duct injection processes. The injection of the particle-laden jet into cocurrent flow further complicates the process. The wetting of the sorbent particles by injecting water spray into the flue duct appears to enhance sorbent utilization. However, the residence time of the droplets (typically about 30 micron in diameter) is about 2.5-3.0 seconds before evaporation and the interaction between the water droplet and the sorbent has to take place before that. The interaction between the water droplet and sorbent has to be improved to enhance wetting and mixing. The scavenging also has to be improved. There is a need for experimental data to validate, develop, and improve water spray and two- and three-phase slurry models, the emphasis being on mixing, dispersion and increasing turbulence. Another issue of interest is the location of the water spray and sorbent injection location relative to one another along the axial direction. Additionally, for certain ratios of spray velocity to flue gas velocity, regions of negative flow appear. There is a need to define the onset of such negative flows to avoid such flow conditions in FGD processes.

Literature Review

A literature survey of research conducted in the area of solid-gas and solid-liquid-gas flows is presented below. The emphasis is placed upon experimental work in the area of two-phase studies for a variety of configurations such as particle-laden flow, concentric jet flow, pipe flow and for spray-cocurrent flow interactions relevant to the in-duct injection process for FGD.

Single Phase Flow Confined Ducted Jets

The injection of a gas-solid jet into a cocurrent stream is somewhat similar to the studies of cocurrent jets reported in the literature. However, in most of these studies the injected fluid is of the same phase as that of the cocurrent flow. Good overall descriptions of how the jet and cocurrent flow interact and develop are given in Abromovich [4].

Mikhail [5] divided the duct diameter of the confined jet flow into three basic regions and obtained a solution for the jet growth and centerline velocity decay from linearized differential-integral equations of motion. Forstall and Shapiro [6] studied the diffusion of mass and momentum between coaxial jets with the central jet velocity larger than that of the outer flow. They concluded that mass diffusion was more rapid than momentum diffusion.

Alexander et al. [7] investigated the relative rates of diffusion of momentum and energy in a turbulent ducted jet, where the velocity of central jet is considerably higher than that of the external stream, and found that energy diffused more rapidly than momentum. Becker et al. [8] experimented with ducted turbulent jets, placing their emphasis on the flow regime in which appreciable recirculation occurs. A mapping of the flow field of mean velocity was made with impact tubes, and concentration profiles were made by a light-scattering technique. Hill [9] predicted the mean flow field for jet mixing using the differential-integral technique.

Free Coaxial Jets

Chigier and Beer [10], Champagne and Wygnanski [11], and Durao and Whitelaw [12] studied coaxial free jets. They observed that the length of the potential core was dependent upon nozzle geometry, turbulence fluctuation at the nozzle exit, and the velocity ratio between the central and annular jets. When the velocity of the annular jet was faster than that of the central jet, the length of the central potential core strongly depended upon the velocity ratio. It was found that the increase of the velocity of the annular jet resulted in a shorter potential core of the central jet. Durao and Whitelaw [12](1973) pointed out that the very low initial velocity of the central jet was responsible for the sudden drop of the central jet velocity near the nozzle exit. This is due to the entrainment of fluids from the potential core into the low pressure region immediately downstream of the nozzle.

Confined Jet Flow with Sudden Expansion

Confined turbulent coaxial jets with sudden expansion were investigated by Guruz and Ilicali [13], Johnson and Bennett [14], Habib and Whitelaw [15], and Freeman and Szczepura [16]. Guruz and Ilicali measured the axial mean velocity and turbulent shear stress with a pitot tube. The velocity ratio between the central and annular jets was equal to two. Johnson and Bennett (1984) measured the mean velocity, turbulent fluctuation and turbulent mass transport to describe the mixing in confined coaxial jets having a velocity ratio of 0.5 between the central and annular jets. The LDA technique was employed to measure the velocity, and laser-induced fluorescence (LIF) was used for concentration measurements.

Habib and Whitelaw [15] employed three different techniques (pressure probe, hot wire, and LDA) to study confined coaxial turbulent jets with two different velocity ratios between the central and annular jets. Measurements of flows having a velocity ratio between central and annular equal to 1 and 0.33 showed that the velocity gradient at the centerline was very steep immediately downstream of the potential core and that high turbulence intensities were observed downstream of the recirculation zone. The flow with a velocity ratio, central to annular, of 0.33 revealed more rapid mixing and faster decay than the flow with a unity velocity ratio. A slightly larger recirculation zone was also observed for the higher annular velocity case.

Freeman et al. [16] measured the mean and turbulent fluctuation with a two-channel LDA system. Three velocity ratios (central to annular: 0.625, 1.0 and 1.25) were considered. It was found that the rapid decay of the centerline velocity was due to strong diffusion and convection transport in the radial direction. The mean velocity distribution was relatively insensitive to the variation in the central jet Reynolds number below a value of 14000 when the velocity of the central jet was higher than that of the annular jet. Also, different velocity ratios appeared to have a qualitatively similar flow.

In the studies reported in the preceding pages, a wide range of flow measurements was obtained by a combination of pressure probe, hot-wire, and LDA. The category that most closely resembles this solid-gas injection process is the injection of gas into a coflowing stream of a similar fluid. In the majority of investigations cited in the preceding pages the velocity measurements have been carried out using a pressure probe and hot-wire anemometer. For example, as indicated by Becker and Brown [17], pressure probes are subjected to errors in regions of significant flow fluctuations. Similarly, the analysis of hot-wire signals become increasingly uncertain as the turbulence intensity increases above 20%. The mixing process for confined, directed, jet flow is turbulent in nature and the measurement technique plays an important role. The measurements of confined, single-phase jet flow will serve as baseline data to compare the particle-laden flows.

Particle-Laden Flows, Turbulent Free Jet

A turbulent free jet is the flow obtained when an inert mixture of particles and incompressible fluid issues from a nozzle into an unbounded region containing quiescent fluid. Several experimental studies of particle-laden free jet flows have been reported in the literature. Some of the earliest experimental work in gas-solid flows was carried out by Laats and Frishman [18] using hot wires in a round, unconfined, turbulent air jet containing 20-60 µm dusty particles. They found that the mean gas velocity profiles are narrower in a two-phase jet than in a single-phase jet. In addition, they found that the maximum velocity (i.e., the centerline velocity) decays slower in a two-phase jet but that the velocity profiles are independent of particle concentration. Unfortunately, they did not measure the initial conditions at the nozzle exit and did not report the material density of the particles.

Abramovich [4] discussed the effects of the dispersed phase on the structure of a turbulent gaseous jet. The two-phase jets showed narrower spreading angles compared to single-phase jets. Another highly referenced data set is that obtained by Hetsroni and Sokolov [19] who used the hot wire technique to study air jets laden with 13 µm cotton seed oil droplets in axisymmetric jets. They observed that the mass flux is at a maximum at the center and decreases across the jet, that it progressively decreases along the centerline of the jet, and that there is a significant change in the fluid mean velocity field compared to that for single-phase jets. The intensity of turbulence is also lower for the two-phase jets. This leads to a decrease in gaseous axial velocity fluctuations in the jet far field.

Goldsmith and Eskinazi [20] also used hot wire anemometry to measure local mean velocities and the concentration of aerosols in a two-phase particle-laden jet. Safflower oil droplets of 3.3 µm mean diameter were used, impacting locally upon a single wire. The hot wire was also used for obtaining the mean and fluctuating velocities in the axial direction. The rate of diffusion of droplets was lower and the width of the two phase jet was slightly narrower than for the single-phase jet. Goldsmith and associates [21], in a later investigation, conducted experiments involving a single-phase air jet and a jet utilizing dibutylphthlate droplets (6-23 µm in diameter). They observed that larger particles dispersed faster than smaller ones. This trend was also observed by Wells and Stock [22] in grid turbulence studies.

Yuu et al. [23] measured particle concentration as well as mean axial velocities for both gas and particle phases in an air jet. A pitot tube, micro manometer, and photoelectric dust counter were the instruments used. The results indicated that particle inertia and large eddies play important roles in particle-laden flows. Melville and Bray [24] examined available experimental data in turbulent, two-phase, axisymmetric jets with the volume fraction of the secondary phase much lower than unity and, using physical arguments and dimensional analysis, they obtained good correlations for the mean fluid velocity and particle mass flux in terms of the initial particle loading.

Wall and co-workers [25] conducted experiments with 180 μm sand particles and measured velocity and concentration distribution using a micro manometer as well as an isokinetic probe in a particle-laden jet in the developing region. They found that, due to particles, the jet expansion angle is less than that of single phase jet, leading to a reduction in the area for entrainment.

PARTICLES AFFECT DIFFUSION CHARACTERISTICS AND ATTENUATION OF TURBULENCE. Subramanian and co-workers [26-28] conducted experiments to measure the velocity and concentration profile in the developing and fully developed regions in a particle-laden jet. They found that, in the developing region, particles reduce the entrainment. In the fully developed region, as particle loading is increased, the decay of velocity is slower and the rate of decay of the concentration profile is also slower. Particles have a tendency to suppress the jet spread and this suppression increases with particle loading. Particles reduce the mixing and this effect progressively increases with loading.

In most of the experiments discussed so far, the experimental facilities contained contractions and the particle velocities at the nozzle exit generally lagged the fluid velocities and could not be accelerated as fast as the fluid. More recent experiments have been conducted with flows exiting from long pipes in an attempt to match the initial fluid and particle velocities. The gas flow originates as a fully developed turbulent flow rather than as a top-hat-shaped potential core with thin shear layers. Another weakness with the earlier experiments was that hot wires proved to be unreliable and less accurate in two-phase flows as well as in highly turbulent or reversing flow regions. Hot wires have generally been replaced with non-intrusive LDA.

Popper et al. [29] reported additional results on a round jet, containing 50 μm diameter oil drops, using a single component LDA system. Downstream of the nozzle exit the droplet velocities decayed more slowly than the velocity of air measured with a hot wire. This resulted from the large inertia of the particles compared to gas. The interaction of particles with fluid is in the form of viscous drag resulting in momentum transfer from particles to the fluid, causing decreased axial fluid velocity. No gas-phase measurements using LDA were reported. McComb and Salih [30] also used one-dimensional LDA to obtain particle velocities and concentration profiles, but did not report any continuous-phase data. The measurements of both Yuu et al. [23] and McComb et al. were limited to very low particle loading; i.e., the loading ratio was much smaller than unity. This implies that, while gas flow influences particle dispersion, the effect of the particles on the structure of the gas flow is small.

Levy and Lockwood [31] used a simple monochromatic two-beam, forward scatter LDA system to study sand-laden turbulent jets. Scattering amplitude discrimination was used to distinguish between small seed particles and the large sand particles. Mean and fluctuating axial profiles for the gas and large sand particles were determined for five different sets of particles. The mass loading ratio of particles to air ranged between 1 and 3. The intermediate size range of sand particles (380-700 µm) exhibited the largest mean spreading while the size range 850-1200 µm caused the greatest increase in the air velocity fluctuations over the level in a single-phase jet. This increase in turbulence was attributed to the increased mean velocity gradients in the gas phase, but again, they did not report the nozzle exit conditions.

Modarress et al. [32] developed a two-component LDA for measurement of particle and gas velocities. This technique was employed in their study for particle-laden jets seeded with glass beads. The glass beads had a diameter of 50 µm, and two initial particle mass loading ratios of 0.32 and 0.85 were considered.

Shuen et al. [33] used LDA to measure velocities of gas and particles. The mass flux was also determined by sampling probes. Particles of 79, 119, and 207 µm in diameter with mass loading of 0.2 and 0.66 were considered in the experiment. It was concluded that the injector conditions are essential in predicting particle-laden flows.

Fleckhaus et al. [34] reported the suppression of turbulence by 64 µm and 132 µm glass particles in a turbulent jet, with a particle mass ratio of m = 0.3. The particle velocity variance reported by them was much lower than that of the fluid. Maeda et al. [35] used a phase doppler analyzer (PDA) in their studies and did not observe any variation in the mean particle diameter measured at different radial locations within the jet.

Tsuji et al. [36] studied the effect of coarse particles on jet turbulence. Measurements in particle-laden jets were made using three devices: a pitot tube, LDA, and a specially designed optical fibre probe. Particles ranging in size from 170-1400 µm were used. The effect of larger particles was qualitatively different from the case of smaller particles (as discussed earlier). In addition, the turbulence intensity is reduced at higher particle loading.

Mostafa et al. [37] and Hardalupas et al. [38] reviewed the previous experiments on particle-laden jets, conducted studies using a two-component PDA, and measured mean and fluctuating velocities. Mostafa et al. made detailed velocity measurements for the near field region of the jet (x/d < 12) using 105 µm glass beads. The results were qualitatively similar to those published previously. Hardalupas et al. [38] studied the region up to x/d = 30 for three different particle sizes. The particle inertia was related to the lagrangian integral time scale of the fluid by developing a turbulent Stokes number. For the largest particle size (200 μm), which had the smallest Stokes number, the flow was found to have little effect on particle velocity characteristics. The flow affected the smaller particles (40 μm and 80 μm), resulting in centerline velocity decay. Hardalupas et al. reported that in the region up to $x/d = 30$, the turbulent Stokes number should still be too small for turbulence to profoundly alter particle paths. Therefore, the initial conditions of the flow, rather than the interactions of the fluid and particles, were dominant in the determination of the level of particle velocity fluctuations in the flow.

Zoltani and Bicen [39] utilized LDA to study dilute ($< 7.1 \times 10^{-6}$ solid concentration) two-phase jet flows. They reported that the velocity profile for the particles was flatter, indicating a turbulent flow field. Fan et al. [40] used a laser diffraction measurement technique to obtain the local particle concentration in silica gel laden jet flows. The solid concentration levels and the particle size ranges were 2.8×10^{-4} and 18.5 μm to 261.6 μm respectively. They observed that larger particles (>160 μm) tend to concentrate around the jet centerline.

Longmire and Eaton [41] have tried to examine the behavior of particles in a jet dominated by vortex ring structures. An axisymmetric air jet with 55 μm glass particles was axially forced with an acoustic speaker to organize the vortex ring structures rolling up in the free shear layer downstream of the nozzle exit. Flow visualization was obtained using a pulsed copper laser. Hot wire anemometry and LDA were the instrumentation used. Instantaneous photographs and videotapes showed that particles are clustered in the saddle region downstream of the vortex rings and propelled away from jet axis by the outwardly moving flow in these regions. Results indicate that the particle dispersion is governed not by diffusion but by convection due to large-scale turbulence structures. Dispersion control mechanisms were demonstrated using single and double waveforms inputted to the speaker for particle-to-air mass loading ratios up to 1.0.

Concentric Jet Flows

One method of contacting particles with gases is to get them as separate jets into a mixing zone. The jets may be free or confined by the walls of the system. Hedman et al. [42] conducted experiments to determine the mixing characteristics of a central particle-laden primary jet with a clean, gaseous, secondary jet. The radial velocity and particle mass flux profiles were measured using a pressure probe and an isokinetic probe, respectively. The coarser particles mixed more rapidly relative to the fine particles.

Subramanian et al. [26,43] have reported on a series of experiments in concentric jets with 150-180 μm sand particles. The particles were injected either in primary or in secondary flows. Special probes were used for making velocity and concentration measurements. They observed that when particles are in the primary stream, mixing improves as particle loading increases. When the particle is in the secondary stream, the annular potential core and entrainment increase as the particle loading is increased.

Pipe Flow

Two-phase flow in pipes has been of interest for a long time in pneumatic conveying. Owen [44] discussed the behavior of fine particles in a shear flow and transport of solid particles by a turbulent air stream in a horizontal pipe. Tsuji et al. [45] provided detailed measurements of an air-solid two-phase flow in horizontal pipe. They found an increase in velocity asymmetry as the loading ratio increased and the air velocity decreased. They observed that the effects of solid particles on air-flow turbulence varies greatly with particle size. Since there is an asymmetric structure of particle concentration due to gravity, they extended this work and investigated flow in a vertical pipe [36]. They observed that large particles increased air turbulence throughout the pipe cross section, while small particles reduce it. Lee et al. [46] conducted experiments on glass particle-air flow and showed that the presence of 800 µm glass particles increased the turbulence over the entire cross section. Theofanous et al. [47] measured mean velocity and turbulent two-phase dispersed flow and their results were in good agreement with other results.

Shahnam et al. [48] conducted an experiment in gas-solid flow in an axisymmetric sudden expansion for different expansion ratios. They obtained axial velocity profiles and turbulent intensities. Flow properties were measured using fiber optics LDA. They observed that the center line velocity decayed at a faster rate for smaller expansion ratios. In a relatively recent study, Liljegren and Vlachos [49] utilized a phase doppler particle analyzer (PDPA) to study two phase particulate air flows in a horizontal pipe. Glass spheres of 50 µm nominal diameter were used. The particle concentration range was 10^{-4} to 10^{-3}. Particle size and velocities were measured. They observed that an increase in loading results in an increase in velocity fluctuations.

Modeling

In addition to the above three categories, modeling of particle-laden jets is also relevant to the project. Various models are discussed by Longmire and Eaton [41]. They concluded that Eulerian as well as Langrangian methods of modeling particle dispersion can be developed based upon available experimental data. The models contain a certain number of adjustable constants, and can also be optimized to produce good fits to existing sets of time averaged data. Longmire and Eaton, in their conclusion, state that none of the models reported account for the organized, large scale ring structures which have been shown to be present in near and far fields of jet flows. They also concluded that the experiments and models available so far provide very useful qualitative information about how particle motion is related to flow structures; but more quantitative data, especially on jet flows, is needed.

Conclusions

Though considerable work has been done in the area of injection of single-phase jets into cocurrent flow, generally acceptable mixing phenomena criteria have not been developed. Confined particle-laden turbulent flow, in which particle-laden primary jet flow is injected into the duct containing secondary gas flow, is even more complex. The complexity of turbulent mixing is increased by the influence of various factors such as particle loading, particle size, and primary and secondary velocities. Understanding of this complex flow field may contribute to the improvement of sorbent utilization and sulfur dioxide capture in FGD processes. This calls for precise and insightful experiments to obtain the characteristics of particle-laden cocurrent jet flows.

Unfortunately, the complex nature of the confined particle-laden turbulent jet flow poses significant problems in making measurements. The gas-particle environment poses numerous problems for both optical and nonintrusive diagnostics. The inability to measure fundamental quantities within the gas-particle flow greatly limits the understanding of the flow characteristics. Detailed information about the behavior of confined particle-laden flow is important for an understanding of mixing characteristics. Improved diagnostics are required to provide (1) the necessary information about the gas-particle flow and (2) detailed data suitable for model development.

A review of the literature in the area of atomizer sprays in cocurrent flow (Kadambi et al. [50]), did not reveal any paper in the area of interaction of water spray and particles in cocurrent flow relevant to dry FGD processes.

This literature review reveals that though there have been some studies of two phase mixing problems and particle-laden jets, so far many of the fundamental questions involved in the two- and three-phase flows, especially relevant to FGD processes, have not been resolved. In addition, the mixing of the sorbent particles in FGD processes can possess different features because of different particle sizes, jet Reynolds number, particle loading, and agglomeration properties. The information required to develop a better understanding of the phenomenon of mixing relevant to FGD processes where a two- or three-phase, gas-solid, gas-solid-liquid, turbulent, horizontal jet issues into a cocurrent flow with relatively smaller diameter particles has to be developed.

The Objectives of Investigation of Transport Processes in FGD

A five-year plan was developed to improve our understanding of such flows involved in dry FGD processes.

The objectives of this five-year plan of study are to experimentally obtain a basic understanding of (1) the turbulent flow structure of the mixing zone and its influence on particle dispersion; (2) the effect of particle loading on turbulent properties and mixing; (3) the effect of jet entrainment; (4) water spray-sorbent interaction, sorbent wetting, and mixing; (5) the flow field where certain ratios of jet velocity to flue gas velocity result in regions of negative flow and define the onset of negative flow; and (6) sorbent reactivity in the immediate mixing zone.

To meet these objectives it was necessary to develop a test facility that would provide the test conditions encountered in flue gas ducts in power plants so that tests could be conducted to improve our understanding of the transport processes involved in FGD.

Design Consideration for the Sorbent Injection Facility

In the design of the test facility, emphasis was placed upon simulating the conditions encountered in in-duct FGD processes. To be able to simulate the conditions encountered during the injection of sorbent (lime) jet into cocurrent flue gas, geometric, dynamic, and kinematic similarity were considered. Dynamic and kinematic similar-ity, particle-laden jet-to-stream momentum flux ratio, the Reynolds number, the Stokes number, the Froude number, and the particle loading and particle/fluid density ratio were considered. The ranges of these nondimensional parameters encountered in the COOLSIDE process are shown in Table 5.1.

Table 5.1. Comparison of Ranges of Nondimensional Parameters

Nondimensional Parameters Processes	Test Section	As Encountered in In-Duct FGD
Reynolds Number	1.5×10^4-1.7×10^4	1.5×10^4-1.8×10^4
Stokes Number	0.26-92	0.35-31
Froude Number	0.6-8.0	0.2-3.0

Table 5.1. Comparison of Ranges of Nondimensional Parameters

Nondimensional Parameters Processes	Test Section	As Encountered in In-Duct FGD
Particle-laden Jet-to-Stream Momentum Flux Ratio	1.4-601	1.49-598
Particle Loading	2.5-10	2.5-10
Particle-Gas Density Ratio	2066	1942

Geometric Similarity

In the COOLSIDE process nine one-inch internal diameter pipes are used for injecting the sorbent laden two-phase flow into the flue gas. The nine pipes are placed symmetrically in a 14 ft by 14 ft cross-section chamber in a 3 by 3 matrix configuration. The pipes are placed with their centers 3 feet apart. The distance from the pipe centers to the outer walls of the chamber is 4 ft. Based on the available experimental data Kadambi et al, [51] the cone angle of the outer edge of the shear layer of the jet was estimated to be 18 degrees [4], which indicates that there might not be any interaction between adjacent jets for an axial distance of 9 feet. Most of the mixing takes place within a distance of 20 diameters axially downstream of the jet exit as indicated from available data [27,28,51,52]. A jet tube size of 7.3 mm diameter and duct size of 88.6 mm by 88.6 mm were selected so that the duct wall would not interact with the jet for at least 35 diameters downstream of the pipe exit, ensuring no effect of the wall on mixing.

Particle-Laden Jet-to-Stream Momentum Flux Ratio

When a particle-laden jet is injected into a cocurrent stream, it entrains fluid from the surrounding cocurrent duct stream. The higher velocity gas-particle jet stream imparts momentum to the lower velocity gas in the duct. It is assumed that the sole force producing the deceleration of the particle-laden jet and acceleration of the surrounding fluid is the tangential shear stress within the mixing region. The development of the mixing layer can be analyzed in terms of (a) spread range, (b) entrainment rate, and (c) shear stress distribution.

The spreading rate of the mixing layer is a function of the velocity ratio and the density ratio of the two streams. The entrainment process occurs due to the effect of small scale turbulence at the interface and also due to the mixing resulting from the jet's potential core engulfing the surrounding volume of fluid in the duct. In addition, the amount of gas-particle jet that is entrained by the duct stream must depend upon the mass flux flow ($\rho_a U_{df}$) and momentum flux

($\rho_a U_{df}^2$). Similarly, the amount of duct stream that can be entrained by the gas-particle jet must depend upon the jet mass flux ($\rho_o U_o$) and the momentum flux $\rho_o U_o^2$. Eddy viscosity is a measure of the momentum exchange and depends upon momentum flux of both streams. Therefore the jet-to-stream momentum flux ratio is an important parameter for this study, as it drives (controls) mixing, entrainment, and jet expansion. The equations for the equivalent jet density, ρ_o, and U_o, the equivalent jet velocity, can be derived as $U_o = U_g (1+\gamma m)/(1+m)$ and $\rho_o = (1+m)^2 \rho_a / \{(1+\gamma m)(1+m\rho_a/\gamma r_p)\}$ where m is the mass loading ratio, U_g is the gas phase velocity, and $\gamma = U_p/U_g$. These equations were obtained utilizing equivalent jet continuity and momentum equations and the solid mass loading and volume fractions.

The particle-laden jet-to-stream momentum flux ratio range for the COOLSIDE processes is 1.49 to 598. The available air flow rates in our laboratory are sufficient to develop a test facility which provides ratios from 1.4 to 601. Thus, all the jet-to-stream momentum flux ratio encountered in the COOLSIDE process can be duplicated in our test facility design.

Reynolds Number

The Reynolds number $\rho_a d U_o/\mu$ represents the ratio of the inertia forces to the viscous forces and as with all flows, dictates the onset of turbulence in gas particle pipe flows. In single-phase flows in pipes, the Reynolds number essentially dictates the structure of the turbulence and thus influences the turbulence time scales, the ratio of the friction velocity to the centerline velocity, and other important macroscopic quantities. In gas-particle flows the Reynolds number has been shown to affect the particle diffusivity in horizontal pipes [53]. The Reynolds number range obtainable in this test facility jet is 0.5×10^4 to 1.7×10^4 as compared to 0.5×10^4 to 1.8×10^4 for the jet in the COOLSIDE process. Thus, the turbulence structure in the particle-laden jet in the test facility is very similar to that encountered in the COOLSIDE jet.

Stokes Number

The Stokes number $\rho_a d_p^2 U_o/18\mu d$ is the ratio of particle relaxation time to the macro-scale eddies of fluid turbulence. The Stokes number provides a measure of the particle's ability to respond to turbulence in different flow conditions [44,49]. The particle relaxation time is the time taken for the particle to accelerate from rest to 63% of the free stream velocity. The Stokes number range for the test facility is 0.26 to 92 as compared to 0.35 to 31 for COOLSIDE. Thus, the COOLSIDE range is more than duplicated in the test facility. The response of the particle to the fluid turbulence in the proposed test facility is similar to that in the COOLSIDE FGD processes.

Froude Number

The Froude number ($\rho_a U_o^2 / \rho_p gD$) represents the ratio of inertia forces to gravity forces. It is the parameter that governs the degree of deposition of the particle phase in a horizontal pipe. The ability of turbulence to suspend the particles against gravity in horizontal tubes is characterized by this number. Again, the range for the test facility of 0.6 to 8.0 as compared with 0.2 to 3.0 for the COOL-SIDE processes indicates that the Froude number is duplicated in the test facility.

Particle Loading and Particle-to-Fluid Density Ratio

The level of particle loading clearly has an important influence on the characteristics of gas-solid particle mixture flow. Particle loading is often described in terms of a mass ratio of particles or a mass loading of the particles. The particle-to-fluid density ratio, in addition to the particle mass ratio, permits the determination of the volume fraction. Boothroyd [54], discussing conventional dimensional analysis, also includes the mass loading ratio and the particle fluid density ratios as relatively important parameters for obtaining similarity in suspension flows. These parameters are important in addition to the Reynolds number, Stokes number and Froude number. For relatively dilute suspension flows where the magnitude of the drag force exerted by the fluid on the particles can be relatively large, Liljegran [49] believes that parameters based upon the relative mass of the particles produce better measures of the ability of the particles to affect the fluid. Again the particle loading is 2.5 to 10 in the proposed test facility, which is the same as that encountered in the COOLSIDE process. The particle-to-gas density ratio is 2066 in the test section as compared to 1942 in the COOL-SIDE process. Thus, the available particle-loading ratio and the particle-to-gas density ratio in the test section duplicate those encountered in COOLSIDE FGD process.

The preceding design considerations were utilized in finalizing the particle-laden jet-injection test facility. The facility and its instrumentation are discussed in the next section.

Sorbent Injection Facility

Sorbent Injection (Particle-Laden Jet Test Facility)

The primary objectives in the design of the flow apparatus were to obtain a steady flow of particles and air at the jet exit and to obtain steady flow in the duct. A schematic diagram of the test facility is shown in figure 1 and the test parameters are provided in Table 5.2. A jet tube of ID 7.3 mm is mounted in the center of the square duct (7) of size 88.6 mm. The length of the duct prior to the test section was chosen such that fully developed turbulent flow was obtained at the location of measurement [55]. A honeycomb was placed at the air inlet in order to get uniform flow. A shop air compressor supplied air to the duct and to the particle-laden jet flow generator. Particles (glass, lime) were fed to the jet tube via venturi inductor (5) as shown in figure 2.

The primary concern with the injection of the particles is the uniformity of delivery. To facilitate the mixing of particles in the jet air stream, a location far upstream (x/d=100) was selected for the addition of particles. Air supplied to the venturi inductor created suction around the body of the inductor, entraining the particles into the jet tube air. The particle feed was regulated using a control valve (4) attached at the bottom of the venturi inductor device. The mean particle feed rate achieved during any individual test was found to be extremely steady, provided that the position of the control valve was not changed. Particle mass loading was varied from 0 to 10 using a control valve. Near the end of the flow system, the particles were separated from the air by a cyclone separator (11) and collected in a container at the bottom of the separator. A filter bag is connected at the end. A load cell arrangement (12) in the container was used to obtain the mass of glass particles collected and the particle flow rate.

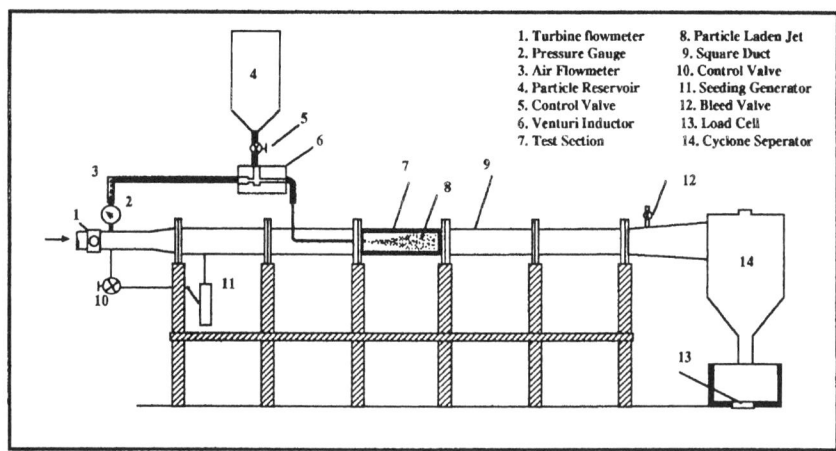

Figure 5.1 Schematic of particle-laden jet flow test facility.

Table 5.2 Test Facility Parameters and Specifications

Jet Diameter	7.3mm
Outer Square Duct	88.6 x 88.6mm^2
Test Section Length	300 mm
Flow Velocity	10 - 20m/s
Jet Velocity	30 - 35m/s
Jet Reynolds Number	$1.5 \times 10^4 - 1.7 \times 10^4$
Particle Size	2 - 100 μm
Particle Density	2450 kg/m^3
Particle Material	Glass
Particle Loading	2.5-10 (mass flow rate of particle/air mass flow rate)

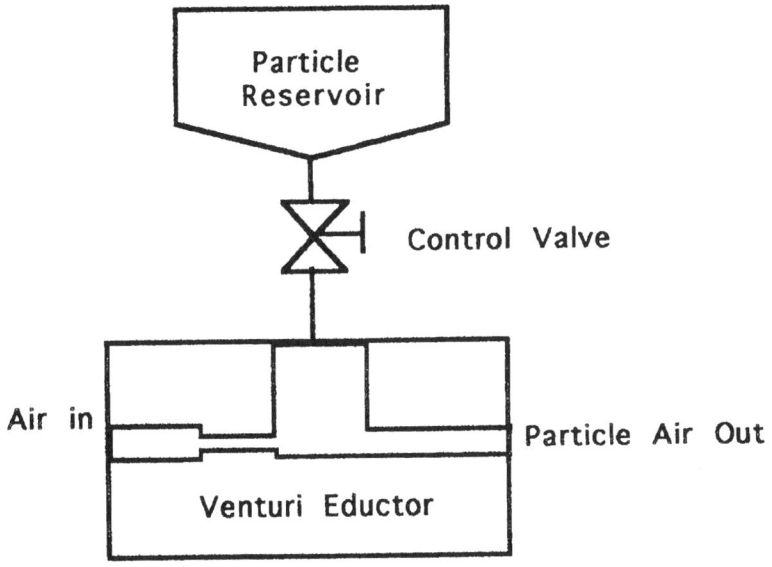

Figure 5.2 Particle feeder.

The mixing of the injected particle-laden jet and cocurrent air flow occurs at the test section (9). The measurement section has an optical (pyrex glass) window that allows the use of laser-based, nonintrusive optical techniques, such as phase doppler interferometry and laser doppler anemometry for flow field measurements. The PDI/LDA instrument is commonly referred to as PDA or PDPA. A seeding system (6), shown in figure 3, was also developed (Chinnapalaniandi [55], Kadambi and Chinnapalaniandi [56]). It was used in this study to measure the velocity profile for single-phase flow and for duct flow. Micronized talc powder was used as the seeding material [57]. The seeder is a cyclone separator in reverse; the powder at the base of the column is entrained by the flow and carried out by the central vortex.

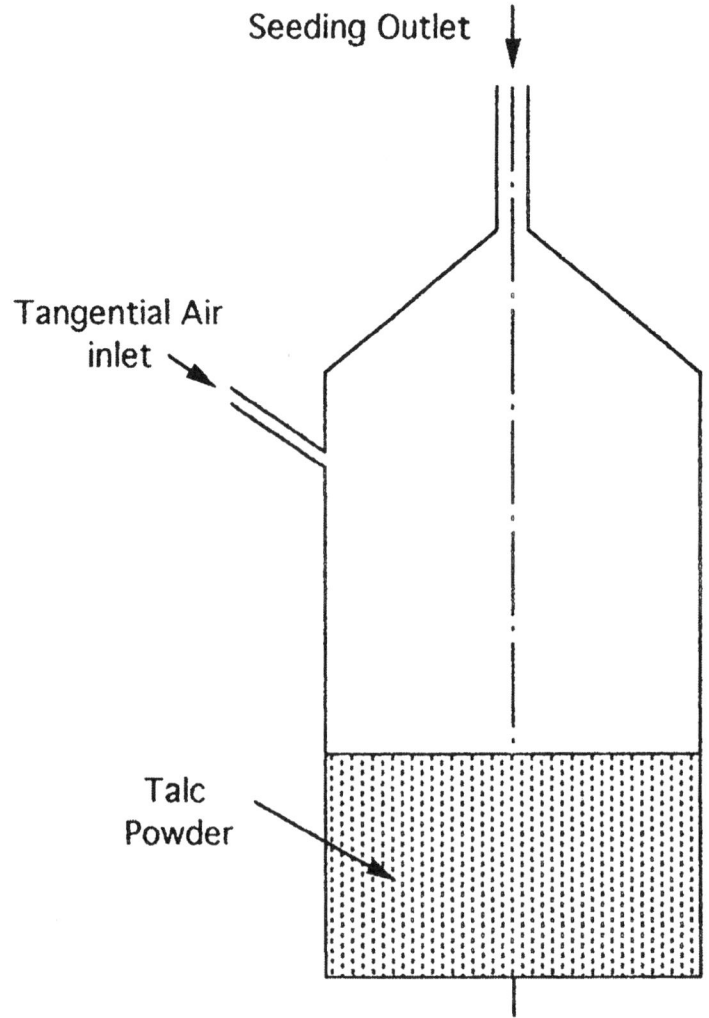

Figure 5.3 Schematic diagram of seeding generator.

Some preliminary tests with single-phase (air) flow were conducted in the test facility to ascertain the effect of the duct wall boundary layer on the mixing and deceleration of the jet in cocurrent flow. Tests were conducted with jet a centerline velocity of 35 m/s and cocurrent flow of nearly 20 m/s. The flow was found to be symmetric about the centerline along the horizontal and vertical plane. The Reynolds number based upon the duct hydraulic mean diameter and the flow properties was 118,000 and the velocity profile was turbulent. The jet diverging angle was found to be 17.3 degrees.

These angles are similar to the diverging angle of a jet in quiescent flow. The review of literature discussed earlier showed that the angle formed by particle-laden jets is relatively smaller than for comparable single-phase air jets. An 18 degree cone angle would reach the walls at x=35d and a 16 degree cone angle would reach the walls at x=39d. This result was useful in indicating that the particles from the jet would not reach the duct walls along the axial distance of interest, which is 5<x/d<25, where nearly all of the mixing takes place. The results of the preliminary tests are reported in detail by Chinnapalaniandi [55].

Characteristics of Particle Phase (Glass Particles, Lime)

Spherical glass particles and hydrated lime particles were used in this study. Glass particles were used since the PDA technique can be used only for spherical particles for the simultaneous measurement of particle size and velocity. This limitation of the PDA was later overcome with our development of transit time laser Doppler velocimetry (TTLDV) and then nonspherical lime particles were used. The characteristics of the two types of particles are provided below.

Glass Particles

The range of diameter of the glass particles used was 2-100 µm. Nearly all of the particles are spherical [55]. The properties of the particles are given in Table 5.3. The average particle density was measured to be 2450 kg/m^3. A Sedigraph 5000E was used to obtain the particle size distribution. The range of particle diameter varies from 1 µm to about 100 µm. Ninety percent of the mass of particles have a size of 3.5-100 µm. The particle size distribution was also measured using a PDA system at the centerline, downstream of the jet at x/d=5, with a mass loading of 10, and plotted along with the Sedigraph data in figure 4. The distribution shapes are comparable.

Table 5.3 Nominal Properties of Glass Particles

Diameter range(microns)	2-100
Density(Kg/m^3)	2450
Stokesian time constant(ms)	0.029-74

Table 5.3 Nominal Properties of Glass Particles

Terminal Velocity(cm/s)	0.024-73
Refractive index	1.5

Lime Particles.

Hydrated lime particles were obtained from Dravo Corporation through the courtesy of Dr. M. Babu. The lime particle size distribution varied from about 3 microns to 100 microns. Microtrac tests conducted at the Pittsburgh Energy Technology Center (courtesy of Dr. D. Wildman) revealed a mean size of 26 microns. The sizes obtained from the microtrac and the TTLDV technique developed for this project compared very well and will be discussed in a later section.

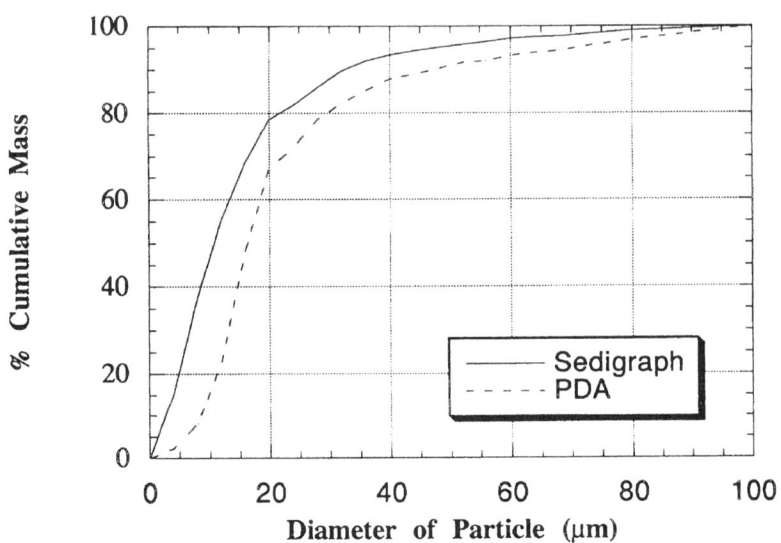

Figure 5.4 Particle size distribution from the Sedigraph and PDA (m=10).

Laser Based Optical Measurement systems

In this section the PDA system and the TTLDV system developed for this investigation will be described.

Phase Doppler Anemometry System. The PDA is the major instrument used in this study. The PDA is manufactured by Dantec Electronics. It utilizes a 5 watt

Argon-ion laser and is equipped with fiber transmission optics, receiving optics, and a model number 58N10 signal processor. The PDA simultaneously measures the size, velocity and concentration of spherical particles and is based upon LDA and PDI principles. The operation of an LDA can be explained by using a simple fringe model (figure 5). The laser beam from a continuous wave laser is split into two parallel beams which then pass through a focusing lens causing the beams to intersect at the focal point, which is also the measurement location. A set of plane parallel interference fringes are formed at the intersection. The fringe spacing is determined by the laser wavelength and the angle between the two beams. A particle passing through the measurement volume scatters light as it passes through the fringe pattern. The scattered light is collected and converted into an electric signal by a high speed photo detector. The frequency of the electrical signal is directly proportional to the particle velocity perpendicular to the fringe direction. This signal is then processed through a signal processor to obtain the velocity. This simple model describes the LDA operation. The fringe pattern produced at the crossing point of the beams will be stationary, resulting in directional ambiguity. This directional ambiguity is removed by utilizing a Bragg cell in one of the beams to provide a frequency shift. Consequently, the fringes are no longer stationary and the directional ambiguity is thus removed.

PDA uses the phase of the scattered light to obtain the information regarding the particle size. As with LDA, PDA can also be explained using the fringe model. Two photo detectors placed at separate locations in space receiving the scattered light from the particle in the measurement volume will see the same frequency but with a relative phase shift. For a spherical particle the phase shift is proportional to its diameter.

The detector position and corresponding phase responses are indicated in figure 5. Selecting a wide detector spacing (U1-U2) gives a phase shift which varies rapidly with particle diameter. However, one can only obtain unambiguous measurements for diameters corresponding to phase shifts less than 360°. Larger particles are indistinguishable from smaller particles and may have a phase shift greater than 360° will fold down providing incorrect results. In order to extend the unambiguous range and retain a high resolution, a three-detector configuration is used. The closely spaced detector pair (U1,U3) with slowly varying phase is used to resolve ambiguity in the measurements from the primary pair (U1,U2). Additionally the third combination can be used for self consistency and validation checks. The Dantec PDA software includes a procedure that attempts to calculate the particle concentration at each measurement point on the basis of the particle data rate, particle diameter measurements, and an estimate of the effective cross-sectional area of the measurement volume. This procedure is fully documented in the Dantec User's manual [58].

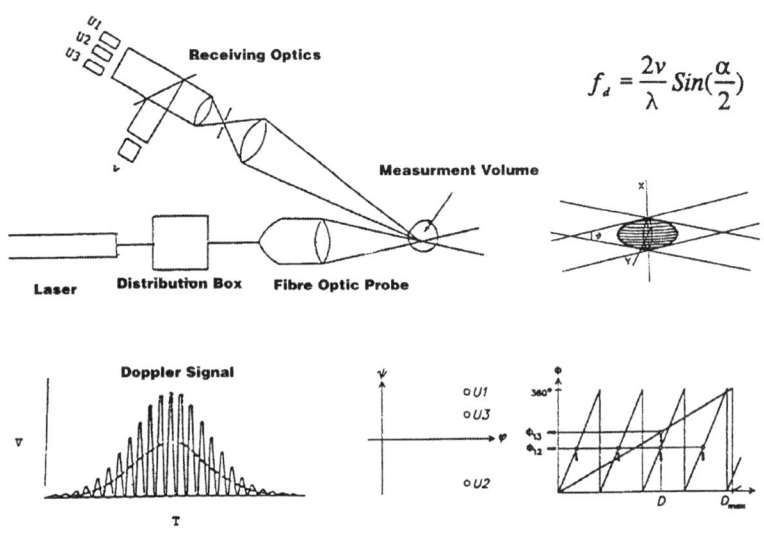

Figure 5.5 PDI/LDA principles.

Details of PDA/LDA theory and operations are provided in References [59,60,61]. The PDA transmitting and receiving optics are mounted on a three-dimensional traverse. This allows the mapping of the particle-laden jet flow in the duct. The PDA was operated in the back scatter second order refraction mode at an angle of 158°, as recommended in Dantec's instruction manual. Each measurement data set consisted of 3000 PDA/LDA measurements.

The PDA allows the simultaneous measurement of size, velocity and concentration of spherical particles, droplets, or bubbles. This measuring technique has been mainly applied to characterize spray systems [62,63], where only the size and velocity of the droplets are of interest. In complex two-phase flows at higher particle loading; however, it is important to measure the velocity of the continuous phase in the presence of the dispersed phase in addition to the size of the particles. Such applications of the PDA are still rare; only particle velocity and size were measured in two-phase pipe flow by Liljegren [49].

The PDA technique is limited to spherical particles. As a result, the technique cannot be relied upon to measure the size of nonspherical particles. This limitation was confirmed experimentally by Alexander et al. [64]. For single phase measurement in jet flows, instead of the glass particle feeding device, another seeder was used for seeding talc powder in the jet and the PDA was operated in the LDA mode.

A distinction between the continuous phase and the discrete phase is necessary in order to determine the gas-phase velocities in the presence of particles. The PDA allows physical discrimination between the phases. By sizing all particles, the size scores from particles used to seed the flow can be extracted and utilized to generate a separate set of statistics for the gas phase. This approach has recently been applied to sprays [65]. Hence, discrimination of phases is inherent in the operation of the instrument. By seeding the flow, sizing all the particles, and then extracting the particles which are small enough to track the flow (particles less than 5 µm were used in the study), the velocity of the gas phase within the particle-laden jet flow field was deduced.

DEVELOPMENT OF TRANSIT TIME LDV(TTLDV) TECHNIQUE. As stated earlier, the PDA technique is useful only for spherical particles and cannot be used to obtain the size of nonspherical, irregular particles, especially lime. Thus, one cannot obtain local concentration of nonspherical solids in such flows using PDA or PDPA. Therefore, to successfully study lime laden solid gas flows, one needs instrumentation that will provide the local particle velocity, size, and concentration as well as fluid velocity in the flow region of interest. Since PDA and PDPA cannot be used for obtaining sizes of nonspherical particles, the transit time LDV technique combination, called the TTLDV technique, was developed and used for lime laden jet flows.

The TTLDV technique utilizes the particle transit time in the measurement volume and the velocity obtained by the LDV technique. The operating principles of LDV have been explained by using a simple fringe model. The particle transit time is obtained by analyzing the intensity of the laser beam light scattered by the particle. For a particle whose dimension in the direction of fringes is d, if the transit time to cross the measurement volume major diameter of length D is t, then we have

$$tv = (d + D)$$

where v is the velocity of the particle obtained from LDV measurement. The length D of the measurement volume is fixed by the laser beam wavelength and the beam crossing angle. Thus, if one knows the transit time and velocity of the particle, one can obtain the linear dimension of the particle crossing the measurement volume. In general, due to shear and drag, irregular shape particles tend to move through the flow with their major principal dimension oriented streamwise in the flow direction. Based upon prior knowledge of the particle dimensions, one can then estimate the other mutually perpendicular dimension. However, in high shear flows where the particle may tend to rotate, this technique may not provide meaningful results. The covariance signal processor (Dantec Electronics model number 58N10) is utilized to simultaneously obtain the transit time and the velocity of the particle crossing the measurement volume. In principle, the technique appears to be simple and easy to apply. However, there are difficulties associated in the application of the technique which have to be overcome.

One major problem area is that if the particle is passing through the edge of the measurement volume, it will not cross the diameter D, but rather a relatively smaller length. This problem can be addressed by either ascertaining that only the data from those particles which cross a fixed number of fringes, whose total spacing corresponds to D, are used or alternatively, by knowing the smaller length (the number of fringes) crossed. The size of the measurement volume relative to the particle size (the linear dimension crossing the measurement volume) is also very important. A substantially larger particle can tend to block the two laser beams and thus not allow the laser beams to intersect at the measurement location. Thus, a judicious selection of the measurement size is very important and has to be based upon the range of the size of particles involved. Another problem stems from the fact that the intensity of the scattered light is a function of the particle surface, the laser beam intensity, the photomultiplier voltage and gain, and the particle size. Of these factors, the laser beam intensity and photomultiplier voltage and gain can be preset and kept constant. Particle surface depends upon the type of particle, and so long as the photodetector can detect the light scattered by the particle, the technique will work. Particle size is, of course, an unknown which we want to obtain. Some knowledge of the statistical variation of particle sizes in the flow will allow one to select a measurement size and number of fringes which will accommodate the range of particle size to be encountered. Initial feasibility calibration tests utilizing known diameter wires, placed radially on the circumference of a rotating wheel and crossing the measurement volume, were conducted. These were followed by tests with known size spherical glass particles and irregular shaped lime particles.

Calibration Tests

ROTATING WIRE TESTS. Feasibility calibration tests were conducted with known size wires which were attached radially to a rotating wheel. Each wire crossed the LDV probe volume once per revolution. Different size wires were mounted in such a way so as to cross the probe volume as close as possible to the probe volume's centerline. Figure 6 shows the rotating wheel, the wires, and the measurement volume. Wires of diameters 1150 microns and 250 microns were used. The LDV set up included a 5 w Argon ion Laser, LDA/PDA receiving and transmitting optics, covariance processor, and a 486DX computer for data acquisition and analysis. Green beams of wave length 514.5 nm were used. For the test configuration, the measurement volume had a major diameter of 90 microns.

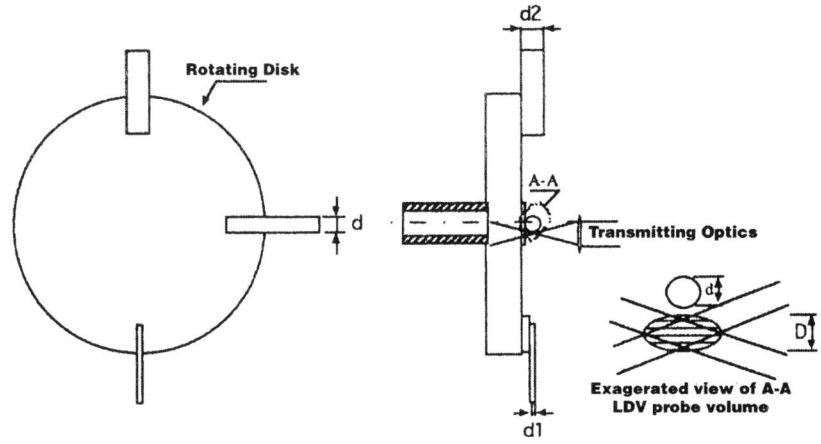

Figure 5.6 The wire disk assembly and the LDV probe volume.

Figures 5.7, 5.8, and 5.9 show the results from a representative calibration test where two wires of diameters 1150 and 250 microns were used. Figure 5.7 shows the histogram of the two wire velocities which are the same (1.7 m/s) and are within ± 0.5% of the velocity obtained from the wheel speed. Figure 5.8 shows the transit time histograms for the two wires. The histogram of the product Vt is shown in figure 5.9. The mean values associated with the histograms for the two wires are 310 and 1130 microns, and if one subtracts the length of the measurement volume (90 microns) the wire sizes one obtains are 220 and 1040 microns, as compared to measured values of 250 and 1150 microns. These values are within 12% and indicate that it may be feasible to use this technique to obtain the particle size and velocities simultaneously in solid-fluid two phase flows.

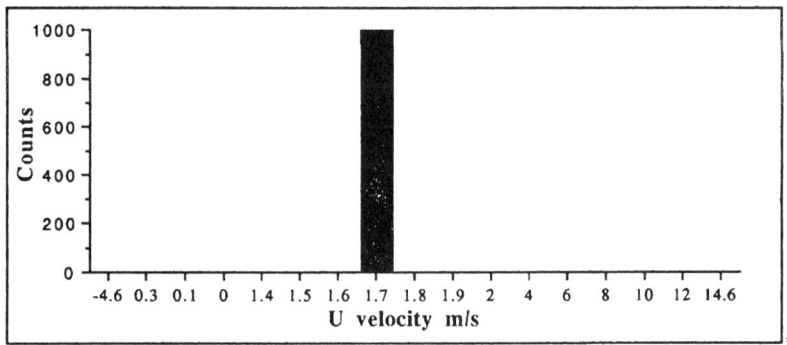

Figure 5.7 Histogram for two rotating wire velocities.

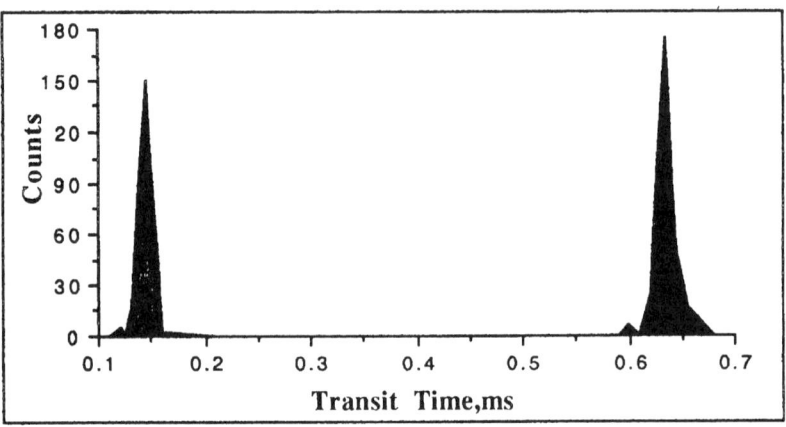

Figure 5.8 Histogram of the two rotating wire transit time. The larger value is for the 1150 micron wire.

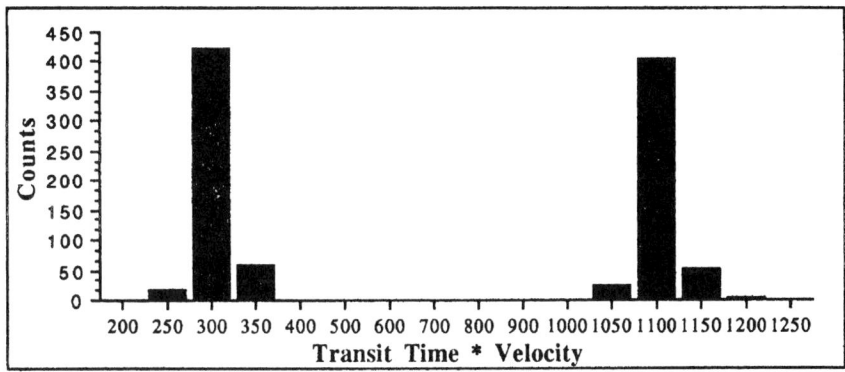

Figure 5.9 Histogram of product vt for the two rotating wires. The larger value is for the 1150 micron wire.

The differences in the wire sizes obtained in these tests and the measured values can be attributed to (i) slight bending of the wires which is not of any consequence in two phase flows, and (ii) the wire not crossing exactly at the center of the measurement volume. Item (ii) has to be addressed in solid-fluid flows. One also has to obtain the optimum measurement volume size for a given range of particle sizes. The issue of particles not crossing in the middle of the measurement volume can be addressed by ensuring that only signals from particles that cross a given number of fringes in the measurement volume are accepted.

GLASS PARTICLE TESTS. The next set of tests was conducted in the sorbent injection test facility using micron size glass beads. The beads were injected into a cocurrent flowing air stream. Particle sizes were obtained using PDA and TTLDV methods. Focusing lenses of focal lengths 310 mm and 160 mm were used in the transmitting optics. The beam separation was 36 mm. For this beam separation distance, the fringe spacing and measurement volume diameter (D) were 4.44 microns and 2.30 microns and 0.15 mm and 0.078 mm, respectively, for the 310 mm and 160 mm focusing lenses. This allowed us to investigate the effect of fringe spacing and measurement volume diameter size on the measurements. Off axis back scattering on the second order refraction mode was used for PDA measurements.

Experiments were carried out at a location 2.5 cm downstream of the jet exit along the jet centerline. Spherical glass beads of mean diameter 40 microns were used. Mean particle velocities were varied from 9.3 m/s to 18 m/s. A very important aspect of the test was the selection of optimal fringe count which ascertained that particles were passing very close to the center of probe volume. For the given LDA parameters, a fringe count of 35 ensures that the measurement volume size is crossed. Gain and voltage settings were set to ensure high data and validation rates. The effect of fringe spacing and the number of fringes (i.e., the measurement volume size) on the TTLDV measurement is shown in Figure 5.10. The number of fringes in the measurement volume stays the same (35) for the two lenses (160 mm and 310 mm focal length) but the fringe spacing (df) is different, resulting in a smaller measurement volume size of 78 microns as compared to 150 microns for the 310 mm lens. One can observe that results with the 160 mm focal length lens compare very well with the PDA data since the measurement volume size of 78 microns is close to the mean glass particle size of 40 microns as compared to 150 microns for the 310 mm focal length lens. One can also observe from the figure that a selection of the minimum number of fringes larger than the optimum 35, obtained from LDA parameters for the measurement volume, results in obtaining larger particle sizes. One can count more than 35 fringes because of the 40 Mhz fringe shift and if the particle is moving at a smaller velocity than the rest of the particles. This is normally true for larger size particles, thus biasing the data towards larger size and not accepting the relatively smaller size particles which crossed the measurement volume along the center but crossed 35 fringes only. However, it should be noted that the measurement volume size considered is the size obtained from the LDA parameters only. The TTLDV method results compare very well with the PDA results for a fringe spacing of 35 and are within 10% of the results obtained from PDA. Even for fringe spacings larger than the optimum (40) the results are within 25% of PDA data. Figure 5.11, a typical TTLDV test with the 160 mm lens, shows the comparison of particle diameter size obtained from PDA and TTLDV. As can be seen from the figure, the mean sizes obtained from PDA (43.2 microns) and TTLDV (41.0 microns) are very close.

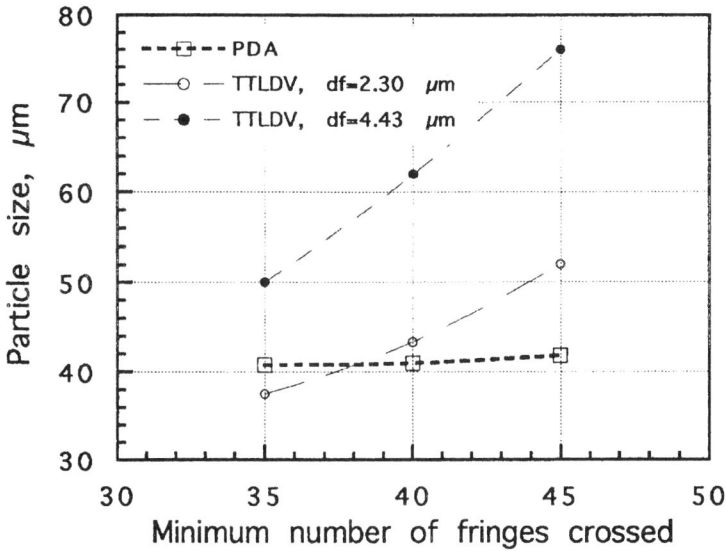

Figure 5.10 The effect of measurement volume size and number of fringes crossed on the TTLDV measurements.

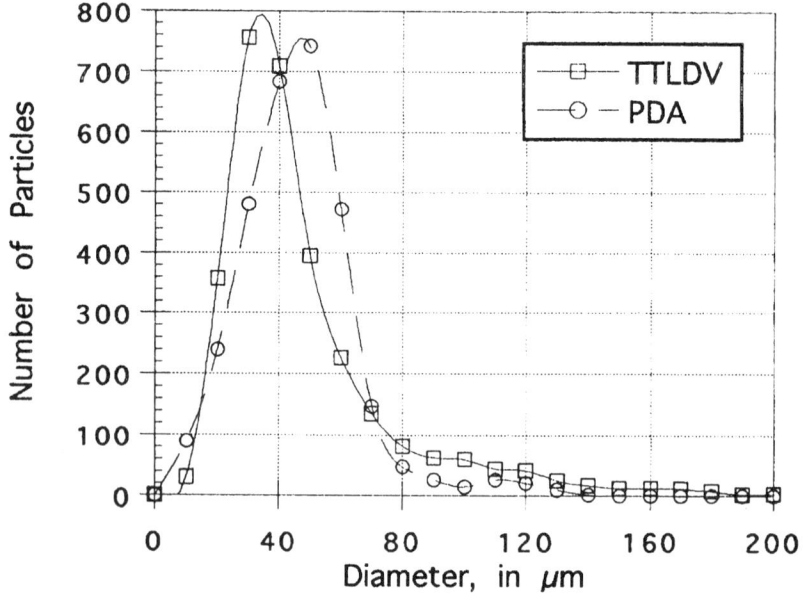

Figure 5.11 Glass particle size distribution obtained from PDA and TTLDV (160 mm lens, TTLD mean = 41.0 µm, PDA mean = 43.2 µm).

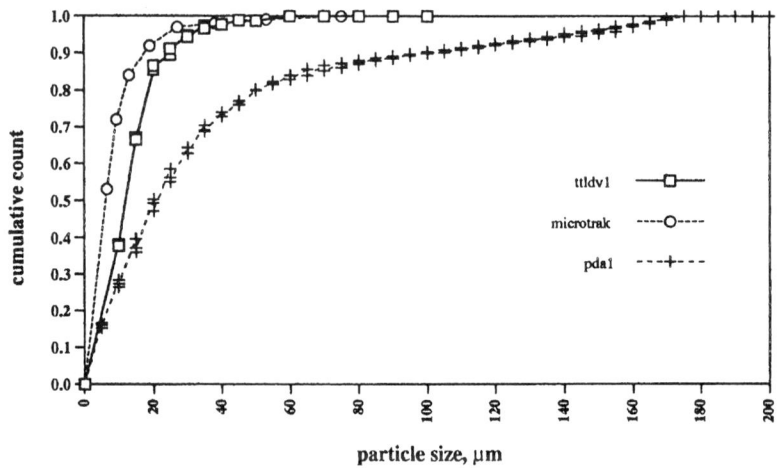

Figure 5.12 Comparison of the lime particle size distribution obtained from TTLDV, PDA, and Microtrac results.

IRREGULAR SHAPED LIME PARTICLE TESTS. Experiments with irregular shaped lime particles of mean size 26 microns were also performed in the sorbent injection test facility. The mean size of 26 microns was obtained by analyzing the lime particles with the Microtrac. The Microtrac utilizes the Fraunhoffer diffraction method to obtain the particle size. As in the case of glass beads, measurements were made 2.5 cm downstream of the jet exit along the center line. The mean particle velocity was 18.0 m/s. A beam separation of 36 mm, as in the glass beads test, was used. A 160 mm focal length lens was used in the TTLDV mode for the lime tests. Data were also taken using the PDA. Figure 12 shows the cumulative size distribution obtained for lime tests using TTLDV, PDA, and Microtrac data.

The agreement between the TTLDV and Microtrac data is good; while the PDA results are off by a substantial margin (a factor of 4). The Sauter mean size of 26 microns obtained from Microtrac compares favorably with TTLDV Sauter mean size of 32 microns. The PDA Sauter mean size of 119 microns is off by more than 4 times indicating the inability of PDA to obtain reliable data for irregularly shaped particle size.

These results show that the TTLDV technique can be used for simultaneous measurement of irregular shaped particle size and velocity in solid-fluid flows. Kadambi et al [66] provides further details of the TTLDV technique.

Test Results and Discussion

The first series of tests was conducted in a single-phase flow, with an air jet exiting into cocurrent duct flow, to provide a baseline data for comparison purposes. The second series of tests were conducted with a particle-laden jet (glass as well as lime) in cocurrent flow. Results of these tests will be discussed in this section.

SINGLE PHASE FLOW MEASUREMENTS. The objective of the single-phase measurements was to verify that the fluid velocity characteristics in the test configuration match those obtained by other investigators and to provide a baseline measurement for comparison to the particle-laden flows. Measurements were made for the flow conditions shown in Table 5.4. The mean and fluctuating fluid velocities were measured for the confined single phase jet flow along the jet axis and also across the jet in the region of $5 < x/d < 20$.

The axial component of the mean and fluctuation velocities at the centerline are similar to that observed in earlier investigations [6,67]. The velocity profile is quite flat near the jet exit in the potential core region for x/d up to 5, and a lengthening of potential core is indicated as Uo/Udf decreases. This is followed by a nearly linear velocity decay in the region of 5 < x/d < 20. A slower rate of velocity decay was exhibited when cocurrent flow velocity was decreased. The magnitude of velocity fluctuation peaked at locations in the region of 5 < x/d < 15 for the flow configurations. A similar trend was observed by Antonia and Bilger [68], but after the potential core the fluctuation velocity was nearly constant up to x/d = 80. Then it increased slightly for x/d > 100.

Table 5.4 Test Conditions

Single Phase (m=0):		Particle-Phase (Glass Beads):		
Centerline Jet Velocity, U_o (m/s)	35, 30	Particle diameter (microns)	2-100	
Mass flow rate of jet flow ms (kg/s)	1.7×10^{-3}, 1.5×10^{3}	Particle density, p_p (kg/m^3)	2450	
Reynolds number of jet $\rho d U_o/\mu$	1.7×10^{4}, 1.5×10^{4}	velocity ratio U_o/U_{df}		mass loading
Cocurrent Velocity, U_{df} (m/s)	20, 15, 10	Case 1:	35/20	2.5, 7.5, 10
		Case 2:	35/15	2.5, 7.5, 10
		Case 3:	35/10	2.5, 7.5, 10
		Case 4:	30/10	2.5, 7.5, 10

Note: PDA measurements were made at axial location of x/d=0.1 of jet tube

Axial Velocity Profile Across the Jet

The exit mean fluid velocity was measured at an axial location of x/d=0.1 downstream of the jet tube exit. The mean velocities for jets at Reynolds numbers of 17033 and 14600, with a constant duct flow Reynolds number of 119117, were compared with the 1/7th power law profile,

$$U/U_\circ = (1 - 2r/d)^{1/7}$$

Jet Reynolds numbers are based on the jet tube diameter, jet velocity at the exit of the tube, and air properties. The duct Reynolds number is based on the duct velocity, hydraulic diameter, and air properties. The measured profiles are in good agreement with an empirical 1/7th power law profile. Modarress [32] measured mean velocity using LDA for a single phase jet, 1 mm downstream of jet exit. He found an experimental value of n = 6.67 which compares well with our data. This implies that the talc powder follows the flow very well when used as a seed for LDA.

Figures 5.13a and 5.13b show the characteristic change in the radial distribution of intensity of axial turbulence $\sqrt{u'^2}/Uc$ and intensity of radial turbulence, $\sqrt{v'^2}/Uc$, respectively, due to the presence of an external stream. In all cases, the relative intensities along the jet axis are reduced and in the outer region of the jet are also reduced. However, in the middle of the distribution, there is a region where turbulent intensity shows its maximum amplification. This region is considered to be a characteristically active turbulent field in the confined jet. Turbulence intensity increases from the stream intersection, the edge of the jet to a maximum near the centerline of the jet, due to the velocity gradient and production of vortices.

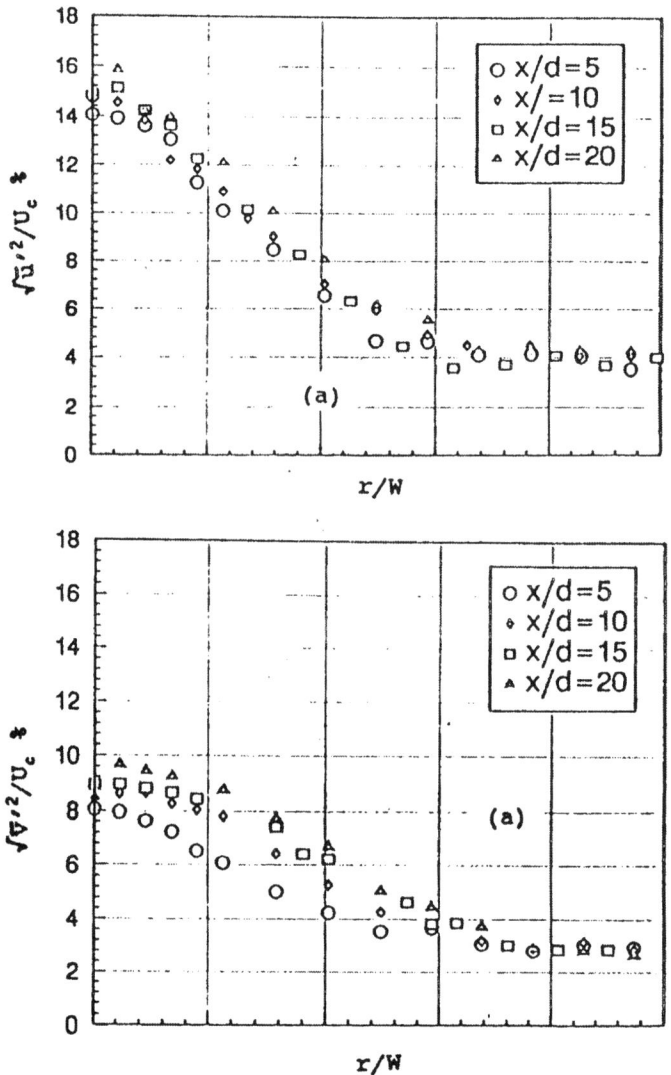

Figure 5.13 a: Intensity of axial turbulence at axial locations of x/d=5, 10, 15 and 20 (case 1). b: Intensity of radial turbulence at axial locations of x/d=5, 10, 15 and 20 (case 1).

Comparison of the profiles of the intensity of axial turbulence with the profiles of intensity of radial turbulence shows that the shapes of the two profiles are similar. The values of the intensity of radial turbulence are, however, approximately one-half the values of the intensity of axial turbulence.

Particle-Laden Flow Measurements

Tests with particle-laden jets were conducted after completion of the single-phase tests. The test conditions are provided in Table 5.3. The conditions reported in the table are immediately downstream (x/d=0.1) of the jet. PDA/LDA measurements were made at axial locations of x/d=5, 10, 15, and 20 downstream of the jet exit. Mean and fluctuating components of the velocities for gas and particle phases, particle concentration, particle size profiles, and Reynolds stresses were obtained from the data. The results of the single-phase tests were used as the baseline data for these flows.

Axial Development Along the Centerline

In the text and in the figures that follow the terms "gas phase" (and the subscript "g") and "particle phase" (and the subscript "p") are used to denote quantities measured for particle-laden flows and "single phase" is used to denote the result of pure air flow.

The presence of particles in the flow has an immediate effect on the gas-phase velocity at the jet exit. The gas-phase velocity was slightly lower than that for the single-phase data at the jet exit. This is due to momentum transfer from the gas phase to particles occurring inside the tube section, since the particles enter the tube at a velocity lower than that of the air flow. As a result, the gas-phase velocity is slowed down and particles are accelerated. At the jet exit, particles have a velocity lower than the gas phase. The difference in velocity is known as the slip velocity. The effect of mass loading in the development of the velocity profile will be discussed in the following section.

Figure 5.14 a shows the effect that the presence of particles has on the gas phase along the centerline of the jet when the mass loading is 2.5 (case 1). At the particle-laden jet exit, the particle velocity is less than the gas-phase velocity of the solid-gas slurry. The gas-phase velocity is less than the single-phase (m=0) air velocity. Figure 5.14(b) shows that the effect of increasing the mass loading from 2.5 to 10 is to decrease the mean axial particle velocity decay relative to the single phase. The behavior of the axial velocity profile is qualitatively similar for the two mass loadings. A general trend, however, can be summarized as follows: An increase in mass loading results in higher mean particle velocities in the region of $5 < x/d < 20$. Near the jet tube, the particle velocity lags behind the fluid velocity and this velocity lag increases with mass loading. In the region of $5 < x/d < 20$, the particle-phase velocity of a particle-laden flow becomes higher than that of single-phase flow. This is due to a reduction in the jet spread in the presence of particles.

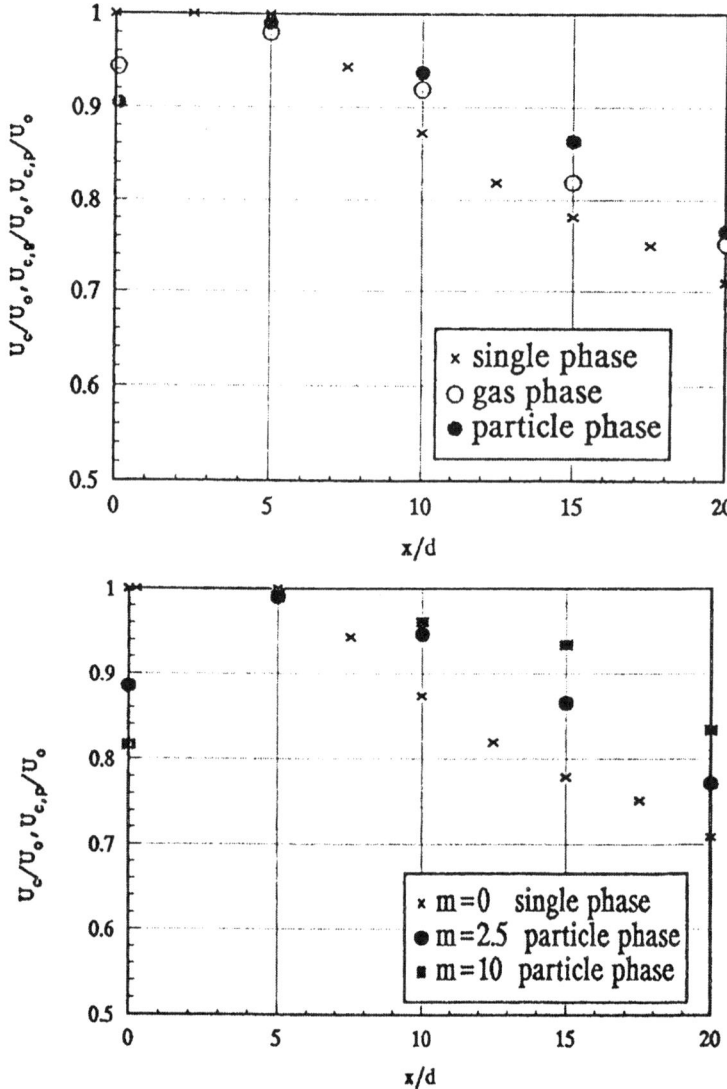

Figure 5.14 a: Mean particle phase and gas-phase velocities for case 1, m=2.5.
b: Mean particle phase-velocity along the centerline for case 1, m=2.5, m=10

It is observed that in the region $0 \leq x/d \leq 5$ the particle and gas phases accelerate until the slip velocity changes signs. The explanation for this behavior was obtained by mapping the flow across the jet and taking the data in the radial direction up to 10 mm from the centerline of the jet at axial locations of x/d=0.1, 3, and 5 for U_o=35 m/s, U_{df}=20 m/s, and m=5. The results are shown in Figure 5.15. The tube radius, r_o, is 3.65 mm. At an axial location of x/d=0.1 we were able to measure the velocity up to the radial location of r=3.5 mm. No data were observed between 3.5<r<3.8, indicating that there were no particles in this region. The tube wall thickness (t) was 1.2 mm, and the tube thickness to radius ratio, t/ro was 33%. The radial region extending from r=3.5 to r=3.8 is therefore in the wake region. It appears that the cocurrent flow pushes the wake region slightly down resulting in a reduction in the flow area at the tube exit. This in turn results in the acceleration of the flow exiting from the tube. The wake effect extended to the axial location x/d=3. There is a velocity defect at radial location $r \approx 6mm$, but at the axial location x/d=5, there is no wake effect from the tube wall, and hence no more acceleration. The particles were, therefore, initially accelerated by the gas phase until the slip velocity changed signs; i.e., until the two phases reached equilibrium. This indicates that, initially, momentum is transferred from the gas phase to the particles. Beyond the potential core, the gas-phase velocity decreased due to the spreading of the jet, but at a reduced rate because of the momentum received from the faster particles. For the region 5<x/d<20, the mean gas-phase velocity along the centerline of the jet became greater than that for single phase flow. This resulted from the reduced spread of the particle-laden jet. The presence of particles has a pronounced effect on the rate of decay of the mean gas phase velocity along the jet axis. The effect of increasing the mass loading of the particles on the axial development of the centerline turbulence, $\sqrt{u}$'p2 and $\sqrt{v}$'p2 are depicted in figures 5.16 and 5.17. The values are again normalized by the centerline jet exit velocity of the single phase U_o.

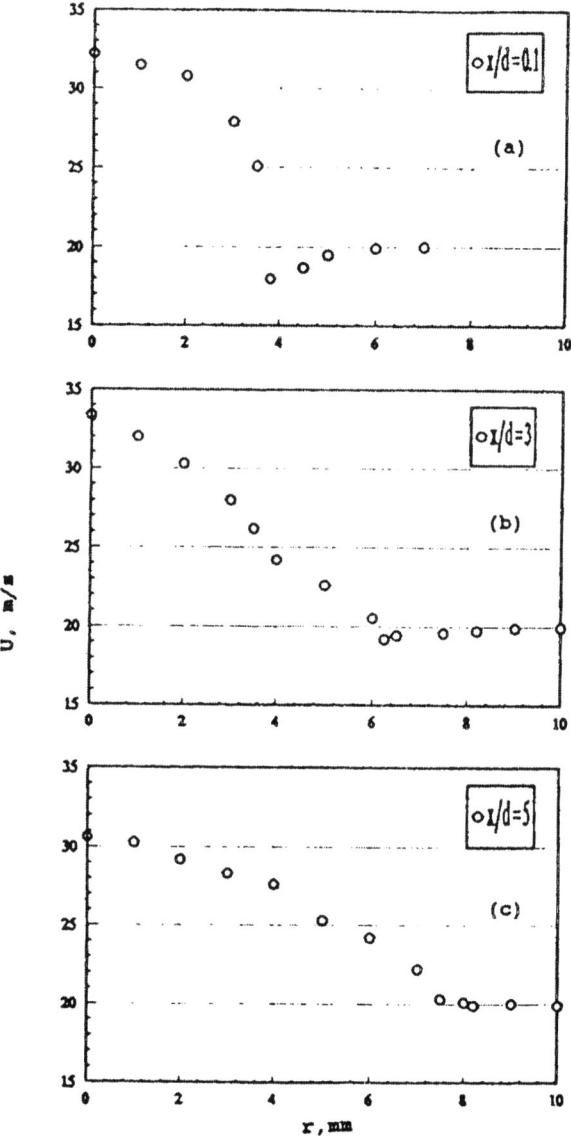

Figure 5.15 Effect of wake due to jet tube wall thickness at (a)x/d=0.1, (b)x/d=3, (c)x/d=5 from $U_0=35$, $U_{df}=20$, m=5.

Turbulent fluctuation velocities for the particle phase are affected more when the particle mass loading is increased. In the region $5 < x/d < 20$, the fluctuation velocity decreases as the particle loading is increased. In the jet exit region x/d < 5, adding particles to the flow increases the fluctuation velocity as compared to single phase fluctuation velocity at the same flow conditions (case 1).

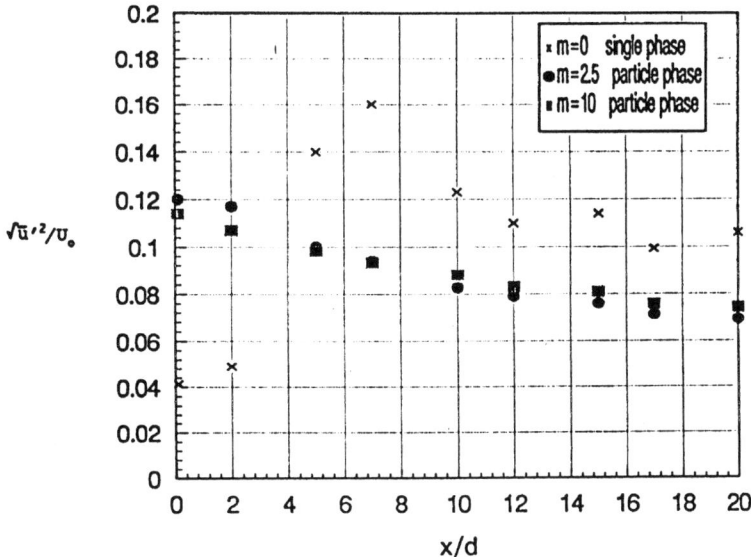

Figure 5.16 Axial fluctuating velocity along the centerline for case 1 at m=2.5 and m=10.

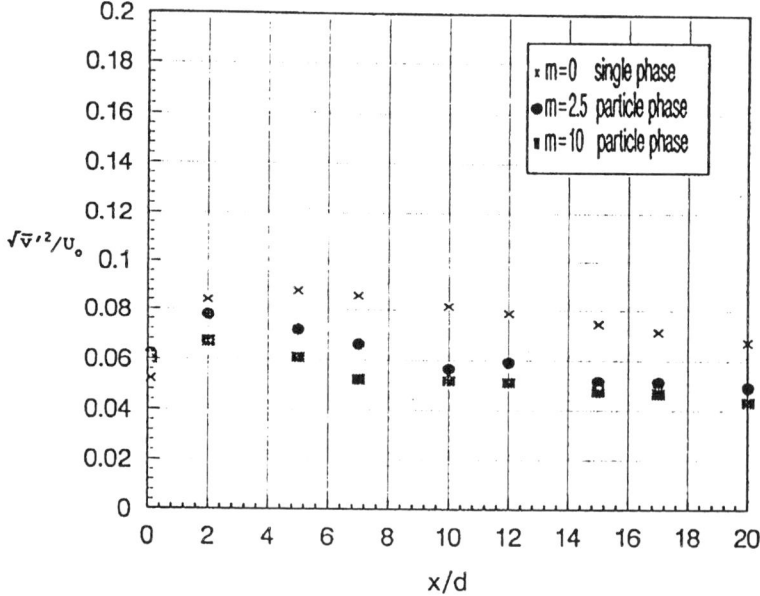

Figure 5.17 Radial fluctuating velocity along the centerline for case 1 at m=2.5 and m=10.

Three notable features of the development are as follows: First, there is a large anisotropy in the turbulence of the particles, largely because $\sqrt{v'}{}_p^2$ is small and does not change with the downstream distance, whereas $\sqrt{u'}{}_p^2$ follows the development of the single phase fluctuating velocity $\sqrt{u'^2}$, at x/d > 5. Secondly, although $\sqrt{v'}{}_p^2$ is always smaller than $\sqrt{v'^2}$ as might be expected, $\sqrt{u'}{}_p^2$ is larger than $\sqrt{u'^2}$ at the exit of the tube and lower beyond x/d ≈ 5. Thirdly, increasing the mass loading results in a slight increase in $\sqrt{u'}{}_p^2$ downstream of x/d =5; in contrast, increasing the mass loading results in a small decrease in $\sqrt{v'}{}_p^2$.

Axial Velocity Profile Across the Jet

The mean axial velocity profiles across the jet exit for particle-laden flows were measured at x/d = 0.1, as shown in figure 18. Single-phase and particle-phase mean axial velocity quantities were normalized with respect to U_o, the jet exit centerline velocity. For the abscissa, the radial distance (r) from the jet axis was normalized by the jet diameter (d). These axial velocity profiles are within the jet tube. The single-phase velocity profile agreed with a power-law of $U/U_o = (1 - 2r/d)^{1/n}$, where the experimental value of n was found to be 6.7, which corresponds to a fully developed turbulent pipe flow. Fan et al. [40] found the experimental value of n to be 6.61 for a fully developed turbulent tube flow at the exit. The particle-phase velocities consistently lagged behind the single phase velocity near the jet exit centerline and led the single-phase velocity near the tube edge for case 1. Since the inlet profiles correspond to a fully developed pipe floHx (the injection tube diameter-to-length ratio was equal to 100) this flow has a thin boundary layer. Therefore, the crossover in the two velocities occurred near r/d = 0.36. Similar behavior for particle phase and single phase were obtained in gas-solid two-phase pipe flow by Liljegren [49]. The profiles change gradually from the single phase to the higher particle-laden jet flows and for the higher particle-laden jet (m = 10), the profile is nearly flat and uniform. As the particle loading increases, the particle phase velocity decreases and its profiles become flat. The origin of the change in the shape of the profile of particle velocity for higher mass loading is not clear, but it may be due to particle-particle interaction, which results in a reduction in momentum.

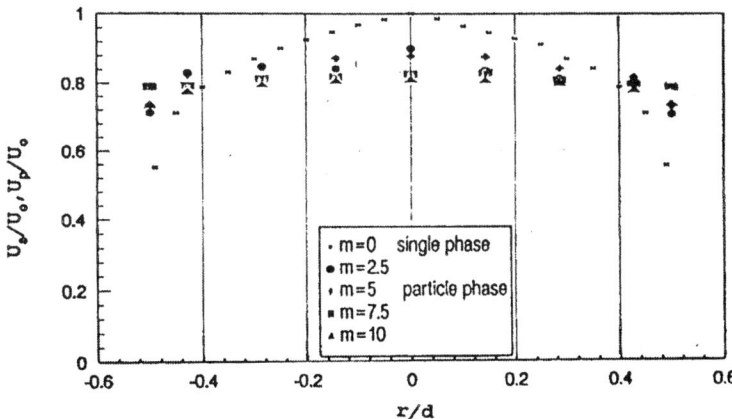

Figure 5.18 Velocity profile at the exit of the jet for x/d=0.1 (case 1).

The addition of particles causes increased difficulty in measuring the necessary quantities due to the need to discriminate between phases, as discussed in earlier. To check the accuracy of the discrimination process, the volume flow rates for the gas phase at the jet exit were calculated via integration of the velocity profiles as

$$Q = 2\pi \int_0^{d/2} U r \, dr$$

Table 5.5 summarizes the results and demonstrates the effectiveness of the discrimination for mass loading of 2.5, 5, 7.5 and 20. The effect of particle loading is shown in figure 5.19.

The radial symmetry of the particle-laden jet in cocurrent flow was also examined. Several tests were conducted by traversing across the jet at different mass loadings and different axial locations. Since the test results show symmetric profiles, the rest of the tests were conducted by traversing along only half the jet width. Figures 5.20a and 5.20b are representative of the data, illustrating the symmetry profiles for velocity and particle number concentration for $m = 2.5$ at $x/d = 20$ and $m = 5$ at $x/d = 20$, respectively, for a jet-to-stream velocity ratio of 35/20. It was also observed that as one moves towards the edge from the jet centerline, the particle concentration falls to nearly zero, and the velocity becomes equal to the cocurrent flow velocity. At that location, the PDA mode was changed to LDA mode in order to obtain the cocurrent flow velocity.

Table 5.5 Discrimination Verification in the Particle-Laden Flows

Mass loading	Volume flow rate (calculated m^3/s)	Volume flow rate (based on rotameter m^3/s)
0	2.37×10^{-3}	2.42×10^{-3}
2.5	2.49×10^{-3}	
5	2.39×10^{-3}	
7.5	2.47×10^{-3}	
10	2.32×10^{-3}	

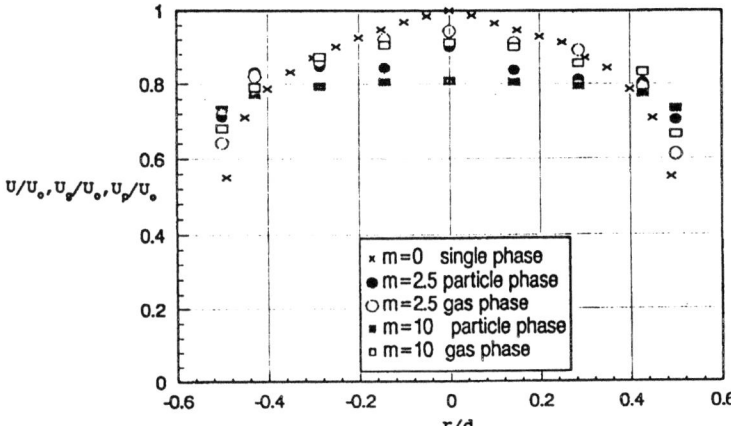

Figure 5.19 Gas-phase and particle-phase velocity profile at the exit of the jet, x/d=0.1 (case 1, m=2.5 and m=10).

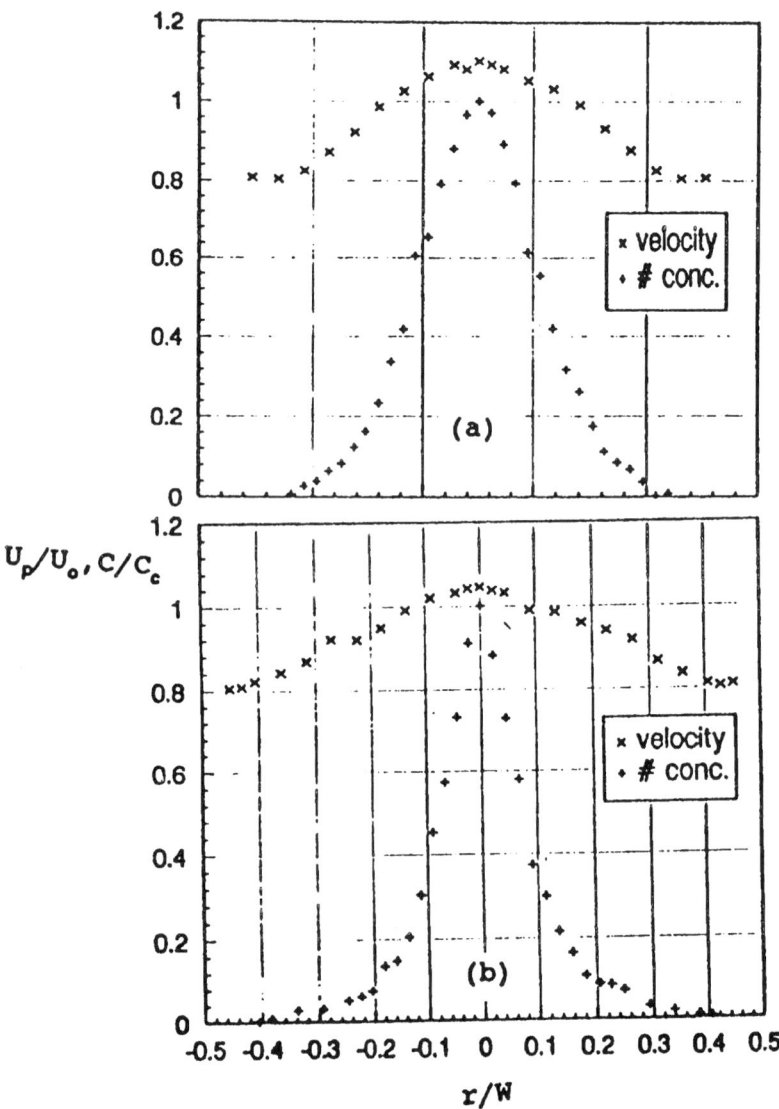

Figure 5.20 a: Particle-phase velocity and particle number concentration profile for case 1, m=2.5, axial location x/d=20. b: Particle-phase velocity and particle number concentration profile for case 1, m=5, axial location x/d=20.

It was observed that the centerline velocity of the dispersed phase at x/d = 20 is about 1.105 times that of the carrier fluid, although the latter is 1.20 times the former at the tube exit for a mass loading of m = 10. This can be explained by the fact that large diameter (> 10 μm) particles do not respond well to the fluid turbulent fluctuations. Thus, the main force that accelerates a particle in the radial direction is the viscous drag exerted on the particle by the fluid radial velocity v_f. Now this drag force is proportional to (v_f-v_p), and since $v_p<v_f$, the resulting force will be directed inward, thus limiting the radial spread of the particles. This is evident in Figure 5.21 where the concentration of the solid particles is nearly zero at a radial distance of r/W ≈ 0.3, while the fluid spreads to at least 1.67 times this distance. Conservation of momentum of each phase then results in the particle-phase axial velocity being higher than that of fluid, and in turn the particles continue to be a source of momentum for the fluid. It is also clear from Figure 5.21 that the single-phase jet is wider than the particle-laden jet. This will be discussed later in this section.

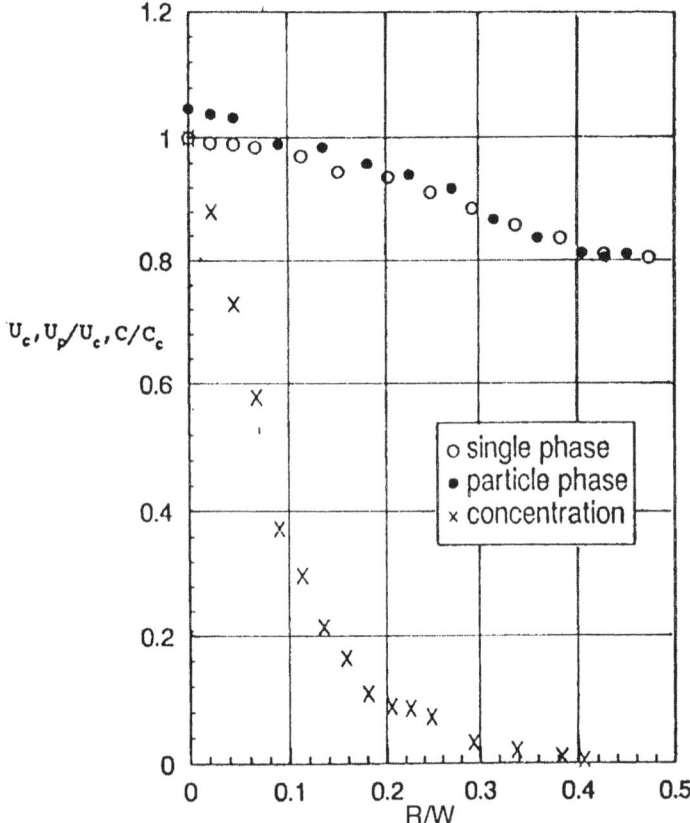

Figure 5.21 Mean axial velocity and concentration profiles at x/d=20 for case 1, m=5.

The variation in the mean velocities of gas phase as a function of particle mass loading is shown in Figure 5.22. The gas phase and particle phase mean velocities increased with an increase in particle loading in a nearly linear fashion. It was observed that the slip velocity is nearly constant. There is an interaction between the particle and gas phases. The additional momentum flux for the gas phase at the higher particle loading relative to the single phase comes from the loss of momentum flux by the particles. Momentum transferred to the gas increases with mass loading.

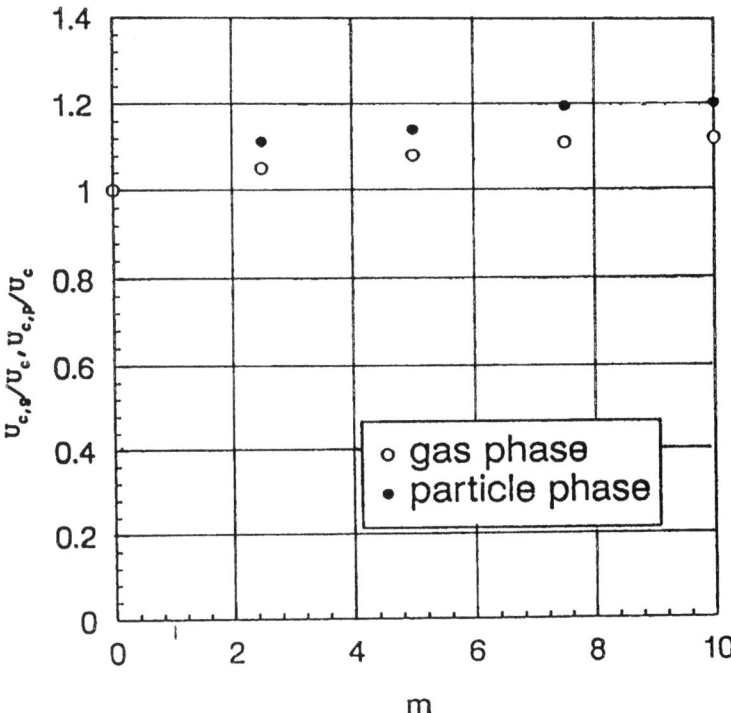

Figure 5.22 Effect of mass loading on particle-phase and gas-phase centerline velocity at axial location x/d=15 (case 1).

The effect of the particles on axial and radial turbulence intensity were studied and their profiles for the single phase and particle-laden (m=5) jet, for case 1 at an axial location of x/d=20, is shown in figures 5.22 and 5.23. It is observed that particles influence the fluctuating axial and radial component velocities. The particle fluctuation velocity is lower than the single phase and gas phase fluctuating velocities. Notice that particle loading also results in a reduction in the peak rms velocity profiles in the radial direction, further reducing any source of spreading. The possible explanation is as follows: The fluid that carries particles experiences a deceleration due to the global drag of the particles, and this, in turn, reduces the fluctuating velocity components of the turbulent flow.

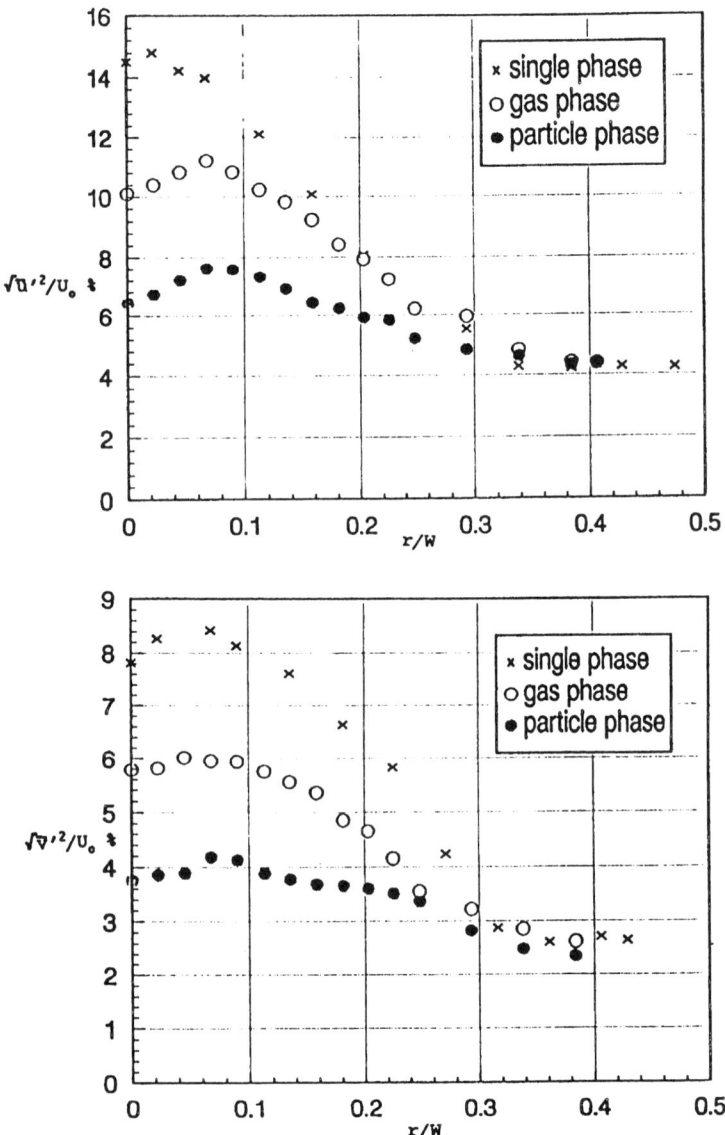

Figure 5.23 a: Intensity of axial velocity fluctuation, x/d=20 (case Hx m=5).
b: Intensity of axial velocity fluctuation, x/d=20 (case 1, m=5).

The axial and radial turbulence intensities ($\sqrt{u'^2}/U_o$, $\sqrt{v'^2}/U_o$) were investigated for (case 1) mass loadings of 2.5 and 10 at an axial distance of x/d=20, for the single-phase and two-phase solid-gas flows, respectively. The gas-phase turbulence intensities are lower than for comparable single phase flows. However, at a lower particle loading (m=2.5) the differences are small. At higher mass loading, the gas-phase turbulence intensities are much lower than for the single phase. The interaction between the particles and the fluid is based on their size (d_p), slip velocity (U_p-U_g), and fluid density and viscosity and can be explained by the particle Reynolds number:

$$Re_p = |U_p - U_g| d_p \rho / \mu$$

For our particle-laden experiment, Re_p was calculated and found to be on the order of 1. For smaller particles, when $Re_p < 10$, the rate of turbulent energy dissipation is increased, compared to single phase flow, by a ratio of $(1+m/r)^{1/2}$. Thus at the sub-energetic range, the particles extract energy from the flow and dissipate it. The intensive interaction between the small particles and the turbulent eddies of the fluid causes a decay of the fluid velocity fluctuations. On the other hand, for larger particles, when Re_p is of the order of 100, considerable turbulence is produced in the wakes of the particles. An examination of the data by Tsuji et al. [36,45,70] reveals that small particles (d_p=200 µm, Re_p on the order of 1 always cause suppression in the turbulence of the mainstream. Larger particles (dp=3000 µm, Re_p on the order of 1000) always cause an increase in the turbulence intensity in the main stream.

The particle-phase fluctuations were generally observed to be lower than those for the gas phase. An increase in the particle loading resulted in a slight decrease in the particle velocity fluctuations and the peak value of intensity was reduced. The radial fluctuating velocity component of particle-laden jet flow is responsible for the growth of the jet and the mixing in the shear layer region. The intensity of radial turbulence is less than that for axial turbulence. The magnitude of $\sqrt{v'^2}$ for the gas phase is less than that for the particle phase and there is only a weak radial dependence of $\sqrt{v'^2}$ for the particle phase. Radial velocity fluctuations are lower for the particle phase for flows with higher loading, indicating a strong damping effect of the particles.

Turbulent shear stress is important in mixing processes. A particle-laden jet with high velocity imparts momentum to the lower velocity cocurrent stream. It is assumed that the sole force producing the deceleration of the particle-laden jet and acceleration of the surrounding fluid is the tangential shear stress within the mixing region. The turbulent shear stress (u'v') was obtained by the simultaneous measurement of the axial and radial fluctuating velocities and is shown as

Transport Processes Involved in FGD

a function of radial location in figure 5.24. The measurements were made for case 1 at an axial location of x/d=20 for a mass loadings of 2.5 and 10. Values of the shear stress in the particle-laden flow are less than that for the single-phase flow in the entire jet region, because larger eddies are decayed by particles. These results show a decrease in shear stress with an increase in mass loading, because an increase in mass loading results in a reduction in the spreading rate of the shear layer.

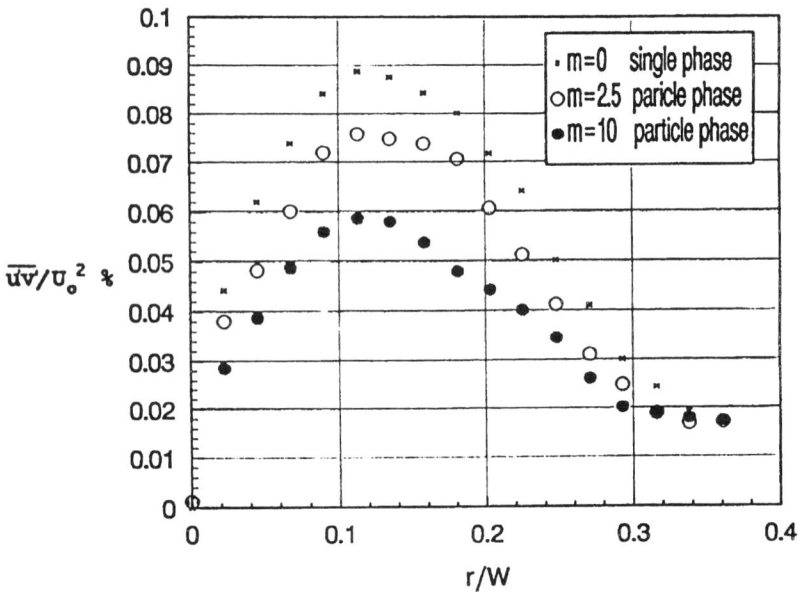

Figure 5.24 Turbulent shear stress at axial location x/d=20, case 1.

Jet Spreading Rate and Entrainment

Spreading of the jet, such as in the mixing of a turbulent jet in coflowing surrounding streams, is of interest from the following viewpoints: First, such a study will lead to a better understanding of free turbulence and the mechanism of turbulent diffusion. Second, although the outer boundary of mixing is the obvious measure of the spreading of the jet, determination of the point where the axial velocity reaches the cocurrent stream velocity is far more difficult than determination of the point where it reaches half its maximum value. Therefore, the spreading characteristics of the jet are commonly represented by the jet half-width, $R_{1/2}$.

The jet half-width increases linearly with increasing axial position. It is also observed that it depends upon the velocity ratio of jet-to-duct stream. When the velocity ratio U_o/U_{df} increases, jet half-width increases. Forstall [6] noted the same trend. The effect of mass loading on the spreading rate of the jet is also studied. For a particle-laden jet, the slope $dR_{1/2}/dx$ is a function of mass loading, and the slope decreases with an increase in mass loading. For mass loadings of 2.5 and 10, the reduction in spreading rate is about 9.5% and 29.5%, respectively, when compared with the single-phase spreading rate.

As a particle-laden jet shoots from an axisymmetric source, the air from the duct stream progressively mixes with the particle-laden jet stream. This is termed entrainment. The entrainment of a particle-laden jet is responsible for cocurrent stream interaction and mixing. The volume flow rate at different axial locations for a particle-laden jet flow is obtained from gas-phase velocity data. The gas-phase velocity is deduced from particle and size statistics as explained earlier. Then the entrainment coefficient for an axisymmetric jet is given by

$$E = Q_e/Q_o = Q/Q_o - 1$$

where

$$Q = 2\pi \int_o^{d/2} Ur\,dr$$

The variable Q_o represents the volumetric airflow rate at the exit of the jet tube. The variable Q represents the airflow rate at an axial position downstream of the jet, while Q_e represents the entrained airflow rate at any axial position x/d, ($Q_e = 0$ at x/d = 0). The integral in the above equation was approximated numerically by means of the trapezoidal rule. Entrainment of the particle-laden jet flow in a cocurrent duct stream was analyzed for effect of mass loading and jet velocity.

Figure 5.25 shows entrainment distributions for the particle-laden jet with mass loadings of 0 (single phase) and 5 and for $U_o = 35$ m/s and $U_{df} = 10$ m/s. In all axial locations the entrainment is reduced for the particle-laden jet. In the particle-laden jet, the jet spreading is reduced; hence, the peripheral area is also reduced. This results in a reduction in the entrainment of the duct stream. As discussed earlier, the slope of the jet edge reduces with increasing particle mass loading. The reduction in the spread reduces the area across which entrainment may occur [71]. This is a consequence of the effect of the particles on turbulence. At an axial location of x/d = 15, the entrainment is reduced by 22% when the mass loading is 5. Subramanian et al. [28] observed a reduction in entrainment of about 11% for m=1, at an axial location of x/d=5 for a jet to coflowing stream ratio of 2. It has been shown by his experiments on entrainment by a round jet of air containing lycopodium powder that particles can either increase or decrease the entrainment. Particles which are very fine (<2 µm) lead to a jet of homogeneous fluid; such particles move with the fluid molecules and the turbulence is enhanced. This is valid when the particle relaxation time is very much smaller than the eddy time scale. Coarse particles (>10 µm) on the other hand suppress the turbulence leading to reduced entrainment. Since, in this experiment the mean particle size used is > 20 µm, a similar trend is observed.

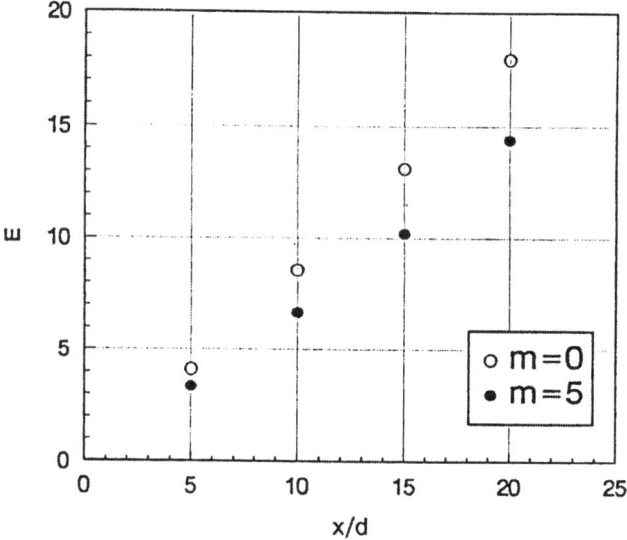

Figure 5.25 Entrainment in single and particle-laden jets (m=5, case 3).

It was observed that, with an increase in mass loading, the entrainment is reduced. The particles modulate the turbulence, as a result of which the transverse velocity of the jet is reduced. This leads to a narrowing of the jet at progressive axial distances and a lessening of the disturbances at the edges, thereby slowing down the mixing process and the entrainment. In summary, the particles are seen to minimize the mixing by suppressing the turbulence, as a result of which the entrainment at the jet edge is lowered and the entrainment is a function of mass loading and jet velocity.

Measurement of Mean Particle Diameter and Particle Concentration

It is observed that for higher mass loadings the particle size is slightly larger at the center of the jet. Therefore, the phenomenon affecting the particle diameter variation across the jet width is closely tied to the variation in the mass loading. It is noted that the particle size measured by PDA is larger than that from the Sedigraph measurement. The explanation is as follows. The Sedigraph particle size analyzer measures the sedimentation rates, according to Stokes' law, of particles dispersed in a liquid and automatically plots these data as cumulative mass percent versus equivalent spherical diameter. But the PDA requires a smooth-surfaced particle to determine a diameter for spherical particles. When PDA determines the particle size, it rejects non-spherical particles. It is observed from SEM pictures and optical microscope, that the glass particles contain a lot of irregularly shaped particles for sizes below 20 µm. Since PDA determines the size for spherical particles only, for the glass particles used in this study it is biased towards larger particles.

The variation in mean diameter was measured along the radial direction of the jet shear layer region for mass loadings of 2.5, 5, 7.5 and 10 at axial location x/d=10. It is observed that, for all mass loadings, the particle diameter is largest at centerline of the jet and decreases towards the edge of the jet. Fan et al. [40] found similar results for a particle-laden turbulent-free jet.

Figure 5.26 shows the profiles of particle concentration (number/cm_3) as a function of radial location at axial positions x/d = 5 and x/d = 20 for mass loadings of 2.5 and 10 (case 1). The volumetric concentration of particles in the shear layer region of the jet was also studied. Figure 5.27 shows the normalized volumetric particle concentration verses radial distance at axial location x/d=15 for case 3. The amount of dispersed phase at a given point in space is frequently the quantity of primary interest in order to understand the sorbent reaction and utilization in the mixing zone. Two observations may be made.

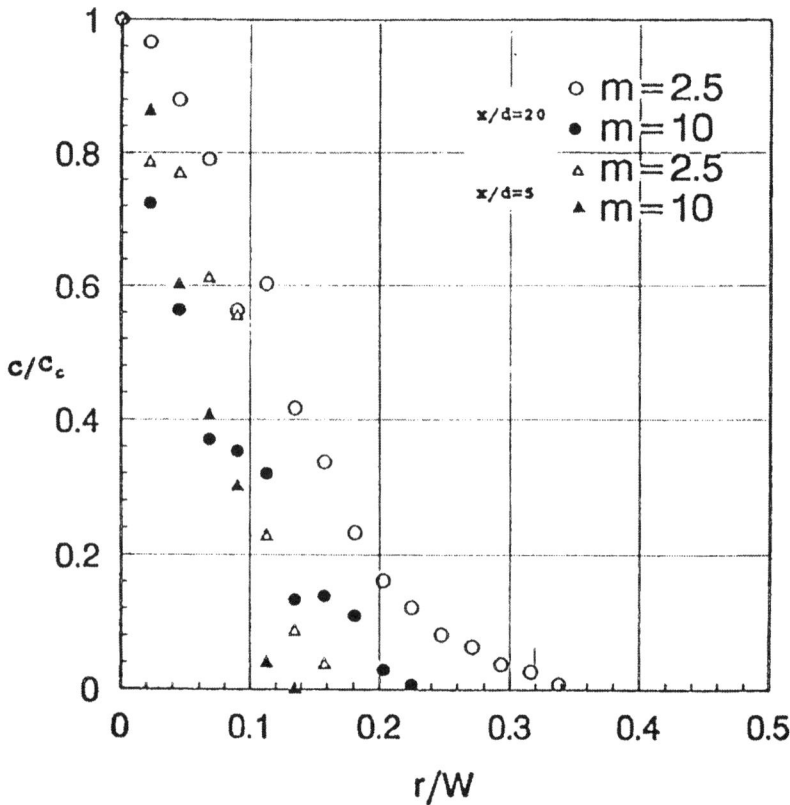

Figure 5.26 Particle number concentration profile for case 1.

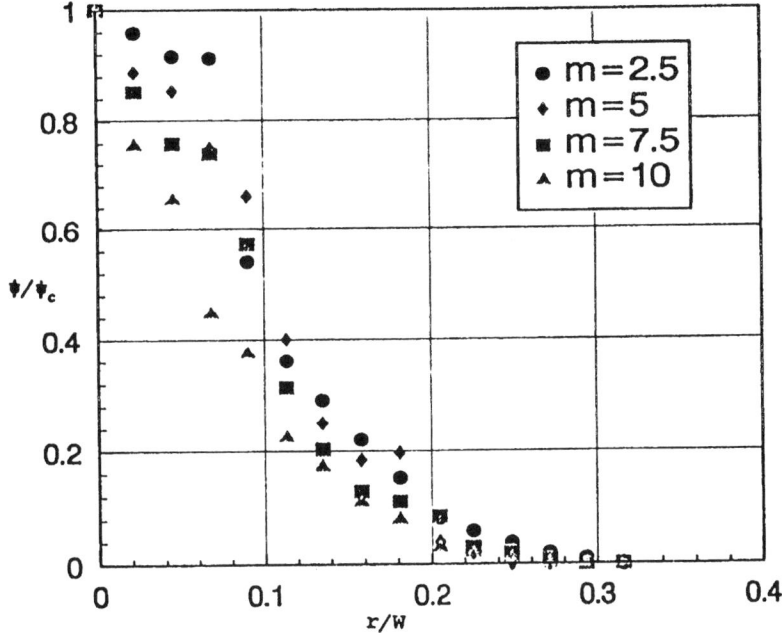

Figure 5.27 Particle volumetric concentration profile, x/d=15, case 3.

First, for the particle-laden jet with higher mass loading, the growth of the jet width and the corresponding growth of the shear layer region is smaller than that for a jet with lower mass loading, thus indicating that for a higher mass loading the jet will stay relatively dense and relatively inactive in terms of particle dispersion and mixing. Second, the jet expansion rate between downstream locations x/d = to 5 and x/d = 20 was lower for the higher particle loading jet than for the lower particle loading jet. Higher particle loading reduces velocity fluctuations in the gas and particle phases, as discussed earlier, and also leads to a reduction in the expansion of the particle-laden jet.

Conclusions

The objective of this research was to characterize the particle and fluid interaction in the mixing zone of particle-laden jets in cocurrent flow. An experimental test facility was designed and constructed to study such flows. Nonintrusive phase Doppler anemometry (PDA/LDA) and TTLDV were used for measuring the mean and fluctuating velocities of single-phase and particulate flows as well as the particle size and concentration. Tests were conducted for different flow conditions and mass loading were varied from 0 to 10. Table 5.3 provides the test conditions. This section summarizes conclusions drawn based on analysis of the test data. The main findings are as follows.

Single-Phase Flow

Measurements of the centerline axial mean and fluctuation velocities indicate the following. After the potential core, a nearly linear velocity decay is observed in the region $5 < x/d < 20$. However, with a decrease in cocurrent stream velocity, a slower rate of velocity decay is observed. The maximum fluctuating velocity appears in the region $5<x/d<10$ for all the data and is greater for a higher jet-to-cocurrent stream velocity ratio.

Similarity of the transverse profiles of axial velocity is observed in the turbulent mixing zone.

Turbulent intensity along the axial direction is higher, as expected, than turbulent intensity in the radial direction and the magnitude is greater for higher jet-to-cocurrent stream velocity ratios.

The width of the jet depends upon the jet-to-cocurrent flowing stream velocity ratio and axial distance. The jet spreading rate is greater for higher jet-to-cocurrent velocity ratios.

Particle-Laden Flows

Phase Doppler anemometry can be successfully utilized for studying particle-laden flows. Particle size and velocity and flow velocities can be obtained allowing us to discriminate between the phases.

The particle velocities are less than the single-phase fluid velocities along the centerline near the exit of the jet. At downstream locations greater than $x/d = 10$ the situation reverses, with the particle velocities being greater than the fluid velocities. This is because the axial velocity decay is faster for a single phase than for a particle phase due to higher spreading of the shear layer in the single phase. This phenomenon is further enhanced with an increase in particle loading due to inertia.

The slip velocity is nearly constant for all mass loadings. Addition of particles to the flow results in an increase in the gas-phase mean velocities.

Particle mass loading has a significant effect on the development of the particle-laden jet. In the radial direction along the jet width, the particle mean velocities and the fluctuating components are dependent upon the mass loading. Larger mass loading results in a higher mean velocity and a smaller fluctuating component. The presence of particles suppresses turbulent fluctuations, since Re_p is on the order of 1. This is also known as turbulence modulation and results in a reduction in the turbulence intensity and shear stress. These effects became more pronounced when mass loading is increased. The jet spreading rate decreases with increasing mass loading. This results from a reduction in the radial fluctuation velocity.

Particle-laden jet entrainment depends upon the jet velocity, mass loading, and axial distance. Entrainment increases with higher jet velocity and lower mass loading.

Larger particles appear at the centerline of the jet. Particle size decreases from the centerline to the edge of the jet.

Since higher mass loading results in a lesser spreading rate, the particles are concentrated in a smaller radial distance and there is a resulting reduction in the particle dispersion and mixing.

Mass loadings greater than 5 result in the reduction in turbulence intensity, Reynolds stress, jet entrainment, and spreading rate. That is, the mixing is reduced. Lower mass loadings, m<5, show better mixing characteristics and therefore may be beneficial for in-duct injection processes.

Investigation of Spray-Concurrent Flow

In the dry FGD process water spray is injected into the flue gas ducts to enhance the reaction between lime and sulfur dioxide. The presence of water spray results in a substantially improved lime utilization and the interaction of water spray and lime is critical in terms of efficient sorbent utilization.

The spray is injected into the cocurrent flowing flue gas in the flue gas duct. The spray expands in the duct and in doing so entrains the cocurrent flowing flue gases. It has been observed that for certain cocurrent flowing conditions the spray begins to stagnate and a flow reversal takes place. The flow reversal results in the separation of the water droplets from the flow and the deposition of the droplets on the duct wall. This results in decreased sorbent particle and droplet interaction and consequently decreased sorbent utilization. An understanding of the flow reversal phenomenon and the ability to predict the onset of flow reversal for given flow conditions is very important in the design and operation of in-duct FGD systems.

The literature review described earlier indicated that little work has been reported on characterization of flow reversal for liquid sprays in ducted jets. Chigier and Beer [72] have described the phenomenon of flow reversal in two cocurrent flowing jets of different densities but of the same phase. Though the study was for single-phase flow only, their findings were of great interest for this work.

The objectives of the current study were to conduct tests to investigate water droplet and cocurrent flow interactions in a duct relevant to the induct FGD processes and develop a criterion to predict the onset of flow reversal.

Experimental Setup

The sorbent injection test facility was modified by replacing the sorbent injection nozzle with a spray atomizing nozzle. A schematic diagram of the facility is shown in Figure 5.1. Spraying System's model SS 1/8 JJ atomizing nozzle was used for injecting the water spray into a 88.6 mm by 88.6 mm horizontal square duct. The spray produced by this nozzle is similar to the spray encountered in FGD processes. The nozzle inner diameter is 1.3 mm. A honeycomb is placed close to the inlet end of the horizontal duct to obtain uniform cocurrent air flow. Air for both the duct and the nozzle is provided by a shop air compressor. Water is supplied to the nozzle from a constant pressure tank. The nozzle is mounted in the center of the duct and carefully aligned to ensure that the spray at the nozzle exit is parallel to the duct walls.

The duct is made up of transparent plexiglass walls except for the 300 mm long test section which is made up of pyrex glass. The nozzle exit coincides with the test section inlet. The glass sides and the square configuration of the duct facilitate the use of a laser based nonintrusive PDA for obtaining droplet size and velocity simultaneously. The interaction between the spray and cocurrent duct flow occurs in the test section. The airflow rate in the duct and in the atomizer are monitored using a turbine flowmeter and a regular flowmeter, respectively. The major instrumentation used is the PDA described earlier. The PDA is a nonintrusive optical instrument which can be used to simultaneously measure the size, velocity, and concentration of the spherical particles. The PDA transmitting and receiving optics are mounted on a three dimensional traverse. This allows for the mapping of the particle-laden jet flow in the duct. The PDA was operated in the back scatter second order refraction mode at an angle of 149°, with parallel polarization, as recommended in the Dantec manual [58]. At each measurement point, 4000 PDA/LDA measurements were made [73, 74].

The uncertainties in the measurements were as follows. For the PDA/LDA measurements the uncertainty in the measurement point location was ± 0.01 mm, the uncertainty in the velocity measurements was ± 1.5%, uncertainty in the particle diameter was ± 8%, and the uncertainty in the droplet concentration was ± 4%. For flows measured using the turbine flowmeter and the regular flowmeter the uncertainties were ± 1% and ± 2%, respectively, and the uncertainty in the particle mass loading was ± 1.0%.

Experimental Procedure

Initial tests were conducted to ensure that the spray was axisymmetric in the duct. Spray axisymmetry can be observed in figure 5.28, which shows the spray velocity profiles along the vertical and horizontal axes at four axial locations. With the establishment of flow axisymmetry, measurements were made along the vertical diameter only for the tests. The test conditions are shown in Table 5.4. Measurements were made along a vertical plane at four axial locations. Spray velocity, droplet size, and concentration measurements were made at axial locations of 5, 10, 20, and 30 cm downstream of the nozzle exit for each test. The air temperature and the water temperature were measured near the inlet to the nozzle.

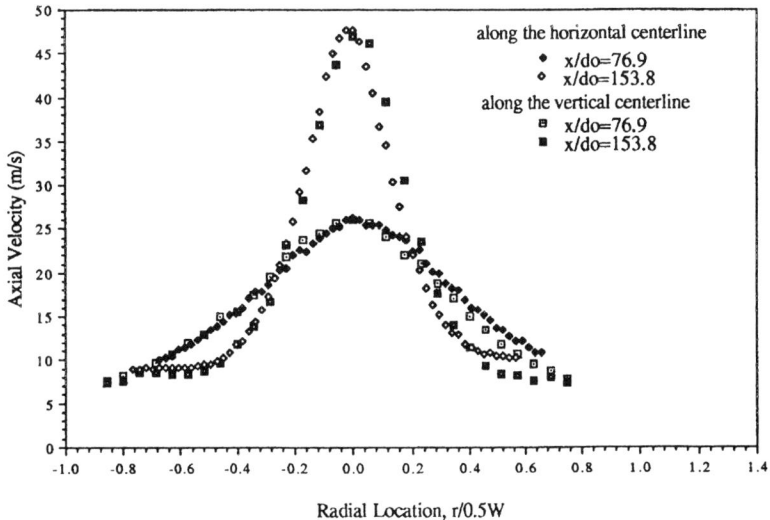

Figure 5.28 Velocity profiles showing axisymmetric spray.

Transport Processes Involved in FGD

The experiment was started at a fixed water flow rate and atomizer air pressure. The cocurrent airflow velocity was set at the higher limit. The spray was traversed at four axial locations, 5, 10, 20, and 30 cm from the nozzle exit. The probe was then returned to the 5 cm axial location and the cocurrent air velocity was reduced. The measurements were made at the same four axial locations. The cocurrent air velocity was then decreased again until flow reversal is observed. This reversal is accompanied by the wetting of the test section window. Once this takes place, it is impossible for the laser beam to penetrate the test section window and no further measurements are possible. The experiment was stopped and the test section was removed and cleaned. The next set of data was taken at a reduced nozzle air pressure as indicated in Table 5.6. The entire process described above was then repeated for the other test parameters.

Table 5.6 Test Conditions

Mass flow rate of water in atomizer $\dot{m}_w(kg/s)$	8.3×10^{-4}, 3.3×10^{-4}, 1.2×10^{-3}
Mass flow rate of air in atomizer, $\dot{m}_o(kg/s)$	1.1×10^{-3}, 8.9×10^{-4}, 7.2×10^{-4}, 5.7×10^{-4}
Axial Measurement Location (cm)	5, 10, 20, 30
Nondimensional Measurement Location (x/d_0)	38.5, 76.9, 153.8, 230.7
Air density, $\rho(kg/m^3)$	1.17
Water density, $\rho f(kg/m^3)$	999.7

Atomizer Conditions		Air Pressure (psig)	Duct Air Flow Conditions
	Water Flow Rate (cc/min)		Duct Air Flow (m/s)
Case 1	50	50	10, 8, 6, 4
Case 2	50	40	10, 8, 6, 4
Case 3	50	30	10, 8, 6, 4
Case 4	20	50	10, 8, 6, 4
Case 5	20	40	10, 8, 6, 4
Case 6	20	30	10, 8, 6, 4
Case 7	20	20	10, 8, 6, 4
Case 8	70	50	10, 8, 6, 4
Case 9	70	40	10, 8, 6, 4
Case 10	70	30	10, 8, 6, 4

Experimental Results

In order to ascertain the accuracy of the PDA measurements, a mass balance was performed at an axial location 5 cm downstream of nozzle exit. The amount of water flow exiting the nozzle was calculated by utilizing the droplet size and the velocity profile inside the spray envelope at the 5 cm axial location. The water flow rate was within 6% of the flowrate measurements upstream of the nozzle. This small amount of difference is attributed to the evaporation of water droplets.

Experimental Velocity Profiles and Prediction

Velocity profiles were obtained for the various conditions shown in Table 5.5 at four separate axial locations. A model based on experimental data was developed to describe the velocity profiles inside the spray envelope. It is given by Equation 5.1. It is a cosine function. A comparison of the experimental velocity profile and the cosine model is shown in Figure 5.29. This approximation is very reliable in the region far downstream from the spray ($x/d_0 > 80$).

Equation 5.1

$$\frac{u_s - u}{u_c - u} + -\frac{1}{2}\left(1 + Cos\frac{\pi r}{2y_{0.5}}\right)$$

Transport Processes Involved in FGD

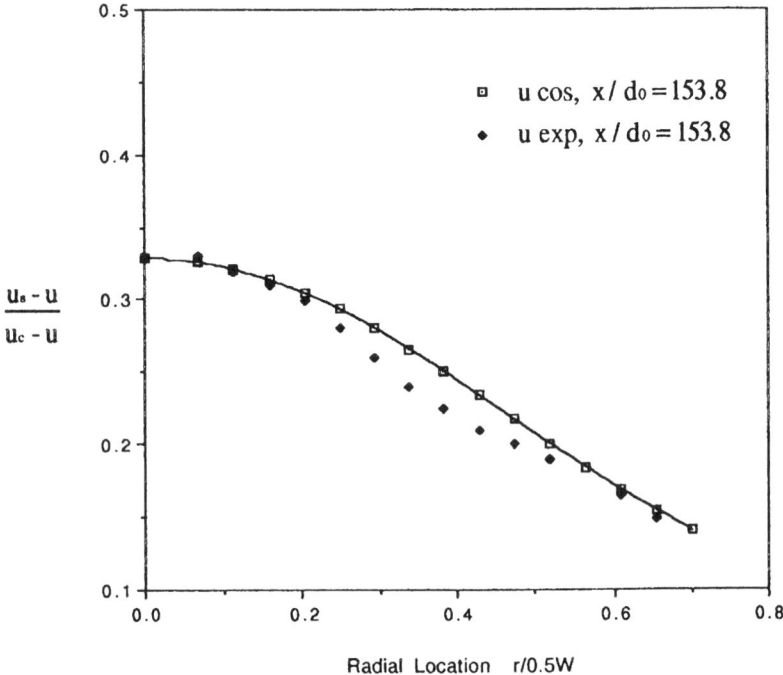

Figure 5.29 Comparison of experimental velocity profile with the cosine model for case 4: uo=681.48 m/s, u=6m/s.

The velocity distributions in this region are very similar to those obtained experimentally. Some ambiguity arises for flow in the region close to the nozzle exit. The ambiguity is attributed to several reasons. Equation 5.1 describes the flow of a free jet in a cocurrent stream but its functional dependence has been adapted to fit experimental data for a ducted jet. For the free jet, the cocurrent velocity remains constant as one moves downstream; however, for the ducted jet, as the spray expands, more of the cocurrent air is entrained in the flow and the flow cross section provides less area for the air stream. Thus, the cocurrent air stream velocity for the experimental conditions is not constant.

It is very important to note that the cosine distribution describes the jet velocity inside the spray envelope only. The largest amount of ambiguity when comparing experimental results in the upstream region is at the edge of the spray. The discrepancies in this region are largely due to the fact that the model assumes the spray boundary to be linear. In the ducted jet, however, the spray boundary swells outward. The amount of swelling depends on the flow rate of the cocurrent air. As the boundary is approached, the model defines an exact point where the spray velocity reaches the gas phase velocity, based on this linear assumption. Discrepancies arise when experimental measurements are made in this region. Since the spray boundary is curved outward, the spray velocity reaches the gas phase velocity at a greater radial distance than that proposed by the model. As one moves downstream, the jet momentum is dispersed during expansion and the curved boundary approaches the linear approximation made by the cosine model. The cosine model is in good agreement with the flow inside the spray envelope.

Flow Reversal

As the spray expands inside the duct, the cocurrent air flow is entrained by the spray. If the mass flow rate of cocurrent air is not sufficiently high enough to feed the entrainment needs of the spray the flow will reverse, as indicated in Figure 5.30. In order to develop an understanding of the flow reversal phenomenon associated with two-phase droplet cocurrent flows, a model for the flow characteristics inside the spray region was developed.

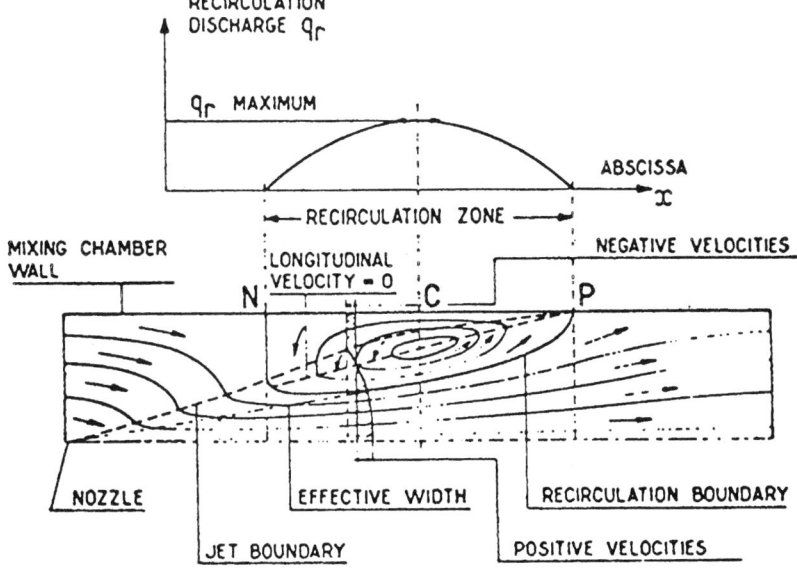

Figure 5.30 Map of flow field showing flow reversal.

Transport Processes Involved in FGD

Chigier and Beer [72] analyzed the interaction of a jet of gas in a coaxial gas stream. The same jet gas as well as different jet gas/coaxial gas combinations were used and the density difference between the jet and coaxial gases was either negligible or relatively very small. Two-phase (droplet-gas) flow, however, was not analyzed. A similar approach was adopted for the spray droplet cocurrent flow model developed in this study. The potential core is a region immediately downstream from the nozzle exit, in which the velocity of the nozzle fluid remains unchanged. The length of potential core is an important parameter in the development of the model. The potential core and other parameters of the jet are shown in figure 5.31. The potential core length, x_p, is obtained from the equation below.

Equation 5.2

$$\frac{x_p}{d_o} - 4 + 55\lambda$$

where λ is the ratio of cocurrent air mass flow rate to atomizer mass flow rate and is defined by the equation below.

$$\lambda - \frac{\rho u}{\rho_o U_o}$$

The denominator in this equation corresponds to the region just outside the nozzle exit. u_0 is the velocity of the nozzle at the exit and r_0 is the density of the mixture of air and water in the nozzle calculated using the relative mass fractions of air and water in the nozzle. The experimental data were utilized in obtaining Equation 5.2. The velocity decay on the centerline in the fully developed region is the given by Equation 5.3.

Equation 5.3

$$\frac{u_c - u}{u_o - u} - \frac{x_p}{x}$$

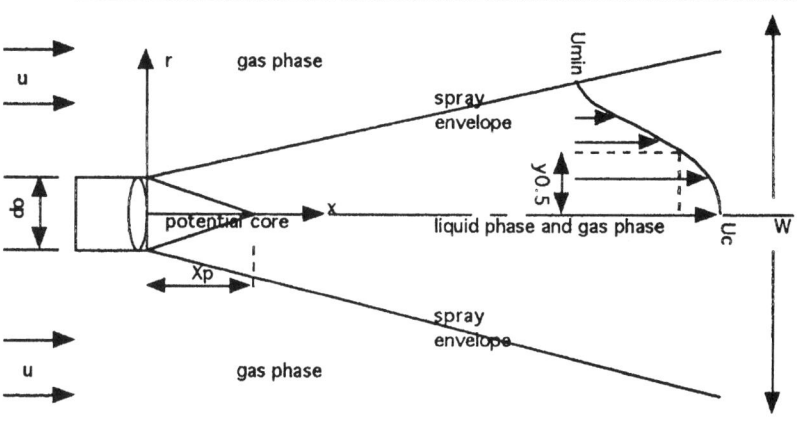

Figure 5.31 Spray parameters.

Here u_c is the centerline velocity at an axial distance of x. Decay of the centerline velocity predicted by this equation is in good agreement with the test data obtained at the four different axial locations. The rate of jet expansion is given by the equation below.

$$\frac{y_{0.5}}{0.5 d_o} - \left(\frac{x}{x_p}\right)^{(1-\lambda)}$$

The radial location $y_{0.5}$ corresponds to the position where the velocity is equal to

$$y_{0.5} - \frac{u_{min} + u_c}{2}$$

Development of Criterion for Prediction of Flow Reversal

As the spray expands in the duct the cocurrent flowing air stream is entrained by the jet. The emergence of flow reversal is based on a critical point in the flow field where all of the cocurrent flowing gas is entrained by the jet but the spray has still more room to expand in the duct. In order for the spray to continue downstream of this point, a recirculation zone develops. This recirculating zone extends to the point where the spray meets the duct wall. This critical point is based on the assumption that the spray boundary is linear (as obtained from the

Transport Processes Involved in FGD

cosine model for spray velocity distribution). The equation below gives the mass flow rate inside the spray envelope during expansion.

Equation 5.4

$$\dot{m}_s - \int p u dA = \dot{m} + \dot{m}_o$$

where m is the mass flow rate of the cocurrent air and m_o is the total mass flow rate issuing from the nozzle. As stated earlier, the critical point is reached when the entire cocurrent flowing air is entrained by the spray. Substituting the cosine model, Equation 5.1, for the velocity in Equation 5.4 results in Equation 5.5, which provides the mass flow rate inside the spray.

Equation 5.5

$$\dot{m}_s - \int \rho \{ u + \frac{u_c - u}{2} \cos\left(\frac{\pi \tau}{2 Y_{0.5}}\right) 2\pi \} r dr$$

To obtain the value of the critical axial locations X_N, we substitute X_N for x in Equation 5.3 on page 408, and with that value of $y_{0.5}$, solve Equation 5.4 and Equation 5.5. The solution is provided in Kadaba [73] and the resulting value of X_N is given by the equation below.

$$X_N = 6.25 q' L$$

where q' is the modified Thring-Newby parameter (Chigier et al.[72] for ducted flows and 2L is the hydraulic diameter of the duct. The modified Thring-Newby parameter is defined by the equation below.

$$\Theta - \frac{\dot{m} + \dot{m}_o}{\dot{m}_o} \frac{d_o}{2L} \left(\frac{\rho_0}{\rho}\right)^{1/2}$$

The mass flow rate of reversed flow is calculated from the equation below

Equation 5.6

$$\frac{\dot{m}_x}{\dot{m} + \dot{m}_o} - \frac{0.47}{\Theta'} - 0.5$$

This development is based on the principle that the flow reversal occurs when all the cocurrent air flow is entrained by the spray before flow reversal begins. Then for the spray to continue to propagate, it requires more entrainment, which is supplied when the flow recirculates.

Based upon the preceding discussion, a criterion for the determination of the onset of flow reversal was developed. A step-by-step procedure for determining the occurrence of flow reversal for a particular test condition is shown in Table 5.7. The first step is to obtain λ, the mass flow rate ratio for the two phases. Then for second step, obtain the potential core length x_p using Equation 5.1 on page 405. Recalling that the model development was based on the assumption of a linear spray boundary, by using similar triangles one can obtain the axial distance required for the spray to reach the duct wall (X_w) in form of the equation below.

$$\frac{L}{X_w} = \frac{1}{4 + 55\lambda}$$

Table 5.7 Procedure for Determination of Flow Reversal

Parameter Calculated	Equation for the Parameter
Mass Flow Ratio for Two Phases	$\lambda = \dfrac{\rho u}{\rho_o U_o}$
Potential Core Length	$\dfrac{x_p}{d_o} = 4 + 55\lambda$
Modified Thring-Newby Parameter	$\Theta = \dfrac{\dot{m} + \dot{m}_o}{\dot{m}_o} \dfrac{d_o}{2L}\left(\dfrac{\rho_o}{\rho}\right)^{1/2}$
Location of Flow Reversal Initiation	$X_N = 6.25 \Theta' L$
Axial Distance from Nozzle Exit to Spray Interaction with Wall	$\dfrac{L}{X_W} = \dfrac{1}{4 + 55\lambda}$
Existence of Flow Reversal	if $X_N < X_W$
Mass Flow of Recirculation	$\dfrac{\dot{m}_o}{\dot{m}_o + \dot{m}} = \dfrac{0.47}{\Theta'} - 0.5$

The next step is to compare the values of X_N and X_W. If X_W is less than X_N, then the spray hits the duct wall before the flow reversal can occur, and there will be no flow reversal. However if X_W is greater than X_N, then the flow reversal will occur as the spray is still away from the wall. Thus if $X_N < X_W$, flow reversal will occur. The recirculating mass flow rate can then be calculated from Equation 5.6 on page 410. This criterion was applied to our experimental results and it correctly predicted flow reversal for the test conditions for which flow reversal was observed.

Conclusions

Experiments were conducted to characterize the interaction of a water spray and cocurrent air flow in a horizontal duct. A nonintrusive phase Doppler interferometry/laser Doppler velocimetry technique was used to obtain the spray droplet sizes and the drop and flow velocities. A wide range of spray conditions and cocurrent flow velocities were investigated. Spray velocity profiles were obtained at various axial locations. A cosine model for predicting the velocity profiles was obtained. The agreement between the experiments and model was within ±5 %. Flow reversals in the duct were observed for certain cocurrent flow conditions. Flow reversal occurs at a location in the duct when all the cocurrent flowing air is entrained in the expanding spray and region of negative flow develops between the spray and the duct wall. A criterion for predicting the flow reversal in spray-cocurrent flow in a duct was developed. The comparison between the experimental results and the predictions obtained using the criterion was very good. All the cases where flow reversals were observed were correctly predicted. An equation to obtain the recirculating mass flow rate was also developed.

Particle-Laden Jet Reversal in Cocurrent Flow

Tests to study particle cocurrent flow interactions resulting in flow reversal were also conducted. Spherical glass particles of average diameter 40 microns were used. PDA measurements of velocity profiles were made at normalized axial distances (x/d) of 0.1, 5.0, 10.0, 15.0, and 20.0. The mass loading was varied up to 2.5. The duct mean velocity U_{df}, (i.e., the cocurrent velocity) was reduced from 10 m/s to a value where flow reversal occurred.

The cosine power law was modified slightly by considering the particle velocity at the edge of jet, i.e., at the boundary for the axial location instead of the average duct velocity. The modified equation is

$$\frac{U - U_{edge}}{U_c - U_{edge}} = \frac{1}{2}\left[1 + Cos\left(\frac{\pi r}{y_{05}}\right)\right]$$

where $y_{05} = \frac{U_c + U_{edge}}{2}$

Here U is the particle velocity, U_{edge} is the velocity at the edge and U^c is the centerline velocity. The comparison between the data and the modified cosine law is good (Figure 5.32). The criteria developed for predicting flow reversal for the spray was also used for predicting flow reversal for the particle laden jet. The criteria successfully predicts the flow reversal for the particle laden jet.

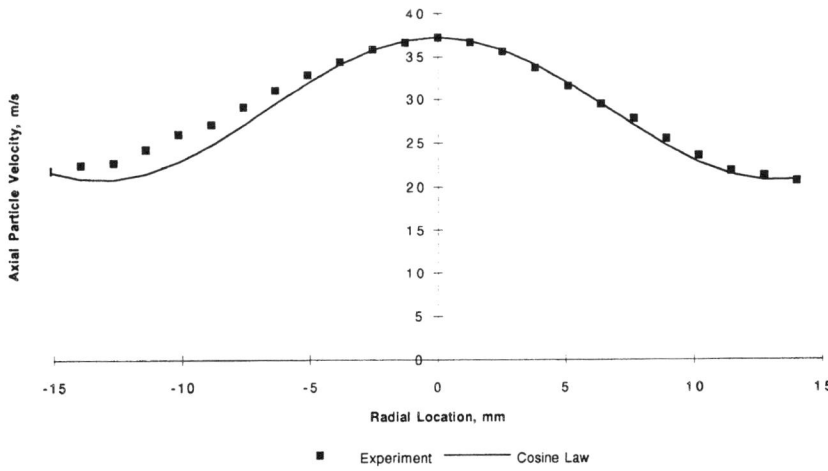

Figure 5.32 Comparison of the experimental velocity profile with the modified cosine law (test number 3, x/d=15, U=6 m/s).

Summary

A thorough literature survey of particle-laden jet/cocurrent flow interactions was undertaken. The survey indicated a lack of information in the area of parameters of interest in FGD processes. A sorbent injection test facility that simulates flue gas flow conditions encountered in power plants was designed and constructed.

Particle-laden jet/cocurrent flow interactions were investigated using nonintrusive optical laser diagnostics techniques (PDA/LDA, TTLDV) in this facility. A new technique for simultaneous measurement of nonspherical particle size and velocity (TTLDV) was successfully developed. The TTLDV technique has potential applications in particle-laden multiphase flows. The conclusions made regarding particle-laden jet-cocurrent flow interactions regarding mixing characteristics may prove helpful in designing such systems. Tests have also been conducted with swirling vanes attached to the jet tubes. This results in generating a higher level of turbulence and improved mixing without requiring any additional power. Reversal of atomizing spray and particle-laden jets in cocurrent flow was studied in the sorbent injection test facility. A criteria was developed to predict the flow reversal in flue gas ducts for spray droplets and sorbent particles. The criteria successfully predicted the cases for which flow reversal occurred in the tests. This criteria will allow the designer to avoid the flow conditions which result in flow reversal and deposition of droplet and sorbent particles on the duct wall and floor leading to poor sorbent utilization.

Nomenclature

a	radius of the jet tube
C	axial particle # concentration
Cc	centerline particle concentration
d	diameter of the jet tube
dp	diameter of the particle
dp,c	diameter of the particle at the centerline
d0	exit (orifice) diameter of nozzle
E	entrainment coefficient
FGD	flue gas desulfurization
L	one half hydraulic diameter of duct
LDA	laser doppler anemometry
LIF	laser-induced fluorescence
m	mass loading
m	mass flow rate of the cocurrent air
ma	mass flow rate of the air in the nozzle
mo	mass flow rate at nozzle exit
mr	mass flow rate of recirculation
ms	mass flow rate inside the spray
mw	mass flow rate of water in atomizer
PDA	particle dynamics analyzer, phase doppler anemometer, both names mean the same instrument
PDI	phase doppler interferometry
PDPA	phase doppler particle analyzer
Q	volume flow rate
Qe	entrainment volume flow rate
Qo	volume flow rate at exit of the tube
r	radial coordinate
R1/2	radial distance at U=(Uc+Udf)/2
Rej	Reynolds number of the jet
u	velocity of gas phase in axial direction
uc	velocity at centerline of spray(5 cm from nozzle exit)
u0	velocity at nozzle exit
us	spray velocity
umin	Spray edge velocity
$\sqrt{u'2}$	ms axial velocity

$\sqrt{u'g2}$	rms axial velocity for gas phase
$\sqrt{u\,p'2}$	rms axial velocity for particle phase
U	axial velocity for single phase
Uc	centerline velocity for single phase
Uc,g	centerline velocity for gas phase
Uc,p	centerline velocity for particle phase
Udf	duct flow velocity
Ug	axial velocity for gas phase
Uo	single phase exit centerline velocity
Up	axial particle phase velocity
u'v'	Reynolds stress
$\sqrt{v'2}$	rms radial velocity
vf	radial fluid velocity
vp	radial particle velocity
$\sqrt{v'2g}$	rms velocity for gas phase
$\sqrt{v'2p}$	rms radial velocity for particle phase
W	half duct width
x	axial distance downstream of nozzle
xp	potential core length
XN	axial location of flow reversal initiation
XW	axial distance required for spray to hit wall
ρa	density of air
ρp	density of particle
P	density of air
Po	density of air at nozzle exit
Pw	density of water
γ	= rp/ra
μ	micron; viscosity of air
ψ	particle volumetric concentration
ψc	centerline particle volumetric concentration
Θ'	Thring - Newby parameter

References

1. Drummond, C., Babu, M., et al."Duct Injection Technologies for SO_2 Control," *Proceedings: First Combined Flue Gas Desulfurization and Dry SO_2 Control Symposium*, EPRI Report Gs-6307, St. Louis,Mo, Oct.(1988)
2. Kadambi, J. R., P. Chinnapalaniandi, and J. S. T'ien, "Investigation of Transport Processes Involved in FGD," *Final Report for First Year*, submitted to OCDO, November 1991
3. Adler, R. J., L.S. Fan, J. R. Kadambi, T. C. Keener, S. J. Khang, M. Prudich, and K. Raghunathan, "Flue Gas Desulfurization and Acid Rain Control," *Report of the Consortium of Ohio Universities to OCDO*, September 1989
4. Abramovich, G. N., *The Theory of Turbulent Jets*, MIT Press, Cambridge, MA (1963)
5. Mikhail, S., "Mixing of CoAxial Streams Inside a Closed Conduct," *J. Mechanics Engineering Science*, 2, 59-68 (1960)
6. Forstall, W., and A. H. Shapiro, "Momentum and Mass Transfer in Coaxial Gas Jets," *J. App. Mech.*, 72, 399-408 (1950)
7. Alexander, L. G., A. Kivnick, E. Commings, and E. D. Henze, "Transport of Momentum and Energy in a Ducted Jet," *AIChE. J.*, 1, 55-73 (1955)
8. Becker, H. A., H.C. Hottel, and G. C. Williams, "Mixing and Flow in Ducted Turbulent Jets," *Ninth Symposium on Combustion*, Academic Press, 892-900 (1963)
9. Hill, P. G., "Turbulent Jets in Ducted Streams," *J. of Fluid Mechanics*, 22, 161-186 (1965)
10. Chigier, N. A., L. M. Beer, "The Flow Region Near the Nozzle in Double Concentric Jets," *J. of Basic Engineering*, 86, 797-804 (1964)
11. Champagne, F. H, and Wygnanski, "An Experimental Investigation of Coaxial Jets, International," *J. of Heat and Mass Transfer*, 14, 1445 (1971)
12. Durao, D. and J. H. Whitelaw, Turbulence Mixing in the Developing Region of CoAxial Jets, *J. Fluids Engineering*, 95, 467-473 (1973)
13. Guruz, K. and C. Ilicali, "An Investigation of the Isothermal Confined Coaxial Jet System," *Comb. Sci. and Tech.*, 25, 193-2088 (1981)
14. Johnson, B. V. and J. C. Bennett, "Statistical Characteristics of Velocity, Concentration, Mass Transport, and Momentum Transport for Coaxial Jet Mixing in a Confined Duct," *J. of Engr. for Gas Turbine and Power*, 106, 121-127 (1984)
15. Habib, M. A. and J. H. Whitelaw, "Velocity Characteristics of a Confined CoAxial Jet," *J. Basic Engineering*, 10, 521-529 (1979)
16. Freeman, A. R. and R. T. Szczepura, "Mean and Turbulent Velocity Measurements in an Abrupt Axisymmetric Pipe Expansion with a Complex Inlet Geometry," *International LDA Symposium*, Lisbon, Portugal, Section 4.5 (1982)
17. Becker, H.A. and A. P. G. Brown, "Response of Pitot Probes in Turbulent Streams," *J. of Fluid Mechanics*, 62, 84-114 (1974)
18. Laats, M. K., and F. Frishman ," Scattering of an Inert Admixture of Different Grain Size in Two-Phase Axisymmetric Jet," Heat Transfer, *Soviet Research*, 2(6)(1970)
19. Hetsroni, G., M. Sokolov, "Distribution of Mass, Velocity and Intensity of Turbulence in a Two-Phase Turbulent Jet," *J. App. Mech.*, 15-327 (1971)
20. Goldschmidt,V., and S. Eskinazi, "Two-Phase Turbulent Flow in a Plane Jet," *J. App. Mech.*, 33, 735-747 (1966)
21. Goldschmidt, V., M. K. Householder, G. Ahmadi, and S. Chuang, "Turbulent Diffusion of Small Particles Suspended in Turbulent Jets," *Progress in Heat and Mass Transfer*, 6 (1972)

Transport Processes Involved in FGD

22. Wells, M. R., and D. E. Stock, "The Effect of Crossing Trajectories on Dispersion of Particles in a Turbulent Flow," *J. Fluid. Mech.*, 136 (1983)
23. Yuu, S., N. Yasulouchi, Y. Hirosawa, and T. Jotaki, "Particle Turbulent Diffusion in a Dust Laden Round Jet," *AIChE.J.*, 24, 509-519 (1978)
24. Melville, W. K., and N. C. Bray,"The Two-Phase Turbulent Jet," *Int. J. Heat Mass Transfer*, 22, 279-286 (1979)
25. Wall, T. F., V. Subramanian, and H. Howley, "An Experimental Study of the Geometry, Mixing and Entrainment of Particle-Laden Jets Up to Ten Diameters from the Nozzle," *Trans. I. Chem. Engr*, 60, 231-239 (1982)
26. Subramanian, V., and G. Raman, *Entrainment by a Concentric Jet with Particles in a Primary Stream*, Heat and Mass Transfer Letters, 9, 277-290 (1982a)
27. Subramanian, V.,and N. Raman, "Measurement of Velocity and Concentration for a Two-Phase Turbulent Jet," *Can. J. Chem. Engr*, 62, 314-318 (1984a)
28. Subramanian, V., and G. Raman, "Influence of Free Stream on the Entrainment by Single Two Phase Axisymmetric Jets," *AIChE, J.* 30,1013 (1984b)
29. Popper, J., N. Abuaf and G. Hetsroni, "Velocity Measurements in a Two-Phase Turbulent Jet," *Int. J. Multiphase Flow*, 1, 715-725 (1974)
30. McComb, W. D., and S. M. Salih, "Comparison of Some Theoretical Concentration Profiles for Solid Particles in a Turbulent Jet with the Results of Measurements Using a Laser Doppler Anemometer," *J. Aerosol Sci.*, 9, 99-313 (1978)
31. Levy, Y. F. C. Lockwood, "Velocity Measurement in a Particle Laden Turbulent Free Jet," *Combustion and Flame*, 40, 333-339 (1981)
32. Modarress, D., H. Tan, and S. Elghobashi, "Two Component LDA Measurement in a Two-Phase Turbulent Jet," *AIAA J.*, 22, 624-630 (1984)
33. Shuen, J. S., A. S. P. Solomon, Q.-F. Zhang and G. M. Faeth, "Structure of Particle-Laden Jets: Measurements and Predictions," *AIAA J.* 23, 396-404 (1985)
34. Fleckhaus, D., K. Hishida and M. Maeda, "Effect of Laden Solid Particles on the Turbulent Flow Structure of a Round Jet," *Exp. Fluids*, 5, 323-333 (1987)
35. Maeda, M., K. Kobashi and K. Hishida, "Application of Phase Doppler Anemometry to Dispersed Two-Phase Flows," *Int. Conf.on Mechanics of Two-Phase Flows*, Taipei, Taiwan, 91-96 (1989)
36. Tsuji, Y., Y. Morikawa and H. Shiomi, "LDV Measurements of an Air-Solid Flow in a Vertical Flow," *J. of Fluid Mech.* 139, 417-434 (1984)
37. Mostafa, A. A., H. C. Mongia, V. G. McDonell and G. S. Samuelsen, "Evolution of Particle-Laden Jet Flows: A Theoretical and Experimental Study," *AIAA J.* 27, 167-183 (1989)
38. Hardalupas, Y., et al. "Velocity and Particle Flux of Turbulent Particle Laden Jets," *Proc.R.Soc.* London A, 426, 31-78 (1989)
39. Zoltani, C. K and A. F. Bicen, "Velocity Measurements in a Turbulent Dilute Two-Phase Jet," *Exp. Fluids*, 9, 295-300 (1990)
40. Fan, J., L. Zhang, H. Zhao and K. Cen, "Particle Concentration and Particle Size Measurements in a Particle-Laden Turbulent Free Jet," *Exp.Fluids*, 9, 320-322 (1990)
41. Longmire, E. K. and J. K. Eaton, *Structure and Control of Particle-Laden Jet*, Dept. of Mechanical Eng., Stanford Univ. Report no. MD-58, September 1990
42. Hedman, P. O.and L. D. Smoot, "Particle-Gas Dispersion Effects in Confined Coaxial Jets," *AIChE J.*, 21, 372-379 (1975)
43. Subramanian, V. and G. Raman, "Entrainment by a Concentric Jet with Particles in the Secondary Stream," *Can. J. Chem. Engr*, 60, 589-592 (1982b)

44. Owen, P. R., "Pneumatic Transport," *J. Fluid Mech,*.39, 407-434 (1969)
45. Tsuji, Y. and Y. Morikawa, "LDV Measurements of an Air-Solid Two-Phase Flow in a Horizontal Pipe," *J. of Fluid Mech*, 120, 385-409 (1982)
46. Lee, S. L. and F. Durst, "On the Motion of Particles in Turbulent Duct Flow," *Int . J. Multiphase Flow*, 8, 125-146 (1982)
47. Theofanous, T. G. and J. Sullivan, "Turbulence in Two-Phase Dispersed Flows," *J. of Fluid Mech*, 116, 343-362 (1982)
48. Shahnam, M.and G. J. Morris, "Gas-Solid Flow in an Axisymmetric Sudden Expansion," *Third International Symposium of Gas-Solid Flows*, FED - 78, ASME, 93-99 (1989)
49. Liljegren, L. M. and N. S. Vlachos, "Laser Velocimetry Measurements in a Horizontal Gas-Solid Pipe Flow," *Exp.Fluids*, 9, 205-212 (1990)
50. Kadambi, J. R., *Investigation of Transport Processes Involved in FGD*, Fourth Year Report, submitted to OCDO, Columbus, OH (1994)
51. Kadambi, J. R., P. Chinnapalaniandi and J. S. T'ien, *Investigations of Transport Properties Involved in FGD*, Final Technical Report for First Year, submitted to OCDO, July 1991
52. Gladnick, P. G., A. C. Enotiadis, J. C. La Rue and G. S. Samuelsen, "Near-Field Characteristics of a Turbulent Co-flowing Jet," *AIAA J.*, 28, 1405-1414 (1990)
53. Ni, S., B. T. Chao and S. L. Soo, "Particle Diffusivity in Fully Developed Turbulent Horizontal Pipe Flow of Dilute Air-Solid Suspensions," *Int . J. of Multiphase Flow*, 16, 43-56 (1990)
54. Boothroyd, R. G., *Flowing Gas-Solid Suspensions*, Chapman and Hall Ltd., London (1970)
55. Chinnapalaniandi, P., *An Experimental Study of Particle-Laden Jet Interactions With Cocurrent Flows*, Ph.D. Thesis, Case Western Reserve University, Cleveland, OH (1992)
56. Kadambi, J. R., P. Chinnapalaniandi and J. S. T'ien, "Investigations of Transport Properties Involved in FGD," *Final Technical Report for Second Year*, submitted to OCDO, July (1992)
57. Lesinnski, J., et al. "Laser Doppler Anemometry Measurements in Gas-Solid Flows," *AIChE J*. 27, 358-364 (1981)
58. Dantec Electronics Inc., *Particle Dynamics Analyzer, User's Manual*
59. Lading, L. and K. Anderson, "A Covariance Processor for Laser Anemometry and Particle Sizing," *Fourth International Symposium on Applications of Laser Anemometry to Fluid Mechanics*, Lisbon, July 1988
60. Bachalo, W. D, "Method for Measuring the Size and Velocity of Spheres by Dual Beam Light Scatter Interferometry," *Applied Optics*, 19, 363 (1980)
61. Saffman, M., "Optical Particle Sizing Using the Phase of the LDA signals," *Dantec Information*, 5, 8-13 (1987)
62. Bachalo, W. D , M. J. Houser and J. N. Smith, "Behavior of Spray Produced by Pressure Atomizers as Measured Using a Phase Doppler Instrument," *Atomization and Spray Technology*, 3, 53-72 (1987)
63. Sankar, S.V. et al., "Liquid Atomization by Coaxial Rocket Injectors," AIAA-91-0691, *29th Aerospace Sciences Meeting*, NV, 1991.
64. Alexander, D., K Wiles, S. Schaub and M. Seeman, "Effect of Non-Spherical Drops on a Phase Doppler Spray Analyzer," *Particle Sizing and Spray Analysis*, 12, 67-72 (1985)
65. McDonell, V. G.and G. S. Samuelsen, "Measurement of the Two-Phase Flow in the Near Field of an Air-Blast Atomizer Under Reacting and Non-reacting Conditions," *Proceedings: Fourth LDA Symposium*, Lisbon, Portugal (1988)

66. Kadambi, J. R., C. Yurteri and V. Kadaba,"A New Approach Using Transit Time For Simultaneous Measurement of Size and Velocity of Non-Spherical Particles," FED-Vol 191, *Laser Anemometry-1994: Advances and Applications*, ASME, NY (1994)
67. Curtet and Ricon, "On the Tendency of Self-Preservation in Axisymmetric Ducted Jet," *J. of Basic Eng*, 766, December 1964
68. Antonia, R. A.and R. W. Bilger, "An Experimental Investigation of an Axisymmetric Jet In a Co-Flowing Stream," *J. of Fluid Mechanics*, 61, 805-822 (1973)
69. Hinze, J. O. *Turbulence*, McGraw-Hill, NY (1975)
70. Tsuji, Y., Y. Morikawa, T. Tanaka and K. Karimine, "Measurement of an Axisymmetric Jet Laden with Coarse Particles," *Int. J. Multiphase Flow*, 14, 565-574 (1988)
71. Townsend, A., *The Turbulent Structure of Shear Flow*, Cambridge University Press (1956)
72. Chigier, N. and L. M. Beer, *Combustion Aero-dynamics*, J. Wiley and Sons Inc., NY (1972)
73. Kadaba, V. P., *Spray Characterization of a Nozzle in Cocurrent Flow*, M.S. Thesis, Dept of Mechanical Engineering, Case Western Reserve University, Cleveland, OH
74. Kadambi, J. R., C.Yurteri, V. Kadaba and M. Assar, *Investigation of Transport Processes in FGD - Fourth Year Final Report* (OCRC93-2.1), submitted to OCDO, September 1994

CHAPTER 6

HIGH TEMPERATURE DESULFURIZATION OF FLUE GAS USING CALCIUM-BASED SORBENTS

L.-S. Fan, A. Ghosh-Dastidar, S. Mahuli, R. Agnihotri
Department of Chemical Engineering
The Ohio State University
Columbus, OH

Abstract

Removal of sulfur dioxide from flue gas is mainly accomplished by contacting the flue gases with calcium-based (limestone, lime and hydrated lime) sorbents which show remarkable sulfur dioxide scavenging abilities. Although the most commonly used industrial practice is wet limestone scrubbing process, dry scrubbing processes, especially dry sorbent injection (DSI), technology offers a more economical and retrofit technology. Application of DSI for flue gas de-sulfurization (FGD) in high temperature range of 800-1200°C (upper-furnace region) involves injection of dry calcium-based sorbents in the above-the-flame regions of a coal-fired furnace. At high temperatures these sorbents undergo calcination resulting in formation of highly reactive CaO which is subsequently sulfated by SO_2 to form $CaSO_4$. Another phenomenon which is typical of high temperature applications is the deactivation of the CaO via thermal sintering. At high temperature, calcination, sulfation and sintering of the sorbent proceed concomitantly. The lack of interest is applying DSI technology for FGD stems from the inherent inefficiencies associated with this process. Under utilization of the sorbent and its inability to meet the required SO_2 removal standards are the main reasons for unacceptance of DSI as a viable FGD process.

A novel entrained flow reactor system is developed with capabilities to study the gas-sorbent reaction kinetics within a few milliseconds. The calcination and sulfation reactions are studied for their inherent characteristics and the influence of internal structural properties on reaction kinetics is also determined. Time resolved kinetic data has revealed that a substantial amount of sorbent sulfation takes place within the first 100 ms of the reaction and at later times, because of very high transport-related resistances, the sulfation reaction is prematurely terminated. The structural studies have clearly shown certain important transformations such as the preferential loss of small pore sizes and the effectiveness of pores of certain optimum size.

An effort is made to provide a mathematical model to describe the experimental findings. Modelling of the overall sulfation process is done in two stages. An independent model for calcination and sintering of the sorbent particles is developed in the first stage. The results obtained from the calcination and sintering model are applied to the second stage of the model to accommodate the sulfation step and provide a comprehensive model.

This research work, with its time resolved kinetic data and insights into role of pore structure on reaction kinetics, has contributed to a more thorough understanding of short-time SO_2/CaO interaction. This chapter has laid the foundation and background for harnessing the sorbent pore structure and tailoring it to develop sorbents with very high reactivity.

Introduction

Sulfur dioxide and other sulfur compounds, such as hydrogen sulfide, have been identified as the main precursors to acid rain. Sulfur in coal is present in both organic and inorganic forms and during any coal combustion process almost all the sulfur present in the coal is transferred into the flue gas, resulting in a significant contribution to the emission of sulfur dioxide by coal-fired utility boilers and coal-fired combustors in power generating facilities. Most of the coal found in the State of Ohio has a high sulfur content (2-3%). More than two-thirds of the emitted sulfur dioxide in the United States comes from the coal-fired boilers in power generating facilities. In the light of overwhelming evidence pertaining to the damaging effects of acid rain to the ecosystem and biosystem, considerable attention is focused on the control and abatement of sulfur dioxide emissions. The recent focus of various environmental acts puts considerable emphasis in making flue-gas desulfurization (FGD) technology an integral part of utility and industrial processes with coal-fired boilers.

Sulfur dioxide control can be effected by either of the two means. Removal of sulfur from the coal prior to combustion is one of the more direct methods. Removal of elemental sulfur from coal can be achieved by either chemical leaching processes or by catalytic reduction. Both of these methods are rather expensive and are associated with other difficulties, such as loss in heating value and generation of other undesirable sulfur products such as hydrogen sulfide. Moreover the heterogeneity of coal itself poses a considerable challenge to its cleaning. The second method of sulfur removal involves the removal of sulfur compounds, mainly the oxides, after coal combustion.

Combustion of coal transfers most of the sulfur present into the flue gas. Prior to venting the waste flue gas it is treated to remove sulfur dioxide. Removal of sulfur dioxide from flue gas is mainly accomplished by contacting it with calcium-based (limestone, lime and hydrated lime) sorbents which have shown remarkable sulfur dioxide scavenging ability. Limestone and lime scrubbers are common installations in coal-fired power generating facilities. In industry, removal of sulfur dioxide from flue gases is usually accomplished by either wet scrubbing or dry scrubbing. Although the most commonly used industrial practice is wet limestone scrubbing processes, dry scrubbing processes, especially dry sorbent injection (DSI) technology offers a more economical retrofit technology.

Sulfur dioxide from the flue gas can be removed at various stages. High-temperature removal of sulfur dioxide would be at the upper-furnace region (800-1200°C) where the temperatures are favorable for rapid reaction. At this temperature only dry scrubbing of the gases by injecting calcium based sorbents (furnace sorbent injection) is possible and the main product formed is calcium sulfate. The reaction scheme for this process is given as

$Ca(OH)_2 \rightarrow CaO + H_2O$

$CaCO_3 \rightarrow CaO + CO_2$

and

$CaO + SO_2 + 1/2 O_2 \rightarrow CaSO_4$

Sulfur dioxide capture in the medium temperature range (375-650°C) or in the economizer section results from the formation of calcium sulfite instead of calcium sulfate because, at this temperature range, calcium sulfite formation is thermodynamically more favorable.

$Ca(OH)_2 + SO_2 \rightarrow CaSO_3 + H_2O$

and

$CaCO_3 + SO_2 \rightarrow CaSO_3 + CO_2$

Furnace Sorbent Injection (FSI)

Application of DSI for FGD in the high-temperature range of 800-1200°C (temperatures typically found in the upper furnace region) leads to injection of sorbent powders in the furnace. Injection of dry calcium-based sorbents ($Ca(OH)_2$ and $CaCO_3$) into the above-the-flame region of a coal-fired furnace is a potential retrofit technology for controlling SO_2 emission from power plants. When Ca-based sorbents are injected into the furnace, they decompose or calcine to give a highly porous, high-surface area CaO.

$$Ca(OH)_2 \rightarrow CaO + H_2O \quad \text{and} \quad CaCO_3 \rightarrow CaO + CO_2$$

The highly reactive CaO then reacts with SO_2 in the presence of oxygen to form solid $CaSO_4$.

$$CaO + SO_2 + 1/2 O_2 \rightarrow CaSO_4$$

In most of the gas-solid noncatalytic reactions, generation of surface area is considered to be an activation step and any process which results in a loss of surface area is a deactivation step. Contrary to calcination, which acts as an activation step, the sulfation reaction is a deactivation phenomenon which results in the buildup of a $CaSO_4$ product layer and hence a loss in available surface area. Along with sulfation, thermal sintering also contributes to deactivation of CaO. Due to thermal sintering, grains coalesce into larger grains reducing the surface area and porosity of reactive CaO [1].

All of the three activation and deactivation steps mentioned above take place within a very short time of sorbent injection. An earlier study [2] measured the calcination rate of hydrates to be even higher than that of carbonates. Their data indicate 70% of calcination of $Ca(OH)_2$ particles after 25 ms at 700°C. Sintering and sulfation also progress with comparable rapidity inside the furnace. Researchers [3] have reported that about 30% of sulfur capture takes place within 30 ms for hydrates at 1090°C. Although time scales for calcination, sintering, and sulfation are different, they may overlap and the degree of overlapping depends on the furnace conditions. Therefore, it is important to understand each of the three reactions independently in order to understand their combined effects.

High Temperature Phenomena: Calcination, Sintering and Sulfation

Contrary to earlier belief, it has been shown recently by researchers [4, 5] that heat and mass transfer do not offer any significant resistance to calcination of small (less than 10 µm) $CaCO_3$ particles. In Powell and Searcy's work [5], chemical reaction was shown to be the rate controlling mechanism. They also showed that the calcination rate was proportional to the surface area of the porous particle, which was later confirmed by other workers [6]. Researchers [1] have also measured the maximum attainable specific surface area of CaO from calcination of limestone and hydrated lime and found that nascent CaO formed from $CaCO_3$ has a surface area of about $100 m^2/g$ and CaO formed from $Ca(OH)_2$ has a surface area of about 80 m^2/g, for 2-10 µm particles at 700°C. This high surface area was available just after calcination before any significant sintering had taken place.

Unfortunately, nascent CaO cannot retain its high surface area and highly porous structure. Sintering causes the BET surface area of the sorbent to decrease rapidly and, in the absence of sulfation, it eventually levels off to an asymptotic value. Studies with pre-calcines suggest that sintering can be correlated by second order kinetics [7,8,9], the rate of surface area reduction being proportional to the square of the instantaneous surface area. Following the work of earlier researchers [10], a two-sphere model for sintering particles was considered and the following generalized expression for sintering phenomenon was proposed [1]

Equation 6.1

$$\left(1 - \frac{S}{S_o}\right)^n = kt$$

where S_o is the initial specific surface area, S is the specific surface area at time t, and k (function of temperature, gas composition, and surface tension) is the sintering rate constant. After obtaining a value of 2.7 for exponent n, it was suggested that the mechanism of sintering was through lattice diffusion [1]. A strong effect of temperature on sintering was shown by other researchers [11] in their simultaneous calcination and sintering studies. An increase in temperature from 1012° to 1152°C resulted in a 35% decrease in the maximum specific surface area attained by CaO, due to a higher degree of sintering. Enhancement of sintering by the presence of CO_2 and H_2O in the flue gas was also reported by several researchers [7,11]. In one of the studies [11] conducted to determine the surface area evolution of $Ca(OH)_2$ during calcination and sintering, it was found that the concentration of H_2O and CO_2 in the gas stream greatly affected the asymptotic surface area produced during sintering. The presence of CO_2 and H_2O in the gas stream would have significant implications on calcium utilization during the sulfation process because sulfur capture is directly influenced by the surface area.

The sulfation process occurs sequentially or consecutively with that of calcination and sintering. The importance of simultaneous calcination and sintering can be best realized by recognizing the proportional relationship between surface area and sulfation of CaO. The mechanistic steps for sulfation include (a) diffusion of gaseous SO_2 to the sorbent particle surface, (b) SO_2 diffusion through the pore structure of the sorbent, (c) diffusion through the buildup $CaSO_4$ product layer, and (d) heterogeneous chemical reaction at the $CaSO_4/CaO$ or $CaSO_4$/gas interface. Since $CaSO_4$, the sulfation product, has a higher volume than CaO (molar volume ratio ≈ 2.86), it can cause pore filling or pore mouth closure, resulting in a dramatic drop in the sulfation rate. A difference in opinion exists on the rate controlling mechanism during sulfation process. In their work, Simon et al. [12] have suggested that suitable conditions can lead to continuous chemical reaction rate control. Braggart et al. [13], on the other hand, have proposed ionic diffusion through the product layer to be the rate-limiting step. However, it has been supposed in their study that SO_4^{2-} ions are formed at the $CaSO_4$/gas interface and then migrate through the $CaSO_4$ product layer to the $CaO/CaSO_4$ interface. Their suggestion is inconsistent with direct experimental findings of Hsia et al. [14], which indicate that Ca^{2+} and O^{2-} ions diffuse through the $CaSO_4$ product layer to the $CaSO_4$/gas interface and the solid state reaction proceeds at a rate limited by such transport of Ca^{2+} and O^{2-}. To identify the dominant diffusing species in the solid-state reaction, they conducted marker experiments with Pt being used as the inert marker. The marker method involves attachment of small inert marker particles to the solid at the original gas/solid interface. As the transport and reaction proceed, the direction of ionic species transport through the product layer influences the relative position of the marker. Results obtained by these researchers [14] indicated that the growth of the product layer takes place by the "outward growth mode," as shown in figure 6.1. Therefore, the reaction proceeds in part by the outward diffusion of Ca^{2+} and O^{2-} ions through the product layer.

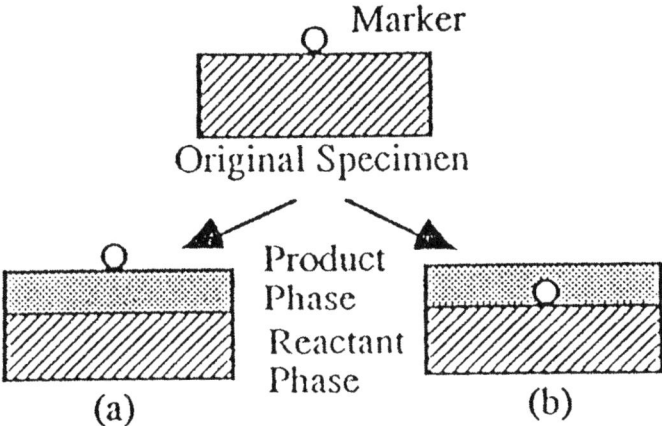

Figure 6.1 (a) Inward growth where reactants move from gas/solid interface to solid/solid interface; (b) outward growth where reactants move from solid/solid interface to gas/solid interface (Hsia et al. [14]).

Significant questions remain to be addressed before a complete understanding of the high temperature CaO/SO_2 reaction will be obtained. As seen in a number of studies, the coupling of the activation and deactivation processes reaches completion within a few hundred ms of injection. Substantial sulfur capture ($\approx 30\%$) is reported within 50 ms after injection into the furnace [15,16]. Unfortunately, there is very little data for sorbent residence times in these times scales. Entrained flow reactors are commonly used for studying reaction kinetics for FSI, but accurate estimation of sorbent residence time and rapid heating and cooling of sorbents remain the most difficult fundamental problems. Moreover, those few studies which have obtained some kinetic data in less than 100 ms time scales, have estimated solid residence time from a knowledge of gas flow pattern. This technique is unsatisfactory, since it requires that the flow patterns are well defined, which is nearly impossible to achieve within the short time scales of interest. As a result, their approach can incur great error, possibly as high as one or two orders of magnitude. Insertion of solid injection and sampling probes also aggravate this problem. In view of this, a great need for an improved technique for solid residence time measurement exists, in order to obtain accurate time-resolved kinetic data for ultrafast phenomena involved in sorbent/SO_2 interaction.

High Temperature Desulfurization of Flue Gas Using Calcium-Based Sorbents

Role of Surface Area and Porosity of the Calcines

Development of extensive internal surface area [17] and concurrent pore structure [18] through calcination are prerequisites for a highly reactive sorbent. In an effort to prove that chemical reaction is not rate limiting during sulfation, work done by Borgwardt and Bruce [17] indicates that the sulfation rate exhibited a greater sensitivity to the specific surface area than expected for chemical reaction control. Their experimental results indicated that the sulfation rate is proportional to the square of specific surface area of the reacting CaO. The appropriate rate expressions and equations illustrating the effect of surface area are given below [17].

Equation 6.2

$$1-3(1-x)^{2/3}+2(1-x)=k_d t$$

Equation 6.2 gives the conversion versus time response under the assumption that diffusion through the product layer surrounding unreacted cores of the spherical grains is the rate-controlling step. Conversion given by X and k_d is the reaction rate constant. In this case, the value of k_d is derived from Fick's law for diffusion into a nonexpanding spherical grain.

$$k_d = \frac{6MD_e C_{SO_3}}{\rho_{CaO}(r_g)^2}$$

where M is the molecular weight of CaO, r_g is the grain radius, D_e is the effective product layer diffusivity, C_{SO_3} is the concentration of SO_3 on the product layer surface and ρ_{CaO} is the density of the CaO grain. The surface area S_g is related to the grain radius by:

$$r_g = \frac{3}{S_g \rho_{CaO}}$$

The effect of sintering is to significantly reduce this surface area while causing only a minor decrease in the total pore volume. However, during sintering, the pore size distribution shifts to higher values, indicating coalescence of small pores into larger ones [19]. The sulfation reaction results in the formation of a larger molar volume product which causes a decline in pore volume and surface area through pore filling or pore mouth plugging. Since small pores fill faster than the larger ones, the pore volume distribution shifts to larger pore sizes as reaction occurs.

Although the role of the surface area and porosity of the calcined sorbent is much clearer than the mechanistic theories of the process, ambiguity remains in the complete understanding of their effects in sulfur capture. Kinetic data for combined calcination and sintering in very short time scales is almost nonexistent [2]. No experimental evidence exists relating the pore size distribution to the conversion for sulfation. In experiments conducted earlier by Gullett and Bruce [19], conversion levels of pre-calcined sorbents were beyond the conversion levels expected by the complete loss of initial porosity. Obviously, the expansion of particle is necessary to account for these conversions. However, a clear picture of the particle expansion phenomenon is not known.

Role of Particle Size

The effect of sorbent particle size upon reactivity has been investigated by a number of researchers, using both pre- and post-reaction sizing methods [20]. The increase in reactivity with smaller sorbent size is likely due to greater surface area to volume ratio as particle size decreases and also due to easier accessibility for diffusing SO_2. Milne and Pershing [15] have also observed that the initial sulfur capture is a strong function of particle size, but after approximately 200 ms, the subsequent sulfur capture is insensitive to particle size. They also reported no particle size effect on sulfur capture below 2.5 μm. Similar observations were made by Cole et al. [8], that below 2-5 μm in size, particle size no longer influences reactivity due to elimination of pore diffusional resistances. However, this result is contrary to the experimental findings of Gullett and Blom [20], who reported no limit to this particle size effect upon higher reactivity and observed constant effect of size over the range of 0.72 to 12.1 μm. Further work is needed to address this question. More importantly, a large amount of reaction data with respect to particle size is needed for short time scales (0-100 ms), in order to gain an insight into the particle size effect.

Role of Sorbent Type

Several past studies have shown the superiority of $Ca(OH)_2$-derived calcines over CaO produced from $CaCO_3$, in terms of reactivity with SO_2 [8, 19]. Three reasons have been suggested for this superior performance of $Ca(OH)_2$: (1) The rate of hydrate calcination greatly exceeds that of carbonates. (2) The inherent particle size of hydrate is much less than the carbonates as a result of the hydration process. This property lessens the mass transfer resistances in $Ca(OH)_2$ and $Ca(OH)_2$ calcines. (3) While CaO derived from $CaCO_3$ possesses cylindrical pores, calcination of hydrates appears to develop a more open, plate-like pore structure, which facilitates easier access of SO_2 to internal surface area. Moreover, this plate-like pore structure of hydrate calcines allows for greater particle expansion, which delays pore filling by product layer build up [19, 21].

However, contradictions still exist between the available studies. It has been claimed by Milne and Pershing [15] that the difference in SO_2 capture between $CaCO_3$ and $Ca(OH)_2$ is primarily attributable to a dramatic difference in prompt SO_2 capture occurring in the first 30 ms after injection, but this superiority of $Ca(OH)_2$ vanishes if the particles are below 5 µm in mean diameter. Variation in sorbent reactivity was also noted among the hydrates or carbonates, depending on the type of parent stone and also the sorbent preparation process [20].

Role of Temperature

Temperature plays an important role in determining the rate of reaction and the rate of sintering. These effects are manifested in the porosity and surface area of the calcine. Sintering is an activated phenomenon; i.e., with increasing temperature, the rate of sintering also increases. In the absence of a more thorough and fundamental knowledge, the rate constant for sintering can be expressed as an Arrhenius type equation

$$k_s = A\exp\left(\frac{E_{as}}{RT}\right)$$

where A is the pre-exponential factor and can be related to the average number of contact points for grains in the particle [1] and E_{as} can be interpreted as the activation energy for sintering. An increase in temperature would increase the rate of sintering resulting in reduction in surface area. Increasing temperature would also increase the extent of calcination and sulfation for the same reaction time.

High-Temperature Entrained Flow Reactor Setup

Introduction

In the upper-furnace region of a coal-fired combustor, all three phenomena, namely calcination, sintering and sulfation, are extremely fast, and substantial sorbent conversion takes place within 200 ms. Although there is no dearth of literature dealing with gas-solid reactions, techniques for accurate kinetic measurement in such a short time scale are virtually nonexistent. To study the ultrafast interaction between different overlapping phenomena, a high-temperature entrained flow reactor system has been developed, which along with its probe systems is capable of performing accurate kinetic measurements for short contact time sorbent-gas reactions, with particular emphasis in the less than 100 ms time scale.

Entrained flow reactor systems are commonly used for studying gas-solid reaction kinetics, because external transport resistances are minimal. However, conventional flow systems can not be readily used for the small time scale of interest associated with sorbent-SO_2 reactions. Because of the high temperature dependence of the reaction, it is crucial to maintain the sorbents at low temperatures until injected into the reactor. Once injected, it becomes equally important to heat up the sorbent particles to the reaction temperature as rapidly as possible. Similarly, particle collection in the sampling probe has to be done with immediate quenching of the gas-solid stream to prevent any further reaction from taking place. Any lapse in fulfilling the above three conditions will result in erroneous kinetic data. Since the particle residence time of major interest ranges within 0–100 ms, it must be estimated to an accuracy of a few ms or better.

In the past, few attempts have been made to obtain time-resolved measurements for SO_2 removal in the 0–200 ms range. Borgwardt et al. [3] employed a phase discrimination probe for sampling solids from the reactor. It appears that the separation of solids, achieved within their probe, may not be complete and some of the fines may be excluded from the subsequent analysis. Literature also cites the use of an axial sampling probe to obtain residence time measurements down to 35 ms [16]. In both the above studies, particle residence time was estimated indirectly from a knowledge of the gas velocity profile, thus requiring an accurate characterization of the fluid motion. This method also required isokinetic sampling. With these systems, measurements at low residence time may not be accurate, because at such short time scales, description of the flow pattern suffers from large uncertainties. Moreover, the presence of the sampling probe is likely to alter the velocity profile inside the reactor.

Entrained Flow Reactor System

The rationale behind the development of a new high-temperature, short-contact time flow reactor is to examine gas-sorbent reaction kinetics. Design considerations for the high-temperature reactor include ironing out the shortcomings and other associated difficulties in earlier designs.

The high-temperature, short-contact time reactor system used to accomplish the kinetic study is shown schematically in Figure 6.2. The reactor setup consists of a furnace, the reactor tubes, a particle feeder, an injection probe, a collection probe and a particle separation/classification system. Various components of the experimental system are discussed below in detail.

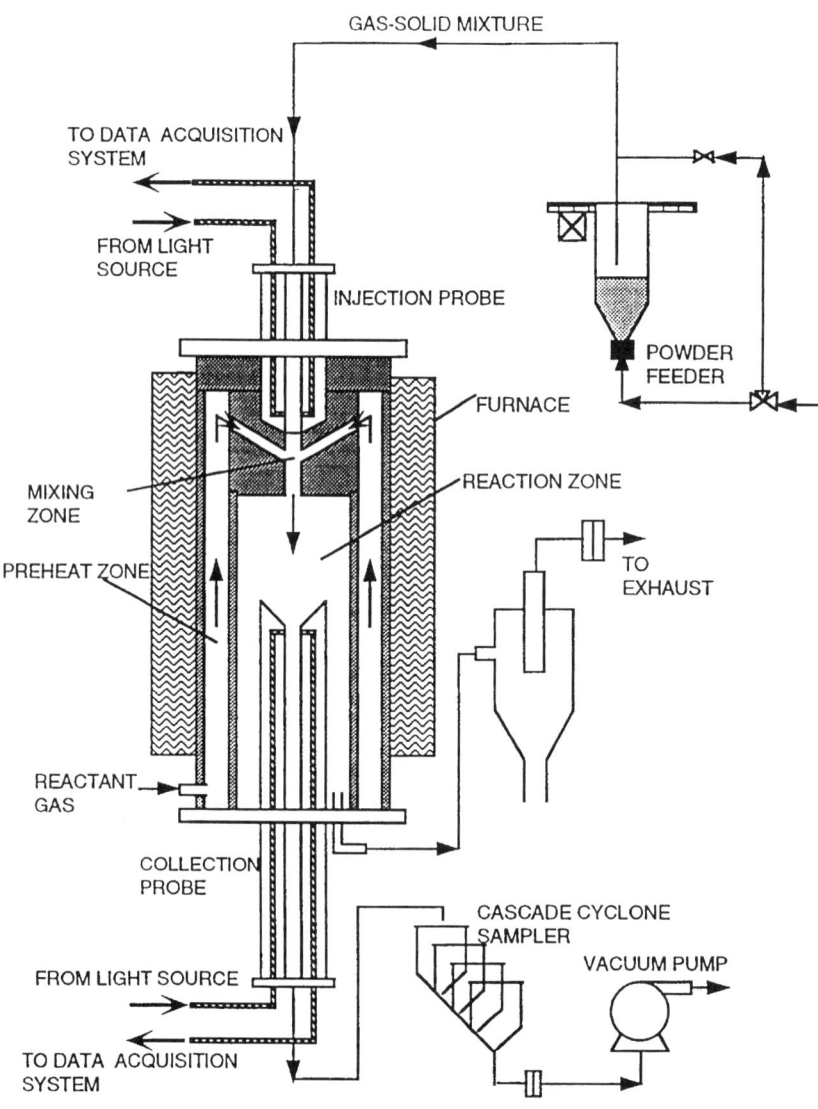

Figure 6.2 Schematic of the entrained flow reactor system.

The Powder Feed System

Sorbent powders used in furnace sorbent injection, due to their small particle size and adhesive nature, belong to Geldart's Group C category. Transportation of these powders to obtain a steady and stable flow is quite a difficult task. Furthermore, the sorbent powders need to be injected in high enough quantity to be easily detected by the optical techniques employed for their residence time measurement. Various feeder designs are available in the literature [22, 23]. While one design fails to produce a stable flow of the powders, the other provides powder concentrations too low to be detected by the optical systems. Considering these requirements and limitations, a novel powder feeder was designed which performs quite satisfactorily in providing steady and low flow of powder when operated under a pulse or continuous injection mode.

The feeder used in the high-temperature reactor system is a plexiglas column with cylindrical top and conical bottom. A schematic of the powder feed system is given in Figure 6.3. At the bottom of the conical section is a 1.2 cm ID tube fitted with a porous filter disc which helps distribute the feeder fluidizing flow. Located above the powder bed is an off-take tube extending outside the column. The position of the off-take tube above the powder bed can be adjusted depending upon the height of the powder bed and the desired rate of powder feeding. The whole powder feed assembly is attached to a vibrator. The entrained sorbent particles from the powder bed are transported by the off-take tube into the reactor. The flow through the porous disc and constant vibration from the vibrator maintain the powder in a "loosened" or locally fluidized state. The vibrator which provides irregular vibrations to the vessel also prevents channeling of fluidizing gas in the powder bed.

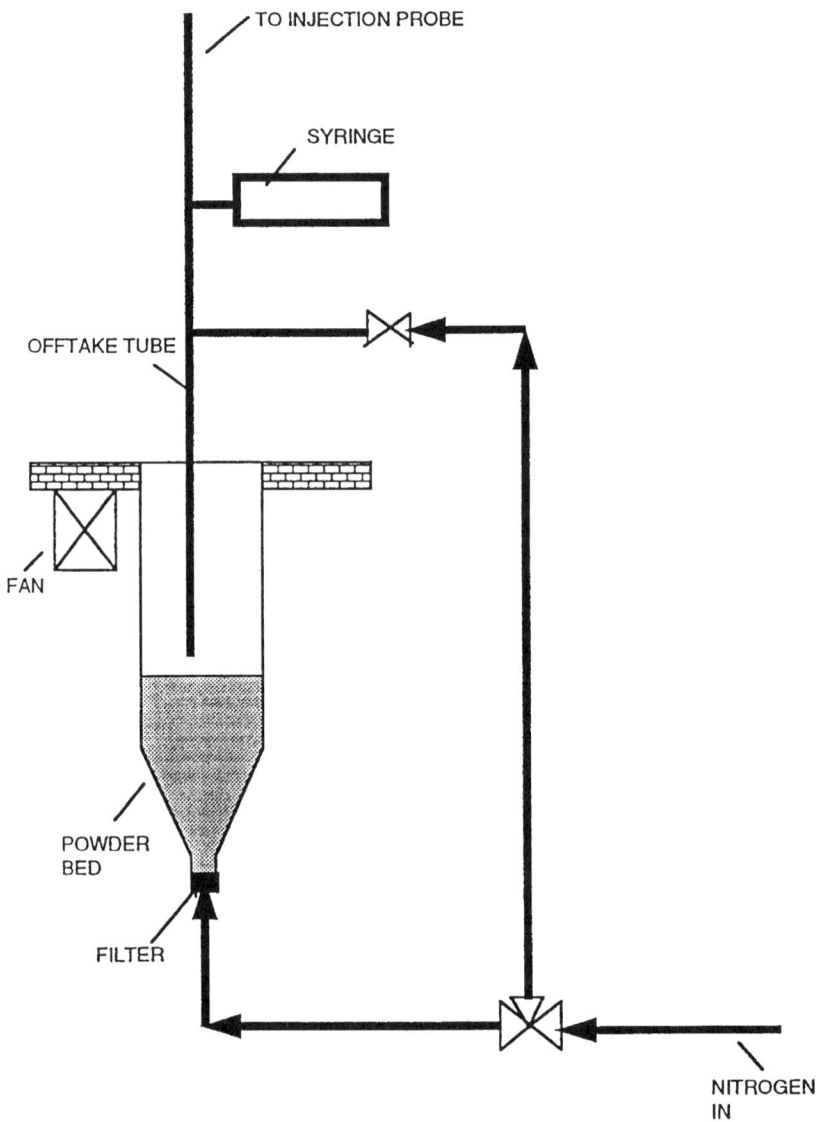

Figure 6.3 Schematic of the continuous powder feed system.

The sorbent particles can be fed continuously or in pulses into the reactor by means of a modified solenoid valve, adopted and modified from the design available in the literature [22]. In continuous operation mode, powder bed is fluidized and the entrained sorbent particles are carried into the reactor through the off-take tube. In pulse feeding mode, the outlet from the solenoid valve is connected to the injection probe of the reactor. When the solenoid is open, the powder is ejected from the feeder into the valve chamber and swept by the transport gas to the reactor. When the solenoid is closed, the transport gas flushes the valve as well as the transport tube connected to the reactor, preventing powder accumulation. At the same time, accumulation in the off-take tube from the feeder is eliminated by the purge line which causes a reverse flow back into the feeder. Thus, as the solenoid opens and closes, a pulse of sorbent powder is injected into the reactor. The powder is injected as a train of pulses during experiments. Duration of a pulse and frequency of pulsing are controlled by a timer connected to the solenoid. In both modes of powder feeding (continuous and pulse), the feeding rate has to be such that the powder hold-up inside the reactor is always less than 0.1%. The powder feeder also has a provision for manual pulse injection of powder, by a syringe, to determine the residence time during continuous powder feeding mode.

The High-Temperature Reactor

The high-temperature reactor consists of two concentric alumina tubes (5.08 cm and 7.62 cm OD respectively) which are housed within a 3-zone, 1500°C, Lindberg vertical furnace. Furnace heating is controlled for each zone separately, by a control console. The reactant gas enters from the bottom into the annular region between the two alumina tubes. While traveling upwards through a heated length of 91.4 cm (36 inches), it is heated up to the desired reaction temperature, and at the top makes a 120° turn to enter into the inner tube as two hot jets. The details of the entrance block design are shown in Figure 6.4. Pulses of sorbent particles are introduced into the reactor by the injection probe, whose tip is located just above the incoming jets. Thus the sorbent stream from the injection probe is impacted by the hot reactant gas jets. The jet impaction technique, successfully employed by earlier researchers [24], causes substantial loss of momentum, creating severe turbulence, which in turn causes rapid heating of the sorbents. Heat transfer calculations indicate that sorbent particles of less than 10μ size heat up within few ms after injection into the reactor, provided the gas and solid phases are well mixed. Optical fibers housed within the injection probe detect a spike for each pulse feeding, and a similar optical fiber system in the collection probe detects another spike as particles are sampled. The collection probe can be moved up and down along the axis of the reactor and, hence, particles can be intercepted at various residence times. The sorbents are cooled down very fast in the collection probe to prevent any further reaction and eventually collected in the particle collection/classification system. A noteworthy design feature of the reactor column is the specially designed flanges for water cooling. All four flanges, sealing both ends of the 7.62 cm and 5.08 cm alumina tubes, must be kept cooled in order to prevent overheating of the viton gasket, thus preserving the seal. In addition to water cooling of the flanges, the alumina tubes are also cooled by several cooling fans at both ends of the reactor column.

To obtain accurate kinetic information it is necessary that the temperature profile in the reaction zone be as uniform as possible. Figure 6.5 shows the axial temperature profile in the reaction zone for two different reaction temperatures. As can be seen from the profiles near the tip of the injection and collection points there is substantial drop in the temperature. This is due to the cooling effects from the specially designed injection and collection probes. It is also evident that away from collection and injection probes the temperature profile is rather stable and uniform. Cooling effects from the injection and collection probe put a restriction on the minimum length of the reaction zone which in turn puts a restriction on the minimum reaction time attainable. Test runs have revealed that reaction zone lengths of less than 3.5 inches don't give a stable uniform temperature profile and kinetic data obtained under these conditions is not reliable.

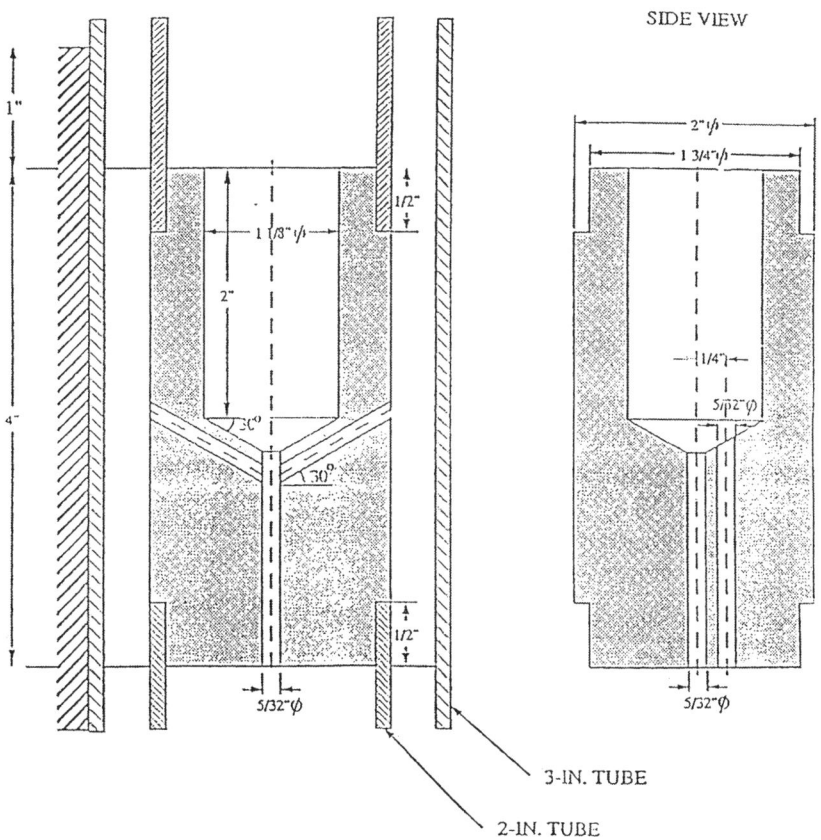

Figure 6.4 Details of the reactor entrance block.

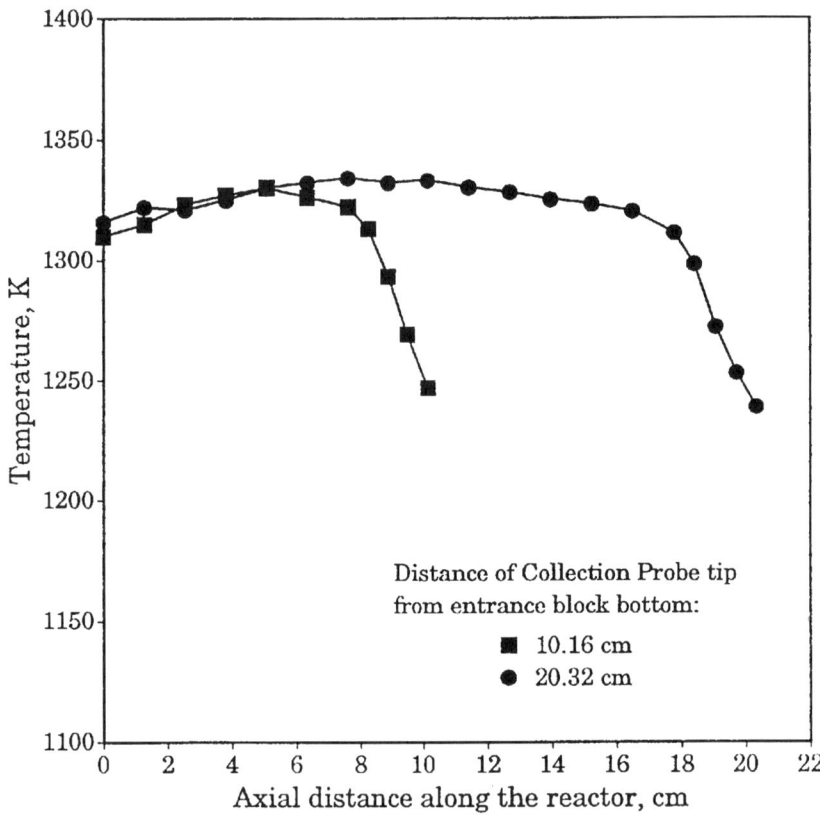

Figure 6.5 Typical axial temperature profile inside the reactor with a nominal reactor temperature of 1323 K.

The Probe System

Design of the probe is critical for obtaining short contact time measurements. The schematic of the collection probe assembly is shown in Figure 6.6. Figure 6.7 and Figure 6.8 show the front view and side view of the collection probe head, the most important component of the probe assembly. Figure 6.9 details the design of the collection probe assembly. The injection probe is similar in design with minor differences needed for probe-specific operation. The front view and side view of the injection probe head are shown in Figure 6.10 and Figure 6.11 respectively. Design details of the injection probe are given in Figure 6.12. Each probe has an OD of 2.5 cm and although the injection probe remains fixed at a particular position, the collection probe can be moved up and down along the axis of the reactor tube. The sorbent particles are injected (injection probe) and sampled (collection probe) through a 3.2 mm hole along the probe center, and pass through the optical path created by an optical fiber assembly as they exit or enter the probe. Two 4.7 mm holes are located diametrically opposite to each other. Each is at a distance of 7.1 mm from the probe central axis and aligned parallel to it. These two holes and the central hole are connected by a 1.6 mm opening radially to create the optical path near the probe tip. Inserted in each 4.7 mm hole is a light guide consisting of a stainless steel tube of 3.2 mm OD, a fiber optic cable inside the steel tube and a miniature 2 mm right angle prism which is located at the top of the fiber optic cable. Each light guide can be withdrawn from the probe in case of any accidental particle deposition on the prism surface. One of the cables is connected to an illumination source, while the other is connected to a high-speed microcomputer data acquisition system via a photomultiplier. The optical guide design is shown in Figure 6.13 in detail. The prisms are assembled in such a way that their vertical sides face each other. The light from the illumination source travels through the fiber optic cable to the base of the prism. The prism causes a 90° turn in the light transmission at its hypotenuse due to total internal reflection, directing the light to the face of the other prism radially, where again a 90° turn directs the light to the photomultiplier-microcomputer data acquisition system. With this optical arrangement, the solid laden gas entering the collection probe crosses the optical path between the two prisms, causing attenuation of the light transmitted.

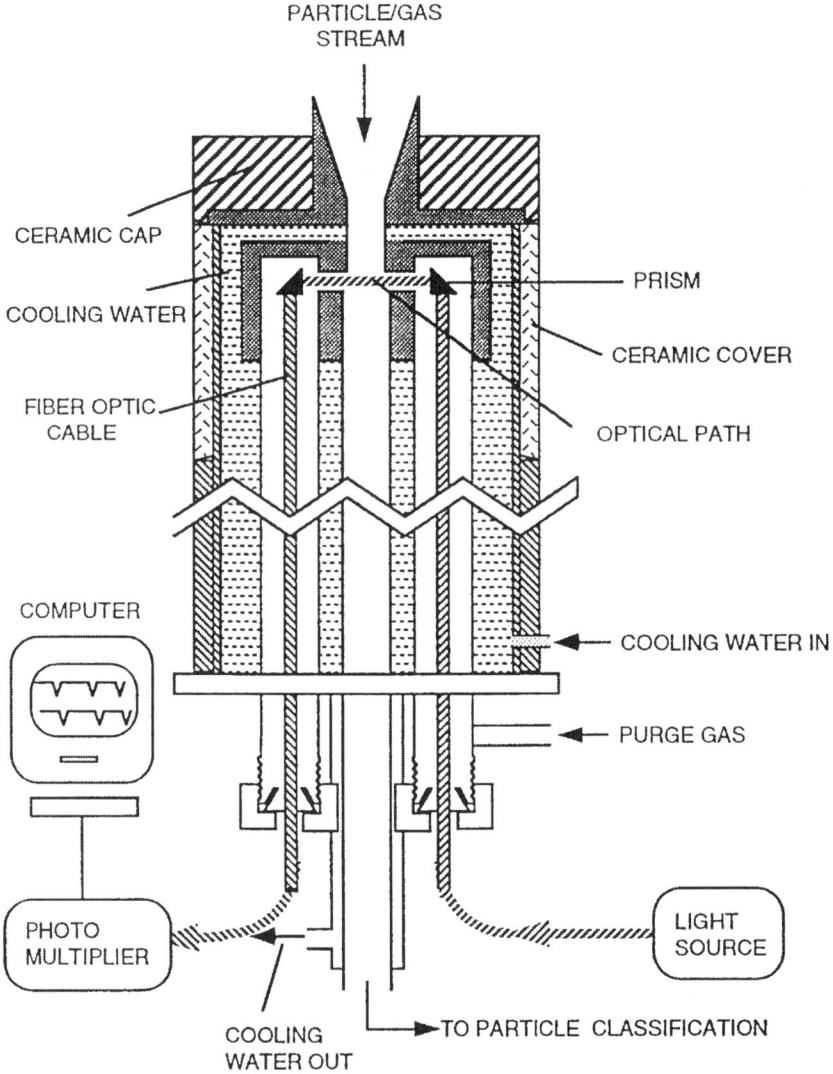

Figure 6.6 Schematic of the collection probe.

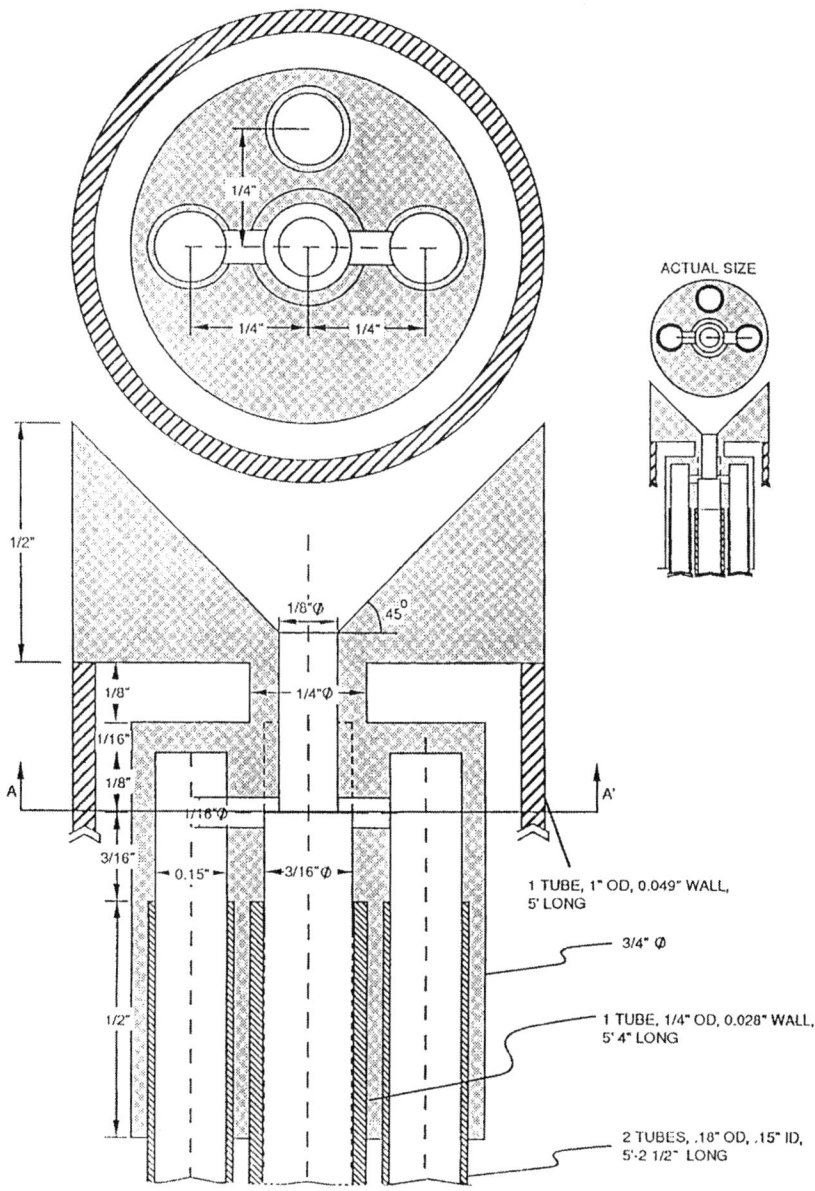

Figure 6.7 Front view of the collection probe head.

High Temperature Desulfurization of Flue Gas Using Calcium-Based Sorbents

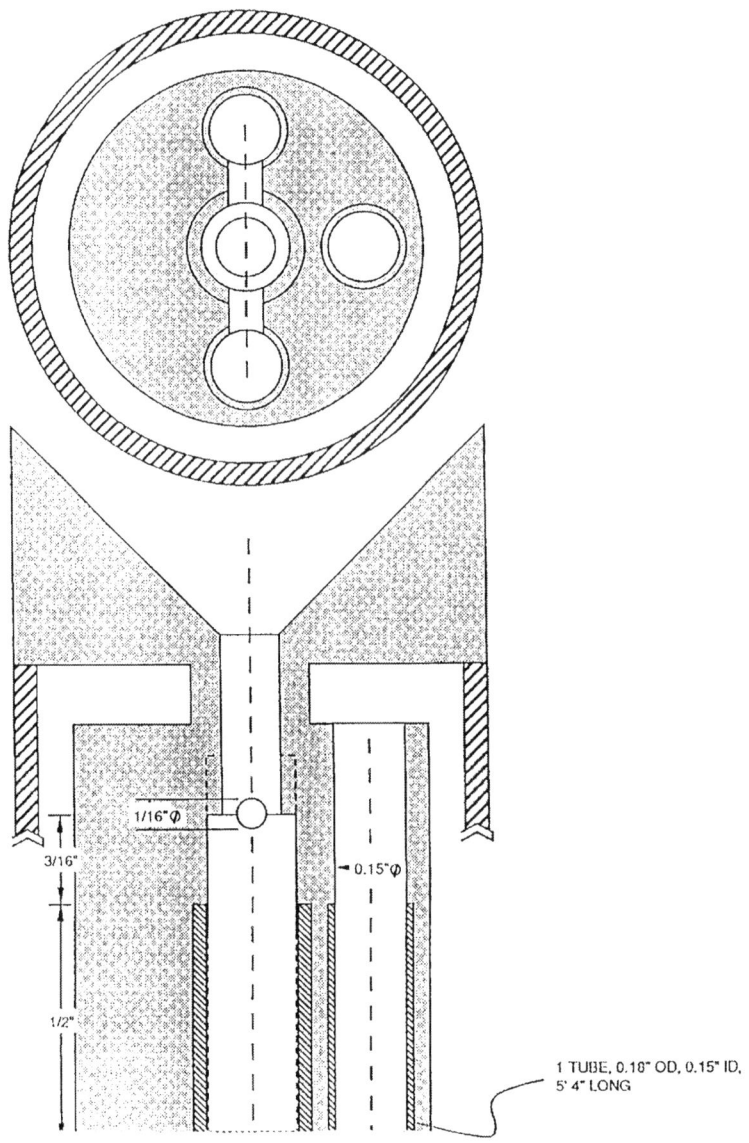

Figure 6.8 Side view of the collection probe head.

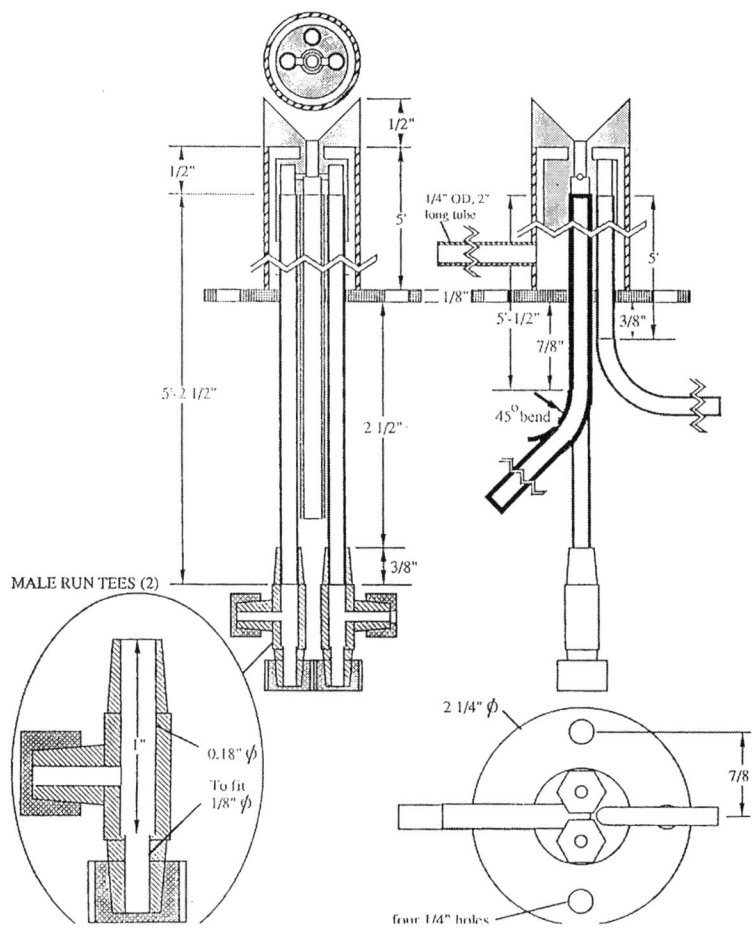

Figure 6.9 Design details of the collection probe assembly.

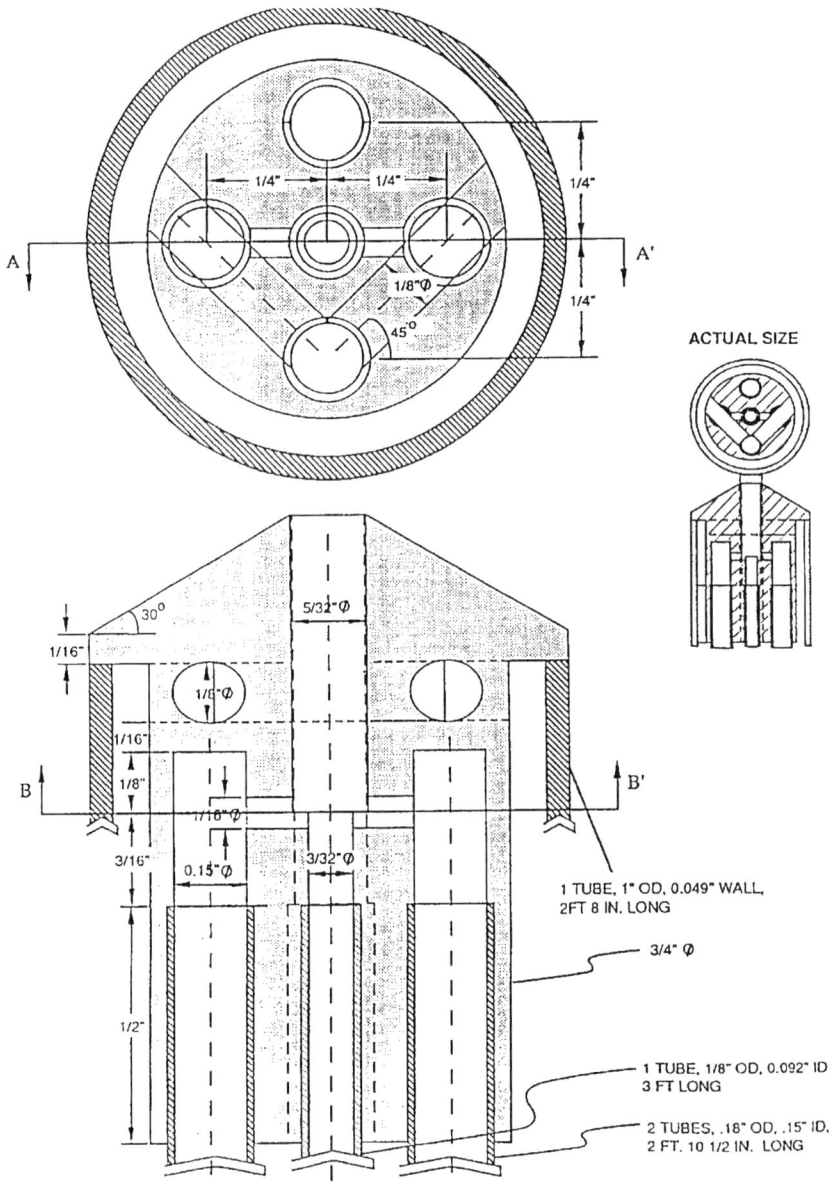

Figure 6.10 Front view of the injection probe head.

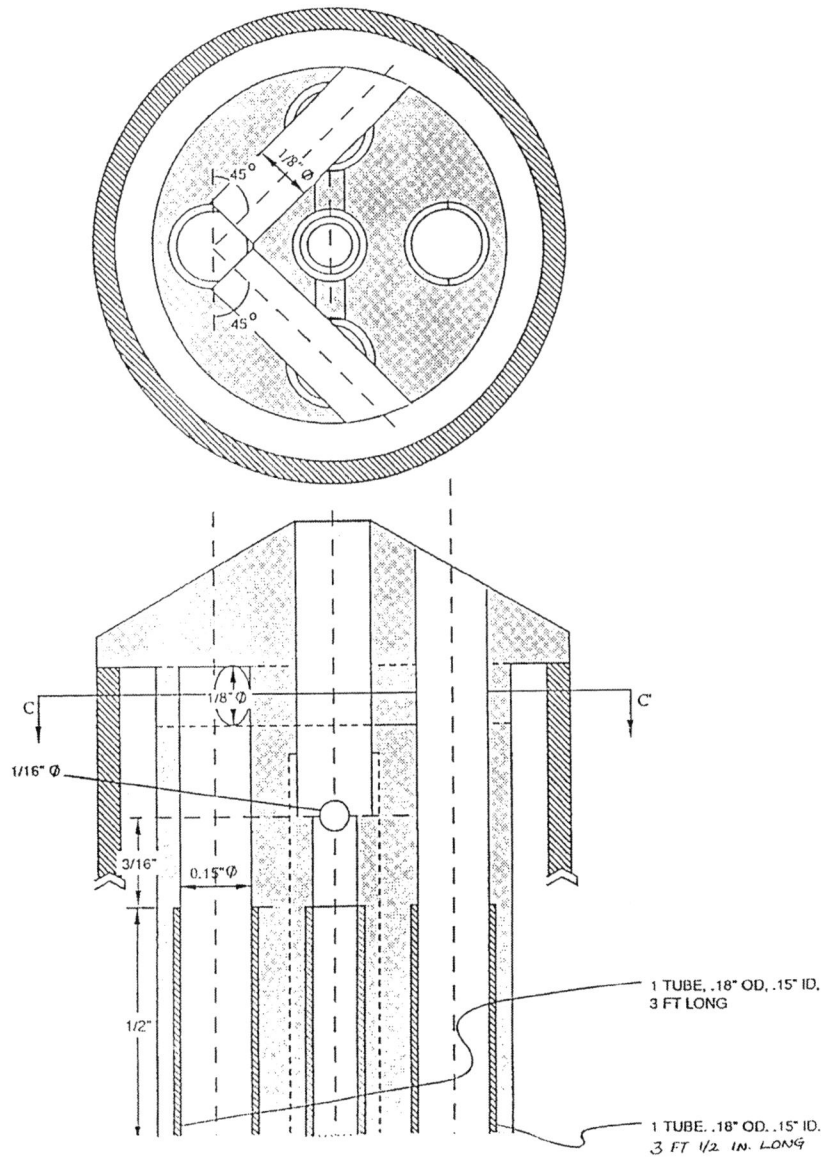

Figure 6.11 Side view of the injection probe head.

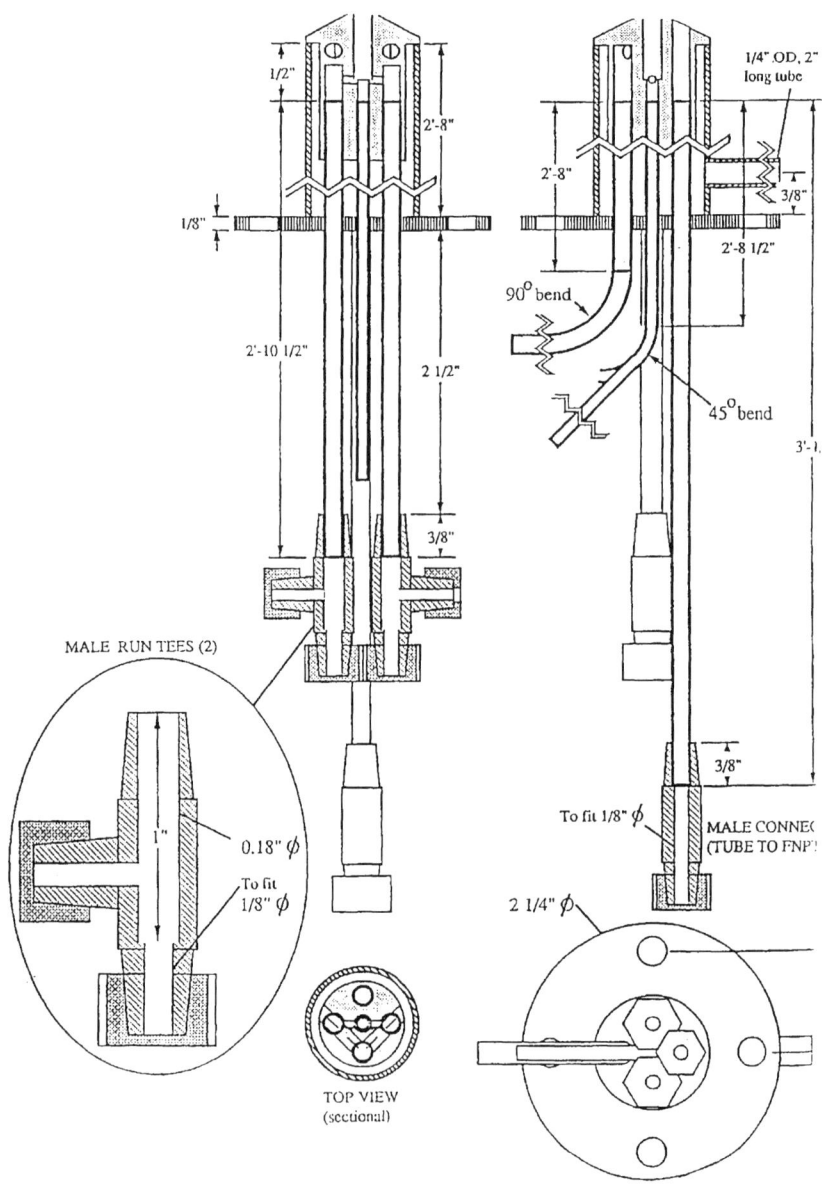

Figure 6.12 Design details of the injection probe assembly.

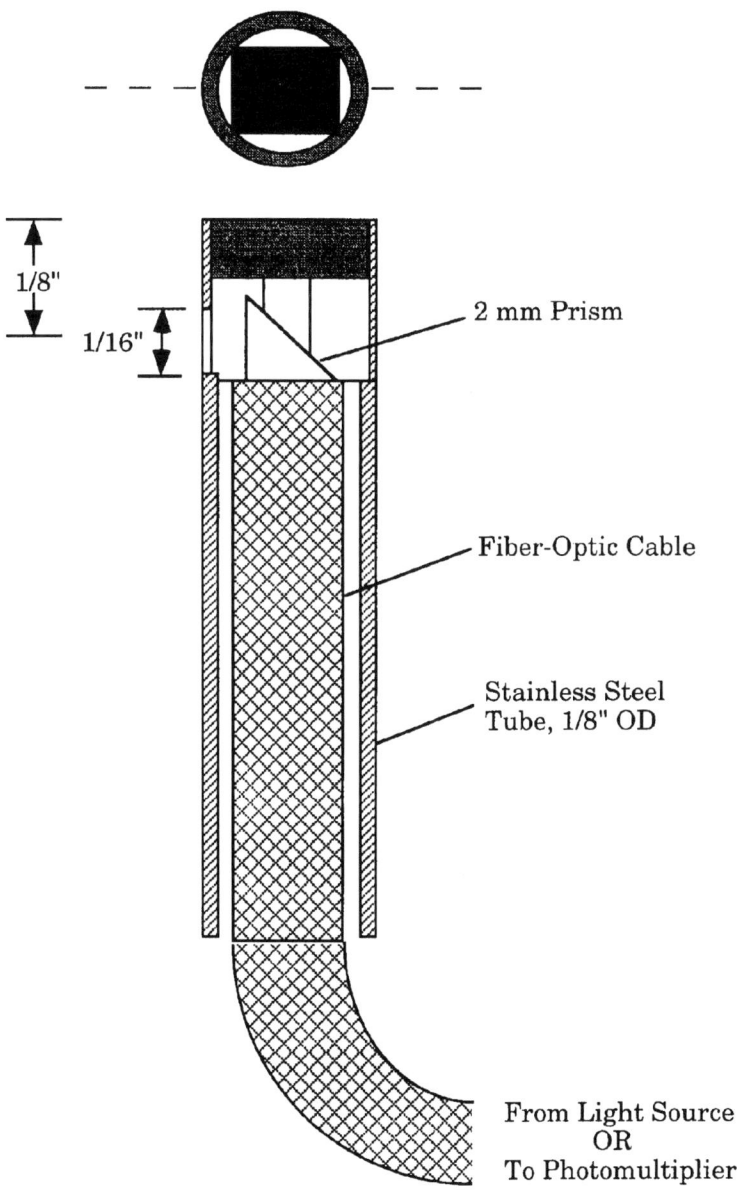

Figure 6.13 Design of optical guide.

High Temperature Desulfurization of Flue Gas Using Calcium-Based Sorbents

Dry nitrogen is purged through the annular space surrounding the fiber optic cable past the prism and mixes with the central gas-solid stream. The purge gas quenches the reaction as soon as the sampled stream enters the collection probe. It also prevents powder from depositing on the prism surface and helps maintain the fiber optic systems at a lower temperature. Cooling water also circulates through the probes to prevent any undesirable heating of the various parts of the probe assembly.

It has been mentioned before that whereas the injection probe is fixed in a particular position, the collection probe is free to move along the axial path of the reactor tube. This provision makes it possible to collect particles at different residence times from the reactor. Among other differences between the two probes is an extra path for thermocouple insertion in the injection probe. The 1.83 m long thermocouple goes through the injection probe head and is free to move up and down inside the reactor, thus making it possible to measure the temperature at different axial locations during the experiments. The tips of the two probes have different configurations. The collection probe tip is made like a funnel to facilitate particle collection, whereas such an arrangement is neither required nor provided in the injection probe head.

The main premise behind the design of injection and collection probe is to limit the reaction to the reaction zone which lies between the tips of injection and collection probe. Particles coming out of the injection probe should be instantaneously heated up to the reaction temperature and the reaction should be quenched as soon as the particles enter the collection probe. Reaction quenching is effected by cooling the particles as well as the gases coming out of the reaction zone.

The heat-up rate of particles is estimated for $Ca(OH)_2$ particles subjected to an increase in the surrounding temperature. A heat balance on a single, spherical $Ca(OH)_2$ particle accounting for convection from the bulk gas, radiation from the wall, and the endothermic heat of the calcination reaction gives

Equation 6.3

$$\frac{\rho_p V_p C_p}{4\pi R_p^2}\frac{dT_p}{dt} = h(T_b - T_p) + (\sigma_{\epsilon p}(T_b^A - T_p^4) - \gamma_{Ca(OH)_2})\Delta H_c$$

The $Ca(OH)_2$ particle is assumed to be initially at room temperature as it is transported by nitrogen gas through the water-cooled injection probe. The hot bulk gas impinges upon the $Ca(OH)_2$ particle and due to the high bulk/transport gas ratio, the resulting gas is assumed to reach the final temperature instantaneously, implying a step increase in the surrounding temperature to T_b. The surrounding gas is assumed to be stagnant and the internal temperature gradients in the particle are assumed to be insignificant, since the Biot number for the small particles of interest is less than 0.1 [25]. Earlier researchers [26] have also justified the assumption of a flat temperature profile for small diameter particles. Theoretical calculations were carried out to determine the heating time for particles of different sizes and the results are presented in Figure 6.14. A simple calcination rate equation [11] is used to estimate the rate of calcination.

Equation 6.4

$$\ln(1-x) = -k_c S_{Ca(OH)_2} t$$

Equations 6.1 and 6.2 are simultaneously solved for a bulk gas temperature of 1000°C, and a particle initial temperature of 27°C. The results indicate that the particles which are less than 10 μm in diameter are heated to the final temperature within 2 ms of injection into the hot gas stream. The cooling effect of the endothermic calcination reaction is observed to be insignificant for such small particle sizes. Calculations done by other researchers [16,26] also estimated negligible heat-up times for such small particles. Additional heat transfer calculations confirm that the assumption of neglecting temperature gradient inside the particle is indeed valid.

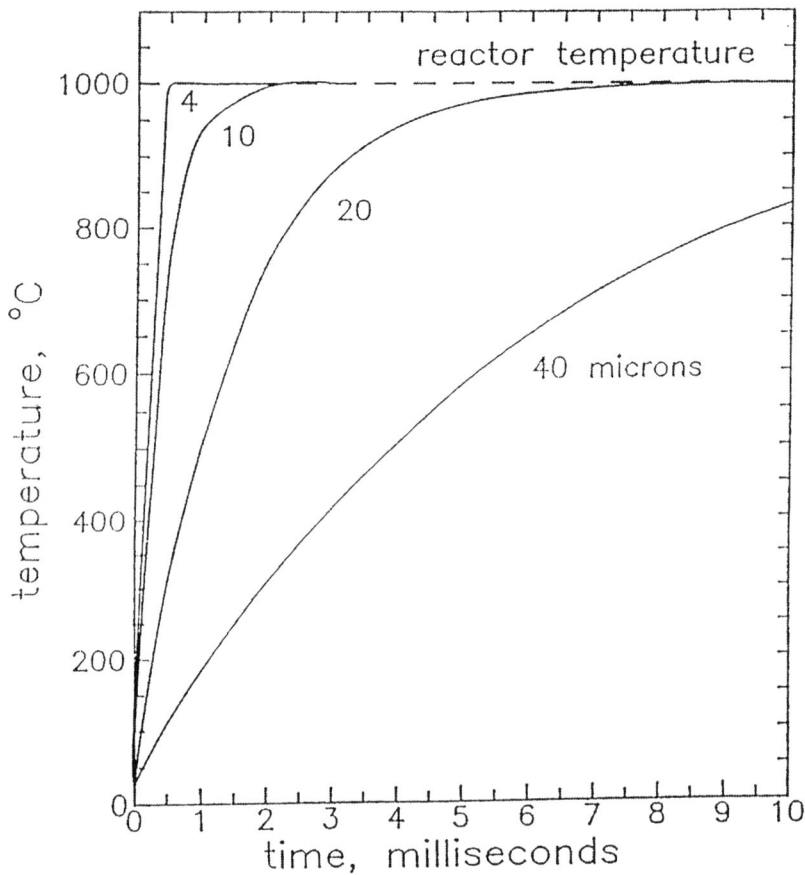

Figure 6.14 Theoretically calculated particle heat-up rate assuming the Nusselt number to be equal to 2.0.

Quenching of the reaction is achieved by rapid mixing of the hot particle/gas stream with cold nitrogen purge and by the water-cooling of the collection probe. Sufficient flow of circulating water is used to maintain the collection probe assembly at about 50°C. Considering a reaction temperature of 1100°C, the temperature of the gas-particle stream drops to about 1000°C at the probe tip, as seen in Figure 6.5. Upon entering the collection probe, the hot stream is mixed with two cold N_2 jets. Heat balance calculations are performed using reactant gas conditions of 12 lpm (STP) and 1000°C, a nitrogen purge of 3 lpm (STP) at 50°C and an inside wall temperature of 50°C. Further simplifying assumptions of instantaneous gas mixing, flat particle temperature, and no entrance effects are similar to heat-up calculations. It is seen that the desired cooling of the gas/particle stream is achieved within 2 ms. Sufficient quenching of the reaction is believed to have been achieved when the gas/solid mixture attains a temperature below 500°C, since no appreciable conversion of CaO to $CaSO_4$ occurs below 800°C, and $Ca(OH)_2$ calcination is reasonably slow at temperatures less than 500°C. Experimental verification of quenching is also carried out by studying the effect of increasing purge flow on calcination. Increasing the flow rate of the cold purge from 3 lpm to 5 lpm results in no significant difference in the extent of decomposition which confirms that adequate quenching of calcination reaction is achieved with 3 lpm of nitrogen flow.

Particle Collection/Classification System

The particle collection assembly is constituted by a filter and a cyclone or a cascade impactor. For pulses with a low solid feed, a filter made of quartz wool is placed after the collection probe. The filter catches all the particles including fines. The filter is subsequently taken out of the filter holder for kinetic measurements and analysis. For pulses with higher solid feed, a cyclone is employed to achieve the preliminary gas-solid separation. In this case, the particles collected in both the cyclone and the filter are analyzed for sorbent conversion. For each experimental condition, a cascade impactor is also used in place of the cyclone to obtain an in situ particle size distribution. The particle size distribution data is useful for investigating the effect of particle size on reactivity. During sulfation experiments, the collection path stretching from the probe outlet to the filter assembly is kept heated to about 350°C to avoid any SO_3 condensation in the line or the solid sample. A high temperature ball valve, placed in between the collection probe outlet and the particle collection device helps in collecting the sample without a need to close off reactant gas flow into the reactor. Instead, the entire gas stream is diverted towards the bulk outlet of the reactor, where it again passes through a cyclone and a filter assembly to separate particles before being vented out through a master exhaust system. An Edwards 2-stage vacuum pump, at the end of the sampling system, facilitates particle collection in the collection probe. Both the bulk outlet and collection probe outlet gas streams are monitored by flow meters to account for the total distribution of flow.

Particles of different sizes can also be collected in situ using a cascade of cyclones. Each cyclone (stage) is designed to capture particles of a specific size range depending upon the overall flow through the cyclone. A knowledge of the overall flow rate through the cyclone can be used to determine the average particle size for the particles captured in that cyclone. Particles are collected by the cascade cyclone sampler according to their aerodynamic and are aerodynamically equivalent in size to the unit density spheres. The aerodynamic size of a particle gives information relating the physical size, shape, and density of the particle and indicates how the particle will behave in any environment. The mean aerodynamic particle size is directly related to the actual mean particle size, assuming spherical particles, as

$$D_{50} = \frac{AD_{50}}{\sqrt{\rho_{solid}}}$$

where D_{50} is the actual mass median particle diameter, AD_{50} is the aerodynamic mass median particle size and ρ_{solid} is the density of the solid.

Rehydration of partially calcined sorbent particles could drastically change the internal structural morphology. Since the collected sorbent particles are constantly exposed to the outgoing gases carrying some amount of H_2O, present as a result of the calcination of $Ca(OH)_2$, it is important to make sure that no significant rehydration of collected sample takes place. In order to verify that no significant rehydration of the collected sorbent occurs in the cascade cyclone sampler, samples of high surface-area CaO are placed in all stages of the sampler. H_2O-laden N_2 gas is passed over the CaO samples for about 30 minutes, which is the typical duration of calcination experiment. The H_2O concentration in the gas is representative of the experimental calcination conditions (based on sorbent feed rate and assuming 100% calcination). A maximum of 3-5% rehydration is observed with this high surface- area CaO, which confirms that during a calcination run, rehydration is not significant in the collection system.

Data Acquisition System

The fiber optic cables are connected to photomultipliers which convert the light intensity to a voltage signal. These photomultipliers are interfaced with Laser 80386 AT computer via a Metrabyte DAS-16 data acquisition system. The typical frequency of sampling of the optical signals is 1000–2000 Hz.

Experimental Approach

Operating Conditions

Calcination experiments are performed using commercial grade calcium hydroxide and calcium carbonate powders (Linwood Mining and Mineral Co.). During calcination experiments, bulk gas contains only nitrogen whereas, during sulfation runs, an air/SO_2 mixture is introduced as the reactant gas. This air/SO_2 mixture can be diluted with dry N_2 to change the SO_2 concentration in the reactant gas. Due to jet impaction at the reactor entrance block, the injected particle stream from the top rapidly mixes with the gas stream and is instantaneously heated to the reaction temperature. After the desired residence time, the partially reacted mixture is sampled by the collection probe, immediately quenched, and passes through the particle separation system. The residence time of the particles in the reactor is estimated on-line from the fiber optics system located at the tip of the injection and collection probes. The residence times can be varied by varying the gas flow rate and/or by moving the collection probe axially.

For all the calcination experiments, 10 lpm (STP) of reactant gas (nitrogen) is preheated by passing it through the annular region of the reactor assembly as illustrated in Figure 6.2 on page 432. 1 lpm (STP) of N_2 gas flow is used for both the feeder fluidizing flow and the injection probe purge. Three lpm (STP) of N_2 is used as the quench and purge gas through the collection probe. The bulk outlet of the reactor is kept closed during the experiment and all of the 15 lpm (STP) of gas is drawn through the sampling line.

For sulfation experiments, instead of using pure N_2, 10 lpm (STP) of an air/SO_2 mixture diluted with pure N_2 is used as the reactant gas. For all the sulfation runs the reactant gas consists of 5.5% O_2, 3900 ppm SO_2, and the balance N_2. An added concern during sulfation experiments is the condensation of H_2SO_4 acid mist in the collection system and on the reacted sorbent sample. Though the formation of acid, if any, is in trace amounts, precaution is taken to prevent any contamination of the sample. This is done by heating the cascade cyclone to about 330°C with a heating mantle during sample collection.

On-Line Estimation of Particle Residence Time

As mentioned before, for the purpose of determining the particle residence time during reactor operation, the sorbent particles are injected into the high-temperature environment as a train of pulses through the injection probe. During continuous powder feeding mode for residence time measurements, the fluidizing flow is diverted through a bypass line, and pulse injection of solids is done manually by a syringe in the feeder line. Just before a pulse of particles enters the reaction zone, it crosses the optical path located at the tip of the injection probe. Since these solids cause attenuation of the transmitted light, the flowing powder pulse produces a spike in the optical signals measured. After a certain residence time in the reactor column, the pulse enters the collection probe, and during entry produces another spike in the signal. The time delay between these two spikes is the residence time of the solids in the reactor.

A statistically sound procedure for estimation of time lag is through cross-correlation functions. The cross-correlation $r_{xy}(\tau)$ between two signals x(t) and y(t+τ) is given by

$$r_{xy} = \int_0^\tau x(t)y(t+\tau)dt$$

The first maximum in $r_{xy}(\tau)$ occurs at the value of τ which is the time lag between the two signals. For signals with a large number of data points, r_{xy} can be computed rapidly via Fast Fourier Transforms (FFT). The fourier transform of the equation above gives

Equation 6.5

$$R_{xy}(f) = X(f)Y^*(f)$$

where S.F., Y(f) and R_{xy} (f) are the fourier transforms of x(t), y(t) and $r_{xy}(t)$, respectively, f is the frequency variable, and Y*(f) is the complex conjugate of Y(f). Thus, from the FFT of x(t) and y(t), $R_{xy}(f)$ is evaluated using Equation 6.5. Inverse FFT of $R_{xy}(f)$ yields $r_{xy}(t)$. Since the first maximum in $r_{xy}(t)$ occurs at the time lag between the two probe signals x(t) and y(t), the residence time of solids in the reactor is known. Computation of cross-correlation using FFT is much faster than direct evaluation using Equation 6.5, hence is preferred for on-line estimation.

Optical signal assembly and data acquisition system are used to collect residence time data for various experimental runs. For two representative runs using the conventional entrance configurations, the two probe signals and their cross-correlation functions are shown in Figure 6.15 and Figure 6.16. The maximum in the cross-correlation function occurs at 4 and 476 ms respectively, measuring the residence time of solids in the reactor for these runs. The accuracy of estimation of the residence time is about 0.5 ms, which can be further improved by increasing the sampling frequency of the signals. As seen from these figures, the time lag between the signals is too small to be estimated from a visual observation of the two signals.

With this approach for short contact-time measurements, isokinetic sampling is not necessary, nor is a precise knowledge of the fluid dynamics in the reactor required. This method is accurate and imposes no restrictions or assumptions regarding the flow patterns.

Post-Reaction Analyses of Data on Reaction Kinetics

The particle collection system has already been discussed in detail in the previous section. By conducting the experiments at a specific particle residence time, enough sample is collected for subsequent kinetic analysis. A thermogravimetric analyzer (TGA) and Ion-chromatograph (IC) are used for obtaining the extent of reaction and other reaction kinetic information. For calcination studies, the partially calcined sample is further calcined to 100% decomposition in a TGA, and the back calculation gives the degree of calcination achieved within the entrained flow reactor. For sulfation runs, the sampled sorbents are tested for sulfate ion content (Ion Chromatography). The extent of conversion and amount of sulfur capture for a particular reaction temperature can thus be estimated. Using the cascade cyclone sampler for each experimental condition, particles of various sizes are collected, and from the analysis of particles of different size groups, information on particle size effect is obtained.

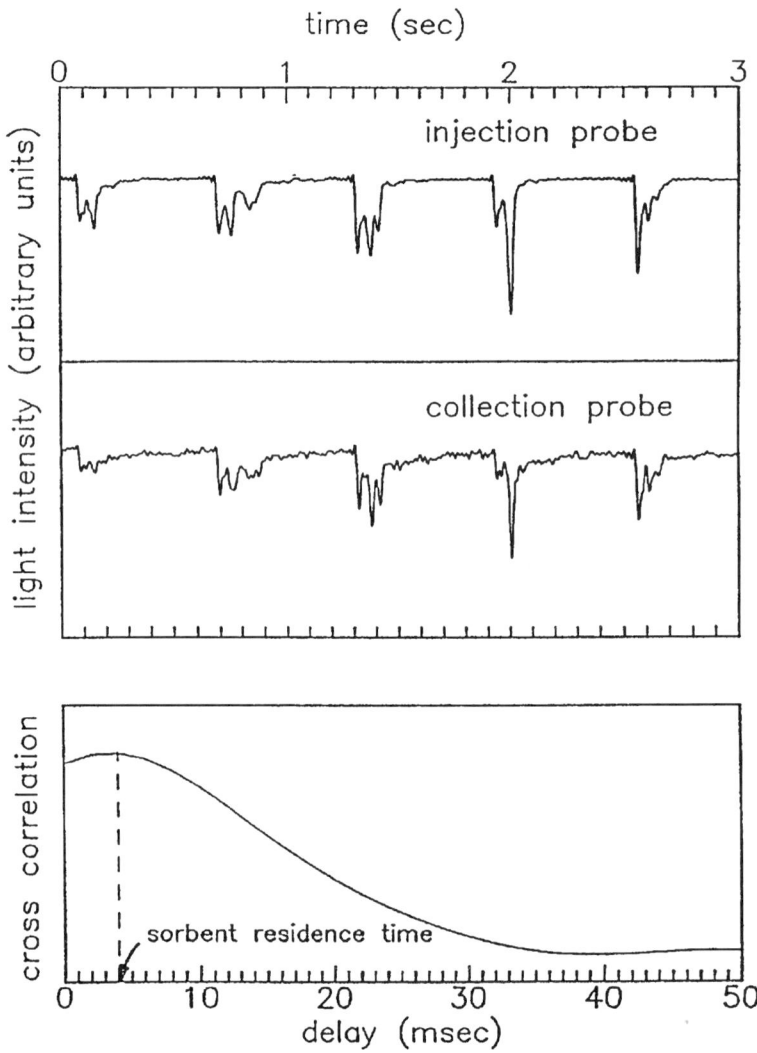

Figure 6.15 Injection and collection probe signals and their cross-correlation. Calculated residence time = 4 ms.

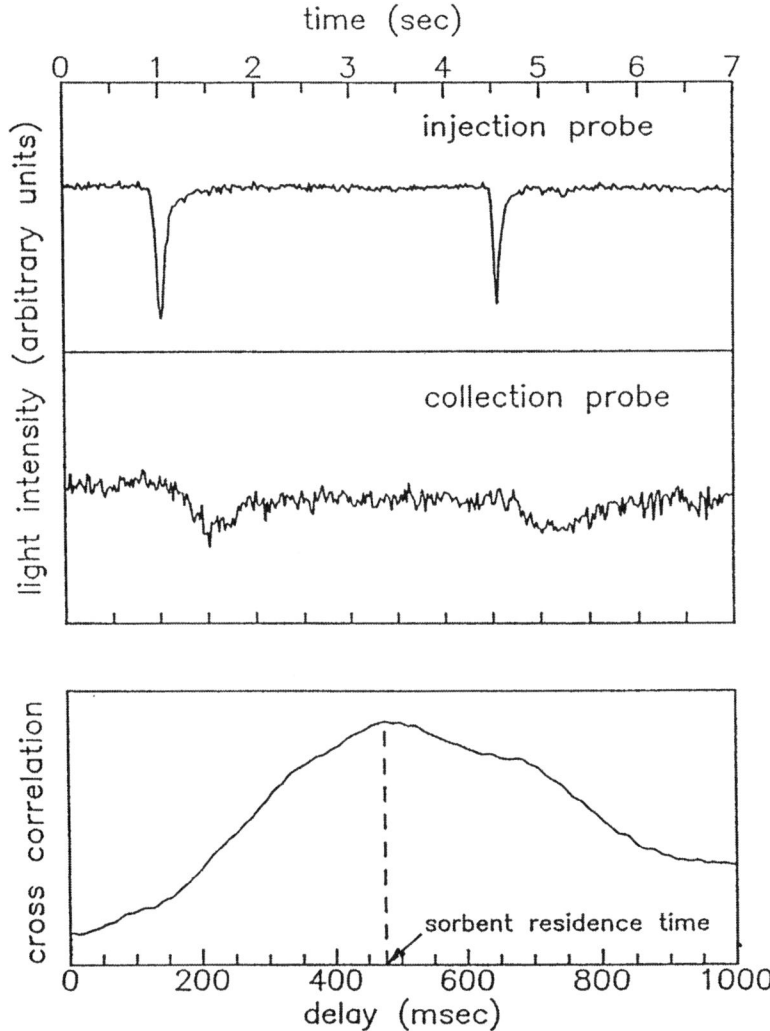

Figure 6.16 Injection and collection probe signals and their cross-correlation. Calculated residence time = 476 ms.

In any gas-solid noncatalytic reaction study, particle morphology plays an important part. Determination of particle structural properties, such as specific surface area of the particle and its porosity and pore size distribution is essential

Dry Scrubbing Technologies for Flue Gas Desulfurization

to a better understanding of the reaction mechanism. Quantitative information regarding structural properties of sorbent particles is obtained by using the BET low-temperature nitrogen adsorption method. Scanning Electron Microscopy (SEM) is used to get qualitative information regarding the sorbent particle structure.

Ca-Based Sorbents Used in High-Temperature Flue Gas Desulfurization

Calcium-based sorbents, typically limestone ($CaCO_3$) and hydrated lime ($Ca(OH)_2$), are the most commonly used sorbents for FGD. Extensive work has been done by earlier researchers to explore the reaction kinetics and other associated phenomena for CaO and SO_2 interaction and it has been well documented that the extent of SO_2 capture varies significantly for different types of limestones depending on their geological origin. This difference is manifested in differences in their structural properties and composition. Chemical composition and structural properties (surface area and porosity) of $CaCO_3$ and $Ca(OH)_2$ (as received) used for our calcination and sulfation studies are presented in Table 6.1. Even though the geological origin of the sorbent influences the SO_2 capture the overall mechanism of capture is generic and applicable to all types of $CaCO_3$ and $Ca(OH)_2$.

Table 6.1. Composition and Initial Structural Properties of Sorbents Investigated

Sorbent	Linwood Hydrated Lime*	Linwood Carbonate*
Composition (wt%)		
$Ca(OH)_2$	94.0	
$CaCO_3$	1.0	97.0
SiO_2	0.9	0.8
Al_2O_3	0.6	0.5
MgO	1.0	
CaO	1.0	
Fe_2O_3	0.5	0.5
K_2O		
Na_2O		
S		

Table 6.1. Composition and Initial Structural Properties of Sorbents Investigated

Sorbent	Linwood Hydrated Lime*	Linwood Carbonate*
$MgCO_3$		1.0
BET surface area (m^2/g)	16.9	3.0
Pore Volume (cc/g)	0.06	0.02

*Linwood Mining and Minerals Co.

As already mentioned, the sulfur capture is effected by the calcined sorbents (CaO) rather than the original sorbent, and the involved reactions are

$$Ca(OH)_2 \rightarrow CaO + H_2O$$

and

$$CaCO_3 \rightarrow CaO + CO_2$$

Calcination of $Ca(OH)_2$ takes place at a temperature of about 450°C, whereas for $CaCO_3$, appreciable calcination doesn't occur till 675°C. Although the structural properties of the calcined sorbent eventually determine the amount of SO_2 capture, initial structural properties of the sorbent do influence the evolution of porosity and surface area of calcined sorbent. $Ca(OH)_2$, due to its crystal structure and method of preparation, is usually more porous and has a higher surface area than $CaCO_3$, which possesses a nonporous structure. Upon calcination, the open structure of the original $Ca(OH)_2$ as compared to that of $CaCO_3$, offers less diffusional resistance to escaping H_2O which in turn leads to a more porous CaO. The relatively nonporous structure of $CaCO_3$ creates temperature and concentration gradients in calcining $CaCO_3$ particles which could retard the calcination process by making the outward diffusion of CO_2 difficult. Researchers [8, 21] have shown that under identical conditions of high temperature and flue gas composition, $Ca(OH)_2$ seems to work better than $CaCO_3$ in scavenging SO_2 from flue gas. The reason put forth to support these results is that the calcination of a $CaCO_3$ particle takes longer than that of a $Ca(OH)_2$ particle due to its closed structure.

Experimental Results and Discussion: Calcination of Calcium Based Sorbents

Introduction

In this study, calcination of $Ca(OH)_2$ was carried out at high temperatures representative of the temperature range in the upper-furnace region of coal-fired combustors, to get an insight into various ultrafast processes which influence the structure of calcined sorbent. As already mentioned, injected sorbents undergo calcination to form CaO and the involved reactions are

$$Ca(OH)_2 \rightarrow CaO + H_2O$$

and

$$CaCO_3 \rightarrow CaO + CO_2$$

At upper-furnace temperatures of 800-1200°C, calcination and the accompanying process of sintering are extremely fast and a substantial amount of calcination takes place within the first 100 ms. In entrained conditions adequate sorbent/gas mixing, particle size, and particle heat-up are important factors in achieving complete calcination.

Effect of Temperature on Calcination

Temperature has a very strong effect on calcination of Ca-based sorbents and it dictates the rate and extent of calcination. Temperature also affects the development of internal structure of the calcined and partially calcined sorbent which eventually effects the total SO_2 capture.

The phenomenon of calcination of $Ca(OH)_2$, along with the accompanying phenomenon of sintering, was studied in a specially designed high-temperature entrained-flow reactor (described in detail earlier). Calcination of $Ca(OH)_2$ was studied at three temperatures (1173, 1223, and 1323 K) covering the upper-furnace temperature range. Figure 6.17 shows the experimental data for calcination of 3.6 μm $Ca(OH)_2$ at these temperatures for residence times ranging from 10 to 300 ms. A similar trend is observed in all the calcination curves. Initially, calcination progresses extremely fast and exhibits a strong influence of temperature. At higher residence times, the reaction rates attenuate considerably and conversion tends to flatten out. Nearly 75% of the final extent of calcination (at 250 ms) is achieved within the first 50 ms at all the temperatures studied.

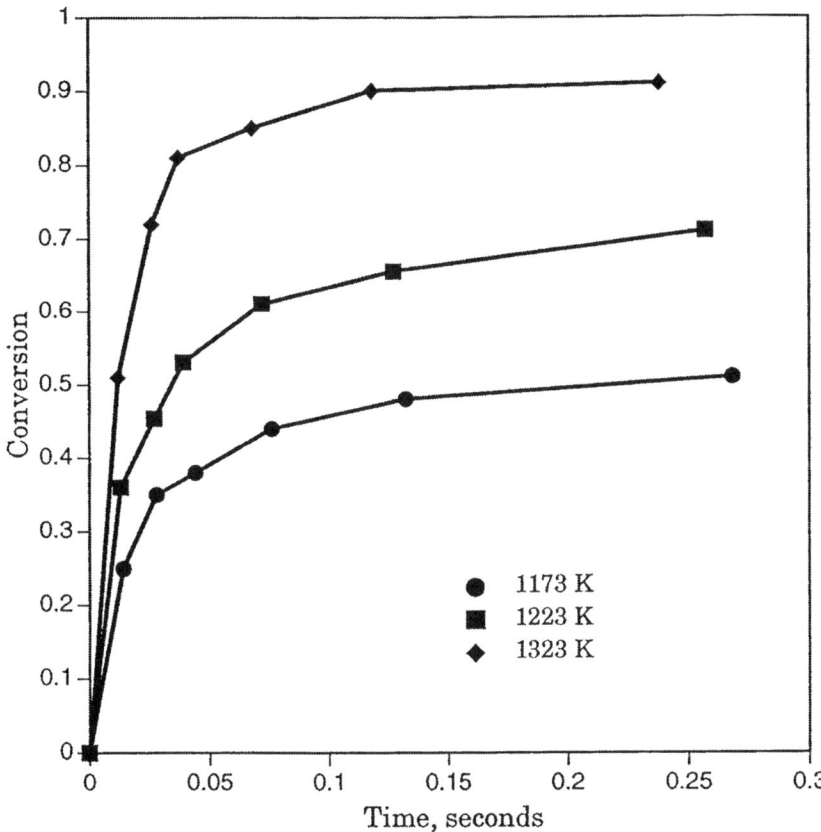

Figure 6.17 Effect of temperature on calcination of 3.6 μm Linwood Ca(OH)$_2$.

The sharp attenuation of reaction rate at higher residence times is seen clearly from the 1323 K data which exhibits virtual cessation of calcination at about 90%. This near "die-off" of the reaction at higher residence times is also observed at the lower temperatures of 1173 and 1223 K. The characteristic behavior of high initial rate followed by attenuation with virtual die-off is closely linked with the structural effects accompanying the reaction. For these high temperature studies, sintering of product CaO holds the key to explaining and modeling these observed behaviors.

The strong influence of temperature during the initial stage of reaction and the high porosity of the nascent CaO suggest that chemical reaction at the CaO/Ca(OH)$_2$ interface may be the rate-limiting step. The product CaO layer rapidly sinters losing surface area and porosity, offering increasing diffusional resistance to the gaseous product, H$_2$O, through the product shell. This leads to an increasing H$_2$O concentration in the product layer and at the interface, which not only enhances the rate of CaO sintering but also retards the calcination rate. All these phenomena eventually result in the overall calcination reaction being limited by outward H$_2$O diffusion through the CaO. At longer residence times, calcination curves for all three different temperatures rise very slowly indicating a diminished effect of temperature and similar overall rate of reaction with, possibly, an identical controlling regime.

Experimental data is used to establish values of the calcination rate constant, k_c, and the sintering rate constant, k_s. The activation energies and the pre-exponential factors are obtained from the Arrhenius-type plots shown in Figure 6.18. The activation energy value for the calcination reaction is calculated to be 22.5 kcal/mole, while that for sintering is 57 kcal/mol. The calcination activation energy obtained in this study compares with 16.4 kcal/mol estimated by Mai and Edgar [11] from calcination data of larger particles (12.5 μm) of Ca(OH)$_2$ in an entrained flow reactor. The sintering activation energy of 57 kcal/mol compares to values of 77 kcal/mol [11] and 58 kcal/mole [2] reported in the literature.

Internal Surface Area Development with Calcination

The overall surface area evolution, at three different temperatures, of the partially calcined Ca(OH)$_2$, shown in Figure 6.19, represents the net effect of the two opposing phenomena of calcination and sintering. For the two lower temperatures of 1173 K and 1223 K, calcination kinetics dominates over the sintering kinetics at short time scales; hence, surface area of the partially decomposed Ca(OH)$_2$ initially rises above the parent hydroxide surface area. The surface area goes through a maximum, then rapidly decreases due to sintering and tends to level off to an asymptotic value at higher residence times. On the other hand, at the higher temperature of 1323 K, the effect of sintering is more pronounced and the observed surface area decreases continuously toward the asymptotic value from the very beginning of the decomposition process. For 1173, 1223 and 1323 K beyond 250 ms, the surface area attains values of about 17.5, 16.0, and 13.5 m^2/g, respectively.

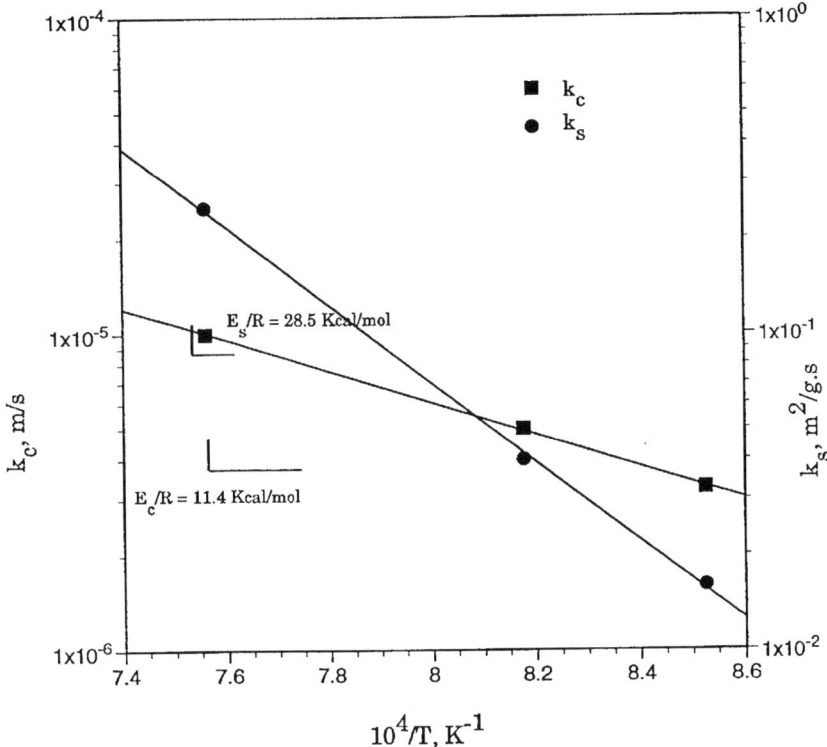

Figure 6.18 Arrhenius plots to estimate activation energies of calcination and sintering.

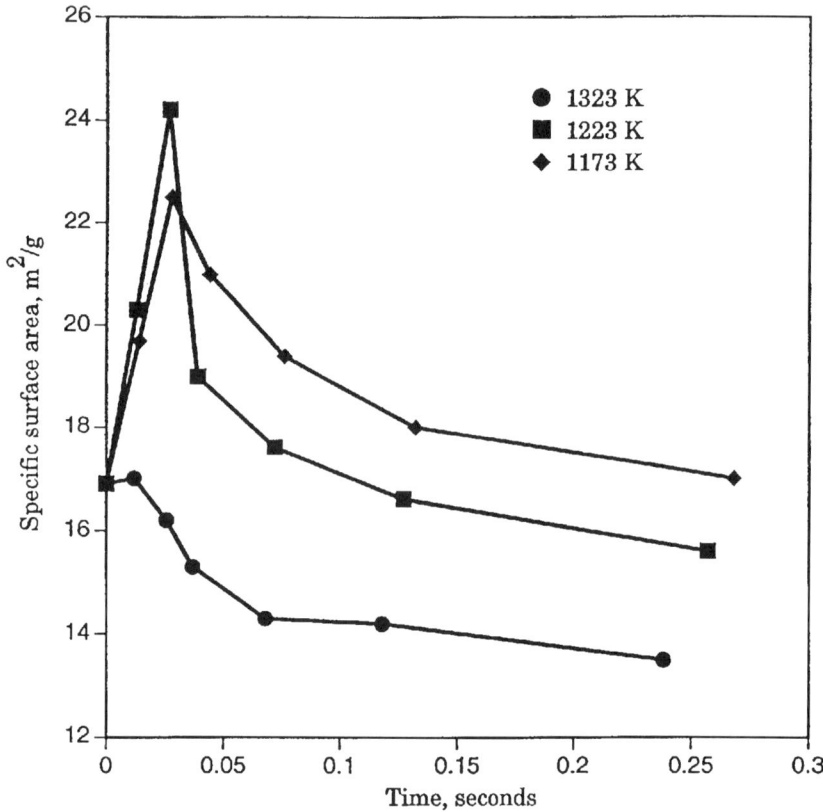

Figure 6.19 Effect of temperature on surface area evolution during calcination of 3.6 μm Linwood Ca(OH)$_2$.

In principle, sintering follows calcination, and since calcination is an activation step; i.e., by generation of surface area, surface area of the partially calcined Ca(OH)$_2$ should initially show an increase over the parent Ca(OH)$_2$ powder surface area and reach a maximum value before dropping to an asymptotic value. A possible explanation for such a behavior could be that at higher residence times loss of surface area of the calcined sorbent due to sintering offsets the surface area generated by further calcination of Ca(OH)$_2$. This trend is quite evident at lower temperatures but the data obtained at 1373 K fails to illustrate this trend. The most probable explanation for this apparent contradiction is that the entrained-flow reactor is limited to sampling partially calcined particles with at least 15 ms residence time. Residence times lower than this value are possible to measure, but because of the nonlinear axial temperature profile in the reaction zone, samples collected at lower residence time (less than 15 ms) would not give accurate information about the kinetics of the reaction. At higher temperatures sintering is extremely fast and its effect is pronounced within 20 ms. The extent of calcination and evolved surface area of partially calcined sorbent particles go hand in hand for a particular reaction temperature. Figure 6.20 shows the calcination and surface area curves for 1323 K in the same plot. The surface areas for various temperatures reported in this study are significantly lower than most of the data available in the literature [1,11]. Borgwardt [1] has reported that after 1 second of calcination at 1000°C, the evolved surface area reaches a value of 32 m^2/g which is substantially higher than the observed surface area value in this study.

The apparent discrepancy between these results and the available data in literature clearly demonstrates a greater effect of sintering encountered in this study. A possible explanation is that in the earlier studies external heat transfer resistance was significant due to poor sorbent/gas mixing. Moreover, for particles of larger sizes there was considerable heat transfer resistance across the particle. As a result, sintering slowed down considerably, but calcination progressed at reasonably high rate. The above argument gains further credibility if the surface area evolution curves for smaller and larger particles are compared. Figure 6.21 shows the surface area evolution of 3.6 μm and greater than 7.8 μm particles at a reaction temperature of 1050°C. The larger size group possibly contains some very big particles, with high internal heat transfer resistance. It is probable that the inner core of these particles never experiences the high temperature at which calcination is being studied. As a result, this size group sinters less and always shows a higher surface area.

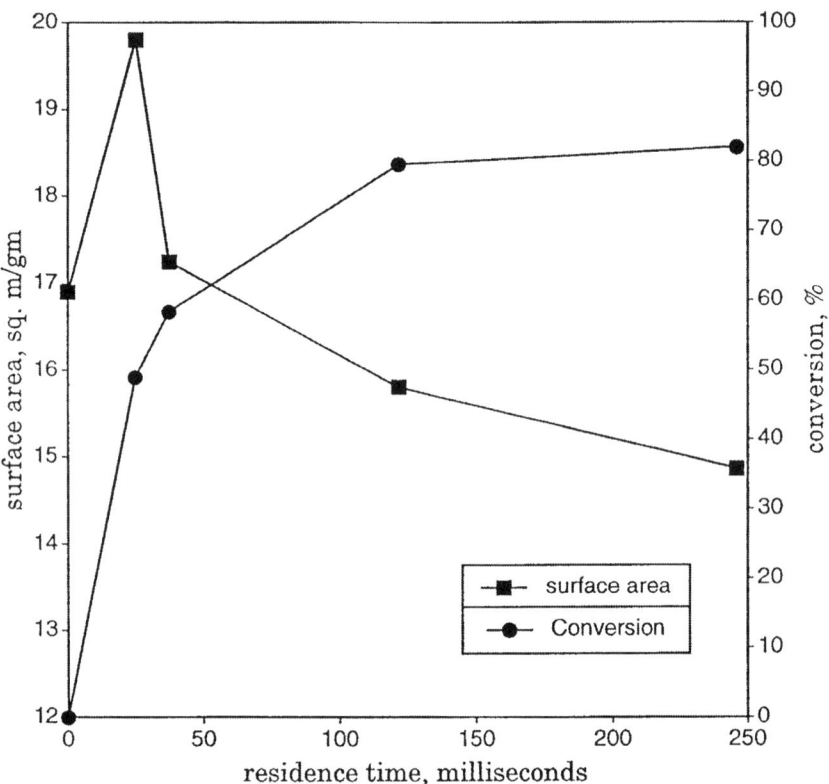

Figure 6.20 Surface area development with calcination of $Ca(OH)_2$ at 1050°C.

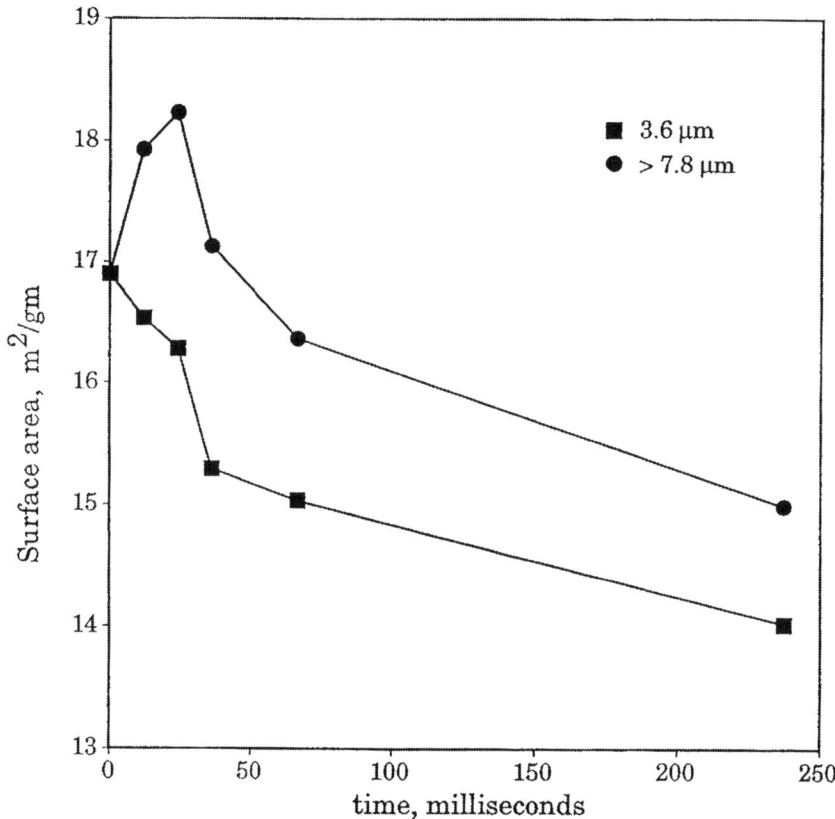

Figure 6.21 Influence of particle size on surface area development during calcination of $Ca(OH)_2$ at 1050°C.

SEM analyses were conducted to complement the studies of structural changes accompanying the simultaneous calcination and sintering. Figure 6.22a shows the structure of 42% calcined 1.5 μm particles after 24 ms at 1223 K. The particles have a surface area of 24 m^2/g. The granular nature of the surface and the presence of uneven cracks are easily observable. Figure 6.22b shows a SEM photomicrograph of 90% calcined powder of the same size after 235 ms at 1323 K. The particles have a much reduced surface area of about 13 m^2/g. From this picture, the surface appears much less granular and considerably glazed with reduced porosity. The grain size appears to be larger than the grains of partially calcined sorbent at 24 ms, due to the sintering process.

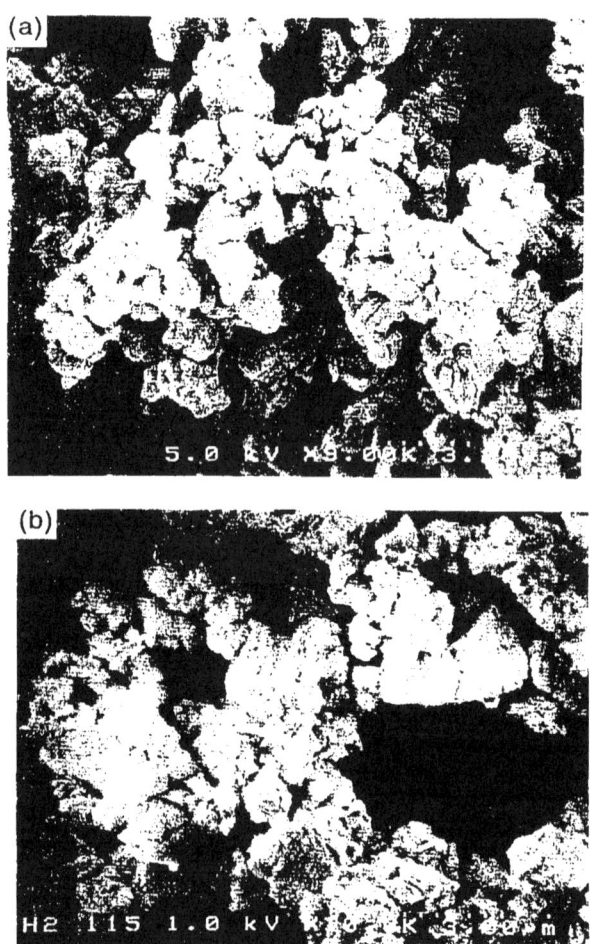

Figure 6.22 SEM photomicrographs of partially calcined $Ca(OH)_2$ particles. (a) 42% calcined, 1.5 µm particles after 24 ms at 1223 K, surface area of 24 m^2/g. (b) 90% calcined, 1.5 µm particles after 235 ms at 1223 K, surface area of 13 m^2/g.

Effect of Particle Size on Calcination

The particle size of the sorbent plays an important part in determining the influence of internal heat and mass transfer resistances on the extent of calcination. Figure 6.23 shows the effect of particle size on $Ca(OH)_2$ calcination at 1050°C (1323 K). For the three different particle sizes of 1.5, 3.6, and greater than 7.8 µm (mean aerodynamic diameter), no effect of particle size is noticed for the smallest particles. These results suggest that for very small particles, diffusional resistance is not significant in the initial stages of the reaction. In this initial period, surface reaction is very rapid and it is the main contributing factor to the overall particle conversion. Even in the later stages of the reaction, no particle size effect is discernible for the two smaller particle sizes. However, for the largest size group, the effect of particle size is quite pronounced from the very onset of the reaction. This can be explained by the fact that significant temperature gradients exist in particles of bigger size. Even though substantial hydroxide conversion takes place on and near the particle surface in the initial stage of the reaction, this accounts for a relatively smaller fraction of the whole particle, compared to the particles of smaller sizes. In the later stages of the reaction, when the intraparticle mass transfer becomes more important, the size effect of the largest size group becomes more visible.

A comparison of surface area evolution for different particle sizes shows a distinct particle size effect on the extent of sintering. Figure 6.24 shows the surface area evolution curves for 3.6 μm and greater than 7.8 μm particles at 1223 K.

The larger particles exhibit a higher maxima in surface area than the smaller particles. The peak surface area is about 24 m^2/g for the 3.6 μm particles while it is nearly 29 m^2/g for the >7.8 μm particles. Such higher peak surface areas observed for the larger particles compares well with the data obtained by Mai and Edgar [11], who observed a peak surface area of about 32 m^2/g for 12.5 μm particles at 1285 K. This strongly suggests a heat-transfer limited sintering for the bigger particles. The larger particles probably experience internal temperature gradients with the inner core requiring more time to reach the reaction temperature and winter to a lower degree. Calcination, on the other hand, progresses at reasonably high rate, though slower than that for the smaller particles.

Development of Porosity and Pore Size Distribution of Calcined Sorbent Particles

Most of the previous efforts involving modeling of short-time sorbent/SO_2 reactions suffers not only from inadequate ultrafast sorbent conversion data, but also from a lack of knowledge on how the internal structure of the sorbent changes with the simultaneous progression of the three ultrafast mechanisms inside the coal-fired furnace. No experimental attempt was made, until this study, to explore the transient nature of all the sorbent physical characteristics, namely surface area, porosity, average pore size, and pore size distribution. However, lack of this much-needed information makes it practically impossible to have a clear understanding of the sorbent/SO_2 interaction phenomena, especially when all the reaction processes greatly affect the resulting sorbent structure and these sorbent structural properties in turn influence the subsequent sulfur capture.

Hence, in order to develop a sophisticated and comprehensive reaction model, the importance of short-time sorbent structural data must be well realized. Rapid changes in sorbent surface area, porosity, and pore size distribution should be investigated for $Ca(OH)_2$ calcination experiments with reaction times of 0 to 250 ms.

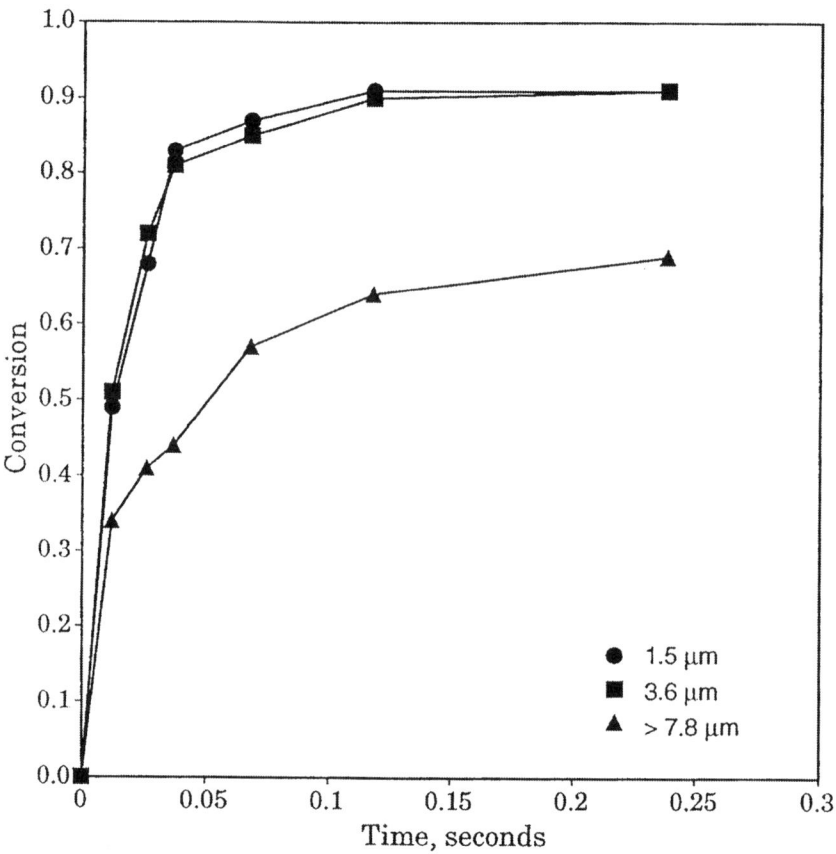

Figure 6.23 Influence of particle size on calcination of Linwood $Ca(OH)_2$ at 1323 K.

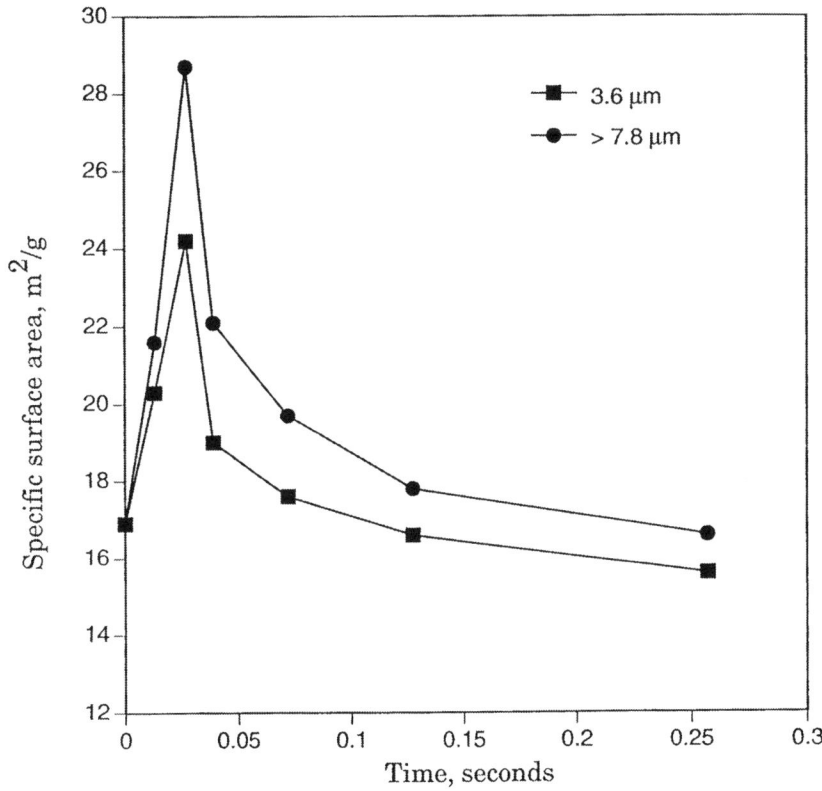

Figure 6.24 Effect of particle size on surface area evolution during calcination of Linwood Ca(OH)$_2$ at 950°C.

The internal surface area for Ca(OH)$_2$ as well as h-CaO (CaO obtained from calcination of Ca(OH)$_2$) lies predominantly in the pores of size 20 to 500 A. Such pores are classified as mesopores. The physical adsorption-desorption isotherm exhibited by mesoporous solids such as Ca(OH)$_2$ and h-CaO is classified as a Type IV isotherm according to the BDDT classification [27]. The exact shape of the pores of mesoporous solids can be cylindrical, parallel plate-shaped, or spherical with a narrow opening. Another characteristic feature of mesoporous solids is the hysteresis loop obtained in the adsorption-desorption isotherm. The exact shape of the hysteresis loop varies from one adsorption system to another and is determined by the shape of the associated pores. The hysteresis loop is classified into Types A, B, and E [27] and these are exhibited by cylindrical, parallel plate, and spherical with narrow opening types of pores, respectively. Earlier researchers [18] have shown that h-CaO possesses a parallel plate or slit-like pore structure; i.e., it exhibits a Type B hysteresis loop. A similar hysteresis diagram for $Ca(OH)_2$ is not available in the literature. Hence, the adsorption-desorption isotherm for $Ca(OH)_2$ was constructed using a BET apparatus and low-temperature nitrogen adsorption, and is shown in Figure 6.25.

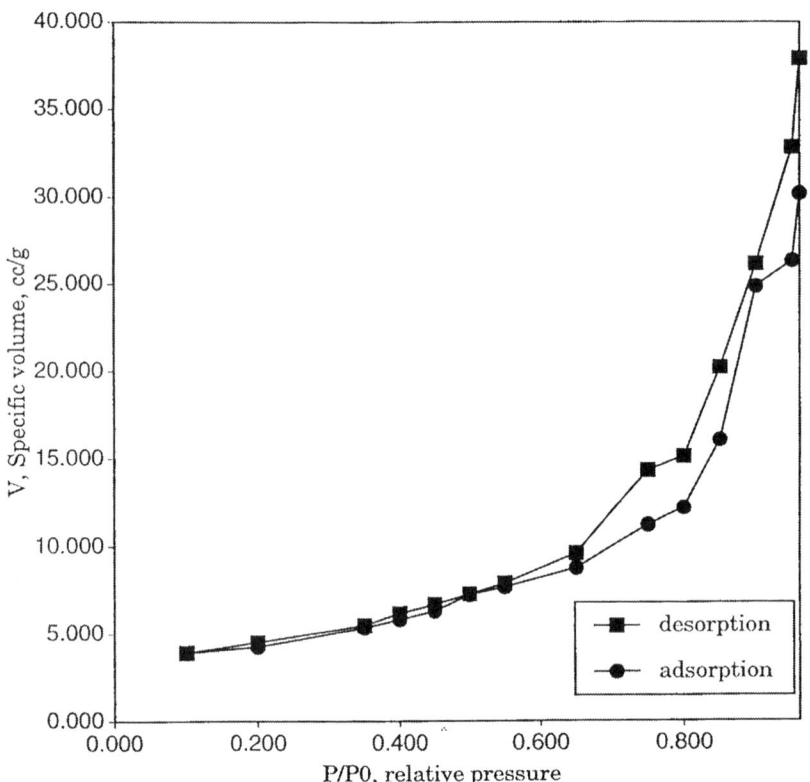

Figure 6.25 $Ca(OH)_2$ adsorption-desorption isotherm.

The experimental desorption isotherm data for the parent sorbent and all the partially reacted sorbents was analyzed using a parallel plate model and the pore-size distribution worksheets were developed as described in the literature [28, 29].

Structural evolution studies with linwood $Ca(OH)_2$ were carried out at two temperatures, 1273 and 1353 K, in order to observe the surface area, porosity, pore volume, and pore size distribution changes with calcination and sintering. Figure 6.26 shows the sorbent conversion with porosity data for the same calcination run. As opposed to surface area, at 1273 K the overall porosity of the sorbent increases with increasing sorbent conversion to CaO, indicating that the effect of sintering is less severe on overall porosity than on overall surface area. This is clearly understandable as sintering causes grain coalescence or coalescence of smaller pores into larger pores. As most of the surface area resides in smaller pores, it is readily lost due to sintering. On the other hand, larger pores primarily contribute to the total sorbent porosity and, consequently, the effect of sintering on this sorbent property is less pronounced. In fact, any loss of porosity due to sintering is hidden by the evolving porosity due to further calcination, and the overall porosity increases with increasing sorbent decomposition. At longer residence times the porosity of the calcined $Ca(OH)_2$ starts to level off, as can be seen clearly in Figure 6.26.

This is because at longer residence times almost complete calcination has taken place and generation of new pores has virtually ceased, whereas deactivation from sintering continues and causes unabated loss in pore volume and surface area.

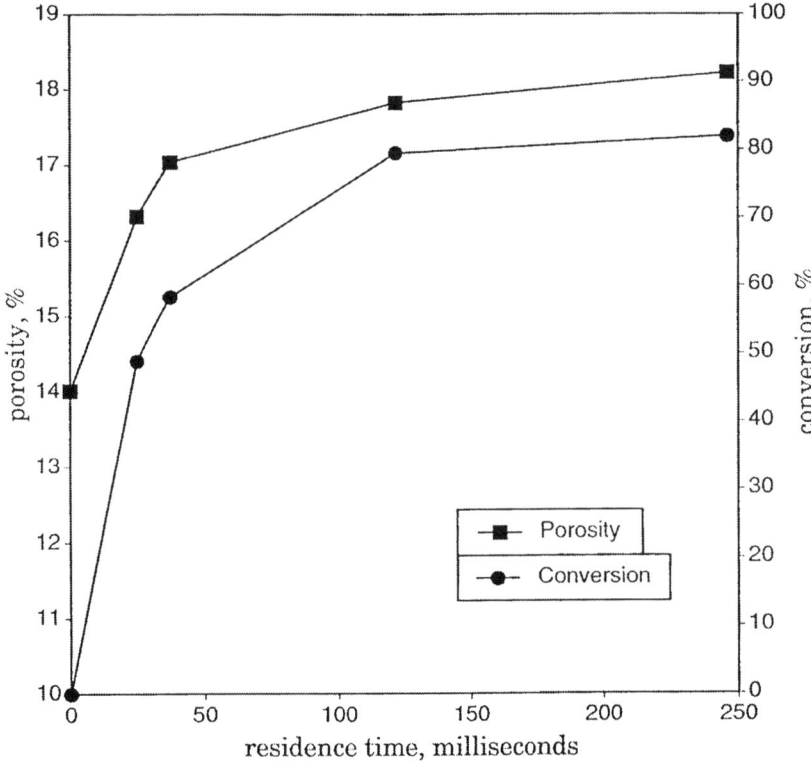

Figure 6.26 Porosity development with calcination of $Ca(OH)_2$ at 1000°C.

The pore size distribution data shown in Figure 6.27 for 1273 K provides some further insight into the pore structure evolution phenomenon. The parent $Ca(OH)_2$ powder exhibits a peak in the pore size distribution at a pore size (plate separation) of 105 A, and a substantial pore volume in a wide size range of 60-300 A. As a result of the calcination reaction, CaO is formed and undergoes rapid sintering to make larger pores. Some small peaks in the size range of 20-60 A for these partially calcined sorbents probably point to the pores of nascent CaO. However, the pore size distribution clearly shifts to the right due to rapid sintering and a careful comparison of 79% and 83% calcined sorbents (not much difference in conversion) indicates a shift to the larger pores with time, induced by sintering.

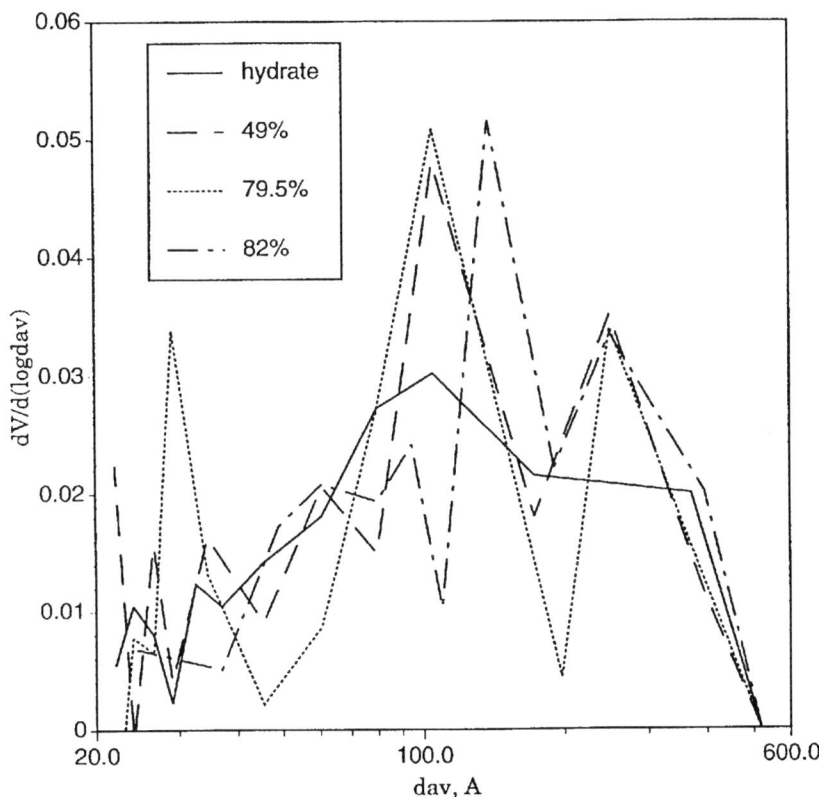

Figure 6.27 Pore size distribution during calcination of Ca(OH)$_2$ at 1000°C.

This shifting of the entire distribution to the right is clearly elucidated in Figure 6.28 from pore size distribution curves for 1353 K. As can be observed, the peak pore size shifts appreciably from about 105 A for parent Ca(OH)$_2$ powder to about 300 A for the 91% calcined sorbent.

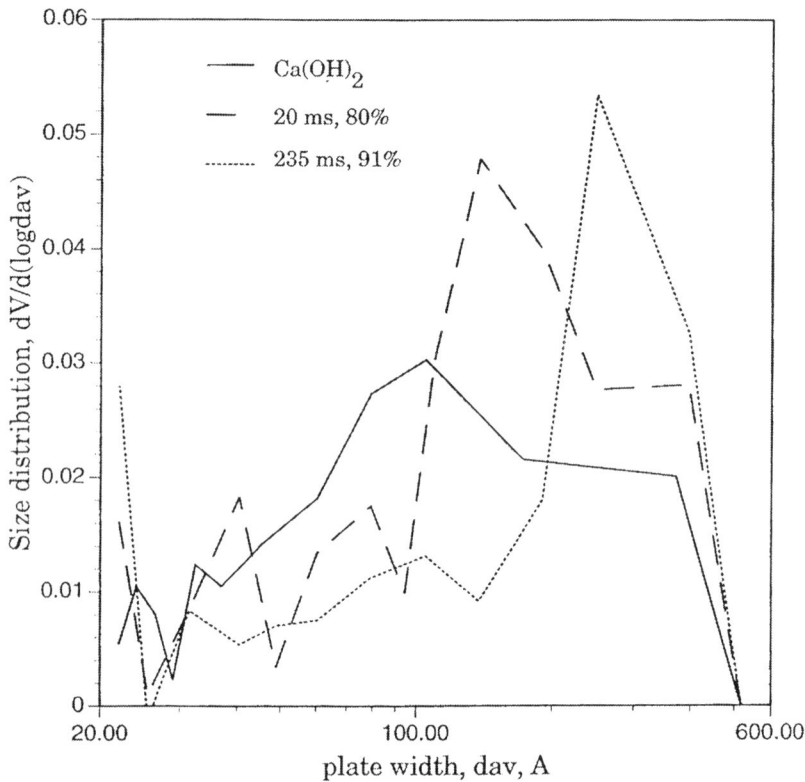

Figure 6.28 Pore size distribution during calcination of $Ca(OH)_2$ at 1080°C.

Experimental Results and Discussion: Sulfation of Calcium Based Sorbents

Introduction

Sulfation studies were carried out in the specially designed high-temperature entrained flow reactor at temperatures representative of the upper-furnace region. The effect of sulfation on structural properties of the sorbent was studied to get an insight into the deactivation mechanism of sulfation. Experiments were carried out with pre-calcined and pre-sintered as well as uncalcined $Ca(OH)_2$ to study the sulfation reaction, independent of calcination and sintering and concurrent with calcination and sintering, respectively. At elevated temperatures in the flue gas environment, calcination, sintering, and sulfation of calcium-based sorbents show considerable overlap. All three phenomena take place at a very rapid rate. Calcined sorbents undergo sulfation according to the following reaction scheme.

$$CaO + SO_2 + 1/2O_2 \rightarrow CaSO_4$$

The sulfation reaction is most favored in the temperature window of 900-1100°C with higher temperatures showing more capture. At temperatures above 1200°C the sulfur capture is reduced because of the thermal decomposition of $CaSO_4$. From the kinetics point of view, higher temperatures in the range of 900-1100°C increase the rate of calcination and sulfation but also increase the rate of sintering. Thus the increased rate of sulfation at higher temperatures is offset, to a certain extent, by a greater loss of surface area due to sintering.

When the sorbent is injected into the furnace it undergoes calcination to produce highly porous and reactive CaO which has a higher surface area than the parent powder. The CaO reacts with SO_2 forming a solid product $CaSO_4$. Calcium sulfate has a higher molar volume than CaO and thereby causes loss in surface area and porosity. Unlike sintering, deactivation of the sorbent due to sulfation occurs because of product layer build up on the surface which leads to pore filling and pore mouth plugging. The steps involved in the mechanism of sulfation reaction include diffusion of bulk gas to the particle exterior, diffusion through the pores to particle interior, diffusion through the solid sulfate layer, and then surface reaction at the CaO interface. Under the chemical reaction control regime the diffusion steps are very fast and the rate is determined only by the surface reaction. As the sulfate product layer begins to build up, the diffusional resistance to SO_2 increases and eventually the reaction shifts to the product layer diffusion control regime.

Effect of Temperature on the Sulfation of CaO and Ca(OH)$_2$

The effect of temperature on sulfation only can be studied by using pre-calcined and pre-sintered sorbent. This effect is shown in Figure 6.29. The average aerodynamic particle size of the sorbent is 3.9 µm with an initial surface area of 3.9 m^2/g. As the sorbent applied in this sulfation study is pre-calcined and pre-sintered, the results show the effect of the independent sulfation reaction without any other simultaneously occurring processes. For all the temperatures the conversion curves follow a similar pattern. The sulfation progresses very rapidly in the first 30-35 ms of the reaction. After this initial period, the sulfation rate experiences a dramatic drop. However, a strong temperature dependence of the reaction can be noticed for both the initial and later stages of the sulfation process.

The drastic drop in sulfation rate after the initial period can be best explained by a shifting controlling regime theory. At the beginning of the sulfation process, the reaction is probably controlled by a surface reaction mechanism. As the reaction product $CaSO_4$ rapidly covers the available CaO surface, product layer diffusion eventually becomes the rate-determining step. A strong temperature dependence observed at these later reaction times supports the past theory that ionic diffusion through the product layer is the rate controlling mechanism rather than gaseous diffusion [3,13]. If gaseous diffusion were rate limiting, then the dependence on temperature would be less pronounced than what is observed in the reported data. A typical activation energy for a gaseous diffusion controlling reaction is around 5 kcal/mole; but calculations with these results show an activation energy as high as 32 kcal/mole, which is typical of an ionic diffusion mechanism.

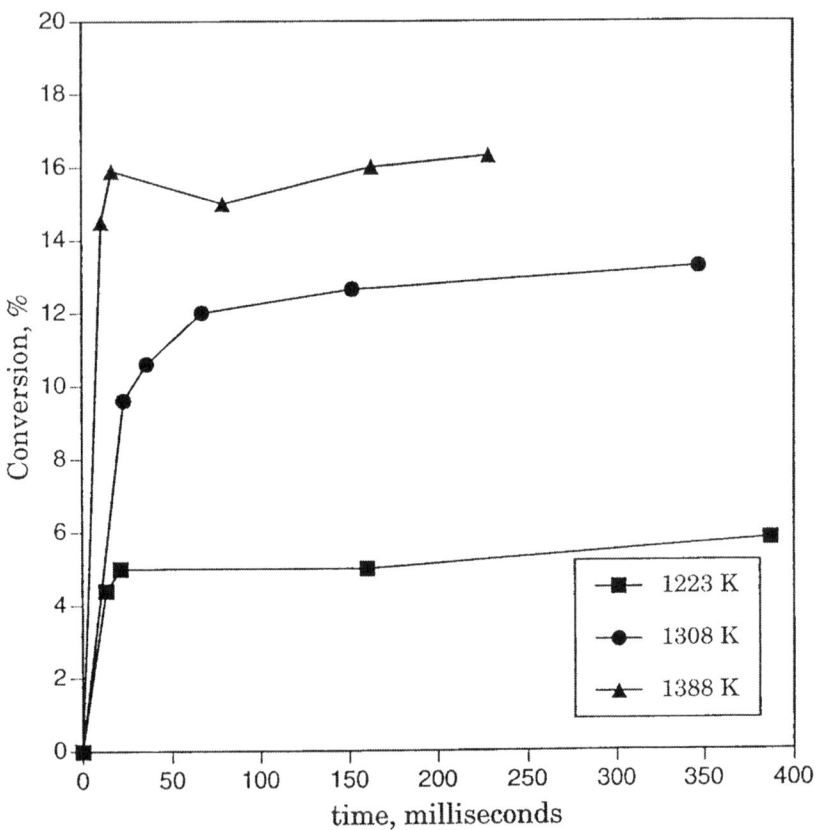

Figure 6.29 Effect of temperature on sulfation of pre-sintered h-CaO (3.9 μm) with an initial surface area of 3.9 m²/gm.

Availability of the low residence time sulfation data for the first time in this field of research helps gain a further insight into the sulfation mechanism. Most of the earlier researchers [13,15,17] suggested that ionic diffusion through the product layer becomes rate controlling from the very onset of the sulfation process. Such a mechanism was accepted for the purpose of modeling even though at the early stages of the reaction, the product layer is either nonexistent or not thick enough to offer substantial diffusional resistance. The reason for ignoring the initial surface reaction was most probably due to the lack of experimental data below 30 ms. On the contrary, this study shows multiple data points below 30 ms which clearly indicate the existence of a much faster rate controlling regime in the initial times. Moreover, the experimental data obtained from times greater than 50 ms distinctly show the takeover of another controlling mechanism at later times.

The effect of temperature on sulfation of $Ca(OH)_2$ is shown in Figure 6.30. The conversion curves here show a combined effect of all the three simultaneously occurring processes: calcination, sintering, and sulfation. Such a situation closely resembles the furnace sorbent injection conditions. The results clearly show a distinct influence of temperature on sorbent conversion from the very outset of the sulfation reaction.

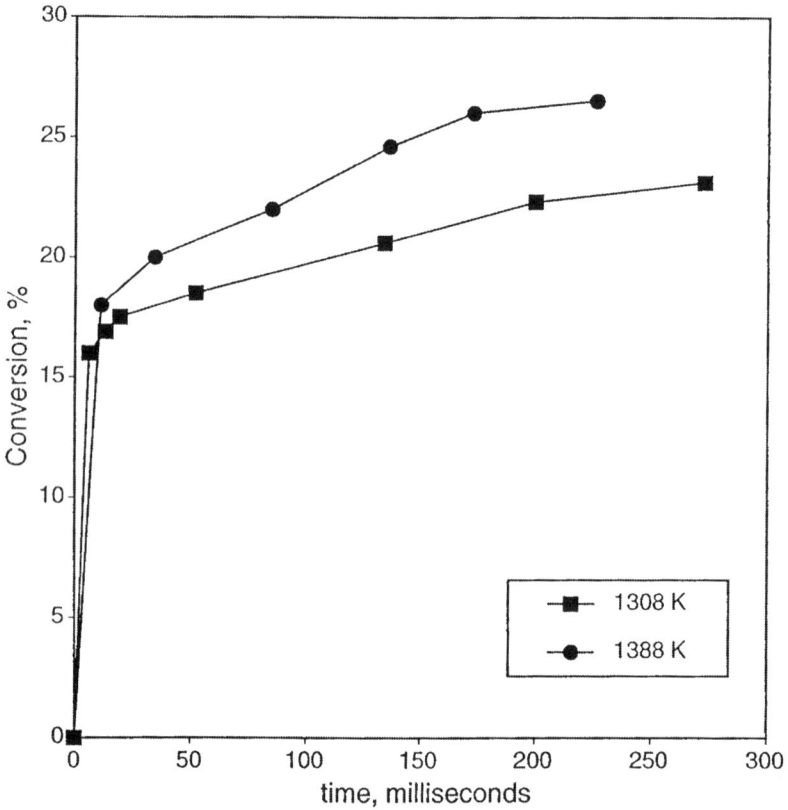

Figure 6.30 Effect of temperature on sulfation of 3.9 μm Linwood Ca(OH)$_2$.

Experimental data for reaction times as low as 6-10 ms show a 16% sorbent conversion at 1035°C (1308 K), whereas for 1115°C (1388 K), the conversion reaches about 18%. Over the entire reaction time period studied (to about 250 ms) sorbent utilization at the higher temperature is always greater than that for the lower temperature. At 0.25 s, a sorbent conversion of 23% is achieved for 1035°C (1308 K), which compares to a conversion of 26% at 1115°C (1388 K). The short-time sulfation data reported in this study are first of their kind for times less than 30 ms. However, a comparison can be made with the available literature data for reaction times greater than 30 ms. Sulfation data, reported by Milne et al. [3], for 2.5 μm raw Linwood hydrates at temperatures of 990 (1263 K), 1095 (1368 K), and 1175°C (1448 K), and a Ca/S ratio of 2 are available in literature. For a comparable temperature of 1095°C (1368 K), they observed

15% conversion in the first 40 ms of the reaction, which compares to a 18% conversion reported here in 11 ms at 1115°C (1388 K). By performing experiments with the same $Ca(OH)_2$ powder and at reaction condition which was differential with respect to SO_2, Gullett et al. [16] obtained a 16% conversion in 50 ms at a temperature of 1000°C (1273 K). However, for longer reaction times of 200-300 ms, the conversion reported in these past studies are in fair agreement with the results shown here.

Effect of Initial Surface Area and Porosity on CaO Sulfation

Calcination of $Ca(OH)_2$ results in the formation of highly porous, high surface-area CaO. As already mentioned, increased surface area results in increased sulfur capture. The nature of the interaction of calcination and sintering with sulfation makes it difficult to study the effects of various structural properties of calcined sorbents on sulfation at elevated temperatures. To study the effect of surface area of the CaO only, the sorbent has to be independently calcined and sintered to the extent that further exposure to high temperatures would not cause any change in the surface area of CaO. Figure 6.31 shows the effect of the initial CaO surface area on the sulfation reaction at a temperature of 1035°C (1308 K).

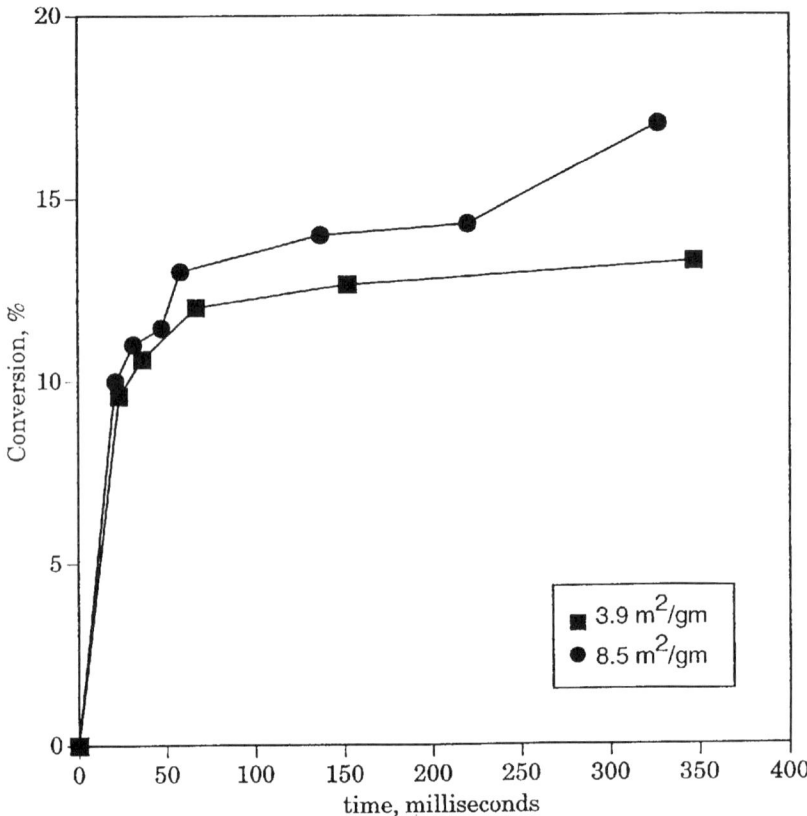

Figure 6.31 Effect of initial surface area of pre-sintered CaO (3.9 μm) on sulfation reaction at 1308 K.

The CaO samples have been sintered long enough prior to the reaction study to eliminate the possibility of any further sintering in the reactor. It can be seen from the conversion plot that the sample with the higher surface area (8.5 m^2/g) reacts faster than the one with the low surface area (3.9 m^2/g) over the entire time period studied. Moreover, the difference in conversion between the two samples is enhanced by increasing the reaction times. In other words, the effect of initial surface area is more pronounced in the later reaction times when product-layer-diffusion is expected to be the controlling mechanism. Such a phenomenon can be explained by either a simple grain model [17] or by a random pore model [30, 31]. For both models, the reaction rate is proportional to the sorbent BET surface area when the surface reaction is rate controlling, whereas for product-layer-diffusion control the rate becomes proportional to the square of the initial surface area. The widening gap between the two conversion curves clearly exhibits a higher sensitivity to surface area, thus supporting the theory of a product-layer-controlling regime.

Effect of Particle Size on Sulfation

The particle size effect during $Ca(OH)_2$ sulfation is presented in Figure 6.32. The reaction temperature is 1388 K, and the particles are classified into four size groups of 1.2, 3.9, 4.9, and greater than 9 μm according to their mean aerodynamic diameters.

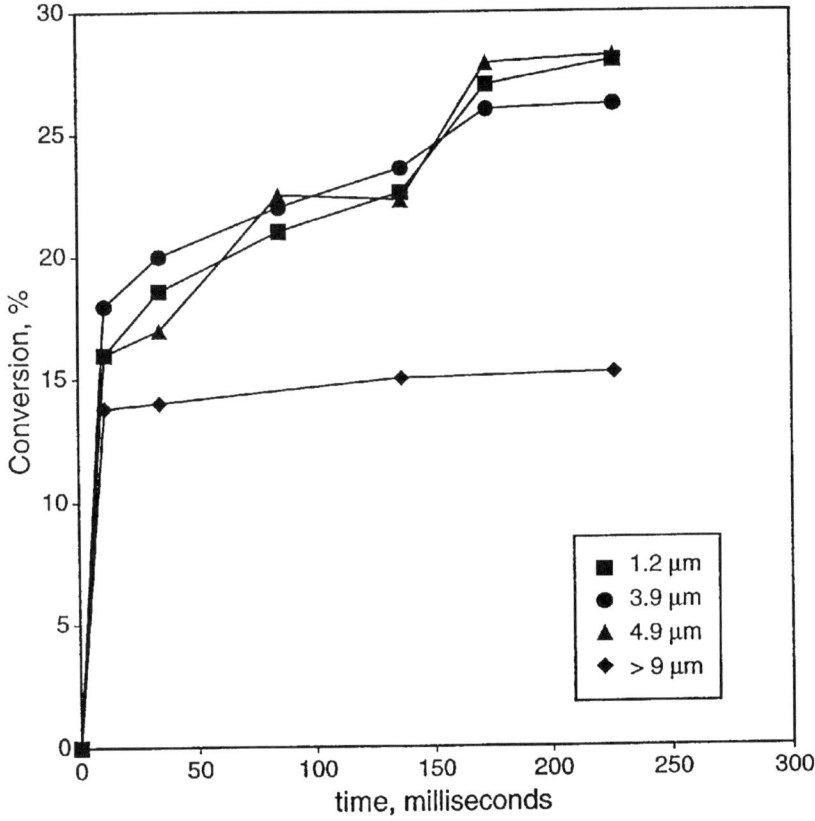

Figure 6.32 Effect of particle size on sulfation of Linwood $Ca(OH)_2$ at 1388 K.

Dry Scrubbing Technologies for Flue Gas Desulfurization

It can be seen from the figure that for the three smaller particle sizes, the effect of sorbent size on conversion is almost nonexistent. On the other hand, for particles larger than 9 μm, a slower sulfation rate can be observed over the entire reaction time. These results agree with the findings by several researchers [8, 15] that below 2-5 μm, sulfation is no longer influenced by particle size. Elimination of pore diffusional resistances below a certain size has been cited as the possible reason. However, this finding is contrary to the observations reported by Gullett and Blom [20], who reported no limit to this particle size effect upon higher reactivity and observed a constant effect of size over the range of 0.72 to 12.1 μm.

Development of Internal Structural Proper-ties During Sulfation

Gas-solid noncatalytic reactions are characterized by the transient nature of internal structural properties of the solid. Unlike calcination, during sulfation a solid product is formed and the $CaSO_4$ product has a higher molar volume than the CaO consumed. The difference in molar volume for the two solids leads to transient internal surface area, porosity, and pore size distribution. The temporal nature of these vital structural properties influences the course of the reaction and the extent of sulfur capture and causes the virtual "die-off" of the sulfation reaction.

The evolution of sorbent structural properties during sulfation of $Ca(OH)_2$ in the entrained flow reactor was studied at a reaction temperature of 1353 K. Figure 6.33 shows a plot of sorbent surface area during sulfation along with the extent of sulfation. As can be seen from figure 6.33, the overall surface area decreases with increasing sorbent sulfation and starts to level off as the extent of sorbent sulfation slows down. All three mechanisms, namely calcination, sintering, and sulfation, progress simultaneously and affect the resulting sorbent structural properties. It is interesting to compare the surface area development for the sorbent with and without sulfation. Figure 6.34 compares the development of surface area for the same $Ca(OH)_2$ powder during calcination only and during simultaneous calcination and sulfation. Interaction with sintering is present in both the cases. The surface area of the sorbent upon sulfation is less than that after calcination for the whole range of reaction time studied. A possible explanation for this observation is that during calcination, sintering is the only deactivation step, whereas during concurrent calcination, sintering, and sulfation, sorbent is subjected to deactivation from both sulfation and sintering.

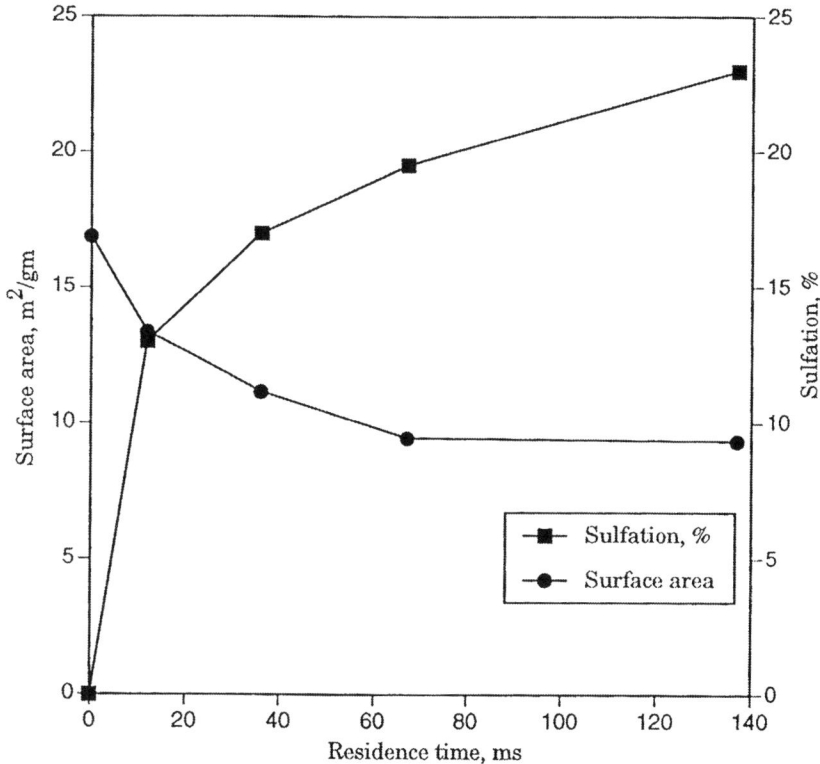

Figure 6.33 Surface area development during sulfation of $Ca(OH)_2$ at 1353 K.

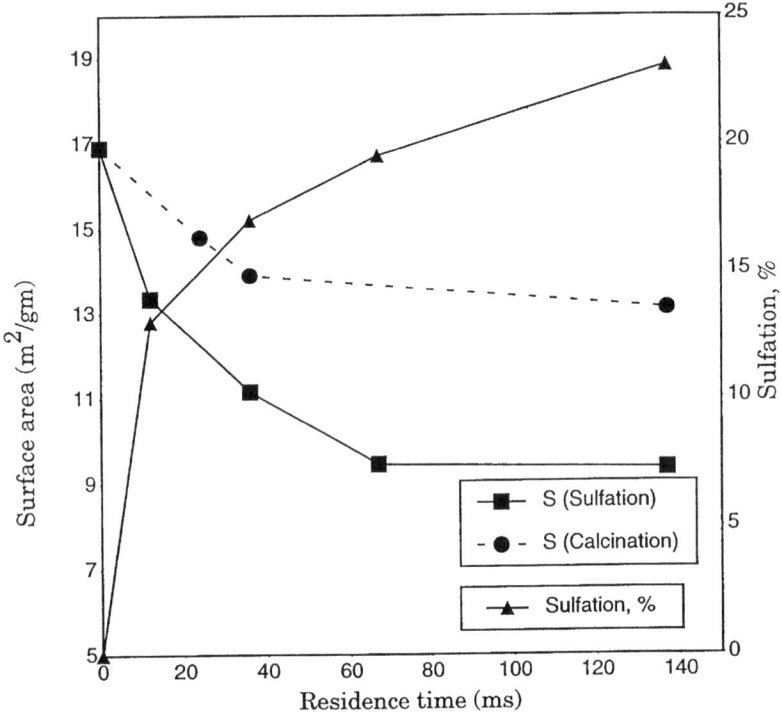

Figure 6.34 Surface area development during sulfation and during calcination only of Ca(OH)$_2$ at 1353 K.

The evolution of porosity with sulfation is illustrated in Figure 6.35, along with its evolution during calcination for the same Ca(OH)$_2$ powder and reaction conditions. Unlike calcination, porosity decreases continuously during sulfation. In addition to sintering, which causes coalescence of smaller pores into larger ones, deposition of the higher volume reaction product (CaSO$_4$) fills up the CaO pores and reduces the overall porosity. The filling up of the CaO pores and the plugging of pore mouths reduces the access of SO$_2$ into the interior of the CaO and retards the rate of the reaction. As the smaller pores get filled up rapidly, the overall pore size distribution shifts to the right.

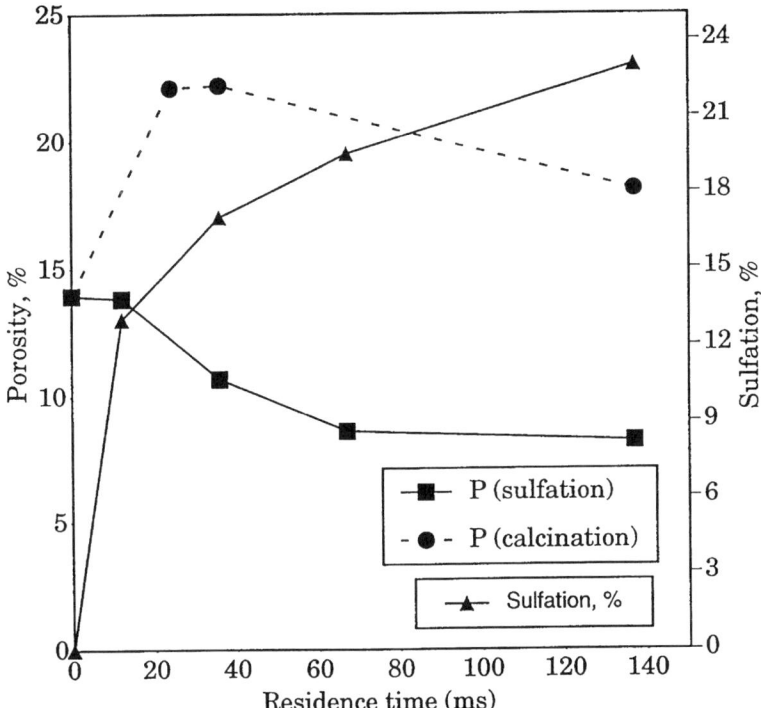

Figure 6.35 Porosity evolution during sulfation and during calcination only of Ca(OH)$_2$ at 1353 K.

Figure 6.36 gives the pore size distribution of partially sulfated sorbent. As the sulfation conversion increases from 13 to 19.5%, the contribution of the pores smaller than 100 A to the overall pore volume gradually decreases. The maximum volume contribution slowly shifts to the pores in the size range between 300-400 A.

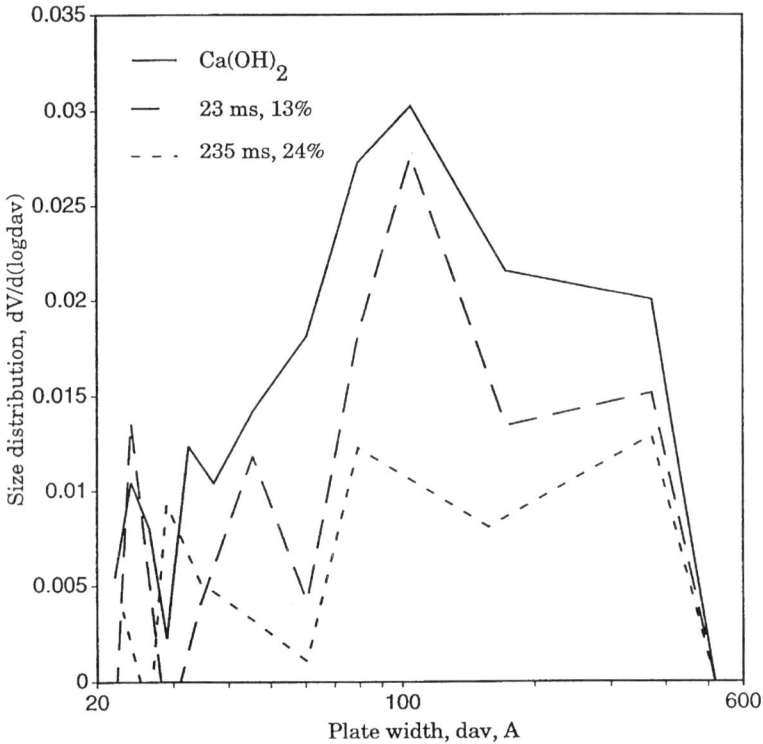

Figure 6.36 Pore size distribution of partially sulfated Ca(OH)$_2$ at 1080°C.

Reaction Modeling for High-Temperature SO$_2$ Sorption

Introduction

Mathematical modeling of gas-solid noncatalytic reactions is an important part of any process or facility which involves such reactions. Mathematical modeling of these systems is important because of design and scale-up requirements and to make the interpretation of experimental data easier. There are various mathematical models available in the literature with the sharp interface model, the volume reaction model, and the particle-pellet or grain model being the most commonly used models. Another class of mathematical models for gas-solid noncatalytic reaction approaches the problem by using the basic porous nature of the solid reactant and the pore size distribution. Such models are generally called pore models [32].

The sharp interface model is the most commonly used model and is well described in literature [33]. The model is mainly applicable to nonporous solids and assumes that the reaction occurs in a topochemical manner at a sharp interface dividing the reacted solid (ash) and unreacted core. This model is mostly applied to gas-solid reactions which have first order kinetics. This model in its unmodified state doesn't take into account the structural changes brought about by the ongoing chemical reaction although various researchers have suggested modification to take into account this shortcoming [34,35,36]. The sharp interface model is relatively simple and can be effectively used for nonporous solids. The important modeling parameters are the surface reaction rate constant, k_s, and the effective diffusivity through the product layer, D_{eG}.

Unlike the sharp interface model where the solid reactant is assumed to be nonporous, in the volume reaction model the reacting solid is porous and the gas can permeate inside the solid. The reaction can take place over the whole volume of the solid. Depending on the size of the pores there could be a radial concentration gradient causing the reaction in the interior to proceed relatively more slowly than at the surface. When diffusion is limiting, the concentration of the gaseous reactant can drop rapidly near the surface (extremely fast reaction) and the volume model approaches the sharp interface model [32].

In the particle-pellet or grain model a porous solid particle can be visualized as consisting of a large number of tiny nonporous spherical grains with the void between the grains giving the particle its porosity. Reactant gas diffuses through these voids to reach the surface of the grains where the reaction can be considered to take place according to the sharp interface model. A detailed description and solution scheme for this model can be obtained from the literature [37,38]. To take into account the structural changes various modifications to the grain model are proposed [28,39]. In these models the grain radius is assumed to vary under the combined influence of chemical reaction and sintering. The main parameters in modified grain model are the reaction rate constant, k_c, grain radius, r_g, sintering rate constant, k_s, effective gas diffusivity, D_{eg}, and grain shape.

As already mentioned, there is another class of gas-solid noncatalytic reaction mathematical models which involves the porous nature of the solid and the pore size distribution of the solid reactant. These models approach the problem in a completely different manner focussing on changes taking place in the pores of the solid. The simplest form of pore model is the single pore model where the solid particle is assumed to have pores of uniform size and changes in any single pore are considered representative of changes taking place in the whole particle. The model parameters are average pore radius, radius of the solid particle, effective pore length, effective diffusivity of the gaseous reactant, and reaction rate constant. A more evolved pore model is the distributed pore model which is similar in concept to single pore model except for the fact that it considers the pore size distribution of the solid reactant particle. The pore models described earlier assume no interaction between the pores. To take into account the intersection between randomly positioned pores inside the solid, a different model is proposed [30,31,32,40]. This model is called the random pore model. It assumes that the actual reaction surface of the solid reactant is formed by a random overlapping of cylindrical pores having a specified size distribution.

All the mathematical models for calcination and sulfation of Ca-based sorbents are lacking because of the fact that they are not applicable accurately to the short-time scale interactions between calcination, sintering, and sulfation of the Ca-based sorbents. The mathematical model developed here describes the kinetics and surface area evolution during simultaneous calcination and sintering of very small (d_p < 5 mm) $Ca(OH)_2$ particles. The overall concept of the model is illustrated in figure 6.37.

A single $Ca(OH)_2$ particle is assumed to be composed of identical, spherical, non-overlapping grains. Calcination takes place on a single grain of $Ca(OH)_2$ according to the sharp interface model. As hydroxide decomposes, smaller grains of CaO are formed surrounding the unreacted $Ca(OH)_2$ core, and the product gas, H_2O, diffuses out through the CaO product layer. The CaO micrograins sinter rapidly, reducing the internal surface area and porosity of the product layer.

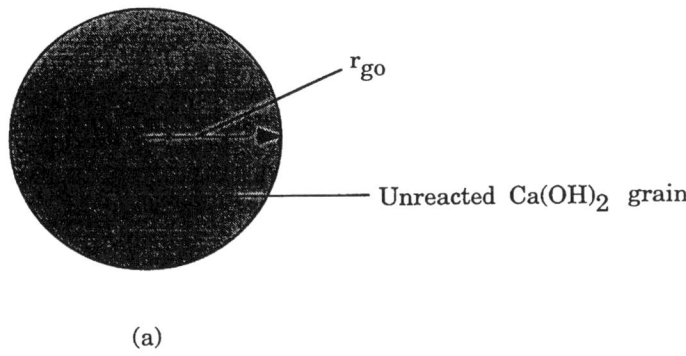

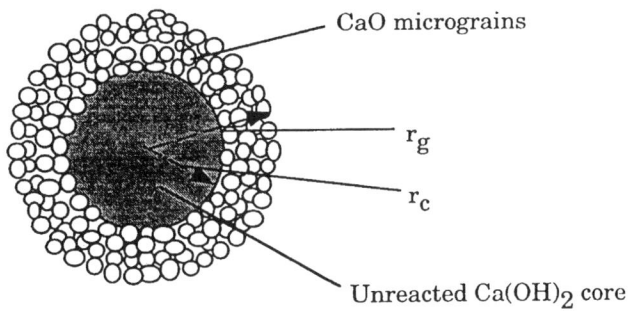

Figure 6.37 Schematic illustration of model. (a) Single, spherical Ca(OH)$_2$ grain. (b) Partially calcined grain with inner Ca(OH)$_2$ core and sintering CaO micrograins.

The overall reaction progresses according to the following steps: decomposition of Ca(OH)$_2$ at the CaO/Ca(OH)$_2$ interface, diffusion of the product gas H$_2$O through the product shell to the grain surface, diffusion of H$_2$O through the inter-grain voids to the bulk gas phase, and subsequent sulfation of CaO. For small diameter Ca(OH)$_2$ particles, intraparticle mass and heat transport are not considered to be rate limiting [4]. All the grains can be considered to be at the same conditions of temperature and gas concentration, and H$_2$O diffusion from the grain surface to the bulk gas is assumed to experience negligible resistance. On the other hand, the product CaO layer offers resistance to the H$_2$O transport. The product shell is composed of CaO micrograins of varying ages with varying degrees of sintering, and hence it can be described as a multilayered CaO shell, each layer corresponding to CaO of certain age. The most recently formed CaO layer, which immediately surrounds the unreacted hydroxide core possesses the highest surface area and porosity.

The sulfation reaction has not only been one of the most extensively studied reactions experimentally, it has also posed an enormous challenge to researchers trying to model its conversion and pore structure behavior. In fact, the sulfation reaction can be said to have served as a benchmarking tool for testing and validation of most of the new gas-solid reaction models developed. The experimental data on sulfation is of two types: (a) $Ca(OH)_2$ sulfation, with all the three phenomena occurring concomitantly or (b) independent CaO sulfation, with precalcined and presintered CaO.

Most of the early efforts to model $Ca(OH)_2$ sulfation assumed the calcination reaction to be instantaneous and neglected the effect of sintering altogether. Most modeled the experimental data of independent $CaO-SO_2$ reaction. Borgwardt and Bruce [17] as well as Bruce et al. [21] applied the limiting case of product layer control with the grain model to fit their $CaO-SO_2$ reaction data. They derived the specific rate constants for product layer control from their best fit of the data. Bhatia and Perlmutter [31] tested their model predictions with experimental $CaO-SO_2$ data. They found that a combination of kinetic and diffusional resistances is needed to explain the experimental data. The specific reaction rate constant and the product layer diffusivity were the two model parameters varied to fit the model. Christman and Edgar [41] also used the surface reaction rate constant and the product layer diffusivity as the two variable model parameters which they estimated from the kinetic data. They estimated the reaction rate constant from the initial rate data and found their value to be in reasonable agreement with Bhatia and Perlmutter [31]. However, they noted that the values of both the parameters varied over a large range in the literature. Simons and Garman [12] modeled the plugging of the smallest pores as the rate-controlling step in leveling off of the conversion-time data. Alvfors and Svedberg [26] modeled the sulfation of CaO using the overlapping grain concept.

Comprehensive $Ca(OH)_2$ sulfation modeling has been attempted by very few researchers due to the complex nature of the various phenomena and the lack of a thorough understanding of their interaction. In $Ca(OH)_2$ sulfation, there exist consecutive reactions where the product of the first reaction (CaO) becomes the reactant for the second, as well as concomitantly undergoes structural changes due to sintering. Mai's [11] work used a pore model for the sulfation and second-order kinetics for sintering. Milne and Pershing [15] presented a combined model using the grain model for sulfation, second-order kinetics for sintering and an empirically modified shrinking-core model for the calcination. Alvfors and Svedberg [26] attempted to model the total process occurring when a sorbent particle is injected into the flue gas atmosphere. The calcination is described by first-order kinetics at the reaction surface, similar to the empirically verified model of Borgwardt [6]. The sintering is based on the German and Munir's [10] formula. The calcination and sintering models together are used to predict the available surface area as a function of time. The predicted area is then used to calculate calcium sulfation. The sulfation itself considers pore diffusion, product layer diffusion and chemical reaction. Alvfors and Svedberg [26] used their PSSM model for predicting the sulfation behavior by adjusting the model parameters. They concluded that qualitative predictions of conversion of limestone to calcium sulfate were possible but quantitative predictions proved to be difficult.

Calcination Modeling

The decomposition reaction takes place at the $Ca(OH)_2$/CaO interface and is assumed to be first order with respect to H_2O partial pressure at the hydroxide core surface [9].

From solid reactant balance, the calcination rate can be expressed in terms of conversion as

$$\Upsilon_{Ca(OH)_2} = k_c \frac{(P_e - P_c)}{RT}$$

and

Equation 6.6

$$\frac{dx}{dt} = \frac{3r_c^2 k_c (P_e - P_c)}{RT r_{go}^3 C_s}$$

where k_c is the calcination rate constant (in m/s) and C_s is the solid reactant concentration (in gmole/m³). The equilibrium dissociation pressure, P_e can be obtained as a function of temperature using a standard thermochemical approach [42].

$$P_e = \exp(-0.23 \ln T - 0.00177T - 46300T^{-2} - 12997.5T^{-1} + 18.2)$$

The species continuity equation for product gas H_2O diffusing through the product CaO assuming the pseudo-steady state approximation can be written as

Equation 6.7

$$\frac{\delta^2 P}{\delta r^2} + \frac{2\delta P}{r\delta r} = 0$$

with the following boundary conditions:

$$-D_{eff}\frac{\delta P}{\delta r}\bigg|_{r=r} = k_c \langle P_c - P_c \rangle$$

and

$$P|_{r=r_g} = 0$$

The latter boundary condition assumes negligible bulk H_2O concentration and no resistance to intraparticle gaseous diffusion.

Although several researchers have investigated the effect of CO_2 pressure on $CaCO_3$ decomposition kinetics [9,43] and the diffusion of gaseous CO_2 through porous lime, similar studies on the effects of H_2O partial pressure and H_2O diffusion are nonexistent. Both Knudsen and ordinary diffusion may be important at different stages of reaction. Hence, Knudsen and ordinary diffusion coefficients are estimated and the effective diffusivity is derived from the following equations:

$$D_{AB} = 8.938E\text{-}6(T/273)^{1.5}P_t^{-1}$$

$$D_k = 0.13\varepsilon T^{0.5} S^{-1}$$

and

Equation 6.8

$$D_{eff} = (D_k^{-1} + D_{AB}^{-1})\varepsilon^2$$

S and ε are the surface area and porosity of the multilayered product shell, respectively. The most sintered CaO layer possesses the lowest surface area and porosity and determines the overall rate of diffusion through the entire product shell. Equation 6.8 approximates the value of the calcine's tortuosity as ε^{-1} [44].

Sintering Modeling

Most of the previous sintering models [9,11] have used second-order kinetics [45] in which the rate of sintering is expressed as

$$\frac{dS}{dt} = -k(S - S_a)^2$$

where S_a is the asymptotic surface area of CaO at the specific sintering temperature and k_s is the sintering rate constant (in $m^2/g\text{-}s$).

The presence of H_2O accelerates the rate of sintering of CaO [11, 17]. In order to account for enhanced sintering in the presence of H_2O, the sintering rate constant is modified according to the following proposed correlation [4]:

$$k_{sm} = k_s(1 + BP^m)$$

Based on the reported work, m and B are taken as 0.17 and 6, respectively, in this correlation.

According to the model, the CaO product shell is divided into multiple layers depending on the time interval in which they are formed. At the end of the I-th time interval, the CaO formed during j-th time interval will have a surface area,

$$S_{j,i} = S_a + \frac{1}{\frac{1}{S_o - S_a} + k_{sm}(t_i - t_j)}$$

where S_o is the surface area of the nascent CaO formed during j-th time interval. The surface area of the entire product shell and the overall specific surface area of the partially calcined particle are calculated as

$$S_{CaO} = \sum_{j=1}^{j=i} S_{j,i} z_j$$

and

$$S_s(x, t) = (x)S_{CaO} + (1 - x)S_{Ca(OH)_2}$$

where z_j is the fraction of CaO formed during the j-th time interval.

A host of previous studies have indicated a linear relationship between surface area and porosity in the low surface area porosity range [3,19,20]. Here it is assumed that $\Delta\varepsilon$ is proportional to ΔS and ε approaches zero as the CaO surface area approaches the asymptotic value. In such a case, the particle porosity results from intergrain voids alone. With such an assumption, the porosity of an individual layer of CaO can be written as a function of its surface area as

$$\varepsilon = \varepsilon_0 \left(\frac{S - S_a}{S_o - S_a} \right)$$

where ε_o is the theoretical porosity of the nascent CaO.

The calcination rate Equation 6.6 on page 493 is coupled with the product H_2O continuity Equation 6.7 on page 494. The core radius r_c is related to the solid conversion as

$$r_c = r_g(1-x)^{1/3}$$

The instantaneous grain radius, r_g, changes as a result of both calcination and sintering which work to oppose each other. Sintering alone causes the grains to grow by combination of adjacent grains. The model stipulates that the unreacted core of each small grain remains at the center of the new grain [41]. Further, the grain size is not a function of the radial position since the model assumes negligible concentration gradients within the particle.

$$r_g = \frac{3}{S_s(x,t)\rho(x,t)}$$

The H_2O partial pressure profile in the product layer can be obtained analytically by solving the continuity Equation 6.7 with the given boundary conditions. The H_2O partial pressure at the interface of the unreacted $Ca(OH)_2/CaO$ can be expressed in terms of r_c, r_g and D_{eff}.

$$P_c = -P \frac{\dfrac{k_c r_c^2}{D_{eff}}\left(\dfrac{1}{r_g} - \dfrac{1}{r_c}\right)}{1 - \dfrac{k_c r_c^2}{D_{eff}}\left(\dfrac{1}{r_g} - \dfrac{1}{r_c}\right)}$$

The calcination rate Equation 6.6 was solved with the initial condition of x=0 at time t=0 using a fourth order Runge-Kutta integration scheme. This local conversion corresponds to the overall particle conversion. The S_s value obtained from the model represents the predicted surface area of the partially calcined particle. The reaction constant, k_c, and the sintering constant, k_s, represent the two specific rate parameters of the model.

Comprehensive Sulfation Modeling

The calcination and sintering model described above used the grain model as the basis but was extensively modified to take into account the sintering of the product CaO layer and the diffusion resistance the porous sintering CaO layer offers to the outward diffusion of H_2O (in case of $Ca(OH)_2$ calcination). The model thus incorporated a number of unique features and was able to fit both the experimental kinetics and the surface area evolution data extremely well. In order to develop the sulfation model, both the grain model formulation as well as the pore model formulation were studied for a number of factors such as ease of adaptability with the existing work, the solution scheme, and the reliability, robustness etc. The random pore model [46] and the distributed pore model [47] have well-refined formulation and a number of capabilities. Furthermore, the random pore model has been applied to the sulfation reaction and has been shown to model the experimental data well. After careful consideration of the various existing models, it was decided to use the random pore model formulation for developing the comprehensive model.

Since the sulfation is restricted to the sintering CaO only, the region of interest is the product CaO shell with its micrograins and associated porosity and surface area. In order to develop the pore model formulation, the reacting solid CaO is now visualized not as grains but rather it is represented as a network of random pores. For the random pore model formulation, the pores are considered to be of the same size and their intersections are taken into account. A schematic illustration of the product CaO layer building during calcination of a $Ca(OH)_2$ grain, according to the random pore model, is shown in figure 6.38.

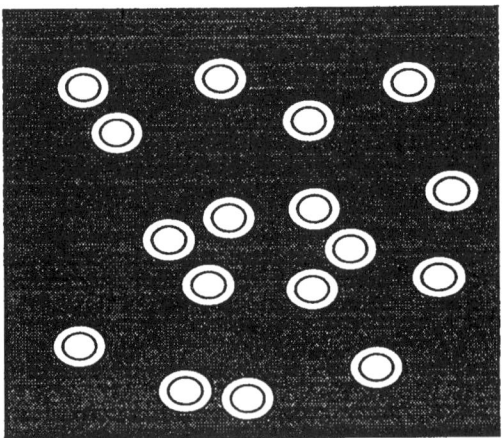

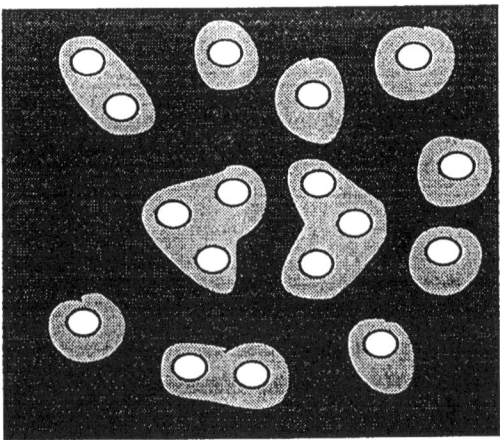

Figure 6.38 Schematic of the overlapping pore structure of the random pore model. (a) Early stage, showing product layer around each pore. (b) Intermediate stage, showing some overlapping reaction surfaces (after Bhatia and Perlmutter [30]).

The various diffusion and reaction steps are (a) SO_2 diffusion through the intergrain voids, (b) diffusion through the intragrain voids of the sintering and sulfating CaO, (c) diffusion through the product layer (after its formation), and finally (d) reaction with CaO.

The inherent assumption involved here is that there is no external diffusion resistance because of the small particle size. Moreover, it is assumed that calcination and sintering proceed unhindered during sulfation. This assumption has been made by several researchers in attempting to model the experimental data. Further, intraparticle (i.e., intergrain) resistance to heat or mass transfer is considered insignificant. This allows us to neglect step (a) from the formulation.

A differential mass balance for diffusion and reaction of SO_2 through the porous CaO product shell of the grain can be written as

Equation 6.9

$$\frac{1}{r^2}\frac{\partial}{\partial r}\left(D_{eff} r^2 \frac{\partial C}{\partial r}\right) = \frac{\rho_{CaO}(1-\varepsilon_0) dx_s}{M_{CaO}} \frac{}{dt}$$

Here, r is the radial distance along the grain and D_{eff} represents the effective diffusivity of SO_2 through the pores of CaO product. C is the concentration of SO_2 inside the porous structure of CaO. dx/dt is the local rate of conversion of CaO to $CaSO_4$. The boundary conditions are based on radial symmetry and on no mass transfer resistance at the grain boundary

Equation 6.10

$$\frac{\partial C}{\partial r} = 0 \text{ at } r = 0$$

and

Equation 6.11

$$C = C_{bulk} \text{ at } r = r_g$$

The local reaction rate, dx/dt at the internal surface of the pores (actually the interface of the CaO and $CaSO_4$), is obtained from the random pore model [30,46] which takes into account the growing product layer and the diffusion of the reactant gas through the layer. The pores are considered to be cylindrical in shape and their random overlapping nature is taken into consideration.

Equation 6.12

$$\frac{dx_s}{dt} = \frac{k_r S_o C(1-x_s)\sqrt{1-\Psi \ln(1-x_s)}}{(1-\varepsilon_0)\left[1 + \frac{\beta Z}{\Psi}(\sqrt{1-\Psi \ln(1-x_s)} - 1)\right]}$$

Ψ is a structural parameter defined as

$$\Psi = \frac{4\Pi L_o(1-\varepsilon_o)}{S_o^2}$$

β can be considered as a Biot modulus and takes into account the ratio of chemical reaction rate compared to the rate of product layer diffusion.

$$\beta = \frac{2k_r \rho_{CaO}(1-\varepsilon_o)}{M_{CaO} D_p S_o}$$

Z is the molar volume ratio of the product $CaSO_4$ to reactant CaO. k_r is the specific reaction rate constant and D_p is the diffusivity through the product layer.

The three parameters, ε_o, S_o, and L_o are the core features of the random pore model. The model considers the actual reaction surface to be the result of a random overlapping of a set of cylindrical pore surfaces of distributed sizes. e_o, S_o and L_o are the total enclosed volume, total surface area and total length of the overlapping cylindrical pore system, respectively. Bhatia and Perlmutter [46] have indicated that these quantities can be determined from measurements of the pore volume distribution, $v_o(r)$. Such pore volume distribution data can be obtained from BET measurements.

$$\varepsilon_o = \int_0^\infty v_o(r)dr$$

$$S_o = 2\int_0^\infty \frac{V_0(r)}{r}dr$$

and

$$L_o = \frac{1}{\pi}\int_0^\infty \frac{V_0}{r^2}dr$$

The conversion-time behavior can be obtained by the simultaneous solution of Equation 6.9 on page 499 and Equation 6.12 with boundary conditions, Equation 6.10 and Equation 6.11. The solution scheme requires the simultaneous solution of two highly nonlinear ordinary differential equations. Further, some of the existing models [30] have made a number of simplifying assumptions and have shown reasonable agreement with the experimental data. In this work, the intragrain mass transfer resistance is neglected, then chemical kinetics and product layer diffusion control the overall reaction rate. Equation 6.12 can be integrated to give the local reaction rate which is also representative of the global reaction rate for the particle due to insignificant intragrain transport resistances. The solution procedure requires the predictions of ε_o, S_o and L_o, which are the structural properties of the porous CaO reactant. The calcination and sintering model calculates the porosity and surface area of the individual layers of the CaO and also the average values of the above by integration over the entire layer structure. In the original random pore model formulation, the properties ε_o, S_o and L_o are calculated from experimental data and are invariant since the reactant is not undergoing any sintering induced structural changes. However, in this case, sintering largely affects the values of the above three parameters and their variation cannot be neglected. ε_o and S_o are calculated from the calcination and sintering model. L_o itself is not available from the model or any experimental data, an average representative nonvariant value is assumed for this parameter. k_r and D_p are the two variable model parameters.

Comparison of Experimental Data and Model Predictions

Figure 6.39 shows the experimental data and model prediction for calcination of 3.6 mm $Ca(OH)_2$ at three different temperatures for residence times ranging from 10 to 300 ms. A similar trend is observed in all the calcination curves.

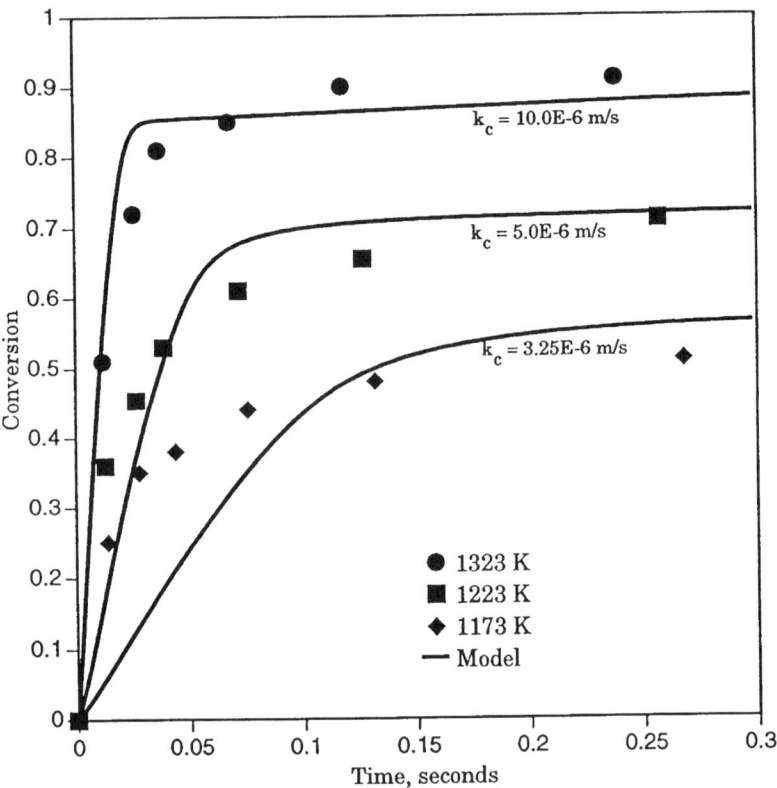

Figure 6.39 Effect of temperature on calcination of 3.6 µm Linwood Ca(OH)$_2$. Experimental data and model predictions.

Initially, calcination progresses extremely quickly and exhibits a strong influence by temperature. At higher residence times, the reaction rates attenuate considerably and conversion tends to flatten out. The sharp attenuation of reaction rate at higher residence times is seen clearly from the 1323 K data which exhibits a virtual ceasing of calcination at about 90%. This near "die-off" of the reaction at higher residence times is also observed at lower temperatures of 1173 and 1223 K. The model discussed earlier assumes that the reaction proceeds on a single Ca(OH)$_2$ grain according to the shrinking core model. The strong influence of temperature during the initial stage of reaction and the high porosity of the nascent product layer suggest that chemical reaction at the CaO/Ca(OH)$_2$ interface may be the rate-limiting step. The product CaO layer rapidly sinters losing surface area and porosity. This offers increasing resistance for the gaseous product H_2O to diffuse through the product shell. This leads to increasing H_2O concentration in the product layer and at the interface which not only enhances the rate of CaO sintering but also retards the calcination rate. All these phenomena eventually result in the overall calcination reaction being dominated by outward H_2O diffusion through the CaO.

Experimental data is used to establish values of the two parameters, the calcination rate constant, k_c, and the sintering rate constant, k_s. The initial surface area of the nascent CaO produced in the dispersed environment is taken to be 100 m²/g [3]. The asymptotic CaO surface area is a function of the temperature of study, and its value is obtained by utilizing the experimental data. S_a is estimated to be 15.5 m²/g at 1173 K and 12.5 m²/g at 1323 K. The rate constants are determined by trial and error procedure to obtain the best fit of the experimental data. The activation energies and the pre-exponential factors are obtained from the Arrhenius-type plots shown in Figure 6.18 on page 463.

The model fits the experimental data fairly well, closely predicting both the initial steep slope and the rate attenuation of the experimental data. However, for the calcination temperature of 1173 K, the model underestimates the initial slope while at higher residence times, its prediction lies above the experimental data. The model fits the higher temperature data of 1323 K very well at both short time scales as well as at higher residence times.

The overall surface area evolution of the partially calcined $Ca(OH)_2$ and model predicted values are shown in Figure 40. For the two lower temperatures of 1173 K and 1223 K, surface area first increases and attains a maximum then rapidly decreases due to sintering and tends to level off to an asymptotic value at higher residence times. At the higher temperature of 1323 K, experimentally observed surface area values show a monotonic decrease indicating a strong effect of sintering.

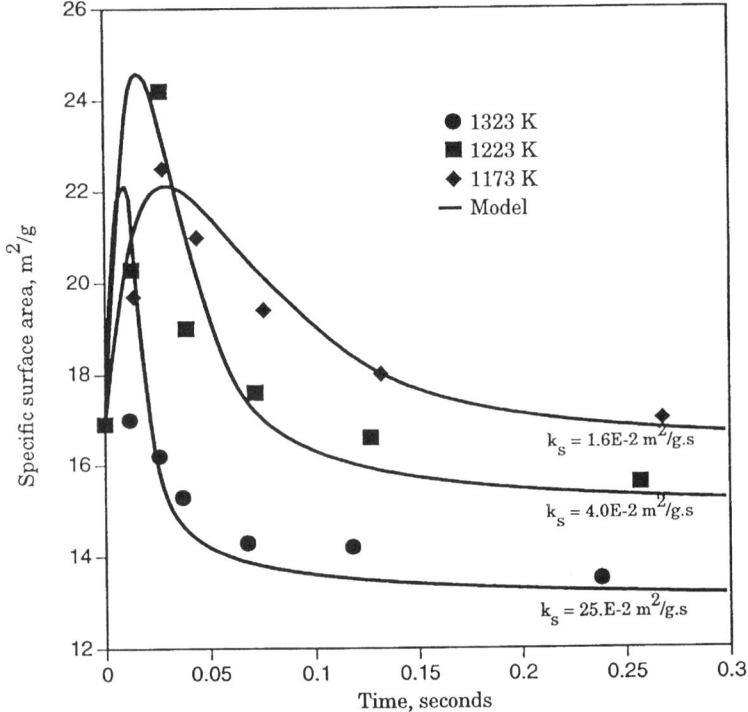

Figure 6.40 Effect of temperature on surface area evolution during calcination of 3.6 µm Linwood Ca(OH)2. Experimental data and model predictions.

The model predictions agree well with the observed trends. At the two lower temperatures, the model predicts the initial rise in surface area and the maxima; the predicted maxima are lower than the observed values for both the temperatures. The rapid reduction and the asymptotic leveling also match with the observed trends. At 1323 K, however, the model predicts a sharp spike in surface area at a very low residence time of about 8 ms before the rapid downward trend toward the asymptotic value. This predicted sharp spike is not observed experimentally, probably due to lack of data at the low residence time of 8 ms.

The model also simulates the buildup of H_2O partial pressure at the CaO-$Ca(OH)_2$ interface with increasing reaction time, as shown in Figure 41. With increasing decomposition temperature, the H_2O partial pressure approaches the equilibrium dissociation value sooner, which results in an early drop in the calcination rate. This prediction is in agreement with the calcination experimental data of Figure 6.39, where attenuation in reaction rate is exhibited earlier at higher temperatures.

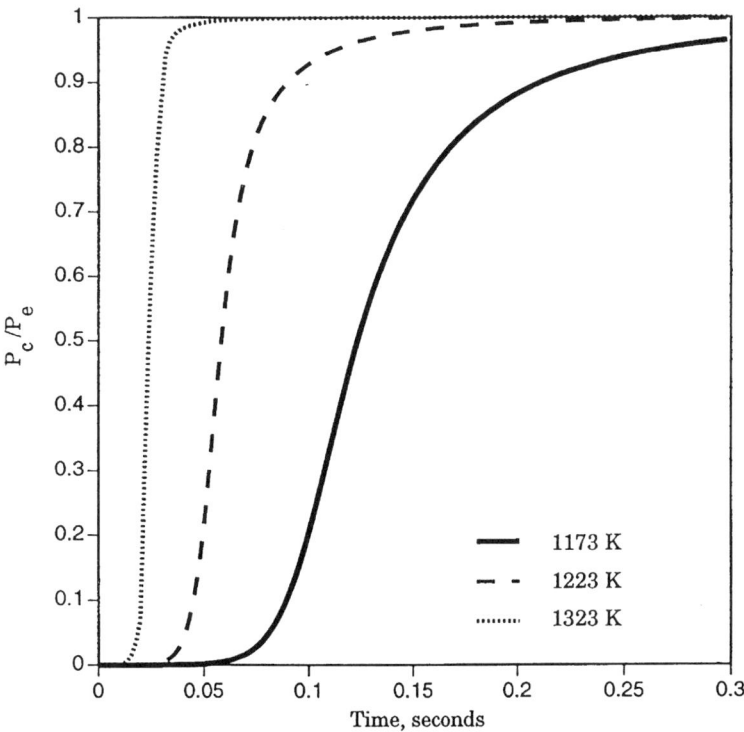

Figure 6.41 Ratio of interface H_2O partial pressure to the dissociation pressure with time. Model simulation.

In order to test the comprehensive sulfation model, the entrained flow reactor kinetic data for $Ca(OH)_2$ sulfation at 1080°C and at other temperatures was utilized. The predictions of the model for three temperatures of 950, 1035 and 1080°C are shown in Figure 6.42. For comparison, experimental data; at 1035 and 1080°C is also shown in the figure. The model predictions agree qualitatively with the data; however, the initial high reaction rate is not predicted well by the model. Moreover, the experimental conversion can be seen to be leveling off even as early as 200 ms, which is not represented well by the model. The variation of the reactant CaO surface area and porosity are shown in Figures 6.43 and 6.44 respectively. As can be seen, the steep changes in both these structural characteristics lead to very drastic changes in the nondimensional parameters, Ψ and β, which influence the sulfation rate. The changes in Ψ and β are shown in Figure 6.45 and Figure 6.46 respectively. As can be seen, both Ψ and β level off asymptotically after the first 50 ms of reaction and sintering. The value of k_r can be computed from the initial rate data [30] or can be obtained together with D_p by fitting the experimental data at specific temperatures. The values of k_r obtained in this work varied from 1.0 @ 10^{-6} at 950°C to 2.6 @ 10^{-6} (in m^4/kmol-s) at 1080°C. These values compare with those reported in the literature. Bhatia and Perlmutter [30] found k_r to be 0.834 @ $10^{-6} m^4$/kmol-s at 980°C. The variation in these values can also be partly attributed to the different reactivities of $CaCO_3$ and $Ca(OH)_2$ as well as compositional variations. The values of D_p were 0.06 @ $10^{-12} m^2$/s at 950°C and 0.22 @ $10^{-12} m^2$/s at 1080°C. Bhatia and Perlmutter [30] calculated D_p as 0.86 @ $10^{-12} m^2$/s at 980°C, while Hartman and Coughlin [18] obtained 0.6 @ $10^{-12} m^2$/s, both of which were for limestone-derived CaO. Figure 6.47 shows the Arrhenius type plots for k_r and D_p used to obtain the activation energies for reaction and product layer diffusion. The activation energy for reaction was found to be 24 kcal/gmol and for product layer diffusion it was found to be 38 kcal/gmol. Borgwardt and Bruce [17] obtained 36.6 kcal/gmol. Bruce et al. [21] obtained 39 kcal/gmol by using the product layer controlled expression of the grain model to fit their CaO sulfation data. Bhatia and Perlmutter [30] obtained 29 kcal/gmol and proposed highly activated solid-state diffusion as the mechanism for product layer diffusion. In fitting the model to the experimental data, it was observed that β has a strong influence on the predictions as well as the stability of the solution scheme. The value of β is seen to level off to about 40 for all the three temperatures studied. The asymptotic value of β was observed to give a better and more robust performance, and the model predictions shown in Figure 42 use the asymptotic β values. Bhatia and Perlmutter [30] found the best fit value of β to be 200 in order to model Hartman and Coughlin's [18] data. They fitted their model such that the reaction ceased beyond a specific time due to surface pore closure. Moreover, the conversion-time curves were found to be not as sensitive to the value of k_r. Considering the large value of β, it can be said that this model represents the entire sulfation data using the limiting case of product layer diffusion control. The initial rapid reaction rate occurs during the period in which the extent of sulfation is quite small (less than 5%) and so is the product layer buildup. During this period, both kinetic and diffusion limitations will need to be considered. However, this period lasts for a very short time (less than 25 ms) and, therefore, the overall behavior is better explained by the diffusion-control mechanism. The main drawback of this scheme is the inability to predict the initial prompt capture.

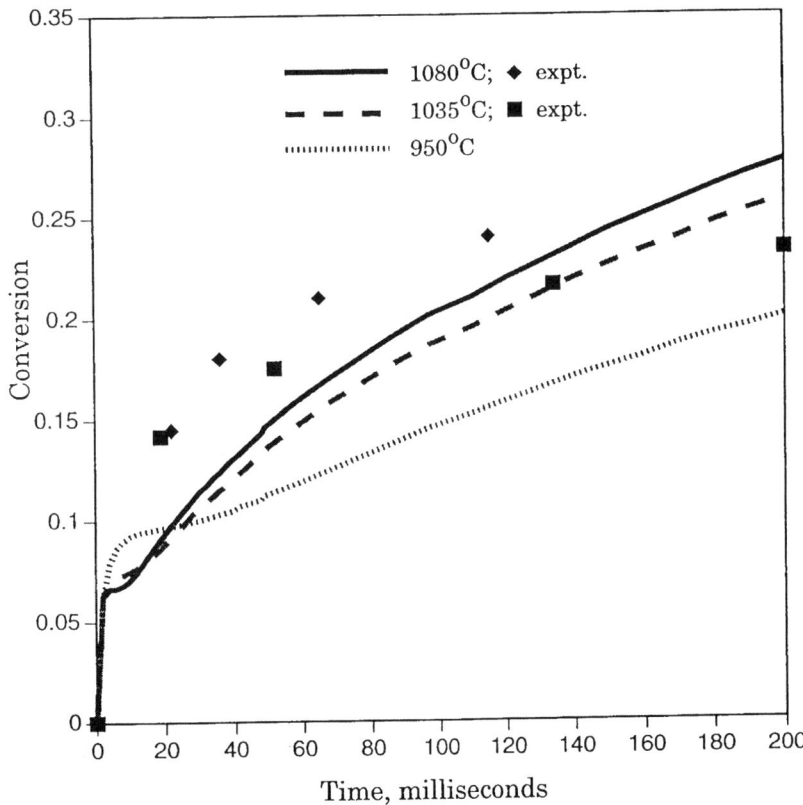

Figure 6.42 Comparison of comprehensive model predictions with experimental sulfation data for 3.9 μm Linwood Ca(OH)$_2$.

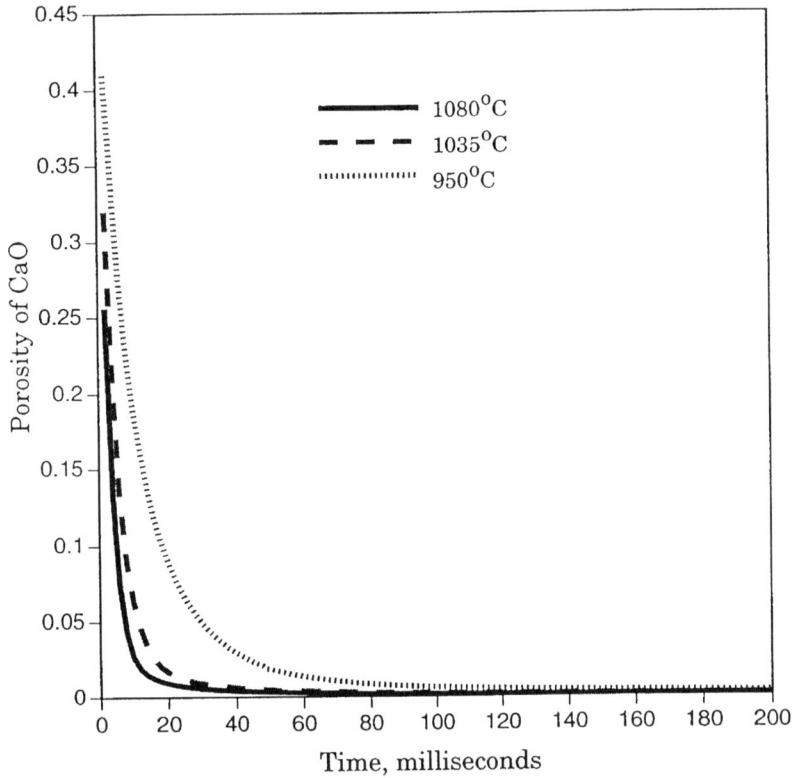

Figure 6.43 Variation of CaO porosity as predicted by the calcination and sintering model and used in the sulfation modeling.

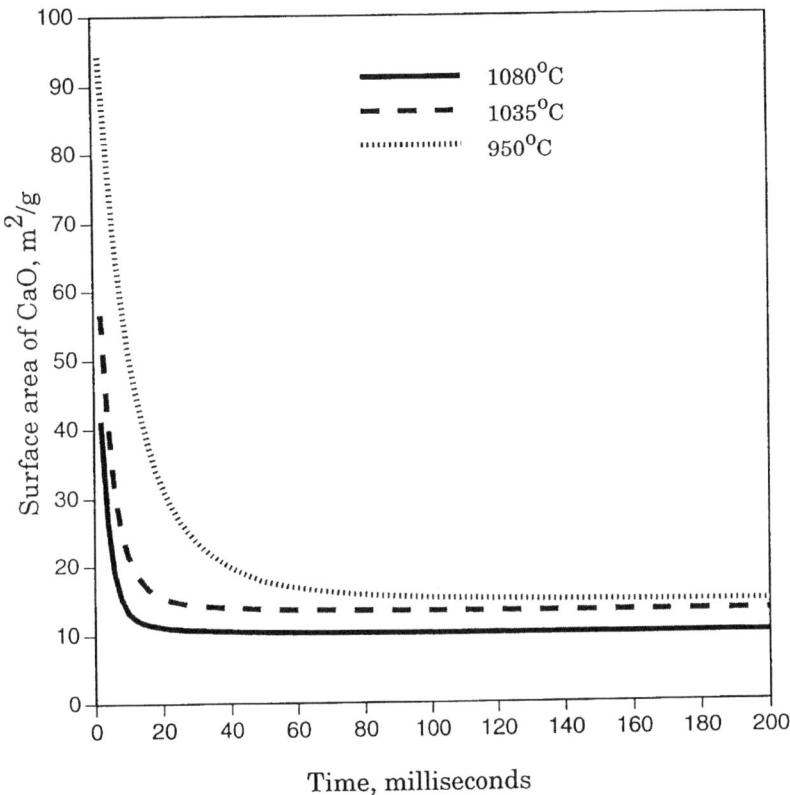

Figure 6.44 Variation of CaO surface area as predicted by the calcination and sintering model and used in the sulfation modeling.

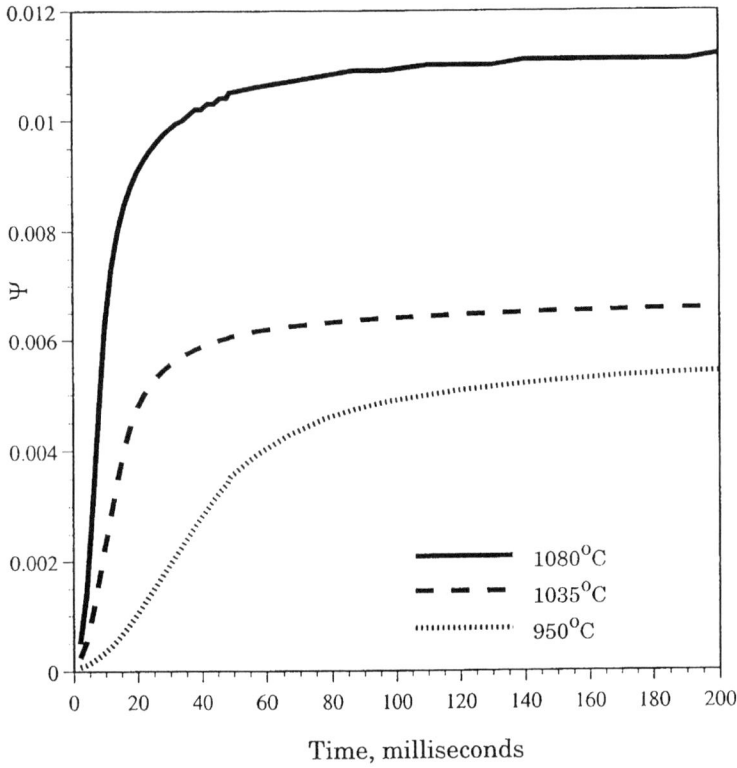

Figure 6.45 Variation of the structural parameter ψ with time; model simulation.

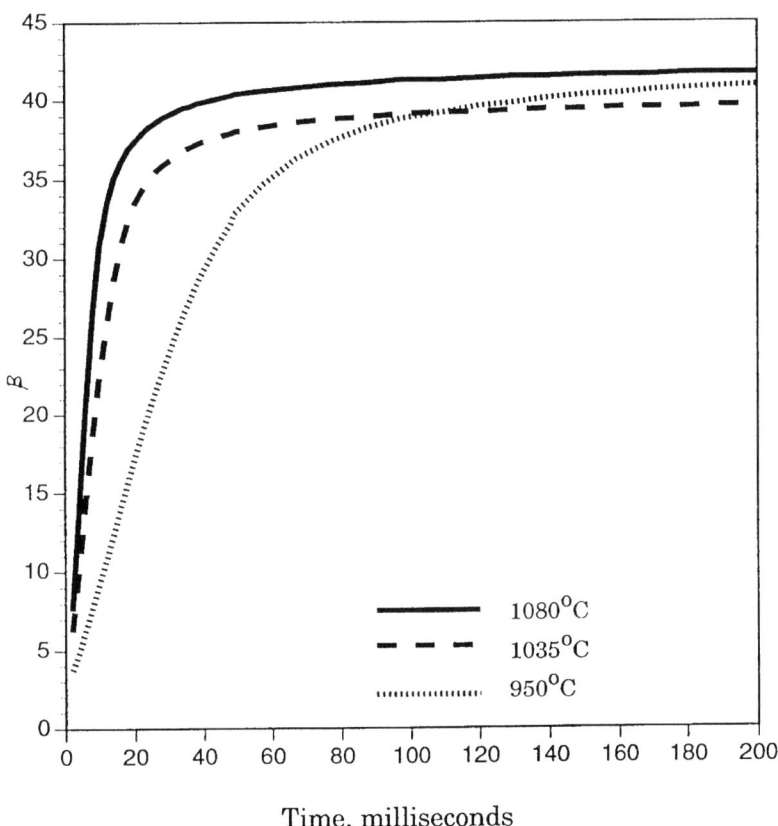

Figure 6.46 Variation of the parameter β with time. Model simulation.

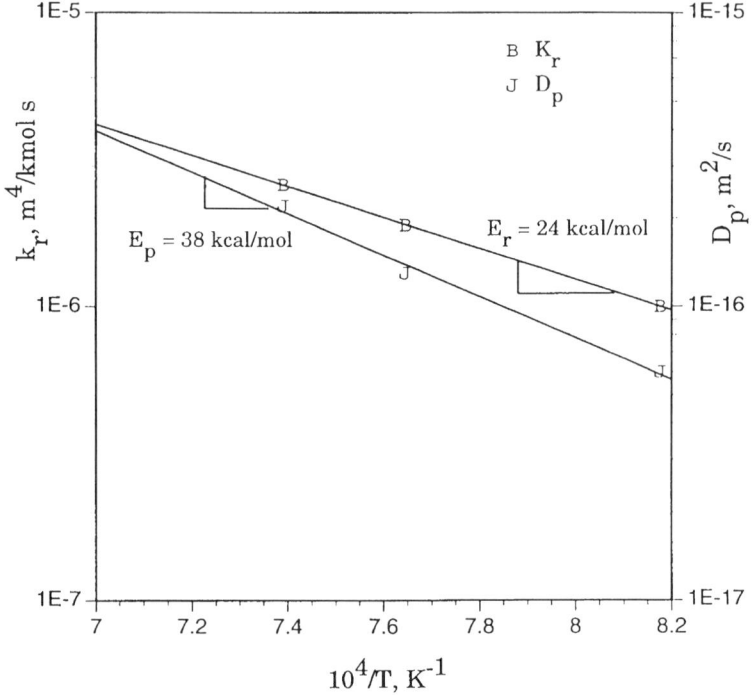

Figure 6.47 Arrhenius plot to estimate activation energies of surface reaction and product layer diffusion.

Modified Calcium-Based Sorbent

Introduction

Even though, from the economic point of view, the furnace sorbent injection (FSI) technique is probably the most viable FGD technology, it is not widely used in industry. The biggest drawback for FSI technology is its inability to meet the standards regulated by the Environmental Protection Agency (EPA). EPA mandates that 95% of all the SO_2 emitted from the coal-fired combustors be removed before venting the flue gas, but SO_2 removal levels achieved by the FSI technique (55-60%) are far below the desired levels of SO_2 removal. Surface area loss, because of pore closure and product layer buildup, is considered to be the main reason for poor sorbent utilization and low sulfur capture. Approximately 60% of the surface area becomes unavailable for the reaction with SO_2 in the first 100 ms of the reaction [48]. One method of increasing the sorbent utilization is by decreasing the particle size [48]. In order to enhance the SO_2 removal by decreasing the particle size and thereby changing its internal structure, it has been suggested that pure Ca-based sorbents should be modified using surface active agents (surfactants).

Surfactant Modifiers

A surfactant is a substance that, when present at low concentration in a system, has the ability to adsorb onto the surface or interfaces of the system and alter the surface or interfacial free energies of those surfaces (or interfaces). Surface active agents have a characteristic molecular structure consisting of a structural group that has very little attraction for the solvent (water), known as a lyophobic (hydrophobic) group, together with a group that has strong attraction for the solvent (water) called the lyophilic (hydrophilic) group. This type of structure is called an amphipathic structure. Depending on the nature of the hydrophilic group, surfactants are classified as anionic if the surface active portion of the molecule bears a negative charge. The catatonic surface active part bears a positive charge, Zwitterionic, if both positive and negative charges are present in the surface active portion of the surfactant, and nonionic if the surface active part of the surfactant bears no charge.

Lignosulfonate Modified Calcium Hydroxide

Kirchgessner and Josewicz [48], reported that addition of calcium lignosulfonate, an anionic surfactant, along with the water of hydration of $Ca(OH)_2$ increases the SO_2 capture of the resulting $Ca(OH)_2$ to up to 65%. They also observed that the principle reason for enhanced SO_2 capture was due to the reduction in particle size of the modified hydroxide and that the main mechanism of size reduction appears to be through deagglomeration of the $Ca(OH)_2$ crystals.

Limited literature results on lignosulfonate modified sorbents indicate that the final concentration of lignosulfonate in the modified sorbent is an important parameter in its ability for sulfur capture. Modified hydroxide sorbents were prepared with varying lignosulfonate concentration, from 0.5 to 2%. The modifier-promoted sorbents were made by calcining $Ca(OH)_2$ at 600°C, mixing a specific amount of lignosite (calcium lignosulfonate) in a known amount of calcined powder, and then adding excess water. The slurry was dried overnight in a box furnace at 60°C which yielded a dried cake of promoted sorbent with specified lignosulfonate concentration. The sorbent was then ground and sieved. Particles of less than 38 μm size were used for reaction studies.

Experimental Results and Discussion: Comparison Between

Modified and Pure Calcium Hydroxide

In this work, the effect of calcium lignosulfonate surfactant on the SO_2 capture ability of calcium hydroxide was tested in the high-temperature entrained flow reactor with special emphasis on short-contact times (less than 100 ms) and development of internal structural morphology. A comparison between pure and lignosulfonate (1.5%) modified hydroxide sorbent is shown in the pore size distribution plot in Figure 6.48. It can be seen that the modified sorbent has a much higher pore volume than that of pure hydroxide. Furthermore, the surface area of the modified sorbent is about 45 m^2/g, as opposed to 16.9 m^2/g for that of pure sorbent. Also, the modified hydrate possesses an overall porosity of 32% as compared to the 14% porosity for the parent hydrate. These differences in the structural properties of the parent sorbent probably suggest that calcination of modified sorbent occurs much faster than for the unmodified one, since calcination rate is proportional to the sorbent BET surface area. In addition to internal structural characteristics, a primary particle size analysis was also performed with the modified and parent $Ca(OH)_2$. Figure 6.49 shows the cumulative particle size distribution curves obtained from Sedigraph analysis. As can be seen, the median primary particle size (at 50% mass finer) of the modified hydrate is about 1.2 microns as compared to almost 2 microns for the parent hydrate.

Kinetic and structural studies were performed with 1.5% lignohydrate because it is documented that at 1.5% weight percent of calcium lignosulfonate calcium hydroxide shows maximum SO_2 capture [48]. Sulfation data (1353 K) of modified sorbent is shown in Figure 6.50 for comparison with the sulfation data for pure hydroxide sorbents. The lignosulfonate modified hydrate shows superior performance in the initial 50 ms. At higher residence times, however, the rate of sulfur capture attenuates considerably. The conversion profiles of both the hydrates (pure and modified) become parallel to each other at higher times, with the lignohydrate maintaining the initial edge achieved in the first 30 ms. In order to understand and explain this characteristic behavior, the internal structure of partially sulfated lignohydrate is studied. Pore size distribution curves for partially calcined ligno-modified hydrate and pure hydrate are shown in Figure 6.51. As can be seen from the figure, for the same calcination time (20 ms) the pore volume for the ligno-modified hydrate is much higher than the pore volume of the pure hydrate.

Figure 6.52 shows a comparison of pore volume between modified and pure hydrate powders during calcination and during sulfation. The lignohydrate initially possesses much higher pore volume which is rapidly lost within the first 20 ms of sulfation to approach very closely to that of the unmodified hydrate. The significant drop in porosity, within 20 ms, for lignohydrate corresponds to the rapid buildup of a $CaSO_4$ layer on the internal surface. This leads to extensive pore closure. Thus, the lignohydrate does not seem to hold any advantage over the pure hydrate (with respect to the internal structure and porosity) beyond 50 ms. Further insight can be obtained into this phenomenon by studying the pore size distribution behavior with sulfation as shown in Figure 6.53. As can be seen, the pore size distribution curves of the partially sulfated samples lie much below the initial lignohydrate and the peak can be seen to shift towards the right indicating closure of smaller pores.

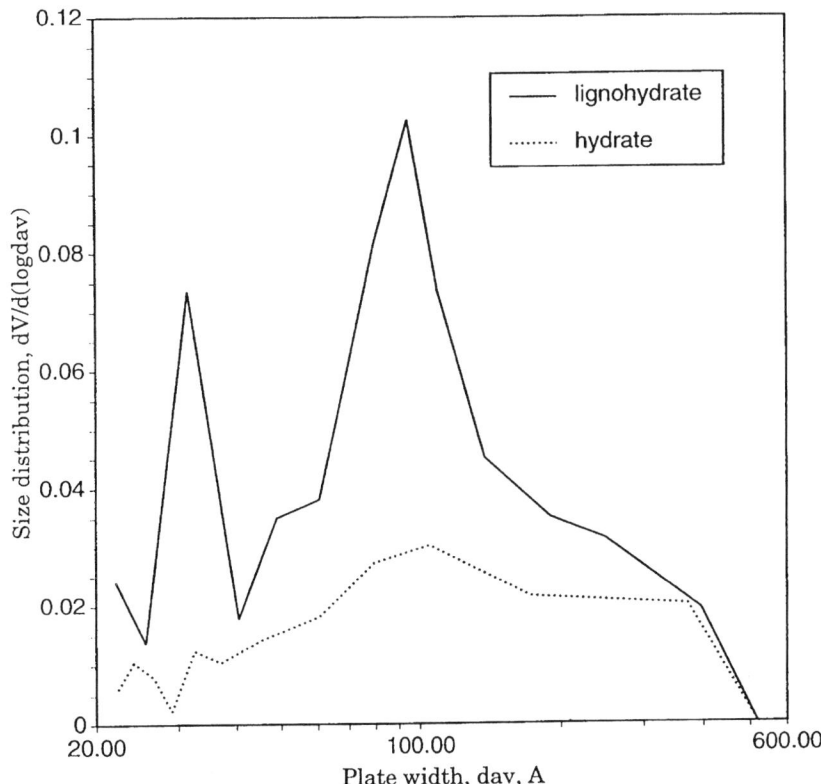

Figure 6.48 Comparison of pore size distribution for pure $Ca(OH)_2$ and 1.5% ligno-$Ca(OH)_2$.

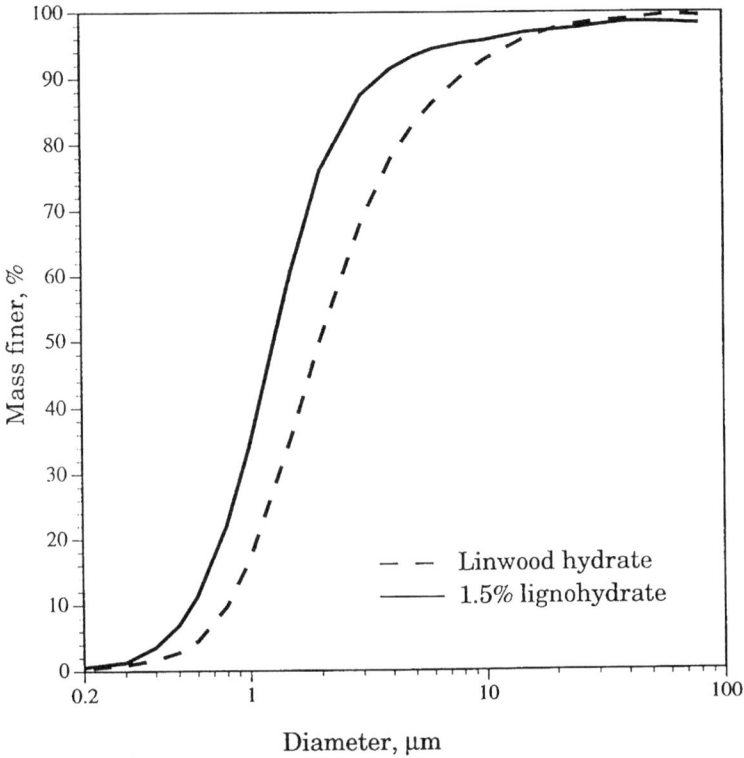

Figure 6.49 Primary particle size distribution from Sedigraph analysis.

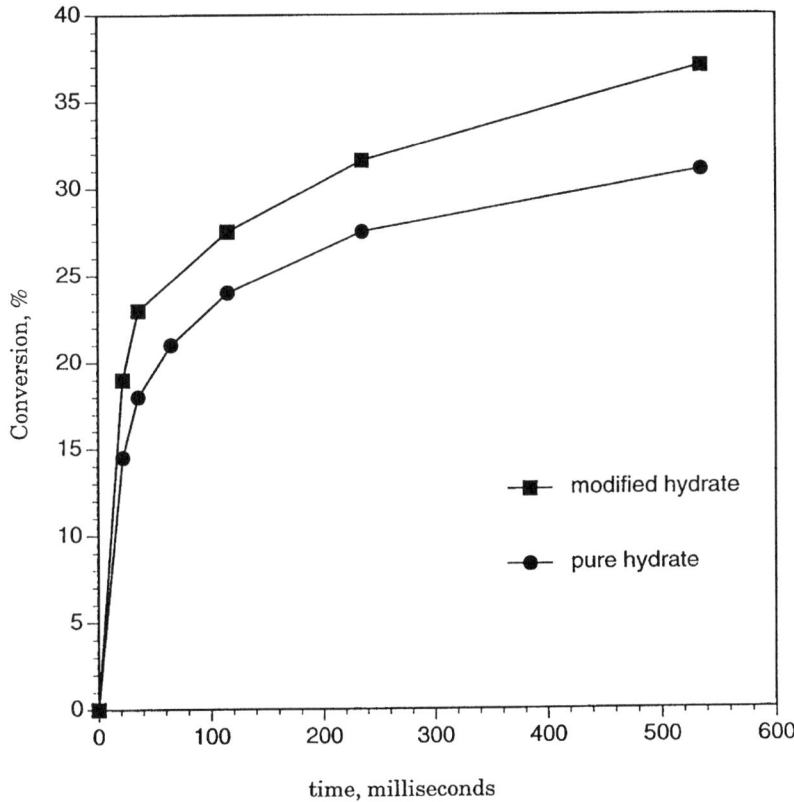

Figure 6.50 Sulfation of 3.9 μm pure hydrate and 1.5% lignohydrate at 1080°C.

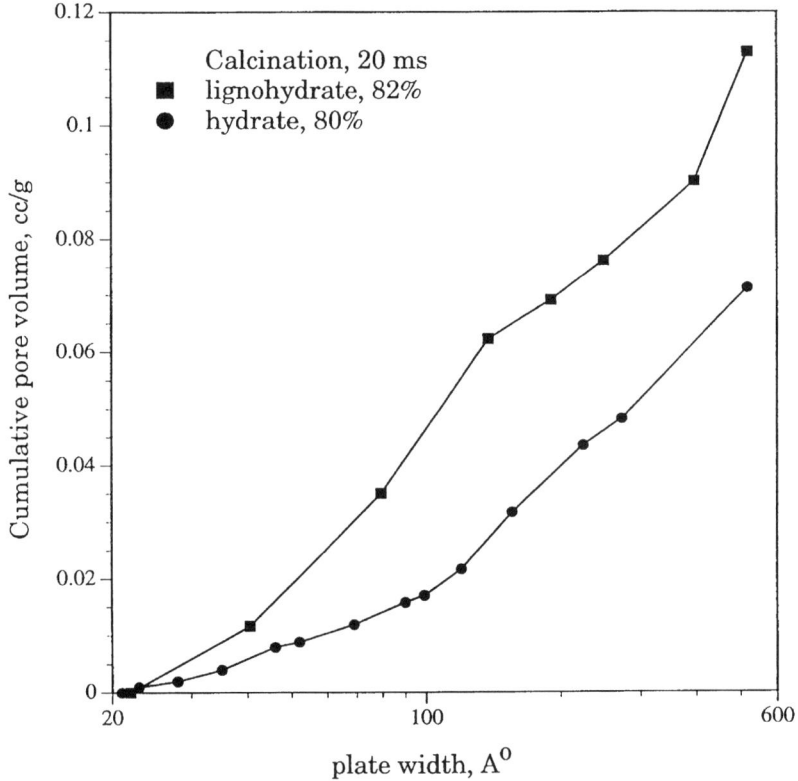

Figure 6.51 Pore volume distribution after 20 ms of calcination at 1080°C. Particle size: 3.9 µm.

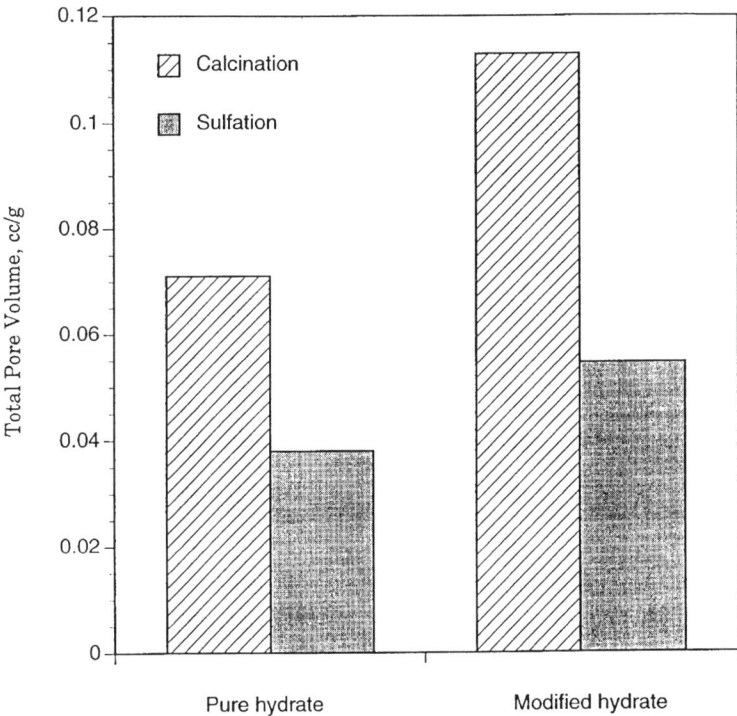

Figure 6.52 Comparison of pore volumes determined experimentally after 20 ms of sulfation and calcination at 1080°C.

Conclusions

The influence of internal structural properties, such as pore structure, surface area and porosity, and its interaction with reaction kinetics has been the main focus of this chapter. The high temperature interaction of SO_2 with calcium-based sorbents, though a subject of numerous investigations, cannot be said to have been completely understood. The three different phenomena of calcination, sintering, and sulfation, each with its own influence and interaction with the internal pore structure, make this a complicated and challenging problem to study. Furthermore, their ultrafast nature and thereby their concomitant occurrence under high temperature conditions adds more complexity. This research has contributed to a more thorough understanding of the various processes.

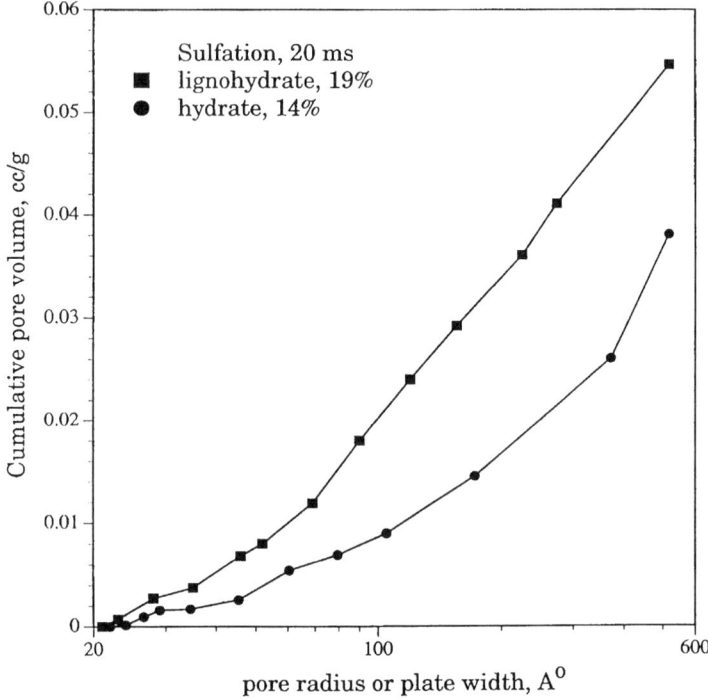

Figure 6.53 Pore volume distribution after 20 ms of sulfation at 1080°C. Particle size: 3.9 μm.

The $Ca(OH)_2$-SO_2 interaction phenomena has been investigated in two stages: first, the calcination and sintering alone in the absence of SO_2; and second, the combined calcination, sintering, and sulfation reaction in the presence of SO_2. While the ultrafast nature of the calcination reaction is well investigated in the available literature, an understanding of the influence of sintering under such short-contact time conditions was lacking. This research has experimentally investigated the enormous effect of sintering and has successfully modeled the kinetics and structural evolution of the calcine during ultrafast calcination and sintering. The mathematical model developed for small $Ca(OH)_2$ particles visualizes the CaO formed as micrograins surrounding the unreacted $Ca(OH)_2$, and takes into account the age distribution and varying levels of sintering of the CaO micrograins formed at various times. The extremely high initial calcination rate and the sudden attenuation and subsequent levelling off has been observed. A mechanism is proposed to explain this behavior based on the increased resistance to outward diffusion of H_2O due to the drastic sintering of the product CaO layers. The model explains and fits the experimental data very well at temperatures of 950°C and higher, but at lower temperatures it under-predicts the initial reaction rate.

In addition to surface area and porosity, the size distribution of the pores plays a crucial role in the kinetics and utilization. The investigations of transformations in the pore size distribution both in the absence and presence of SO_2 were carried out. During calcination and sintering, there is initially a generation of pores in the smallest size range of less than 50 A° (due to calcination). This is more prominent at temperatures lower than 1000°C, where calcination initially dominates sintering leading to increased porosity and generation of pores. However, as time progresses, sintering leads to continuous loss of pores, starting with the smallest pores first. This is enhanced at higher temperatures where the porosity reduces drastically and the small pores formed cannot sustain themselves due to an increased sintering rate. Moreover, sintering is shown to be more sensitive to temperature than the calcination reaction (due to its much higher activation energy). In the presence of SO_2, the preferential filling of the smallest pores by the higher molar volume product, $CaSO_4$, has been shown to take place. Further, the enhanced loss of overall pore volume over the entire size range is clearly demonstrated. This study has shed more light on the evolution of pores of various size ranges and the implications of the loss of the smallest pores.

The investigation and development of modified and improved (high surface area and porosity) calcium-based sorbents has been an important part of this research towards improving the sorbent reactivity and utilization. Investigations performed with the high surface area hydrate and a critical comparative investigation with unpromoted hydrate have been able to correlate the pore structure and reactivity. Previous researchers have acknowledged the importance of the optimum size range of pores but a definite knowledge of the "best" range and the advantage to be derived had been missing. Pores smaller than 50 A diameter are likely to be quickly filled or plugged with sulfation and pores larger than 200 A in diameter possess much smaller surface area to contribute significantly to sulfation rate. The modified hydrate is shown to be superior to the low surface area hydrate but only in the initial 50 ms or so. The modified hydrate is observed to have a very high sintering rate and to lose most of its high surface area and porosity within the first 50 to 100 ms, after which it behaves very similarly to the unpromoted hydrate. However, the advantage derived in the first 50 ms or so is retained by the modified hydrate, leading to its higher ultimate utilization. The results indicate about a 20% improvement in utilization of the modified hydrate compared to the pure. This chapter has laid the foundation and background for harnessing the sorbent pore structure and tailoring it to develop sorbents with very high reactivity.

This research work, with its time-resolved kinetic data and on-line measured residence time, has contributed to a more thorough understanding of the short-time SO_2/Ca interaction. However, there most certainly remain a number of uncomplicated tasks, shortcomings and improvements that warrant further investigation.

The calcination and sulfation reactions were studied for their inherent characteristics; however, under actual conditions, CO_2 and H_2O are major constituents of the flue gas. The presence of CO_2 and H_2O can not only influence the calcination of the sorbent thermodynamically (due to its reversible nature), they can also influence and accelerate the sintering characteristics. Hence, in order to accurately predict and model the reaction kinetics and pore structure transformations under actual boiler conditions, studies with simulated flue gas will be very useful.

The structural studies have clearly shown certain important transformations such as the preferential loss of small pore sizes and the effectiveness of pores of certain optimum size. However, an exact quantitative analysis of pore structure transformations has not been attempted. Hence, a necessary step towards that ultimate objective would be to gather more data on the pore structure transformations, while the model development is in progress. Such a model and its experimental validation would not only represent an important advancement in this field, it could also serve in the theoretical predictions of the sorbent tailoring work that is outlined below.

A simplified form of the sulfation model scheme has been used for prediction of experimental data. The sulfation model itself utilizes the predictions of the calcination and sintering model as its input. The combination of the sulfation model with the calcination and sintering model needs to be improved upon. These shortcomings may be largely responsible for the mediocre model prediction ability. Incorporating these improvements will be very useful in making the model more robust and comprehensive. Further, application of the model to predict the high reactivities of the high surface area carbonates could yield some interesting theoretical insights for future work.

Nomenclature

B	Constant to incorporate effect of H_2O partial pressure on sintering
C_p	Specific heat capacity of particle
C_s	$Ca(OH)_2$ concentration
C_{SO3}	Concentration of SO_3 on the product layer surface, in Equation 6.2
D_{AB}	Molecular diffusivity of H_2O in N_2
D_{eff}	Overall effective product layer diffusivity
D_k	Knudsen diffusivity
^aH_c	Heat of $Ca(OH)_2$ calcination reaction
h	Convective heat transfer coefficient
k_e	Specific reaction rate constant for sulfation during product layer
k_d	Specific reaction rate constant for sulfation during product layer diffusion control regime, in Equation 6.1
k_s	Specific rate constant for CaO sintering
k_{sm}	Modified specific rate constant for CaO sintering
m	Constant to incorporate effect of H_2O partial pressure on sintering
P	Partial pressure of H_2O in the CaO product layer
P_c	Partial pressure of H_2O at the $CaO/Ca(OH)_2$ interface
P_e	Equilibrium H_2O dissociation pressure for $Ca(OH)_2$ calcination
P_t	Total pressure
R	Universal gas constant
R_p	Particle radius
r_c	Instantaneous unreacted core radius
r_g	Instantaneous grain radius
r_{go}	Initial grain radius
S	Specific surface area of product CaO layer
S_a	Asymptotic specific surface area of CaO at a particular temperature
S_{CaO}	Specific surface area of CaO product shell
$S_{Ca(OH)2}$	Instantaneous specific surface area of $Ca(OH)_2$
S_o	Specific surface area of nascent CaO

$S_{j,i}$	Specific surface area of CaO formed in the j-th time interval at the end of i-th interval
S_s	Overall specific surface area of the particle
T	Reaction temperature
T_b	Bulk gas temperature
T_p	Particle temperature
t	Time
V_p	Particle temperature
x	Fractional conversion of $Ca(OH)_2$ to CaO
x_s	Fractional conversion of CaO to $CaSO_4$
Z_j	Fraction of total CaO formed in the j-th time interval

Greek Letters

$\Upsilon_{Ca(OH)}$	Rate of $Ca(OH)_2$ calcination reaction
ε	Product layer porosity
ε_o	Initial product layer porosity
ε_p	Emissivity factor of the particle
ρ	Overall solid density of the grain
ρ_p	Density of the particle
ρ_{CaO}	Density of CaO
$\rho_{Ca(OH)23}$	Density of $Ca(OH)_2$
σ	Stefan-Boltzman constant

Reference

1. Borgwardt, R. H., "Sintering of Nascent Calcium Oxide," *Chem. Eng. Sci.*, 44(1), 53 (1989)
2. Bortz, S. J. and P. Flamen, "Recent IFRF Fundamental and Pilot Scale Studies on the Direct Sorbent Injection Process," *Proceedings: First Joint Symposium on Dry SO_2 and Simultaneoustaneous SO_2/NO_x Control Technologies*, 1, EPA-600/9-85/020a, (NTIS PB85-232353) (1985)
3. Milne, C. R., G. D. Silcox, D. W. Pershing and D. A. Kirchgessner, "High-Temperature, Short-Time Sulfation of Calcium-Based Sorbents. 2. Experimental Data and Theoretical Model Predictions," *Ind. Eng. Chem. Res.*, 29(11), 2201 (1990)
4. Beruto, D. and A. W. Searcy, "Use of Langmuir Method for Kinetic Studies of Decomposition Reactions: Calcite ($CaCO_3$)," *J. Chem. Soc., Faraday Trans.*, 7(2), 145 (1974)
5. Powell, E. K. and D. W. Searcy, "The Rate and Activation Enthalpy of Decomposition of $CaCO_3$," *Metall. Trans.*, 11b, 427 (1980)
6. Borgwardt, R. H., "Calcination Kinetics and Surface Area of Dispersed Limestone Particles," *AIChE J.*, 31(1), 103 (1985)
7. Bortz, S. J., V. P. Roman, R. J. Yang, P. Flament and G. R. Offen, "Precalination and its Effect on Sorbent Utilization During Upper Furnace Injection," *Proceedings: Joint Symposium on Dry SO_2 and Simultaneous SO_2/NO_x Control Technologies*, 1, EPA-600/9-86-029a, (NTIS PB87-120465) (1986)
8. Cole, J. A., J. C. Kramlich, W. R. Seeker, G. D. Silcox, G. H. Newton, D. J. Harrison and D. W. Pershing, "Fundamental Studies on Sorbent Reactivity in Isothermal Reactors," *Proceedings: Joint Symposium on Dry SO_2 and Simultaneous SO_2/NO_x Control Technologies*, 1, EPA-600/9-86-029a, (NTIS PB87-120465) (1986)
9. Silcox, G. D., J. C. Kramlich and D. W. Pershing, "A Mathematical Model for the Flash Calcination of Dispersed $CaCO_3$ and $Ca(OH)_2$ Particles," *Ind. Eng. Chem. Res.*, 28(2), 155 (1989)
10. German, R. M. and Z. A. Munir, Surface Area "Reduction During Isothermal Sintering," *J. Am. Ceram. Soc.*, 59, 379 (1979)
11. Mai, M. C. and T. F. Edgar, "Surface Area Evolution of Calcium Hydroxide During Calcination and Sintering," *AIChE J.*, 35(1), 30 (1989)
12. Simons, G. A., A. R. Garman and A. A. Boni, "The Kinetic Rate of SO_2 Sorption by CaO," *AIChE J.*, 33(2), 211 (1987)
13. Borgwardt, R. H., K. R. Bruce and J. Blake, "An Investigation of Product-Layer Diffusivity for CaO Sulfation," *Ind. Eng. Chem. Res.*, 26, 1993 (1987)
14. Hsia, C., G. R. St. Pierre, K. Raghunathan and L.-S. Fan, "Diffusion Through $CaSO_4$ Formed During the Reaction of CaO with SO_2 and O_2," *AIChE J.*, 39(4), 698 (1993)
15. Milne, C. R. and D. W. Pershing, "Time Resolved Sulfation Rate Measurements for Sized Sorbents," *Proceedings: Fourth Annual Pittsburgh Coal Conference*, 109 (1987)
16. Gullett, B. K., J. A. Blom and G. R. Gillis, "Design and Characterization of a 1200°C Entrained Flow, Gas/ Solid Reactor," *Rev. Sci. Instrum.*, 59(9), 1980 (1988)
17. Borgwardt, R. H. and K. R. Bruce, "Effect of Specific Surface Area on the Reactivity of CaO with SO_2," *AIChE J.*, 32(2), 239 (1986)
18. Hartman, M. and R. W. Coughlin, *AIChE J.*, 22, 490 (1976)
19. Gullett, B. K. and K. R. Bruce, "Pore Distribution Changes of Calcium-Based Sorbents Reacting with Sulfur Dioxide," *AIChE J.*, 33, 1719 (1987)

20. Gullett, B. K. and J. A. Blom, Calcium "Hydroxide and Calcium Carbonate Particle Size Effects on Reactivity with Sulfur Dioxide," *Reactivity of Solids*, 3, 337 (1987)
21. Bruce, K. R, B. K. Gullet and L. O. Beach, "Comparative SO_2 Reactivity of CaO Derived from $CaCO_3$ and $Ca(OH)_2$," *AIChE J.*, 35(1), 37 (1989)
22. Hamor, R. J. and I. W. Smith, *Fuel*, 50(4), 374 (1971)
23. Gullett, B. K. and G. R. Gillis, *Powder Tech.*, 52, 257 (1987)
24. Sonnet, D., S. Afara, C. L. Briens, J. F. Large and M. A. Bergougnou, "Circulating Fluidized Bed Technology II," *Proceedings: Second International Conference on Circulating Fluidized Beds,* Compiegne, France, 565 (1988)
25. Holman, J. P., *Heat Transfer*, McGraw-Hill, NY (1972)
26. Alvfors, P. and G. Svedberg, "Modelling of the Simultaneous Calcination, Sintering and Sulphation of Limestone and Dolomite," *Chem. Eng. Sci.*, 47(8), 1903 (1992)
27. Gregg, S. J. and K. S. W. Sing, *Adsorption, Surface Area and Porosity*, Academic Press (1982)
28. Ranade, P. V. and D. P. Harrison, "The Variable Property Grain Model Applied to the Zinc Oxide-Hydrogen Sulfide Reaction," *Chem. Eng. Sci.*, 36, 1079 (1981)
29. nnes, W. B., "Use of Parallel Plate Model in Calculation of Pore Size Distribution," *Analytical Chemistry*, 29(7), 1069 (1957)
30. Bhatia, S. K. and D. D. Perlmutter, "A Random Pore Model for Fluid-Solid Reactions: II. Diffusion and Transport Effects," *AIChE J.* 27(2) (1981)
31. Bhatia, S. K. and D. D. Perlmutter, "Unified Treatment of Structural Effects in Fluid-Solid Reactions," *AIChE J.*, 29(2) (1981)
32. Ramachandran, P. A. and L. K. Doraiswamy, "Modeling of Non-Catalytic Gas-Solid Reactions," *AIChE J.*, 28(6), 881 (1982)
33. Szekely, J., J. W. Evans and H. Y. Sohn, *Gas-Solid Reactions*, Academic Press (1976)
34. Shen, J. and J. M. Smith, "Diffusional Effects in Gas-Solid Reactions," *Ind. Eng. Chem. Fund.*, 4, 293 (1963)
35. Rehmat, A. and S. C. Saxena, "Multiple Non-Isothermal Noncatalytic Gas-Solid Reactions: Effect of Changing Particle Size," *Ind. Eng. Chem. Proc. Des. Dev.*, 16, 502 (1977)
36. Evans, J. W., J. Szekely, W. H. Ray and Y. K. Chaung, "On the Optimum Temperature Progression for Irreversible Non-Catalytic Gas-Solid Reactions," *Chem. Eng. Sci.*, 28, 683 (1973)
37. Calvelo, A. and J. M. Smith, "Intrapellet Transport in Gas-Soild Non-Catalytic Reactions," *Chemeca Proceedings*, 3, 1 (1970)
38. Szekely, J. and J. W. Evans, "A Structural Model for Gas-Solid Reactions with a Moving Boundary-II. The Effect of Grain Size, Porosity and Temperature on the Reaction of Porous Pellets," *Chem. Eng. Sci.*, 26, 1901 (1971)
39. Georgakis, C., C. W. Chang and J. A. Szekely, "A Changing Grain Size Model for Gas-Solid Reactions, *Chem. Eng. Sci.*, 34, 1072 (1979)
40. Bhatia, S. K. and D. D. Perlmutter, "The Effect of Pore Structure on Fluid-Solid Reactions: Application to the SO_2-Lime Reaction," *AIChE J.*, 27(2) (1981)
41. Christman, P. G. and T. F. Edgar, "Distributed Pore-Size Model for Sulfation of Limestone," *AIChE J.*, 29(3) (1983)
42. Hartman, M. and A. Martinovsky, "Thermal Stability of the Magnesian and Calcareous Compounds for Desulfurization Processes," *Chem. Eng. Commun.*, 111, 149 (1992)
43. Darroudi, T. and A. Searcy, "Effect of CO_2 Pressure on the Rate of Decomposition of Calcites," *J. Phy. Chem.*, 85, 3971 (1981)
44. Smith, J. M., *Chemical Engineering Kinetics*, 3rd ed., McGraw-Hill, NY (1981)

45. Nicholson, D., "Variation of Surface Area During the Decomposition of Solids," *Trans. Faraday Soc.*, 61, 990 (1965)
46. Bhatia, S. K. and D. D. Perlmutter, "A Random Pore Model for Fluid-Solid Reactions: I. Isothermal, Kinetic Control," *AIChE J.*, 26(3) (1980)
47. Christman, P. G. and T. F. Edgar, "Distributed Pore-Size Model for Sulfation of Limestone," *AIChe J.*, 29(3) (1983)
48. Kirchgessner, D. A. and W. Josewicz, "Enhancement of Reactivity in Surfactant-Modified Sorbents for Sulfur Dioxide Control," *Ind. Eng. Chem. Res.*, 28(4), 413-418 (1989)

CHAPTER 7

KINETIC STUDIES ON THE MEDIUM TEMPERATURE $Ca(OH)_2$ SORBENT INJECTION FGD PROCESS

Soon-Jai Khang, Timothy C. Keener*
Anbo Wang, Zhenwei Wang
Department of Chemical Engineering
*Department of Civil &
Environmental Engineering
University of Cincinnati

Abstract

Recent progress in dry sorbent injection flue gas desulfurization (FGD) has resulted in a medium-temperature sorbent injection process. Compared to conventional furnace injection, this new process exhibits equivalent SO_2 removal performance at a much lower operation temperature, allowing the injection point to move to a lower temperature region downstream of a furnace.

Considering dehydration and carbonation reactions would take place simultaneously with sulfation reaction when a $Ca(OH)_2$ particle is injected into a flue gas stream at medium temperatures, the experimental and kinetic analysis of each of the possible reactions were conducted independently over a broad spectrum of reaction conditions. Reaction conversions for the solitary carbonation and the solitary sulfation reactions in a temperature range of 600-1100°F and with residence times within 1000 milliseconds were determined. Mathematical models simulating the solitary carbonation and the solitary sulfation processes were developed based on the experimental data obtained. The activation energies and frequency factors for these reactions have been calculated.

Following the solitary reaction studies, experimental tests on the simultaneous dehydration, sulfation, and carbonation reactions of $Ca(OH)_2$ were also conducted. The carbonation and sulfation conversions with a simulated flue gas in the reaction temperature range of 600-1100°F and residence time within 100 ms are presented.

Based on the solitary reaction models, a comprehensive mathematical model was developed to describe the medium-temperature $Ca(OH)_2$ sorbent injection process. This model incorporates the simultaneous dehydration, sulfation, and carbonation of $Ca(OH)_2$, as well as the further sulfation of the newly formed $CaCO_3$. The deactivation of sorbent caused by the loss of porosity for reactant gas diffusion was considered by introducing time-dependent activities which are commonly used to describe a catalyst deactivation process. The model was validated for both the sulfation and carbonation conversion data at the studied reaction temperatures. The model-predicted conversions were shown to be in good agreement with the experimental data.

It was found that the competing carbonation reaction was the major factor affecting the SO_2 removal performance of a medium-temperature sorbent injection process. The simulation results showed that the carbonation reaction affected the sulfation reaction behavior by means of two important mechanisms. First, it adds additional diffusion resistance to SO_2 gas because of the formation of $CaCO_3$. The second mechanism is the further sulfation of the newly formed $CaCO_3$, which could partially compensate for the negative effect of the competing carbonation reaction.

Introduction

Background

A large number of flue gas desulfurization (FGD) processes have been developed and are expected to play an important role in reducing SO_2 emission from power plants. One new development is the dry scrubbing process. The process has been established as an alternate approach for achieving moderate levels of SO_2 reduction in coal-burning utility boilers. It is attractive for retrofit of existing boilers since the capital equipment requirements and overall sulfur reduction costs per ton of sulfur removed are less than for most other options such as wet FGD.

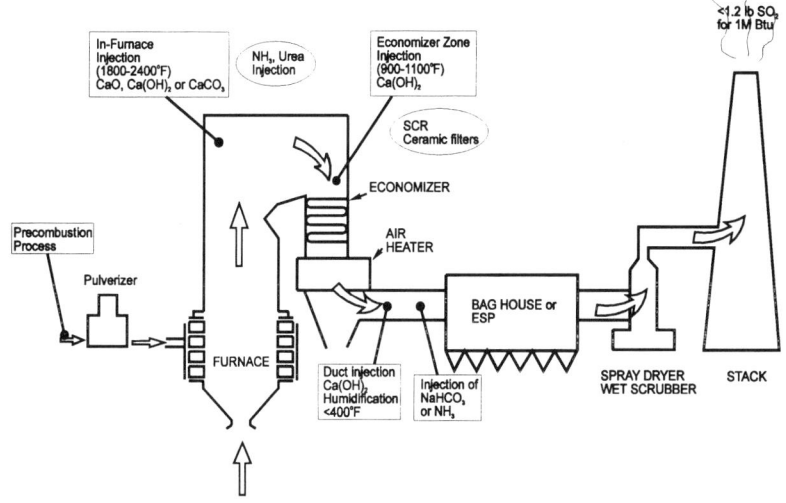

Figure 7.1 Dry SO$_2$ removal processes.

In dry SO$_2$ removal processes, the sorbent, usually calcium hydroxide or lime, is injected and reacts with SO$_2$ to form a dry product which is removed in the particulate collector. As shown in Figure 7.1, three general approaches are available or under development for bringing the sorbent into contact with the SO$_2$: furnace sorbent injection, economizer sorbent injection, and duct sorbent injection.

These three approaches correspond to three temperature windows that are available for SO$_2$ removal by dry sorbent injection. They are furnace injection at a temperature around 2000-2300°F (1100-1250°C) using a variety of calcium-based sorbents, economizer injection between 800 and 1000°F (450-550°C) using hydrates, and duct injection into the gas passage leading to the particulate collector, either at temperatures approaching within 9-27°F (5-15°C) of adiabatic saturation for hydrates or at 270-360°F (130-180°C) for sodium sorbents.

Furnace Sorbent Injection. Furnace sorbent injection has been established as an alternate approach for achieving moderate levels of SO$_2$ reduction on coal-burning utility boilers. The process is attractive for retrofit of existing boilers since the capital equipment requirements and overall sulfur reduction costs per ton of sulfur removed are less than for other options such as some wet scrubbing FGD processes. In the furnace sorbent injection process, sorbent, usually calcium hydroxide or limestone (CaCO$_3$), is injected above the flame zone and mixes with the combustion products containing SO$_2$. The SO$_2$ reacts chemically with lime (CaO), formed from the calcination and dehydration of sorbent, to produce solid calcium sulfate ($CaSO_4$). The $CaSO_4$ is removed with fly ash in particulate collection equipment.

Furnace sorbent injection technology has been tested since the late 1960s. Most recently, the U.S. EPA initiated research on a limestone injection multistage burner (LIMB) process. In the LIMB process, dry sorbent is injected, along with the coal, directly into the multistage burners. The process was demonstrated at Ohio Edison's Edgewater Station on a 105-MW unit in 1987 [1-3]. Although the LIMB process achieves removal efficiencies lower than other flue gas scrubbers, it is easily retrofitted to existing boilers and its capital investment is relatively low. With a calcium-to-sulfur ratio of approximately two, SO_2 removal in the furnace ranging from 55% to 72% was achieved. The removal was dependent on the specific sorbent utilized and the degree of humidification employed. The potential market for furnace sorbent injection includes all pre-1971 coal burning utility boilers, primarily those burning eastern and midwestern bituminous coals.

DUCT SORBENT INJECTION. Similar to furnace injection systems, duct injection systems use the duct between the boiler outlet and dust collector inlet to capture SO_2 with either lime- or sodium-based compounds. The alkali can be injected dry or as a slurry or solution, but in any case, requires humidification to be effective. It also requires either a baghouse or relatively large precipitator to collect the increased particulates.

Duct injection systems have undergone extensive testing; however, there is no large commercial installation currently in operation in the U.S. Testing has been done with both sodium salts and lime, with the former being much more effective for SO_2 removal. Under normal operating conditions, the duct injection system is capable of approximately 50% SO_2 removal with a medium-sulfur coal. Again, the system is highly dependent upon the amount of humidification and how close one reduces the gas temperature towards saturation. With a very close approach to saturation, the SO_2 removal efficiency can be increased to as high as 70%. The various duct injection processes under development are summarized in Table 7.1.

To use duct injection technology on an existing unit, it is essential either to have a long straight run of flues, or to modify the flows and install hoppers as necessary, as well as to have an oversized precipitator or baghouse to accommodate the increased dust loading. From a capital standpoint, this system is similar to a furnace injection system except that it eliminates the expense of furnace modifications, and makes all of the modifications necessary downstream of the boiler. It tends to be less capital-intensive than a furnace injection system. It must, however, use either lime or sodium salts as a reagent, since limestone cannot be used.

Table 7.1. Duct Injection Processes

Organization	Process	Configuration
Conoco [5]	Coolside	Hydrate injection into humidified flue gas (downstream of water sprays with additives).
Bechtel [6,7]	Confined zone dispersion (CZD)	Injection of a finely atomized slurry of lime and water into duct through dual-fluid atomizing nozzles. An outlet temperature of 160°F (71°C) is maintained with an approach to saturation of 35°F (19.4°C).
General Electric Environ. Service [8,9]	Lime based in-duct	Injection of lime slurry with a rotary atomizer into the duct. In a 50,000 acfm pilot plant burning high-sulfur coal, a 50% SO_2 removal was obtained with a Ca/S ratio of 1.5.
Dravo [8,10]	Hydrate addition at low temperature (HALT)	Hydrate injection with direct contact of the water spray and the hydrate. For Ca/S ratios near 2, 50% removal was achieved for an approach to saturation temperature of 20-30°F (11-16.7°C).
EPRI	Hybrid pollution abatement system (HYPAS)	Initial removal of fly ash, humidification, and sorbent injection ahead of a second particulate collector.
EPA	ESOX	Use of the first field of an ESP as a spray dryer.

ECONOMIZER SORBENT INJECTION. Recently, a temperature window around 1000°F (550°C) was discovered that yielded SO_2 capture rates similar to furnace injection with commercial hydrates and significantly better performance with specially prepared sorbents [4]. This encouraging finding established a new process, the economizer sorbent injection process. In this process, hydrated sorbent is injected into a temperature envelope centered around 1000°F (550°C), which is generally found prior to the economizer section, to react with SO_2. The desulfurization product is removed from the flue gas stream in a particulate collector. Calcium hydroxide injected into the economizer reacts with SO_2 as follows.

$$Ca(OH)_2(solid) + SO_2(gas) \rightarrow CaSO_3(solid) + H_2O(gas)$$

In contrast to furnace injection processes where the reaction temperature is around 2000°F (1100°C), $Ca(OH)_2$ reacts directly with SO_2 in an economizer injection process since the temperature in the process is too low for a complete dehydration reaction. In this temperature range, the main product is $CaSO_3$ instead of $CaSO_4$.

From the standpoints of capital investment and operating cost, an economizer injection process is similar to a furnace injection process. The sulfation reaction rate in this temperature range is comparable or even higher than that at 2000°F (1100°C). Since sulfite formation is very fast (<250 ms) and the window is approximately 180°F (100°C) wide, the process is compatible with the high quench rates, typically 900-1080°F/s (500-600°C/s), found through economizers.

This intermediate temperature process appears to have a prospect for further increases in utilization. Since sintering is not an issue at such low temperatures, it should be possible to achieve a higher sorbent utilization by optimizing the process. In terms of engineering considerations, there are two other advantages of the new process over furnace injection process.

First, the sorbent injection location at 1000°F (550°C) is typically just above or in the economizer section, eliminating any possible negative impact of sorbent injection on the radiant heat exchange.

Second, the relatively low injection temperature should allow the insertion of simple sorbent injectors into a flow stream and good mixing should be more easily achieved.

In spite of these advantages, there is no large commercial installation currently in operation in the U.S. Further extensive testing is needed in search of approaches that further enhance sulfur dioxide capture, and sufficient process understanding to prepare design guidelines.

OBJECTIVE. The objective of this project is to study the fundamental mechanism of the interaction between flue gas and calcium sorbent during a medium temperature sorbent injection process (or an economizer injection process).

In order to understand the interaction between flue gas and sorbent when reactions occur simultaneously, experimental and kinetic studies for each of the possible reactions have been conducted independently over a broad spectrum of reaction conditions. It is expected that a better knowledge of the relationship between these reactions will lead to a design and an operation procedure of a more efficient dry FGD process.

This research is divided into six parts. The first part describes the design, construction and operation of two entrained-flow reactors. The following parts investigate each of the possible reactions which may occur in medium-temperature range. In an economizer injection process, the major chemical reactions which may occur are

$Ca(OH)_2(solid) + SO_2(gas) = CaSO_3(solid) + H_2O(gas)$

$Ca(OH)_2(solid) + CO_2(gas) = CaCO_3(solid) + H_2O(gas)$

$Ca(OH)_2(solid) + heat = CaO(solid) + H_2O(gas)$

$CaO(solid) + SO_2(gas) = CaSO_3(solid)$

and

$CaO(solid) + CO_2(gas) = CaCO_3(solid)$

As soon as hydrated lime is injected into flue gas, $Ca(OH)_2$ will immediately react with SO_2 and begin to dehydrate simultaneously. The CaO formed from the dehydration reaction may react further with SO_2. In addition, the $Ca(OH)_2$ and CaO may also react with CO_2 in the flue gas to form $CaCO_3$. The experimental results and the kinetic studies of the dehydration, carbonation, sulfation and simultaneous carbonation-sulfation reactions are presented in the later parts.

Entrained Flow Reactor

Introduction

Under a typical medium-temperature injection condition, the reactions between flue gas and sorbent particles are extremely fast, on the order of milliseconds. Short contacting time, entrained flow reactors were developed in this research to study these fast gas/solid reactions.

Entrained flow reactor systems are commonly used for studying gas/solid reaction kinetics due to their advantage of minimizing interparticle effects. However, the design of an entrained flow gas/solid reactor is often difficult due to the need to fulfil the global kinetic study requirements.

Because of the dramatic temperature effects on reactions, solid particles must be maintained at low temperature until injected into the reactor. Once injected, the sorbent particles should be heated to the reactor temperature and equilibrate with the gas phase in the reactor very rapidly.

In addition, after a short period of reaction the collected particles have to be cooled rapidly in an inert gas environment to prevent any possible additional reactions. For higher temperatures, isothermal conditions are difficult to obtain. Besides, incomplete mixing of solid and gaseous reactants and problems in sampling the solid product may also compound the operational problems. Finally, because of the sticky nature of fine sorbent powders, additional complications are encountered in smooth feeding and dispersion of the powder into a reactor.

A few studies in the past have been attempted by using an entrained flow reactor to obtain time-resolved measurements for SO_2 removal. The tests were performed at a high temperature in a short reaction time within a ms scale. These entrained flow reactors designed and used for research include electrically heated drop-tube furnaces with natural gas precombustors [11-13], electrically heated furnaces [14], and natural-gas-fired furnaces [15]. Problems with these reactors include difficulty in maintaining constant temperatures or the direct uses of combustion gases as the reactant gas may affect the accuracy of the measurement for the gas/solid reaction of interest.

Two kinds of entrained flow reactors were designed and constructed in this research to investigate the short time interaction between flue gas and sorbent in the medium temperature range. The first setup, referred to as entrained flow reactor I in this research, was used to study reactions with residence times over 500 ms. The second setup, referred to as entrained flow reactor II, was used to conduct the investigation on the reactions for times less than 100 ms. Entrained flow reactor II was constructed by modifying entrained flow reactor I.

Entrained Flow Reactor Construction and Operation

The system is schematically shown in Figure 7.2. It consists of six parts: a reaction chamber, a preheater, a heat exchanger, a sorbent feeder, a liquid N_2 quencher, and a product sorbent collecting system.

The reaction chamber is a 3-foot long by 1.5 inch inside diameter quartz tube heated in a three-zone Lindberg tube furnace. With three independently controlled heating zones, the tube furnace is capable of maintaining a uniform temperature up to 2200°F (1204°C). Sorbent is fed to the top of the reaction chamber and entrained by reactant gas to move downward to the bottom. As soon as sorbent particles leave the reaction chamber, they are quenched quickly by a liquid N_2 quencher to about 350°F (177°C). The mean residence time of sorbent particles in the reaction zone can be controlled to a range of 0.5 to 1.5 seconds by varying the reactant gas velocity.

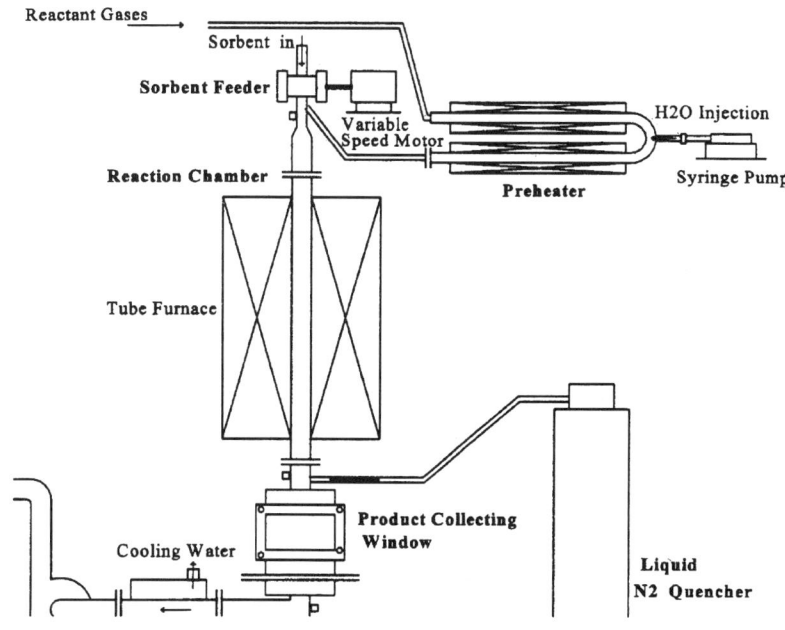

Figure 7.2 Schematic diagram of entrained flow reactor I.

A preheater is used to preheat the reactant gases to a temperature around 350°F (177°C). Pre-mixed multicomponent gas cylinders are used as the reactant gas to eliminate any possible concentration change during tests. Water is injected by a syringe pump in the middle section of the preheater where the gas temperature is above the boiling point of water. A heat exchanger is used to cool product gases before entering the outlet pump. Temperatures inside the reaction chamber, the preheaters, the product particle collector and the heat exchanger are monitored by eight type-K thermocouples which are connected to a computer-based data acquisition system.

Sorbents are fed into the reactor by a rotary solid particle feeder. The detailed structure of the sorbent feeder is shown in Figure 7.3. The sorbent feeder consists of two parts: a four-vane, closed-end rotor and the rotor house. The rotor is driven by a computer-controlled variable speed motor which controls the sorbent feed rate during tests. A gradually divergent section connects the sorbent feeder and reaction chamber. This section prevents back-mixing but allows maintenance of a uniform velocity profile and ensures that the sorbent is well dispersed in the reaction chamber.

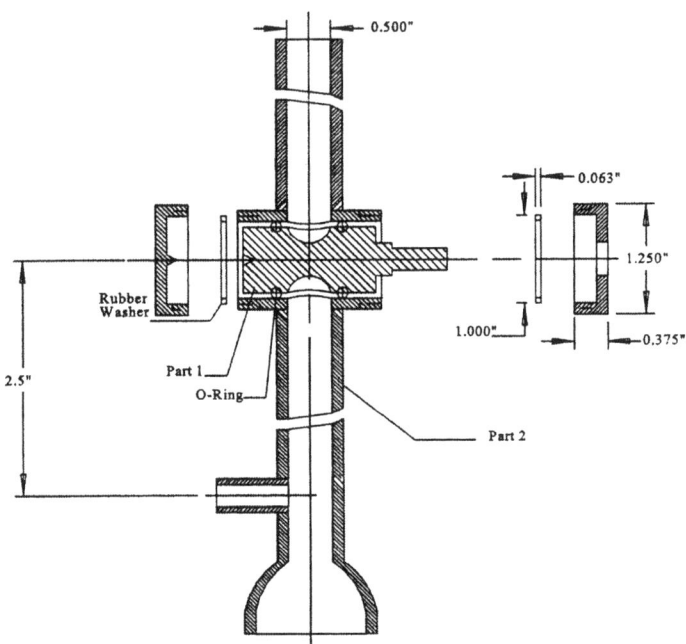

Figure 7.3 Detail structure of the sorbent feeder for entrained flow reactor I.

The collecting filter, which is made up of a paper filter on a stainless steel screen, is located between two parts of the sample collecting chamber. The unloading window on the sample collecting chamber makes it convenient to unload sample products after each test. The product gas through the collection section is exhausted by a 2-3/4 horsepower exhaust fan.

The objective of the above experimental setup is to obtain a short residence time for the sorbent particles and a quick quench by using the liquid nitrogen to stop the reaction at any stage. To avoid a significant reactant gas concentration change through the reaction chamber caused by the gas-solid reaction, the sorbent particle concentration in the gases is kept very low (less then 3 g/m^3). An isothermal condition throughout the reaction chamber was also maintained. As illustrated in Figure 7.4, the temperature measurement inside the reactor with three type-K thermocouples showed uniformity of temperature within ± 2°C throughout the reaction chamber.

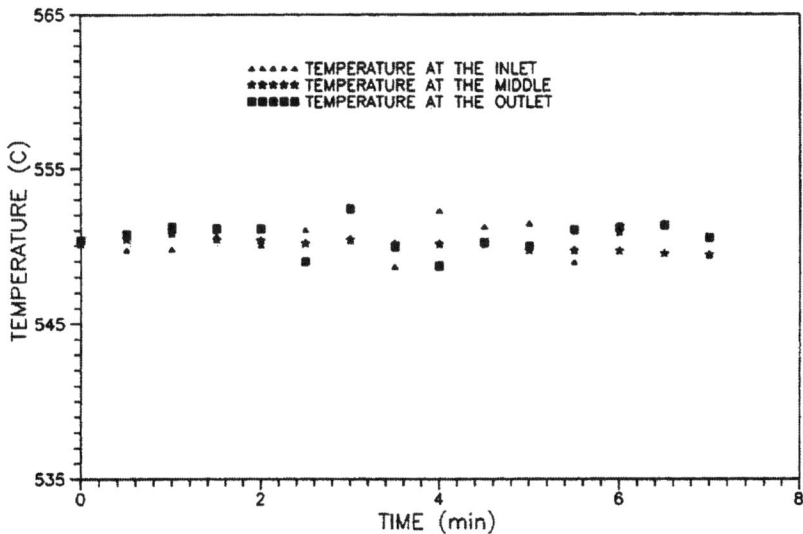

Figure 7.4 Temperature in the reaction chamber of entrained flow reactor I.

Entrained Flow Reactor II

Test results in entrained flow reactor I revealed that the reaction between flue gas and sorbent did not progress significantly after 500 ms. This emphasized the importance of the reaction mechanism within the first 500 ms.

Since the diameter of the reaction chamber of entrained flow reactor I used for the first part of research was relatively large, testing below 500 ms required a prohibitively large amount of reactant gas. To rectify this situation, modifications were made to entrained flow reactor I.

In the entrained flow reactor II system, the diameter of the reaction chamber was reduced from 1.5 inches, requiring on 1/9 of the previous reactant gas. Due to the smaller gas flow, the prehater and sorbent feeder were modified as well.

The entrained flow reactor II system is show schematically in Figure 7.5. It consists of a furnace, a reactor with jacketed preheating and reaction tubes, a heat exchanger, a sorbent feeder, a N_2 gas quenching and diluting units, and a sorbent collecting unit.

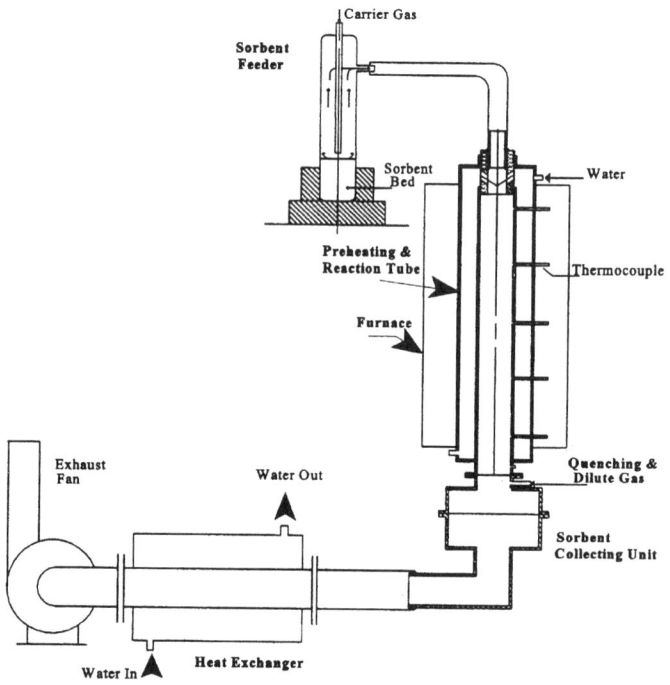

Figure 7.5 Schematic diagram of the entrained flow reactor II.

The reactor consists of two concentric stainless steel pipes (0.5 and 1.5 inch I.D. respectively) which are housed within a tube furnace. The furnace is a 3-feet-long, 3-inch-I.D. Lindberg model 55247-4, with three independently controlled heating zones, capable of maintaining a temperature up to 1200°C. Reactant gas is preheated in the annular space between the outer (1.5 inch I.D.) shell and the inner (0.5 inch I.D.) reactor tube.

The reactant gas enters the bottom of the reactor and travels upward through the preheating zone to the top of the reactor and then makes a 120° turn to enter the inner tube, forming two hot jets. The details of the entrance block design are shown in Figure 7.6. At the top of the reactor, sorbent particles are injected into the reactor just above the incoming hot jets by means of the sorbent feeder described below. Thus, the sorbent stream is impacted by the hot reactant gas jets. This jet impaction technique, successfully employed by Sonnet et all [16], causes a substantial loss of momentum which creates severe turbulence and in turn causes rapid heating of the sorbents. Heat transfer calculations indicated that sorbent particles of less the 10 μm would be heated within a few ms after injection into the reactor, providing a well-mixed isothermal mixture of the gas and solid particles. The gas and the entrained particles travel downward through the isothermal reactor tube and as soon as the sorbent particles leave the reaction chamber, they are quickly quenched and diluted by a large amount of nitrogen gas (in a ratio of dilute gas/reactant gas up to 10) supplied from an N_2 cylinder. The particles are collected with a paper filter. The temperature of the exhausted gas through the paper filter is maintained around 45°C and continuously monitored.

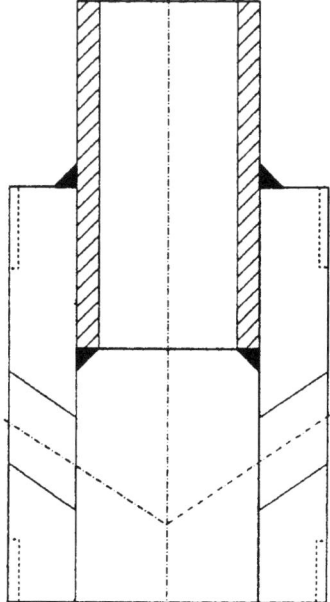

Figure 7.6 Detailed structure of the entrance block design.

In a fashion similar to that in entrained flow reactor I, water is injected by a syringe pump at the stop of the preheater where the gas temperature is kept above the boiling point of the water. To monitor the temperature profile inside the reactor, five type-K thermocouples were inserted into the reactor chamber through the stainless steel thermocouple ports. The signals from the thermocouples are sent to a computer data acquisition system.

The sorbent particles used in the study, due to their small particle size (about 10 μm) and adhesive nature, are very difficult to feed and transport steadily. A modified sorbent feeder was designed to provide a reliable feeding condition.

The schematic diagram of the sorbent feeding system is shown in Figure 7.7. The sorbent particles are entrained by the carrier gas, which flows over the surface of a sorbent bed and discharges out of a side-branch tube. The sorbent feeding rate is controlled by changing the carrier gas flow rate and the position of the carrier gas tube. A series of preliminary tests indicated that a stable flow of the sorbent particles can be obtained with this type of sorbent feeder for small sorbent particles. During the tests, the particle feeder was capable of injecting sorbent as small as 2 μm at a rate of 250 mg/min. The gas flow through the reactor was varied from 15 to 50 L/min to control particle residence time from 50 to 100 ms. The reactor temperature was controlled in the range 500 to 1100°F.

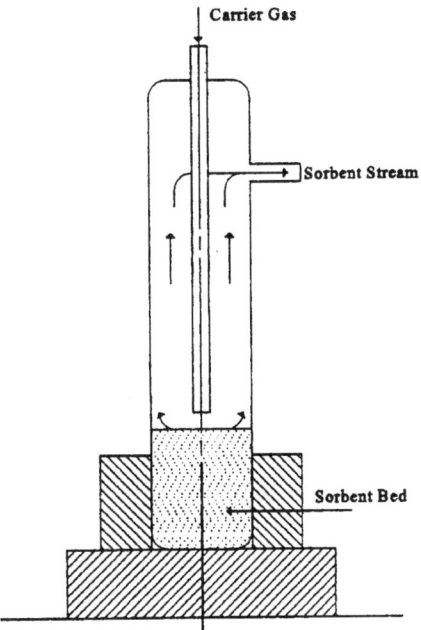

Figure 7.7 .Schematic diagram of the sorbent feeder for entrained flow reactor II.

Particle Heat Transfer Rate Determination

A knowledge of the transient temperature increase of the sorbent particle after injection into the entrained flow reactor is important for creating an appropriate design, selecting operating conditions to ensure rapid heating, and guaranteeing the accuracy of data. With simplifying assumptions, heat transfer calculations were carried out on sorbent particles subjected to a step increase in temperature.

Let us consider a $Ca(OH)_2$ practice initially at room temperature T_o. At time t=0, this particle is introduced into hot nitrogen gas where the bulk gas temperature is T_b. The sorbent particle is assumed to be spherical and its surrounding gas is taken to be stagnant. Heat is transferred from the bulk gas to the particle mainly by convection as soon as the particle is injected into the hot gas stream. Radiation from the reactor wall (assumed to also be at T_b) to the particle also contributes to the overall heat transfer. All the heats of reaction during the initial heating period are neglected.

The internal temperature gradients in the particle are assumed to be insignificant based on the fact that the Biot number for the particles of interest are small ($B_{io}=hr_p/k_p=0.17$). A heat balance accounting for radiation and convection results in a nonlinear differential equation.

Equation 7.1

$$\alpha = \frac{dT_p}{dt} = h(T_b - T_p) + \sigma\varepsilon(T_b^4 - T_p^4)$$

where

$$\alpha = \frac{\left(\frac{4}{3}\pi r_p^3\right)\rho C_p}{\left(\frac{4}{3}\pi r_p^2\right)} = \frac{r_p \rho C_p}{3}$$

T_p, r_p, h, σ,ε,C_p and ρ are the particle temperature (K), the heating time (s), the particle radios (m), the convective heat transfer coefficient (W/m²K), the Stefan-Boltzman constant (5.669•10⁻⁸ W/m²K⁴), the emissivity of $Ca(OH)_2$, the heat capacity of $Ca(OH)_2$, and the density of particle, respectively.

Equation 7.1 was solved using Gear's method to determine the temperature history of sorbent particles after injection. The fixed parameters needed to solve the equation have been chosen from the literature and are summarized in Table 7.2. The results for T_b=800°F (427°C) and T_b=1000°F (438°C) are shown in Figure 7.8 and Figure 7.9. The results indicate that for d_p=5 and 10 µm, the particles reach 99% of the reactor temperature within one ms. The typical particle size used in this research is 10 µm or less.

Table 7.2 Heat Transfer Parameters Obtained from the Literature

Parameter	Value	Sources
Thermal conductivity of bulk gas (k_g)	0.0647 (W/m•K)	[17, 18]
Thermal conductivity of particle (k_p)	0.751 (W/m•K)	[7]
Heat transfer coefficient (h)	h-k_gNu/r_p Nu=2	[8,19]
Heat capacity of particle (C_p)	1208.0 (J/kg•K)	[7]

Table 7.2 Heat Transfer Parameters Obtained from the Literature

Parameter	Value	Sources
Particle density (ρ)	$2.2 \cdot 10^3$ (kg/m^3)	[7,20]
Emissivity of particle (ε)	0.5	[21,22]

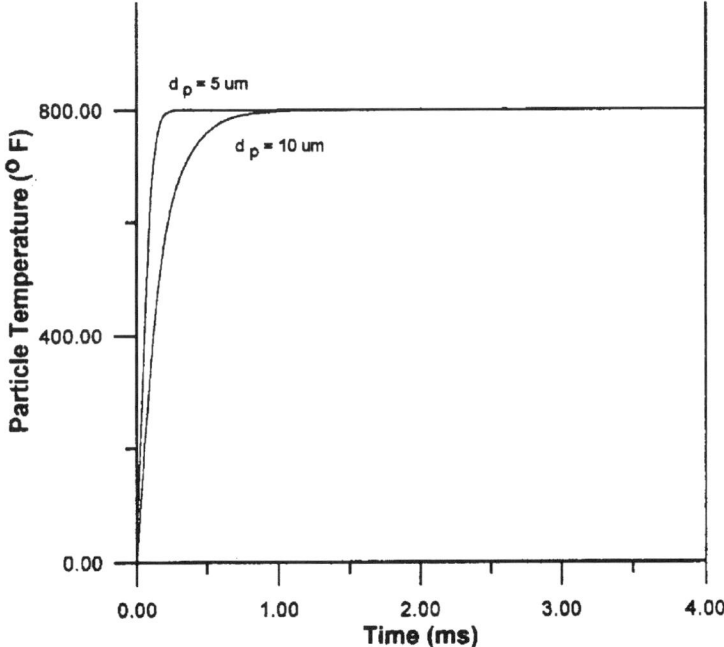

Figure 7.8 Temperature increase of a sorbent particle injected into a furnace at 800°F (427°C).

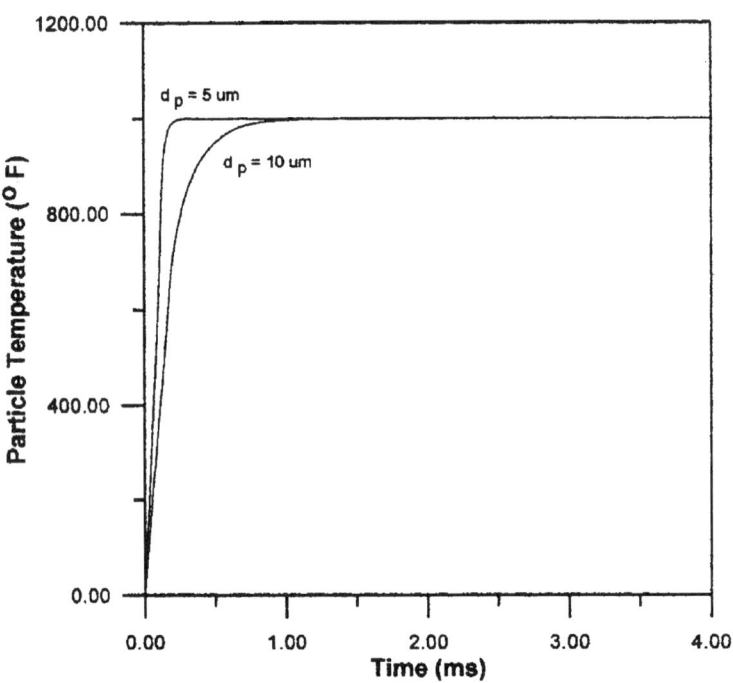

Figure 7.9 Temperature increase of a sorbent particle injected into a furnace at 1000°F (538°C).

In practice, the heat transfer rate through the gas film is much higher due to the fact that the surrounding gas is not stagnant. Also, for a given mass of sorbent particle, the surface area is at a minimum for a spherical geometry and the sorbent particles are far from being spherical. The simplifying assumptions cause the heating rate to be underestimated. On the other hand, ignoring the reaction effects during heat up may tend to cause the heating rate to be overestimated. On the whole, the above calculations indicate that as long as the gas phase in a gas jet entering the reactor is mixed rapidly with the hot reactant gases, the particle attains the reactor temperature quickly, and kinetic data measured for residence times of 10 ms or more are not influenced by the particle heat-up time.

After leaving the reaction chamber, sorbent particles are quickly quenched and diluted by a large amount of nitrogen gas (in a ratio of dilute gas/reactant gas up to 10) supplied from N_2 cylinder. Calculations for cooling rates are similar to those of heating rates and, hence, quenching should also be achieved within one ms.

Particle Residence Time Determination

The residence time of the sorbent particles inside the reaction chamber was estimated by a fluid dynamics model. For the particle size of entrained-flow reactor test (d_p<10 μm), the calculation results show that the velocity difference between the gas and the particle was negligible. Therefore, the particle residence time was safely assumed to be practically equal to the gas residence time.

Dehydration Reaction of Ca(OH)$_2$

Dehydration is an important step in a hydroxide sorbent injection process, especially under medium temperature conditions where the dehydration of Ca(OH)$_2$ takes place simultaneously with the sulfation reaction. The dehydration reaction produces a more porous material, CaO, which may change the activity of the original sorbent to SO$_2$.

Literature Review

The dehydration of calcium hydroxide is an endothermic decomposition reaction which takes place when solid Ca(OH)$_2$ is subjected to heat. The reaction products are solid calcium oxide and water vapor. It can be represented by the following equation.

$$Ca(OH)_2(s) \rightarrow CaO(s) + H_2O(g)$$

Some research has been conducted to study the dehydration of calcium hydroxide. Hartman et al. [23,24] studied the equilibria of the dehydration reaction of Ca(OH)$_2$. Using a standard thermochemical approach, the equilibrium constant, or equilibrium partial pressure, was expressed algebraically as a function of temperature as

Equation 7.2

$$Ln P_{H_2O} = -\frac{13,151.1}{T} + 20.7912$$

where P_{H_2O} (kPa) is the equilibrium partial pressure of water vapor and T(K) is the temperature of the calcium hydroxide particle.

The removal behavior of the desulfurization process depends on the initial state of the dehydration. According to the equilibria data shown in Equation 7.2, the dehydration of calcium hydroxide starts up at medium temperatures, from 660°F to 1110°F (350 to 600°C), depending on the concentration of water vapor in the gas phase. The typical flue gas composition, where the water vapor concentration is about 8%, the decomposition temperature calculated from Equation 7.2 is around 806°F (430°C).

The kinetics of the dehydration reaction of calcium hydroxide have been studied by other researchers. Criado and Morales [25] studied the mechanism of the dehydration of calcium hydroxide and Mu and Perlmutter [26] reported the thermal dehydration kinetics of a high-purity calcium hydroxide (98% by weight) under nonisothermal conditions. The results were extremely sensitive to heating rates and decomposition temperatures (617-779°F or 325-415°C). Dehydration kinetics were obtained at the very slow heating rate of 1.8°F/min (1°C/min).

Mu and Perlmutter's results were fitted to a reaction control shrinking core model [26] shown in the equation below. The activation energy derived from their test data is 180.0±1.25 kJ•mol^{-1}. The frequency factor is 1.94•10^{11}s^{-1}. The standard error of the model fit is 4.7%.

$$\frac{dX}{dt} = k_0 \exp\left(\frac{-E}{RT}\right)(1-X)^{\frac{2}{3}}$$

Criado and Morales [25] suggested a pseudo homogeneous model for their dehydration tests.

$$\frac{dX}{dt} = k_0 \exp\left(\frac{-E}{RT}\right)(1-X)$$

The obtained activation energy (E=116 kJ•mol^{-1}) was close to the heat of decomposition, which is 109.28 kJ•mol^{-1}.

The kinetics of the dehydration reaction of calcium hydroxide in temperature range 625°-824°F (330-450°C) was recently reported by Irbien et al. [27,28]. A Perkin-Elmer TGA-7 Thermogravemetric Analyzer was used to study two types of calcium hydroxide: the commercial calcium hydroxide which contains 78–88% Ca(OH)$_2$ and a large-surface-area calcium hydroxide reagent that was obtained under laboratory conditions. The test conditions were isothermal temperatures in range of 625-842°F (330-450°C), sample sizes of 5-25 mg, particle sizes of 1-60 µm, gas flow rates of 10-100 ml/min, and heating rates below 50°C/min. The tests showed a strong influence of temperature on the dehydration rate. The kinetics tests were conducted with the maximum heating rates (90°F/min or 50°C/min), a small sample size (5 mg) and a large flow rate (100 ml/min). The maximum heating rate was used to achieve the isothermal condition as quickly as possible. The small sample size and large flow rate were used to minimize the mass transfer resistance.

A genernalized grain model originally developed by Szekely et al. [29] was used to regress the kinetics data by assuming that the gaseous products within the pellet are at pseudosteady state, the initial solid structure is uniform, the solid structure is not affected by the reaction, and the product gas (water vapor) concentration is low. The best fit was obtained when the structure parameter ψ was set to zero, which corresponds to a pseudohomogeneous model.

$$\frac{dX}{dt} = k_{s0}\exp\left(-\frac{E_a}{RT}\right)\frac{p_o*\exp\left(-\frac{\Delta H}{RT}\right)}{RT}S_0(1-X)$$

In the above equation, the second exponential term represents the equilibrium water vapor pressure-temperature relationship, which was originally reported by Halstead and Moore [30]. The regression results were used to find the fitted parameters: $k_{s0}=1.8 \cdot 10^{20}$ cm/s, $E_a=280.4$ kj/mol, $p^*_o=1.834 \cdot 10^8$ atm, $-\Delta H=138.5$ kJ/mol.

The influence of water vapor was also studied by conducting two experiments at 400°CX with a 22.6 mg sample in dry nitrogen and in humidified nitrogen ($p_{H_2O} = 0.0168$ atm), The result showed that the water vapor inhibited the dehydration process. After 50 minutes of reaction, a conversion of 80% was achieved under the dry N_2, while a conversion of only 65% was achieved under the humidified nitrogen.

It was also observed that the initial solid sample size and gas flow rate had some influence on the dehydration rate indicating the existence of a significant mass transfer resistance during their dehydration test. Recently Zhang [31] studied the dehydration kinetics of calcium hydroxide and the influence of humidity with a Dupont Thermogravemetric Analyzer. Two series of tests were conducted: dehydration tests in dry N_2 and dehydration tests in humidified N_2 (8% water vapor balanced by N_2). The test temperature covered the whole medium temperature range from 572 to 1112°F (350–600°C). The results indicated that the humidity had a moderate influence on the lower temperature dehydration tests, but little influence on the higher temperature dehydration rates. Based on the TGA data, a homogeneous dehydration model was developed. The activation energy for this model is relatively large, E=144.5kJ/mol for the temperature range 626 to 878°F (330–470°C) and E=60.24kJ/mol for the temperature range 878 to 1136°F (470 to 630°C), indicating a strong temperature dependence. Zhang [31] commented that it was very difficult to obtain the isothermal kinetic data with a TGA furnace due to its limited heating rate (for a DuPont-951 TGA, 100°C/min; for a Perkin Elmer TGA, 50°C/min). For instance, with an isothermal temperature and heating rate set at 806°F (430°C) and 180°F/min (100°C/min) respectively, 40% of the tested $Ca(OH)_2$ already reacted before the temperature reached the isothermal point. This problem became more serious for higher temperature tests.

All the dehydration research reviewed above was performed in fixed-bed reactor and thermal gravimetric analyzers. In an investigation conducted by Mai and Edgar [32] however, an entrained-flow reactor was used to replicate the entrained-flow conditions which particles would experience in a furnace sorbent injection application. They studied the concomitant dehydration and sintering of $Ca(OH)_2$ at high temperatures (1002°C and 1152°C). Dehydration runs were performed under entrained-flow conditions to directly observe rates of dehydration for calcium hydroxide. A pseudohomogeneous model was used to fit the dehydration results. The apparent activation energy was 68.3 Kj•mol^{-1}. Measurable dehydration of the $Ca(OH)_2$ particle was not observed under residence times of 0.1 s. The observation implied that a significant heat-transfer limitation existed in the experimental system. The heat up time lag was subtracted from measured residence times.

In conclusion, some research has been reported on the equilibria and chemical kinetics of the calcium hydroxide dehydration reaction; however, no investigation involved the dehydration at medium temperatures under the entrained-flow conditions.

Equilibrium of the Dehydration Reaction

To verify the equilibrium data of the dehydration reaction, the decomposition temperatures of $Ca(OH)_2$ in N_2 under different humidities were studied in a Dupont 951 Thermogravemetric Analyzer. Tests were performed in dry nitrogen and humidified nitrogen with water vapor percentage of 8% and 15%, respectively.

As shown in Figure 7.10 and 7.11, dry nitrogen was passed through an evaporator system, which was immersed in a temperature-controlled water bath to set the water concentration in a gas phase. A Cole-Parmer immersion circulator, consisting of an immersion heating element, temperature sensor and a built-in circulating pump, was immersed in the water bath. To prevent any initial condensation of water vapor in the TGA chamber at the beginning of a test, the following experimental procedure was used: Loaded a sample on the sample holder, but did not insert the balance assembly into the furnace as shown in Figure 7.10. Allowed the furnace to come up to a temperature of 200°C, which was much higher than the dew point of the humidified nitrogen and lower than the decomposition temperature of calcium hydroxide, the equilibrated for 3–5 minutes. Pressed the start button and immediately slid the balance into the furnace.

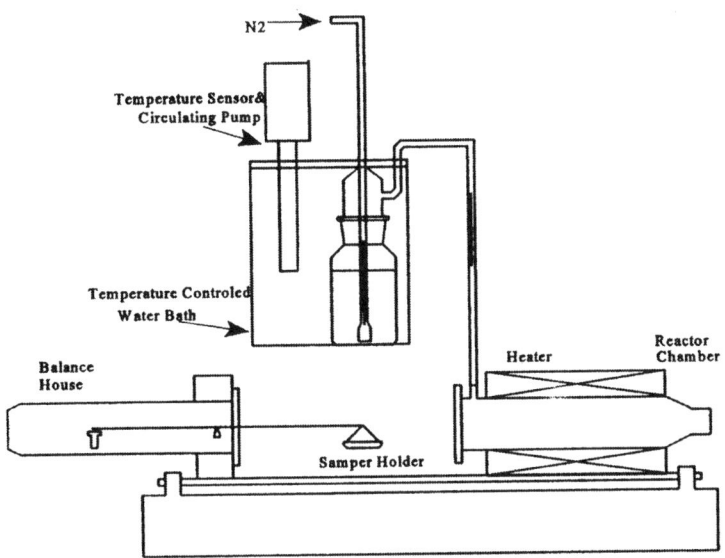

Figure 7.10 TGA setup at preheated stage.

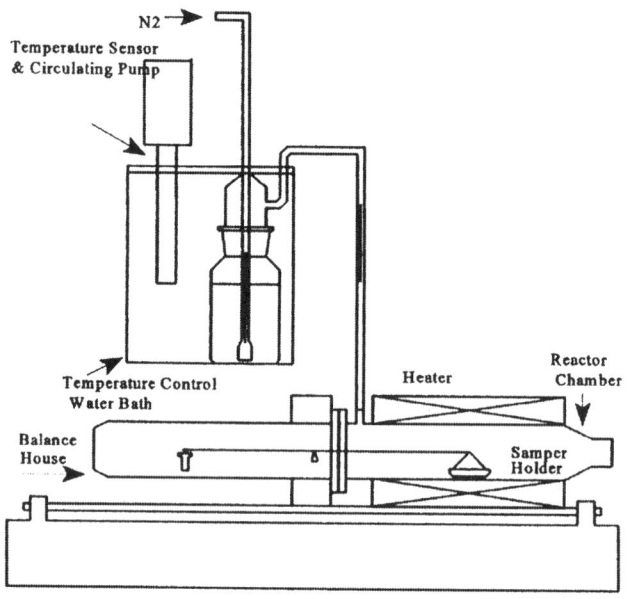

Figure 7.11 TGA setup during test.

A sample size around 10 mg and heating rate of 10°C/min were used in the TGA tests. As listed in Table 7.3, the experimental results are very close to the value calculated from Equation 7.2 indicating its validity for estimating the decomposition temperature of calcium hydroxide in humidified nitrogen.

Table 7.3 Comparison of the Decomposition Temperature Measured by TGA and the Decomposition Temperature Calculated by Equation 7.2

	Dry	8% H_2O	15% H_2O
TGA Result	316°C	420°C	456°C
Equation 7.2	320°C	430°C	455°C

Dehydration Kinetics in the Entrained Flow Reactor

The dehydration tests were conducted in the entrained flow reactor system under the following experimental conditions: temperature of 600-110°F (315.5-593.3°C); reaction times of 35, 40, 50, 75 and 100 ms; pure N_2 reactant gas; Ca$(OH)_2$ technical powder sorbent; and a gas flow rate of 15-50 l/min.

A LECO elemental analyzer was used to measure the hydrogen and carbon contents in product sorbents. In LECO analyzer, high temperature combustion of a nominal 2 mg sample was used as the means of removing the elements from the material. The products of the combustion were CO_2, H_2O, N_2 and So_x. They were carried through the system by a helium carrier. The hydrogen and carbon content were measured by a nondispersive infrared absorption detection system. The results are listed in Table 7.4.

Table 7.4 Dehydration Reaction of $Ca(OH)_2$

		H wt%			Conv %	
RxTime	35 ms	50 ms	75 ms	35ms	50 ms	75 ms
600°F	---	2.4300	----	----	3.152	----
750°F	---	2.3480	----	----	6.395	----
900°F	2.3530	2.2835	2.1830	4.909	10.087	15.704
1000°F	---	1.8735	----	----	29.804	----
1100°F	---	1.7440	----	----	35.698	----

The experimental data were analyzed and converted to reaction conversion. For the dehydrated reaction, the conversion in terms of $Ca(OH)_2$ is defined as

$$X_{Ca(OH)_2} = \frac{F_{Ca(OH)_2} - F_{oCa(OH)_2}}{F_{oCa(OH)_2}}$$

where $F_{oCa(OH)_2}$ (=94.1%) is the initial mole fraction of $Ca(OH)_2$ in $Ca(OH)_2$ technical powder and F_{CaCO_3} is the mole fraction of $CaCO_3$ in the product. The calculated conversions are listed in Table 7.4 and shown in Figure 7.12 and 7.13.

As shown in Figure 7.12, the dehydrating conversion increased with reaction temperature from 3% at 600°F to 36% at 1100°F. The dehydration conversion increased rapidly over 900°F. Figure 7.13 shows the dehydration conversion versus sorbent residence time. The dehydration conversion increases with residence time continuously from 5% at 35 ms to 16% at 75 ms.

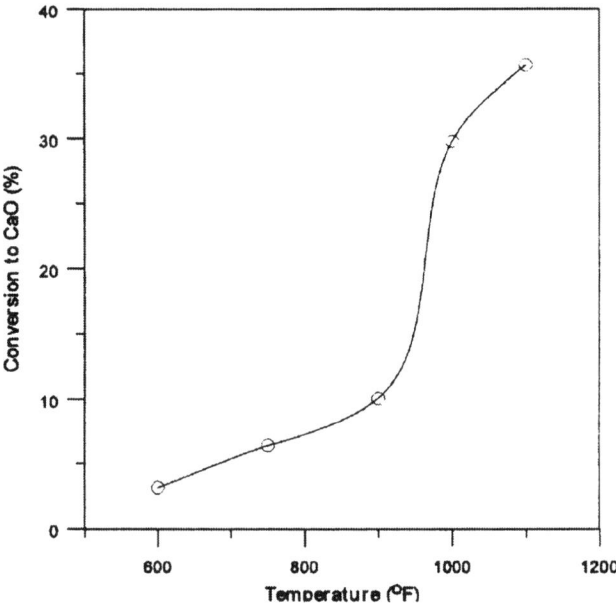

Figure 7.12 Dehydrated reaction of $Ca(OH)_2$ under the medium temperature range (residence time 50 ms).

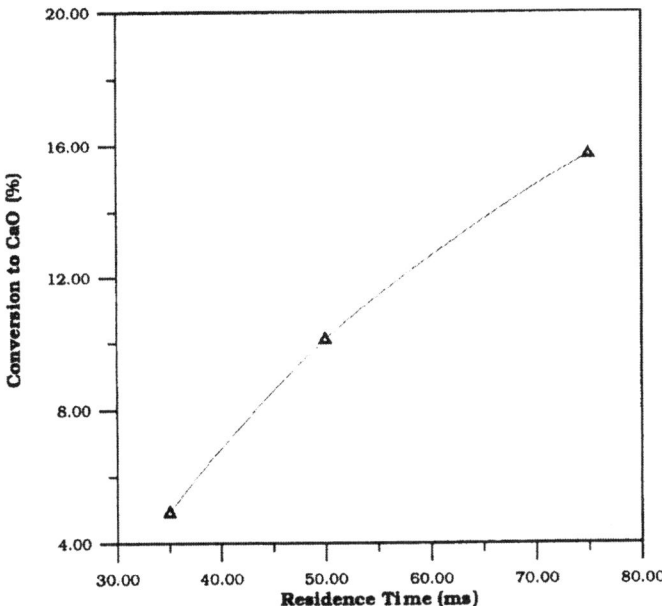

Figure 7.13 Dehydration reaction of $Ca(OH)_2$ in the residence time from 35 to 75 ms (reaction temperature 900°F).

Dehydration Model

The dehydration of $Ca(OH)_2$ involves three potential rate-controlling steps. The first is the heat transfer to the surface of the particle and the second is through the product layer (porous CaO) to the reaction interface. The third is the mass transfer of water vapor from the reaction interface to the surrounding gas phase.

The time for the temperature at the surface of the particle to attain a temperature close to the temperature in the surrounding gas was estimated by assuming a flat temperature profile inside the particle. For a particle size less the 20 μm, the time required to reach 95% of the total temperature change was on the order of a few ms. The internal temperature gradients in the particle were assumed to be insignificant based on the fact that the Biot number for the particles of interest is small enough ($b_{io} < 0.2$). The calculated temperature profile in the particle as a function of time (Carslaw and Jaeger [33]) justified the assumption of a flat temperature profile. Thus, heat transfer to the particle was not considered to be rate controlling for small particles in an entrained flow reactor.

Kinetic Studies on the Medium Temperature Ca(OH)₂ Sorbent Injection FGD Process

A shrinking-core model and pseudohomogeneous model were used to model the calcium hydroxide decomposition process. The shrinking core model assumes that the dehydration reaction takes place at the exterior surface of the particle first. As the dehydration reaction proceeds, the reaction zone moves into the interior of the solid, leaving behind a layer of CaO product. Three successive steps during the dehydration reaction are listed in what follows for the irreversible decomposition reaction: the decomposition reaction of $Ca(OH)_2$ at the surface of the unreacted $Ca(OH)_2$ core, penetration and diffusion of water vapor through the CaO product layer to the surface of the particle and diffusion of water vapor from the surface of the particle through the film surrounding the particle.

Since the magnitude of the resistances of the above steps usually differ greatly, it is possible to simplify the analysis by considering the step with the highest resistance to be rate controlling. The conversion versus reaction time for these respective situations are given by following equation.

Equation 7.3

$$\frac{t}{\tau} = X$$

and

$$\tau = \frac{\rho d_p}{6 k_g C_{eq}}$$

where ρ is the density of particle, d_p is the diameter of particle, k_g is the mass transfer coefficient between the gas phase and the particle, and C_{eq} is the water vapor concentration on the surface of particle which is assumed equal to the equilibrium water vapor concentration of $Ca(OH)_2$ at the given temperature.

Equation 7.4

$$\frac{t}{\tau} = 1 - 3(1-X)^{\frac{2}{3}} + 2(1-X)$$

and

$$\tau = \frac{\rho d_p^2}{24 D_e C_{eq}}$$

where D_e is the effective diffusion coefficient of water vapor in the ash layer and C_{eq} is the water vapor concentration on the reaction surface which is assumed equal to the equilibrium water vapor concentration of $Ca(OH)_2$ at the given temperature.

Equation 7.5

$$\frac{t}{\tau} = 1 - (1-X)^{\frac{1}{3}}$$

and

$$\tau = \frac{\rho d_p}{2k_s}$$

where k_s is the zero order rate constant for the dehydration reaction.

The Pseudohomogeneous Model

The pseudohomogeneous model has been used to describe the calcination process of $CaCO_3$ [34,35] and the dehydration process of $Ca(OH)_2$ [10] in previous investigations. The pseudohomogeneous model is based on the following assumptions: the dehydration reaction takes place homogeneously throughout the particle (i.e., the temperature profile throughout the particle is uniform and the gas diffusion resistances are negligible), the dehydration reaction is irreversible, and the intrinsic dehydration reaction rate constant follows Arrhenius' law.

According to these assumptions, a particle is converted continuously and progressively to CaO throughout the particle. Therefore, the rate of reaction may be expressed as

Equation 7.6

$$-\frac{dN_{Ca(OH)_2}}{dt} = K_d N_{Ca(OH)_2}$$

where $N_{Ca(OH)_2}$ is the moles of calcium hydroxide in the particle. The dehydration rate constant, K_d, may be expressed using a Arrhenius form.

$$K_d = K_{o,d} e^{-\frac{E}{RT}}$$

An expression relating conversion and time can be obtained by integrating Equation 7.6.

$$Ln(1 - X) = -k_d t \qquad \text{Equation 7.7}$$

The appropriate rate expression was determined by fitting the data from Figure 7.11 to Equation 7.3, 7.4, 7.5, and 7.7 with the results summarized in Table 7.4.

As shown in Table 7.5, the regression result indicates that the diffusion of water vapor through the product layer is not the controlling step. This is consistent with the conclusions reached by Borgwardt [40] and Silcox et al.[36] during their research of the thermal decomposition of $CaCO_3$ using a small particle size. Even though the product layer is apparently a major resistance to mass transfer in large spheres, its effect should diminish as the size of the $Ca(OH)_2$ is reduced. Since the particle size in this research is very small (less than 10 μm), the product layer becomes vanishingly thin, and the effect of chemical reaction becomes more prominent.

Table 7.5 Comparison of the Dehydration Reaction Model Fits

Equation Number	Model	Deviation in X, Conversion	Correlation Coefficient
7.3	Diffusion through gas film	0.0220	0.9168
7.4	Diffusion through product layer	0.0525	0.6806
7.5	Chemical reaction	0.0237	0.9074
7.7	Pseudohomogeneous	0.0246	0.9025

The gas film diffusion control model also fits the dehydration data very well. However, since hard porous solid CaO forms during the dehydration reaction, the resistance of the gas-phase product through this "ash layer" must be much greater than through the gas film surrounding the particle. Hence, in the presence of this nonflaking CaO layer, the film resistance can safely be ignored.

The comparisons in Table 7.5 show that the regression results are very close for Equation 7.5 and Equation 7.7. However, as it has been proved above that the temperature profile throughout the sorbent particle is uniform and the gas diffusion resistance is negligible, it can be concluded that the pseudohomogeneous model corresponds more closely to what really takes place during a dehydration process.

The dehydration data from this research were analyzed to establish rate parameters. The Arrhenius temperature dependency of the rate constant was used to calculate the activation energy, E, and the frequency factor, $K_{o,d}$. The rate constants were established by minimizing the sum of the squares of the error between fitted and experimental values. The activation energy and frequency factor were then calculated based on these determinations.

In Figure 7.14, the effect of temperature is shown as an Arrhenius plot of the reaction rate constants which were calculated based on the pseudohomogeneous model. A change in the activation energy at about 900°F (483°C) was observed. The calculated activation energy was 27 kJ/mol for temperature less than 900°F (482°C) and 70 kJ/mol for temperature over 900°F (482°C). The frequency factors estimated from this study were 0.152 l/s and 189 l/s for dehydration temperatures less than and over 900°F (482°C), respectively.

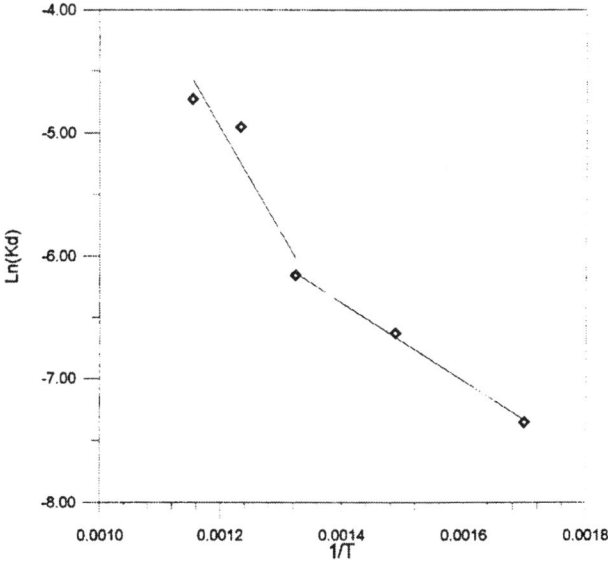

Figure 7.14 Arrhenius plot of dehydration data.

The activation energy estimated from this study, 70kJ/mole for temperatures over 900°F (482°C), is in good agreement with the value of 68.3 kJ/mole (at

1002°C and 1152°C) established by Mai and Edgar and greatly extends the applicable temperature range. In addition, the agreement between the two sets of data, in spite of the differences in reactor design, entrainment atmosphere, and sampling technique, indicates that the experimental results of this study are consistent with the existing data for $Ca(OH)_2$ dehydration in a dispersed system.

The fact that the activation energies obtained in this research shift to a higher level when the temperature is over 900°F indicates that the dehydration reaction is more temperature-sensitive at higher temperatures. This could be due to the fact that at a higher temperature the dehydration reaction rate increases dramatically by producing a more porous structure inside the particle (the molar volume is reduced from 23.4 cm^3/mole of the initial $Ca(OH)_2$ to 16.7 cm^3/mole of the product CaO). This significant porosity increase makes the diffusion resistance of water vapor through the particle completely vanish and the dehydration reaction become fully controlled by the chemical reaction itself.

Conclusions

A pseudohomogeneous model was proposed for the dehydration of $Ca(OH)_2$ which assumes that the dehydration proceeds at substantially equal rates in all sections of the particle. The model agrees well with the previous experimental data for dispersed dehydration systems.

The results show that the homogeneous chemical reaction throughout the particle is the most important mechanism especially when the temperature is over 900°F. The activation energy obtained in this temperature range is 70 kJ/mole.

Carbonation Reaction

Literature Review

Unlike the high-temperature sulfation window around 2000°F (1093°C), where $CaSO_4$ is the only thermodynamically stable compound, the medium-temperature window (800–1200°F, or 427–649°C) allows both $CaCO_3$ and $CaSO_3$ to be stable. As shown in previous sections, the amount of SO_2 capture initially depends on the rate of five competing reactions

$$Ca(OH)_2 \rightarrow CaO + H_2O$$

$$Ca(OH)_2 + SO_2 \rightarrow CaSO_3 + H_2O$$

$$CaO + SO_2 \rightarrow CaSO_3$$

$$Ca(OH)_2 + CO_2 \rightarrow CaCO_3 + H_2O$$

and

$$CaO + CO_2 \rightarrow CaCO_3$$

For these reasons, the carbonation reaction is an important mechanism in the medium-temperature sorbent injection process. Little research has been reported in the literature in this area. This section presents experimental tests and the kinetic analyses of the carbonation reaction of calcium hydroxide.

Carbonation of Ca(OH)$_2$ in Entrained Flow Reactors

The carbonation tests for $Ca(OH)_2$ were conducted in the entrained flow reactors I and II for particle residence times from 500–1000 ms, and 40–100 ms respectively. Test conditions were a temperature of 600°F to 1200°F; reaction time of 50, 75, 100, 500 and 1000 ms; 14% CO_2 with the balance N_2; $Ca(OH)_2$ technical powder sorbent; and a gas flow rate of 15-50 l/min.

A LECO elemental analyzer was used to measure the carbon content of the reacted sorbents. The measured carbon percentage in the product under the different reaction conditions is listed in Tables 7.6, 7.7, and 7.8.

Table 7.6 Conversion of the Ca(OH)$_2$ Carbonation Reaction from 600°F to 1100°F for Residence Times from 50 ms to 100 ms

	Rxn Temperature	50 ms	75 ms	100 ms
C wt%	600 °F	1.454	1.408	1.593
		1.459	1.386	2.067
	900 °F	2.426	2.666	2.690
		2.401	2.670	2.656
	1100 °F	4.617	4.952	5.004
		4.742	5.010	4.944
Conv.%	600 °F	3.5	3.16	4.40
		3.54	2.04	7.74
	900 °F	10.19	11.65	11.76
		9.94	11.71	11.39
	1100 °F	25.00	26.84	26.56
		25.96	27.28	26.28

Table 7.7 Conversion of the Ca(OH)$_2$ Carbonation Reaction Below 50 ms at 600°F to 1100°F

	C%		Conv.%	
600° F	1.454	1.495	3.5	3.537
900°F	2.426	2.401	10.19	9.95
1000°F	4.479	NA	24.92	NA
1100°F	4.617	4.742	25.0	26.0

Table 7.8 Conversion of the Ca(OH)$_2$ Carbonation Reaction at Temperatures of 600 to 1200°F and at the residence times of 500 ms and 1000 ms

Rx Time	C wt%		Conv%	
	0.5 s	1 s	0.5 s	1 s
600°F	2.367	2.705	10.74	12.3
750° F	2.756	2.916	12.66	12.71
900° F	2.924	3.267	12.86	14.11
1000° F	5.077	5.825	26.5	31.75
1100° F	5.140	4.224	26.57	NA
1200° F	5.588	6.097	29.85	33.86

Tests were also conducted to study the reactivity of CaO with CO_2 within the medium-temperature range. The results revealed that the newly-formed CaO from the dehydration of Ca(OH)$_2$ was very active and the carbonation rate observed was almost the same as that of Ca(OH)$_2$. Details of these tests and results are presented in the previous section.

Figure 7.15 shows the carbonation conversion change with residence time under different reaction temperatures. The carbonation conversion increases slightly with the residence time from 75 ms to 1000 ms. It appears that the carbonation conversion reaches its upper limit after 75 ms and the reaction becomes very slow after this point. In addition, as revealed in Figure 7.15, even at a temperature as high as 1100°F, this upper limit of carbonation conversion is only about 26%. This incomplete conversion to $CaCO_3$ could be due to pore blocking during the formation of $CaCO_3$, which has a larger molar volume than that of either $Ca(OH)_2$ or CaO.

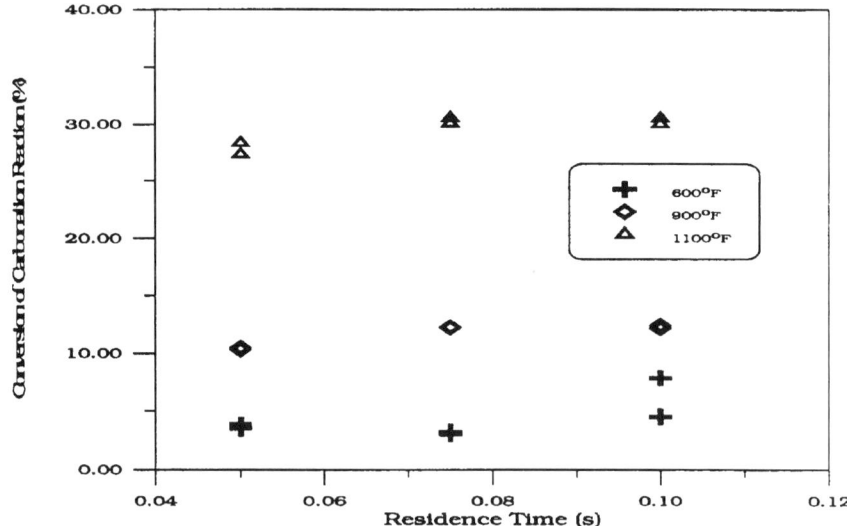

Figure 7.15 Effects of residence time on carbonation conversion at various temperatures.

The effect of reaction temperature on the carbonation reaction is shown in Figure 7.16. Carbonation conversion increases with reaction temperature, and just like the situation in the dehydration reaction, the carbonation conversion increases dramatically over 900°F. Comparing the conversion-temperature curves in Figure 7.12 and Figure 7.15 for the dehydration and carbonation reactions respectively, the same inflection points were observed at a temperature around 900°F. A possible explanation is that the carbonation reaction starts on the fresh surface of the $Ca(OH)_2$ particles at a high rate initially; however, the reaction product, $CaCO_3$, accumulates on the surface and block pores, which causes a rapid increase in diffusion resistance as the reaction proceeds. This in turn impedes further carbonation reaction. In the medium-temperature range the dehydration reaction takes place simultaneously with carbonation. When the temperature is over 900°F, the dehydration reaction becomes significant and produces more porous structure inside the $Ca(OH)_2$ particle. Such increases in porosity, not only provide more fresh surface for reaction, but also effectively reduce the diffusion resistance through the particle and enhance the carbonation reaction dramatically.

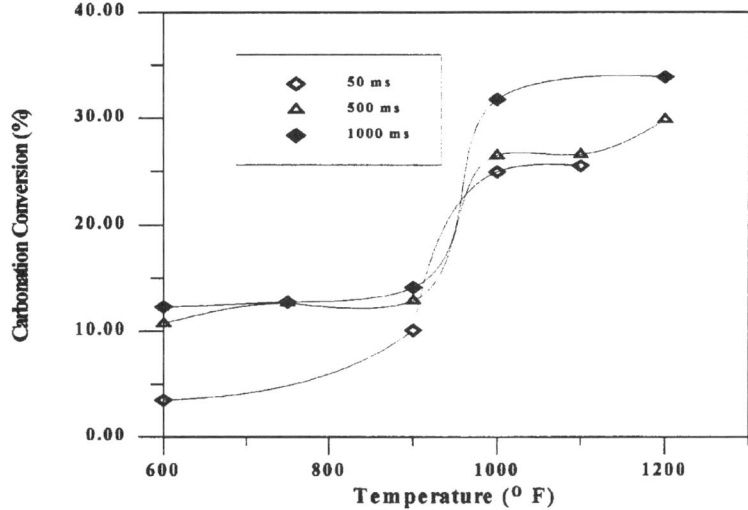

Figure 7.16 Effect of temperature on the carbonation reaction.

To verify the pore-blocking theory, the surface area of original $Ca(OH)_2$ and the reaction products were measured with a BET analyzer. The results are shown in Table 7.9, Figure 7.17 and Figure 7.18.

Table 7.9 BET surface area from the carbonation reaction

Rxn Temperature	Area (m²/g) Original 10.1 (m²/g)					
	600°F		900°F		1100°F	
50 ms	5.774	5.782	5.168	4.380	2.370	2.294
75 ms	NA	NA	3.900	3.815	2.616	2.281
100 ms	NA	NA	4.056	NA	2.303	NA

Dry Scrubbing Technologies for Flue Gas Desulfurization

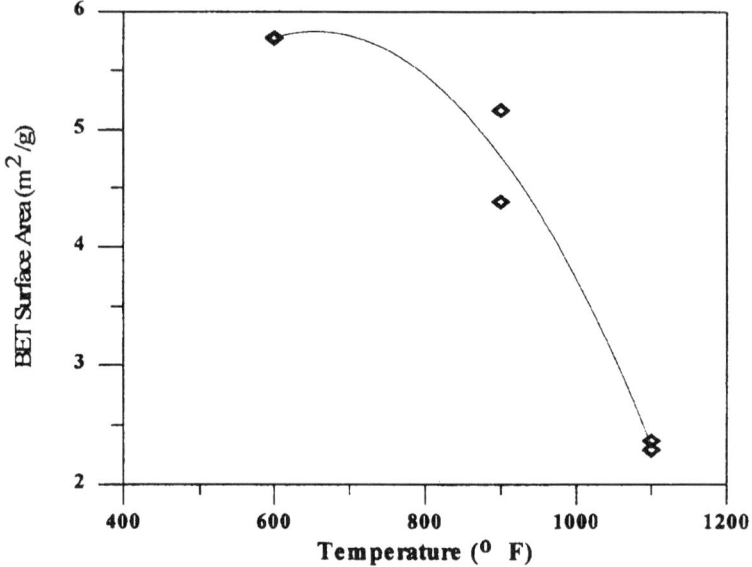

Figure 7.17 BET surface area versus carbonation temperature (residence time 50 ms).

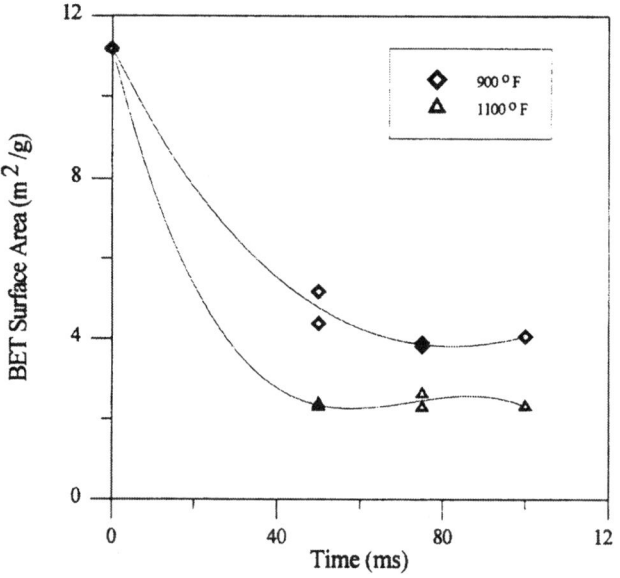

Figure 7.18 BET surface area versus reaction time.

As shown in Figure 7.17, the surface area decreased as the reaction temperature increased. The surface area dropped significantly from the initial surface area of 10.1 m^2/g to 2.3 m^2/g at a reaction temperature of 1100°F, which corresponds to significant carbonation at this temperature, supporting the pore-blocking theory. The relationship between surface area and residence time is shown in Figure 7.18. The surface area decreased with residence time and the surface area reached its lower limit after 75 ms, corresponding to the apparent reaction slow-down after this point.

Reactivity of CaO with CO_2

It has been widely believed that, compared to the direct $Ca(OH)_2$ carbonation reaction, the rate of reaction of CaO with CO_2 is very small in the medium temperature range [40-43]. Consequently, in a medium-temperature hydrated sorbent injection process, the carbonation of CaO formed due to the dehydration of $Ca(OH)_2$ is traditionally ignored.

However, as described in previous sections, the experience in handling the dehydration product indicates that the newly formed CaO is very active even at room temperature. Since the reactivity of newly formed CaO to CO_2 is important in understanding the whole picture of the carbonation of $Ca(OH)_2$, it is necessary to verify experimentally the reactivity of the fresh CaO.

A series of tests was conducted in the Dupont 951 Thermogravimetric Analyzer. In order to compare the solitary carbonation rate of $Ca(OH)_2$ with that of CaO, all the tests were conducted at a temperature around 300°C, where the dehydration of $Ca(OH)_2$ does not take place during the duration of experiment.

To investigate the reactivity of fresh CaO, the $Ca(OH)_2$ was first dehydrated in N_2 in a TGA. After the dehydration was completed, the newly formed CaO was cooled down to room temperature in pure N_2 in the TGA reaction chamber. Following this, the TGA reactor was set to a temperature of 300°C. After ensuring an isothermal condition in the reaction chamber, the inlet gas to the TGA was switched quickly from pure N_2 to 14% CO_2 gas balanced with N_2. The weight change history of the CaO after the introduction of CO_2 gas was recorded. A similar procedure was also used for $Ca(OH)_2$, except that the dehydration step was omitted.

The results of these two tests produced carbonation conversions of 5.47% for CaO and 4.85% for $Ca(OH)_2$ in the first minute. This observation infers that the freshly dehydrated CaO is as reactive as $Ca(OH)_2$ carbonation reaction is accompanied by or involves the simultaneous dehydration of calcium hydroxide.

$$Ca(OH)_2 \rightarrow CaO + H_2O$$

The CaO product in the reaction above may then react further with CO_2 to form $CaCO_3$. Due to the molar volume difference among $Ca(OH)_2$ (23.4 cm^3/mol), CaO (16.9 cam^3/'mol) and $CaCO_3$ (36.9 cm^3/mol), the formation of $CaCO_3$, whether by carbonation of $Ca(OH)_2$ or carbonation of CaO is accompanied by a net decrease in porosity. Although it is possible that the simultaneous dehydration makes calcium hydroxide more porous, subsequent carbonation causes large decreases in porosity. On a molar basis, the pore volume consumed by the formation of $CaCO_3$ during the carbonation of CaO is much greater than the pore volume developed by dehydration of $Ca(OH)_2$. This analysis suggests that owing to a gradual decrease in porosity, considerable diffusion resistances can develop within the calcium hydroxide particle during exposure to CO_2-containing reactant gas. As a result, the activity of calcium hydroxide to CO_2 gas will decrease continuously.

Since this process of deactivation of $Ca(OH)_2$ is analogous to the catalyst-poisoning process, where the deactivation of catalyst is also caused by deposition and physical blocking of the surface, a general-order catalyst deactivation model, developed originally by Levenspiel [37], is used here to describe the process of gradual reduction in reactivity of $Ca(OH)_2$ to CO_2. The most commonly used models, First- and the third-order deactivation kinetics are briefly described based on plausible physical mechanisms, although their application does not necessarily require accurate physical interpretations.

First-Order Deactivation Model

In this model, the reaction is assumed to be proportional to the remaining amount of reactant and is assumed to take place homogeneously throughout the particles (no diffusion resistance) because the particle is porous and the size is very small. The other assumptions are a first-order reaction (proportional to the remaining amount), that the partial volume does not change during the reaction, and that the reaction rate of $Ca(OH)_2$ is equal to that of CaO for each reactant gas (as shown by the TGA results).

Based on the above assumptions, the carbonation reaction rate, -r_c (mol/m^3•s), is defined as

Equation 7.8

$$-r_c = k_c C_{Cg} a_c$$

where k_c is the carbonation reaction rate constant based on the initial volume of $Ca(OH)_2$ (1/s) and C_{Cg} is the concentration of CO_2 in the bulk gas phase (mol/

m³). The activity of the pellet, a_c, is assumed to follow a linear "poisoning" relation frequently used in the petroleum industry.

Equation 7.9

$$a_c = \frac{N_{R_o} - k_{dC} N_c}{N_{R_o}}$$

where N_R, N_{R_o}, and K_{dC} are the number of moles of $CaCO_3$ produced (mol), the original number of moles of $Ca(OH)_2$ (mol) and the intrinsic deactivation constant, respectively.

From Equation 7.8 and Equation 7.9 we obtain

Equation 7.10

$$\frac{dX_c}{dt} = \frac{k_c C_{Cg}}{\rho_R} a_c = K_C C_{Cg} a_c$$

and

Equation 7.11

$$-\frac{da_c}{dt} = \frac{k_{dC} k_{Cg} C_{Cg}}{\rho_R} a_c = K_{dC} C_{Cg} a_c$$

where ρ is N_{ro}/V_o, the molar density of solid product $CaCO_3$(mol/m³), V_o is the original volume of particle, which has been assumed kept constant during the reaction; K_C is k_c/ρ_r (m³/(mol•sec)); and K_{dC} is $k_{dC} K_C/\rho_R$, the apparent deactivation constant, (m³/(mol•sec)).

The term "first-order deactivation" comes from the fact that the decreasing rate of activity is linearly related to the activity itself as expressed in Equation 7.11.

Third-Order Deactivation Model

In this model, the reaction is assumed to take place only at the surface of the particles. Since the molar volume of the reaction product is larger than the original material, the volume of the particle increases and the product layer becomes thicker during the reaction. Figure 7.19 illustrates the microscopic flat geometry where the reactant core shrinks and the product layer grows as reaction proceeds.

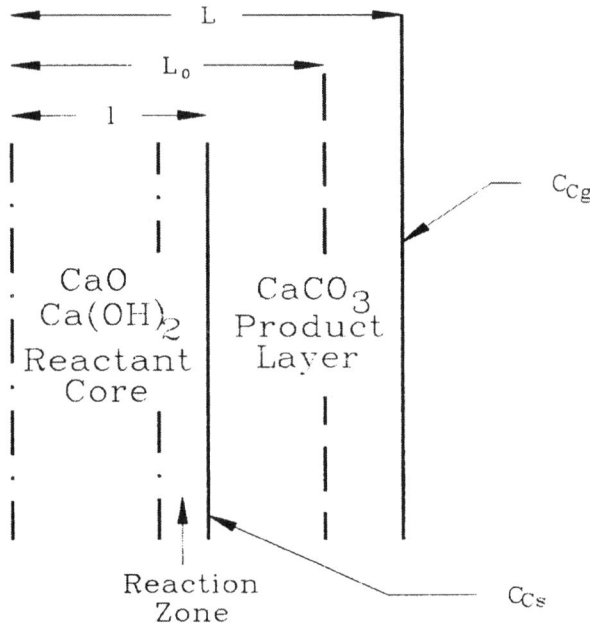

Figure 7.19 Flat geometry of the reactant core and the product layer for the carbonation reaction.

Both reaction and diffusion take place through the product layer; therefore the rate equation is expressed in terms of mass transfer first.

Equation 7.12

$$-r_C = D_{CO_2} A \frac{(C_{Cg} - C_{Cs})}{L - l} \cdot \frac{1}{AL_0}$$

where A, C_{Cs}, and D_{CO_2} are the total surface area (m²), the concentration of CO_2 at the reaction surface of the reactant core (mol/m³) and the diffusivity of CO_2 through the product layer (m²/sec), respectively.

The rate equation for the reaction is also expressed in terms of the bulk kinetics.

Equation 7.13

$$-r_c = k_c C_{cs} \varepsilon_C$$

where ε_C is the effectiveness factor, defined as

$$\frac{1}{\varepsilon_C} = \sqrt{\frac{k_C}{D_C}} L_0$$

and D_C is the diffusivity of CO_2 through the reactant core.

After rearranging Equation 7.12 and Equation 7.13, the reaction rate equation becomes

$$-r_C = \frac{C_{Cg}}{\dfrac{(L-l)L_0}{D_{CO_2}} + \dfrac{1}{\varepsilon_C k_C}} = \frac{k_C C_{Cg} \varepsilon_C}{\dfrac{k_C \varepsilon_C (L-l)L_0}{D_{CO_2}} + 1}$$

By defining the pellet activity, a_c, as

Equation 7.14

$$a_C = \frac{1}{\dfrac{k_C \varepsilon_C L_0 (L-l)}{D_{CO_2}} + 1} = \frac{1}{C_1(L-l) + 1}$$

the thickness of the product layer is related to the activity as

Equation 7.15

$$L - 1 = \frac{1}{C_1}\left(\frac{1}{a_C} - 1\right)$$

where the constant C_1 combines all the constants appearing in Equation 7.14.

$$C_1 = \frac{k_c \varepsilon_c L_0}{D_{CO_2}} = \frac{\sqrt{k_C D_C}}{D_{CO_2}}$$

Finally the growth rate of product layer becomes

Kinetic Studies on the Medium Temperature Ca(OH)$_2$ Sorbent Injection FGD Process

$$\frac{dN_C}{dt} = \frac{dA(L-l)}{\rho_C dt} = (-r_C)AL_0 = k_{Cg}\varepsilon_C C_{Cg} a_c AL_0$$

Equation 7.16

where ρ_C is the molar density of CaCO$_3$.

Substituting Equation 7.15 into Equation 7.16 with the assumption $D_C \approx D_{CO_2}$ and performing a series of algebraic manipulations, the following equation of activity is obtained:

$$-\frac{da_C}{dt} = \frac{k_C C_{Cg} \varepsilon_C}{\rho_C} a_C^3 \sqrt{\frac{k_C}{D_C}} L_0 = \frac{k_C C_{Cg}}{\rho_C} a_C^3$$

Similar to the first-order deactivation model, all the above equations can be summarized into two equations for the third-order deactivation case.

Equation 7.17

$$-r_C = k_C C_{Cg} \varepsilon_C a_C = k_C C_{Cg} a_C$$

and

$$-\frac{da_C}{dt} = \frac{k_C C_{Cg} \varepsilon_C}{\rho_C} a_C^3 = K_{dC} C_{Cg} a_C^3$$

The right-hand side of Equation 7.17 indicates the third-order dependency of the deactivation rate on the activity.

General dth-order Deactivation Model

The actual geometry may not match the simple flat geometry of the third-order deactivation model. A dth-order deactivation model is proposed to accommodate the non-ideal geometry of the third-order model:

$$-r_C = K_C C_{Cg} a_C$$

and

$$-\frac{da_C}{dt} = K_{dC} C_{Cg} a_C^d$$

The order of deactivation, d, generally lies between 1 and 3.

Model Application and Results

For the carbonation reaction experimental runs performed by the authors, the first order deactivation model appears to be the best suited among the three models mentioned above.

After calculating the CO_2 concentrations at the top and the bottom of the reactor, there is very little change in CO_2 concentration during the reaction (from 14% to 13.5%). Therefore, we can assume the CO_2 concentration to be constant in the reactor. The differential equations (Equation 7.10 and 7.11) can be easily solved and manipulated to provide the carbonation conversion as a function of the reaction time.

$$X_C(t) = \frac{K_C}{K_{dC}}(1 - e^{(-K_{dC}C_{cg}t)})$$

The conversions for the carbonation reaction have been previously measured at residence times from 0.05 s to 1 s at temperatures from 600 to 1100°F. By using these conversion data and a statistical method using the least-sum-of-squares-of-error criterion, K_c and K_{dC} values can be obtained for each temperature listed in Table 7.10. Using these calculatated parameters, simulated kinetic curves at different temperatures have been obtained. Figure 7.20 shows the comparison between the simulated curves and the experimental data. It can be seen that for the carbonation reaction, the first order model fits the experimental data well.

Table 7.10 Rate Constants for Carbonation Reaction

Temperature	Kc	K_{dC}
600°F	0.235	1.883
900°F	1.307	8.140
1100°F	6.537	20.556

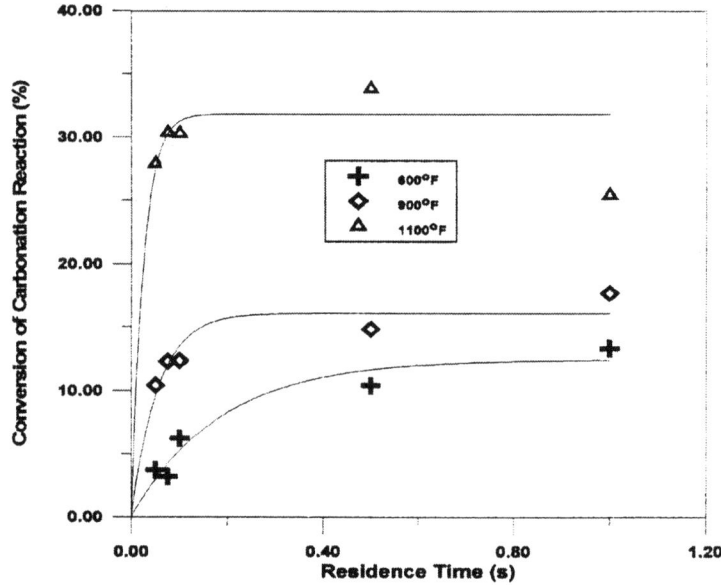

Figure 7.20 Model validation for the carbonation data using the first-order deactivation model.

As shown in Table 7.10, the reaction rate constant for the carbonation reaction increases with temperature, as does the apparent deactivation rate constant. From Figure 7.20, we can also see that the carbonation conversion increases quickly and reaches its highest level in a very short time. During this time the activity of the particle decreases significantly due to the pore closure. This prevents the particle from reacting further with the reactant gas.

Figure 7.21 gives the relationship between reaction rate constant and temperature. A straight line fit is obtained. From the slope and the intercept of the line, the activation energy (E) and the frequency factor (k_0) are calculated. These values for the carbonation reaction are $E=4.856 \cdot 10^4$ (J/mol) and $K_{CO}=4317.5$ (sec^{-1}).

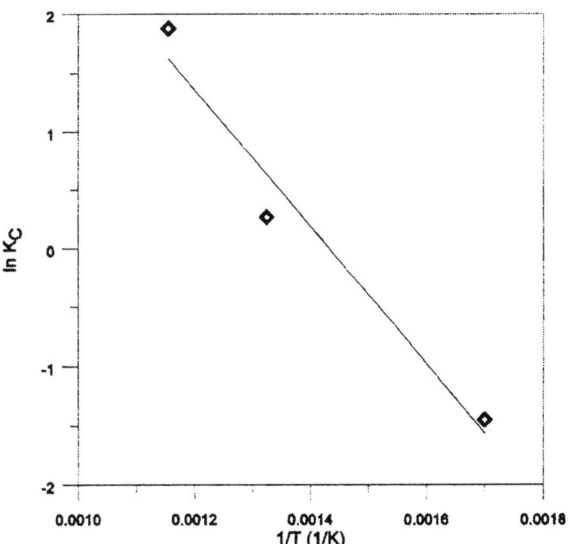

Figure 7.21 Arrhenius equation fit for K_c.

Figure 7.22 shows the relationship between the deactivation constant and temperature. It also follows the Arrhenius law. From the fit, the deactivation energy and the frequency factor are $E_a=3.573 \cdot 10^4$ (J/mol) and $K_{dCO}=2710.6$ (sec^{-1}).

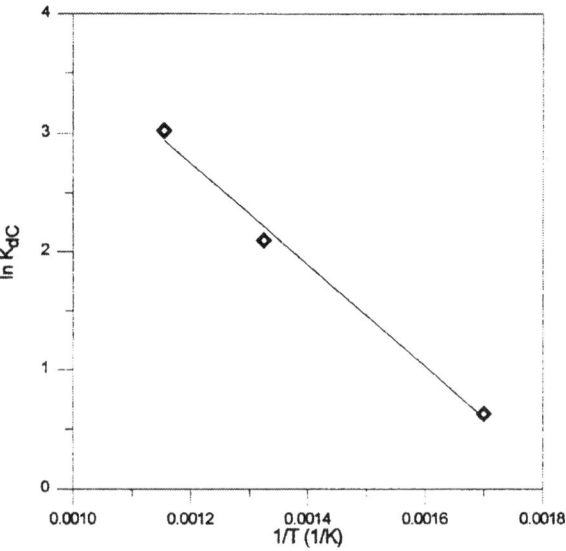

Figure 7.22 Arrhenius equation fit for K_{dC}.

Conclusions

A kinetic study on the carbonation reaction of $Ca(OH)_2$ was conducted in two entrained-flow reactors. Carbonation conversions with a reaction temperature range of 600–1100°F and residence time between 50 and 1000 ms were presented. The results indicate that at the initial state, the carbonation of $Ca(OH)_2$ is extremely fast. At 1100°F, more than 25% conversion is reached within the first 50 ms. However, the reaction slows down after that and the conversion value reaches its upper limit (about 26%) in about 75 ms. The reason for such a lower upper limit for the conversion value is believed to be due to the combination of surface deposition and pore blocking by the product, $CaCO_3$, because it has a much larger molar volume than that of the reactant. The observed decrease in BET surface area in the product confirms this pore blocking deactivation theory.

It is found that at the initial state, the newly formed CaO has an equivalent reactivity to that of $Ca(OH)_2$.

A first-order deactivation model was developed to simulate the carbonation process of $Ca(OH)_2$ particle. The model was validated for the carbonation test data with different temperatures. The activation energies and frequency factors for both carbonation reaction and the deactivation process were obtained. Using this kinetic data, the conversion predicated by the model is in good agreement with the conversion obtained experimentally.

Sulfation Reaction

Introduction

The sulfation of $Ca(OH)_2$ is the major reaction which contributes to the removal of SO_2 from flue gas. In order to fully understand the mechanism of medium temperature desulfurization process, experimental tests were conducted in this research to investigate the independent sulfation reaction of $Ca(OH)_2$. This section presents the test results and their kinetic analyses to understand the mechanism of medium temperature sorbent injection process.

Sulfation in an Entrained Flow Reactor

To investigate the individual sulfation reactions of $Ca(OH)_2$, sulfation tests were conducted in the entrained flow reactor. The experimental conditions were temperatures of 600-1100°F; reaction times of 50-75 ms; a reactant gas with 3000 ppm SO_2, 3% O_2, and the balance N_2; $Ca(OH)_2$ technical powder sorbent; and pure N_2 (99.99%) quenching gas.

The sulfation conversion of $Ca(OH)_2$ for each test was obtained by measuring the total sulfur content in the reacted sorbent. A LECO-517 total sulfur analyzer was used to measure the sulfur content in the product. The method is based on the ASTM method E30-47. The method includes heating the sample at a high temperature (2600-2700°F) in an oxygen environment to drive the sulfur out and oxidize it to SO_2, and then determining the amount of SO_2 by titration.

Tests were also performed to study the reactivity of CaO to SO_2 within the medium-temperature range. The results revealed that the newly formed CaO from the dehydration of $Ca(OH)_2$ is very active, and just as in the case of the carbonation reaction, the observed sulfation rate is almost the same as that of $Ca(OH)_2$ in this temperature range. The details of these tests and results are presented in the next section.

Based on the assumption that $Ca(OH)_2$ and newly formed CaO from dehydration has the same sulfation rates under medium-temperature conditions, the experimental data were analyzed and converted to reaction conversions. The calculated sulfation conversions under different sulfation conditions are listed in Table 7.11.

Table 7.11 Sulfation Reaction of $Ca(OH)_2$ at 600-1100°F for Residence Times from 50 ms to 100 ms

	Rxn Temperature	50 ms	75 ms	100 ms
S wt%	600 °F	0.784	0.606	0.8203
		0.891	0.786	0.871
	900 °F	NA	2.371	2.533
		2.106	2.189	2.216
	1100 °F	3.042	NA	3.659
		2.884	3.059	3.805
Conv.%	600 °F	1.822	1.405	1.902
		2.071	1.818	2.007
	900 °F	NA	5.458	5.779
		4.896	5.017	5.057
	1100 °F	6.686	NA	7.658
		6.341	6.518	7.967

Similar to the situation of carbonation, the sulfation conversion reaches its upper limit at 75 ms and the reaction becomes slow after this point. The upper limit of sulfation conversion at 1100°F is only about 8%. This incomplete conversion to $CaSO_3$ is believed due to pore blocking during the formation of $CaSO_3$, which has the larger molar volume than that of either $Ca(OH)_2$ or CaO.

Reactivity of CaO with SO_2

In previous research on the medium-temperature injection process conducted by Bortz et al. [40], in order to confirm that $Ca(OH)_2$, and not CaO, is the reactant species, Longview hydroxide was dehydrated in a muffle furnace to between 50 and 60 mole percent CaO with the remainder being hydroxide. This material was then injected into a reactor at temperatures ranging from 800 to 1200°F, and solid samples were collected. Utilizations were considerably reduced relative to the raw Longview and were independent of temperature. The authors believed that the 10 percent utilization that did occur with the precalcines was probably due to the remaining hydroxide present in the dehydrated Longview and, consequently, the reaction between CaO with SO_2 could be ignored relative to the reaction of $Ca(OH)_2$

Since then, it has been widely accepted that compared to the direct sulfation reaction of $Ca(OH)_2$, the reaction rate of CaO with SO_2 is small enough so that the further sulfation reaction of CaO formed by the dehydration of $Ca(OH)_2$ can be ignored. However, the carbonation test results presented in the previous section indicate that newly formed CaO is as active as $Ca(OH)_2$. Since the reaction between CaO with CO_2 is very similar to that with SO_2, it is possible that the reactivity of the newly formed CaO to SO_2 cannot be ignored either.

To verify this, a series of tests was conducted in the Dupont 951 Thermogravimetric Analyzer. In order to compare the solitary sulfation rate of $Ca(OH)_2$ with that of CaO, all the tests were conducted at a temperature around 300°C, which is low enough so that little dehydration of $Ca(OH)_2$ will take place. The same procedures used in the investigation presented in the previous section were also used to study the reactivitites to SO_2. Just as in the case of the carbonation reactions, the experimental sulfation conversion in the initial 1.5 minutes the two cases were very close, 3.17% for CaO and 3.09% for $Ca(OH)_2$.

Based on the above tests it was concluded that the further sulfation reaction of the newly formed CaO is as important as that of $Ca(OH)_2$ in the medium-temperature range. This contradicts previous researchers' observations [39,40] where partially predehydrated hydroxide was used that could have been precarbonated before the sorbent was injected into the reactor. Their CaO sorbents were not formed in situ, and were thus less active than the newly formed CaO.

Sulfation Model

Similar to the carbonation reaction, the sulfation reaction includes three different reactions.

$$Ca(OH) \rightarrow CaO + H_2O \uparrow$$

$$Ca(OH)_2 + So_2 \rightarrow CaSO_3 + H_2O \uparrow$$

and

$$CaO + SO_2 \rightarrow CaSO_3$$

In the sulfation reaction, the concentration of SO_2 is very low and changes significantly from the top to the bottom of the reactor. Therefore, we can no longer assume a uniform concentration of SO_2 throughout the reactor. CSO_2 is considered a variable in the calculation of the kinetic parameters for the sulfation reaction. Thus, the parameter estimation for the sulfation reaction is somewhat more complicated than that for the carbonation reaction.

Kinetic Studies on the Medium Temperature Ca(OH)₂ Sorbent Injection FGD Process

Similar to the carbonation reaction, the first order deactivation model was found to be most adequate for the sulfation reaction. The rate equation for sulfation and its activity become

Equation 7.18

$$-r_A = \frac{dX_s}{dt} = K_A C_{Ag} a_A$$

and

Equation 7.19

$$-\frac{da_A}{dt} = K_{dA} C_{Ag} a_A$$

where the parameters have similar meanings as those for the carbonation reaction except that the subscript A indicates the reactant, SO_2. The concentration of SO_2 in the gas phase is related to SO_2 uptake in the sorbent as shown below.

Equation 7.20

$$\frac{dC_{Ag}}{dt} = (-r_A)\alpha = K_A C_{Ag} a_A \alpha$$

where α is the volume fraction of sorbent particles in the gas stream. After comparing Equation 7.18 and 7.20, the following linear relationship is obtained between the gas-phase concentration of SO_2 and the sulfation conversion of sorbent.

Equation 7.21

$$C_{Ag} = C_{Ag,0} - \alpha X_s$$

where $C_{Ag,0}$ is the initial concentration of SO_2 in the feed stream. By dividing Equation 7.18 by Equation 7.19, and solving the obtained differential equation, we have the relationship between X_s and a_A

Equation 7.22

$$a_A = 1 - (K_{dA}/K_A)X_s = 1 - k_{dA}X_s$$

Substituting Equation 7.21 and 7.22 into Equation 7.18, the following differential equation is obtained for sulfation conversion.

$$\frac{dX_s}{dt} = (C_{Ag,0} - \alpha X_s)(K_A - K_{dA} X_s)$$

After solving the above differential equation, the sulfation conversion is obtained as a function of time.

Equation 7.23

$$X_s = \frac{K_A C_{Ag,0}(1 - e^{(C_{Ag,0} K_{dA} - \alpha K_A)t})}{\left(K_A \alpha - e^{(C_{Ag,0} K_{dA} - \alpha K_A)t}\right) C_{Ag,0} K_{dA}}$$

Model Application and Results

The experimental values of sulfation conversion were used to fit Equation 7.23 and the rate constants were estimated.

Table 7.12 summarizes the rate constants. Figure 7.23 shows the comparison of the simulated curves with experimental data for the sulfation reaction. For the sulfation reaction, the first order deactivation model fits the experimental data well.

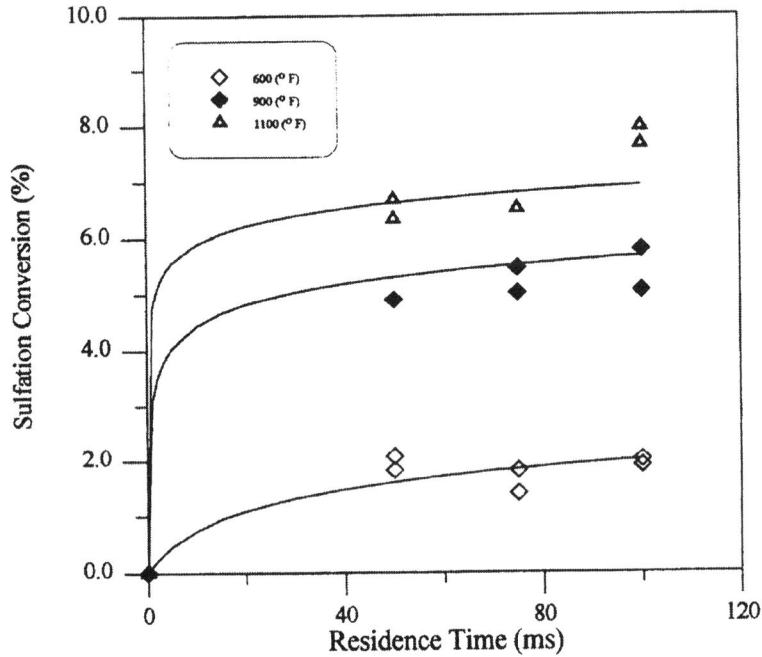

Figure 7.23 Validation of the sulfation model.

Table 7.12 Rate contents for the sulfation reaction

Temperature	K_A	K_{dA}
600°F	20.895	1017.23
900°F	84.095	1301.36
1100°F	124.20	1485.896

Figures 7.24 and 7.25 illustrate the reaction rate constant and the deactivation constant as a function of time. We see that both constants follow the Arrhenius law. From the linear fit, we obtained the following results for sulfation reaction: $K_o=6311.3$ (s^{-1}), E=2.776x10^4 (kJ/kg-mol), $K_{a0}=3266.13$ (s^{-1}), and $E_a=5716.0$ (kJ/kg-mol).

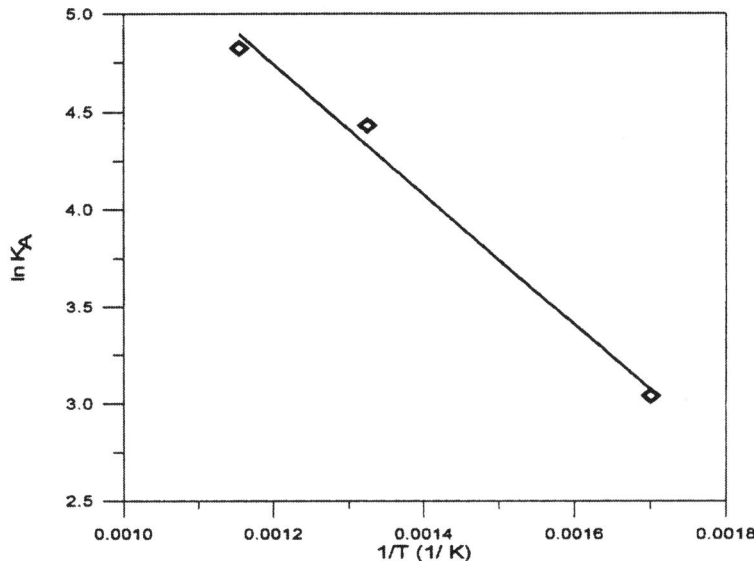

Figure 7.24 ln K_A versus 1/T plot for determination of the activation energy for the sulfation reaction.

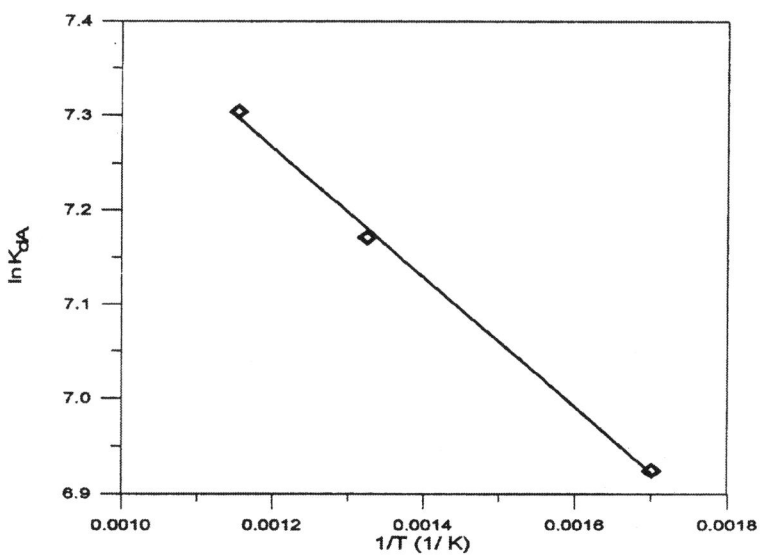

Figure 7.25 ln KdA versus 1/T plot for determination of the activation energy for the sulfation deactivation.

Comparing this set of results with that of a carbonation reaction, we can see that the frequency factor for the sulfation reaction is higher than that for the carbonation reaction, and the activation energy is lower than that for the carbonation, indicating the particle is more reactive with SO_2. This has been shown by the reaction results. Although the conversion of the sulfation reaction is lower than that of the carbonation reaction due to the much higher concentration of CO_2 (about 50 times larger than the concentration of SO_2), the reactivity between $Ca(OH)_2$ and SO_2 is still larger.

When comparing the deactivation of the carbonation and sulfation reactions, it can be seen that the deactivation energy and the corresponding frequency factor for the deactivation of sulfation reaction are smaller and larger, respectively, than those for the deactivation of carbonation reaction. This means that $Ca(OH)_2$ deactivates more easily by the sulfation reaction than by the carbonation reaction, while it is more reactive with SO_2 than with CO_2. This finding is consistent with the fact that the molecular weight of the product $CaSO_3$ is larger than that of $CaCO_3$, making it easier for $CaSO_3$ to block the pores of the $Ca(OH)_2$ particle.

Conclusions

The kinetic study on the solitary sulfation reaction of $Ca(OH)_2$ in the medium temperature range was conducted in an entrained-flow reactor. The sulfation conversion in a reaction temperature range of 600-1100°F and residence times from 50–100 ms were presented. The results indicate that similar to the carbonation reaction, the sulfation reaction is extremely fast at the initial stage and the conversion reaches an upper limit within the last 75 ms. However, this upper limit of sulfation conversion is low, less than 8% for the reaction at 1100°F, because of pore blocking and surface deposition of $CaSO_3$. This upper limit of sulfation conversion is much lower than that for the carbonation reaction reported in the previous section. The much lower SO_2 concentration in the reaction gas and relatively higher molar volume of product are believed to be the reasons for the sulfation limitation.

The sulfation reactions of $Ca(OH)_2$ is also accompanied by dehydration at medium temperatures. A series of TGA tests are conducted to determine if the fresh dehydration product, CaO would react with SO_2. By comparing the reactivities of CaO and $Ca(OH)_2$ in the TGA, it was found that the newly formed CaO has an equivalent reactivity to that of $Ca(OH)_2$. This result differs from the conventional assumption widely used in previous research that the further sulfation reaction of CaO is insignificant. The dehydration reaction not only provides more porous sorbent for SO_2 to diffuse through but also produces an equally reactive reactant.

A first-order deactivation model was developed to simulate the sulfation process. The model considers simultaneous dehydration and sulfation as well as gradual sorbent deactivation due to the pore blocking and reaction surface deposition which occurs as the reactions proceed.

Simultaneous Sulfation and Carbonation Reaction

Literature Review

As mentioned in previous sections, the interaction between flue gas and $Ca(OH)_2$ sorbent in the medium temperature range is a very complex process which involves simultaneous carbonation, sulfation, and dehydration of the injected sorbent. Upon injection of hydrated lime into the flue gas, the $Ca(OH)_2$ immediately reacts with SO_2 and also begins to dehydrate. The dehydration product, CaO, may further react with SO_2. The $Ca(OH)_2$ and CaO may also react with the abundant CO_2 present in the flue gas to form $CaCO_3$. In addition, the carbonation product, $CaCO_3$, may react further with SO to form $CaSO_3$.

$$Ca(OH)_2 s + SO_2(g) \rightarrow CaSO_3(s) + H_2O(g)$$

$$Ca(OH)_2(s) + \text{heat} \rightarrow CaO(s) + H_2O(g)$$

$$CaO(s) + SO_2(g) \rightarrow CaSO_3(s)$$

$$Ca(OH)_2(s) + CO_2(g) \rightarrow CaCO_3(s) + H_2O(g)$$

$$CaO(s) + CO_2(g) \rightarrow CaSO_3(s)$$

and

$$CaCO_3 + SO_2 \rightarrow CaSO_3 + CO_2$$

The overall SO_2 removal performance is affected by sorbent characteristics and operating conditions such as the injection temperature, residence time, and flue gas composition.

The ability of $Ca(OH)_2$ to react rapidly with SO_2 at medium temperature (600-1200°F) was first found in the EPRI-sponsored Dry Sorbent Emission Control program in 1985[4]. Tests conducted under this program showed that SO_2 capture removal with pressure-hydrated dolomite decreased as the injection temperature was gradually reduced from 1800°F to approximately 1200°F and then again increased with further reductions in injection temperature. This prompted a short study at the Southern Research Institute [38,39] where Genstar pressure-hydrated dolomite was injected at temperatures ranging from 200–2400° F.

These initial encouraging results led to a series of bench- and pilot-scale tests on the medium temperature hydrated sorbent injection process performed by S. J. Botz and G. R. Offen [3,40] from 1986 to 1988. The majority of their experiments were conducted in an electrically heated, isothermal plug flow reactor. Sorbent was fed with a screw feeder and injected with a radial sprayer to achieve rapid mixing at the top of the reactor. The reactant gas composition was maintained at 3000 ppm SO_2, 10% CO_2, 4% O_2, 15% H_2O, and the balance N_2. Solid samples were collected with an air quench probe. Tests conducted to assess possible reaction between the sorbent and SO_2 in the sampling system showed that although reaction could occur, the conversion levels in the reactor are large compared to those occurring in the probe and filter. Therefore, further reaction after the reactor could be ignored.

The tests focused on investigating the effects of injection temperature and initial SO_2 concentration. Effort was also made to optimize the hydrated sorbents. The following results were obtained:

The reaction rate of calcium hydroxide with SO_2 and CO_2 at around 1000°F was significantly faster than the reaction rate of CaO with SO_2 and CO_2 near 2000°F.

The initial SO_2 concentration has a direct effect on the final sorbent sulfation conversion. The sulfation conversion increased when SO_2 concentration increased.

Calcium hydroxide reacted directly and very rapidly with both SO_2 and CO_2. The optimum injection temperature was around 1000°F.

The sulfation rate in this temperature range was very fast. At a temperature around 1000°F, the final conversion level was achieved in 250 ms or less.

Surface area was the most important hydrate characteristic for sorbent reactivity. Total sorbent conversion to $CaSO_3$ and $CaCo_3$ increased dramatically with surface area up to about 40 m^2/g and more slowly thereafter.

A proof-of-concept demonstration of an integrated dry injection process for coalfire boiler control was conducted by a team from Research-Conttrell and Reiley Stoker sponsored by DOE and EPRI [41, 42]. The process consists of dry injection of dehydrated lime at the economizer for primary capture of SO_2, combustion modification using low NO_x burners to reduce NO_x emissions, and dry injection of a commercial grade sodium bicarbonate at the air heater exit for additional SO_2 and NO_x removal. The SO_2 removal by dry injection of hydrated lime at the economizer was independently evaluated. To identify the best calcium for the proof-of-concept demonstration, pilot-scale tests were performed that involved the injection of calcium hydroxide at various points in the flue gas system downstream of a $7.237 \cdot 10^4$ W coal-fired combustor. The flue gas flow from the furnace was approximately 0.0264 Nm^3/s, and the gas residence times, cooling rates, and temperatures were comparable to those found for full-scale utility boilers. Sorbents were injected by means of a compressed air-driven eductor.

Recently Babcock & Wilcox has developed and patented a process known as the SNRB process which can reduce SO_2, NO_x and particulates from the flue gas simultaneously. In this process, hydrated lime is injected into the convection pass of a boiler upstream of the economizer, where the flue gas temperatures may range from 900-1100°F. Simultaneous dehydration and sulfation of the sorbent begins immediately upon injection and continues as the flue gas passes through the economizer and into the baghouse.

A series of pilot-scale tests was performed in a SNRB laboratory pilot facility to evaluate the effect of economizer injection on SO_2 removal. The facility consisted of a pilot test furnace (0.5 MW equivalent), insulated ductwork, a sorbent injection system, and a heat exchanger to simulate a utility boiler economizer section. A medium-sulfur Ohio No. 8 coal (about 2.0-2.5% sulfur) was burned in the pilot test furnace to produce flue gas containing approximately 1500-2500 ppm SO_2, 600-800 ppm NO_x, and 3-4% of O_2.

To evaluate SO_2 removal performance over a range of temperature typical of utility boiler economizers, the sorbent injection temperature was varied from 800–1100°F, resulting in baghouse temperatures of 600–850°F. Over the range of temperatures tested, SO_2 removal is essentially independent of injection temperature. While the majority of SO_2 removal occurs in the duct, additional removal in the baghouse was also observed.

Simultaneous Dehydration, Sulfation and Carbonation in the Entrained Flow Reactor

A series of simultaneous carbonation-sulfation tests was carried out at reaction temperatures from 600-1100°F; reaction times of 50 ms, 75 ms and 100 ms; reactant gas composition of 3000 ppm SO_2, 3% O_2, 8% H_2O, 14% CO_2, and the balance N_2; $Ca(OH)_2$ technical powder sorbent; and a gas velocity of 15-50 l/min.

The direct experimental results are listed in Table 7.13. The weight percentages are converted to fractional conversions of carbonation and sulfation. The calculated carbonation and sulfation conversions under the different reaction conditions are listed in Table 7.14.

Table 7.13 Experimental Results of the Simultaneous Carbonation and Sulfation Reaction of $Ca(OH)_2$

	Rxn Temperature	50 ms	75 ms	100 ms
C wt%	600 °F	NA	1.263	1.231
		NA	1.252	1.190
	900 °F	2.231	2.113	2.11
		2.23	2.104	2.197
	1100 °F	4.049	3.684	3.503
		4.583	3.586	3.843
S wt %	600 °F	NA	1.197	1.251
		NA	1.266	1.192
	900 °F	1.667	2.530	2.363
		2.045	2.684	2.561
	1100 °F	2.213	2.993	2.900
		1.801	3.356	3.417

Table 7.14 Conversions in the Simultaneous Carbonation and Sulfation Reaction

	Rxn Temperature	50 ms	75 ms	100 ms
Carbonation Conversion (%)	600 °F	NA	2.305	2.062
		NA	2.238	1.769
	900 °F	9.134	8.328	8.120
		9.223	8.300	8.799
	1100 °F	21.87	18.76	16.84
		25.87	18.19	19.70

Table 7.14 Conversions in the Simultaneous Carbonation and Sulfation Reaction   cont.

		Rxn Temperature	50 ms	75 ms	100 ms
Sulfation Conversion (%)		600 °F	NA	3.047	3.174
			NA	3.225	3.019
		900 °F	4.316	6.554	6.042
			5.327	6.969	6.586
		1100°F	5.765	7.624	7.172
			4.736	8.584	8.657

Figures 7.26-7.28 show the carbonation and sulfation conversions at temperatures of 1100°F, 900°F, and 600°F, respectively. The sulfation conversion rapidly increases with residence time from 0–75 ms. The sulfation conversion remains relatively constant after 75 ms, which suggests that the majority of the sulfation is completed within the first 75 ms.

In contrast to sulfation conversion, the carbonation conversion decreases with residence time when the residence time is over 50 ms. To identify the reason for this reduction, the carbonation conversions in a CO_2 environment and a CO_2/SO_2 mixture environment are compared in Figures 7.29-7.31. It is revealed that under the temperature range investigated in this research, the reduction on the carbonation conversion is not observed under the CO_2 environment. It indicates that the equilibrium conversion is not the factor that leads to a reduction in the conversion. It is implied that the carbonation reduction observed in a CO_2/SO_2 mixture is due to the further reaction between the newly formed $CaCO_3$ and the SO_2 in the reaction gas.

Kinetic Studies on the Medium Temperature Ca(OH)$_2$ Sorbent Injection FGD Process

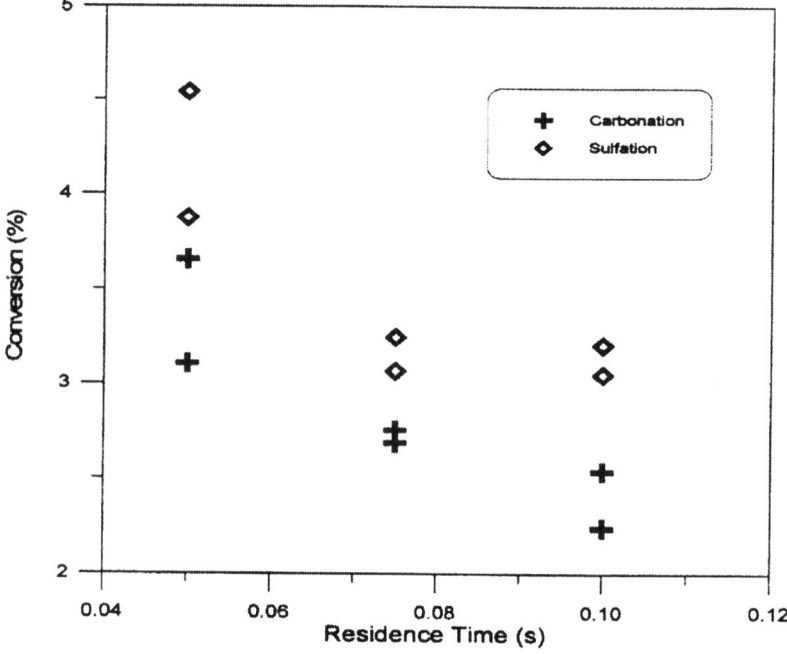

Figure 7.26 Sulfation and carbonation conversions at 600°F in 14% CO$_2$ and 3000 ppm SO$_2$.

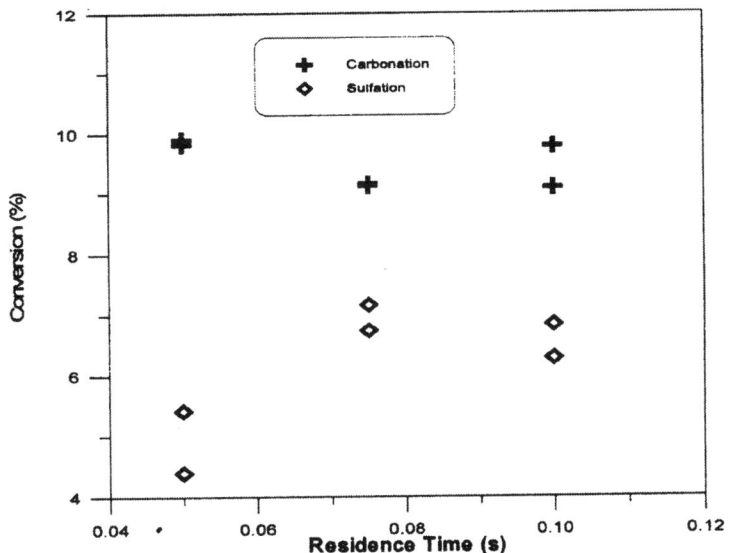

Figure 7.27 Sulfation and carbonation conversions at 900°F in 14% CO_2 and 3000 ppm SO_2.

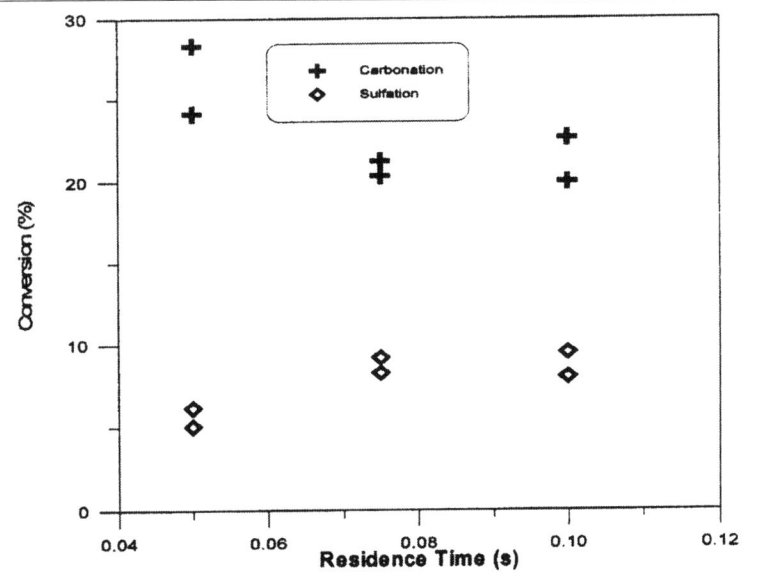

Figure 7.28 Sulfation and carbonation at 1100*F in 14% CO_2 and 3000 ppm SO_2.

Dry Scrubbing Technologies for Flue Gas Desulfurization

Kinetic Studies on the Medium Temperature Ca(OH)$_2$ Sorbent Injection FGD Process

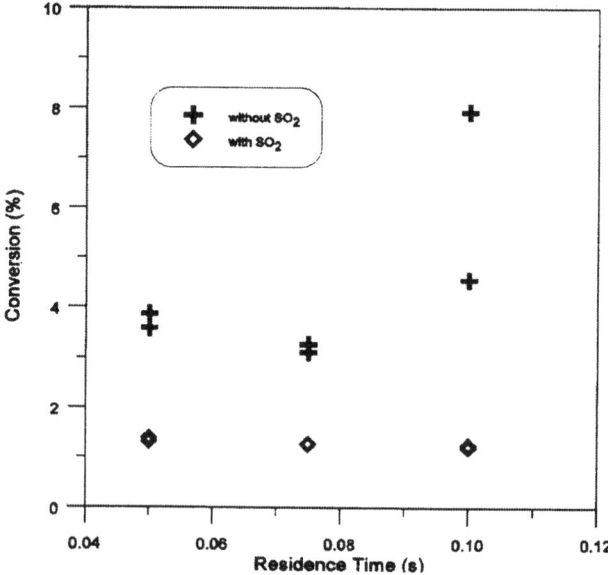

Figure 7.29 Influence of SO$_2$ on carbonation conversion at 600°F.

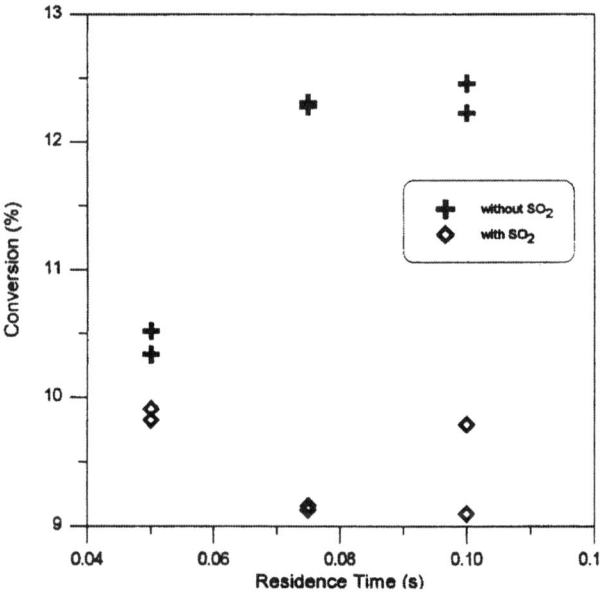

Figure 7.30 Influence of SO$_2$ on carbonation conversion at 900°F.

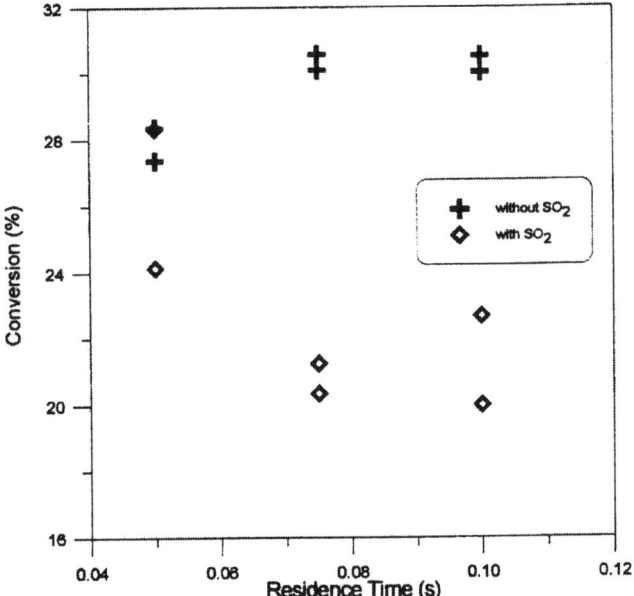

Figure 7.31 Influence of SO_2 on carbonation conversion at 1100°F.

An additional sulfation of the newly formed $CaCO_3$ will reduce the negative effect of the competing carbonation reactions and finally increase the ratio of $CaSO_2/CaCO_3$ in the used sorbent, leading to increased sorbent utilization.

Kinetic Model for Simultaneous Dehydration, Sulfation and Carbonation

In a medium-temperature sorbent injection desulfurization process, the $Ca(OH)_2$ sulfation reaction is accompanied by simultaneous carbonation and dehydration reactions. The dehydration product, CaO, may then react further with CO_2 and SO_2 in the flue gas to form $CaCO_3$ and $CaSO_3$. In addition, the newly formed $CaCO_3$ is also reacting with SO_2. As is the case in solitary carbonation or sulfation processes, due to the molar volume difference among $Ca(OH)_2$ (23.4 cm^3/mol), CaO (16.9 cm^3/mol), $caCO_3$ (36.9 cm^3/mol) and $CaSO_3$ (45.8 cm^3/mol), the formation of $CaSO_3$ as well as $CaCO_3$, whether by reaction with $Ca(OH)_2$ or reaction with CaO, is accompanied by a net decrease in porosity of the reactant particle.

Similar to the case of solitary carbonation or sulfation, this rate of simultaneous sulfation and carbonation is mainly determined by the deactivation rate of $Ca(OH)_2$ due to the pore-blocking during the above-mentioned reactions. Since this pore-blocking deactivation process is analogous to a catalysts-poisoning process, in which the deactivation of catalysts is also caused by deposition and physical blocking of the reactant surface, the deactivation model derived from the catalyst reaction described in the previous sections is used here to describe the process of gradual reduction in reactivity of $Ca(OH)_2$ to CO_2 and SO_2.

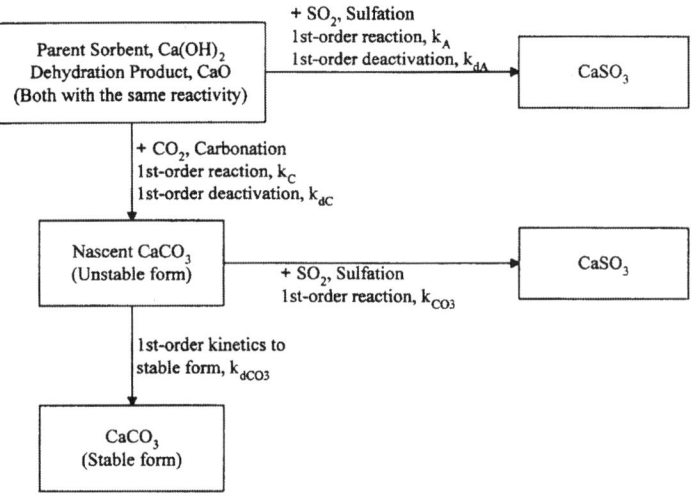

Figure 7.32 Overall kinetic scheme of the simultaneous sulfation and carbonation of $Ca(OH)_2$.

The deactivation model developed to describe the simultaneous sulfation and carbonation process is based on the following assumptions:

1. The carbonation and sulfation reactions are parallel and do not interfere with each other in terms of the reaction rate per mole of available $Ca(OH)_2$. Consequently, the carbonation and sulfation models and their reaction rate constants obtained in the previous two sections can be used here to calculate the carbonation conversion.

2. The carbonation and sulfation reactivity of newly formed CaO are the same as those of the parent sorbent $Ca(OH)_2$.

3. The carbonation reaction first produces nascent $CaCO_3$ which is unstable and either reacts with SO_2 to produce $CaSO_3$ or becomes a stable form of $CaCO_3$. The nascent $CaCO_3$ decays to produce a stable form of $CaCO_3$ according to first-order kinetics.

Figure 7.32 illustrates the overall kinetic scheme. In the following subsections, each step is described in detail. The reaction between nascent $CaCO_3$ and SO_2 is described first.

Reaction with nascent $CaCO_3$

For the rate of reaction between $CaCO_3$ and SO_3 ($-r_{AC}$), it is assumed that the newly formed CaCO, (nascent $CaCO_3$) from the carbonation reaction has the highest potential to react with SO_2 at the time of its formation, and that the reaction rate decays with time due to the fact that the nascent $CaCO_3$ slowly becomes a stable, hardened form of carbonate. So the fraction of nascent $CaCO_3$ formed between time t' and t'+dt' is

$$x_C'(t')dt'$$

and is the differential amount of $CaCO_3$ formed and not the total conversion of $CaCO_3 x_c$, which is affected by further reaction with SO_2.

Assuming a first-order decay during the time 0 to t-t', the rate of reaction with SO_2 decays due to the reduced amount of nascent $CaCO_3$ available. The rate constant is then

$$k_{co_3} e^{-k_{dco_3}(t-t')}$$

where k_{CO3} is the initial rate constant for the $CaCO_3$-SO_2 reaction when the $CaCO_3$ is fresh. The decay rate constant, k_{dco_3}, is the first-order rate constant for the decay of nascent $CaCO_3$ to stable $CaCO_3$, which reduces the amount of available solid sorbent for the sulfation reaction. The overall reaction rate at time t for the fraction $x_c'(t')$ dt' is

$$k_{co_3} e^{-k_{dco_3}(t-t')} C_{Ag}(t) x_C'(t')dt'$$

The total reaction rate at time t is obtained by integrating the above expression.

Kinetic Studies on the Medium Temperature Ca(OH)₂ Sorbent Injection FGD Process

Equation 7.24

$$-r_{AC} = k_{CO_3} C_{Ag}(t) \int_0^t x_C(t') e^{-k_{dCO_3}(t-t')} dt'$$

Defining the activity of the CaCO₃-SO₂ reaction as

Equation 7.25

$$a_{CO_3} = \int_0^t x_C(t') e^{-k_{dCO_3}(t-t')} dt'$$

can be simplified to the following general rate-deactivation form.

$$-r_{AC} = k_{CO_3} C_{Ag} a_{CO_3}$$

Similar to the solitary carbonation or sulfation reactions, the reaction rate of the simultaneous reaction is mainly determined by the deactivation rate of Ca(OH)₂. The assumptions are similar to those mentioned above for the individual reaction. However, the conversion by carbonation and sulfation are modified by the reaction from CACO₃ to CaSO₃. (The amount of carbonate is reduced and the amount of sulfite is increased to the same extent.) Therefore, the reaction rates for the carbonation and sulfation reactions become

$$-r_C = (-r'_C) - (-r_{AC}) = k_C C_{Cg} a_C - k_{CO_3} C_{Ag} a_{CO_3}$$

and

$$-r_A = (-r'_A) + (-r_{AC}) = k_A C_{Ag} a_A + k_{CO_3} C_{Ag} a_{CO_3}$$

where -r_C' is the rate of reaction between Ca(OH)₂ and CO₂, -r_A' is the rate of reaction between Ca(OH)₂ and SO₂, and -r_{AC} is the rate of reaction between CaCO₃ and SO₂.

The rate constants, k_C and k_A, and are the same values as defined for the solitary carbonation and sulfation reactions in Equation 7.10 and 7.11.

Activity

The rate of change in activity can be obtained by differentiating Equation 7.25.

$$\frac{da_{CO_3}}{dt} = \frac{d}{dt}\int_0^t x_C'(t')e^{-k_{dCO_3}t-t'}dt' = x_C' - k_{dCO_3}\int_0^t x_C'(t)e^{-d_{dCO_3}(t-t')}dt'$$

Recognizing that the last integration is the activity itself, the above equation is simplified to the following form:

$$\frac{da_{CO_3}}{dt} = \frac{dx_C}{dt} - k_{dCO_3}a_{CO_3}$$

where k_{dCO_3} is the deactivation constant for $CaCO_3$-SO_2 reaction.

The activities of the carbonation and sulfation reactions are defined as linear functions of their respective poisoning substances.

$$a_C = \frac{N_0 - k_{dA}n_C - k_{dCaSO_3}n_S}{N_0} = 1 - k_{dA}x_C - k_{dCaSO_3}x_S$$

and

$$a_A = \frac{N_0 - k_{dA}n_S - k_{dCaSO_3}n_C}{N_0} = 1 - k_{dA}x_S - k_{dCaSO_3}x_C$$

where n_s and n_c are the numbers of moles of $CaSO_3$ and $CaCo_3$ produced from $CA(OH)_2$ without considering the amount modified by the nascent $CaCO_3$ reaction respectively, and x_s and x_c are the corresponding conversions. These expressions are analogous to those of heterogeneous catalysts where the deactivation is caused by poisoning species gradually occupying the available active sites. The deactivation constants, k_{dA} and k_{dC}, are the same values as defined for the solitary sulfation and carbonation reactions listed in Equation 7.22 and 7.9. The two additional constants, k_{dCaCO_3} and k_{dCaSO_3} define the additional deactivation of the sulfation and carbonation reactions due to the deposition of $CaCo_3$ and $CaSO_3$ respectively These constants, therefore, may be understood as the "enhancement" constants for the respective deactivations.

In summary, the following equations are obtained for the simultaneous reaction

$$\frac{dX_c}{dt} = \frac{1}{\rho_R}(-r_C) = K_C C_{Cg} a_C - K_{CO_3} C_{Ag} a_{CO_3}$$

Kinetic Studies on the Medium Temperature Ca(OH)₂ Sorbent Injection FGD Process

$$\frac{dX_s}{dt} = \frac{1}{\rho_R}(-r) = K_A C_{Ag} a_A + K_{CO_3} C_{Ag} a_{CO_3}$$

$$-\frac{da_c}{dt} = k_{dC}\frac{dx_C}{dt} - k_{dCaSO_3}\frac{dx_S}{dt} = K_{dC} C_{Cg} a_C - K_{dCaSO_3} C_{Ag} a_A$$

$$-\frac{da_A}{dt} = k_{dA}\frac{dx_S}{dt} - k_{dCaCO_3}\frac{dx_C}{dt} = K_{dA} C_{Ag} a_A - K_{dCaCO_3} C_{Cg} a_C$$

and

$$\frac{da_{CO_3}}{dt} = \frac{dx_C}{dt} - k_{dCO_3} a_{CO_3} = K_C C_{Cg} a_C - K_{dCO_3} a_{CO_3}$$

In this model, four additional kinetic constants, K_{CO_3}, k_{dCO_3}, K_{dCaSO_3} and K_{dCaCO_3} are added in addition to the constants already needed for solitary sulfation and carbonation reactions.

Model Application

The above equations describe the overall simultaneous carbonation and sulfation process. By assuming that the carbonation and sulfation reactions are parallel and not interfering with each other in terms of the reaction rate mole of available Ca(OH)₂, reaction rate constants and deactivation constants, K_c, K_A, K_{dC}, and K_{dA} which are obtained from the individual carbonation and sulfation reactions, can be used in the above equations. However, K_{dCaCO_3}, K_{dCaSO_3}, K_{CO_3}, and k_{dCO_3} are unknown and need to be determined. A computer program was written to evaluate these parameters. The computer program first solves the differential equation by fourth-order Runge-Kutta method, then optimizes the parameters until the error function, the equation below, reaches a minimum value

$$F = \sum [X_c(t_i) - X_{Ct}]^2 + \sum [X_S(t_i) - X_{St}]^2$$

Figure 7.33 and 7.34 show the comparison of fitted curves and experimental data for conversions of carbonation and sulfation. It can be seen that the simulation can represent the simultaneous reaction process well.

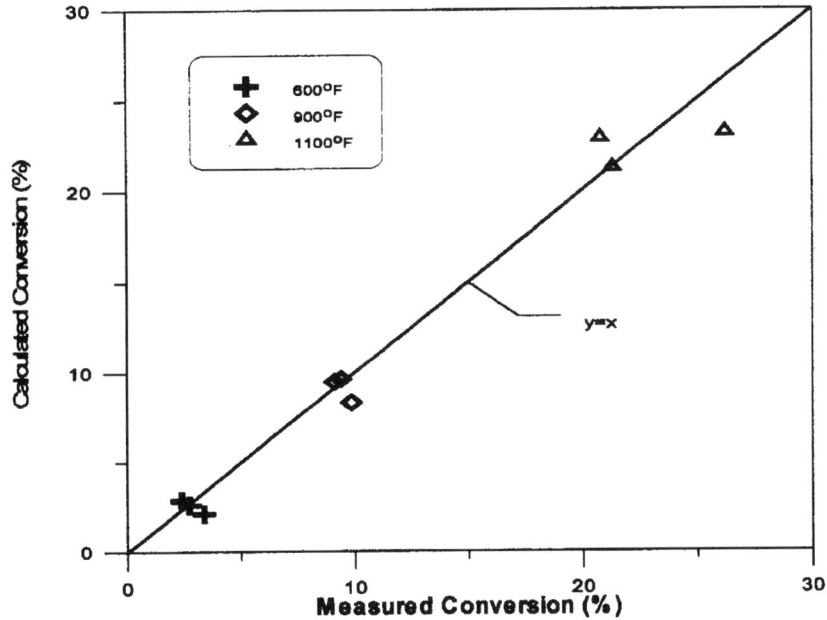

Figure 7.33 Model validation for carbonation conversion with the simultaneous reaction.

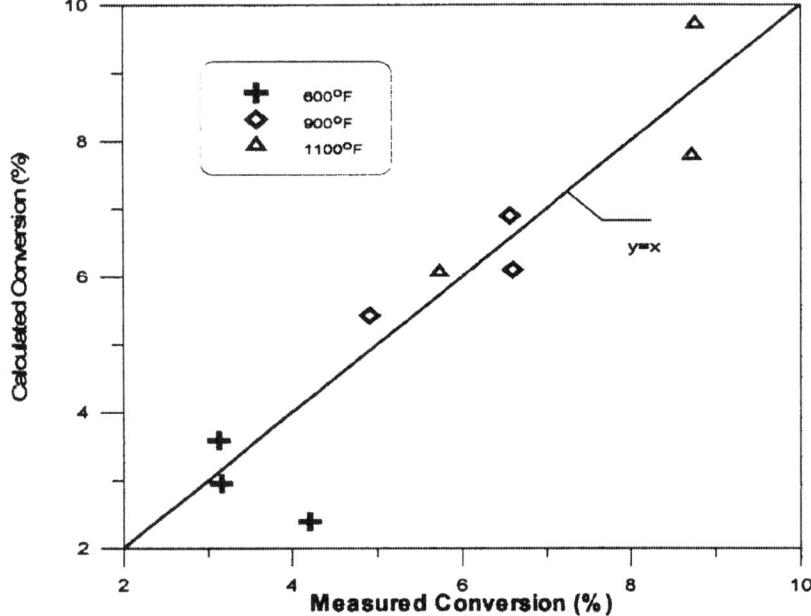

Figure 7.34 Model validation for sulfation conversion with the simultaneous reaction.

Figure 7.35 shows the relationship between the reaction rate constant, k_{CO_3} and temperature. From this Arrhenius plot, the activation energy and frequency factor for the reaction of $CaCO_3$ with SO_2 to form $CaSO_3$ can be calculated giving E_{CO3}=4511.2 kJ/kg mol and $K_{co3,0}$=285 s^{-1}.

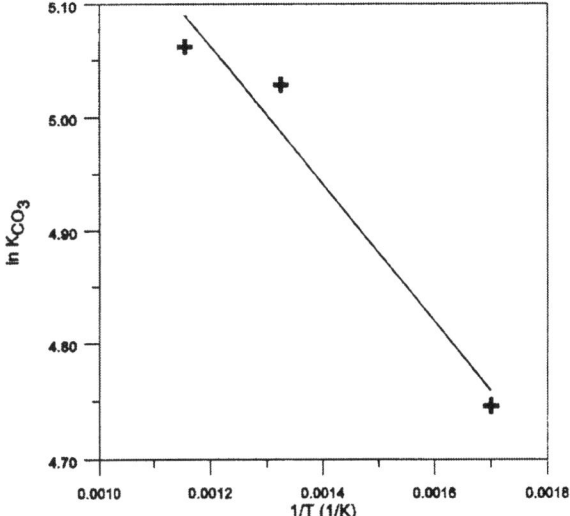

Figure 7.35 ln K_{CO3} versus 1/T.

Conclusions

The kinetic studies on the medium-temperature $Ca(OH)_2$ sorbent injection FGD process has been studied in an entrained-flow reactor. The carbonation and sulfation conversions in a simulated flue gas were presented for a reaction temperature range of 600-1100°F and residence times below 100 ms.

The results show that both sulfation and carbonation conversions strongly depend on the reaction temperature. For a typical economizer temperature range (600-1100°F), both sulfation and carbonation conversions increase with reaction temperature.

The residence time also plays an important role in the medium temperature sorbent injection process. Unlike the solitary sulfation reaction presented in the previous section, the observed sulfation conversion increases continuously with residence time even though the rate slows down quickly after the short internal period of reaction. The effect of residence time on the carbonation conversion, however, is more complicated. The carbonation reaction is extremely fast at the initial stage and the carbonation conversion reaches its upper limit quickly. After the initial period, the carbonation conversion begins to decrease with residence time. This reduction in carbonation conversion is believed to be due to the further reaction of the newly formed $CaCO_3$ with SO_2.

It has been believed in previous research that the carbonation of $Ca(OH)_2$ is a competing and undesired reaction which consumes the available $Ca(OH)_2$ wanted for the sulfation reaction. This observed further sulfation, however, compensates the negative effect of the competing carbonation reaction and finally increases the ratio of $CaSO_2$ to $CaCO_3$ in the used sorbent. Consequently, it is possible that, by increasing the residence time of sorbent in the ductwork, the sorbent utilization could also be increased.

A mathematical model was developed to simulate the simultaneous sulfation and carbonation process of calcium hydroxide particle. The model describes the simultaneous sulfation and carbonation of $Ca(OH)_2$ as well as the further sulfation of the newly formed $CaCO_2$. The first-order deactivation of sorbent is considered by introducing time-dependent deactivation coefficients which are commonly used to describe a catalyst deactivation process in other fields.

Based on the assumption that the intrinsic sulfation and carbonation reactions are parallel and not interfering with each other, the activation energies and frequency factors previously obtained in solitary reactions for both the carbonation and sulfation reactions were used in this simultaneous sulfation and carbonation model. The model was validated using the sulfation and carbonation conversion data in the temperature of 600–1100°F. The activation energy and frequency factor for the deactivation process were obtained. Using these kinetic data, the predicted conversions were found to be in good agreement with the experimentally obtained conversion.

The simulation results show that the carbonation reaction affects the sulfation reaction behavior by means of two important mechanisms. First, it adds additional diffusion resistance for SO_2 gas because of the formation of $CaCO_3$, consequently accelerating the deactivation of calcium hydroxide sorbent. The second mechanism is the further sulfation of the newly formed $CaCO_3$ which compensates the negative effect of the competing carbonation reaction.

Recommendations

Based on the framework of this simultaneous reaction model, modifications may be made to better represent the realistic situation of a calcium hydroxide particle undergoing the simultaneous dehydration, sulfation and carbonation reactions. This includes the introduction of the effect of the initial surface area and the initial pore size distribution into the model. In addition, the sorbent deactivation due to the further sulfation of the newly formed $CaCO_3$ may be included, especially for longer residence times when this deactivation may become significant.

In order to supply sufficient data for model modification, further experiments on the effect of sorbent properties, such as initial surface area and initial pore size distribution are recommended.

References

1. Nolan, P. S., "Results of the EPA LIME Demonstration at Edgewater," presented at the EPA/EPRI 1990 SO_2 Control Symposium, New Orleans, LA, May 8-11, 1990
2. Goots, T. R., M. R. Gogineni, T. J. Purdon, P. S. Nolan, J. L. Hoffmann and T. W. Arrigoni, "Results From Lime Extension Testing at the Ohio Edison Edgewater Station," presented at the 1991 EPRI/EPA/DOE SO_2 Control Symposium, Washington, D.C., December 3-6, 1991
3. Bolli, R. E., "Ohio Edison Clean Coal Projects, CIRCA: 1991", presented at the 1991 EPRI/EPA/DOE SO_2 Control Symposium, Washington, D.C., December 3-6, 1991
4. Bortz, S. J., "SO_2 Removal through Hydroxide Injection at Economizer Temperatures," *Proceedings: 1986 Joint EPRI/EPA Symposium on Dry SO_2 Simultaneous SO_2/NO_x Control Technologies*, Volume 2, EPRI CS4966, December 1986
5. Yoon, H., "Sorbent Improvement and Computer Modeling Studies for Coolside Desulfurization," *Proceedings: 1986 Joint EPRI/EPA Symposium on Dry SO_2 and Simultaneous SO_2/NO_x Control Technologies*, 2, EPRI CS4966, December 1986
6. Abrams, J. Z., "Development of the Confined Zone Dispersion Process for Desulfurization of Flue Gas," *Coal Technology '86 Conference Papers*, 1 and 2, 253-264 (1986)
7. Cooch, J. P., E. B. Dismukes and R. S. Dahlin, "Scale-up Tests and Supporting Research for the Development of Duct Injection Technology," *Topical Report No.1-Literature Review*, Presented to DOE, Southern Research Institute (1989)
8. Shilling, N. Z., "Technical Status Report-Development of Lime Based In-Duct Scrubbing-A Cost Effective SO_2 Control Technology," presented at the American Power Conference, Chicago, IL, April 1986
9. Murphy, R., "Status of the DOE/GEESI In-Duct Scrubbing Pilot Study," *JAPCA*, 36, No. 8, 953-958 (1986)
10. Babu, M., "Results of 1 MMBtu/hour Testing of Halt (Hydrate Addition at Low Temperatures) for SO_2 Control," *Proceedings: 1986 Joint EPRI/EPA Symposium on Dry SO_2 and Simultaneous SO_2/NO_x Control Technologies*, 2, EPRI CS4966, December 1986
11. Roman, V. P., L. J. Muzio, M. W. McElroy, K. W. Bowers and D. T. Gallaspy, "Flow Reactor Study of Calcination and Sulfation," *Proceedings: First Joint Symposium on Dry SO_2 and Simultaneous SO_2/NO_x Control Technologies*, EPA-600/9-85-020a (NTIS PB85-232353), 1, 7-1 (1986)
12. Cole, J. A., J. C. Kramlich, G. S. Samuelsen, W. R. Seeker and G. D. Silcox, "Fundamental Studies of Sorbent Reactivity in Isothermal Reactors," *Proceedings: First Joint Symposium on Dry SO_2 and Simultaneous SO_2/NO_x Control Technologies*, EPA-600/9-85-020a (NTIS PB85-232353), 1, 10-1 (1986)
13. Bortz, S. and P. Flament, "Recent IFRF Fundamental and Pilot Scale Studies on the Direct Sorbent Injection Process," *Proceedings: First Joint Symposium on Dry SO_2 and Simultaneous SO_2/NO_x Control Technologies*, EPA-600/9-85-020a (NTIS PB85-232353), 1, 17-1 (1986)
14. Gullett, B. K., G. C. Snow, J. A. Blom and D. A. Kirchgessner, "Design of Graphite Element Drop-Tube System for Study of SO_2 Removal by Injected Limestone Sorbent," *Rev. Sci. Instrum.* 57, 2599 (1986)
15. Newton, G. H., D. J. Harrison, G. D. Silcox and D. W. Pershing, "Control of SO_x Emissions by In-Furnace Sorbent Injection: Carbonates vs. Hydrates," *Environ. Prog.* 5, 140 (1986)

16. Sonnet, D., S. Afara, C. L. Briens, J. F. Large and M. A. Bergougnou, "Circulating Fluidized Bed Technology II," *Proceedings: Second International Conference on Circulating Fluidized Beds*, Compiegne, France, 565 (1988)
17. Perry, R. H. and C. H. Chilton, *Chemical Engineers Handbook*, 6th edition, McGraw-Hill Inc. (1984)
18. Bird, R. B., W. E. Stewart and E. N. Lightfoot, *Transport Phenomena*, John Wiley & Sons, Inc. (1960)
19. Rowe, P. N., K. T. Claxton and J. B. Lewis, "Heat and Mass Transfer from Signle-Sphere in Extensive Flowing Fluid," *Trans. Instn. Chem. Engrs.*, 43, No. 1, 14 (1965)
20. Boynton, R. S., *Chemistry and Technology of Lime and Limestone*, Wiley-Interscience Publication (1980)
21. Hedin, R., "Investigation of the Lime-Burning Process," *Handlinger Proceedings*, NR32, Stockholm (1961)
22. Rao, T. R., D. J. Gunn and J. H. Bowen, "Kinetics of Calcium Carbonate Decomposition," *Chem. Eng. Res. Des.*, 67, No. 1, 38 (1989)
23. Hartman, M. and Martinovsky, "Thermal Stability of the Magnesian and Calcareous Compounds for Desulfurization Processes," *Chem. Eng. Commun.*, 111, 149 (1992)
24. Hartman, M. and O. Trnka, "Reactions between Calcium Oxide and Flue Gas Containing Sulfur Dioxide at Lower Temperatures," *AIChE J.*, 39, No.4, April 1993
25. Criado, J. M. and J. Morales, "On the Thermal Decomposition Mechanism for the Dehydroxylation of alkaline earth hydroxides," *J. Thermal Anal.*, 10, 103-110 (1976)
26. Mu, J.and, D. D. Perlmutter, "Thermal Decomposition of Carbonates, Carboxylates, Oxalates, Acetates, Formates and Hydroxides, "*Thermochim. Acta*, 49, 207-218 (1981)
27. Irabien, A., J. R. Viguri, and I. Ortiz, "Thermal Dehydration of Calcium Hydroxide. 1. Kinetic Model and Parameters," *Ind. Eng. Chem. Res.*, 29, 1599-1606 (1990)
28. Irabien, A., J. R. Viguri, F. Cortabitarte and I. Ortiz, "Thermal Dehydration of calcium Hydroxide. 2. Surface Area Evolution," *Ind. Eng. Chem. Res.*, 29, 1606-1611 (1990)
29. Szekely, J., J. W. Evans and H. Y. Sohn, "Porous Solids of Unchanging Overall Sizes," *Gas-Solid Reactions*, Academic Press: NY (1976)
30. Halstead, P. E. and A. E. Moore, "The Thermal Dissociation of Calcium Hydroxide," *J. Chem. Soc.*, 3873-3875 (1957)
31. Zhang, C. X., *Experimental Kinetic Studies on the Simultaneous Sulfation/Carbonation/ Dehydration Reactions in Medium Temperature Sorbent Injection Process*, M.S. Thesis, University of Cincinnati (1991)
32. Mai, M. C. and T. F. Edgar, "Surface Area Evolution of Calcium Hydroxide During Calcination and Sintering," *AIChE J.* 35, 30-36 (1989)
33. Carslaw, H. S. and J. C. Jaeger, *Conduction of Heat in Solids*, Oxford University Press, Oxford (1959)
34. Beruto, D. and A. W. Searcy, "Use of Lagmuir Method for Kinetic Studies of Decomposition Reaction: Calcite($CaCO_3$)," *J. Chem. Soc., Farad. Trans.*, 7, 2145 (1974)
35. Borgwardt, R. H., "Calcination Kinetics and Surface Area of Dispersed Limestone Particles," *AIChE J.*, 31 (1), 1 (1985)
36. Silcox, G. D., J. C. Kramlich and D. W. Pershing, "A Mathematical Model for the Flash Calcination of Dispersed $CaCO_3$ and $Ca(OH)_2$ Particles," *Ind. Eng. Chem. Res.*, 28, 155 (1989)
37. O. Levenspiel, *Chemical Reaction Engineering*, Chapter 15, Second Edition, John Wiley, NY (1972)

38. Beittel, R., "Dry Sorbent Emission Control" *Southern Research Institute Monthly Report No. 18*, Contract No. 195-83-003, October 1985
39. Beittel, R., "Effects of Injection Temperature and Quench Rate on Sorbent Utilization," *Proceedings: 1986 Joint Symposium on Dry SO_2 and Simultaneous SO_2/NO_x Control Technologies*, June 2-6, 1986
40. Bortz, S. J., V. Roman and G. R. Offen, "Hydrate and Process Parameters Controlling SO_2 Removal During Hydroxide Injection near 1000°F," *Proceedings: First Combined FGD and Dry SO_2 Control Symposium*, October 25-28, 1988
41. Helfritch, D., S. Bortz, R. Beittel, P. Bergman and B. Toole-O'Neil, "Combined SO_2 and No_x Removal by Means of Dry Sorbent Injection," *Environmental Progress*, 11, No.1, February 1992
42. Helfritch, D., S. Bortz, R. Beittel, P. Bergman and B. Toole-O'Neil, "ASO_2 and NO_x Control by Combined Dry Injection of Hydrated Lime and Sodium Bicarbonate," *1991 EPRI/EPA/DOE SO_2 Control Symposium*, Washington, D.C., December 3-6, 1991

CHAPTER 8
ADVANCES IN SPRAY DRYING DESULFURIZATION FOR HIGH-SULFUR COALS

Tim C. Keener*, Jun Wang*, and Soon-Jai Khang**
*The Department of Civil and Environmental Engineering
**The Department of Chemical Engineering
The University of Cincinnati
Cincinnati, Ohio

Abstract

Advances in spray drying desulfurization for high-sulfur coals were investigated, with a focus on two major areas where improvements can be made: the spray dryer technology itself and the chemical process involving the absorption of SO2. This chapter documents (1) measuring lime dissolution rates at different conditions and with additives in a bench-scale apparatus; (2) direct tests of additive mechanisms in a pilot spray dryer absorber; (3) tests of hydrated fly ash and lime mixtures as new sorbents in a pilot spray dryer; and (4) theoretical studies on the mechanisms of the additive effects.

The mechanisms responsible for improvements in performance with the use of additives were investigated and discussed. Changes in the hygroscopicity of the sorbent slurry were incorporated into a spray dryer mathematical model that examines the drying of slurry droplets, mass transfer, and reaction in the droplet. This model describes the improvements measured in experimental tests with reasonable accuracy.

The lime dissolution rate was measured in the presence of a variety of inorganic and organic additives. The effects of additives were studied directly in a spray dryer. Hygroscopic additives had significant effects on improving lime utilization and SO2 removal in spray dryers; buffer additives had little or no effect. Experiments showed that over 90% removal of the SO2 concentration from a high sulfur coal flue gas can be achieved in spray dryers with the appropriate use of additives and operating conditions.

Introduction

Spray drying flue gas desulfurization (FGD) is considered a well-established commercial technology for utility applications with low-sulfur coal. This SO_2 scrubbing technology is backed by 15 years of operating experience since the first full-scale demonstration system on a coal-fired utility boiler began operation in 1980. Compared to wet lime/limestone scrubbing technology, the spray dryer has the reported advantages of fewer major equipment items and thus lower capital costs, high reliability, lower space requirements, lower potential for corrosion, potential for lower energy consumption, absence of a wastewater stream, lower water consumption, and less sensitive and simpler process chemistry. The greatest disadvantage of spray dryer technology is perhaps the higher cost of lime reagents used in relation to limestone used for the dominant wet scrubbing processes. Since reagent requirements increase with increasing coal sulfur content, the higher operating costs with medium-to-high sulfur coal cut into the capital cost advantages held by spray drying FGD systems. This situation is further aggravated by a lower reagent utilization as the coal sulfur level increases. Therefore, there is a demand for continuing the development of techniques that enhance SO_2 removal efficiency, improve lime utilization, and further reduce capital costs to make spray dryers more economically attractive.

Two major fields exist in which improvements can be made. One is the spray dryer technology itself (including drying process), and the other is the chemical process involving the absorption of SO_2. Research areas that may yield higher spray dryer desulfurization performance are improvements on spray dryer and atomizer designs, use of additives to enhance sorbent (lime) utilization and increase SO_2 removal efficiency, fly ash hydration with lime to form new sorbents, integrated effects of a spray dryer and baghouse combination, analytical investigation on the characteristics of additives to find common properties that could improve sorbent utilization, and theoretical studies and simulation of the physical and chemical mechanisms in the spray drying process to shed light on future directions.

The difficulty in reaching a theoretical understanding of the desulfurization spray drying process lies in the simultaneous mass transfer, heat transfer and reaction of $Ca(OH)_2$ and SO_2. Since the reaction is most effective in the liquid phase, the drying rate controls the sorbent utilization. The use of additives adds more complications to this study.

A comparison between drying applications in food and other industries and desulfurization drying is helpful for fully understanding the SO_2 absorption process. This comparison is shown in Table 1. The SO_2 mass transfer is considered to occur in steps. First SO_2 undergoes gas phase mass transfer, then it changes phase at the droplet surface according to Henry's Law. Finally it undergoes liquid phase mass transfer and reacts with dissolved $Ca(OH)_2$. Chemical reactions between SO_2 and $Ca(OH)_2$ can be as simple as the following:

$Ca(OH)_2 + SO_2 \rightarrow CaSO_3 + H_2O$

$2CaSO_3 + O_2 \rightarrow 2CaSO_4$

This study focused on the effects of additives on sorbent utilization compared with baseline results. Various additives have been tested based on their functions in spray dryer desulfurization-related processes. The properties of additives include hygroscopicity (NaOH, NaCl, $NaHCO_3$, and $CaCl_2$), oxidation of S(IV) to S(VI) (H_2O_2), buffering capacity (benzoic acid, formic acid), dissolution rate enhancement (sucrose, NH_4Cl etc.), and the use of silica (fly ash hydrated with lime) to form calcium silicate or aluminate compounds.

Experimental results showed that the inorganic sodium compounds that exhibit hygroscopicity significantly increase the SO_2 removal efficiency compared with that of the baseline. Hydrogen peroxide can also significantly enhance sorbent utilization and increase SO_2 removal by means of its oxidation potential. Organic acids that act as buffers show little effect on enhancing sorbent utilization. Sugar, which can increase $Ca(OH)_2$ dissolution rates, has some positive effects on sorbent utilization with a small amount of addition. Studies on the dissolution of $Ca(OH)_2$ are few and the dissolution magnitude under the influence of a variety of additive conditions has not been previously known. A bench-scale dissolution rate experimental system was constructed to study the lime dissolution rate. This understanding should lead to a better method of predicting and optimizing spray dryer performance for FGD.

Table 8.1 Comparison between Product Drying and Desulfurization Drying

Item	Product Drying	Desulfurization Drying
Main Mechanism	Evaporation from atomized product solution droplet	Evaporation from atomized $Ca(OH)_2$ slurry droplets + SO_2 diffusion into the droplets + SO_2 and $Ca(OH)_2$ reaction
Macro Balances	Moisture balance and enthalpy or heat balance	Moisture balance and enthalpy or heat balance
Micro Processes	Combined heat and mass transfer (vapor from the droplets and heat into the droplets)	Combined multi-component mass transfer and heat transfer (vapor from the droplets, heat and SO_2 into the droplet)
Design Parameters	Total evaporation time for the mean size droplet and product moisture content	Total evaporation time for the mean size droplet, product moisture content, and SO_2 removal efficiency
Droplet Size Effect	Smaller size will reduce the total evaporation time	Smaller size droplets will be dried up more quickly and they have more surface area which is positive for SO_2 absorption, but the time interval for SO_2 diffusing into the liquid phase of the slurry will be greatly reduced which has a negative effect
Solution Effect	Droplets containing dissolved solids will reduce their surface vapor pressure. Therefore, they evaporate at lower rates than pure liquid droplets of equal size.	Same as the left column, which is a possible explanation for the hygroscopic additive effect. (For many salts, the water vapor pressure difference between pure liquid and solution is significant).

Fly ash was slurried with quicklime at elevated temperatures to enhance spray dryer performance. Bench-scale experimental results showed that this hydration process greatly increased the total surface area of the solids. Pilot-scale tests in the spray dryer revealed that the heating step significantly increases calcium utilization and SO_2 removal of these fly ash/quicklime sorbents.

Additive effects have been extensively investigated and have been helpful in improving spray dryer desulfurization performance. Although this research was focused on spray dryer applications, these additive properties are ubiquitous and can be applied to many other fields.

Literature Review

General Spray Dryer Operations for Desulfurization

The application of spray dryer technologies to the FGD process turned the spray dryer as a basic product drying device into a combination of both a drying device and a chemical reactor. Industrial-scale applications of spray drying started in the milk and detergent industries in the 1920s [1], while applications of spray dryers to FGD use was commercialized in the 1980s.

The overall process for spray drying FGD includes four consecutive operations: reagent preparation, spray drying desulfurization, particulate collection, and solid waste treatment and disposal. The reagent, which is usually lime, is slaked with water to form a slurry which is less than about 25% solids by weight to simplify pumping. The slurry is then pumped through a strainer before reaching the spray nozzles or rotary atomizer. Nozzles or rotary atomizers atomize the slurry into tiny droplets in the spray dryer, and the droplets contact the hot flue gas. While the SO_2 in the flue gas diffuses into the droplets reacting with the alkaline material, the water contained in the droplets evaporates by the heat transferred from the flue gas. Before the droplets reach the outlet of the spray dryer, they become solid particles which constitute both reaction product and unreacted reagent. These dried particles, along with the fly ash, are carried by gas flow to a particulate collector, typically a fabric filter, and collected in the dust hopper. The dust is then removed from the dust hopper for further waste treatment and disposal.

Reagent Properties and Preparations

Though sodium carbonate has also been commercially used as a reagent, lime is the preferred reagent because of its lower cost, widespread availability, and the fact that its reaction product is not highly soluble in water, which simplifies disposal.

There are about 40 different definitions of lime [2]. Usually, the term lime connotes only the burned form of lime, usually quicklime. It may be calcitic (CaO), magnesian (MgO), or dolomitic, with dolomitic being the cheapest. The quicklime used in spray drying desulfurization is usually calcitic. Quicklime must be slaked into a lime slurry which by definition is a form of lime hydrate in aqueous suspension that contains considerable free water. The reactivity of the lime slurry depends not only on the type of lime but also on the sizes of the solid lime hydrate suspensions. Therefore, the slaking process itself has a significant impact on the reactivity of the lime slurry. Good slaking can make the solid particles of lime hydrate very small to maximize the available surface area. The parameters for determining the quality of the lime slurry are [3,4] quicklime quality, makeup water quality, slaking temperature, water-to-lime ratios, equipment type, and operation and addition of additives.

The most commonly used slakers are ball mill slakers and paste slakers. Though ball mill slakers produce finer particles, which have more surface area, paste slakers produce a less abrasive slurry and have lower capital costs and lower power requirements [5].

The solubility of lime is another factor that is believed related to lime reactivity. Solubility of lime hydrate is inversely proportional to temperature [2]. This relationship is shown in Figure 8.1. Though $Ca(OH)_2$ is about 100 times more soluble than CO_2-free $CaCO_3$, it is still considered as slightly soluble. Therefore, methods of increasing $Ca(OH)_2$ solubility have been widely investigated. It was calculated that commercial hydrated limes are about 7% more soluble on an average than chemically pure $Ca(OH)_2$ [6], and that this slight deviation is due to traces of highly soluble alkalis (K_2O and Na_2O) that exist in the commercial hydrates. It was also calculated that a freshly slaked quicklime with fine micro particles was about 10% more soluble than a coarse, crystalline type of hydrate [7].

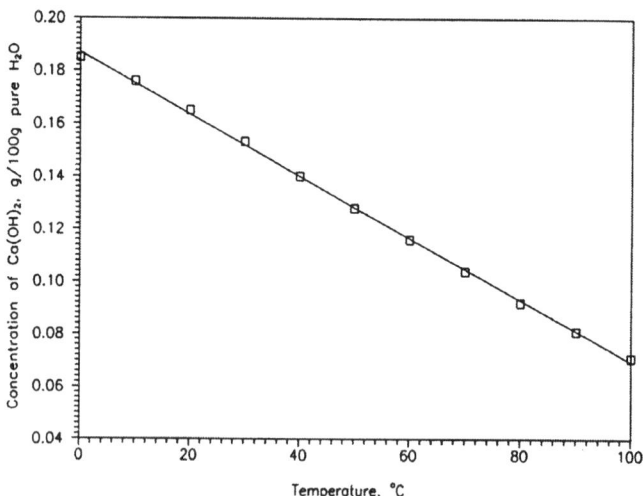

Figure 8.1 Solubility of lime expressed as $Ca(OH)_2$ versus temperature [2].

Solubility of lime under the effects of other factors has been summarized as follows [2]: MgO and such impurities as silica, alumina, and CO_2 (as $CaCO_3$) do not affect the solubility of $Ca(OH)_2$, but may retard its dissolution rate; lime solubility decreases sharply in alkaline solutions and increases in acid salt solutions (some organic substances can increase $Ca(OH)_2$ solubility by forming a new chemical compound between the dissolved organic material and $Ca(OH)_2$; agitation increases both dissolution rate and solubility; the dissolution rates of K_2O and Na_2O, as impurities in commercial lime, are far greater than that of $Ca(OH)_2$.

For high solubility sorbents, such as NaOH or Na_2CO_3, the reagent preparation is simply making a solution of the desired concentration in a suitable mixer.

Spray Drying Desulfurization

The spray dryer absorber is the main piece of equipment for FGD in this process. Large commercial spray drying chambers may have a cylinder diameter of 50 feet (about 15 meters), a height of 45 feet (about 14 meters), and a conical bottom. The temperatures of flue gases entering the spray dryer typically range from 250°F to 350°F. Rotary atomizers can produce reagent slurry droplets much less than 100 micrometers; usually the drops are about 50-75 micrometers. The slurry is pumped to the atomizer by a slurry pump specifically designed for this type of service. The droplets contain $Ca(OH)_2$, water, and any recycled products. The residence time for the droplets in the spray dryer chamber is empirically chosen as 10 seconds, and within this residence time two processes simultaneously take place. First, SO_2 contained in the flue gas diffuses to the surface of the droplets and, after being absorbed by the droplets, reacts with the alkaline reagent in the droplets to form sulfite or sulfate salts. Then, water contained in the droplets evaporates vigorously as heat is transferred to the droplet surfaces from flue gases. As a result, the droplets are dry when leaving the spray dryer chamber, and generally the dry powder has less than 1% free moisture [4,8].

Currently two types of atomizers are in use: rotary atomizers and two-fluid nozzle atomizers. The rotary atomizer creates droplets by rotating a high speed disk or wheel. When slurry is fed to the disk, it flows outward by the influence of centrifugal force, and breaks into droplets when leaving the edge of the disk [1]. The atomizer disk is usually 12-14 inches in diameter and is spun at speeds ranging from 3000 to 50,000 rpm [9]. The droplet sizes depend on the slurry properties and disk rotating speeds. Nozzle atomizers can also be called pneumatic nozzle atomizers since the atomization process involves impacting liquid slurry with high velocity air. The mechanism of atomization is that high-velocity air creates high frictional forces over the liquid slurry surface and causes the liquid slurry to disintegrate into droplets [1]. Droplet size depends on the slurry properties, the liquid flow rate, and the velocity of the air or the air pressure.

According to the direction of the initial contact velocity between flue gas and droplet spray, spray drying mixing can be divided into two types: cocurrent (codirection) and countercurrent (counter-direction). For rotary atomizers, the counter\current mixing pattern has advantages such as reducing wall impingements, decreasing radial trajectory, and increasing spray droplet suspension time in the flue gas. Nevertheless, the advantage is offset by the problem of droplet deposits on the atomizer surfaces [1].

The atomized lime-water slurry droplets contain a large number (for 20% slurry each droplet would contain about 105 particles) of discrete alkali $Ca(OH)_2$ particles. The reactions involving gaseous SO_2 and a lime-slurry droplet have been postulated as sequential [10,11] and consisting of the following four steps:

Step 1, ionization of water

$H_2O \leftrightarrow H^+ + OH^-$

$K_w = 10^{-14}$ (T=298 K);

Step 2, two-step ionization of dissolved SO_2,

$SO_2(aq) + H_2O \leftrightarrow H^+ + HSO_3^-$

$K_1 = 1.29 \times 10^{-2}$ (T=298 K)

$HSO_3^- \leftrightarrow H^+ + SO_3^{2-}$

$K_2 = 6.01 \times 10^{-8}$ (T=298 K);

Step 3, two-step dissolution and ionization of $Ca(OH)_2$,

$Ca(OH)_2(s) \leftrightarrow CaOH^+ + OH^-$

$K_{so} = 1.55 \times 10^{-4}$ (T=298 K),

$CaOH^+ \leftrightarrow Ca^{2+} + OH^-$

$K_3 = 3.24 \times 10^{-2}$ (T=298 K);

Step 4, precipitation of dissolved Ca^{2+} and SO_3^{2-} ions,

$Ca^{2+} + SO_3^{2-} + 1/2\, H_2O \leftrightarrow CaSO_3 * 1/2 H_2O(s)$

$K_4 = 2 \times 10^7$ $(T=298\ K)$.

The equilibrium constants change little within the temperature range of interest in the spray dryer (80-300°F, or 300-423 K).

At high pH values, the calcium ions are converted to $CaOH^+$ complex. In the lime-slurry droplet, the pH will vary from acidic at the gas-liquid interface to alkaline at the liquid-solid interface. At low pH values, sulfite ions are depleted by the reverse rate of Step 2. Steps 1, 2 and 4 are considered fast while Step 3 is considered slow [11]. The rate-controlling mechanisms may be considered, therefore, to be the diffusion of reactants to the reaction zone and the solid dissolution rate [12,13].

SO_2 removal increases with increasing sorbent/SO_2 molar ratios and decreases with increasing approach to saturation temperature. As the sorbent stoichiometry is increased, the efficiency of sorbent utilization will be decreased leading to higher reagent costs. Operating the spray dryer at lower approach to saturation temperatures generally has the effect of greatly increasing the SO_2 removal efficiency and sorbent utilization. However, the possibility of water condensation and wet droplet deposition on the walls increases as the operating temperature approaches closer to the saturation temperature.

Particulate Collection

The devices for particulate collection are usually baghouses (fabric filters) and electrostatic precipitators (ESP).

There are three basic types of baghouses: reverse-air, shake/deflate, and pulse-jet units. Reverse-air and shake/deflate types of baghouses have low air-to-cloth ratios, while the pulse-jet type baghouse has a higher air-to-cloth ratio. Utilities generally use reverse-air baghouses while industrial boilers typically have pulse-jet baghouses [4,14]. The bag filter performance is rated in terms of its permeability, cleanability, and durability.

It has been shown that fabric filters play a major role in desulfurization if installed downstream of the spray dryer [14]. Once collected on the bag surface, the dust forms a dust cake which can still absorb more SO_2. SO_2 removal may reach as high as 25% in the baghouse alone [14]. Therefore maintaining a thick dust layer on the bags without causing a substantial pressure drop is an objective during operation of the baghouse. The baghouse has the advantage of being insensitive to the higher particulate loadings. However, it can be damaged by water and acid condensation since the spray dryer products mixed with fly ash can form cementatious substances which can 'blind' the bags. Acids formed from

SO_2 can 'eat' holes in the bags. By maintaining the spray dryer at higher approach to saturation temperatures and, perhaps, by using flue gas reheat between the spray dryer and baghouse, the condensation problem can be avoided.

Two types of ESPs exist: pipe-type precipitators and duct-type precipitators. Generally the pipe-type precipitators are used for small gas flows, for collection of mists and fogs, and frequently for applications requiring water-flushed electrodes. Duct-type precipitators are used for larger gas flows, dry collection, and sometimes also for water-flushed service [15]. The performance of an ESP is affected by many factors which cover the properties of the gas flow, the electrical properties of the particles, and the energization forms of corona discharge.

The ESP can also contribute to additional SO_2 removal if installed downstream of spray-dryer absorbers, but the additional removal is less than that by a baghouse. This baghouse advantage is believed due to the intimate contact between the solids and flue gas, which reduces the SO_2 gas-phase mass transfer resistance [16]. While the additional SO_2 removal in ESPs does not strongly depend on the approach to saturation temperature, the additional SO_2 removal in the baghouse is very sensitive to the approach to saturation temperature [17,18]. In both ESPs and baghouses the removal decreases as the approach to saturation temperatures increases. The calcium-sulfur molar ratio also affects the SO_2 removal [14,19]; the removal increases with increasing molar ratios for both devices.

Recycle, Solid Waste Treatment and Disposal

Since there is a certain amount of unreacted $Ca(OH)_2$ in the spray dryer product, most spray drying systems today recycle part of their waste solids back into the feed slurry. Recycle has been shown to increase sorbent utilization by allowing partially reacted sorbent to react further with the flue gas. In addition, recycling fly ash with $Ca(OH)_2$ has been shown to be more beneficial toward SO_2 removal than recycling $Ca(OH)_2$ alone [20].

The spray dryer waste product has also been used to detoxify hazardous wastes such as metal plating wastes, oil sludges, and heavy metal processing sludges. Heavy metal leaching characteristics of cadmium waste, chromium plating waste, and aluminum can waste were reduced from hazardous to nonhazardous by fixating with dry products of spray dryers. Oil sludge leaching characteristics were not affected by spray-dryer waste fixation [21].

Spray-dryer waste products are typically disposed with fly ash in landfills. The waste products are dampened in rotary unloaders preparatory to their trip to a suitable site to be used as fill material. Tests such as optimum moisture, compressive strength, permeability, and chemical composition are used to decide suitability for landfilling [22,23]. The stabilization of lime spray dryer waste from treating orimulsion fuel flue gases has been shown to reduce the leaching of Ca, V, and SO_4^{2-} in landfills [24]. Orimulsion is a new fuel consisting of small amorphous particles of a naturally occurring bitumen, emulsified into water.

Methods of Increasing SO_2 Removal and Sorbent Utilization

Spray dryer systems have cost advantages over the wet scrubbing systems when used for low sulfur coal flue gas treatments. With increasing coal sulfur content, the reagent requirements increase, and reagent utilization decreases. Therefore, for medium– to high-sulfur coal flue gases, a need exists to enhance spray dryer SO_2 removal efficiency and improve lime utilization.

ADDITIVES FOR ENHANCING SORBENT UTILIZATION. Much work related to additives has been done to enhance wet and dry desulfurization processes. The types of additives that can increase sorbent utilization in desulfurization have been classified according to the following [10]: deliquescent inorganic salts and related inorganic salts, sodium-containing basic compounds, organic compounds, and oxidation catalysts.

In wet processes, inorganic species such as magnesium or sodium (by holding alkaline sulfite in solution) and organic acids (by buffering the pH of the bulk solution through a series of equilibrium reactions among various species in the solution) have been used to assist in SO_2 removal [25-33]. Forced oxidation (air sparging) and, more recently the oxidation inhibition with thiosulfate (inhibition of the natural oxidation of sulfite to sulfate) have been the means to reduce gypsum scaling [31,32,35-40].

In high temperature (around 2000°F) dry processes, sodium compounds such as NaOH, Na_2CO_3, $NaHCO_3$, and NaCl have all been tested to enhance the sulfur-removing capability of the sorbent, typically calcitic hydrates or dolomitic hydrates [41-44]. Muzio et al [44] have also investigated other inorganic compounds including Li_2SO_4, $LiNO_3$, K_2CO_3, Cs_2SO_4, $Fe(NO_3)_3$, and $FeCl_3$. They found for experiments at 2100°F, Ca/S=2, and additive/CaO weight ratio of 0.03 the use of additives resulted in the following increase in SO_2 capture (efficiency increases are shown by parenthesis) when compared with the baseline CaO: Cs(33%), K (29%), Na(20%), Li(0%), and Fe(0%).

In the low-temperature (about 300°F) dry and spray drying desulfurization process, Ruiz-Alsop and Rochelle [45,46] tested the effectiveness of 18 additives toward improving calcium utilization in dry/dry SO_2 capture. The additives tested included two buffers (5 wt% glycolic acid, and 1 wt% adipic acid), three organic deliquescent substances (5 wt% monoethanolamine, 5 wt% ethylene glycol, and 5 wt% TEG) and thirteen inorganic deliquescent substances (5 mole%

Na$_2$SO$_4$, 5 mole% Na$_2$SO$_3$, 5 mole% CaCl$_2$, 10 mole% NaCl, 10 mole% NaOH, 5 mole% Ca(NO$_3$)$_2$, 10 mole% NaNO$_2$, 10 mole% NaNO$_3$, BaCl$_2$•2H$_2$O, Na$_2$S$_2$O$_3$, KCl, NaBr•2H$_2$O, and LiCl). Experiments were done in a sand-bed reactor at 54-74% relative humidity, which was set to simulate the conditions in bag filters downstream of a spray dryer. It was found that the inorganic deliquescent substances enhanced the SO$_2$ removal efficiency, but the two buffers and the three organic deliquescent substances caused the SO$_2$ removal to decrease. The most effective inorganic deliquescent materials were LiCl, KCl, NaBr, and NaNO$_3$. In dry process, the sodium deliquescent materials such as NaOH, NaHCO$_3$, Na$_2$CO$_3$, NaCl, NaNO$_3$, Na$_2$SO$_3$, and Na$_2$SO$_4$ have been widely shown to be able to enhance the SO$_2$ removal and Ca(OH)$_2$ utilization [17,47-52].

Mechanisms for the enhancement of SO$_2$ capture by using additives in Ca(OH)$_2$ slurries have been suggested to consist of three categories [47]. These are to increase or retain moisture at the sorbent surface, to enhance the basicity of the sorbent, and to change the sorbent particles' physical properties, particularly the surface area of the hydrate.

There are two other possible mechanisms by which the additives can enhance SO$_2$ removal. One is that the additives could increase the solubility or dissolution rate of the solid Ca(OH)$_2$ particles [2]. Another is that oxidation enhancers could oxidize S(IV) in CaSO$_3$ to S(VI) in CaSO$_4$ [53] and prevent CaSO$_3$ from precipitating onto the sorbent surface and blocking the solid from dissolving into the solution. Little evidence has been offered in the literature to support any of these proposed mechanisms.

FLY ASH HYDRATION WITH LIME TO INCREASE SORBENT UTILIZATION. Calcium silicates and calcium aluminates that are available in fly ashes have been shown to be very reactive toward SO$_2$. Fifty-three single and binary metallic oxides that include aluminates, ferrates, carbonates, and titanates have been screened both thermodynamically and kinetically for reactivity with SO$_2$ [54]. Barium titanate and calcium aluminate were the most promising [55]. Kinetic studies have also been performed on two forms of CaSiO$_3$, two forms of β-Ca$_2$SiO$_4$, γ-Ca$_2$SiO$_4$, and Ca$_3$SiO$_5$. The results showed that CaSiO$_3$ and β-Ca$_2$SiO$_4$ have higher overall rates and capacities for sulfation as compared with CaO on a molar basis. The other two silicate forms, γ-Ca$_2$SiO$_4$ and Ca$_3$SiO$_5$, are not as reactive as CaO [56]. The formation of the di- and tri-calcium silicates in fly ashes requires temperatures to be above 900°C. In the sulfation product analysis (X-ray diffraction and infrared), it was found that in the sulfated CaSiO$_3$ products, both silica and SO$_4^{2-}$ are chemically bonded to calcium, while in the sulfated Ca$_2$SiO$_4$, silica is not chemically bonded to calcium sulfate [56].

Extensive efforts have been made on improving the methods of hydrating fly ashes with $Ca(OH)_2$ to extract the useful silicate and aluminates from fly ashes, and the treated fly ash and $Ca(OH)_2$ mixtures have also been widely applied to different FGD systems. Most of these applications, showed that fly ash hydrated with $Ca(OH)_2$, which forms calcium silicates and aluminates, is more reactive than the $Ca(OH)_2$ alone. It is recognized that hydration of fly ashes with $Ca(OH)_2$ at elevated temperatures results in a more reactive sorbent [57-64]. The surface areas of the hydration mixtures are increased with increasing hydration time and temperature. However, surface area alone does not necessarily make the mixtures more reactive [65]. The specific type of silica in the ash also has a significant influence on reactivity [65].

It has been postulated that the limiting step for the hydration process (the reaction of silica from fly ashes with $Ca(OH)_2$) is the dissolution of silica from the fly ash [66,67]. Several approaches have been used to increase the silica dissolution rate. These have included the use of additives (sodium hydroxide, ammonium phosphate, and phosphoric acid) [59,68-70], more reactive forms of silica (diatomaceous earth, bentonitic clays) [60,61,69,70], pressure hydration [60], and fly ash grinding before hydration [58,63]. Although some positive effects arise from the use of the additives and pressurized hydration, these methods of increasing silica dissolution significantly increase the cost of sorbent preparation. The wet grinding of both high calcium and high silica fly ashes can generate very reactive sorbent due to the pozzolanic reaction between the calcium and the silica of the fly ash in the wet grinder [58].

Sorbents prepared from ground fly ash are always more reactive than sorbents made from unground fly ash at the same hydration conditions. Pilot-plant results have verified that initially grinding the fly ash before hydration with $Ca(OH)_2$ decreases the optimal hydration time from 12 hours to 3 hours [63].

Previous Lime and Limestone Dissolution Rate Studies

The prediction of lime and limestone dissolution rates is important in determining scrubber performance for FGD systems. A wide range of studies has been conducted to investigate limestone dissolution rates in typical wet scrubber systems. These studies can be divided into three categories: limestone dissolution rate studies by varying parameters such as temperature, and pH, etc.; limestone dissolution rate studies under the influence of magnesium, sulfite, and sulfate ions; and limestone dissolution rate studies under the influence of other additive ions.

For the first category, the limestone dissolution rates were extensively tested by changing such variables as pH, temperature, particle size, magnesium content, and aqueous sulfite and sulfate concentration [71], and the experimental data were correlated by a best-fit model equation as:

Equation 8.1

$$ln(DR) = a + b_1(TP) + b_2(pH) + b_3ln(STIR) + b_4ln(TSA) + b_5ln(Mg)$$

where
DR is the dissolution rate of calcium or magnesium (mg/g-min),
TP is the inverse of temperature (K-1),
pH is the solution pH,
STIR is the stirring rate in the reactor (rpm),
TSA is the total surface area of the test material (cm^2/g),
Mg is the magnesium concentration in the reactor feed solution (mg/l), and
a, b_1, b_2, b_3, b_4, and b_5 are constants.

Tests showed that the presence of sulfite produced a significant increase in the dissolution rate of $CaCO_3$.

At low pH values (<4), the dissolution rate is approximately proportional to the hydrogen ion concentration [72-77]. This means that at low pH, the dissolution rate is controlled by the diffusion of hydrogen ions to the crystal surface.

In the middle pH range (4.5-7), the limestone dissolution rate was originally found to vary linearly with pH, and the rate is considered essentially independent of the CO_2 partial pressure [74,78]. Recent studies have shown that the effect of partial pressure of CO_2 on the dissolution rate varies between pH 4 and 5 [79-81]. At a pH above 5, the dissolution rate is controlled by diffusion of OH^-, HCO_3^-, and other species, and by the finite rate of CO_2 reaction. The partial pressure of CO_2 influences the dissolution rate to a lesser extent in the lower pH range. In this range most dissolved CO_2 exists as H_2CO_3 and the dissolution rate is considered controlled by the hydrogen ion transfer. However, at higher pH the dissolved CO_2 is converted to HCO_3^- and the solubility of CO_2 in molecular plus ionic form is very high compared with that at lower pH. Besides controlling hydrogen ion transfer, the effect of the other components, resulting from dissolved CO_2, on the dissolution rate seems significant.

At higher pH levels, the dissolution rate is much lower and occurs via poorly defined mechanisms [74,82]. Due to the lack of applicability of the dissolution rate at higher pH, the dissolution of limestone in higher pH (>8) has not received much study.

Advances in Spray Drying Desulfurization for High-Sulfur Coals

Though temperature is an important factor affecting limestone dissolution rate, very little effort has been made in determining correlations under different conditions. Equation 8.1 shows the dissolution rate to be exponentially related to temperature in the form of $DR = c\ EXP(b_1/T)$ where c is a constant dependent on the other parameters, and b_1 is also a constant related to the type of stone. Values of b_1 ranged from -1700 for Fredonia calcium carbonate to -5670 for Aragonite magnesium carbonate.

The smaller the particle size for a given mass, the higher the dissolution rate. The particle size effect is correlated to total surface area [71], and this is shown in Equation 8.1. Dissolution rate is directly proportional to the total surface area of the limestone particles. For a given weight of limestone, since the total surface area is inversely proportional to particle diameter, the dissolution rate is inversely proportional to particle diameter.

The magnesium ion has been found to be an inhibitor for limestone dissolution. It has been shown that the presence of magnesium ions reduces the dissolution rate of $CaCO_3$ [83-85]. One explanation for this phenomenon is that magnesium ions may absorb onto the limestone surface and slow the dissolution process much the same as many inhibitors. A dissolution study by the Electric Power Research Institute [86] shows that at pH values of 5.0 and 5.8, the limestone dissolution rate is independent of the magnesium content in the solids until the $MgCO_3$ concentration reaches 4–5%. Beyond 5%, the magnesium concentration increase results in a decrease in the overall limestone dissolution rate but an increase in the magnesium dissolution rate.

Sulfite ions have been found to have a dual effect on the dissolution of limestone [87]. The sulfite/bisulfite pair acts as a buffer at the limestone interface by providing hydrogen ion which enhances the dissolution reaction. However, sulfite also inhibits the dissolution by adsorbing onto the limestone as calcium sulfite. The presence of sulfite has been shown to increase the dissolution rate of $CaCO_3$ significantly [71]. The most affected, Fredonia stone, was increased from 1.8 (without sulfite) to 11.7 mg/g-min (with sulfite). Chan and Rochelle [79] quantified inhibition by sulfite in concentrations greater than 1 mM at conditions far from equilibrium. Measured rates for reagent calcite were much less than values predicted by a mass transfer model that accounted for the buffering effect of sulfite/bisulfite. The experimental data also showed that at low concentrations of sulfite and for a given pH, the dissolution rate is enhanced over the rate observed without sulfite. Jarvis et al. [88] studied FGD solution effects on ten different limestone types. Of the species studied, magnesium and adipic acid were found to be minor inhibitors while sulfite and aluminum fluoride complexes were noted to be significant inhibitors to the limestone dissolution rate. The magnitude of sulfite inhibition was found to be dependent on limestone type. The effect of sulfite on the limestone dissolution rate was modeled as mass transfer controlled with the calcite surface concentrations controlled by surface kinetics involving sulfite at the surface [87]. The dissolution rate in the presence of sulfite was found to be controlled by a combination of surface kinetics and mass transfer.

Sulfate ions have been shown to increase limestone dissolution rate. One study with limestone from Warner Co., in Bellefonte, Pennsylvania (97.15% $CaCO_3$, 1.10% $MgCO_3$, 1.35% insolubles) showed that the presence of dissolved sulfate ions, at pH values below 5.0, increased the dissolution rate slightly [76]. The limestone dissolution rate increase was also observed when such salts as Na_2SO_4 and $MgSO_4$ were added to the solution [89]. An increase in the dissolution rate is thought to be the effect of bisulfate ions coexisting in the solution in equilibrium, by providing an additional means of diffusing acidity (bisulfate ions) to the limestone surface. Sulfate is a typical species present in scrubbing systems and is a known inhibitor for calcium sulfite crystallization [87]. The presence of sulfate in solution reduces both the dissolution and crystallization rates [90,91].

Depending on the type of ions formed, other additives can affect limestone dissolution rates. Jarvis et al [88] studied the effects of solution species on limestone dissolution rates. Solution species that were tested included pH, Ca^{2+}, Mg^{2+}, Na^+, SO_3^{2-}, SO_4^{2-}, CO_3^{2-}, Fe^{3+}, Al^{3+}, F^-, and adipic acid. Of the species studied, magnesium and adipic acid were minor inhibitors while sulfite and aluminum fluoride complexes were noted to be significant inhibitors to the limestone dissolution rates. Dissolution rates decreased with increasing Ca^{2+} and CO_3^{2-} since these species decrease the solubility of $CaCO_3$ at the limestone surface. No significant effects were noted with changes in the concentrations of Na+, SO_4^{2-}, and Cl⁻, and it was confirmed that sulfate does not inhibit limestone dissolution in a manner analogous to sulfite. The trace metals Fe2+, Fe3+, and Mn2+ (at a concentration of 5 ppm) also had no effect on the dissolution rate. Recent studies [89] showed that in chloride solutions such as $CaCl_2$, $MgCl_2$ and NaCl, the limestone dissolution rate decreases when the salt concentration increases. The presence of barium and/or strontium [84] and benzoic acid [78] has been shown to increase the limestone dissolution rate. Barium and strontium only slightly increase the dissolution rate, whereas 0.2% benzoic acid increased the dissolution rate by 10% at pH 5.5. In the studies of crystal habit [92], it was found that several crystal habit modifiers, mostly organic acids, significantly reduce crystal growth rate and particle size.

Spray Dryer FGD Models

A spray drying desulfurization model has been developed by Damle [16] and later expanded by Partridge [12]. Partridge's model of the constant-rate drying period is based on film theory and treats the atomized slurry droplet as a sphere of discrete sorbent particles with the fluid phase uniformly distributed around individual sorbent particles. The absorption rate is considered a process of absorption of the SO_2 into the slurry drop accompanied by an instantaneous chemical reaction. The rate is governed by the size of four transfer coefficients: the gas-film mass transfer coefficient, the liquid side SO_2 mass transfer coefficient, the liquid $Ca(OH)_2$ mass transfer coefficient and the solid dissolution rate constant. The overall rate is assumed to depend on these resistances in series. The model estimates these coefficients from correlations and determines if one resistance is much larger than the others. If so, that portion of the transfer process is assumed to control the overall rate.

Lime Dissolution Rate Studies

Introduction

Spray drying processes have different physical and chemical features compared with wet scrubbing operations. The dissolution rates of lime and limestone could become a critical limiting step for desulfurization. Spray drying processes contain two phenomena that are totally different from wet scrubbing. One is the short residence time of the liquid phase in droplets, and the second is the highly concentrated ions in droplets and the ion interactions accompanying the water evaporation. According to the literature review, the second phenomenon has a relationship to the lime dissolution rate, and because of these phenomena, the lime dissolution rate needs special consideration and study. Since the dissolution rate can become the limiting step for the overall SO_2 removal in spray dryers (as seen from using different sorbents such as NaOH, $Ca(OH)_2$, and $CaCO_3$), a higher dissolution rate could mean higher sorbent utilization. Based on the above reasons, the objectives of these studies are as follows: The first is to conduct experiments to measure the dissolution rate of lime, and study the mechanisms of the sorbent ($Ca(OH)_2$) dissolution rates and the factors that have the greatest effects on the dissolution rate. The second is to study the influence of additives on lime dissolution rates. These additives may exist as trace compounds in the sorbents, the products of the SO_2, or other gaseous pollutants reacted with the sorbents or added as artificial additives.

Experimental Approach and Limestone Test Results

The experimental setup was designed and constructed, and is schematically shown in . It contains three major blocks. The first block is the sorbent solution container mounted on a temperature controlled hot-plate stirrer. The sorbent solution in the container is agitated by a magnetic mixer, and its pH and temperatures are monitored through a pH probe and a thermocouple, respectively. A gas sparger is immersed into the solution for saturating the solution with a specific gas. The second block consists of a combination of pH meter and pH stat, a data acquisition system, a cylinder for containing acid solution, a balance for monitoring the weight of acid, and a titration liquid pump. This block works to decrease the solution pH by injecting acid whenever the pH value in the solution gets higher than the upper pH limit value preset on the pH stat. The pH stat controls the titration pump. The third block is composed of a rotation motor and its speed controller. The rotating disk is attached to one end of the shaft, the other end of which is clamped to the motor. The speed of rotation can be changed by the motor speed controller.

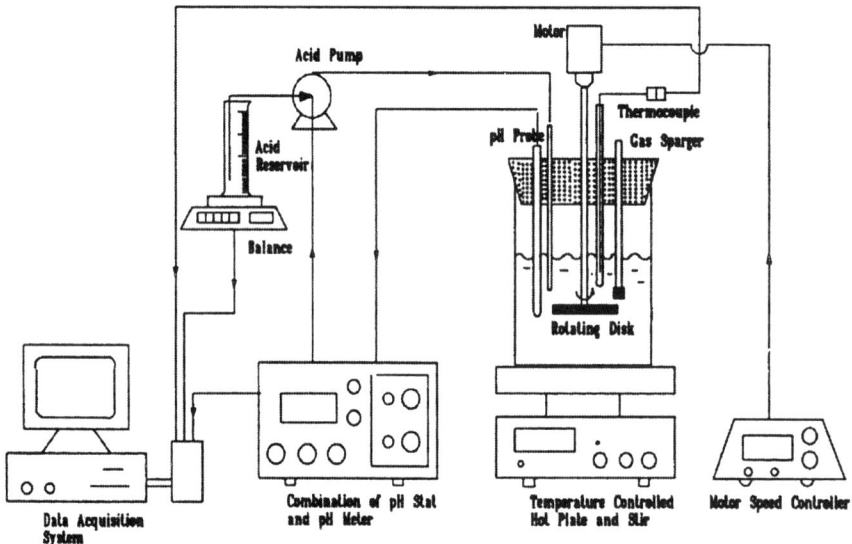

Figure 8.2 Schematic diagram of experimental setup.

A test of limestone dissolution rates was used to verify the experimental system so the results could be compared with published data. Fisher reagent grade limestone was chosen for the initial dissolution rate test. The overall test results are shown in Figures 8.3 and 8.4.

Advances in Spray Drying Desulfurization for High-Sulfur Coals

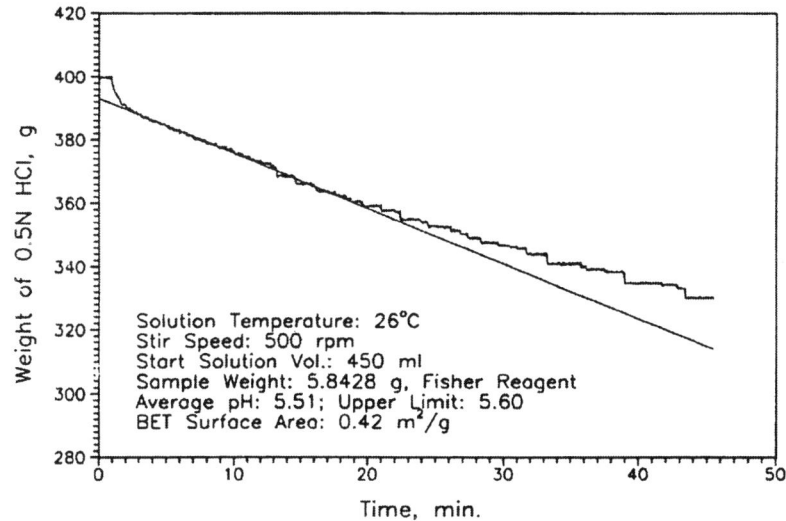

Figure 8.3 Acid HCl weight in the reservoir versus time at pH upper limit 5.6 in the Fisher reagent limestone dissolution test.

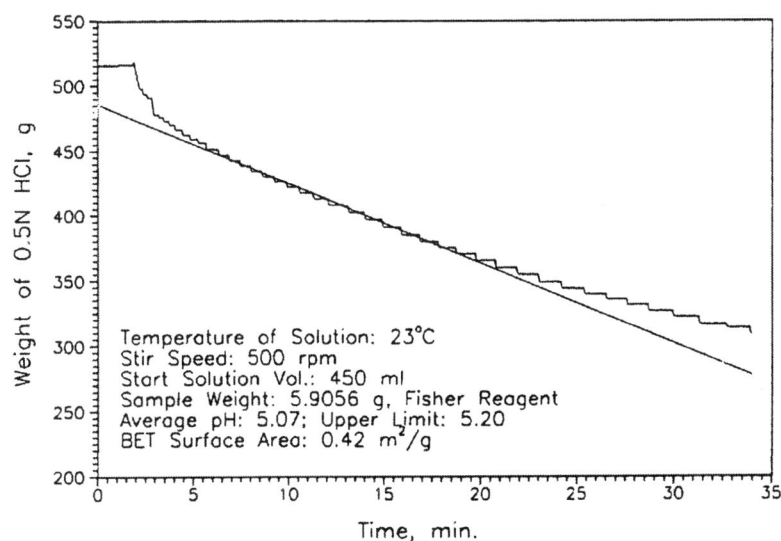

Figure 8.4 Acid HCl weight in the reservoir versus time: pH upper limit 5.2; Fisher reagent limestone dissolution test.

Figures 8.3 and 8.4 give acid (HCl) weight versus time. The pH upper limit was set manually in the pH stat, and for these two cases were 5.6 and 5.2, respectively. The acid consumptions were monitored continuously by the balance.

Both the disappearance of smaller particles and the size reduction of larger particles will contribute to the reduction of the overall sorbent surface area. This phenomenon is more clearly seen in Figures 8.3 and 8.4 in which the acid consumption rates deviate from a straight line as time passes; i.e., less acid is demanded to keep the limestone dissolution at constant pH.

The dissolution rate can be calculated by knowing the slope of the acid versus time relationship and the limestone BET surface area. The relationship is based on the following:

ΔW (g) is the acid weight change.
Δt (min) is the time change.
$\Delta W/\Delta t = L$ (g/min) is the acid solution consumption rate.
C_{acid} (mole/liter) is the acid concentration.
W_s (g) is the initial weight of sorbent (limestone).
BET (m^2/g) is the BET surface area of the sorbent to be tested.
ρ (g/cm^3) is the density of the acid solution (the dilute acid density is assumed to be the same as the density of water).
d is the sorbent/acid complete reaction molar ratio (for $CaCO_3$/HCl d=1/2).

The acid consumption rate R' in gmole/s is

$$R' = \frac{LC_{acid}}{60 \cdot 10^3 - \rho}$$

The sorbent dissolution rate (flux) Rs in gmole/cm^2-s is

$$R_s = \frac{LC_{acid} d 10^{-8}}{60 \cdot BET \cdot W_s \cdot \rho}$$

The BET surface area of Fisher Scientific reagent limestone was measured as 0.42 m^2/g. The dissolution rates of the limestone under different pH values were obtained and compared with the literature [88]. It can be seen in Table 8.2 that the measured dissolution rates are comparable.

Rotating Disk Technique and Lime Dissolution Rate Experimental Studies.

Rotating Disk Technique. In the lime dissolution rate study, it was found that the lime solution pH value is very difficult to control. This is because the reagent lime particles dissolve much faster than the limestone particles. To measure the dissolution rate of lime, the surface area of lime has to be reduced to counter this effect.

Table 8.2 Comparison of Results for Fisher Reagent Limestone Dissolution Rate

Solution Treatment	pH	Temperature C	Rate x10^9 gmol/cm^2-s
With N_2 Sparging [88]	4.5	25	4.78
	5.0	25	1.36
	5.5	25	0.74
	5.75	25	0.54
	6.0	25	0.36
Without Sparging (This Test)	5.07	23	1.03
	5.51	26	0.30

Therefore, the idea of using a rotating disk was proposed. A 4 cm diameter die was designed to press the $Ca(OH)_2$ into a solid disk. The die, which was made of stainless steel, consists of a cylinder and two pistons. The two pistons have a small clearance with the cylinder. When making the $Ca(OH)_2$ disks, the bottom piston is fit into the cylinder, and then an amount of $Ca(OH)_2$ is poured into the cylinder. Finally, the upper piston is carefully slid into the cylinder.

The die thus mounted is ready to be pressed in a hydraulic compressor. The optimal press force for forming good $Ca(OH)_2$ disks is about 6 tons which is equivalent to a pressure of about 48 MPa [93]. After reaching this pressure, the force is released by turning off the hydraulic valve on the press. The die is then taken out of the press, and the bottom piston is slowly turned and removed. Then the die is put back into the press again to have the disk pushed out by the upper piston. When the disk is out, it is wrapped in a wax paper for further processing. The upper piston is also taken out of the cylinder, and all the die components are carefully cleaned using lens papers and methanol.

The disk thus made is not very smooth on the surfaces, especially on its edges. Therefore, surface polishing is necessary. The first step is to use abrasive sandpaper 320 (approximately 33 micro grit) to get rid of major defects on the disk surfaces. The second step is to use abrasive sandpaper 600 (approximately 14 micro grit) to make the disk surface more smooth. The final step is to use polishing cloth with diamond spray of 1 micrometer. The disks are put into glass bottles filled with nitrogen gas for future use.

The compressed lime disk was tested for its chemical composition changes. The major concern was that the pressure might change the water content in the lime.

Three TGA dehydration tests showed that from zero to 48 MPa the lime water content did not change. The TGA results are shown in Figures 8.5, 8.6 and 8.7. Figure 8.5 is a dehydration test with lime powder; Figure 8.6 is for a pressed disk lime with 24 MPa pressure, and Figure 8.7 is for a pressed disk lime with 48 MPa.

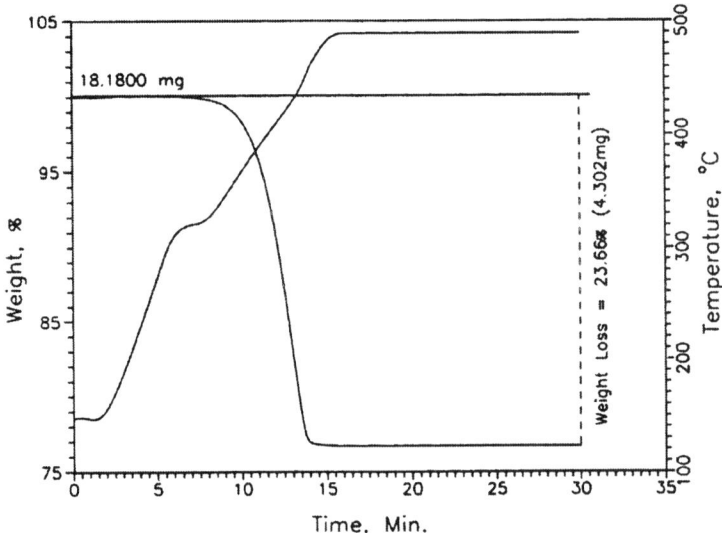

Figure 8.5 TGA dehydration test for lime powder.

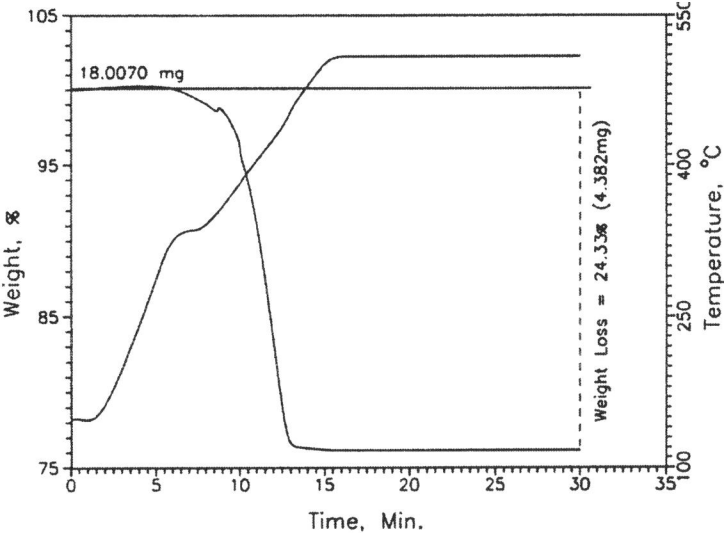

Figure 8.6 TGA dehydration test for 24 MPa pressed.

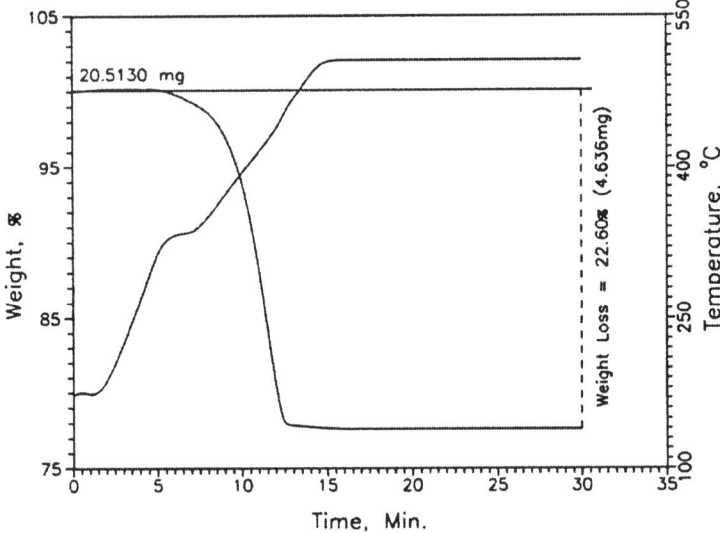

Figure 8.7 TGA dehydration test for 48 MPa pressed lime.

Before the dissolution rate experiment, a 3mm diameter hole is drilled in the center of disk and it is attached to the tip of a 3/8" diameter shaft by super glue (the diameter of the tip of the shaft is 2mm). The thickness of the disk is measured by a micro calliper, and the surface area of the disk is calculated by adding the two disk face surface areas and the rim surface area, and subtracting the surface area occupied by the hole in the disk center. The disk is then lowered into 450 ml of super-Q water for the dissolution experiment. The pH upper limit is set according to the experiment. The solution temperature is controlled by a temperature-controlling hot-plate stirrer. The acid used was from Fisher Scientific as standardized concentrated acid.

Lime Dissolution Rate At Different Conditions. The dissolution rate of lime is shown in Figure 8.8 versus disk rotating speeds at room temperature. It can be seen that when the disk rotating speed is above 300 rpm, the dissolution rate is almost independent of the disk rotation rate. Therefore, a disk rotational speed of 300 rpm was chosen as the speed for subsequent experiments.

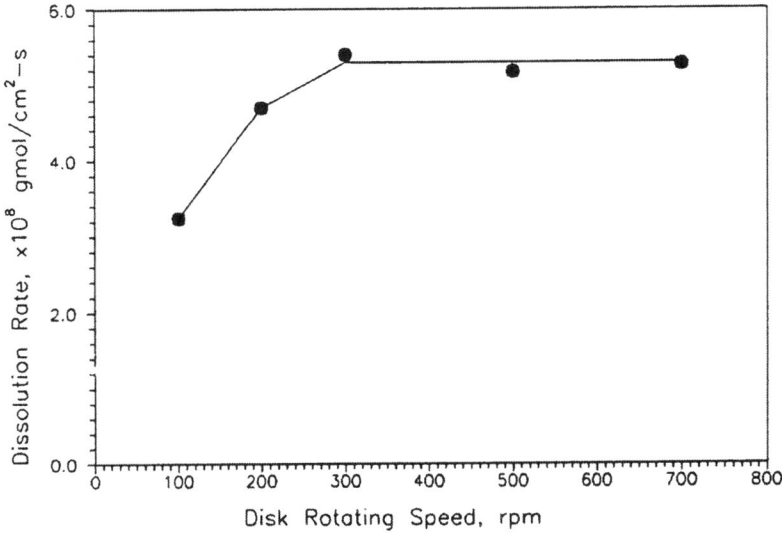

Figure 8.8 Lime dissolution rate versus disk rotation speed.

A series of experiments was conducted to test lime dissolution rates under different solution pH values. These experiments were conducted by keeping the solution temperature constant at 50°C to simulate the conditions in a spray dryer. Figure 8.9 shows the dissolution rate versus pH. Below pH 6, the dissolution rate increases with decreasing pH. For a typical measurement, the variation of pH, temperature, and consumption of acid are shown in Figures 8.10 and 8.11, respectively. From Figure 8.11, the pH of the solution changed over a wider range than that in limestone tests. This is probably because the lime rotating disk is a surface source for dissolution while the limestone particles are well mixed in the solution. Maintaining a uniform pH value is easier for the limestone solution. Figure 8.10 gives the solution temperature versus time, and it can be seen that the temperature was held reasonably constant by the hot plate. Figure 8.11 shows the acid consumption rate versus time. Though the consumption rate is not strictly a straight line, it can be considered a straight line between 5 and 20 minutes. This time choice is based on two considerations. One is that initially the disk surface may not be smooth and free of dust that can make the initial dissolution rate greater. Second is that after some time has elapsed the solution may have a high ion concentration that may result in some deposition on the disk surface to make the dissolution rate decrease.

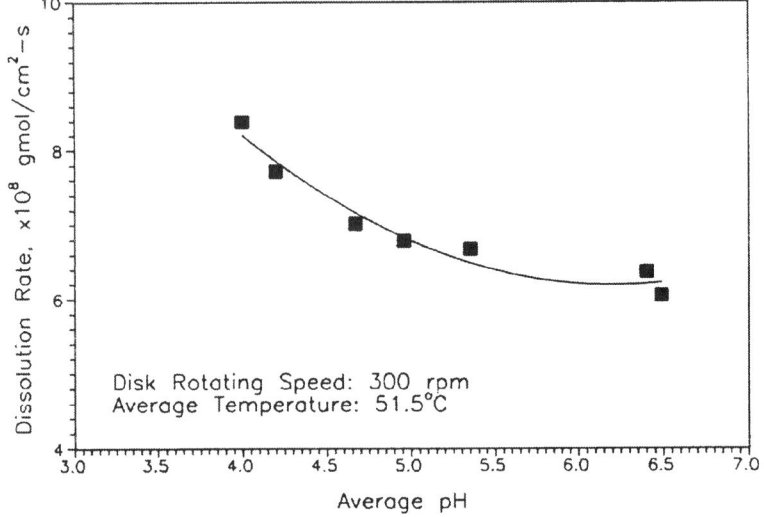

Figure 8.9 Lime dissolution rate versus solution pH.

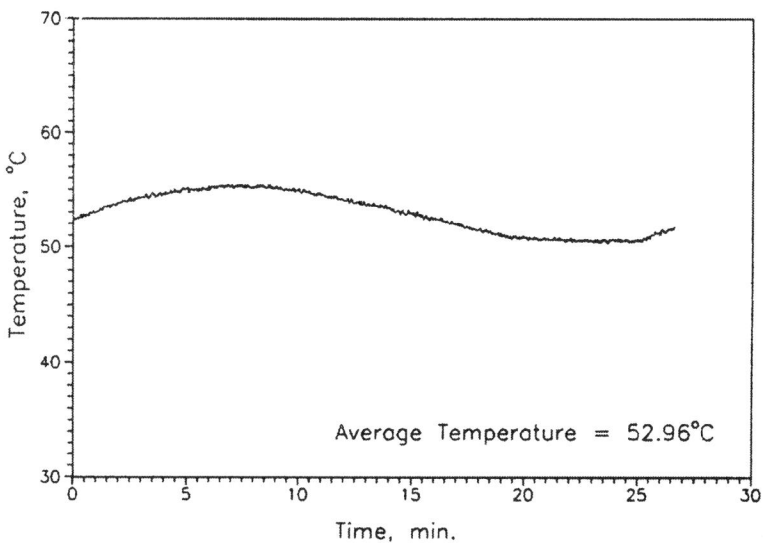

Figure 8.10 Solution temperature versus time for lime dissolution study.

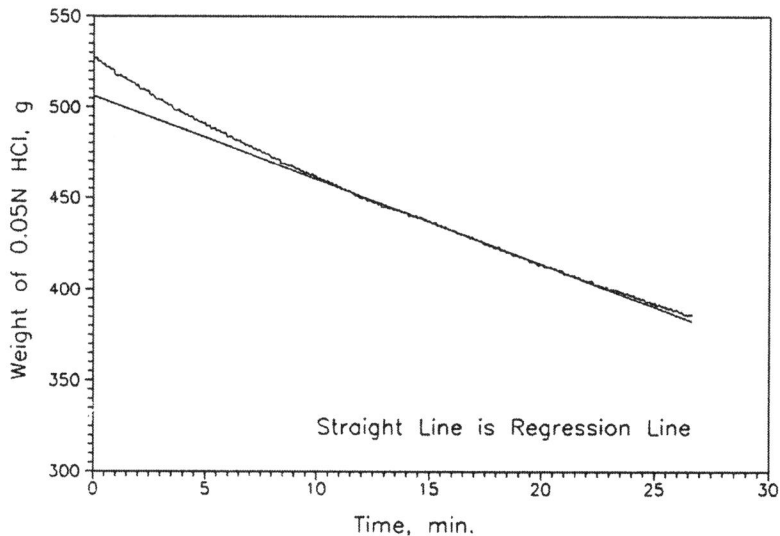

Figure 8.11 Acid HCl weight in reservoir versus time for lime dissolution study.

EFFECTS OF ADDITIVES ON LIME DISSOLUTION RATE. From solubility studies [2], it was found that most inorganic salts increase $Ca(OH)_2$ solubility by 10-15% in 0.1-0.2% salt solutions, but several soluble organic substances markedly increase $Ca(OH)_2$ solubility far greater than any inorganic chemicals. Glycerol at 25°C yields a steady increase in solubility up to a glycerol concentration of 69%. $Ca(OH)_2$ solubility at the maximum glycerol level is 3.55g CaO/100g H_2O, which is about 22 times greater than $Ca(OH)_2$ solubility in pure water. Phenol produces an even greater rise in $Ca(OH)_2$ solubility. At 23°C and 30% phenol concentration $Ca(OH)_2$ solubility can reach 8.67g CaO/100g H_2O, which is an increase of 75 times over pure water. Sugar exerts the greatest influence on lime solubility. At 25°C and 35% sugar concentration, $Ca(OH)_2$ solubility can reach a maximum of 10.1g CaO/100g H_2O, which is 100 times more than $Ca(OH)_2$ solubility in pure water. The solubility of $Ca(OH)_2$ with respect to the sugar concentration is shown in Figure 8.12 where 35% sugar weight concentration corresponds to the maximum $Ca(OH)_2$ solubility.

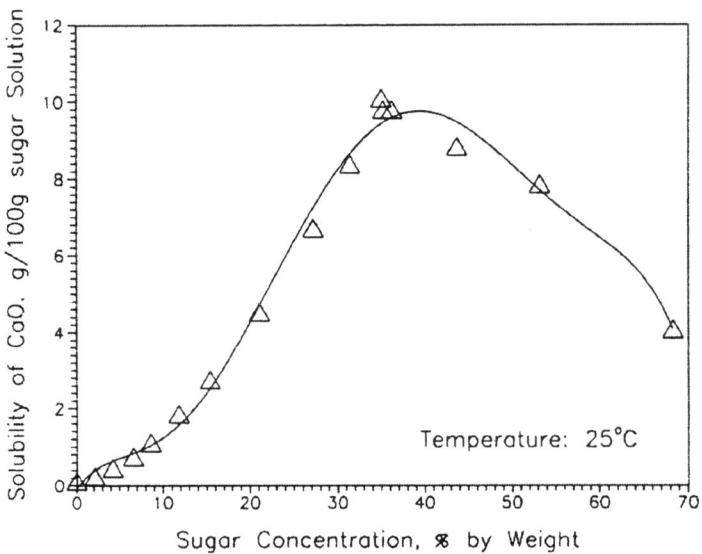

Figure 8.12 $Ca(OH)_2$ solubility in sugar solution [2].

The dissolution rates of lime with some organic additives are shown in Figure 8.13. It can be seen in Figure 8.13 that in the additive concentration range of less than 10% by weight phenol has the greatest effect on enhancing lime dissolution rates. When the phenol concentration gets to 20%, the lime dissolution rate declines because not all the phenol dissolves in water. Sugar enhances the dissolution rate of lime steadily. The dissolution rate of lime can be related to the sugar weight concentration and fitted as an exponential regression line.

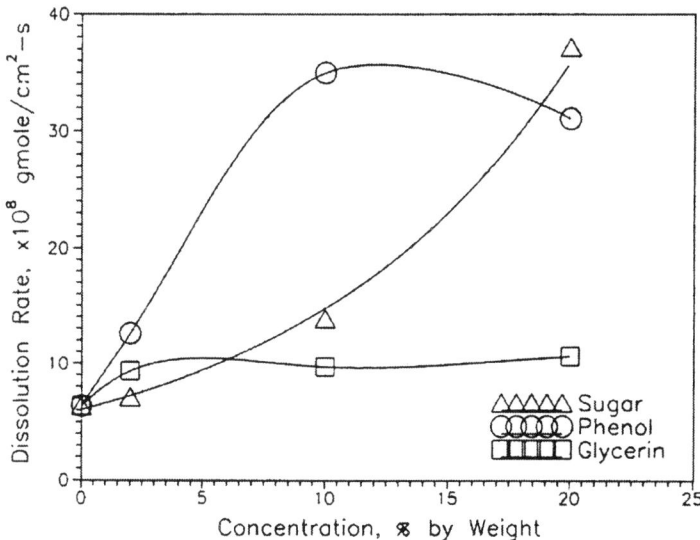

Figure 8.13 Lime dissolution rate for different additive solutions

Adding glycerin to the solution results in little improvement on lime dissolution rates over pure water, and an increase in glycerin concentration leads to no further improvement in lime dissolution rates. Tap water was tested for the effect of impurities on lime dissolution rates. Using laboratory tap water, it was found that tap water (dissolution rate $6.39 \cdot 10^{-8}$ gmole/cm^2-sec at average temperature 51.05°C) did not change the lime dissolution rate much compared with the lime dissolution rate in pure super-Q water (dissolution rate $6.05 \cdot 10^{-8}$ gmole/cm^2-sec at average temperature 49.04°C). The test results for organic additives are shown in Table 8.3.

Table 8.3 Results of Lime Dissolution Rate Affected by Organic Additives

Additive Name	Additive % Weight in Solution	pH Stat Upper Limit	Average pH of Solution	Average Temp of Solution °C	Dissolution Rate of Lime x10^8 gmol/cm^2-sec.
Pure Water	0%	7.00	6.49	49.04	6.05
Tap Water	0%	7.00	6.95	51.05	6.39
Sucrose	2%	7.00	6.21	55.52	7.11
	10%	7.00	6.61	51.68	13.86
	20%	7.00	7.18	48.63	37.19
Phenol	2%	7.00	6.71	50.75	12.58
	10%	7.00	6.91	46.99	34.99
	20%	7.00	6.91	44.24	31.07
Glycerin	2%	7.00	6.13	50.97	9.36
	10%	7.00	6.28	51.75	9.72
	20%	7.00	6.42	50.97	10.64
$NH_4C_2H_3O_2$	2%	7.00	7.02	47.28	21.32
Urea	2%	7.00	4.52	45.60	7.52
Alcohol	10%	7.00	5.89	54.12	4.98

Inorganic additives were also tested for their effects on lime dissolution rates. It was found that most inorganic ammonium compounds can increase lime dissolution rates tremendously. The results from tests on dissolution rates versus temperature showed that NH_4Cl accelerated the mass transfer process. These results are summarized in Figure 8.14 where an Arrhenius plot of dissolution rate versus inverse temperature is given. The activation energy from this figure was calculated to be 3.63 kcal/g-mole, which is much less than that of a chemical reaction. The experimental results also showed that ammonium sulfate does not produce the same effect on lime dissolution enhancement. Further studies found that the insignificant effect by ammonium sulfate is due to the lower solubility of calcium sulfate that could be back deposited onto the disk surface to block calcium hydroxide from further dissolution into the solution. With the other ammonium compounds, all the calcium compounds formed by the cations and calcium ions have high solubilities. The solubilities of corresponding calcium compounds are shown in Table 8.4. The effects of inorganic additives on lime dissolution rates are shown in Table 8.5. Figure 8.15 shows some of the most effective ammonium compounds.

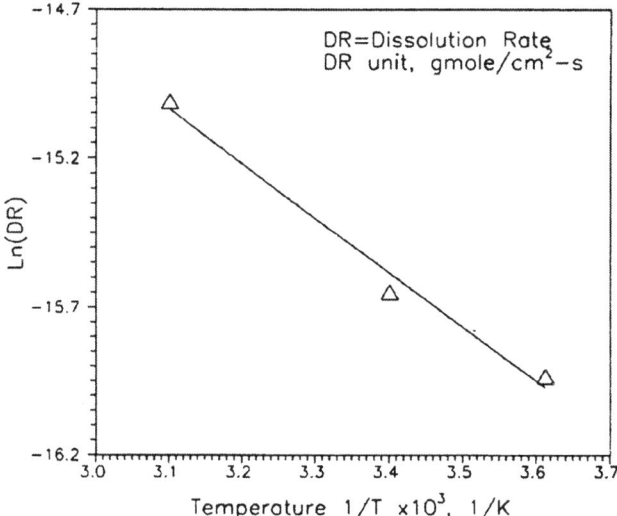

Figure 8.14 Lime dissolution rate versus temperature as derived from activation energy formula.

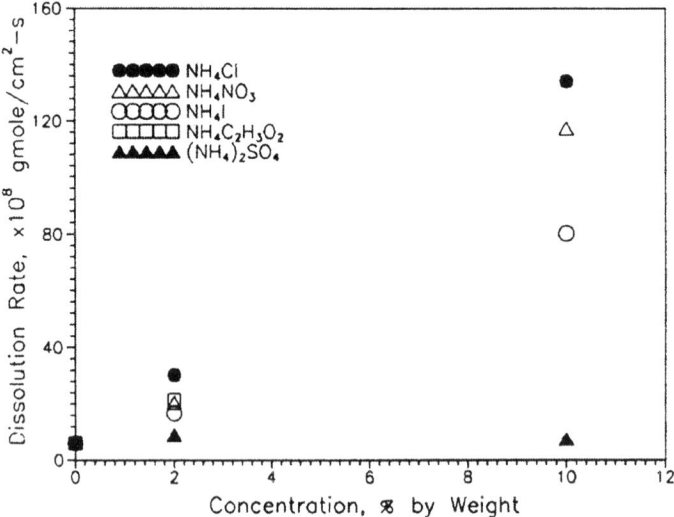

Figure 8.15 Effect on lime dissolution rate by ammonium compounds.

Table 8.4 Solubilities of Calcium Compounds (by replacing ammonium ion with calcium ion from the ammonium compounds) [94]

Name	0 °C	10 °C	20 °C	30 °C	40 °C	50 °C	60 °C	70 °C	80 °C	90 °C	100 °C
Acetate •$2H_2O$ $Ca(C_2H_3O_2)_2$	37.4	36.0	34.7	33.8	33.2	……	32.7	……	33.5	……	……
Acetate •$1H_2O$ $Ca(C_2H_3O_2)_2$	……	……	……	……	……	……	……	……	……	31.1	29.7
Chloride $CaCl_2$ •$6H_2O$	59.5	65.0	74.5	102	……	……	……	……	……	……	……
Chloride $CaCl_2$ •$2H_2O$	……	……	……	……	……	……	136.8	141.7	147.0	152.7	159
Nitrate •$4H_2O$ $Ca(NO_3)_2$	102.0	115.3	129.3	152.6	195.9	……	……	……	……	……	……
Nitrate •$3H_2O$ $Ca(NO_3)_2$	……	……	……	……	237.5	281.5	……	……	……	……	……
Nitrate $Ca(NO_3)_2$	……	……	……	……	……	……	……	……	385.7	……	363.6
Sulfate •$2H_2O$ $CaSO_4$	0.176	0.193	……	0.209	0.210	……	0.205	0.197	……	……	0.162
Hydroxide $Ca(OH)_2$	0.185	0.176	0.165	0.153	0.141	0.128	0.116	0.106	0.094	0.085	0.077

Table 8.5 Results of Lime Dissolution Rate Affected by Inorganic Additives

Additive Name	Additive % Weight in Solution	pH Stat Upper Limit	Average pH of Solution	Average Temp of Solution °C	Dissolution Rate of Lime $\times 10^8$ gmol/cm^2-sec.
Pure Water	0%	7.00	6.49	49.04	6.05
Tap Water	0%	7.00	6.95	51.05	6.39
NH_4Cl	2%	7.00	6.97	49.43	30.29
	10%	7.00	7.02	48.56	133.97
	20%	7.00	7.04	51.02	888.84
NH_4NO_3	2%	7.00	6.94	46.81	20.41
	10%	7.00	6.98	50.81	117.26
NH_4I	2%	7.00	6.95	47.98	16.74
	10%	7.00	7.02	51.15	80.18
$(NH_4)_2SO_4$	2%	7.00	6.89	47.43	9.04
	10%	7.00	6.95	51.59	7.44
$NH_4C_2H_3O_2$	2%	7.00	7.02	47.28	21.32
$CaCl_2$	2%	7.00	6.22	53.91	7.56
	10%	7.00	6.38	53.49	9.19
NaCl	10%	7.00	5.20	53.44	6.79

Pilot Spray Dryer Experimental Approach and Results

Spray Dryer and Sampling System

A spray drying absorption pilot plant has been designed, constructed, and operated for direct tests on additive effects. The purpose of the experiments was to verify bench scale experimental results with additives on operational variables such as hygroscopicity and dissolution rate The pilot plant was designed with an air flow capacity of 160 scfm and a residence time in the spray dryer chamber of about 10 seconds. A schematic system flow chart is given in .

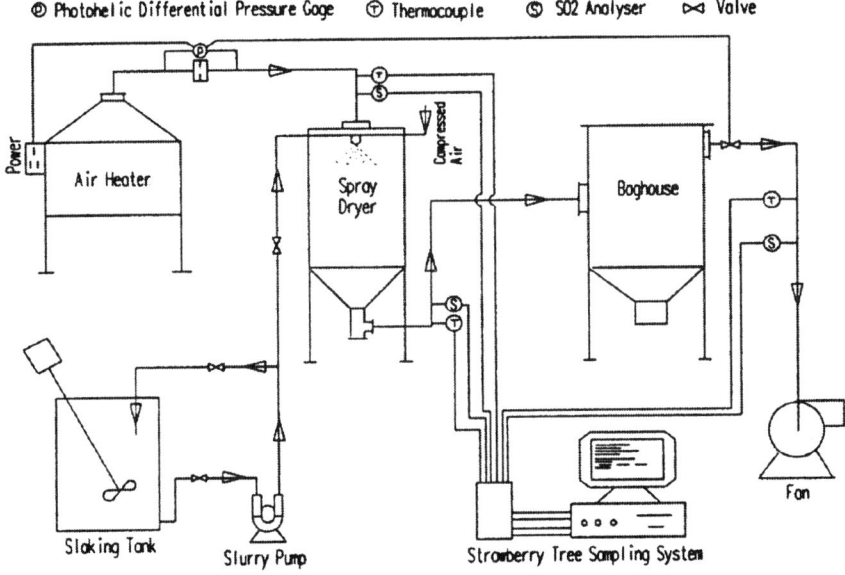

Figure 8.16 Schematic diagram of pilot spray dryer system.

As shown in Figure 8.16, five major portions comprise the pilot plant. They are an electric heater, a slaking system, the spray dryer and baghouse, and the sampling system. The heater has a total capacity of 10kw that can increase ambient air (60°F, 16°C) with a flow rate of 150 cfm to 300°F (150°C). The slaking system consists of two tanks, two stirring motors, and a slurry pump. The quicklime (CaO) is slaked in one tank and then diluted in another. The spray dryer was designed for a nominal gas flow rate of 4.2 m^3/min (150 cfm) with an average gas residence time in the dryer of approximately 11 seconds.

The baghouse is a single-compartment, continuous, automatic, self-cleaning cloth pulse-jet dust collector which has 42 individual bags with a total surface area of 98 square feet (air-to-cloth ratio of 1.6 ft/min). The sampling system consists of three temperature monitors (thermocouples), two SO_2 concentration monitors (Horiba Model 2000 NDIR), and a data acquisition system. The temperatures are monitored by thermocouples mounted at the inlet and outlet of the spray dryer, and at the outlet of the baghouse. The spray dryer inlet and outlet SO_2 sampling ports are located right at the top and bottom of the spray dryer, respectively. The sampled gas is transported first through a particle removal filter, then through an electric cooler to condense the water vapor (the temperature of the cooler is set to 35°F to prevent an ice block), then into a flask to remove any liquid water, and finally passed through the analyzer. The concentration of SO_2 is shown on the meter and the data are simultaneously sent to the computer through the data acquisition system. The sampling tube from the sampling port to the cooler is heated and insulated; this prevents precooling and SO_2 absorption by H_2O. The SO_2 concentration meter is calibrated by 1840 ppm standard SO_2 gas. The macro-parameters that are most important in the operation of the spray dryer are approach to saturation temperature at the outlet of the spray dryer (measured by means of psychrometry, and system dry bulb temperatures), the calcium/sulfur molar ratio, and the residence time of the droplet spray. The droplet average residence time is taken as the hydraulic residence time of the air flow in the spray dryer. The calcium/sulfur molar (Ca/S) ratio is difficult to measure because lime slurry will block normal flow meters. Also, for tests of short duration (<60 minutes), the change of the liquid level in the slurry holding tanks cannot be used with good accuracy to determine the total amount of injected slurry. The Ca/S can be calculated by means of a heat balance to determine the amount of water addition. The amount of water addition can be obtained by using the psychometric chart by knowing the ambient air relative humidity, the spray dryer inlet SO_2 concentration, and the spray dryer outlet temperature. In addition, the amount of water retained in the product must be measured. This calculation neglects the moisture bound by any chemical reactions. The equation for calculating the Ca/S molar ratio is shown below.

Equation 8.2

$$\frac{Ca}{S}(\text{molar ratio}) = 10^6 \cdot \frac{\alpha M_a C(H_{3_o} - H_2)(t_3 - t_2)}{C_s(1-C)M_c(t_{w2} - t_2)}$$

where
α is a correction factor to account for moisture in the product (experimentally, 1.5 for this system),
M_a is the molecular weight of flue gas, (g/mole),
M_c is the molecular weight of CaO, 56 g/mole,
C (%) is the slurry concentration of CaO by weight,

C_s (ppm) is the inlet SO_2 concentration to the spray dryer,
t_2 is the inlet temperature of the spray dryer, any units,
t_{w2} is the saturation temperature by assuming adiabatic system, same unit as t2,
t_3 is the outlet temperature of the spray dryer, same unit as t2,
H_{30} is the humidity ratio at t_{w2}, and
H_2 is the humidity ratio at t_1 or t_2.

The nozzle used is a two-fluid external mixing nozzle. It is shown in and compared with the internal mixing nozzle. Due to a spray nozzle limitation, the slurry solids concentration was limited to a 10% concentration of CaO (by weight). The gas flow rate and inlet temperature were kept constant and equal to 150 cfm (4.2 m3/min) and 300°F (150°C), respectively. The inlet SO_2 concentration was around 2500 ppm. The outlet temperature was changed by the variation of the sprayed slurry volume, and the stoichiometric ratio was calculated by using . At the above condition, the Ca/S stoichiometric ratio was around 1.5.

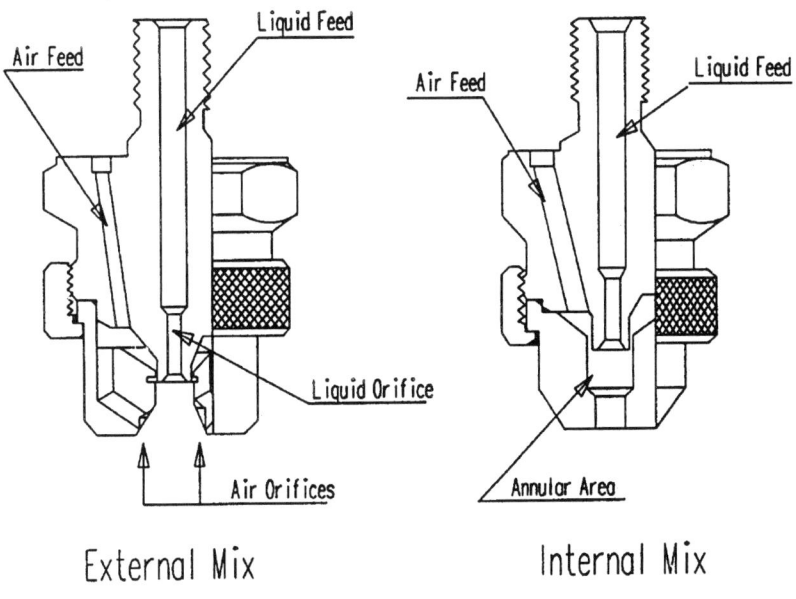

Figure 8.17 Schematic diagram of spray dryer nozzles.

Baseline Test and Parameters that Affect Spray Dryer Desulfurization Processes

The spray dryer inlet air flow rate is monitored by a differential pressure gauge located in the inlet duct. The differential pressure gauge was calibrated by a standard pitot tube. In order to obtain the maximum mixing effect, the SO_2 injection nozzle was mounted directly behind the differential gauge.

Water was first sprayed into the spray dryer, and there was negligible SO_2 concentration difference between the inlet and outlet of the spray dryer when the approach to saturation temperature was around 20°F. Since the water was completely evaporated, it cannot absorb SO_2. Initial tests were performed with $NaHCO_3$. The chemical properties of the $NaHCO_3$ (Type 2 was used in this test) are shown in Table 8.6. The gas flow rate, inlet temperature, and SO_2 concentration were kept constant and equal to 130 cfm, 253°F, and 2783 ppm, respectively. The outlet temperature was changed through the variation of the sprayed solution volume, and the stoichiometric ratio was calculated by the use of the psychometric chart. The SO_2 removal efficiency (%) versus the $NaHCO_3/SO_2$ molar ratio is shown in Figure 8.18.

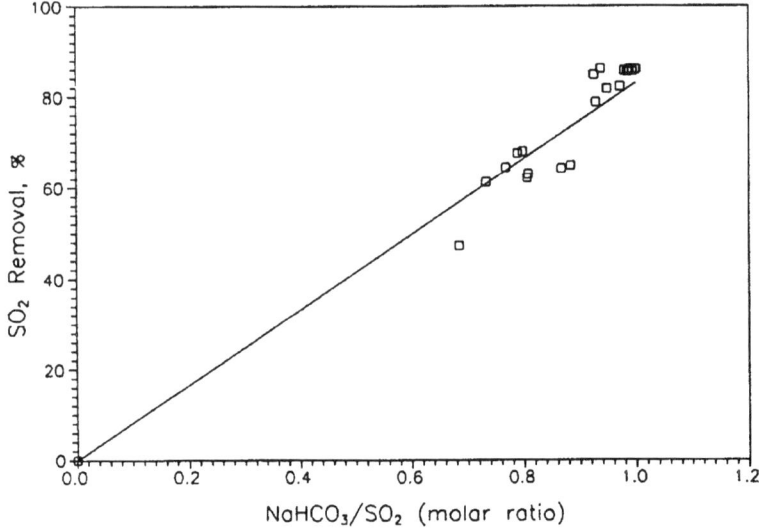

Figure 8.18 $NaHCO_3$ baseline SO_2 removal in pilot spray dryer.

A slurry of 5% CaO by weight could be applied to the internal mixing SU J16 - SS type nozzle. Later a 10% concentration CaO (by weight) slurry was tested in the external mixing SU 1A - SS type nozzle. Both nozzles were from Spray System Co. The following experiments are all based on the external mixing nozzle.

The quicklime used in the experiments is from Dravo Company, at the Black River Plant located in Kentucky. The chemical and physical analyses are listed in Table 8.7.

Table 8.6 The Chemical Properties of NaHCO$_3$ (from Church & Dwight Co., Inc.)

Component	CO$_2$ Calculated (%)	Na$_2$O Total (%)	NaHCO$_3$ Assay (%)	Arsenic (%)	Heavy Metals (%)
Type 2	52.28	36.82	99.8	< 3	< 5
Type 5	52.44	36.93	100.0	< 3	< 5

The SO$_2$ removal efficiency versus the Ca/S molar ratio is shown in Figure 8.19 which shows the SO$_2$ removal efficiency only across the spray dryer. Results from the literature [95] are shown as a comparison. In Figure 8.19, the straight line is the theoretical utilization line, and the notation Dt stands for the spray dryer outlet temperature minus the saturation temperature; i.e., the approach to saturation temperature. From Figure 8.19, it can be seen that the SO$_2$ removal efficiency increases dramatically with a decrease in the approach to saturation temperature.

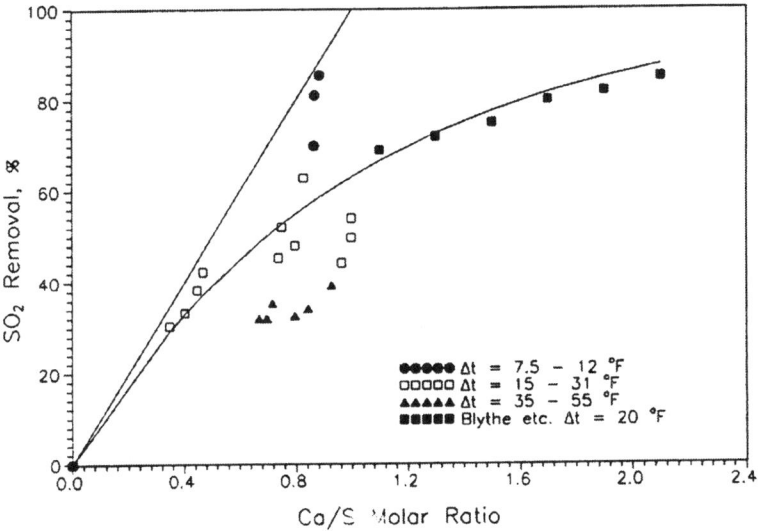

Figure 8.19 Ca(OH)$_2$ baseline SO$_2$ removal in pilot spray dryer.

Dry Scrubbing Technologies for Flue Gas Desulfurization

Table 8.7 Dravo Black River Plant Lime Analysis (Dravo Lime Co.)

Chemical Analysis	3/4 x 1/8 Pebble Quicklime Quicklime	3/4 x 1/4 Double Screen Pebble
Total CaO	93.0	93.0
Available CaO	87.5	88.5
MgO	2.75	2.65
Total Oxides	95.75	95.65
SiO_2	2.05	1.95
R_2O_3	0.86	0.86
Fe_2O_3	0.16	0.16
Al_2O_3	0.7	0.7
Sulfur	0.050	0.045
P	0.004	0.004
Loss on Ignition	1.60	1.50
CO_2	1.20	1.10
H_2O	0.40	0.40
Physical Analysis Screen Size	**% Passing**	
3/4	100	100
1/2	54.0	48.0
3/8	19.8	18.0
1/4	4.5	4.5
1/8	1.3	1.2
No. 20	1.2	1.0
No. 30	1.0	0.8
No. 50	0.9	0.7

Data Reported by Keith Bingham Effective 4/20/88

When the solids are collected, they are not completely dry. Their ultimate moisture content depends on the equilibrium condition between the air and the particulate moisture.

Additives to Increase Ca(OH)$_2$ Utilization.

HYGROSCOPIC ADDITIVES. Four hygroscopic additives were evaluated for their effect on sorbent reactivity. These were NaOH, NaHCO$_3$, NaCl, and CaCl$_2$. These additives were tested by adding them at varying concentrations from 100 to 700mg/liter, to the calcium hydroxide slurry. The results were compared with those obtained in the baseline test program. It was shown that the sodium ion can significantly increase the SO$_2$ removal efficiency and enhance sorbent utilization, owing largely to the hygroscopicity of these sodium compounds.

Based on their alkalinity, the four additives are NaOH (strong base), NaHCO$_3$ (weak base), NaCl (neutral salt), and CaCl$_2$ (the moderate chlorine content in coal reportedly improves SO$_2$ uptake in spray dryers). The test results for the four additives with 500 mg/liter addition are shown in Figure 8.20 where the baseline results are kept as a comparison.

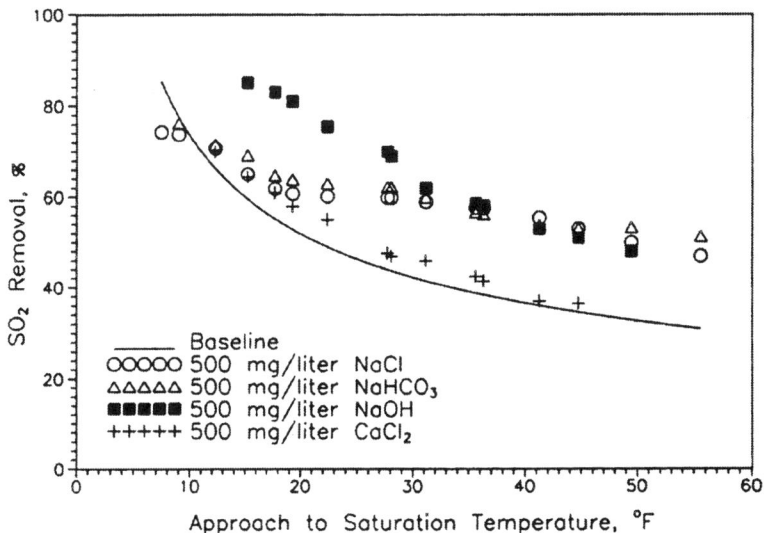

Figure 8.20 SO$_2$ removal efficiency versus approach to saturation temperature.

From Figure 8.20, it can be seen that NaCl and NaHCO$_3$ have almost the same effectiveness in improving the SO$_2$ capture by Ca(OH)$_2$, if the same amount is added. The improvement by NaOH is much higher if the approach to saturation temperature is less than 30°F, while the enhancement by $CaCl_2$ is relatively small. All these additives have the characteristics of hygroscopicity. NaOH is the only one that can increase the basicity of the saturated $Ca(OH)_2$ solution, as can be seen from Figure 8.21. NaCl slightly reduces the pH and $NaHCO_3$ reduces it more. $CaCl_2$, as one of the most frequently investigated additives, has been shown [86] to have less affinity for water vapor in combination with $Ca(OH)_2$ than it does as a pure compound. The explanation for this anomaly is the reaction between $CaCl_2$ and $Ca(OH)_2$ to produce a basic salt, $Ca(OH)_2 \cdot CaCl_2 \cdot H_2O$, which does not have a strong affinity for water vapor.

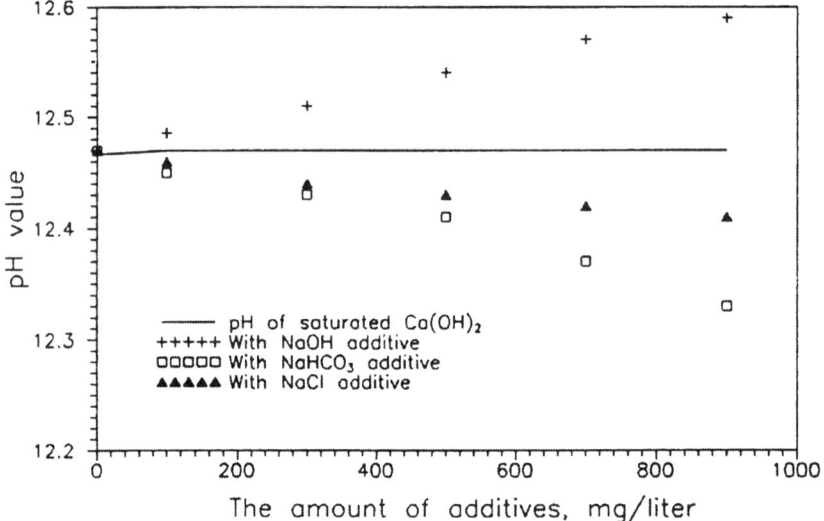

Figure 8.21 The pH of saturated $Ca(OH)_2$ solution after additivie addition.

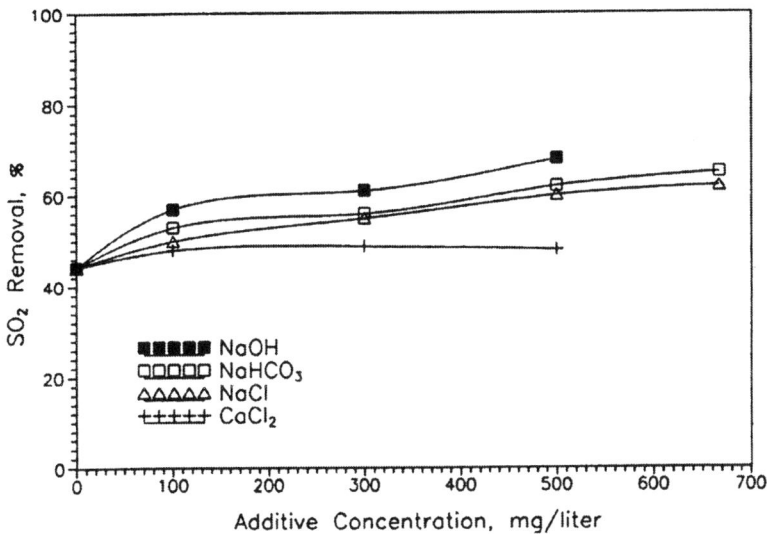

Figure 8.22 SO₂ removal efficiency versus additive concentration.

Figure 8.22 shows that the SO₂ removal efficiency varies with different amounts of additives added around an approach to saturation temperature of about 28°F and a Ca/S molar ratio of about 1.5. From Figure 8.22 we can see the tendency of efficiency changes with additive concentrations.

Table 8.8 Solubilities of Ca2+ and Na+ substances (g per 100g H_2O) [94]

Substances	Temperature				
	20°C	30°C	40°C	60°C	80°C
$CaCl_2 \cdot 6H_2O$	74.5	102	-	-	-
$CaCl_2 \cdot 2H_2O$	-	-	-	136.8	147.0
$Ca(OH)_2$	0.165	0.153	0.141	0.116	0.094
$CaSO_4 \cdot 2H_2O$	-	0.2090	0.2097	0.2047	-
$NaHCO_3$	9.6	11.1	12.7	16.4	-
NaCl	36.0	36.3	36.6	37.3	38.4
$NaOH \cdot 1H_2O$	109	119	129	174	-
Na_2SO_4	-	-	48.8	45.3	43.7

Knowing what additive concentration is desirable in the slurry is important. The data in Figure 8.22 show that when the additives are less than 200 mg/liter, there is a steep portion of the curve where large improvements result from small increases in additive concentrations. This is followed by a region characterized by a less-steep slope, beyond which there should be a flat region which means the absorption process has reached the gas-phase limitation. The alkalinity available to neutralize absorbed SO_2 is no longer a rate-limiting factor. Higher additive concentration levels are of little benefit once the gas-phase limitation is approached.

The hygroscopicity of the additives can be found from the following theory. Since feeds of insoluble solids from slurries and pastes (called suspensions) have negligible vapor pressure lowering effects, the total drying time of the lime slurry droplets containing insoluble and dissolved solids can be evaluated as the drying time of the liquid containing dissolved species. Since some salts (especially the sodium salts) have high solubilities (Table 8.8) and vapor-pressure lowering effects, the dissolved salts can increase the liquid phase residence time of the droplets. Since most of the reactions occur in the liquid phase, this prolonged residence time has a significant effect on SO_2 removal.

The salt concentration in a droplet will increase with decreasing droplet size if the salt is highly soluble because the mass of the salt remains constant and only the water evaporates. Thus, a given mass of dissolved salt serves to reduce the vapor pressure at the droplet surface to a greater degree as droplet size decreases. In competition with this trend is the Kelvin effect that causes an increase in vapor pressure as droplet size decreases. The relationship between saturation ratio and droplet (containing dissolved material) size is given by Hinds as the following:

Equation 8.3

$$\frac{p}{p_s} = \frac{1}{1 + \frac{6imM_w}{M_s \rho \pi d_p^3}} \cdot \exp\left(\frac{4\gamma M_w}{\rho RT d_p}\right)$$

where
P is the partial pressure of water vapor at the droplet surface, any units;
P_s is the saturation pressure at the plane surface of pure water, same units as P;
M_s is the molecular weight of dissolved salt, g/mole;
m is the weight of dissolved salt in the droplet, g;
i is the number of ions each salt molecule forms when dissolves (2 for NaOH);
ρ is the density of the solvent (water), g/cm^3;
M_w is the molecular weight of the solvent (water), g/mole;
γ is the surface tension of the solvent (water), dyn/cm;
d_p is the droplet diameter, cm;
R is the ideal gas law constant, dyn-cm/K; and
T is the absolute temperature of the droplet, K.

In the spray dryer, the Kelvin effect may be small because of the relatively large droplet size (the Kelvin effect is prominent when the droplet sizes are less than 0.1um). From the above equation we can see how the additive salts affect the partial water vapor pressure at the droplet surface. If other variables are held constant, the salts change the partial pressure through im/M_s; i.e., the moles of additive in a droplet times the ion number. From Equation 8.3, higher values of the im/Ms term results in greater reduction of the droplet surface partial pressure.

A comparison of the im/Ms term for different substances is shown in Table 8.9. As shown in Table 8.9, NaOH is the most hygroscopic.

Table 8.9 Comparison of the im/M$_s$ Term for Different Substances

Substance	NaOH	NaHCO$_3$	NaCl	CaCl$_2$	Na$_2$SO$_4$
i	2	3	2	3	3
M$_s$	40	84	58.5	111	142
im/M$_s$	0.05m	0.036m	0.034m	0.027m	0.021m

ADDITIVES WITH ABILITIES TO OXIDIZE S(IV) TO S(VI). There are at least two advantages of oxidizing S(IV) into S(VI) in a spray dryer. One is that since the diffusion of H+ from a droplet surface into a droplet core might become the limiting step (liquid phase mass transfer resistance), the pH decrease on the droplet surface could lead to the halt of $SO_2(aq)$ dissolution into ion forms [$SO_2(aq)$ + $H_2O \leftrightarrow H^+ + HSO_3^-$]. If there exists a way to oxidize the $SO_2(aq)$, HSO_3^-, and SO_3^{2-} into $SO_3(aq)$, HSO_4^-, and SO_4^{2-} respectively (SO_2, HSO_3^- and SO_3^{2-} are known as S(IV) and SO_3, HSO_4^-, and SO_4^{2-} are known as S(VI)), then due to the high dissolution rate of $SO_3(aq)$ and H_2SO_4 at relatively low pH (pH > 4.0, (see Figure 8.23 [96]), the $SO_2(aq)$ concentration will not be nearly as much affected by the liquid side equilibrium and pH. Therefore, the limiting step will be the liquid phase mass transfer of H^+ ions.

Another advantage can be seen from Figure 8.23. S(IV) form can be crystallized easily. On the contrary, S(VI) form is known as a crystallization inhibitor; i.e., S(IV) has a lower solubility than S(VI) in the typical FGD pH operating range. If S(IV) can be oxidized to S(VI), this implies less S(IV) crystallization and more available sorbent surface area for dissolution since the crystal of the S(IV) form will deposit onto the surface of the sorbent and block its contact with the liquid. In the wet scrubbing process where gypsum scaling (due to the S(VI) forms) needs to be controlled, the natural sulfite oxidation (oxidation by oxygen from the flue gas, and from other sources) is inhibited by using thiosulfate to the point where gypsum scaling does not occur [29].

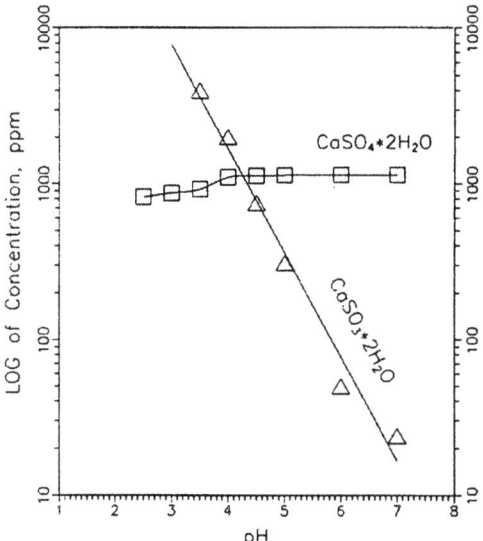

Figure 8.23 Calcium sulfite and calcium sulfate solubilities versus pH.

During $SO_2(g)$ diffusion into the droplet certain amounts of oxygen will simultaneously diffuse into the droplet, but lacking catalysts the reactions between dissolved O_2 and S(IV) are negligible. Certain dissolved metal ions can act as catalysts in the oxidation reaction [54], and metal ions such as Fe^{2+}, Mg^{2+}, K^+, Cs^+, etc. could be the catalyst. H_2O_2 (hydrogen peroxide) is a more powerful oxidizing agent. Considering that H_2O_2 has a larger Henry's Law constant than O_2, the oxidation of S(IV) into S(VI) can be more effective if H_2O_2 is used. Henry's Law constants are listed below in Table 8.10.

Table 8.10 Henry's Constants for Certain Dissolved Gases and Liquids

Gas	H, Matm^{-1} (298 K)
O_2	1.3×10^{-3}
O_3	9.4×10^{-3}
SO_2	1.24
H_2O_2	7.1×10^4

The H_2O_2 used in the additive tests was from Fisher Scientific in a 30% liquid form. It was added directly to the slurry tank. The spray dryer test results using H_2O_2 as an oxidizing additive are shown in Figure 8.24. The H_2O_2, as an additive, has a significant effect in enhancing $Ca(OH)_2$ sorbent utilization and SO_2 removal. At a 20°F approach to saturation temperature, the SO_2 removal efficiency under the H_2O_2 additive is about 20% higher than that of the baseline. From Figure 8.24, the additive concentration ranges from 1 ml/liter to 5 ml/liter. An increase in additive concentration leads to the increase of SO_2 removal efficiency.

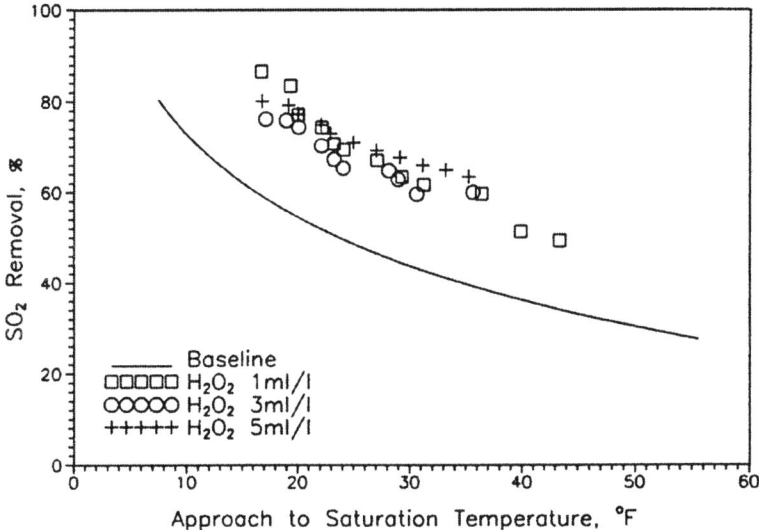

Figure 8.24 SO$_2$ Removal for Ca(OH)$_2$ with H$_2$O$_2$ additive.

SUCROSE - AN ADDITIVE TO INCREASE CA(OH)2 SOLUBILITY AND DISSOLUTION RATE. The dissolution rate of Ca(OH)$_2$ particles is a measure of how fast the alkalinity can be delivered to the solution. Faster dissolution rates of Ca(OH)$_2$ particles result in less liquid phase resistance for SO$_2$ reaction, and therefore higher SO$_2$ removal efficiency.

Sugar has been shown to enhance the dissolution rate of Ca(OH)$_2$, as discussed previously. The spray dryer test results for sugar as an additive are shown in Figure 8.25. It can be seen that with a small amount of sugar added to the slurry, the removal efficiency is increased over the baseline. However, with further increases of sugar (from 100mg/l to 500mg/l) a decrease in the SO$_2$ removal efficiency is seen. The removal efficiency improvement is about 10% at a 20°F approach to saturation temperature.

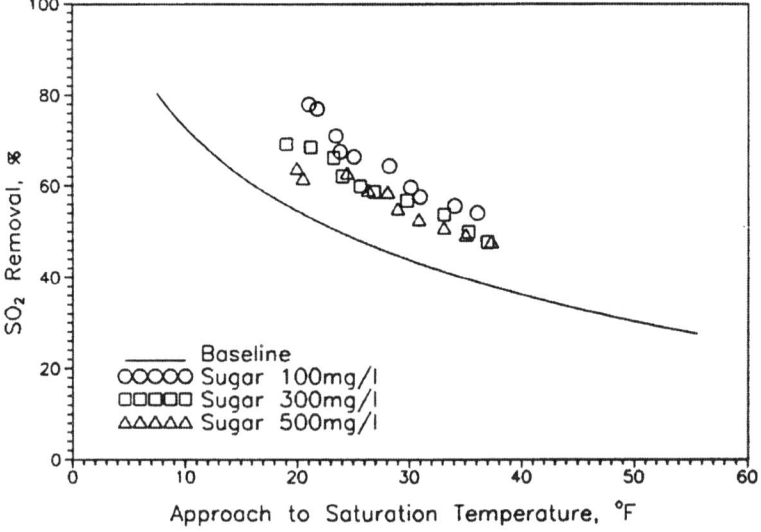

Figure 8.25 SO_2 removal for $Ca(OH)_2$ with sugar additive.

FORMIC AND BENZOIC ACIDS AS BUFFER ADDITIVES. Depending on the dissolution rate of sorbents, the pH at the surface of the droplets in a spray dryer may drop below 5 as SO_2 is absorbed. This low pH value will inhibit further absorption of SO_2. Organic acids, having been widely used in wet scrubbing systems, can be effective as a buffer for preventing the pH from falling too low. In the experiments, benzoic acid and formic acid were used as the organic acid buffer. The test results are shown in Figures 8.26 and 8.27. Little improvement in SO_2 removal was observed compared with baseline results. The benzoic acid used was in crystal form, and the amount added was from 100mg/l to 500mg/l. The formic acid used was in a liquid form (88% concentration). The amount added was from 1ml/l to 5ml/l. The test results showed that the organic acid buffer, which is effective in wet scrubber systems, has little effect in spray dryer systems. This is probably because $Ca(OH)_2$ particles can dissolve faster than $CaCO_3$ particles, and because the pH in the droplets will not drop to as low as 5 before the droplets are dried.

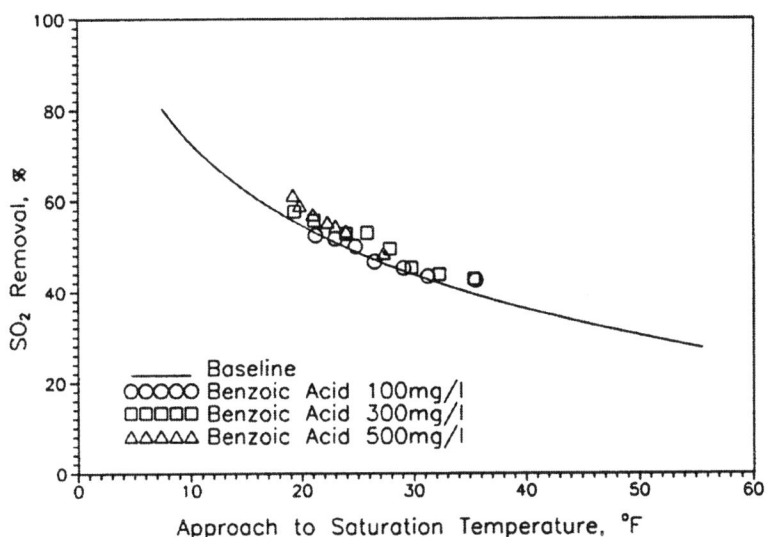

Figure 8.26 SO_2 Removal for $Ca(OH)_2$ with benzoic acid.

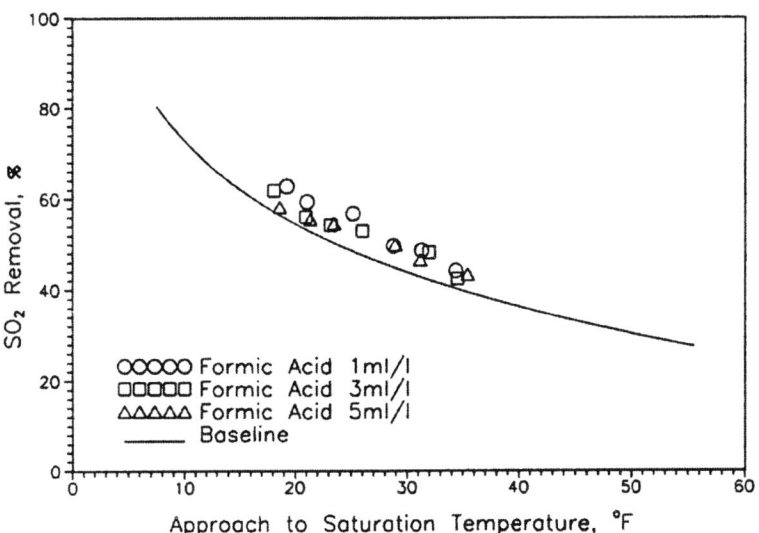

Figure 8.27 SO_2 Removal for $Ca(OH)_2$ with formic acid.

Dry Scrubbing Technologies for Flue Gas Desulfurization

Hydration of Fly Ash with Ca(OH)$_2$ to Increase Sorbent Reactivity and Utilization

Two fly ashes have been used in these tests. They were generated from the #7 and #8 boilers of Cincinnati Gas and Electric Company's Miami Fort (MF) Station. The MF7 ash was generated from a 2.1% sulfur coal, and the MF8 ash was generated from a 0.7% sulfur coal. An analysis of the ashes can be found in Table 8.11. The surface areas of the MF7 and MF8 ashes were measured to be 0.5 m^2/g and 1.2 m^{2}g, respectively.

Table 8.11 Mineral Analysis of Ashes on a Percent Ignited Basis (Courtesy of Standard Laboratories, Inc., South Charleston, W.V., U.S.A.)

Fly Ash	MF7	MF8
SiO$_2$	49.77	56.51
Al$_2$O$_3$	23.76	28.53
Fe$_2$O$_3$	17.86	4.62
K$_2$O	2.21	2.39
TiO$_2$	1.65	2.08
CaO	1.59	1.49

The Ca(OH)$_2$ used in the bench scale tests was reagent grade Ca(OH)$_2$ from Fisher Scientific. The surface area of this material was measured to be 13.9 m^2/g. The hydration solids consist of 4 parts fly ash to 1 part Ca(OH)$_2$. The water/solids ratio was kept at 15:1. The samples were then heated in a Fisher Scientific water bath at 80°C or 95°C for varying periods. After heating, the samples were filtered, and part of the solids was vacuum-desiccated over night. The solids surface area was then measured with a Quantachrome Monosorb BET analyzer.

Figure 8.28 shows the surface areas of the fly ash/Ca(OH)$_2$ hydration mixtures, which were obtained from hydrating the MF7 and MF8 ashes with Ca(OH)$_2$ at 95°C. The fly ash/Ca(OH)$_2$ ratio is 4:1. The surface areas of both mixtures developed markedly with hydration time, owing to the formation of high surface area calcium silicates and aluminates. The MF8 ash and Ca(OH)$_2$ mixtures developed, on average, higher surface areas than the MF7 ash and Ca(OH)$_2$ mixtures. The apparent reason is that MF8 ash has more available silica and alumina than the MF7 ash has, which can be seen from Table 8.11.

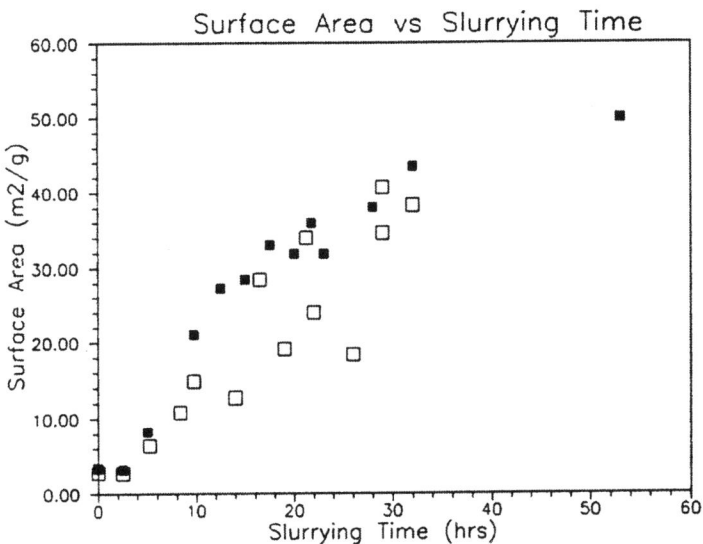

Figure 8.28 Surface area development in MF7 sorbents (□) and MF8 sorbents (■).

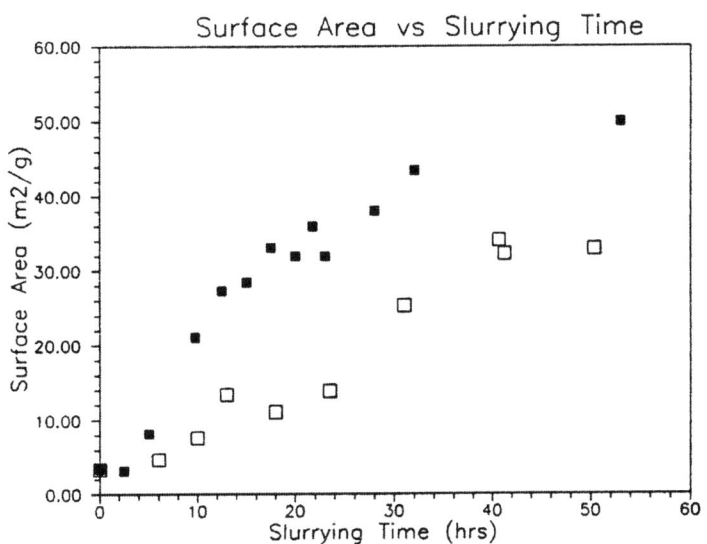

Figure 8.29 Surface area development in MF8 sorbents heated at 80°C (□) and 95°C (■).

Dry Scrubbing Technologies for Flue Gas Desulfurization

Figure 8.29 displays the surface area development from heating 4:1 MF8/ $Ca(OH)_2$ mixtures at two different temperatures, 80°C and 95°C. The surface areas developed at 95°C were much greater than those developed at 80°C.

Sorbents to be spray dried were made with varying ratios of fly ash and quick-lime mixtures, but mostly 4:1 and 1:1 (corresponding to fly ash: Calcium hydroxide ratios of 4:1.32 and 1:1.32). The fly ashes were passed through a 250 µm sieve to remove any potential nozzle-plugging particles. The mixtures were heated in an insulated 150 L stainless steel drum by allowing steam to flow through the outer jacket. A steam solenoid valve and temperature controller kept the slurry temperature constant at 95°C.

The results of the spray dryer tests with the fly ash sorbents are in Figures 8.30, 8.31, and 8.31. Figure 8.30 compares the spray dryer performance of the mixtures of MF7 and MF8 ash hydrated with $Ca(OH)_2$. Both mixtures consisted of 4 parts ash and 1.32 parts $Ca(OH)_2$, and were hydrated at 95°C for 15 h before testing. The MF8 ash and $Ca(OH)_2$ mixtures were more effective at removing SO_2 probably due to the reason that MF8 ash has more available silica to form reactive calcium silicates.

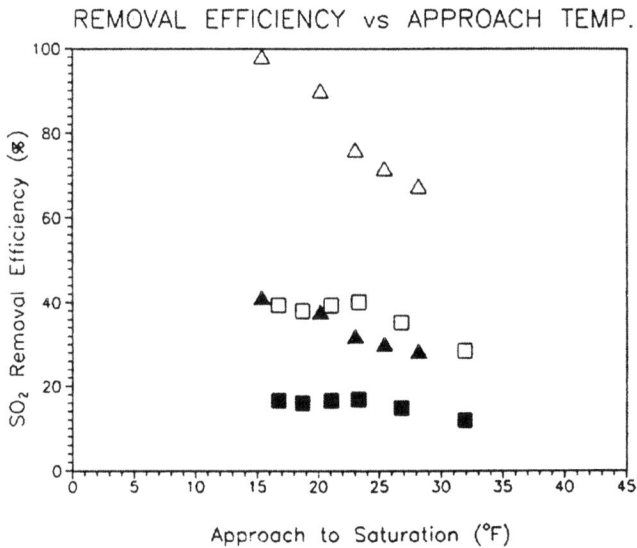

Figure 8.30 SO_2 removal for MF7/Ca(OH)$_2$ (■) and MF8/Ca(OH)$_2$ (▲). Calcium ultilization for MF7/Ca(OH)$_2$ (□) and MF8/Ca(OH)$_2$ (△).

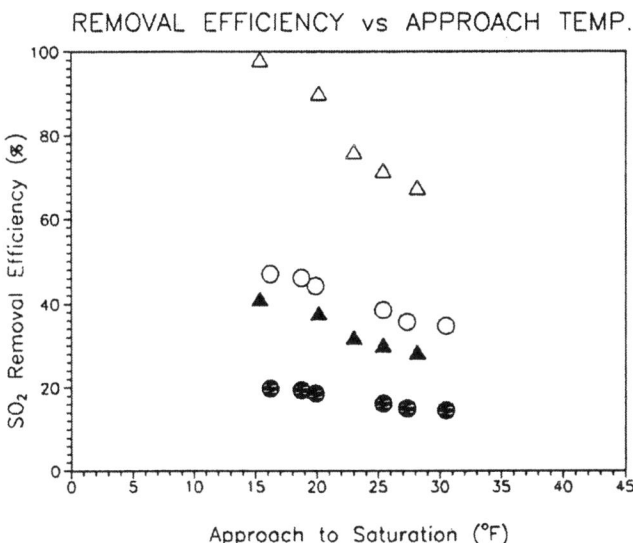

Figure 8.31 SO_2 removal for MF8/Ca(OH)$_2$: heated (▲) and unheated (●). Calcium utilization: Heated (△) and Unheated (○)

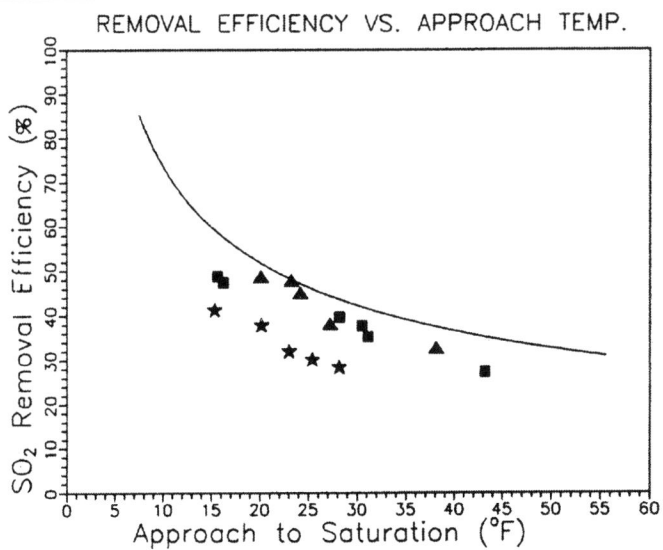

Figure 8.32 SO_2 removal for baseline Ca(OH)$_2$ slurry (-), 1:4 MF8/CaO mixture (▲), 1:1 MF8/CaO mixture (■), and 4:1 MF8/CaO mixture (★).

Dry Scrubbing Technologies for Flue Gas Desulfurization

Figure 8.31 displays the effect of heating the 4:1.32 MF8/Ca(OH)$_2$ slurry for 15 h at 95°C has upon SO$_2$ removal in the spray dryer. The removal increased 20 percentage points at an 11°C (20°F) approach to saturation, which translates into an increase in calcium utilization from 45% to 95%. From heating the mixtures of quicklime and fly ash, the calcium utilization is increased. This suggests that fly ash could be used as a sorbent additive in the spray drying process.

Figure 8.31 compares the SO$_2$ removal of a pure 12.8% Ca(OH)$_2$ solution (10% CaO before hydration) to three heated mixtures with different MF8 ash/Ca(OH)$_2$ ratios. The fly ash mass fractions in the solid content of the slurries were 20%, 50%, and 80%; the heating time was again 15 h, and the temperature was 95°C. None of the slurries removed as much SO$_2$ as the pure CaO slurry, although the 80% and 50% fly ash slurries were within 10 percentage points at an 11°C (20°F) approach. Table 8.12 summarizes the SO$_2$ removal efficiencies and calcium utilizations of each slurry. The Ca utilization in the 0% and 20% fly ash slurries was only 30%, but rose dramatically as more fly ash was substituted. The 50% fly ash slurry had a Ca utilization of 45%, and the 80% fly ash slurry had a 95% Ca utilization. However, the presence of a large percentage of inert material in the fly ash plus the limit of 10% solids loading in the slurries imposed by the two fluid atomizer prevented the testing of a 4:1.32 fly ash/Ca(OH)$_2$ slurry with heavier solids loading.

Table 8.12 Removal Efficiency and Ca Utilization of Heated MF8/Ca(OH)$_2$ Slurries with Different Fly Ash/Quicklime Mass Ratios at 11°C Approach

Fly Ash/CaO	Removal Efficiency (%)	Ca/S Ratio	Ca Utilization (%)
0:1	52	1.75	30
1:4	49.5	1.65	30
1:1	44	1.0	45
4:1	38	0.4	95

Additional sulfur uptake tests with hydrated MF7/Ca(OH)$_2$ and MF8/Ca(OH)$_2$ slurries were performed to help verify the results obtained in pilot scale tests. Both ashes were hydrated with Ca(OH)$_2$ at a 4:1 ratio. The total solids weight percentage was 10% in each case. Samples were heated at 100°C for 2 hours, cooled to room temperature, and then sparged with 2800 ppm SO$_2$ for 5 min, and 30 min separately. After having been sparged, the samples were filtered, dried overnight in a heated vacuum desiccator, and measured for sulfur content in a Leco Total Sulfur Analyzer.

The results of SO_2 sparging tests are shown in Figure 8.33. The hydrated MF8/$Ca(OH)_2$ mixture was more reactive than the MF7/$Ca(OH)_2$ mixture. The unreacted MF8/$Ca(OH)_2$ mixture (corresponding to zero sparging time) contained approximately 0.3% more sulfur than the unreacted MF7/$Ca(OH)_2$ mixture. However, the amount of additional sulfur picked up by the MF8 sorbent during the reaction was much greater. The MF8/$Ca(OH)_2$ mixture had 0.9% more sulfur than the MF7/$Ca(OH)_2$ mixture when reacted for 5 minutes, and it had 2.8% more sulfur when reacted for 30 minutes. These results verify the greater reactivity of the MF8/$Ca(OH)_2$ mixture observed in the spray dryer. The sparging tests were designed to verify the SO_2 capture capabilities of these two mixtures. It was not intended to simulate the spray dryer processes. However, the sparging test results did show some similarity to the spray dryer results. The major SO_2 capture mechanisms between the two tests are the same. They are the mass transfer from the gas phase toward the liquid phase, the interface absorption, and liquid phase mass transfer and reaction. The differences are that the spray dryer involves the heating of the droplets and the evaporation of the droplet water. The evaporation of the droplet water may hinder the mass transfer of SO_2 in the gas phase. So as for sorbent SO_2 capture abilities, the two processes behave the same.

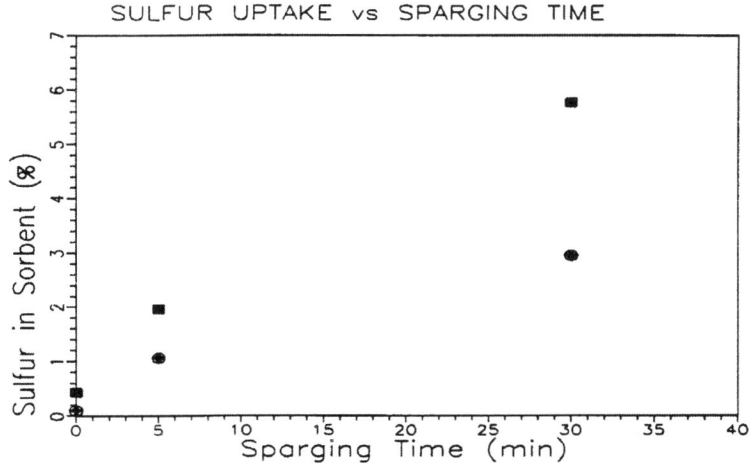

Figure 8.33 Sulfur capture in solution from heated MF7/$Ca(OH)_2$ mixture (●) and heated MF8/$Ca(OH)_2$ mixture (■).

Spray Dryer Modeling

Introduction to "SPRAYMOD" and its Application to Baseline Test Results

A spray dryer mathematical model was used to predict spray dryer performance. This model was initially developed by Damle [16] and later modified by Partridge [12].

The initial model (SPRAYMOD) was observed to be less sensitive to the effects of slurry concentration and SO_2 concentration than the obtained data [12]. SPRAYMOD tended to under-predict at low slurry concentrations and over-predict at high slurry concentrations. The model neglected the liquid phase mass transfer resistance and calculated the SO_2 removal during the evaporation/reaction process considering only the individual resistance that was dominant.

By considering the mass transfer approach in SPRAYMOD to be in error, the modified model aimed at developing a comprehensive model for the constant-rate period using a mechanistic approach that combines the individual resistances that affect the SO_2 absorption rate into a single relationship. The modified model (SPRAYMOD-M) borrowed the material and energy balance calculations from the program SPRAYMOD. Both the gas-phase mass transfer and liquid-phase mass transfer coefficients were included along with a relationship to predict the resistance to lime dissolution. The model was based on film theory and treated the atomized slurry droplet as a sphere with the fluid phase uniformly distributed around the discrete individual sorbent particles. This approach was said to allow predictions of the mass transfer coefficients and to model the enhancement due to increasing solid concentration as evaporation proceeds. Efficiency predictions using the modified model had been compared with the pilot-plant data taken from the pilot spray dryer/baghouse facility using a slipstream of the University of Tennessee's stoker fired boilers.

The development of a spray dryer mathematical model can be generalized as follows: Two processes simultaneously take place in the desulfurization with spray dryers. One is the drying of the slurry droplets, and the other is the SO_2 mass transfer and reaction with the sorbents contained in the slurry droplets.

DRYING. The drying process can be roughly divided into two major periods. The first period is the constant-rate drying period. In this period, the solid particle concentration in the slurry droplet does not affect the water evaporation rate. This period continues until the moisture of the droplets falls below the critical moisture content that is about 30%, by weight, water in the droplets. The second period directly follows the first period, and is named the falling-rate period. In this period, the solid particles contained in the droplets begin to touch each other, and the moisture has to diffuse through the solid matrix. This period continues until the moisture in the dried droplets achieves an equilibrium with the moisture in the surrounding gas. The moisture content in the dried droplets is called the equilibrium moisture content. It is around 5%, by weight, water in the dried droplets.

THE CONSTANT-RATE PERIOD. The mathematical descriptions of the drying processes also follow the above division. In the constant-rate drying period, heat is transferred from the gas phase to the droplets, and water vapor is transferred from the droplet surface to the gas phase. The transfer coefficients are determined by the following correlations [97].

$$Sh = 2 + \alpha Re^{\beta} Sc^{\gamma}$$

and

$$N = 2 + \alpha Re^{\beta} Sc^{\gamma}$$

where
$Sh = k_d d_d/D$ is the Sherwood Number for mass transfer,
$Re = d_d v \rho/\mu$ is the Reynolds Number based on droplet diameter,
$Sc = \mu/\rho D$ is the Schmidt Number,
$Nu = h d_d/k$ is the Nusselt Number for heat transfer,
$Pt = C_p \mu/k$ is the Prandtl Number,
k_d is the gas-phase mass transfer coefficient,
d_d is the droplet diameter,
D is the water vapor diffusivity in gas phase,
v is the relative velocity between gas and droplet,
ρ is the gas density,
μ is the gas kinetic viscosity,
h is the gas phase heat transfer coefficient,
k is the gas phase thermal conductivity,
C_p is the gas specific heat,
$\alpha = 0.6$, $\beta = 0.5$, and $\gamma = 0.33$.

By assuming a quiescent droplet-gas system, a rigorous analysis of the evaporation process gives the rates of mass and heat transfer [98]:

Equation 8.4

$$N_w = 2\pi D c d_4 Ln\left[\frac{1-x_g}{1-x_d}\right]$$

and

$$H = 2\pi k d_d (T_g - T_d) \cdot \left[\frac{\varepsilon}{e^\varepsilon - 1}\right]$$

Equation 8.5

where ε is the dimensionless heat-transfer rate factor, and for water is given by:

$$\varepsilon = \frac{9}{\pi} \cdot \frac{N_w C_{p,l}}{k d_d}$$

Equation 8.6

where

N_w is the molar rate of water evaporation,
c is the mean molar density of gas phase,
x_g is water vapor mole fraction in the bulk gas phase,
x_d is the equilibrium water vapor mole fraction at the droplet surface,
T_g is the temperature in bulk gas phase,
T_d is the temperature of the droplets, and
$C_{p,l}$ is specific heat of water vapor.

The heat transferred from gas phase to the droplets is used to evaporate the moisture in the droplets, and the heat and mass transfer are related by

$$H = N_w \lambda + \frac{\pi d_d^3}{6} \rho_1 C_{p,l} \frac{dT_d}{dt}$$

Equation 8.7

where

ρ_1 is the density of the droplets,
λ is the latent heat of vaporization at T_d, and
t is time.

The water vapor at the droplet surface is constantly in equilibrium with the water in the droplet. Therefore, x_d is given by the water vapor partial pressure at T_d.

Equation 8.8

$$x_d = f(T_d)$$

Equations 8.4, 8.5, 8.7 and 8.8 are combined to solve the drying processes. There are seven unknowns (N_w, x_d, x_g, H, T_d, T_g, and t) contained in these four equations. Additional relationships must be found to solve this problem. The additional equations are found from the gas and droplet mass and energy balances.

In the gas-phase mass and energy balances, the gas temperature, T_g; the gas phase water vapor mole fraction, x_g; and SO_2 mole fraction, x_{SO2}, are solved by the following equations.

$$C_s \frac{dT_g}{dt} = -HTN$$

$$\frac{1}{MW(1-x_g)} \frac{dx_g}{dt} = N_w TN$$

and

Equation 8.9

$$\frac{1}{MW(1-x_g)} \frac{dx_{SO_2}}{dt} = -N_s TN$$

where C_s is the humid heat, and is given by

$$C_s = 0.24 + 0.446 \frac{18 x_g}{MW(1-X_g)}$$

TN is the number of droplets per gram of dry inlet gas, which is known,
MW is the molecular weight of dry inlet gas, and
N_s is the SO_2 molar transfer rate from gas phase to droplets.

The above three equations introduce two additional unknowns, which are xs and N_s. N_s which can be found by considering the SO_2 removal processes.

The droplet energy balance is given by Equation 8.6. In the droplet mass balances, the weight of moisture in a droplet, W; the weight of active sorbent (unreacted sorbent) in a droplet, S; and the total weight of a droplet, WD, are given by the following equations.

$$\frac{dW}{dt} = -18 N_w,$$

Equation 8.10

$$\frac{dS}{dt} = -MW_s \cdot N_s$$

and

$$\frac{dWD}{dt} = -18N_w + 64N_s$$

where MW_s is the molecular weight of the active sorbent.

The above three equations introduce three new unknowns which are W, S, and WD. WD is related to W and S; therefore, there are only two unknowns. The equations listed above are sufficient to solve the drying problem, except that Ns should be obtained by considering the SO_2 mass transfer or absorption process.

THE FALLING-RATE PERIOD. When the moisture in the droplet reaches the critical moisture content at which the solid particles in the droplets begin to touch each other and form a continuous phase [99], the solid concentration starts to influence the drying rate, and the falling-rate period begins. The falling-rate drying period continues until the moisture content in the droplets reaches the equilibrium moisture content. There are no simple correlations to describe this drying process. A linear drying rate between the critical moisture content and the equilibrium moisture content is assumed in the model.

Equation 8.11

$$\text{Drying Rate} = \left(\frac{\text{Drying Rate Based on Gas}}{\text{Phase Resistance Alone}}\right)\left(\frac{X - X_e}{X_c - X_e}\right)^n$$

where
X is moisture content at time t,
X_e is equilibrium moisture content,
X_c is critical moisture content, and
n is a dimensionless parameter. For a linear relation, n=1.

SO_2 Absorption Process

The SO_2 removal in spray dryer absorbers is also modeled according to the drying periods [12]: the constant-rate period and the falling-rate period.

CONSTANT-RATE PERIOD. The conceptual model which served as a basis for model development of the constant-rate period has the following characteristics: Instantaneous reaction between SO_2 and $Ca(OH)_2$ occurs in ion forms in the liquid phase. Droplets possess spherical symmetry. Elementary $Ca(OH)_2$ particles in the atomized slurry droplets are stagnant due to the high apparent slurry viscosity. CO_2 is taken as an inert gas. Reactant sorbent $Ca(OH)_2$ first undergoes dissolution, followed by diffusion toward the reaction zone. The overall rate is controlled by whichever is lower. The effect of product precipitation on sorbent dissolution can be approximated by considering the dissolution rate to be proportional to the lime fraction remaining in the solid phase.

The schematic diagram for the constant-rate period model is shown in . In
x_g is the mole fraction of SO_2 in the bulk gas phase,
x_d is the mole fraction of SO_2 at the droplet surface (gas-liquid interface),
C_{Ai} is the concentration of SO_2 at the interface (in gmole/cm3),
C_B is the concentration of $Ca(OH)_2$ in the bulk liquid phase (gmole/cm3),
C_S is the saturation concentration of $Ca(OH)_2$ at the liquid phase (gmole/cm3),
x is the distance from the gas-liquid interface to the reaction zone (cm), and
x_0 is the liquid film thickness (cm).

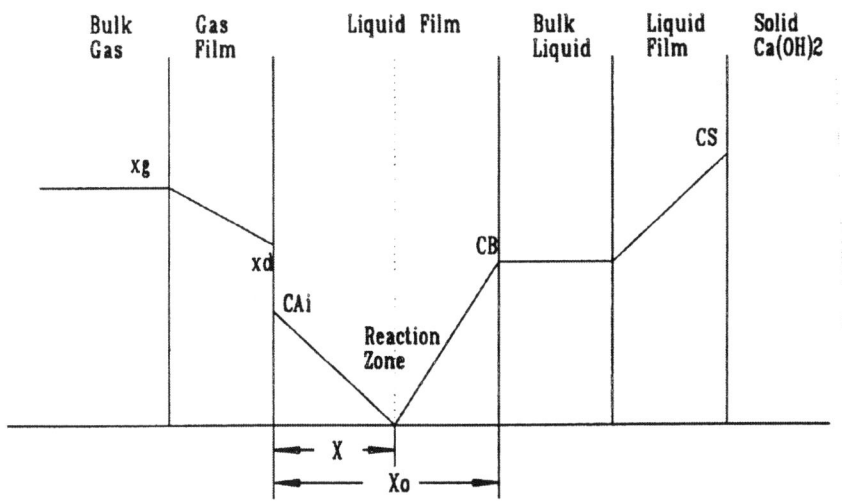

Figure 8.34 Schematic diagram for SO_2 absorption into a slurry drop.

A general description of the SO_2 absorption process follows. The SO_2 absorption process can be considered as four separate steps. SO_2 diffuses from the bulk gas phase to the droplet surface, absorbs into the droplets and diffuse from the droplet surface to the reaction zone. $Ca(OH)_2$ dissolves from the solid phase into the liquid phase, then diffuses to the reaction zone where it reacts with SO_2.

The rate of SO_2 mass transfer through the gas film can be given by

$$N_s = k_g \pi d_d^2 c (x_g - x_d)$$

The rate of SO_2 mass transfer through the liquid film can be given by

$$N_s = k_l \pi d_d^2 (C_{Ai} - 0) \frac{X_0}{X}$$

The sorbent (lime) dissolution rate is given by

$$R_d = k_s \pi d_d^2 (C_S - C_B) N_p$$

The sorbent mass transfer rate in liquid film is given by

$$R_p = k_{Bl} \pi d_d^2 (C_B - 0) \frac{x_0}{x_0 - x}$$

where
k_g and k_l are mass transfer coefficients for SO_2 at gas side and liquid side (cm/s),
R_d is the rate of lime dissolution,
k_S is the mass transfer coefficient for sorbent dissolution (cm/s),
k_{Bl} is the liquid side $Ca(OH)_2$ mass transfer rate (cm/s),
d_p is the sorbent solid particle diameter (cm), and
N_p is the number of sorbent solid particles in a droplet.

The equation for N_s was found (by considering other relationships besides the above four) to be

$$N_s = \frac{\pi d_d^2 \left[\frac{C s_g}{H_A} + \frac{D_{Bl} C_S}{D_t} \right]}{\frac{1}{k_s H_A} + \frac{1}{k_t} + \frac{d_d^2 \delta R_L}{d_d^2 N_p D_t}}$$

where

D is the diffusivity,
H_A is Henry's Law constant for SO_2, and
δ is the film thickness surrounding the sorbent solid particles of diameter d_p.

Modification of the Model by Hygroscopic Additive Effects: The Mechanism of Hygroscopicity

As mentioned previously, the relationship between the saturation ratio and droplet (containing dissolved material) size is given by [100]

Equation 8.12

$$x_d = \frac{p}{p_s} = \frac{1}{1 + \dfrac{6imM_w}{M_s \rho \pi d_d^3}} \cdot \exp\left(\frac{4\gamma M_w}{\rho R T d_d}\right)$$

where
P is the partial pressure of water vapor at the droplet surface,
P_s is the saturation pressure at the plane surface of pure water,
M_s is the molecular weight of the dissolved salt,
m is the weight of dissolved salt in the droplet,
i is the number of ions each molecule of salt forms when it dissolves (2 for NaOH),
r is the density of water (or solvent),
M_w is the molecular weight of water,
γ is the surface tension of water,
d_d is the droplet diameter,
R is the ideal gas law constant, and
T is the temperature of the droplet.

From Equation 8.12 we can see that x_d is directly related to the partial water vapor pressure on the droplet surface. The partial water vapor pressure is affected by the additive through the im/M_s term which is a property of the additives. One possible way to reduce the water evaporation from the droplets is to reduce the driving force for water vapor mass transfer from the droplet surface to the bulk gas phase (i.e., reduce $x_d - x_g$). The hygroscopic additive effect reduces the water vapor pressure, P, on a droplet surface, thus reducing xd and xd-xg. The enhancement of sorbent utilization due to the hygroscopicity of additives can be explained as follows. When water evaporates from the droplets in the spray dryer, the salt concentration in a droplet increases with decreasing droplet size if the salt is highly soluble. This is because the mass of the salt remains constant and it is only the water that evaporates. Thus, a given mass of dissolved salt serves to reduce the vapor pressure at the droplet surface to a greater degree as droplet size decreases. The size change is reflected in the term πd_d^3.

The water vapor pressure-lowering effects by NaOH can be seen from Figure 8.35 where the water vapor pressure is measured above a NaOH solution plane surface [94]. As the NaOH concentration is increased, the partial water vapor pressure is greatly reduced.

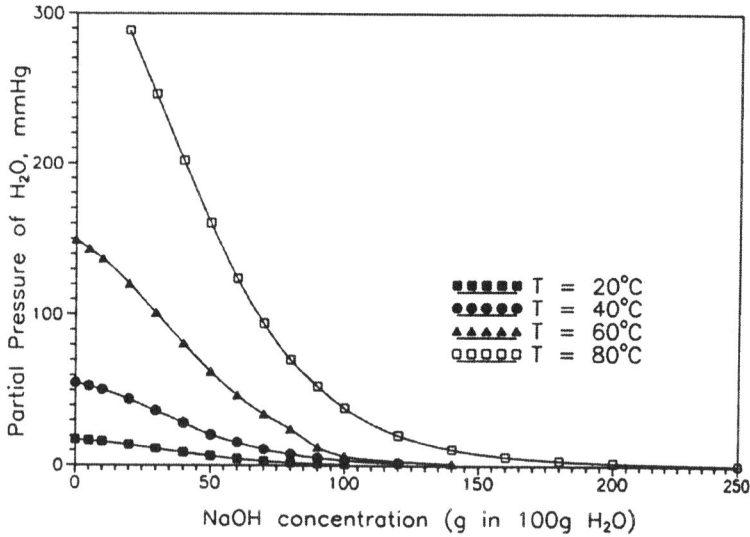

Figure 8.35 Water vapor partial pressure over NaOH solution plane surface.

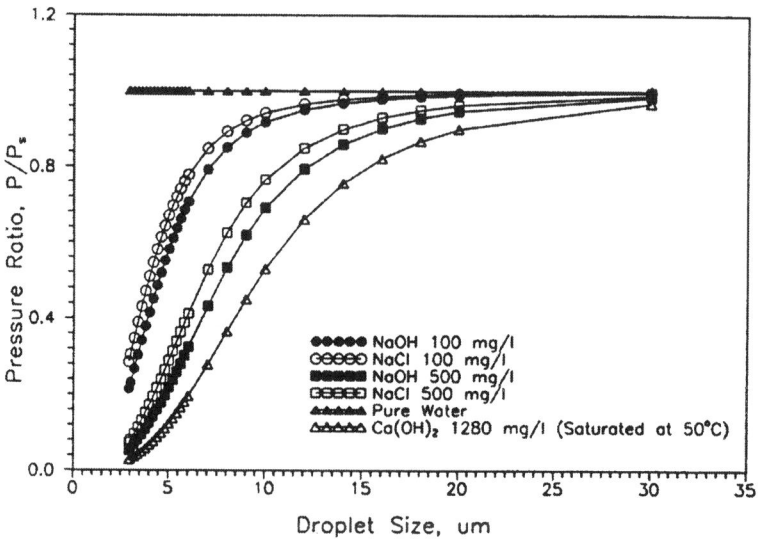

Figure 8.36 Ratio of pressure of droplet surface to pure water plane surface versus droplet size.

Figure 8.36 gives pressure ratios versus droplet size under the influence of additives. The pressure ratio is the ratio of the water vapor pressure on the droplet surface to the water vapor pressure on a plane surface of pure water. Calculations have been made from Equation 8.12. The saturated concentration of $Ca(OH)_2$ is 1280 mg/l at a temperature of 50°C.

Combined with the equation for the calculation of water vapor pressure on the droplet surface in the original model, Equation 8.12 was used to modify the water vapor pressure on the droplet surface to predict the hygroscopic additive effects. The original model was essentially unchanged in terms of the basic principles for drying and SO_2 removal.

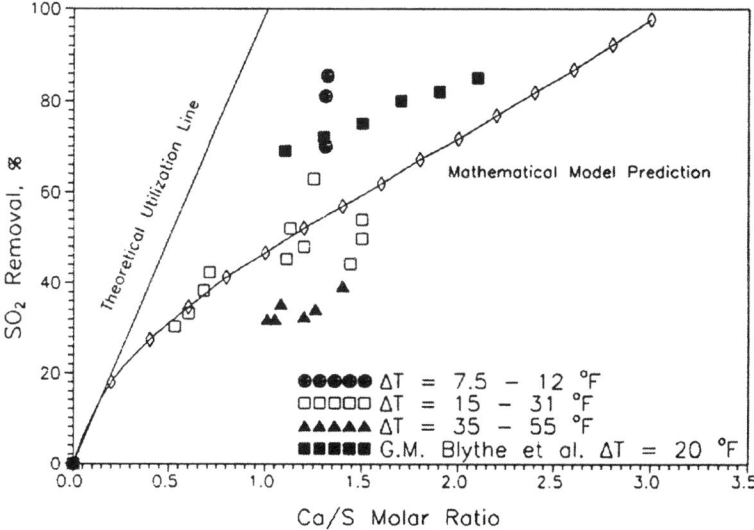

Figure 8.37 SO_2 removal versus Ca/S molar ratio for baseline $Ca(OH)_2$.

The baseline prediction of SO_2 removal was based on the available SPRAY-MOD-M model, and the results were compared with the baseline experimental data. As shown in Figure 8.37, the model prediction of SO_2 removal efficiency versus Ca/S molar ratio fits the SO_2 removal from those experiments described here, but under-predict the SO_2 removal from EPRI's High Sulfur Test Center. This is because the $Ca(OH)_2$ particles in the slaked lime used in this study have a larger particle size. As shown in Figure 8.38, the mode of the particle size used in these experiments is about 6.08 micrometers.

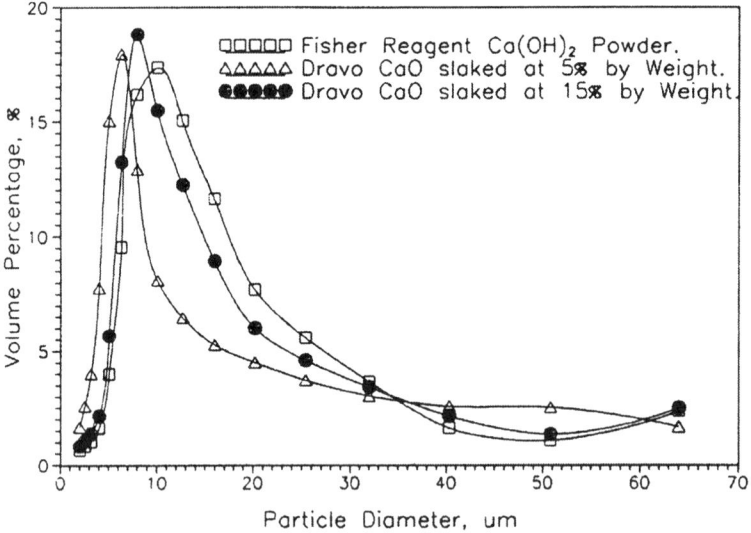

Figure 8.38 Ca(OH)$_2$ particle size distribution (measured by Coulter Counter).

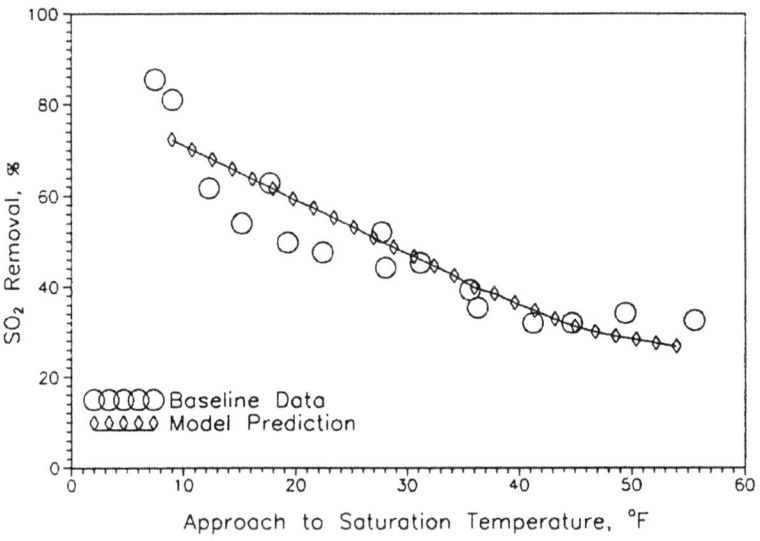

Figure 8.39 Baseline SO$_2$ removal versus approach to saturation temperature.

Table 8.13 gives a list of the parameters and standard operating conditions that are typical in the pilot plant. In the model, the stoichiometric Ca/S molar ratio was varied from zero to 3.0 (in the experiment, the ratio was kept at 1.5 which was used as the standard condition in the model), and the approach to saturation temperature was varied from 5°C (9°F) to 30°C (54°F) (an approach to saturation temperature of 20°F (11°C) was used as the standard condition). When either of them was changed, the other was kept constant at the standard condition.

Table 8.13 SPRAYMOD-M Operating Parameters and Standard Conditions

List of Parameters	SI Units	Eng. Units
Inlet Gas Conditions		
Inlet Gas Temperature	149°C	300°F
% Water in Inlet Gas	1%	1%
Inlet SO_2 Concentration	2500 ppm	2500 ppm
Molecular Weight of Dry Inlet Gas	28.9	28.9
Operating Parameters		
Approach to Saturation Temperature	11°C	20°F
Stoichiometric Ca/S Molar Ratio	1.5	1.5
Mass Fraction of Sorbent in Fresh Solids	1.0	1.0
Recycle Solids Ratio by Mass	0.0	0.0
Mass Fraction of Sorbent in Recycle	0.0	0.0
Residence Time of Gas Phase	10 sec	10 sec
Flow System in Spray Dryer	Plug	Plug
Sorbent Properties		
Inlet Droplet Diameter	100 um	100 um
Inlet Droplet Temperature	15.6°C	60°F
Form of Sorbent	Slurry	Slurry
Sorbent Particle Diameter	6.08 um	6.08 um
Molecular Weight of Sorbent	74	74
Density of Solid Sorbent	2.24 g/cm^3	140 lb/ft^3
Critical Moisture Content	29.2%	29.2%
Equilibrium Moisture Content	6%	6%
Dry Sorbent Reaction Rate Coefficient	0.0	0.0

Table 8.13 SPRAYMOD-M Operating Parameters and Standard Conditions cont.

List of Parameters	SI Units	Eng. Units
Assumed Spray Dryer Efficiency	50%	50%
Program Parameters		
Maximum Allowable Time Step	0.1 sec	0.1 sec
Time Between Printouts	0.5 sec	0.5 sec
Time Step Control Parameter	0.01 sec	0.01 sec

Figure 8.39 gives the comparison between the model predictions and experimental data for baseline SO_2 removal versus approach to saturation temperature. It can be seen from Figure 8.39 that the model fits the data well.

With the addition of NaOH (500 mg/liter) into the 10% $Ca(OH)_2$ slurry, Figure 8.40 shows the model prediction for SO_2 removal versus approach to saturation temperature. As can be seen from Figure 8.40, if the approach to saturation temperature is below 30°F, the model under-predicts the removal.

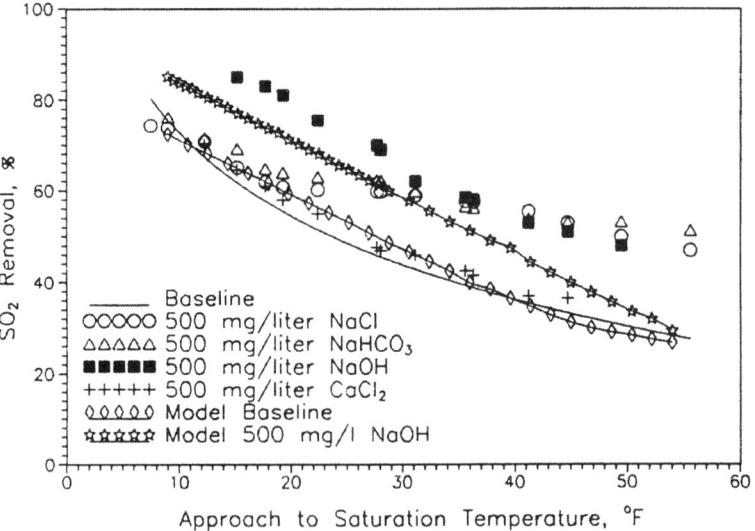

Figure 8.40 Comparison of model prediction and experimental data on SO_2 removal versus approach to saturation temperature.

The model predictions for the effects of the other two additives (NaCl and NaHCO$_3$) are shown in Figure 8.41. The model over-predicts the removal if the approach to saturation temperature is below 28°F, but under-predicts if the approach to saturation temperature is above 28°F. The are no significant differences between the two additives, in terms of SO$_2$ removed, based on the model prediction.

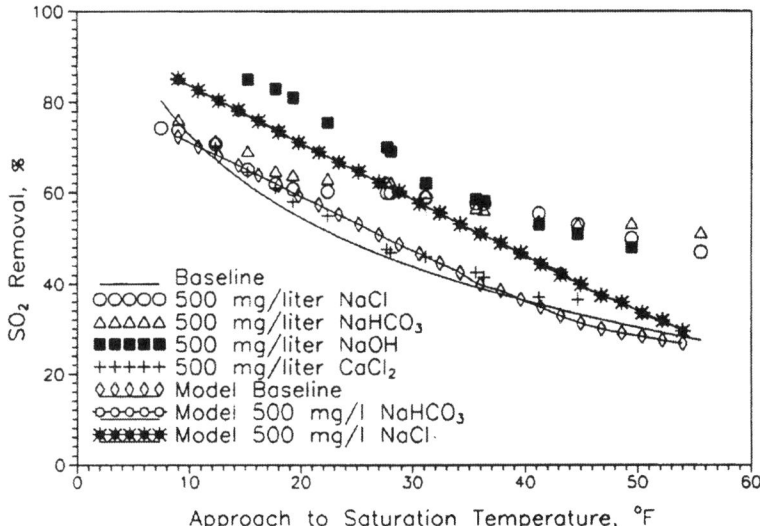

Figure 8.41 Comparison of model prediction and experimental data on SO$_2$ removal versus approach to saturation temperature.

Derivation of a Simplified Model for SO$_2$ Removal in Spray Dryers and Discussion of the Hygroscopic Additive Effects

If the gas phase SO$_2$ mass transfer resistance is assumed to be the limiting step for the overall process in the constant-rate period, a simple correlation can be derived between the SO$_2$ removal efficiencies and the residence time of the liquid phase. Although this hypothesis is simplified, the results may show a clearer picture of the spray dryer desulfurization process.

A schematic of the spray dryer chamber (flue gas flow rate Q, m^3/h) and the spray dryer inlet SO$_2$ concentration is shown in Figure 8.42. The Ca(OH)$_2$ slurry volumetric flow rate is N (m^3/h). The origin of the x coordinate is at the nozzle. At position x the concentration of SO$_2$ is C, and at position x+dx the concentration of SO$_2$ becomes C - dC. The SO$_2$ mass transfer coefficient in the air is kg, the initial droplet diameter (at the birth of the droplet) averaged over the distribution is d_{d0}, and the droplet diameter averaged over the constant-rate period (over time) is d_d.

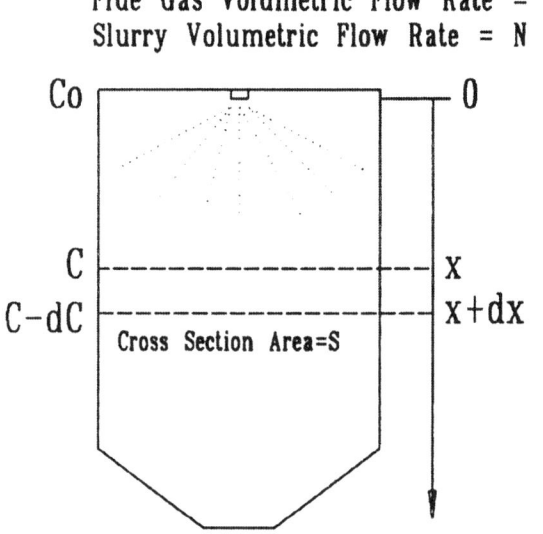

Figure 8.42 Schematic drawing for derivation of spray dryer desulfurization model.

When flue gas moves down from point x to point x+dx, the SO_2 concentration is reduced from C to C-dC. This concentration reduction is due to SO_2 overcoming the gas phase resistance and diffusing to the droplets. Therefore, this concentration reduction is equal to the mass transfer flux times the overall droplet surface area in the volume S•dx, times the flue gas residence time in the volume S•dx. Since N/Q is the slurry volume per volume of air, the number of droplets per volume of air can be determined and is given by the equation below.

$$\frac{N}{Q\frac{1}{6}\pi d_{d0}^3}$$

The overall droplet surface area in the volume S•dx is

$$-dC = \frac{k_g C \cdot \dfrac{6Nd_d^2 S dx}{Qd_{d0}^2} \cdot dt}{S dx}$$

The flux of SO_2 in terms of gas phase resistance is $k_g C$ (here the concentration of SO_2 at the droplet surface is assumed to be zero). The residence time, dt, of the flue gas in the differential volume S•dx is S•dx/Q. Integrating the right-hand side of the equation above from 0 to t and the left-hand side from C_o to C gives

$$\frac{N}{Q\frac{1}{6}\pi d_{d0}^3} \pi d_d^2 (S \cdot dx) = \frac{6Nd_d^2 S dx}{Qd_{d0}^3},$$

$$\frac{dC}{C} = -\frac{6Nd_d^2 k_g dt}{Qd_{d0}^3}$$

The removal efficiency is $h = 1 - C/C_0$; therefore,

$$\frac{C}{C_0} = e^{-\frac{6Nd_d^2 k_g t}{Qd_{d0}^3}}$$

and

$$\eta = 1 - e^{-\frac{6Nd_d^2 k_g t}{Qd_{d0}^3}}$$

From the equation above, an understanding of the hygroscopic additive effects is straightforward. Since the hygroscopic additives cause the retention of moisture on the droplet surface, the constant-rate period will be prolonged, and therefore the drying time will be longer than that without additives in the slurry droplets. If the drying time is increased, the efficiency will be increased assuming other conditions to be the same. An illustration of the efficiency versus time for the typical conditions in the pilot spray dryer is shown in Figure 8.43 where the air flow rate used is 4.25 m³/min, $d_d = d_{d0} = 100$ micrometers, and $k_g = 5.884 \cdot 10^{-5} T^{1.75}$ where T=70°C. The typical slurry input in the spray dryer is about 4 gallons/hour if the approach to saturation temperature is kept around 11°C (20°F). The upper limit for the slurry input is about 6 gallons/hour. The slurry volumetric input depends on the approach to saturation temperatures if the slurry concentration is constant.

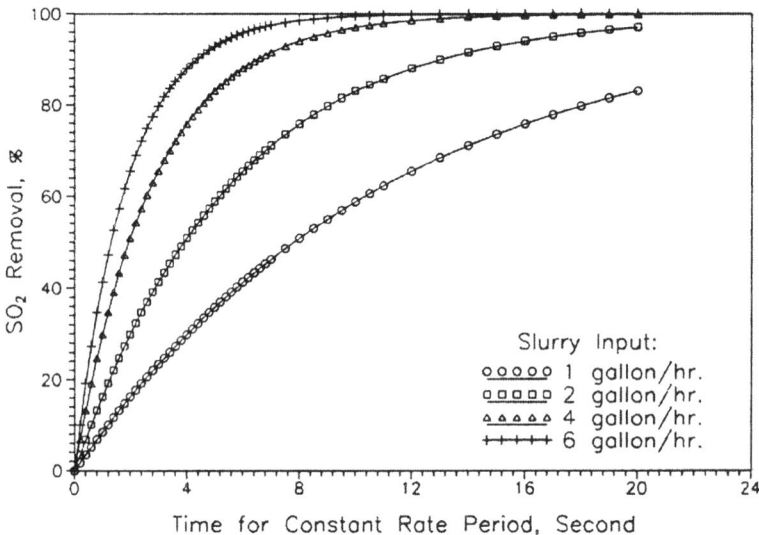

Figure 8.43 SO_2 removal as a function of constant-rate drying period from the simplified model prediction.

From Figure 8.43, we can determine that the values of SO_2 removal per second per gallon of slurry are greater for higher slurry volumetric input rates. This indicates that slurries that have lower solids concentrations (more water per mass of solid sorbent) have higher rates of reaction. Hygroscopic additives produce the same effect by slowing the drying rate for a given slurry concentration.

Conclusions

This study and experiments have shown that over 90% removal of the SO_2 concentration from a high sulfur coal flue gas can be achieved in spray dryers with the appropriate use of additives and operating conditions. The research has attempted to explain the measured results in terms of fundamental chemical and physical phenomena which affect the SO_2 reaction.

The experiments have investigated methods of improving sorbent utilization in spray dryer desulfurization processes. The methods investigated in the study have included: measuring lime dissolution rates at different conditions and with additives in a bench scale apparatus; direct tests of other additive mechanisms in a pilot spray dryer absorber; tests of hydrated fly ash and lime mixtures as new sorbents in a pilot spray dryer; and theoretical studies on the mechanisms of the additive effects.

The mechanisms responsible for improvements in performance with the use of additives have been investigated and discussed. Changes in the hygroscopicity of the sorbent slurry has been mathematically formulated and incorporated into a spray dryer mathematical model. This model has been shown to describe the improvements measured in experimental tests with reasonable accuracy.

The lime dissolution rate has been measured in the presence of a variety of inorganic and organic additives. Among the inorganic additives, NH_4Cl, NH_4NO_3, and NH_4I are very effective at enhancing the lime dissolution rate. Among the organic additives, $NH_4C_2H_3O_2$, sucrose, and phenol are the most effective.

The effects of other additives have been studied directly in a spray dryer. The additives tested include hygroscopic additives, oxidation additives, dissolution rate enhancement additives, buffer additives, and fly ash. Hygroscopic additives such as NaOH, NaCl, $NaHCO_3$, an oxidation additive (H_2O_2), and additives to increase dissolution rate (such as NH_4Cl and Sucrose) have significant effects on improving lime utilization and SO_2 removal in spray dryers. Buffer additives (formic and benzoic acids) have little or no effect.

Two types of fly ashes have been hydrated with lime at elevated temperatures and evaluated. It was found that the treated mixtures developed high surface areas (after 20 hours hydration, the mixture surface area can reach as high as 30 m^2/g, as compared to about 3 m^2/g prior to hydration), and one type of fly ash (MF#8) could greatly enhance lime utilization in spray dryer. This has been attributed to the type of hydrated material produced, and the significant increase in surface area.

Nomenclature

BET	Specific surface area (m^2/g)
c	Molar density of gas phase (mole/cm3)
C	Concentration, percent by weight
C_{acid}	Acid concentration (mole/liter)
$C_{p,l}$	Specific heat of water vapor (J/g-°C)
C_p	Specific heat (J/g-°C)
C_s	Humid gas specific heat (J/g-°C)
C_s	SO_2 concentration (ppm)
d	Sorbent/acid complete reaction molar ratio (for $CaCO_3$/HCl d=1/2)
D	Molecular diffusivity (cm2/s)
d_d	Droplet diameter (mm)
d_d	Droplet diameter averaged over the constant-rate period (mm)
d_{d0}	Initial droplet diameter (mm)
d_p	Sorbent solid particle diameter (cm)
h	Heat transfer coefficient (J/s-cm^2-°C)
H	Heat transfer rate (J/s)
H_2	Humidity ratio at t_1 or t_2 ($g_{H_2O}/g_{dry\ air}$)
H_{30}	Humidity ratio at t_{w2} ($g_{H_2O}/g_{dry\ air}$)
H_A	Henry's Law constant (appropriate units)
i	Number of ions each salt molecule forms when dissolves (2 for NaOH)
k	Thermal conductivity (J/s-cm-°C)
k_{Bl}	Liquid side $Ca(OH)_2$ mass transfer rate (cm/s),
k_d	Gas-phase mass transfer coefficient,
k_g	Gas side mass transfer coefficient (cm/s)
k_l	Liquid side mass transfer coefficient (cm/s)
k_S	Mass transfer coefficient for sorbent dissolution (cm/s),
L	Acid solution consumption rate (g/min)
m	Weight of dissolved salt in the droplet (g)
M_a	Molecular weight of flue gas (g/mole)

M_c	Molecular weight of CaO, 56 g/mole
M_s	Molecular weight of dissolved salt (g/mole)
M_w	Molecular weight of the solvent (g/mole)
MW	Molecular weight of dry inlet gas (g/mole)
MW_s	Molecular weight of the active sorbent in Equation 8.10 (g/mole)
n	Dimensionless parameter in Equation 8.11
N	$Ca(OH)_2$ slurry volumetric flow rate (m³/h)
N_p	Number of sorbent solid particles in a droplet
N_s	SO_2 molar transfer rate from gas phase to droplets (mole/s)
Nu	Nusselt Number, hd_d/k, for heat transfer, (dimensionless)
N_w	Molar rate of water evaporation (mole/s)
P	Partial pressure of water vapor at the droplet surface (atm)
P_s	Saturation pressure at the plane surface of pure water (atm)
P_t	Prandtl Number, $C_p\mu/k$, (dimensionless)
Q	Gas flow rate (m³/h)
R	Ideal gas law constant, (dyn-cm/K)
R'	Acid consumption rate (mole/s)
R_d	Rate of lime dissolution (mole/s-drop)
Re	Reynolds Number, $dd vr/m$, based on droplet diameter (dimensionless)
R_s	Sorbent dissolution rate (mole/cm²-s)
S	Mass of active sorbent (unreacted sorbent) in a droplet (g)
Sc	Schmidt Number, $\mu/\rho D$, (dimensionless)
Sh	Sherwood Number, $k_d d_d/D$, for mass transfer (dimensionless)
t	Time (s)
T	Absolute temperature of the droplet (K)
t_2	Inlet temperature of the spray dryer (°C)
t_3	Outlet temperature of the spray dryer (°C)
T_d	Temperature of the droplets (°C)
T_g	Temperature in bulk gas phase (°C)
TN	Number of droplets per gram of dry inlet gas (g⁻¹)
t_{w2}	Saturation temperature by assuming adiabatic system (°C)

v	Relative velocity between gas and droplet (cm/s)
W	Mass of moisture in a droplet (g)
WD	Total mass of a droplet (g)
W_s	Initial weight of sorbent (limestone) (g)
X	Moisture content ($g_{H_2O}/g_{sorbent}$)
X_c	Critical moisture content ($g_{H_2O}/g_{sorbent}$)
x_d	Equilibrium water vapor mole fraction at the droplet surface (dimensionless)
X_e	Equilibrium moisture content ($g_{H_2O}/g_{sorbent}$)
x_g	Water vapor mole fraction in the bulk gas phase (dimensionless)
x_g	Gas phase water vapor mole fraction (dimensionless)
x_{SO2}	SO_2 mole fraction in Equation 8.9 (dimensionless)

Greek Letters

α	Correction factor to account for moisture in the product in
γ	Surface tension of the solvent (dyn/cm)
δ	Film thickness surrounding the sorbent solid particles (mm)
η	Removal efficiency (dimensionless)
λ	Latent heat of vaporization at Td (J/g)
μ	Viscosity (g/cm^{-s})
ρ	Density (g/cm^3)
ε	Heat-transfer rate factor (dimensionless)
ΔW	Acid weight change (g)

References

1. Masters, K., *Spray Drying Handbook*, Fourth Edition, NY, John Wiley & Sons Inc. (1985)
2. Boynton, R. S., *Chemistry and Technology of Lime and Limestone*, Interscience Publishers, a Division of John Wiley & Sons, Inc., February 1967
3. Beals, J., J. Cannell and J. Hengel, "How FGD Reagent Quality Affects System Performance," *Power*, 128(3), March 1984
4. Kelly, M. E., J. D. Kilgroe and T. G. Brna, "Current Status of Dry SO_2 Control Systems," *Proceedings: Symposium on Flue Gas Desulfurization*, 2, EPRI, Palo Alto, CA, March 1983
5. Reinauer, T. V., J. P. Monat and M. Mutsakis, "Reducing Plant Pollution Exposure: Dry FGD on an Industrial Boiler," *Chem. Eng. Prog.*, 79(3), March 1983
6. Haslam, R., G. Calingaert and C. Taylor, *J. Am. Chem. Soc.*, 46, 308 (1924)
7. Bassett, H. Jr., *J. Chem. Soc.*, 1270 (1934)
8. Stearns Catalytic Corp., *Economic Evaluation of Dry-Injection FGD Technology*, EPRI, Palo Alto, CA., January 1986
9. Midkiff, L.A., "Spray-Dryer System Scrubs SO_2," *Power*, 123(1), January 1979
10. Gooch, J. P. et al. *Humidification of Flue Gas to Augment SO_2 Capture by Dry Sorbents*, Topical Report to Environmental Protection Agency, EPA Cooperative Agreement No. CR811683 and CR812811 (Draft), August 1988
11. Getler, J. L., H. L. Shelton and D. A. Furlong, "Modelling the Spray Absorption Process for SO_2 Removal," *JAPCA*, 29(12), 1270 (1979)
12. Partridge, G. P. Jr., *A Mechanistic Spray Dryer Mathematical Model Based on Film Theory to Predict Sulfur Dioxide Absorption and Reaction by A Calcium Hydroxide Slurry in the Constant-Rate Period*, Ph.D. Dissertation, The University of Tennessee, Knoxville, December 1987
13. Uchida, S. and C. Y. Wen, "Rate of Gas Absorption into a Slurry Accompanied by Instantaneous Reaction," *Chem. Eng. Sci.*, 32, 1277-1281 (1977)
14. Makansi, J., "Particulate Control: Optimizing Precipitators and Fabric Filters for Today's Power Plants, Special Report," *Power*, December 1986
15. White, H. J., *Industrial Electrostatic Precipitation*, Addison-Wesley Publishing Company, Inc., MA (1963)
16. Damle, A. S., *Modelling of SO_2 Removal in Spray Dryer Flue Gas Desulfurization System*, EPA-600/7-85-038, December 1985
17. Yoon, H., M. R. Stouffer, W. A. Rosenhoover and R. M. Statnick, "Laboratory and Field Development of Coolside SO_2 Abatement Technology," *Second Annual Pittsburgh Coal Conference*, Pittsburgh, PA, September 1985
18. Klingspor, J., H.T. Karlsson and I. Bjerle, "A Kinetic Study of the Dry SO_2-Limestone Reaction at Low Temperature," *Chem. Eng. Comm.*, 22, 88 (1983)
19. Babu, M. et al., *5-MW Toronto HALT Pilot Plant Testing - Final Testing Results*, DOE Contract DE-AC22-85PC81012, December 1988
20. Felsvang, K. S., O. E. Hansen and E. I. Rasmussen, *Process for Flue Gas Desulfurization*, US Patent No. 4279873, July 21, 1981
21. Weeter, D. W., "Utilization of Dry Calcium Based Flue Gas Desulfurization Waste as a Hazardous Waste Fixation Agent," *Proceedings: Waste Treatment & Disposal Aspects: Combustion & Air Pollution Control Progresses*, South Atlantic Section, Air Pollution Control Association, February 1981

22. 22.Donnelly, J. R., R. P. Ellis and W. C. Webster, "Dry Flue Gas Desulfurization End-Product Disposal Riverside Demonstration Facility Experience," *Proceedings: Symposium on Flue Gas Desulfurization*, 2, EPRI, Palo Alto, CA, March 1983

23. Vuceta, J., J. P. Woodyard and D. W. Weeter, "Disposal Techniques for Dry Flue Gas Desulfurization Process Residues," *Proceedings: Waste Treatment & Disposal Aspects: Combustion & Air Pollution Control Progresses*, South Atlantic Section, Air Pollution Control Association, February 1981

24. Kuchibotla, S. et al. "The Stabilization of Orimulsion Spray Dryer Waste for Landfill Disposal," *Proceedings: EPRI 1993 SO_2 Control Symposium*, Session 8B, Boston, MA, August 24-27, 1993

25. Jarvis, J. B., Farmer, R. W. and D. A. Stewart, "Description and Mechanisms of Limestone FGD Operating Problems Due to Aluminum/Fluoride Chemistry," 7/79-7/83, *Proceedings: Tenth Symposium on Flue Gas Desulfurization*, 1, EPRI CS-5167, May 1987

26. Chang, J. C. S. and T. G. Brna, "Enhancement of Wet Limestone Flue Gas Desulfurization by Organic Acid/Salt Additives," paper 6b, *Tenth Symposium on Flue Gas Desulfurization*, Atlanta, GA, November 18-21, 1986

27. Wang, S. C. and D. A. Burbank, "Adipic Acid Enhanced Lime/Limestone Test Results at EPA Alkali Scrubbing Test Facility," in *Flue Gas Desulfurization*, J.L. Hudson and G.T. Rochelle, eds., Amer. Chem. Soc. Symp. Ser. 188, 243-265, Washington (1982)

28. Rochelle, G. T. et al. "Buffer Additives for Lime/Limestone Slurry Scrubbing," in *Flue Gas Desulfurization*, J. L. Hudson and G. T. Rochelle, eds., Amer. Chem. Soc. Symp. Ser. 188, 243-265, Washington (1982)

29. Moser, R. E. and D. R. Owens, "Overview on the Use of Additives in Wet FGD Systems," *Proceedings: 1991 SO_2 Control Symposium*, EPRI TR-101054, 1, 3A-1, 3A-21, November 1992

30. Stohs, M. et al. "Results of Formate Ion Additive Tests at EPRI's High Sulfur Test Center," *EPRI 1991 SO_2 Control Symposium*, Washington, D.C., Dec. 3-6, 1991

31. Burke, J. et al. "Results of Sodium Format Addition at EPRI's HSTC and AECI's Thomas Hill Unit 3 FGD System," *Proceedings: 1990 SO_2 Control Symposium*, Vol. 2, 5-23 to 5-44, September 1990

32. Moser, R. E., J. M. Burke and S. M. Gray, "Results of Wet FGD Testing at EPRI's High Sulfur Test Center," *Proceedings: EPA/EPRI First Combined FGD and Dry SO_2 Control Symposium*, EPRI CS-6307, RP 982-41, St. Louis, MO, October 1988

33. Weilert, C .V. and R. D. Norton, "Compliance Strategy for Future Capacity Additives: The Role of Organic Acid Additives," *EPRI 1991 SO_2 Control Symposium*, Washington, D.C., December 3-6, 1991

34. Moser, R., and F. Meserole, U.S. Patent No. 4,994,246, February, 1991

35. Maller, G., F. Meserole and R. Moser, "Use of Thiosulfate and EDTA to Inhibit Sulfite Oxidation in Wet Limestone FGD Processes: Results of Laboratory and Bench-scale Testing," *Proceedings: EPRI 1990 SO_2 Control Symposium*, 3, 7B53-74, September 1990

36. Blythe, G., T. Slater and R. Moser, "Full-scale Demonstration of EDTA and Sulfur Addition to Control Sulfite Oxidation," *Proceedings: EPRI 1991 SO_2 Control Symposium*, EPRI Report TR-101054, November 1992

37. Moser, R., D. Owens and D. Colley, "Control and Reduction of Gypsum Scale in Wet Lime/Limestone FGD by Addition of Thiosulfate: Summary of Field Experiences," *Proceedings: First Combined FGD and Dry SO_2 Control Symposium*, 1, 5-335 to 5-355, April 1989

38. Owens, D. et al. "Inhibited Sulfite Oxidation by Thiosulfate in Wet Lime/Limestone FGD Processes: Results of Laboratory Studies and Testing at EPRI's HSTC Mini-Pilot," *Proceedings: First Combined FGD and Dry SO_2 Control Symposium*, 1, 5-310 to 5-334, April 1989

39. Moser, R. E. and D. R. Owens, "Overview on the Use of Additives in Wet FGD System," *EPRI 1991 SO_2 Control Symposium*, Washington, D.C., December 3-6, 1991

40. Bailey, M. et al. "Results of an Investigation to Improve the Performance and Reliability of HL&P's Limestone Station FGD System," *Proceedings: EPRI 1991 SO_2 Control Symposium*, Washington, D.C., December 3-6, 1991

41. Teixeira, D. P. T. A. Lott, and L. J. Muzio, "Dry Sorbent SO_2 Control for New Power Plants Burning Low Sulfur Western Coals," paper 7, *Proceedings: 1986 Joint Symposium on Dry SO_2 and Simultaneous SO_2/NO_x Control Technologies*, 1, EPRI CS-4966, December 1986

42. Weber, G. F., M. E. Collings and M. H. Bobman, "Enhanced Utilization of Furnace Injected Calcium-Based Sorbents," paper 9, *Proceedings: 1986 Joint Symposium on Dry SO_2 and Simultaneous SO_2/NO_x Control Technologies*, 1, EPRI CS-4966, December 1986

43. Snow, G. C., J. M. Lorrain and S. L. Rakes, "Pilot Scale Furnace Evaluation of Hydrated Sorbents for SO_2 Capture," paper 6, *Proceedings: 1986 Joint Symposium on Dry SO_2 and Simultaneous SO_2/NO_x Control Technologies*, 1, EPRI CS-4966, December 1986

44. Muzio, L. J., A. A. Boni, G. R. Offen and R. Beittel, "The Effectiveness of Additives for Enhancing SO_2 Removal with Calcium Based Sorbents," paper 13, *Proceedings: 1986 Joint Symposium on Dry SO_2 and Simultaneous SO_2/NO_x Control Technologies*, 1, EPRI CS-4966, December 1986

45. Ruiz-Alsop, R.N. and G.T. Rochelle, "Effect of Delquescent Salt Additives on the Reaction of Sulfur Dioxide with Dry $Ca(OH)_2$," *Amer. Chem. Soc. Div. Fuel Chem. Prepr.*, 30(2), 88 (1985)

46. Ruiz-Alsop, R. N. and G. T. Rochelle, "Fossil Fuels Utilization: Environmental Concerns," *Am. Chem. Soc.*, 209-222 (1986)

47. Yoon, H. et al. "Sorbent Improvement and Computer Modeling Studies for Coolside Desulfurization," *1986 Joint Symposium on Dry SO_2 and Simultaneous SO_2/NO_x Control Technologies*, Raleigh, NC., June 2-6, 1986

48. Stouffer, M. F., H. Yoon and F. P. Burker, "The Mechanism of CaO Sulfation in Boiler Limestone Injection," 194th American Chemical Society National Meeting, August 30, New Orleans, LA (1987)

49. Yoon, H. et al., "Pilot Process Variable Study of Coolside Desulfurization," *Environmental Progress*, 7(2), 104-111 (1988)

50. Staudinger, G., P. Melcher and K. Eckersdorfer, "Austrian Experience with Furnace Limestone Injection," presented at the First Combined FGD and Dry SO_2 Control Symposium, St. Louis, MO (1988)

51. Bjerle, I., J. Klinspor and H.T. Karsson, *Ind. Environ. Protec.*, 4(1), (1984)

52. Cunill, F. et al. "Influence of Different Additives on the Reaction Between Hydrated Lime and Sulfur Dioxide," *Environmental Progress*, 10, 4, 273-277, November 1991

53. Seinfeld, J. H., *Atmospheric Chemistry and Physics of Air Pollution*, John Wiley & Sons, Inc. (1986)

54. Lowell, P. S. and T. B. Parson, *Identification of Regenerable Metal Oxide SO_2 for Fluidized-Bed Coal Combustion*, Report by Radian Corporation, Austin, Texas, EPA-650/2-75-065 (1975)

55. Ruth, L. A. and G. M. Varga, Jr., *Regenerable Sorbents for Fluidized-Bed Combustion*, Exxon Research and Engineering Co., Linden, N.J., Quarterly Progress Report Nos. 4-5 to Nat. Sc. Found. under Contract No. AER 75-16194 (1977)
56. Yang, T. R. and M. S. Shen, "Calcium Silicates: A New Class of Highly Regenerative Sorbents for Hot Gas Desulfurization," *J. AIChE*, 25, 811-819 (1979)
57. Jozewicz, W. and G. T. Rochelle, "Fly Ash Recycle in Dry Scrubbing," *Environmental Progress*, 5, 2, 219-224 (1986)
58. Petersen, T., J. Peterson, H. T. Karlsson and I. Bjerle, "Physical and Chemical Activation of Fly Ash to Produce Reagent For Dry FGD Processes," *Proceedings: First Combined FGD and Dry SO_2 Control Symposium*, St. Louis, MO, October 1988
59. Peterson, J. R. and G. T. Rochelle. "Aqueous Reaction of Fly Ash and $Ca(OH)_2$ to Produce Calcium Silicate Absorbent for Flue Gas Desulfurization," *Environmental Science and Technology*, 22, 11, 1299-1304 (1988)
60. Jocewicz, W., C. Jorgensen, J. C. S. Chang, C. B. Sedman and T. B. Brna. "Development and Pilot Plant Evaluation of Silica-Enhanced Lime Sorbents for Dry Flue Gas Desulfurization," *JAPCA*, 38, 6, 796-805 (1988)
61. Jocewicz, W., J. C. S. Chang, C. B. Sedman and T. B. Brna, "Silica-Enhanced Sorbents For Dry Injection Removal of SO_2 From Flue Gas," *J. APCA*, 38, 8, 1027-1034 (1988)
62. Martinez, J. C., J. F. Izquierdo, F. Cunill, J. Tejero and J. Querol. "Reactivation of Fly Ash and $Ca(OH)_2$ Mixtures for SO_2 Removal of Flue Gas," *Ind. Eng. Chem. Res.*, 30, 9, 2143-47 (1991)
63. Hall, B. W., C. Singer, W. Jozewicz, C. B. Sedman and M. A. Maxwell. "Current Status of the ADVACATE Process for Flue Gas Desulfurization," *J. A&WMA*, 42, 1, 103-110 (1992)
64. Ho, C. S. and S. M. Shih. "$Ca(OH)_2$/Fly Ash Sorbents for SO_2 Removal," *Ind. Eng. Chem. Res.*, 31, 4, 1130-35 (1992)
65. Sanders, J., T. C. Keener and J. Wang, "Heated Fly Ash/Hydrated Lime Slurries for SO_2 Removal in Spray Dryer Absorbers," *Ind. Eng. Chem. Res.*, 34, 1, 302-307 (1995)
66. Costa, V. and F. Massazza, "Natural Pozzolanas and Fly Ashes," in S. Diamond, ed., *Effects of flyash Incorporation in Cement and Concrete*, Materials Research Society, Boston, MA, 134 (1981)
67. Tognon, G. and P. Ursella, "Combined Lime and Specific Surface Area of Lime-Pozzolana and Lime-Flyash Mixes," in S. Diamond, ed., *Effects of Fly Ash Incorporation in Cement and Concrete*, Materials Research Society, Boston, MA, 145 (1981)
68. Jorgensen, C. and J. C. S. Chang, "Evaluation of Sorbents and Additives for Dry SO_2 Removal," *Environmental Progress*, 6(2), 26 (1987)
69. Jozewicz, W. and J. C. S. Chang, *Evaluation of FGD Dry Injection Sorbents and Additives, Vol. 1, Development of High Reactivity Sorbents*, EPA-600/7-89-006a (NTIS PB89-208920), May 1989
70. Chang, J. C. S. and C. Jorgensen, *Evaluation of FGD Dry Injection Sorbents and Additives, Vol. 2, Pilot Plant Evaluation of High Reactivity Sorbents*, EPA-600/7-89-006b (NTIS PB89-214314), May 1989
71. Meserole, B. F., B. M. Eklund, K. W. Luke and L. J. Holcombe, *Limestone Dissolution Studies*, Final Report, EPRI CS-4845, Project 1031-4, November 1986
72. Barton, P. and T. Vatanatham, *Env. Sci. Tech.*, 10, 262 (1976)
73. Barton, P. and T. Vatanatham, *Env. Sci. Tech.*, 13, 1420 (1979)
74. Berner, R. A. and J. W. Morse, "Dissolution Kinetics of Calcium Carbonate in Sea Water-Theory of Calcite Dissolution," *American J. of Science* 274(2), 108-34 (1974)

75. Plummer, L. N., D. L. Parkhurst and T. M. L. Wigley, *Amer. Chem. Soc. Symp. Ser.*, 93, 538-573 (1979)
76. Wentzler, H. T. and F. Aplan, "Kinetics of Limestone Dissolution by Acid Wastewaters," *Proceedings of Environmental Control Symposium*, 1972. ASME Metallurgical Society, New York, 513-523 (1972)
77. Uchida, S., C. Y. Wen and W. J. McMichael, "Dissolution Rate of Limestone into Acid Solution," *J. Chin. Inst. Chem. Eng.* 5, 111-114 (1974)
78. Kim, K. Y., M. E. Deming and J. D. Hatfield, "Dissolution of Limestone in Simulated Slurries for Removal of Sulfur Dioxide from Stack Gases," *Env. Sci. Tech.* 9(10), 949-5, (1975)
79. Chan, P. K. and G. T. Rochelle, "Limestone Dissolution - Effects of pH, CO_2, and Buffers Modeled by Mass Transfer," *Amer. Chem. Soc. Symp. Ser.* 188: 75 (1982)
80. Toprac, A. J. and G. T. Rochelle, *Environmental Progress*, 1, 52 (1982)
81. Noda, K., S. Uchida and M. Miyazaki, "Limestone Neutralization of Acid Solutions Containing Dissolved Iron," *J. of Chem. Eng. of Japan*, 22, 3, 253-57 (1989)
82. Sjoberg, L. E., "A Fundamental Equation for Calcite Dissolution Kinetics," *Geochimica et Cosmochimica Acta*, 40, 441-447 (1976)
83. Evans, J. T. and M. G. Moseley, "Suitability and Availability of Texas Limestone for Flue-Gas Desulfurization," *Proceedings of the Gulf Coast Lignite Conference: Geology, Utilization, and Environmental Aspects* (1978)
84. Pesret, F., *Kinetics of Carbonate-Seawater Interactions*, AD-747976. University of Hawaii, Hawaii Institute of Geophysics, Honolulu, HI, August 1972
85. Pearson, H. F. and A. J. McDonnell, "Use of Crushed Limestone to Neutralized Acid Wastes," *ASCE, J. Env. Eng. Div.*, 101(EEI), 139-158 (1975)
86. Gooch, J. P. et al. "Sorbent Development and Production Studies," paper 11, *Proceedings: 1986 Joint Symposium on Dry SO_2 and Simultaneous SO_2/NO_x Control Technologies*, Vol. 1, EPRI CS-4966, December 1986
87. Gage, L. C. and G. T. Rochelle, "Limestone Dissolution in Flue Gas Scrubbing: Effect of Sulfite," *J. A&WMA*, 42, 7, 962-935, July 1992
88. Jarvis, J. B., B. F. Meserole, J. T. Selm, G. T. Rochelle, L. C. Gage and E. R. Moser, "Development of a Predictive Model for Limestone Dissolution in Wet FGD Systems," *EPA/EPRI Combined FGD and Dry SO_2 Control Symposium*, St. Louis, MO (1988)
89. Ukawa, N., S. Okino, M. Oshima and T. Oishi, "Effects of Salts on Limestone Dissolution Rate in Wet Limestone Flue Gas Desulfurization," *J. of Chem. Eng. of Japan*, 26, 1, 112-113 (1993)
90. Tseng, P. C. and G. T. Rochelle, "Dissolution Rate of Calcium Sulfite Hemihydrate in Flue Gas Desulfurization Processes," *Environmental Progress*, 5, 35-40 (1986)
91. Tseng, P. C. and G. T. Rochelle, "Calcium Sulfite Hemihydrate: Crystal Growth Rate and Crystal Habit," *Environmental Progress*, 5, 5-11 (1986)
92. Kelly, B., B. Keough and A. D. Randolph, "Some Effects of Crystal Modifiers on $CaSO_3$.1/$2H_2O$ Crystal habit, Growth and Nucleation," AIChE National Meeting, Houston, March 27-31, 1983
93. Giles, E. D., I. M. Ritchie and B. A. Xu, "The Kinetics of Dissolution of Slaked Lime," *Hydrometallurgy*, 32, 119-128 (1993)
94. Perry, R. H. and D. Green, *Perry's Chemical Engineer's Handbook*, 6th Edition, McGraw-Hill, Inc. (1984)
95. Blythe, G. M. et al. "Results of EPRI's High Sulfur Test Center Spray Dryer/Pulse-Jet Fabric Pilot Test," *EPRI Proceedings: 1990 SO_2 Control Symposium*, Volume 2, 4c-47, 4c-64 (1990)

96. Fellman, R. T. and P. N. Cheremisineff, "A Survey of Lime/Limestone Scrubbing for SO_2 Removal," *Air Pollution Control and Design, Part 2*, Edited by Paul N. Cheremisineff and Richard A. Young, Maral Dekker, Inc. (1977)
97. Ranz, W. E. and W. R. Marshall, Jr., "Evaporation From Droplets," *Chem. Eng. Prog.*, 48, 3, 141-46, 173-80 (1952)
98. Bird, R. B., W. E. Stewart and E. N. Lightfoot, *Transport Phenomena*, John Wiley and Sons, NY (1960)
99. Parti, M. and B. Palancz, "Mathematical Model for Spray Drying," *Chem. Eng. Sci.*, 29, 355-362 (1974)
100. Hinds, W. C., *Aerosol Technology*, John Wiley & Sons, Inc. (1982)

CHAPTER 9

LOW TEMPERATURE DRY SCRUBBING/LEC PROCESS SUPPORT

K.W. Appell, M.J. Visneski, S. Reddy, M. Maldei, K.J. Sampson
M.E. Prudich
Department of Chemical Engineering
Ohio University
Athens, OH

Abstract

Dry scrubbing that takes place after the air preheater (<350°F) is the mode of operation that is the primary focus of this chapter. At the relatively low temperatures that occur in this region of operation, the rate of the gas-solid reaction that drives SO_2 capture in the convective (800-1200°F) and the combustion (1600-2400°F) zones is too slow to be significant. At lower temperatures, the presence of either liquid-phase or vapor-phase water is required in order to mediate SO_2 capture and to produce reasonable capture rates.

More specifically, this chapter focuses on work performed in support of the Limestone Emission Control (LEC) process. The LEC process is a unique system employing standard quarry-sized limestone to remove SO_2 from coal-fired boiler flue gases. In the LEC process, hot flue gases (<350°F) are contacted with a bed of 1/32 to 1/4 inch limestone granules covered with a thin film of water. Sulfur dioxide is absorbed from the flue gas into the water film where it subsequently reacts with dissolved limestone. A layer of reaction products, primarily calcium sulfate and calcium sulfite, forms on the surface of the limestone as the reaction proceeds. The LEC process has demonstrated the ability to remove in excess of 90% (and under some conditions in excess of 99%) of the SO_2 found in coal-combustion flue gases.

This chapter includes a description of a mechanistic model for SO_2 capture by a wetted limestone particle, process models for both fixed-bed and moving-bed LEC reactors, and support studies dealing with limestone solubilities and dissolution rates.

Literature Review

General Background

Flue gas desulfurization (FGD) processes can be categorized as either recovery or throwaway systems. Recovery systems produce a salable by-product of sulfur dioxide removal such as elemental sulfur or sulfuric acid. The by-product produced depends on the location of the plant and the marketability of the product.

Throwaway systems, as their name implies, discard the reaction product of sulfur dioxide removal. The reaction product is usually calcium based and is discarded as a slurry or solid depending on the process used. Throwaway systems can further be categorized as wet/wet, wet/dry, or dry/dry systems. The first descriptor tells whether liquid water is present (wet) or not (dry) during the removal stage. The second descriptor tells whether liquid water is present in the discarded waste product.

In wet/wet systems a sorbent is contacted with the flue gas as a slurry or solution. After reaction with sulfur dioxide the reaction product, usually calcium sulfate and/or calcium sulfite, is removed and discarded as a slurry. Figure 9.1 shows a generic wet/wet scrubbing system. Most of the throwaway systems in use today are of the wet/wet variety. The popularity of this mode of FGD operation is attributed to the fact that this technology is proven and it is currently cost effective. There are also disadvantages. Problems arise from the handling and disposing of large amounts of wet sludge and from equipment scaling, plugging, and corroding.

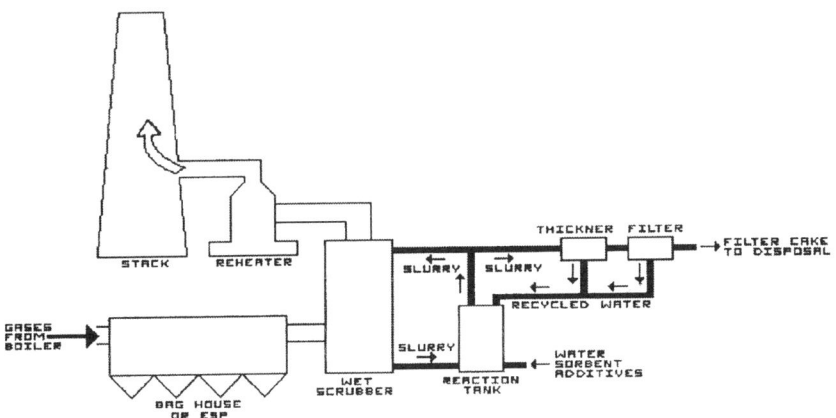

Figure 9.1 Generic wet/wet flue gas scrubbing system.

Recently, dry scrubbing methods have drawn increased attention. These methods tend to have lower capital and operating costs as compared to the wet methods. Also, since the waste products are a dry powder, their handling and disposal are more easily accomplished. Because of the low capital costs, dry scrubbing methods are favored for retrofitting older plants which still have a useful life expectancy but need to comply with stricter SO_2 emission limits.

Dry scrubbing techniques use finely divided calcium-based sorbents (either lime or limestone) and are divided into three categories depending upon the point at which the flue gas is treated. These categories are (1) in-furnace scrubbing where the flue gas temperature ranges from 1800-2400°F, (2) economizer zone scrubbing where the flue gas temperature ranges from 900-1100°F and (3) after the air heater (post-furnace) scrubbing where the temperature is typically below 350°F. Figure 9.2 shows the typical scrubbing sites. Post-furnace scrubbing is the primary concern here and no further comment will be made concerning the other two scrubbing sites.

Dry Scrubbing Technologies for Flue Gas Desulfurization

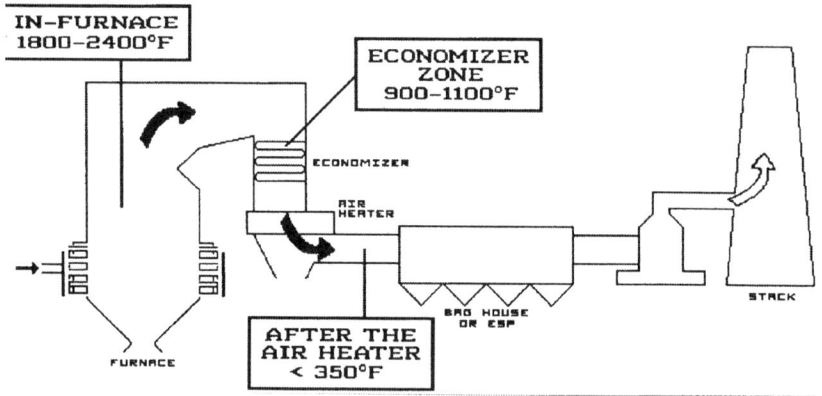

Figure 9.2 Typical injection locations for calcium-based sorbents.

Post-furnace injection involves injecting either a slurry of the sorbent and water or just the sorbent into the flue gas duct between the air heater and the particulate removal device (bag house or ESP). This process is often referred to as in-duct injection or in-duct spray drying. In order for the SO_2 capture reaction to take place, water must be present either in the vapor phase (dry/dry systems) or in the liquid phase (wet/dry systems). When the sorbent is injected as a dry solid the flue gas must be humidified or a bulk water phase must be present for the reaction to take place.

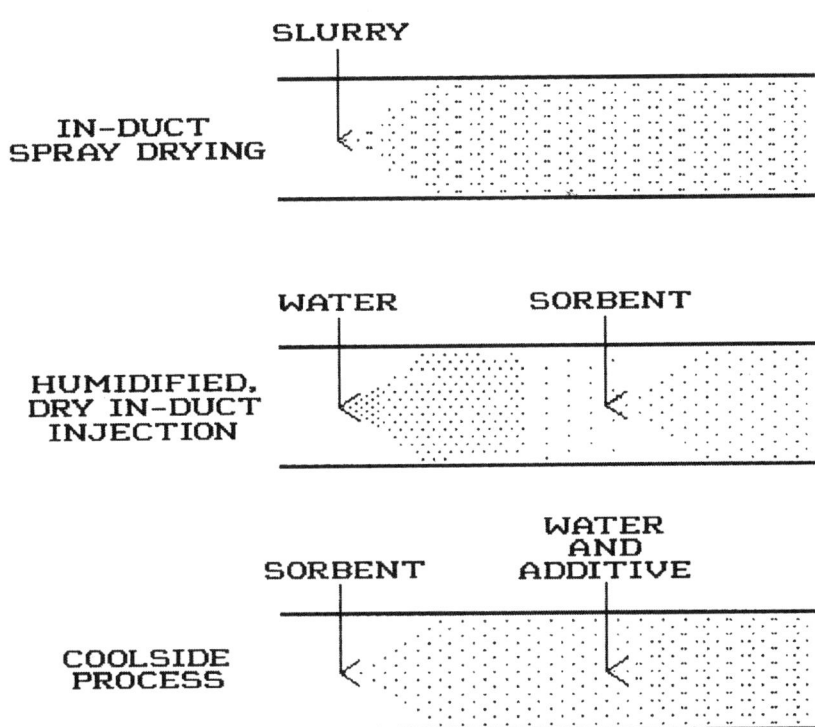

Figure 9.3 Spray nozzle arrangements for in-duct injection processes.

Figure 9.3 shows the most popular arrangements of water/sorbent/slurry injection sites. The first arrangement is in-duct spray drying in which a slurry is directly introduced into the duct. Examples of this wet/dry process are the Bechtel Confined Zone Dispersion (CZD) process and the G.E. In-Duct Scrubbing (IDS) process [1,2]. In both processes the residence time in the duct required to evaporate the water is about one to two seconds and 50% of the sulfur dioxide is removed.

The Dravo Hydrate Addition at Low Temperature (HALT) process is an example of humidified, dry in-duct injection [1,2]. In this dry/dry process, water is injected upstream of the sorbent. Almost all of the water evaporates before reaching the sorbent. SO_2 removal is typically in the 50-60% range.

The third process is the Consol Coolside process [1,2]. In this wet/dry process, the sorbent is injected upstream of the water spray. An additive, such as NaOH, is typically added to the water spray to enhance the SO_2 removal. The residence time for evaporation is one to three seconds and up to 70% of the SO_2 can be removed.

Another wet/dry post-furnace desulfurization process is ETS' Limestone Emission Control (LEC) system [3]. This process does not involve sorbent injection but relies on a fixed or moving bed of quarry-sized limestone. Experimental pilot plant results indicate that the fixed-bed LEC system is capable of removing up to 90% of the SO_2 present in coal-fired boiler flue gas. Recently, a moving-bed configuration of the LEC process has been shown to remove in excess of 90% (and in some cases as high as 99.9%) of the SO_2 entering the dry scrubber [4].

The modeling and experimental work concerning the dry scrubbing processes are the main focus here. Studies on the dry/dry techniques can be divided into low temperature studies (less than 212°F) and high temperature studies (212°F–400°F). The modeling work done for the wet/dry in-duct processes focuses on modeling the slurry droplet. Modeling work on the LEC process focuses on a single particle covered with a thin film of water. (This is a subset of the slurry droplet. The slurry droplet is a collection of particles surrounded by a film of water.) The wet/dry modeling has been primarily concerned with the gas- and liquid-phase resistances to mass transfer. Little work has been done concerning dissolution of the sorbent in wet/dry processes. Because of the chemical similarity between the wet/wet and the wet/dry processes, dissolution studies done for wet/wet scrubbing may be applicable to the wet/dry processes.

An in-depth study of the kinetics of the low-temperature reaction of SO_2 with limestone has been performed by Klingspor et al. [5]. In that study the effect of temperature, SO_2 concentration, oxygen concentration, CO_2 concentration, relative humidity, and particle diameter on the reaction rate was investigated. The kinetic data were determined using a sand bed reactor. In addition to kinetic data, water-limestone sorption curves were obtained and the surface of the solids was characterized by BET surface areas.

Using specially designed sorption equipment, Klingspor et al. [5] determined the adsorption isotherm of water on limestone to be type II of the BET classification. From this information it was further determined that the amount of water adsorbed is directly proportional to the BET surface area and independent of the particle radius.

Klingspor et al. [5] give the reaction scheme for SO_2 and limestone as

$$SO_2(g) + CaCO_3(s) \xrightarrow{O_2, H_2O} CO_2(g) + \begin{cases} CaSO_3(s) \\ CaSO_3 \bullet 2H_2O(s) \\ CaSO_4(s) \\ CaSO_4 \bullet o/ooH_2O(s) \\ CaSO_4 \bullet 2H_2O(s) \end{cases}$$

The amount of oxygen present during the reaction will determine the relative amounts of sulfite and sulfate in the product. From the reaction scheme it would be suspected that O_2 would have an effect on the overall rate of reaction. The results of Klingspor's study showed that oxygen has a slight suppressing effect on the initial reaction rate. As the oxygen content is increased, the reaction rate decreases until about a 2% O_2 content when the rate levels out.

Again, upon examination of the reaction scheme, it would appear that the presence of CO_2 in the flue gas would affect the reaction rate. Klingspor et al. [5] determined that carbon dioxide in their simulated flue gas had no effect on the SO_2 removal rate.

Relative humidity was found to have the most dramatic effect on the SO_2 removal rate. At a relative humidity of less than 24% (which by Klingspor's calculations is below one monolayer of water adsorbed on the surface of the limestone) the reaction rate is virtually zero. Above this point the rate increases exponentially.

Klingspor et al. [5] also found a weak temperature dependence in the temperature range of 104°F to 176°F. An increase in temperature in this range resulted in a slight increase in the SO_2 removal rate.

Results from varying the particle diameter indicated the reaction rate to be solely a function of the external BET surface area. This result suggests that the reaction occurs on the external surface of the particle and no SO_2 penetrates into the bulk of the unreacted limestone.

Upon analysis of the kinetic data, Klingspor et al. [5] determined the reaction order with respect to the SO_2 to change from zero to one as the relative humidity was increased. At relative humidities less than 24% (i.e., one monolayer of water adsorbed) the rate was difficult to determine. From 24–69% relative humidity (1 to 2.3 monolayers of water adsorbed) the order with respect to SO_2 was zero. Increasing the relative humidity above 69% increased the order to 0.8 at a relative humidity of 92% (4.6 monolayers of water adsorbed). Relying on this kinetic analysis a reaction mechanism was devised to explain the increasing order of reaction with increasing relative humidity.

$$H_2O(g) \Leftrightarrow H_2O(ad)$$

and

$$SO_2(g) \Leftrightarrow SO_2(ad)$$

The first step in the mechanism is adsorption of the gaseous species. Next the adsorbed H_2O and SO_2 form a complex.

$$SO_2(ad) + H_2O(ad) \Leftrightarrow Z*$$

The SO_2/H_2O complex ($Z*$) reacts with the solid to form another complex.

$$Z* + CaCO_3(s) \Leftrightarrow Y*$$

Finally the SO_2/carbonate complex ($Y*$) decomposes to form sulfite or combines with oxygen to form sulfate.

$$Y* \rightarrow CaCO_3(s) + CO_s(g) + H_2O(ad)$$

and

$$Y* + (o/ooO_2 \rightarrow CaSO_4(s)) + CO_2(g) + H_2O(ad)$$

The order of the reaction with respect to SO_2 is determined in the adsorption and complexing-with-water steps. As the relative humidity increases the number of water layers adsorbed on the surface increases and the SO_2 must penetrate into the adsorbed water layers. This process then becomes similar to SO_2 dissolution into water which is a first-order reaction with respect to the gas-phase water concentration. Hence the order of reaction gradually changes from zero to one as the relative humidity increases.

Using a sand-bed reactor Jorgensen et al. [6] evaluated several sorbents and additives for dry/dry SO_2 removal. The sorbents were hydrated lime, sodium bicarbonate and limestone and the additives were various inorganic salts.

The reactivity of hydrated lime ($Ca(OH)_2$) was investigated over a temperature range of 130–166°F and a relative humidity range of 15–85%. The reactivity was found to be a strong function of the relative humidity. As the humidity was increased the reaction rate increased. An increase in temperature gave a slight increase in the reaction rate.

Inorganic salts were then mixed with the $Ca(OH)_2$. The salts used were $CaCl_2$, NaCl, NaOH and $NaNO_3$. These additives increased the amount of SO_2 removed with $CaCl_2$ giving the most noticeable increase. The $CaCl_2$, having a high affinity for water, is believed to increase the amount of water adsorbed by the particle, thus increasing the reaction rate.

The outlet concentration of SO_2 from the sand-bed reactor was the same as the inlet concentration after approximately one hour of operation. Upon analysis, Jorgensen et al. [6] determined that 60–90% of the $Ca(OH)_2$ was unreacted. This was attributed to the blinding of the sorbent surface by the reaction product.

Sodium bicarbonate, $NaHCO_3$, was examined at 148°F and at 15–75% relative humidity. An increase in the relative humidity resulted in only a slight increase in the SO_2 reaction rate, although the rates were higher than those observed for $Ca(OH)_2$. A very noticeable difference between $NaHCO_3$ and $Ca(OH)_2$ was the lack of blinding of the surface. After one hour of run time using $NaHCO_3$ the rate of reaction was constant and no sign of blinding was apparent. The authors suggested a mechanism involving the release of CO_2 during the reaction to explain the high reactivity without blinding.

Finally, Jorgensen et al. [6] tried $CaCO_3$ in their sand-bed reactor. They found no reactivity of $CaCO_3$ with SO_2 in a relative humidity range of 40–70%. Mixing the $CaCO_3$ with $CaCl_2$ in a ratio of 1 gram additive to 10 grams of sorbent resulted in an SO_2 removal efficiency similar to that of pure $Ca(OH)_2$. Although the removal efficiencies were similar, the blinding effects were not. It was found that the $CaCO_3$ with the additive showed no signs of blinding after one hour of run time. The shape of the removal curve was similar to that of $NaHCO_3$ where the reaction is constantly proceeding with the rate decreasing only with the depletion of the sorbent.

On the other end of the dry/dry temperature scale, Garding and Svedberg [7] did some modeling work on the reaction of sodium bicarbonate with sulfur dioxide at 392°F. The first step in this reaction is the thermal decomposition of sodium bicarbonate to sodium carbonate.

$$2NAHCO_3(s) \rightarrow Na_2CO_3(s) + H_2O(g)$$

The sulfation reaction occurs next.

$$NA_2CO_3(s) + SO_2(g) \rightarrow NaSO_3(s) + CO_2(g)$$

And finally, in the presence of oxygen, the sulfite is oxidized to sulfate.

$$Na_2SO_3(s) + o/ooO_2(g) \rightarrow Na_2SO_4(s)$$

In formulating their model Garding and Svedberg [7] assumed the thermal decomposition reaction to be fast when compared to the sulfation reaction. It was further assumed that the sulfation reaction is irreversible and first order with respect to the partial pressure of SO_2. The reaction process was then represented by an unreacted core model.

According to this model, the reaction first takes place on the outer surface of the particle. Then, as the reaction proceeds, the reaction zone moves toward the center of the solid particle leaving behind a completely converted material. Therefore, as the reaction proceeds the gaseous reactant must diffuse through the converted material covering the unreacted core. This gas-solid reaction model incorporates three resistances which are (1) external film diffusion, (2) diffusion through the reacted material, and (3) chemical reaction.

The kinetic model was incorporated into a system flow model. The system being investigated was the cocurrent flow of solid and gas phases (that is, injection of the dry sorbent into a gas stream flowing through a duct). This flow model is mathematically similar to the solids being suspended in a finite bath of gas. An analytical solution to the problem was found and the results were compared to experimental findings.

Because of fluctuating operating conditions in the experimental apparatus, Garding and Svedberg [7] were not able to determine any of the kinetic parameters. Therefore they had to rely on previously published data. The mathematical model predictions and the experimental results were in fair agreement.

Rice and Bond [8] constructed a pilot-scale in-duct FGD system to investigate the effect of humidity, temperature and residence time on the SO_2 capture efficiency of $Ca(OH)_2$. When varying the absolute humidity from 0–0.18 lb H_2O/ lb of dry gas the SO_2 removal efficiency increased from 0–40%. The temperature in these runs was held constant so the relative humidity also increased. Since the adsorption of water onto the surface of the solid is dependent on the relative humidity, more water was adsorbed on the surface of the particle at the higher gas-phase water concentrations.

Low Temperature Dry Scrubbing/LEC Process Support

As the temperature was increased from 257 to 464°F the SO_2 removal efficiency decreased from 30 to 5%. In these runs the absolute humidity was held constant; therefore, the relative humidity decreased. As the relative humidity decreased, the water adsorbed of the surface decreased and the SO_2 removal rate decreased because of the strong dependency of adsorbed water to the rate of SO_2 removal. From these results the effect of temperature on the reaction cannot be clearly determined.

Rice and Bond [8] found no affect of residence time on SO_2 removal. The base residence time for the experiments was 0.8 seconds. The results from decreasing the residence time to 0.5 seconds and from increasing the residence time to 1.6 seconds were not statistically different from the base case. These results lead the researchers to conclude that the conversion of SO_2 is not very sensitive to moderate changes in residence time.

As is the usual case for in-duct injection, the solids and gases flow cocurrently so gas-film mass transfer plays an important part in the overall rate of reaction. In an attempt to increase the gas-film mass transfer coefficient Rice and Bond [8] tried injecting the sorbent countercurrently. In this arrangement the solids would be flowing countercurrently to the gas for a short period of time until they would be slowed by the oncoming gas, and then eventually would flow cocurrently. Since the reaction occurs in a very short time, and the solids would be flowing countercurrently for a short period of time an increase in the removal might be achieved. The increase in the rate would result due to the decrease in the gas-film mass-transfer resistance since the relative velocities of the solids and gas would be greater than in cocurrent flow. This increased effect was not observed. The results of countercurrent flow were the same as those for cocurrent flow.

Two other sorbents were tried by Rice and Bond [8] in their experimental apparatus. These were sodium sesquicarbonate and calcium carbonate. At the base operating conditions sodium sesquicarbonate was found to have a higher reactivity than $Ca(OH)_2$ despite the fact that it has a smaller specific surface area than $Ca(OH)_2$. Negligible SO_2 removal was obtained with $CaCO_3$ at the same conditions.

Unlike the dry/dry processes, the wet/dry processes contain liquid-phase water. In the case of in-duct spray drying the water appears in the slurry droplets that dry out as they pass through the duct. The dry particles are then removed in a bag house or electrostatic precipitator (ESP). While residing in the particulate collection device the particles will, to some extent, still remove SO_2 from the gas stream. Droplet evaporation and SO_2 absorption occur simultaneously in the process. Although the drying process is independent of the SO_2 absorption process, the SO_2 absorption process is not independent of the drying process because either a bulk water phase or an adsorbed water phase must be present in order for the SO_2 capture to occur. The closer the gas is to saturation the more slowly the droplet dries and the greater the removal efficiency.

Damle [9] modeled both the SO_2 removal of the slurry droplet and the SO_2 removal of the solids residing in the particulate collection device for a spray dryer using $Ca(OH)_2$ as the sorbent. The first task was to model the drying of the slurry droplet. The drying of the slurry droplet occurs in two stages, a constant-rate drying period and a falling-rate drying period. Both of these processes were modeled. In the constant-rate drying stage the concentration of the solids does not affect the evaporation rate. This stage continues until the moisture level falls below a critical moisture content. During the constant-rate drying period the evaporation rate is solely determined by the resistance of the gas film surrounding the droplet. In the falling-rate drying stage the concentration of the solids reduces the drying rate because the water has to diffuse through the solid matrix. During this stage the droplet changes from water being the continuous phase to the solid constituting the continuous phase. The falling-rate drying stage ends when the moisture level of the solid agglomerate reaches its equilibrium moisture content. During the constant rate drying period the solid particles are free to move within the droplet, but during the falling-rate drying stage the solids begin to touch each other and become immobile.

The rate of drying in the constant-rate period is determined by the simultaneous heat transfer from the gas phase to the droplet and mass transfer of water from the droplet to the gas phase. The transfer coefficients are determined from previously published correlations. One important parameter in determining these transfer coefficients is the Reynolds number which includes the velocity of the droplet. The velocity of the droplet changes from entry to exit. Initially the droplets have a very high velocity when leaving the atomizer. Due to the resistance of the surrounding gas the particles are decelerated to a steady-state free-fall velocity. The deceleration time depends on the diameter of the droplet and is usually very short. Since the deceleration time is short, the free-fall velocity is used to simplify the calculations.

Because no simple correlations are available describing the falling-rate drying stage, a simplified form was proposed as

$$\text{Drying Rate} = (\text{Drying Rate based on Gas-Phase Resitance}) \left(\frac{X - X_e}{X_c - X_c} \right)^n$$

where X is the moisture content at any time, X_e is the equilibrium moisture content, X_c is the critical moisture content and n is an adjustable parameter. For simplicity this rate was assumed to be linear (i.e., n=1).

Like the drying of the droplet, the kinetics of the reaction must be divided into two stages, the wet-particle stage (constant-rate drying period) and the dry-particle stage (falling-rate drying period). During the wet-particle stage the lime particle is surrounded by a thin film of water. A resistance-in-series model is used to describe the mass transfer of SO_2 across the gas film, the diffusion of the dissolved sulfur and calcium species to the reaction sight, and the dissolution of the lime. The model was simplified by assuming the reaction to be instantaneous and by using film theory to describe the concentration profiles in the water phase. A simple film model was also used to describe the dissolution of the lime particle. This model was

$$R_p = \frac{D_{lime}}{\delta}(C^*_1 - C_1)$$

where R_p is the rate of dissolution, D_{lime} is the diffusivity of lime in water, δ is the thickness of the liquid film, C_1^* is the equilibrium solubility concentration of lime in water, and C_1 is the bulk concentration of lime in the water film. Assuming the reaction to be instantaneous forces the concentration of the reactants to be zero at the reaction site. Upon analysis of the three resistances, Damle [9] determined the reaction to be either gas-phase controlled or lime-dissolution controlled. During the dry-particle stage the diffusion of SO_2 into the solid matrix becomes the controlling mechanism. For modeling this stage an unreacted core model was used.

Damle [9] also compared the removal efficiencies of the dry "droplets" while residing in the particulate collection device. The major factor affecting the reactivity of the solids is the equilibrium moisture content. The higher the relative humidity the higher the equilibrium moisture content, and thus the reactivity. The major difference between a baghouse and an ESP is that a baghouse offers no resistance to gas-phase mass transfer of the SO_2 to the surface of the lime particle whereas this is a significant factor in the ESP. In a baghouse the flue gas passes through the layer of solids at a relatively high velocity. In an ESP the flue gas passes over the layer of solids at a relatively low velocity. This intimate contact between the gas and solids at a high velocity in a baghouse leads to the absence of a significant mass-transfer resistance. Thus for similar operating conditions (i.e., moisture content of solids, dust layer thickness, total surface area and concentration of unreacted lime) a baghouse is more effective than an ESP in removing SO_2.

Harriott and Kinzey [10] modeled the gas- and liquid-phase resistances in the spray drying process using the assumption of an infinite rate of $Ca(OH)_2$ dissolution. Both the constant-rate drying period and the falling-rate drying period were modeled. The effect of adding an inorganic salt, $CaCl_2$, was considered in the analysis of the falling-rate drying stage. The addition of the salt decreases the rate of evaporation and therefore increases the overall SO_2 removal efficiency.

Maibodi et al. [11], in modeling the slurry droplet, considered the dissolution of the lime particle. In addition to the dissolution resistance, the gas- and liquid-phase diffusional resistances were also considered. The value of the dissolution rate constant was determined from the experimental data. The value chosen was that which forced the model to fit the data. For the conditions used in their study the predominant resistance was found to be the dissolution of the lime.

Newton et al. [12], in much the same way as Damle [9] and Harriott and Kinzey [10], modeled the drying and SO_2 absorption efficiency of a $Ca(OH)_2$ slurry droplet. The major difference in their work was a more extensive description of the dissolution process. Although the dissolution rate was considered in the model formulation, it was not considered in the simulations. The lime dissolution rate was set at an artificially high value (i.e., infinite dissolution rate) so as not to have any effect on the model predictions. The authors attribute this to a lack of fundamental data on dissolution occurring under typical in-duct conditions.

In all of the wet/dry process models described above the dissolution of the solid phase is either ignored or given a cursory treatment. And as Newton et al. [12] state,

> "Although the model includes a term for the dissolution rate at the sorbent surface, no fundamental data taken under typical duct conditions are available that could be used to set their values. The model predictions shown here were made by setting its value so high that it had no influence on model calculations."

Since the wet/dry and the wet/wet systems are chemically similar and many studies have been done on dissolution in the wet/wet process, these studies can provide a starting point for modeling the dissolution process and determining preliminary performance values for the wet/dry processes.

A sampling of the modeling work done for the dissolution of limestone include papers by Chang et al. [15], Chan and Rochelle [16], Gage and Rochelle [17] and Jarvis et al. [18]. Typically the dissolution is modeled using traditional mass-transfer models. Factors that affect limestone dissolution are limestone type and grind, temperature, pH, partial pressure of CO_2, buffer additives, and impurities.

Fixed-Bed Limestone Emission Control

In the LEC process hot flue gases are contacted with a bed of 1/32 to 1/4 inch limestone granules covered with a thin film of water. Sulfur dioxide is absorbed from the flue gas into the water film and subsequently reacts with the limestone. A layer of reaction products, primarily calcium sulfate and calcium sulfite, precipitates on the surface of the limestone granules throughout the course of the reaction. Particulate matter such as fly ash may also be deposited in the water film.

A small pilot unit was constructed in order to demonstrate the technical feasibility of this fixed-bed limestone flue gas desulfurization process. The pilot unit was designed to handle 400 acfm of flue gas and was installed on a slipstream taken from a 70,000 lb/hr stoker boiler burning a 2.5% sulfur coal and providing steam to Ohio University's Athens campus. The slip stream was taken from the duct leading from the ESP to the stack. The entire LEC small pilot unit project has been described by Prudich et al. [3]. The LEC project conducted over 100 experimental runs on flue gases containing 500 to 3500 ppm SO_2. The program established that the LEC is capable of up to 99% SO_2 removal. An average of greater than 90% removal was realized over extended operational periods.

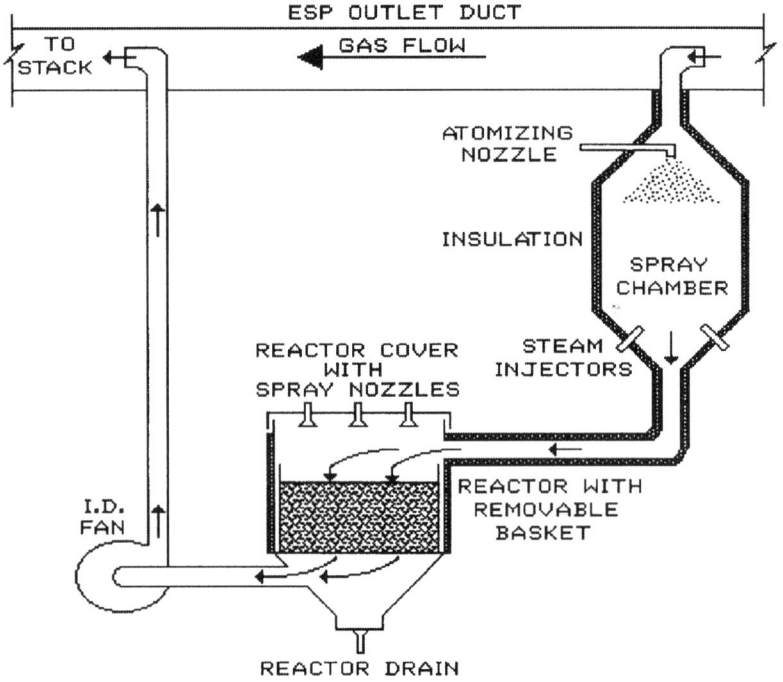

Figure 9.4 Simplified flow schematic of the LEC pilot unit.

Figure 9.4 shows a simplified flow schematic of the LEC pilot unit. The flue gas first entered a spray chamber where the flue gas was conditioned to the desired temperature and humidity. The conditioning was achieved by adding water via an atomizing nozzle and/or by injecting live steam into the chamber. The conditioned gas then entered the reactor, passing downward through the fixed bed of limestone. The reactor contained a removable basket filled with limestone. The basket was 22 inches square and the depth of the limestone was adjustable up to 18 inches. The cleaned gases were then drawn into the induced draft fan and returned to the duct leading to the stack.

Two factors were identified which limit the LEC operation. These were (1) drying of the limestone surface and (2) build up of the reaction product on the reactive limestone surface. A dry limestone surface was effectively nonreactive for LEC process conditions. The build up of reaction product on the surface of the limestone decreased the reaction rate and eventually rendered the limestone inactive.

Low Temperature Dry Scrubbing/LEC Process Support

Three methods were used to supply water to the LEC process. One method was to prewet the bed of limestone before starting the process. This was accomplished by pouring water onto the bed before sealing the reactor. The life of the bed was determined by the amount of conditioning of the incoming flue gas. The higher the relative humidity of the gas the longer the time before the limestone bed became dry and therefore unreactive. The second method was by condensation. Normally the limestone bed was cooler than the incoming flue gas. As the gas contacted the cooler bed, water would condense from the gas phase onto the limestone. The amount of water deposited by condensation would depend upon the conditioning of the flue gas. As the relative humidity increased the amount of water deposited onto the limestone would increase. Condensation would slow and eventually cease as the bed was heated by the incoming flue gas. After condensation stopped, the life of the bed was once again determined by the condition of the incoming flue gas. The final method of water introduction was by spraying water directly onto the limestone bed. As shown in Figure 9.4, spray nozzles were positioned directly over the bed. The life of the bed was determined by the rate of the water spray as well as the condition of the flue gas.

One of two operational modes was used to overcome deactivation by drying. Operating the process with a saturated flue gas would halt evaporation and therefore the bed would not dry out. The other mode was the use of an over-bed spray. In order to prevent the bed from drying, the spray rate would have to be equal to or greater than the rate of evaporation.

Deactivation due to surface blinding could not be prevented but it could be overcome. After the limestone became nonreactive due to the blinding of the surface the solids still retained a core of unreacted limestone. The layer of reaction products was removed by mild attrition. After attrition the stone regained its reactivity and was returned to service. When the surface of the stone was again blinded this process could be repeated until all of the limestone had been utilized.

Figure 9.5 shows experimental data taken from the LEC pilot unit that illustrates both deactivation due to drying and surface blinding. Runs 870714A', 870716A, 870716B, 870717A, 870720A and 870720B represent consecutive runs using the same 12 inch deep bed of limestone. The superficial gas velocity remained constant from run to run at 1.5 ft/s but the SO_2 concentrations ranged from 620 to 870 ppm. The inlet flue gas relative humidity was approximately 50% but varied from run to run. The variation of SO_2 concentrations and gas humidities was due to the fluctuating conditions of the incoming flue gas. Run 870714A' started with a prewet bed of limestone. After a little over 1.5 hours the bed became unreactive. The same bed of limestone was then rewet before starting run 870716A. The bed regained its initial SO_2 removal efficiency before once again losing its reactivity due to drying. The process of rewetting the bed before starting the next run continued for the remaining four runs. Starting with run 870717A, rewetting the bed failed to return its activity to its original level. This decrease in activity was attributed to the blinding of the surface due to the buildup of the reaction products on the reactive surface. Deactivation due solely to surface blinding is illustrated in Figure 9.6. These data show the performance of three different 6-inch-deep beds of limestone. The inlet flue gas was saturated and therefore drying of the bed was not a factor in the decline of the SO_2 removal efficiency of the limestone. Run 871217C is a continuation of 871216B and uses the same bed of limestone. Run 871105A continued using the same bed of limestone started by run 871103A. As can be seen from Figure 9.6 each bed lost activity at different times. The life of the bed was attributed to the SO_2 concentration of the incoming flue gas. Run 871103A/ 871105A with an inlet SO_2 concentration of 344 ppm lasted the longest. Next was run 871214B with an SO_2 concentration of 702 ppm. Run 871216B/871217C with an inlet SO_2 concentration of 736 ppm had the shortest life.

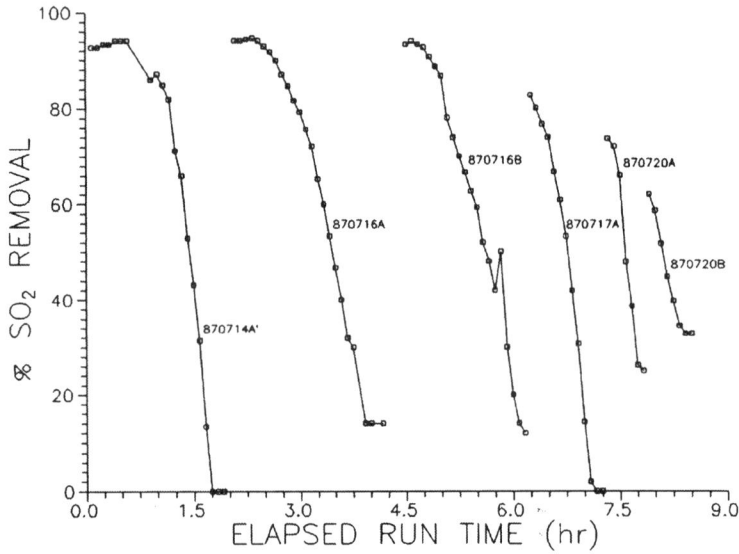

Figure 9.5 Deactivation of an LEC bed due to surface drying and surface blinding (Runs are from Prudich et al. [3] with a limestone bed depth of 12 inches and a superficial gas velocity of 1.5 ft/s.)

Low Temperature Dry Scrubbing/LEC Process Support

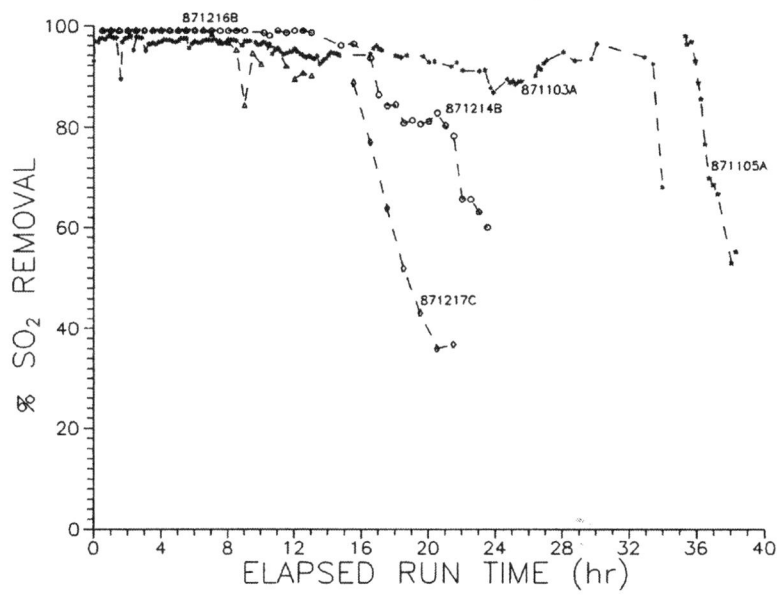

Figure 9.6 Deactivation of an LEC bed due to surface blinding. (Runs from Prudich et al. [3] with a limestone bed depth of 6 inches and a superficial gas velocity of 1.0 ft/s.)

In addition to the difference in operating conditions, the limestone used for the runs shown in Figure 9.5 differs from the limestone used for the runs shown in Figure 9.6.

Moving-Bed Limestone Emission Control

A continuous moving-bed LEC pilot plant sized to handle 5000 acfm was installed on the slip stream of the 70,000 lb/hr stoker boiler providing steam to the Ohio University campus. A simplified flow scheme of the moving-bed LEC process is shown in Figure 9.7. A detailed schematic of the pilot plant is given in Figure 9.8.

Limestone is added to the bed via a 32-foot bucket elevator which is supplied with limestone from the feed hopper, the make-up hopper, the vibrating screen, or the recycle chute by means of a feed screw. The make-up hopper is used to compensate for limestone lost due to reaction with SO_2 and its subsequent removal as reaction product. The limestone bed itself has dimensions of 14 x 36 x 128 inches (Figure 9.9). The depth of the bed in the direction of the flue gas flow is 14 inches. The LEC reactor internals consist of inlet and outlet louvers, overhead bed sprays, and outlet screening. The inlet and outlet louvers (as originally designed) are four inches long and are inclined at a 75° angle with a 3/8 inch overlap between louvers. The flue gas inlet and outlet plenums on the sides of the reactor are sized such that there is uniform distribution of the gas across the whole bed.

The flue gas is drawn from the duct leading to the stack from the ESP. The superficial velocity of the flue gas through the LEC reactor can be varied from 0.3 to 1.6 feet per second. The live hopper at the top of the bed is provided so as to prevent the leakage of ambient air into the outlet gas plenum. An induced draft fan is used to draw the processed flue gas from the outlet plenum. The processed flue gas is directed to the stack. Limestone moves vertically downward through the bed with a velocity sufficient to prevent bridging of the limestone, which may occur when the flue gas dries up the bed. The limestone removed from the bottom of the bed is sent to the discharge screw via a 19 foot bucket elevator. The discharge screw has three outlets, any one of which may be opened. The first outlet directs the limestone to the recycle chute, which leads directly to the feed screw and back to the reactor. The second outlet leads to the reactivation system. The buildup of reaction products on the limestone surface tends to retard the rate of reaction per unit granule surface area. To prevent this surface blinding from rendering the limestone inactive, the limestone granules are passed through a reactivation device that removes the reaction product layer by either dry or wet abrasion. This abrasion exposes a fresh layer of reactive limestone. This mixture of reactivated limestone and abraded material is then passed through a vibrating screen where the abraded material is removed as a waste. The reactivated limestone is then directed to the feed screw and returned to the top of the LEC bed. The third discharge screw outlet is directed to the waste hopper. Normal LEC operation directs the limestone through the reactivation system.

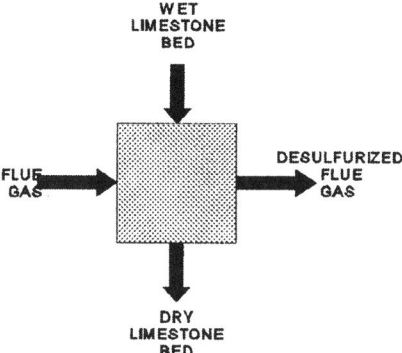

Figure 9.7 Simplified schematic of the LEC process.

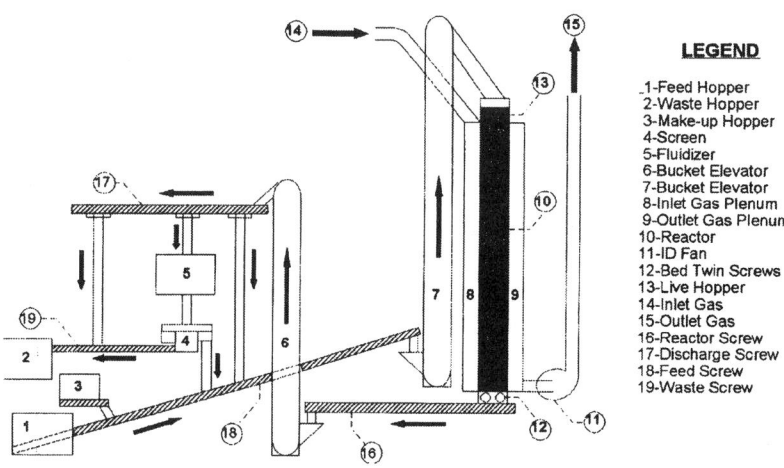

Figure 9.8 Schematic of the moving-bed LEC pilot plant.

Dry Scrubbing Technologies for Flue Gas Desulfurization   **711**

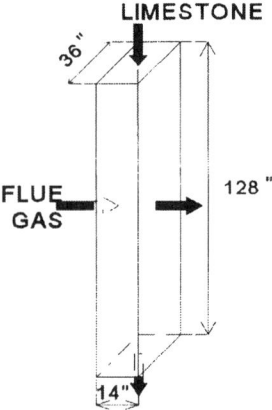

Figure 9.9 Limestone bed geometry for moving-bed LEC pilot plant.

More than 90 experimental trials were performed using the pilot-scale moving-bed LEC dry scrubber with run times ranging up to 125 hours (the results of a typical trial are shown in Figure 9.10). The primary process variables studied were superficial flue gas velocity (0.5 to 1.6 ft/s), limestone bed velocity (4.4 to 21.6 ft/hr), inlet SO_2 concentration (400 to >3000 ppmdv), and water addition rate (0 to 2 gallons/min including both over-bed sprays and humidification water added to the flue gas). SO_2 removal efficiencies as high as 99.9% were achievable for all experimental conditions studied during which sufficient humidification was added to the LEC bed. The periods during which the SO_2 removal efficiency dipped below 90% in Figure 9.10 were caused by cessation of flue gas and limestone bed humidification. These drying times were a result of the research nature of the program and would not be part of a commercial operation strategy.

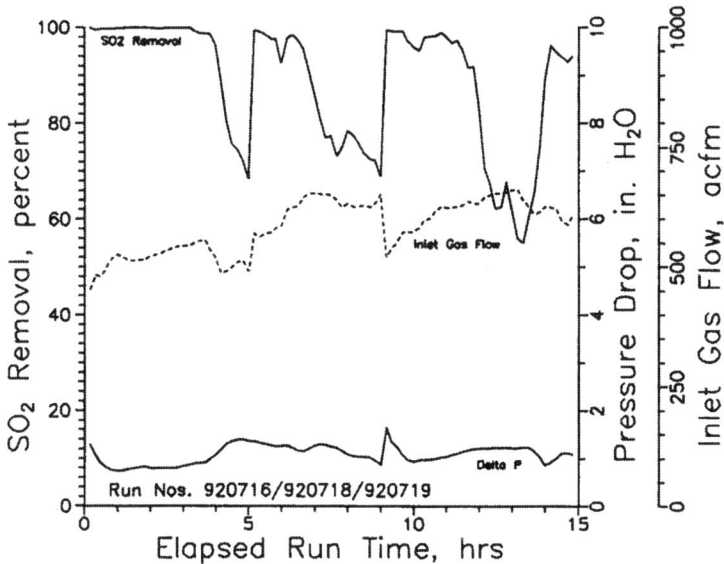

Figure 9.10 LEC SO_2 removal efficiency at high inlet SO_2 concentrations. AASHTO No. 9 limestone, Inlet SO_2: 3500-3900 ppmdv[4].

Limestone Solubilities and Dissolution Rates

Early research in the determination of calcite and aragonite solubilities and dissolution rates has primarily been performed by geochemists and oceanographic researchers. These investigators determined calcite and aragonite equilibria in open seawater as well as under laboratory conditions. In seawater, the first direct studies of calcium carbonate dissolution were conducted by Peterson [19] and Berger [20]. Peterson suspended spheres of Iceland Spar calcite at various depths in the central Pacific Ocean, while Berger studied the dissolution rate of biogenic calcite and aragonite using the same method. Under laboratory conditions, Weyl [21] examined the calcite dissolution rate by directing a jet of water at a crystal of Iceland Spar and found the rate of dissolution to be transport controlled. This phenomenon has been confirmed by other researchers investigating dissolution rates [22].

Dissolution rates have been commonly measured by two different methods, the 'pH-stat' and the 'free drift' method. The pH-stat technique is based on the maintenance of a constant degree of disequilibrium in the carbonate/carbonic acid system. This technique has been applied by Morse and Berner [23], Berner and

Morse [24], Morse [25], Plummer et al. [26], Chan and Rochelle [16], Toprac and Rochelle [27], and Gage [28]. The alternative method, the free drift technique, is an approach to establish equilibrium between the carbonate species from an initial disequilibrium and an under- or super-saturation of the solute. The free drift method has been used by Erga and Terjesen [29], Sjoeberg [30], Plummer and Wigley [22], and Plummer et al. [26] to determine dissolution rates of calcite and aragonite, which are polymorphic forms of calcium carbonate.

Dissolution rate and solubility measurements have been conducted for a variety of biogenic, synthetic, and geologic materials. These materials were either calcitic, aragonitic, or contained other proportions of calcium and magnesium in a single phase. The kinetic behavior of multiple phases which are often present in biogenic materials has also been studied. Biogenic samples were used by Berger [20], Morse and Berner [23], and Walter and Hanor [31]. Synthetic calcite found application in the experiments of Ingle et al. [32] and Sjoeberg [30]. Both synthetic calcite and geologic materials have been examined by Toprac and Rochelle [27]. They collected limestones like Ash Grove, Brassfield, Fredonia, Georgia Marble, Longview, Maysville, Pfizer, and Stoneman to study their respective dissolution rates in a prepared solution. These dissolution rates were reported to be independent of limestone type or source.

Different solvents have been used by researchers to determine dissolution rates and solubilities. In seawater, the solubility of aragonite was measured by Berner [33]. Since the solubility determination of calcite in seawater is very complex, Berner [33] performed additional experiments dissolving calcite and aragonite in distilled water and subsequently related their difference in solubility to the natural solvent. Pseudo-seawater, a solution of $NaCl/CaCl_2$, was selected by Berner and Morse [24]. This solution contained the same amount of calcium ions as present in seawater and was prepared to have an identical ionic strength. Since K^+ and Cl^- ions were not observed to form significant complexes with Ca^{2+}, Sjoeberg [30] applied KCl in distilled water as a solvent.

Solvents containing an initial amount of calcium ions were believed to have an effect on the dissolution rate [30]. Sjoeberg quantified this effect by adding known amounts of analytical grade $CaCl_2$ to the reactant solution. The reaction rate was found to decrease in proportion to the square root of increasing calcium ion concentration. This discovery led to the identification of a disadvantage of the free drift method, the continuous change in the state of solvent saturation.

A factor which has an influence on the dissolution rate of calcite was found to be the surface area [30]. In experiments using pure analytical $CaCO_3$ of different particle sizes, the dissolution rate per unit mass increased proportionally to the increase in calcite surface area. Similar results were achieved by Rochelle and Toprac [27] in 1982. They determined the particle-size distribution of naturally occurring limestones to be an influential parameter on their respective dissolution rates.

Other parameters that are significant in controlling the chemistry of limestone dissolution are the ambient temperature, the solution pH, and the atmospheric partial pressure of carbon dioxide. The ambient temperature determines the equilibrium constants between the chemical species in solution while the solution pH relates to the amount of hydronium ions which can react in the carbonate/carbonic acid system. The partial pressure of CO_2 in the atmosphere indirectly influences the solution pH and defines the equilibrium pH under free drift conditions. Experiments leading to the quantification of these effects were performed by Plummer et al. [26] in 1978.

Limestone dissolution rates and solubilities are related to the solution pH and to the equilibria of the carbonate species. In those equilibria, the ions involved are specified in terms of their activities rather than their concentrations. The activities of electrolytes are dependent on the ionic strength of the solution and can be calculated from the Debye-Hueckel formula. The application of the Debye-Hueckel theory of activity coefficients was reviewed by Garrels and Christ [34]. They stressed the combined effects of the activities of several electrolytes in solution of a given electrolyte in regard to mineral equilibria.

Limestone dissolution in aqueous systems can be manipulated by introducing a small amount of certain chemicals into the system. Those additives either enhance or inhibit limestone dissolution and have attracted the interest of several investigators. Chan and Rochelle [16] studied the influence of the additives adipic-, acetic-, acrylic- and sulfosuccinic acid on limestone dissolution rates. Their experimental results indicated an enhanced mass transfer of calcite due to a promoted acidity transport to the limestone surface. Other carboxylic acids like glutaric, maleic or formic acid were tested to compare their effectiveness in regard to SO_2 removal [35]. The effects of the ions $Na+$, SO_3^{2-}, SO_4^{2-}, CO_3^{2-}, Fe^{3+}, Mn^{2+}, thiosulfate and adipic acid on limestone dissolution were investigated by Jarvis et al. [18]. An extension of this work includes the effect of sulfite ions in solution [28].

Dissolution rates are also influenced by impurities in the limestones and by the magnesium in solid solution with calcium carbonate. Research focused on these factors for the selection of limestone reagents in wet FGD systems was performed by Jarvis et al. [36]. They concluded that limestones with lower concentrations of carbonate may be less desirable for reasons of transportation and waste disposal cost. Low-quality limestones also have higher concentrations of potential inhibitors like iron, silicon, magnesium, dissolved sulfite and aluminum complexes which could adversely affect scrubber operation. Magnesium availability as a solid solution with calcium carbonate and dolomitic content of limestones was also analyzed. The dissolution rates of solid solution magnesium and the magnesium in dolomite were determined. Dolomite in limestones was determined to be inert under normal FGD system operating conditions whereas magnesium, when present as a solid solution, should improve FGD system performance.

However, since limestone properties like solubility or dissolution rate vary among limestones, dependent on their source, there is a need for further research on the kinetics of dissolution of local Ohio limestones, which is the subject of the present research. The data generated on physical and chemical properties of local Ohio limestones will be used in an existing computer program which models a granular limestone dry scrubbing reactor.

Process Theory and Model Development

In the LEC process a thin film of water on the limestone particles is required for SO_2 removal. A simplified representation of the sorbent particle is shown in Figure 9.11. In the SO_2 removal process, the SO_2 is absorbed by the thin film of water on the limestone particle and is then transported in the form of ions to the reaction front. The limestone dissolves in the water layer and is transported to the reaction front where it reacts with the absorbed SO_2. This ionic reaction produces calcium sulfite/sulfate which precipitates onto the limestone particle and then inhibits limestone dissolution.

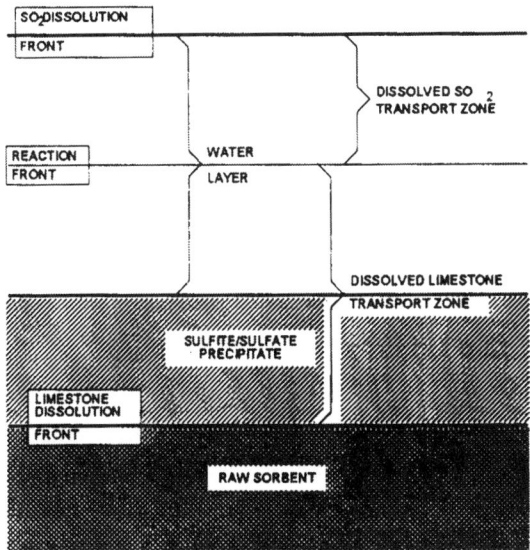

Figure 9.11 Conceptual representation of the sorbent particle in the LEC process.

Visneski [14] has discussed in detail the film theory representation of the concentration profiles of SO_2 and limestone. For the development of this model it has been assumed that the calcium sulfite/sulfate precipitate layer has no effect on SO_2 removal. A schematic for the concentration profiles of SO_2 and limestone is given in Figure 9.12.

Dry Scrubbing Technologies for Flue Gas Desulfurization 717

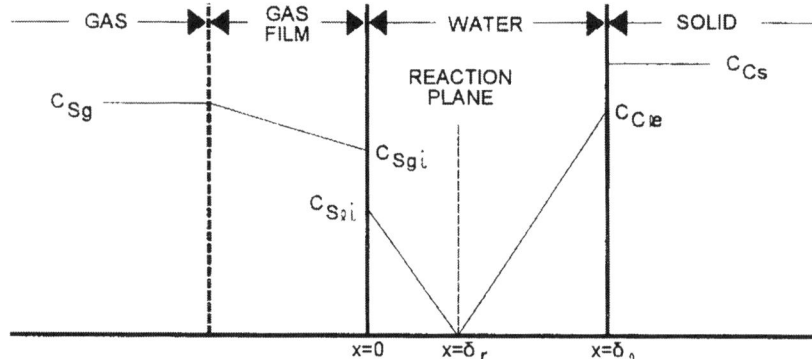

Figure 9.12 Film theory representation of the concentration profiles of sulfur dioxide and llmestone during reaction.

Hence, the film theory model of SO_2 capture kinetics show that the possible ratecontrolling steps are (1) mass transfer of SO_2 across the gas film in contact with the water layer; (2) absorption of SO_2 into the water layer; (3) ionic dissociation of SO_2 in water; (4) transport of SO_2 through the water layer to the reaction front; (5) dissolution of the limestone into the water layer; (6) ionization of the limestone; (7) transport of the dissolved limestone through the water layer to the reaction front; and finally (8) the rate of reaction between the dissolved limestone and the dissociated SO_2.

The rate-limiting step is determined by various factors such as the relative speeds of the flue gas and the limestone bed, the temperature of the reaction, the water content of the limestone bed, and the humidity of the gas stream. The reaction rate is taken to be instantaneous since it is known that ionic reactions occur with great rapidity. It is also assumed that SO_2 and limestone ionize instantaneously. Hence, the rate-controlling steps for SO_2 removal are reduced to the mass transfer rate of SO_2 across the gas film (gas-phase control), the diffusion rate of the species through the water layer to the reaction front (liquid-phase control), and the dissolution rate of limestone (solid-phase control).

Assumptions

In the first-generation modeling, done by Appell [13], of the fixed-bed LEC process, one aspect which was not considered is that the water does not necessarily drain off the bed instantaneously. Water trickles down the bed at finite velocities when the water content of the bed exceeds its saturation concentration. This effect is taken into consideration in the development of this model. Even though low flow rates are conducive to partial wetting, it is assumed for the sake of simplicity that the limestone granules are uniformly wetted.

Several of the assumptions used in the development of the flow model for the moving-bed LEC process have been used in the fixed-bed LEC process modeling work done by Appell [13] and Visneski [14].

The overall assumptions on which the flow model is based are (1) the system is isobaric at atmospheric pressure; (2) the ionic dissociation of the species and the ionic reaction of the species is instantaneous; (3) all limestone particles are the same size and if the particles are not spherical, a sphericity factor is used to describe the sphericity; (4) the limestone particles are completely wetted when water is present on the particles; (5) the water layer on the limestone bed has a uniform thickness within any given differential element; (6) the water flow is in the trickling flow regime; (7) the temperatures of the liquid film and the trickling water are equal to the temperature of the solids; (8) the superficial gas velocity is constant; (9) the flows of the gas phase, solid/liquid phase and the trickling water phase are not deflected by the force of the other flows; (10) the gas phase, solid/liquid phase, and the trickling water move in plug flow with no axial dispersion; (11) calcium sulfite/sulfate precipitation does not affect the mass and enthalpy balances; and (12) the heat- and mass-transfer areas are equal.

Removal Rate of Sulfur Dioxide

The general reaction scheme can be expressed as

$$A(g) \rightarrow A(l)$$

$$B(s) \rightarrow B(l)$$

$$A(l) + bB(l) \rightarrow \text{products}$$

For the reaction of SO_2 and $CaCO_3$, b equals 1 (a 1:1 stoichiometry) so this reaction scheme can be expressed as

$$SO_2(g) + CaCO_3(s) \rightarrow \text{products}$$

The flux of SO_2 (N_{SO_2}) is determined after first writing the mass-transfer equations for each of the four transport steps involved in the reaction.

The mass transport across the gas film is given as

Low Temperature Dry Scrubbing/LEC Process Support

Equation 9.1

$$N_{SO_2} = k_{Ag}(p_A - p_{Ai})$$

where k_{Ag} is the mass-transfer coefficient of SO_2 in the gas film, p_A is the partial pressure of SO_2 in the bulk gas, and p_{Ai} is the partial pressure of SO_2 at the gas-liquid interface.

Using film theory, the transport through the liquid is

Equation 9.2

$$N_{SO_2} = \frac{D_{Al}}{\lambda}(C_{Ai} - 0)$$

and, since the molal flux of SO_2 is equal to the negative of the molal flux of $CaCO_3$,

Equation 9.3

$$N_{SO_2} = -N_{CaCO_3} = \frac{D_{Bl}}{\delta - \lambda}(C_{Bl} - 0)$$

where D_{Al} and D_{Bl} are the diffusivities of SO_2 and $CaCO_3$ in water, respectively, C_{Ai} is the concentration of SO_2 at the gas-liquid interface, C_{Bl} is the concentration of $CaCO_3$ at the liquid-solid interface, λ is the distance from the gas-liquid interface to the reaction plane, and δ is the thickness of the water layer.

The dissolution of $CaCO_3$ can be expressed as

Equation 9.4

$$N_{SO_2} = -N_{CaCO_3} = k_{Bs}(C_B^* - C_{Bl})$$

where k_{Bs} is the dissolution mass-transfer coefficient and C_B^* is the equilibrium concentration of $CaCO_3$ in water.

Interfacial Concentrations

The concentrations of SO_2 and $CaCO_3$ at the interfaces can easily be obtained using Equations 9.1 through 9.4 and Henry's law, which is expressed as

Equation 9.5

$$P_{Ai} = H_A C_{Ai}$$

where H_A is the Henry's law coefficient for SO_2 in water.

Equating Equations 9.1 and 9.2, substituting Equation 9.5 for p_{Ai}, and then rearranging gives

Equation 9.6

$$C_{Ai} = \frac{\frac{P_A}{H_A}}{1 + \frac{D_{Al}}{\lambda H_A k_{Ag}}}$$

Equating Equations 9.3 and 9.4 yields

Equation 9.7

$$C_{Bl} = \frac{C^*_B}{1 + \frac{D_{Bl}}{(\delta - \lambda) k_{Bs}}}$$

Reaction Site and Determination of Control

The location of the reaction plane ($x=\lambda$) can easily be determined. Equating Equations 9.2 and 9.3, substituting Equations 9.6 and 9.7 for the interfacial concentrations, and solving for λ gives

Equation 9.8

$$\lambda = \frac{\delta + \frac{D_{Bl}}{k_{Bs}} - \frac{C^*_B D_{Bl}}{P_A k_{Ag}}}{1 + \frac{C^*_B D_{Bl} H_A}{D_{Al} P_A}}$$

Not only does Equation 9.8 give the location of the reaction plane, it also determines the controlling mechanism for the reaction. A negative value for l is a physical impossibility because the reaction must take place in the water layer. If the conditions are such that Equation 9.8 gives $\lambda \leq 0$ then the reaction takes place at the gas-liquid interface. Therefore the controlling mechanism under these conditions is the mass transport through the gas film.

If $0 < \lambda < \delta$, then the reaction occurs in the water layer at the distance λ from the gas-liquid interface. For this case all three resistances play a part in determining the rate of SO_2 removal but the largest resistance is the diffusion in the liquid layer.

Another impossibility is $\lambda \geq \delta$. In this case the reaction takes place at the liquid-solid interface and the controlling mechanism is the dissolution of the limestone.

GAS-PHASE CONTROL. If $\lambda \leq 0$ (as given by Equation 9.8), then the reaction is controlled by the transport of SO_2 across the gas film. Since the reaction occurs at the gas-liquid interface, the interfacial concentration of SO_2 (C_{Ai}) is equal to zero. The rate of reaction of SO_2 is given by Equation 9.1. Substituting in Henry's law (Equation 9.5) with $C_{Ai} = 0$ gives the rate of reaction for gas-phase control as

Equation 9.9

$$N_{SO_2} = k_{Ag} p_A$$

LIQUID-PHASE CONTROL. If $0 < \lambda < \delta$ (as given by Equation 9.8), the reaction occurs within the water layer. All three resistances contribute to the overall resistance with the diffusion through the liquid layer as the major contributor. The rate of reaction of SO_2 is given by Equation 9.1. Substituting in Henry's law (Equation 9.5), then C_{Ai} (Equation 9.6), and finally λ (Equation 9.8) gives the rate of reaction for liquid-phase control as

Equation 9.10

$$N_{SO_2} = \frac{\dfrac{p_A}{H_A} + \dfrac{C_B^* D_{BI}}{D_{AI}}}{\dfrac{1}{H_A k_{Ag}} + \dfrac{\delta}{D_{AI}} + \dfrac{D_{BI}}{k_{Bs} D_{AI}}}$$

SOLID-PHASE (DISSOLUTION) CONTROL. If $\lambda \geq \delta$ (as given by Equation 9.8), then the reaction is controlled by the limestone dissolution. Since the reaction occurs at the liquid-solid interface the interfacial concentration of $CaCO_3$ (C_{BI}) is equal to zero. The rate of reaction of SO_2 is given by Equation 9.4. Substituting in $C_{BI} = 0$ gives the rate of reaction for dissolution control as

Equation 9.11

$$N_{SO_2} = k_{Bs}C^*_B$$

The dissolution mass-transfer coefficient (k_{Bs}) and the equilibrium concentration of $CaCO_3$ (C_{BI}) are constants. Therefore for the case of dissolution control the reaction rate is a constant. It is not a function of the SO_2 concentration in the bulk gas as are gas-phase control and liquid-phase control.

Fixed-Bed Process Model

Mass and Enthalpy Balances

The flow scheme for the packed-bed reactor is shown in Figure 9.13a. Flue gas enters the reactor at the top (z=0), passes downward through the limestone bed of constant cross-sectional area (A), then exits at the bottom ($z=z_T$). In order to model the bed, a differential element, shown in Figure 9.13b, is analyzed.

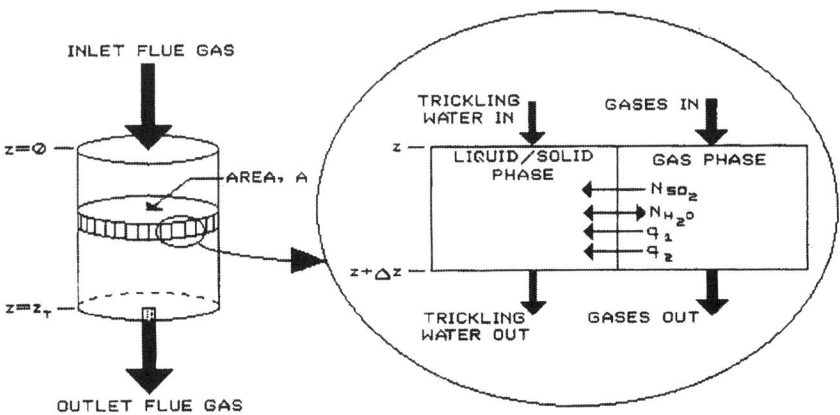

Figure 9.13 Flow scheme (a) and differential element (b) for down flow in a packed bed.

The analysis is simplified by "lumping" the solid (limestone) and liquid (water) phases into a combined liquid/solid phase. This results in a two-phase differential element that consists of (1) the gas phase and (2) the liquid/solid phase. This is the same analysis that is used by Appell [13].

The flue gas (gas-phase) enters the differential element at an axial position z and exits at z+Dz, while the liquid/solid phase remains stationary except for trickling water, due to oversaturation of the liquid/solid phase. (Water does not begin to flow through the bed until the water concentration exceeds the saturation concentration.) The mass transfer of sulfur dioxide (N_{SO_2}) is unidirectional, from the gas phase to the liquid/solid phase. Water (N_{H_2O}) can be transported between the phases by condensation or evaporation. The enthalpy transfer between the two phases is the sensible heat (q1), due to the temperature difference between the two phases, and the latent heat (q2), due to the evaporation or condensation of water. The heat content of the trickling water is assumed to be negligible.

All balances are written in the usual form.

Equation 9.12

IN-OUT=ACCUMULATION

SULFUR DIOXIDE MASS BALANCE. Sulfur dioxide enters the differential element by convection and leaves the element by convection and by transport into the liquid/solid phase. These processes are expressed as follows. The rate of transport in by convection is

Equation 9.13

$$A v_g C_{SO_2}\big|_z$$

The rate of transport out by convection is

Equation 9.14

$$A v_g C_{SO_2}\big|_{z+\Delta z}$$

The rate of transport into the liquid/solid phase is

Equation 9.15

$$N_{SO_2} a_m A \Delta z$$

The rate of accumulation is

Equation 9.16

$$\varepsilon_g A \Delta z \frac{\partial C_{SO_2}}{\partial t}$$

In Equations 9.13 through 9.16 A is the bed cross-sectional area, v_g is the gas-phase superficial velocity, C_{SO_2} is the concentration of SO_2 in the gas phase, N_{SO_2} is the flux of SO_2 into the liquid/solid phase (as expressed by Equations 9.9, 9.10, or 9.11), a_m is the specific area for mass transfer, and ε_g is the bed voidage.

Substituting these equations into Equation 9.12, dividing by $A\Delta z$, then taking the limit as $\Delta z \to 0$ yields

$$v_g \frac{\partial C_{SO_w}}{\partial z} + \varepsilon_g \frac{\partial C_{SO_2}}{\partial t} = -a_m N_{SO_2}$$

GAS-PHASE WATER MASS BALANCE. The gas-phase water enters the differential element by convection and leaves by convection and by condensation to the liquid/solid phase. These processes are expressed as follows. The rate of transport in by convection is

Equation 9.17

$$A v_g Y \rho_g \big|_z$$

The rate of transport out by convection is

Equation 9.18

$$A v_g Y \rho_g \big|_{z+\Delta z}$$

The rate of transport out by condensation is

Equation 9.19

$$N_{H_2O} = \rho_g a_m k_{wg}(Y - Y^*) A \Delta z$$

The rate of accumulation is

Equation 9.20

$$\varepsilon_g \rho_g A \Delta z \frac{\partial Y}{\partial t}$$

In Equations 9.17 through 9.20 Y is the molal humidity, ρ_g is the molal gas density, k_{Wg} is the gas-side water/air mass-transfer coefficient, and Y* is the equilibrium saturation humidity.

Putting these equations into the form of Equation 9.12, dividing by $A\Delta z \rho_g$, then taking the limit as $\Delta z \rightarrow 0$ yields

Equation 9.21

$$v_g \frac{\partial Y}{\partial z} + \varepsilon_g \frac{\partial Y}{\partial t} = -a_m k_{Wg}(Y - Y^*)$$

GAS-PHASE ENTHALPY BALANCE. Enthalpy in the gas phase is carried into and out of the differential element by convection. Enthalpy is transferred from the gas phase to the liquid/solid phase due to the latent heat of vaporization and intrinsic enthalpy of water transferred and due to convective heat transfer between phases. These processes are expressed as follows. The rate of transport in by convection is

Equation 9.22

$$v_g A \rho_g H|_z$$

The rate of transport out by convection is

Equation 9.23

$$v_g A \rho_g H|_{z+\Delta z}$$

The rate of transport out by condensation (enthalpy from phase change) is

Equation 9.24

$$N_{H_2O} \lambda_0 A \Delta z$$

The rate of transport out by the intrinsic enthalpy of water is

Equation 9.25

$$N_{H_2O}(C_{P_{H_2O}}(T_g - T_0))A\Delta z$$

The rate of transport out by sensible heat loss is

Equation 9.26

$$h_g a_H (T_g - T_1) A \Delta z$$

The rate of accumulation is

Equation 9.27

$$\varepsilon_g \rho_g A \Delta z \frac{\partial H}{\partial t}$$

In Equations 9.22 through 9.27 H is the enthalpy of the gas phase, N_{H_2O} is the rate of water transfer from the gas phase, λ_0 is the latent heat of vaporization for water at the reference temperature T_0, $C_{P_{H_2O}}$ is the heat capacity of water, T_g is the temperature of the gas phase, T_1 is the temperature of the liquid/solid phase, h_g is the air/water heat-transfer coefficient, and a_H is the specific heat-transfer area.

The rate of water transfer, N_{H_2O}, can be found by assembling Equations 9.17 through 9.23 and rearranging to give

Equation 9.28

$$N_{H_2O} = v_g A Y \rho_g \big|_z - v_g A Y \rho_g \big|_{z+\Delta z} - \varepsilon_g A \Delta z \rho_g \frac{\partial Y}{\partial t}$$

Assembling Equations 9.22 through 9.27 in the form of Equation 9.12, dividing by $\rho_g A \Delta z$, and taking the limit as $\Delta z \to 0$ yields

Equation 9.29

$$-v_g \frac{\partial H}{\partial z} + (C_{P_{H_2O}}(T_g - T_0) + \lambda_0)\left(v_g \frac{\partial Y}{\partial z} + \varepsilon_g \frac{\partial Y}{\partial t}\right) - \frac{h_g a_H}{\rho_g}(T_g - T_1) = \varepsilon_g \frac{\partial H}{\partial t}$$

The enthalpy of the water/air mixture, H, can be defined as

Low Temperature Dry Scrubbing/LEC Process Support

Equation 9.30

$$H = C_{P_{air}}(T_g - T_0) + Y(C_{P_{H_2O}}(T_g - T_0) + \lambda_0)$$

where $C_{P_{air}}$ is the heat capacity of the air.

Taking the partial derivatives of Equation 9.30 with respect to z and t, then substituting into Equation 9.29 and rearranging gives

Equation 9.31

$$-v_g \frac{\partial T_g}{\partial z} + \varepsilon_g \frac{\partial T_g}{\partial t} = \frac{h_g a_H}{\rho_g (C_{P_{air}} + YC_{P_{H_2O}})}(T_g - T_l)$$

LIQUID/SOLID-PHASE WATER MASS BALANCE. Water enters the liquid/solid phase by condensation and convection and exits by evaporation and convection. The rate of transport in by convection (trickling water) is

Equation 9.32

$$A v_1 \rho_1 \big|_z$$

The rate of transport in by condensation from the gas phase is

Equation 9.33 (the same as Equation 19, see above)

$$N_{H_2O} = \rho_g a_m k_{wg}(Y - Y^*) A \Delta z$$

The rate of transport out by convection (trickling water) is

Equation 9.34

$$A v_1 \rho_1 \big|_{z + \Delta z}$$

The rate of accumulation is

Equation 9.35

$$A \Delta z \frac{\partial C_{H_2O}}{\partial t} = A \Delta z \rho_1 \frac{\partial \varepsilon_{lT}}{\partial t}$$

In Equations 9.32 through 9.35 C_{H_2O} (or $\rho_l \varepsilon_{lT}$) is the concentration of the water occupying the reactor volume, ε_{lT} is the total liquid holdup, ρ_l is the density of the liquid phase, and v_l is the superficial velocity of the trickling water.

Once again substituting into Equation 9.12 and rearranging gives

Equation 9.36

$$\rho_l \frac{\partial v_l}{\partial z} + \frac{\partial C_{H_2O}}{\partial t} = \rho_g a_m k_{Wg}(Y - Y^*)$$

LIQUID/SOLID-PHASE ENTHALPY BALANCE. As mentioned above the convective terms in the enthalpy balance for the liquid/solid phase are ignored. The enthalpy gain for the liquid/solid phase is from the sensible heat, from the gas phase, and from condensation. These can be expressed as follows. The rate of transport in by sensible heat gain from the gas phase is

Equation 9.37

$$q_1 = h_g a_H (T_g - T_1) A \Delta z$$

The rate of transport in by condensation of water from the gas phase is

Equation 9.38

$$q_2 = N_{H_2O}(C_{P_{H_2O}}(T_g - T_0) + \lambda_0)) A \Delta z$$

The rate of accumulation is

Equation 9.39

$$\rho_s C_{P_s} A \Delta z \frac{\partial T_1}{\partial t} + C_{H_2O} C_{P_{H_2O}} A \Delta z \frac{\partial T_1}{\partial t}$$

In Equations 9.37 through 9.39 ρ_s is the density of the solid phase (limestone) and C_{P_s} is the heat capacity of the solid phase.

Placing Equations 9.37 through 9.39 into Equation 9.12, substituting for N_{H_2O}, dividing by $A\Delta z$, and rearranging yields

Equation 9.40

$$\frac{\partial T_1}{\partial t} = \frac{h_g a_H}{(C_{H_2O} C_{P_{H_2O}} + \rho_s C_{P_s})}(T_g - T_1) + \frac{\rho_g a_m k_{Wg}(C_{P_{H_2O}}(T_g - T_1) + \lambda_0)}{(C_{H_2O} C_{P_{H_2O}} + \rho_s C_{P_s})}(Y - Y^*)$$

Model Simulations

The model simulations presented herein are based on the fixed-bed LEC operating conditions used for experimental run 870714A' from Prudich et al. [3]. The experimental data for the SO_2 removal performance of this run is shown in Figure 9.5 and the operating conditions are given in Table 9.1.

Table 9.1 Simulation Conditions.

Superficial Gas Velocity	1.5 ft/s
Limestone Bed Depth	12 in
Inlet Gas Conditions	130°F 53% Relative Humidity 750 ppm SO_2
Initial Bed Conditions	77 °F 0.0423 lb water/lb stone

Blinding of the surface by the precipitation of the reaction products is not considered for these simulations. The decrease in the SO_2 removal efficiency is solely due to the drying of the reactive surface. The amount of SO_2 removed during the operating time of the bed, about 105 minutes, has a negligible effect on the removal efficiency. Evidence for this is also shown in Figure 9.5. A subsequent reactivation of the dry bed by re-wetting returns the bed to its original activity. No sign of surface blinding is apparent at the conditions given in Table 9.1 until 375 operational minutes have elapsed.

Figures 9.14 and 9.15 compare the simulation results to the experimental data of run 870714A' from Prudich et al. [3]. Figure 9.14 compares the outlet bed temperature (i.e., the temperature of the bottom of the limestone bed). Both the model and experimental results show a sharp increase of the bed temperature very early in the run time. Then as the evaporation front moves through the bed the temperature is fairly constant until the drying front reaches the end of the bed. When this occurs the temperature quickly rises to the temperature of the inlet gas. Although there is fairly good quantitative agreement between the model predictions and the experimental data, only the qualitative results can be compared. The LEC data were taken from a pilot unit supplied by flue gas from a coal-fired boiler. Therefore there was no control over the raw flue gas conditions, specifically inlet temperatures and humidities, and only marginal control was obtainable in the conditioning stage. Quantitative results will have to wait until highly controlled experiments have been performed.

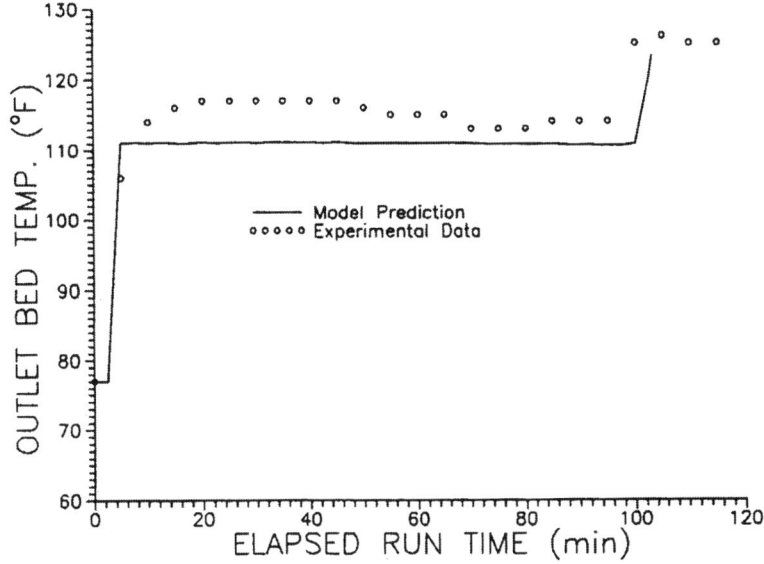

Figure 9.14 Comparison between the outlet bed temperature predicated by the model and experimental data (Data is from Prudich et al.

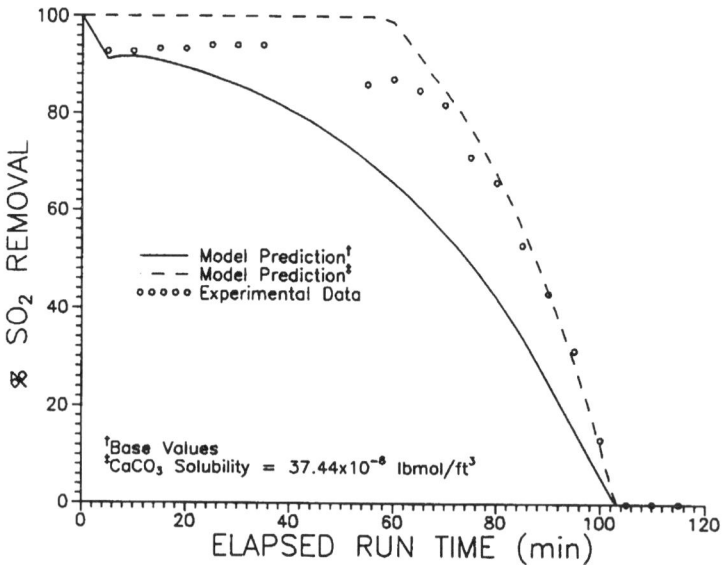

Figure 9.15 Comparison between the SO_2 removal efficiency predicted by the model and experimental data [Data is from Prudich et al. (1988), run 870714A'.]

Figure 9.15 compares the SO_2 removal efficiency of the model prediction to those from run 870714A'. As mentioned above only a qualitative analysis is appropriate. Using the base set of parameter values, the removal efficiency is underestimated by the model. Using a high $CaCO_3$ solubility, the model overestimates the efficiency. The comparison with the high solubility is made because it appears that this value is very important in fitting the model to the experimental data. Increasing the dissolution rate coefficient will not achieve better agreement with the experimental data. The curve would not be much different from the base value curve shown in Figure 9.15. Also increasing the liquid-phase diffusivities will not improve the fit.

The solubility is the one parameter that does not have a precise estimate. The solubility values found in the literature range over three orders of magnitude. Some fundamental work needs to be done to quantify limestone dissolution under the conditions found in the LEC process.

Moving-Bed Reactor Process Model

Model Development

The procedure followed in developing the steady-state flow model of the moving-bed LEC process is similar to that followed by Appell [13]. A full account of the development of this model is given by Reddy [37]. The flows of the flue gas and the limestone bed are unidirectional with the limestone bed moving vertically down and the flue gas moving horizontally across the limestone bed in a cross-flow pattern. The differential element used to develop the model is shown in Figure 9.16.

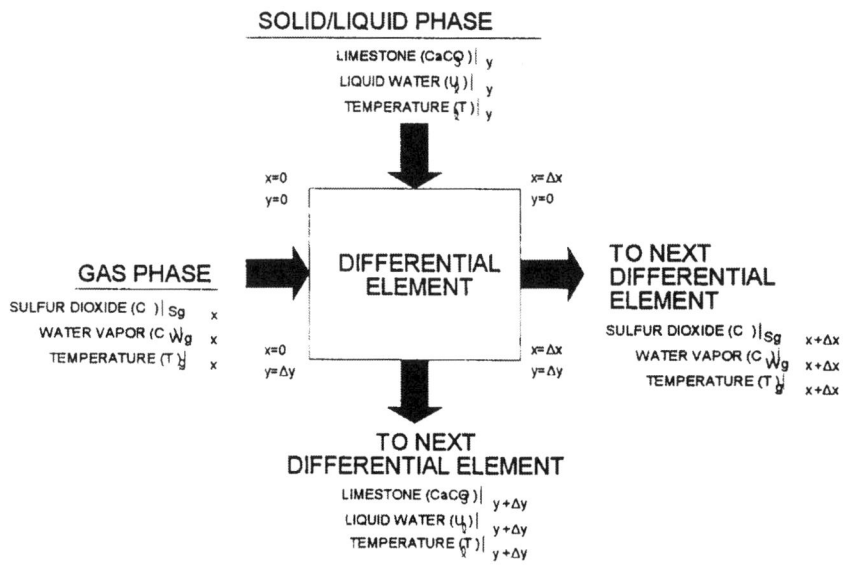

Figure 9.16 Differential element used to develop the model.

The solid phase (limestone), the liquid phase (water adsorbed on the bed) and the trickling water phase (if present) are lumped into one solid/liquid phase and hence the mass and energy balances are based on only two phases, viz. the gas phase and the lumped solid/liquid phase.

The direction of sulfur dioxide mass transfer is from the gas phase to the solid/liquid phase in the presence of adsorbed water. In the absence of water there is no change of SO_2 concentration in the gas phase. Mass transfer of water can occur in either direction (from gas phase to solid/liquid phase or vice versa)

Dry Scrubbing Technologies for Flue Gas Desulfurization

depending on whether the water is condensing from the gas phase or is evaporating from the solid/liquid phase. The transfer of water between the two phases adds to the sensible heat transfer between the two phases.

Model Simulations

The baseline conditions chosen for the computer simulation of the moving-bed LEC process are based on runs made at the LEC pilot plant installed on the slipstream of the stoker boiler providing steam to the Ohio University campus. These baseline conditions for the computer simulation are listed in Table 9.2.

Table 9.2 Simulation Conditions

Limestone Bed Geometry		
Bed Depth	Bed Height	Bed Width
14 inches [direction of flue gas flow]	128 inches [direction of limestone flow]	36 inches [perpendicular to both flows]
Flue Gas Flow Rate		
3000 ACFM		
Inlet Gas Conditions		
201°F	2% Water Vapor (by volume)	750 ppm SO_2
Superficial Limestone Velocity		
14.23 ft/hr		
Inlet Limestone Conditions		
77°F	0% liquid water saturation (Dry Bed)	0.5 GMP overhad spray

The superficial limestone velocity was chosen on the basis of the tachometer reading indicating the speed of the bed twin screws (see Figure 9.8). Low limestone speeds were found to create operational problems such as caking. The maximum limestone flow which the screws and bucket elevators could handle was 14.23 ft/hr of dry limestone.

One of the primary topics of interest for this chapter is the drying and the average SO_2 removal of the bed. As has been explained earlier, the presence of liquid water is essential for the removal of SO_2. It is also desirable that the limestone be dry at the bottom of the bed since this ameliorates the handling of the stone.

Figure 9.17 gives the water content of the bed and the outlet SO_2 concentration profile under the simulation conditions. It can be seen that as the limestone bed moves down, the drying front moves laterally across the bed. Correspondingly, due to the lack of water for SO_2 removal, the outlet SO_2 concentration goes up as the limestone moves down through the reactor. The profile of the drying front appears to be linear. The outlet SO_2 concentration profile is not linear. This is due to the fact that the rate of SO_2 removal has a nonlinear relationship with the water content of the bed. This is more clearly depicted in Figure 9.18 which gives the contours for various SO_2 concentrations throughout the bed. The gap between the contours is pretty regular above the 150 ppm contour, but there is a larger gap between the 150 ppm and 0 ppm contours. This is due to the fact the SO_2 removal follows an S-shaped curve and at lower SO_2 concentrations the rate of SO_2 removal approaches zero.

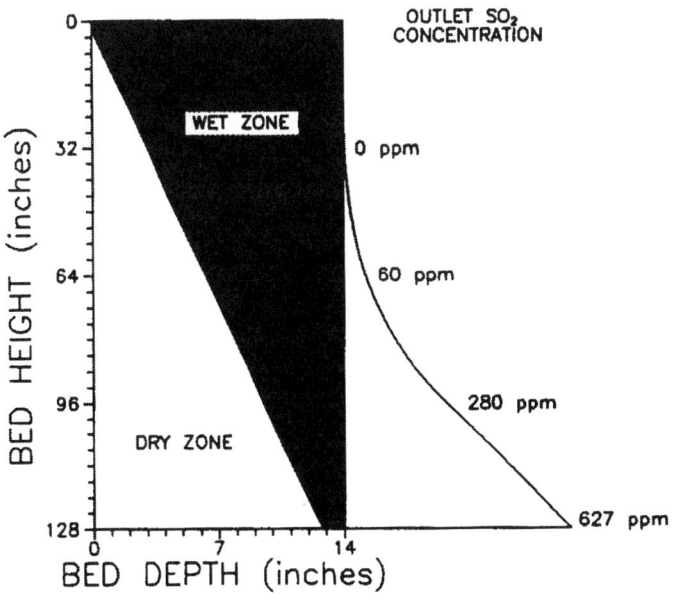

Figure 9.17 Drying front and outlet SO_2 concentration profiles.

Dry Scrubbing Technologies for Flue Gas Desulfurization

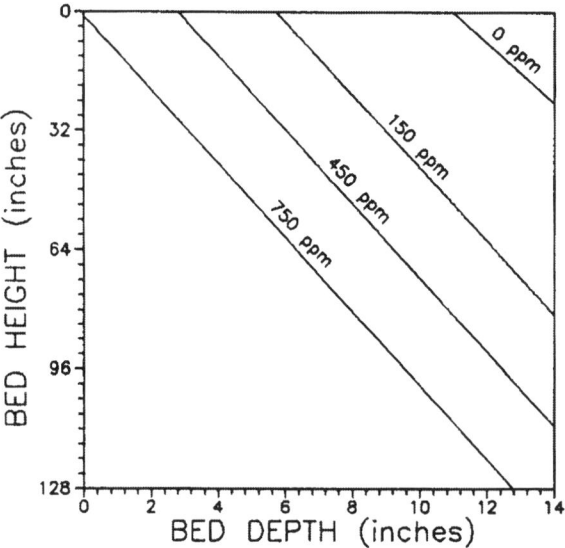

Figure 9.18 SO_2 contours through the bed.

It should be noted that although the SO_2 removal at the bottom of the bed is very poor, what matters is the overall SO_2 removal capacity of the bed. The variation of the average SO_2 concentration (over the entire bed height of 128 inches) with bed depth is given in Figure 9.19. As can be seen from Figure 9.17 the first few inches of the bed in the direction of the gas flow are primarily dry and hence the SO_2 removal is very low. At greater bed depths, the liquid water concentration increases and hence SO_2 removal goes up. The overall SO_2 concentration curve shown in Figure 9.19 is in the form of an S-shaped curve. This would have been more evident for bed depths greater than 14 inches.

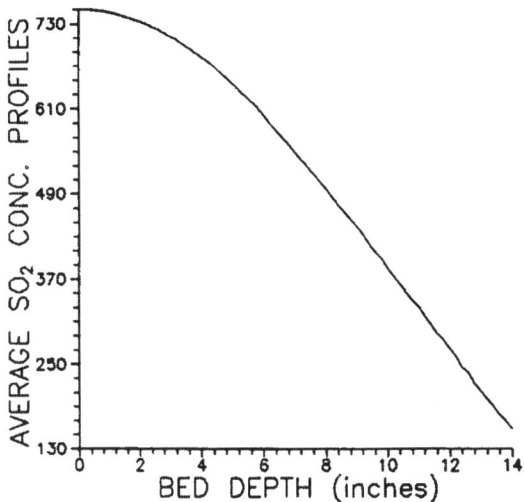

Figure 9.19 Average SO₂ outlet concentration profile.

Comparison of the results of the pilot plant with the computer simulation is not possible at this time. Due to the lack of time it was not possible to use the pilot plant to generate enough experimental data so as to accurately compare the two results. Furthermore there still is a large uncertainty of the actual value for certain empirical correlations used in the model. For example, the equilibrium concentration of limestone has been cited [14] with values several orders of magnitude apart. The use of different values for this correlation can give varied rates of SO_2 removal.

Limestone Solubilities and Rates of Solubilization

Theory

CARBONIC ACID/CARBONATE CHAIN. When water is exposed to the atmosphere, carbon dioxide from the air dissolves in it. One molecule of aqueous carbon dioxide reacts with a molecule of water to form undissociated carbonic acid.

$CO_2(g) \leftrightarrow CO_2(aq)$

and

$CO_2(aq) + H_2O \leftrightarrow H_2CO_3(aq)$.

Since carbonic acid is known to be a weak acid and is mostly present in its undissociated form, the carbonate and the bicarbonate ions react with hydronium ions in two hydrolytic equilibrium reactions.

$$H_2CO_3(aq) + H_2O \leftrightarrow H_3O^+ + HCO_3^-$$

and

$$HCO_3^- + H_2O \leftrightarrow H_3O^+ + CO_3^{2-}.$$

Hydronium ions from the equations above are in equilibrium with neutral water molecules as described by the ionic product of water.

$$2\,H_2O \leftrightarrow H_3O^+ + OH^-.$$

The absorption and association of CO_2 can be expressed by the law of mass action.

Equation 9.41

$$K_{CO_2} = \frac{a_{H_2CO_3(aq)}}{a_{CO_2(g)}}$$

in which the activity of gaseous carbon dioxide can be replaced using Henry's law.

Equation 9.42

$$a_{CO_2(g)} = \frac{p_{CO_2}}{H_{CO_2}}$$

where H_{CO_2} denotes the Henry constant and p_{CO_2} is the partial pressure of CO_2 in air.

Similarly for the reactions with hydronium ions,

Equation 9.43

$$K_{H_2CO_3} = \frac{a_{HCO_3^-} \cdot a_{H_2O^-}}{a_{H_2CO_3(aq)} \cdot a_{H_2O}}$$

Equation 9.44

$$K_{HCO_3^-} = \frac{a_{CO_3^{2-}} \cdot a_{H_3O^+}}{a_{HCO_3^-} \cdot a_{H_2O}}$$

and

Equation 9.45

$$K_{H_2O} = \frac{a_{H_3O^+} \cdot a_{OH^-}}{a^2_{H_2O}}$$

From Equations 9.43 through 9.45, the mole fractions of H_2CO_3, HCO_3^- and CO_3^{2-} can be calculated for a closed system depending on the solution pH. The mole fractions are obtained using the simplifications $a_i \approx c_i$ assuming that the activity coefficients are close to unity. Another simplification, $a_{H_2O} \approx 1$ a, can be obtained from the product of the activity coefficient and the mole fraction of water. Figure 9.20 shows the results of those calculations at constant temperature. The mole fractions of CO_3^{2-}, HCO_3^- and H_2CO_3 at equilibrium have been normalized by dividing the amount of moles of each species by the sum of moles of the three chemical species. From Figure 9.20 it can be shown that pure distilled water which has an initial pH of 7, when exposed to atmospheric air containing carbon dioxide, will have a lower pH at equilibrium.

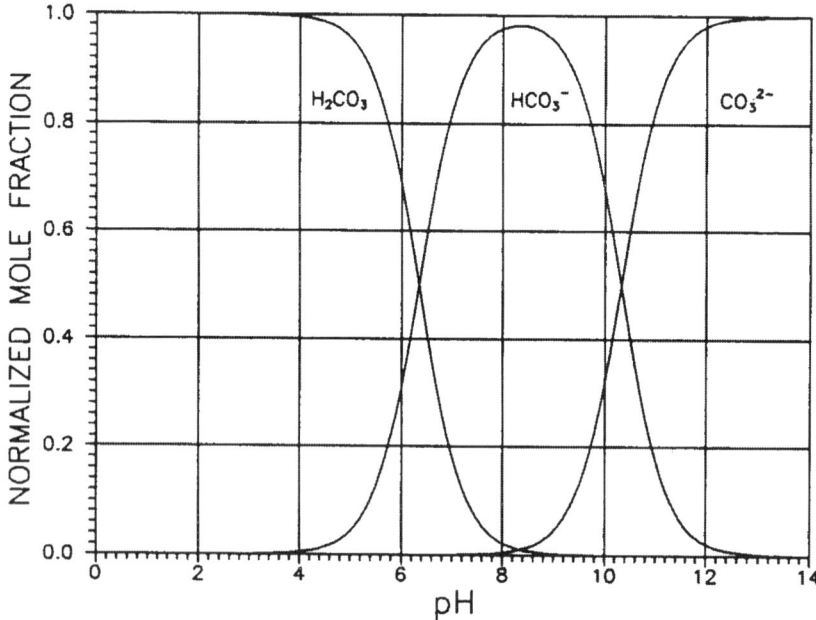

Figure 9.20 Bicarbonate-carbonate equilibria as a function of solultion pH.

At pH 7 the normalized mole fraction of carbonic acid is approximately 0.2, while the value for the bicarbonate ion is 0.8. The normalized mole fraction for the carbonate ions is negligible. According to the CO_2 absorption reaction, dissolving CO_2 from the atmosphere disturbs the equilibrium by increasing the H_2CO_3 concentration. The system responds to this disequilibrium by shifting the equilibrium of the first hydronium ion producing reaction to the right, thereby lowering the solution pH. The decrease of the solution pH will finally come to a halt when a dynamic equilibrium between CO_2 entering and leaving the solution is reached.

HYDRATION OF SULFUR DIOXIDE. In order for a reaction between sulfur dioxide and limestone to occur in a limestone scrubber, sulfur dioxide has to be dissolved into water. At the gas/liquid interface, SO_2 from the flue gas is hydrated by water molecules to undissociated H_2SO_3.

$$SO_2(g) \leftrightarrow SO_2(aq)$$

and

$$SO_2(aq) + H_2O \leftrightarrow H_2SO_3$$

Undissociated sulfurous acid subsequently reacts in two equilibrium reactions to bisulfite or sulfite ions thereby lowering the solution pH. The absorption rate is dependent on the pH of the liquid.

$$H_2SO_3 + H_2O \leftrightarrow HSO_3^- + H_3O^+$$

and

$$HSO_3^- + H_2O \leftrightarrow SO_3^{2-} + H_3O^+$$

Combining the SO_2 absorption and association reactions and applying the law of mass action yields

Equation 9.46

$$K_{H_2SO_e} = \frac{\gamma_{H_2SO_3(aq)} \cdot c_{H_2SO_3(aq)}}{\frac{p_{SO_2}}{H_{SO_2}}}$$

Equation 9.47

$$K_{HSO_3^-} = \frac{\gamma_{HSO_3^-} \cdot c_{HSO_3^-} \cdot \gamma_{H_3O} \cdot c_{H_2O}}{\gamma_{H_2SO_3(aq)} \cdot c_{H_2SO_3(aq)} \cdot \gamma_{H_2O} \cdot c_{H_2O}}$$

and

Equation 9.48

$$K_{SO_3^{2-}} = \frac{\gamma_{SO_3^{2-}} \cdot c_{SO_3^{2-}} \cdot \gamma_{H_2O^+} \cdot c_{H_2O}}{\gamma_{HSO_3^-} \cdot c_{HSO_3^-} \cdot \gamma_{H_2O} \cdot c_{H_2O}}$$

The equilibrium curves for undissociated sulfurous acid, bisulfite and sulfite ions in the equations above, are given in Figure 9.21. The dependency is shown in terms of normalized mole fractions versus liquid phase pH. The normalization was achieved by dividing the amount of moles of each species of SO_3^{2-}, HSO_3^- and H_2SO_3 by their sum of moles. From Figure 9.21 and the equilibrium reactions listed above, it can be recognized that the absorption of SO_2 in water is promoted at a high pH.

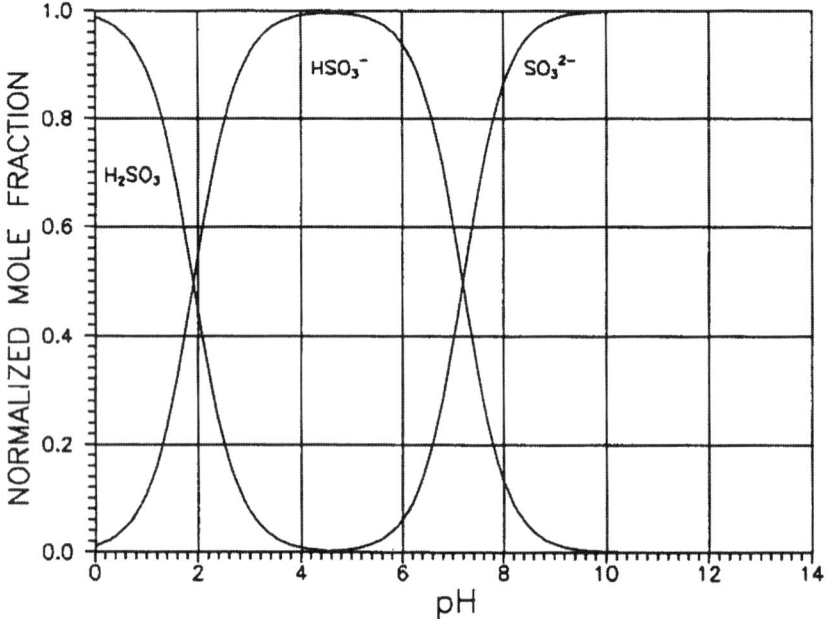

Figure 9.21 Bisulfite-sulfite distribution as a function of pH.

The actual reaction chemistry is more complicated, because the bisulfite ion enters into still another equilibrium reaction. Two equally charged HSO_3^- ions react via a condensation step to form the disulfite or 'pyrosulfite' ion, $S_2O_5^{2-}$ according to

$$HSO_3^- + HSO_3^- \leftrightarrow S_2O_5^{2-} + H_2O$$

LIMESTONE DISSOLUTION. The two major constituents of limestone are calcitic and magnesium carbonates. Magnesium carbonate is often referred to as dolomite. Magnesium may also be present in a solid solution with calcium carbonate. Those mixed crystals make up a true homogeneous solution and can be expressed by

$$Ca_{(1-x)}Mg_{(x)}CO_3 \leftrightarrow (1-x)Ca^{2+} + x\,Mg^{2+} + CO_3^{2-}$$

However, biogenic calcite can contain multiple phases of those mixed crystals which may vary in composition.

It is therefore practical to express the chemical reactions involved in the dissolution of limestones in terms of those major constituents. The solvation of limestone is described by

$$CaCO_3(s) \leftrightarrow CaCO_3(aq)$$

and

$$MgCO_3(s) \leftrightarrow MgCO_3(aq)$$

The dissociation reactions are described by

$$CaCO_3(aq) \leftrightarrow Ca_2^+ + CO_3^{2-}$$

and

$$MgCO_3(aq) \leftrightarrow Mg^{2+} + CO_3^{2-}$$

The law of mass action yields

Equation 9.49

$$K_{CaCO_2} = \frac{a_{Ca^{2+}} \cdot a_{CO_3^{2-}}}{A_{CACO_3(s)}}$$

and

Equation 9.50

$$K_{MgCO_3} = \frac{a_{Mg^{2+}} \cdot a_{CO_3^{2-}}}{a_{MgCO_3(s)}}$$

For the case of

Equation 9.51

$$a_{CaCO_3(s)} = a_{MgCO_3(s)} = 1$$

Dry Scrubbing Technologies for Flue Gas Desulfurization

the solubility products of calcium and magnesium carbonate are defined by

Equation 9.52

$$L_{CaCO_3} = a_{Ca^{2+}} \cdot a_{CO_3^{2-}}$$

and

Equation 9.53

$$L_{MgCO_3} = a_{Mg^{2+}} \cdot a_{CO_3^{2-}}$$

Equations 9.49 and 9.50 should be written in terms of their activity coefficients.

Equation 9.54

$$K_{CaCO_3} = \frac{\gamma_{Ca^{2+}} \cdot c_{Ca^{2+}} \cdot \gamma_{CO_3^{2-}} \cdot c_{CO_3^{2-}}}{a_{CaCO_3(s)}}$$

and

Equation 9.55

$$K_{MgCO_3} = \frac{\gamma_{Mg^{2+}} \cdot c_{Ca^{2+}} \cdot \gamma_{CO_3^{2-}} \cdot c_{CO_3^{2-}}}{a_{MgCO_3(s)}}$$

In these equations, the activity of the solid phase has to be inserted. The activity of the solid phase would be 1, by definition, if the limestone is assumed to be a pure solid. This is seldom true for natural limestones. Therefore, the $CaCO_3$ in a calcitic limestone which has a composition of $(Ca_{0.95}Mg_{0.05})CO_3$ deviates from an activity of 1. Based on the assumption of an ideal solution, the activity of $CaCO_3$ in this limestone is equal to its mole fraction.

The connection between limestone dissolution and the carbonic acid/carbonate chain is the carbonate ion. When limestone dissolves, an excess of carbonate ions is released. These ions have to be in equilibrium with the bicarbonate ions and undissociated carbonic acid. To reestablish equilibrium, hydronium ions are necessary. During limestone dissolution, the solution pH thus increases.

The need of hydronium ions mentioned above is related to another parameter, which is important for the free drift method. This parameter is the 'stoichiometric factor,' which defines how many hydronium ions will be necessary to reestablish an equilibrium state after initial disequilibrium caused by the dissolution of one calcium ion. The stoichiometric factor is the product of the following theory.

From Figure 9.20 it can be noted that, at a pH of 4, the mole fraction of undissociated carbonic acid is nearly 1, while the fraction of bicarbonate and carbonate ions is negligible. For each calcium carbonate molecule dissolved in this environment, one carbonate ion is released and transferred into undissociated acid. Therefore two hydronium ions are necessary for this transfer.

The other extreme case is at a pH of approximately 13, where most of the calcium is present as carbonate ions. Under these conditions, a carbonate ion released from the dissolution of limestone would not affect the equilibrium at all. Therefore no hydronium ion is necessary to maintain the equilibrium of the solution.

From those extreme cases, it can be seen that the value for the stoichiometric factor lies between 0 and 2 for the pH interval from 0 to 14. The factor is calculated by finding first the normalized mole fractions of bicarbonate ions and carbonic acid in solution at a fixed pH. Here, the mole fractions are expressed in terms of concentrations rather than activities, as a means of simplification:

Equation 9.56

$$\psi_{HCO_3^-} = \frac{1}{1 + \frac{\tilde{K}_{HCO_3^-}}{c_{H^+}} + \frac{c_{H^+}}{\tilde{K}_{H_2CO_3}}}$$

and

Equation 9.57

$$\psi_{H_2CO_3} = \frac{1}{1 + \frac{\tilde{K}_{H_2CO_3}}{c_{H^+}} + \frac{\tilde{K}_{H_2CO_3} \cdot \tilde{K}_{HCO_3^-}}{(c_{H^+})^2}}$$

where

Equation 9.58

$$\tilde{K}_{H_2CO_3} = K_{H_2CO_3} \cdot c_{H_2O}$$

Equation 9.59

$$\tilde{K}_{HCO_3^-} = K_{HCO_3^-} \cdot c_{H_2O}$$

and, of course,

Equation 9.60

$$c_{H^+} = 10^{-pH}$$

From the mole fractions of bicarbonate and carbonic acid, the stoichiometric factor can be found by

Equation 9.61

$$f_s = \psi_{HCO_3^-} \cdot 1 + \psi_{H_2CO_3} \cdot 22$$

This factor is shown in Figure 9.22 as a function of pH and temperature.

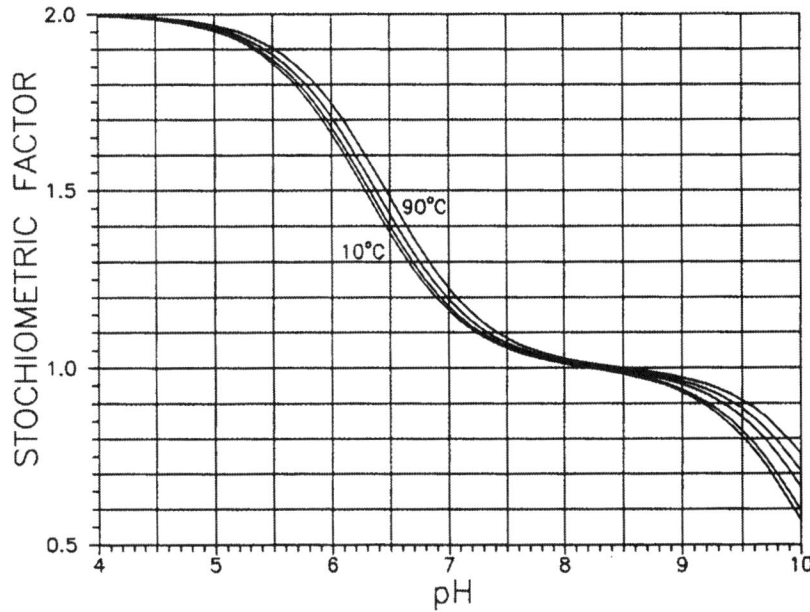

Figure 9.22 Stoichiometric factor, f_s, as a function of pH and temperature.

Looking at Figure 9.22 it is also obvious that a low pH shifts the equilibrium of the carbonate species towards undissociated carbonic acid. Therefore, the amount of carbonate ions in the liquid phase is very small. Since the solubility product has to remain constant, a higher solubility of calcium and magnesium carbonate is expected.

REACTION BETWEEN LIMESTONE AND SULFUR DIOXIDE. In the liquid phase, dissolved calcium and sulfite ions are reacted to form calcium sulfite. Simultaneously, calcium ions can also react with bisulfite ions to form $Ca(HSO_3)_2$.

$$Ca^{2+} + SO_3^{2-} \leftrightarrow CaSO_3(aq)$$

and

$$Ca^{2+} + HSO_3^- \leftrightarrow Ca(HSO_3)_2$$

The bisulfite and the sulfite ions might be oxidized, resulting in the presence of sulfate ions.

$$SO_3^{2-} + \tfrac{1}{2} O_2 \leftrightarrow SO_4^{2-}$$

and

$$HSO_3^- + \tfrac{1}{2} O_2 + H_2O \leftrightarrow SO_4^{2-} + H_3O^+$$

These sulfate ions can react with dissolved calcium ions to form aqueous calcium sulfate.

$$Ca^{2+} + SO_4^{2-} \leftrightarrow CaSO_4(aq)$$

Finally, the aqueous calcium sulfite and sulfate are precipitated in a reaction involving water molecules.

$$CaSO_3(aq) + \tfrac{1}{2} H_2O \leftrightarrow CaSO_3 \cdot \tfrac{1}{2} H_2O(s)$$

and

$$CaSO_4(aq) + 2 H_2O \leftrightarrow CaSO_4 \cdot 2 H_2O(s)$$

As shown by the preceding equations, the pH in the liquid phase surrounding a limestone particle is not constant. The pH is maintained at a high value directly adjacent to the limestone surface because the equilibrium reactions involved in the dissolution of limestone are decreasing the hydronium ion concentration. With increasing distance from the limestone surface, the pH value decreases because the solution of SO_2 in water increases the hydronium ion concentration.

THEORETICAL APPROACH TO ESTIMATE LIMESTONE SOLUBILITIES. Limestone solubility in an aqueous solution can be estimated if most of the ionic species present are known qualitatively. From this knowledge, the equilibrium reactions in which those ionic species interact are also usually understood. The equilibrium constants can be obtained from

Equation 9.62

$$K_{eq.} = e^{-\frac{\Delta G°}{R \cdot T}}$$

where R is the ideal gas constant and $\Delta G°$ is the free energy change of the reaction when all of its reactants and products are in their standard states. The equilibrium constant varies with temperature as described by,

Equation 9.63

$$\ln K_{eq.} = -\frac{\Delta H°}{R} \cdot \frac{1}{T} + \frac{\Delta S°}{R}$$

if the reaction enthalpy $\Delta H°$ and the reaction entropy $\Delta S°$ are reasonably independent of temperature.

For the equilibrium reactions, the law of mass action provides the mathematical equations for subsequent solubility calculations. In regard to the dissolution of limestone, the equations necessary to estimate limestone solubilities were already mentioned in Equations 9.54 and 9.55.

Also, the activity coefficient of a chemical species I in solution is defined as:

Equation 9.64

$$\gamma_i = \frac{a_i}{c_i}$$

Reviewing Equation 9.54 and writing Equations 9.41 and 9. through 9.45 in terms of their activity coefficients leads to

Equation 9.65

$$K_{CaCO_3} = \frac{\gamma_{Ca^{2+}} \cdot c_{Ca^{2+}} \cdot \gamma_{CO_3^{2-}} \cdot c_{CO_3^{2-}}}{a_{CaCO_3(s)}}$$

Equation 9.66

$$K_{CO_2} = \frac{\gamma_{H_2CO_3(aq)} \cdot c_{H_2CO_3(aq)}}{\frac{p_{CO_2}}{H_{CO_2}}}$$

Dry Scrubbing Technologies for Flue Gas Desulfurization

Equation 9.67

$$K_{H_2CO_3} = \frac{\gamma_{HCO_3^-} \cdot c_{HCO_3^-} \cdot \gamma_{H_3O^+} \cdot c_{H_3O^+}}{\gamma_{H_2CO_3(aq)} \cdot c_{H_2CO_3(aq)} \cdot \gamma_{H_2O} \cdot c_{H_2O}}$$

Equation 9.68

$$K_{HCO_3^-} = \frac{\gamma_{CO_3^{2-}} \cdot c_{CO_3^{2-}} \cdot \gamma_{H_3O^+} \cdot c_{H_3O^+}}{\gamma_{HCO_3^-} \cdot c_{HCO_3^-} \cdot \gamma_{H_2O} \cdot c_{H_2O}}$$

and

Equation 9.69

$$K_{H_2O} = \frac{\gamma_{H_3O^+} \cdot c_{H_3O^+} \cdot \gamma_{OH^-} \cdot c_{OH^-}}{(\gamma_{H_2O} \cdot c_{H_2O})^2}$$

The activity coefficients in those equations can be determined by computing the ionic strength of the aqueous solution according to

Equation 9.70

$$I = \frac{1}{2} \cdot \sum c_i \cdot z_i^2$$

where c_i represents the concentration of each species in solution and z_i represents its respective charge. The activity of water can be approximated by the relation

Equation 9.71

$$a_{H_2O} = \gamma_{H_2O} \cdot c_{H_2O} = 1 - 0.017 \cdot \sum c_i$$

where c_i stands for the concentration of each dissolved cationic, anionic or neutral species.

The ionic strength is inserted in the Debye-Hueckel expression to determine the activity coefficients of each chemical species. The Debye-Hueckel theory considers the long-range electrostatic forces of charged ions on each other, resulting in lower activity coefficients. This expression is valid for dilute solutions only, which is the case for the experimental work performed.

Equation 9.72

$$\gamma_i = \exp\left(-\frac{A \cdot z_i^2 \cdot \sqrt{I}}{1 + å_i \cdot B \cdot \sqrt{I}}\right)$$

In this case, A and B are characteristic constants dependent only on the dielectric constant, the density, and the temperature of water. $å_i$ is a value dependent upon the 'effective diameter' of the ion in solution, while z refers to the ionic charge.

The calculation of limestone solubility is an iterative method. When the activity coefficients of the chemical species have been calculated, the concentrations of each ion can be found from Equations 9.54 and 9.66 through 9.69. These concentrations have to be inserted into the equation for the ionic strength [Equation 9.70] and subsequently into the Debye-Hueckel equation to get updated activity coefficients. With these new coefficients, the calculations have to be repeatedly performed, until the values of the coefficients and the various ion concentrations converge. A BASIC program for this iterative method of estimating limestone solubilities has been written and is listed in Reference 39.

Experimental

EQUIPMENT. The experimental setup used for the determination of solubilities and dissolution rates is shown in Figure 9.23. The central piece of equipment used during these experiments was the jacketed reactor which contained a liquid volume of up to 1.5 dm^3. The temperature of the liquid contents in the reactor was varied between 30 and 90°C by means of a constant temperature water bath which provided the water flow for the outer mantle of the reactor.

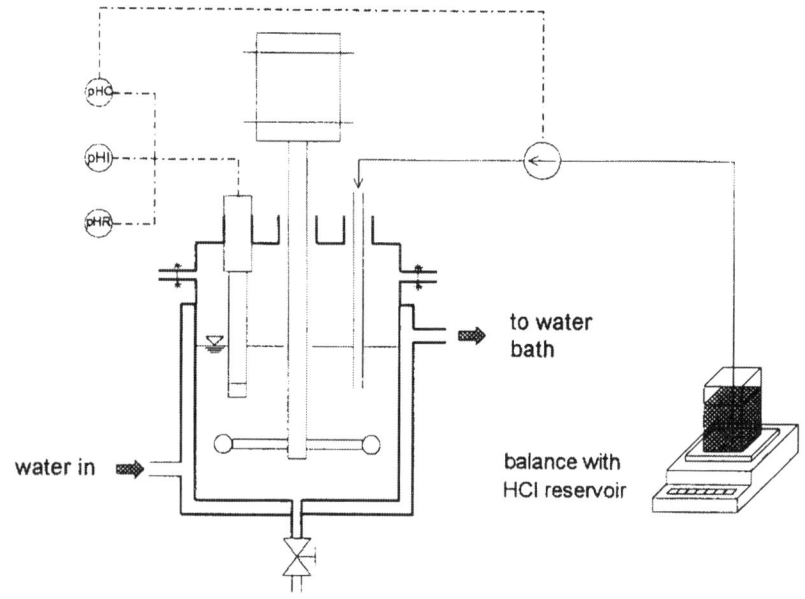

Figure 9.23 Experimental setup.

The basket stirrer, a pH electrode, and a glass tube for the addition of HCl were inserted through the reactor cover. The pH electrode was connected to a pH indicator, recorder, and controller. Depending on the desired pH in solution, the pH controller activated the peristaltic pump which in turn delivered diluted hydrochloric acid through a glass tube into the reactor. The amount of HCl titrated per unit time was determined by weight loss from a reservoir of HCl placed on top of a balance.

The dimensions of the basket stirrer, employed in the free-drift and pH-stat experiments are shown in Figure 9.24. The basket stirrer consisted of two threaded pipe nipples with end caps, which were in horizontal position and opposite to each other. These pipe nipples were welded with short connection rods to the vertical axis of the stirrer. The opening of each end cap was covered by a fine mesh screen to keep the limestone particles inside the pipe nipples.

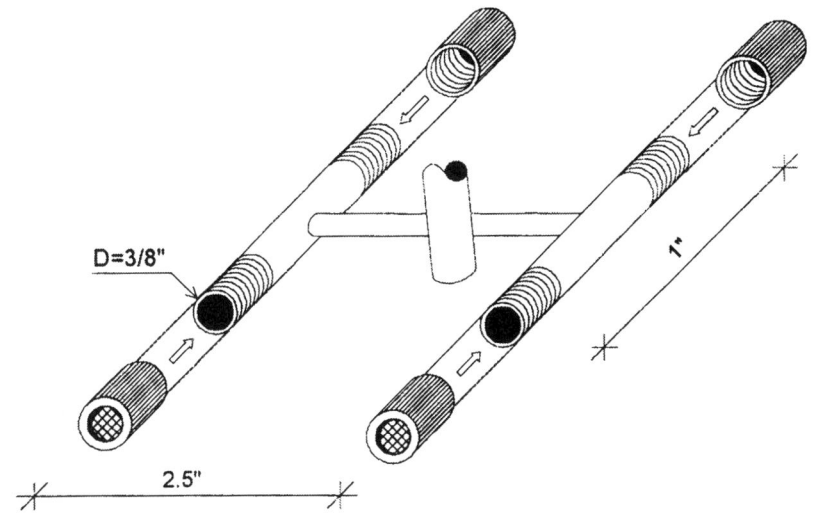

Figure 9.24 Basket stirrer.

LIMESTONES. Five different limestones, including four local Ohio limestones, were used to obtain kinetic data for this study. They are (1) Vanport limestone, (2) Maxville limestone, (3) Mississippi limestone, (4) Bucyrus limestone, and (5) Carey limestone.

Vanport limestone was obtained from the Waterloo Coal Company, Jackson, Ohio. Maxville limestone was collected from the Maxville Quarries Inc., Logan, Ohio, while the Mississippi limestone was obtained from the Mississippi Lime Company, Alton, Illinois. Samples of Carey and Bucyrus limestones were ordered from the quarries of the National Lime & Stone Company at Carey and Bucyrus, Ohio, respectively. Limestones in the form of powder were used in the experiments for the determination of solubilities. Experiments to obtain dissolution rates were performed with particle sizes between 1.18 mm (16 mesh) and 1.00 mm (18 mesh).

Preliminary elemental analyses of some of these limestones provide information about the amount of calcium and magnesium present in representative samples. These analyses were performed according to ASTM, Part 05.05, Method D4326-84. The content of calcium and magnesium was reported according to the amount of oxide present after ignition (Table 9.3).

Table 9.3 Chemical Analysis of Limestones

Source of limestone	Mass-% of CaO	Mass-% of MgO	Ca/Mg mole ratio
Vanport limestone	72.62	0.87	60.0
Maxville limestone	61.50	9.65	4.6
Carey limestone	56.44	39.93	1.0

Additional chemical analyses of the Ca/Mg content were available from the distributors of the limestones. Table 9.4 shows the results of these analyses and compares the Ca/Mg mole ratios. Since limestone composition is known to vary within each sample and within every delivery, an analysis can only be considered as an estimate for the given limestone source.

Table 9.4 Represenstative Analyses from Manufacturer

Source of limestone	Mass-% of $CaCO_3$	Mass-% of $MgCO_3$	Ca/Mg mole ratio
Carey limestone	54.5	45.0	0.73
Bucyrus limestone	80.0	17.0	2.85
Mississippi limestone	97.8 to 98.9	0.4 to 0.95	62.5 to 149.7

However, the close agreement of the data for Carey limestone in Tables 9.3 and 9.4 indicates that this estimate of the composition is typical for this particular limestone.

Another feature of the designated limestones is the amount of magnesium available in a soluble form. No analysis at Ohio University has been performed, thus far, to determine either the amount of soluble magnesium in solid solution or the amount of magnesium in relatively insoluble dolomite.

The five limestones identified were chosen because they represent two important stable equilibria among calcium and magnesium carbonates. The stable phases are referred to as calcite and dolomite, according to their Ca^{2+} and Mg^{2+} content in the mineral. Figure 9.25 distinctively shows these phases represented in a double-logarithmic diagram, where the partial pressure of CO_2 in the air is the horizontal axis and the Ca^{2+}/Mg^{2+} concentration in the limestone is the vertical axis.

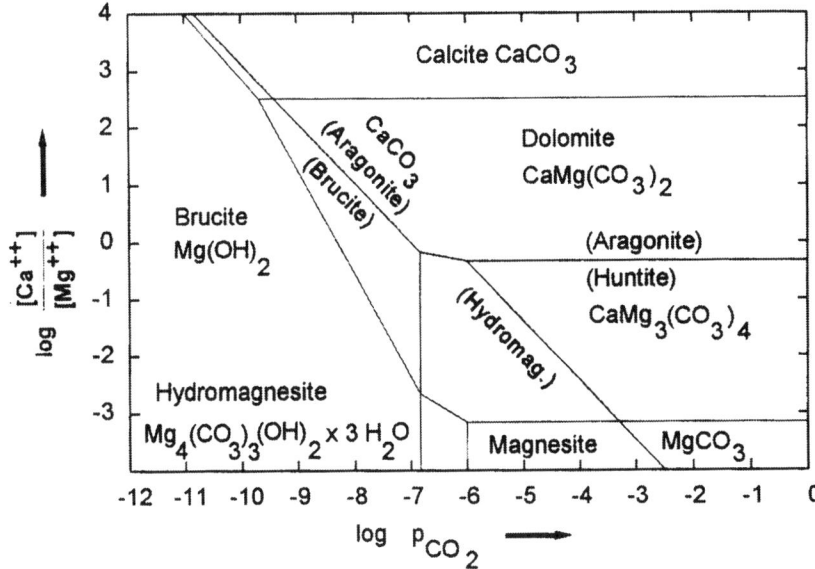

Figure 9.25 Equilibria among calcium and magnesium carbonates [34].

It should be mentioned that limestones occur naturally not only in phases but in polymorphic forms. Three forms of $CaCO_3$ are known. They are calcite, aragonite, and vaterite. For a given temperature and pressure, the prevailing polymorphic form is that which is the most thermodynamically stable. In addition to these polymorphic forms there are two hydrates of calcite, Trihydrocalcite, $CaCO_3 \cdot H_2O$, and Ikaite, $CaCO_3 \cdot 6 H_2O$, which occur naturally.

PROCEDURE. Five different limestones, Maxville, Carey, Bucyrus, Vanport and Mississippi limestone were employed in experiments to determine their solubility and to obtain their dissolution rates by the free-drift and the pH-stat method. In all experiments, distilled water was used as the solvent. The effects of three potential additives, glycine, citric acid, and sodium sulfite, on solubilities and dissolution rates were investigated. The additives were mixed with the solvent prior to an experiment.

For solubility and dissolution rate experiments, two different limestone particle sizes were prepared. The solubility experiments were performed with a powder of each limestone having particle diameters smaller than 40 mesh (< 0.01 mm). Dissolution rate experiments were conducted with particle diameters between 1.18 (16 mesh) and 1.00 mm (18 mesh).

The pH meter used for all experiments was calibrated prior to each experiment. Selective buffers at pH values of 4, 7 and 10 were chosen to cover the pH range of interest. With the calibration of the pH meter, the chart recorder used in the free-drift method was also adjusted.

INTRINSIC SOLUBILITY EXPERIMENTS. The reactor used for the determination of the limestone solubilities was filled with 1.25 liters of distilled water and was surrounded by an outer glass jacket. Water was circulated through the glass jacket in order to control the reactor temperature. Thereby, the temperature at which the solubility of a particular limestone was measured could be set using the temperature controller for the water bath.

A plastic stirrer was mounted in the reactor through one of the four openings at the reactor lid. The pH and temperature electrodes were plugged into the top of the reactor and occupied two additional openings. The fourth opening allowed the exchange of carbon dioxide with the atmosphere. A thin polyethylene tube for the addition of hydrochloric acid or sodium hydroxide was inserted through the fourth opening and just penetrated the solvent surface in the reactor. The acid or base in the tube was transported by a peristaltic pump which in turn was operated by a pH controller connected with the pH meter. The solubility was measured at different pH values of the solvent. The pH value remained constant during an experiment and was set at the pH controller.

The procedure below was followed for each experiment. Upon reaching the desired temperature in the reactor, 0.625 g of limestone powder was added to the reactor contents. The pH value of the solution was adjusted using the pH controller and the system was operated for three hours. After this time, a sample was removed and filtered through a 0.1 micron filter. Subsequently, the sample was analyzed with EDTA (1,2-diaminoethanetetra-acetic acid) and EGTA ({Ethylenebis-[oxyethylenenitrilo]}tetra-acetic acid) to determine the concentration of calcium and magnesium ions present. These quantitative, titrimetric analytical techniques were documented by Jeffery et al. [38].

These solubility determinations were carried out using a matrix of 15 experiments for each limestone. The temperature was varied between 30 and 90°C in steps of 15°C, while the pH was changed between pH 4 and pH 8 in steps of one unit. Additional experiments were performed with the additives glycine, citric acid, and sodium sulfite. The additive concentrations were 5, 10, 15, and 25 mmol/l of the substance for a specified pH and temperature.

DISSOLUTION RATE EXPERIMENTS: FREE-DRIFT METHOD. The free-drift method made use of the same reactor already applied to solubility type experiments, although instead of the plastic stirrer, a spinning basket stirrer was used. The two opposite baskets of the stirrer were filled with 2.25 g of limestone particles. 1.25 liters of distilled water were poured into the reactor. During these experiments, the pH was allowed to drift from an initially low value of approximately pH 4, to its equilibrium value. Each experimental trial lasted between 6 and 12 hours. The pH of the liquid phase was continuously monitored and recorded using a chart recorder which was connected to the pH meter.

Before an experiment was initiated, the pH value of the liquid phase was adjusted to a value of close to pH 4 by the addition of diluted hydrochloric acid. Upon starting an experiment at a desired temperature, the spinning basket stirrer was lowered into the liquid phase. The revolution speed of the stirrer was set to a predetermined value. At the completion of the experiment, the recorded pH versus time curve was evaluated and the dissolution rates determined by the use of a BASIC program [39].

Free-drift experiments were conducted without additives at 30, 60, and 90°C for each of the five limestones. The influence of the three additives was studied with Maxville limestone at 60°C. The additive concentrations used were 5, 10, 15, and 25 mmol/l.

DISSOLUTION RATE EXPERIMENTS: PH-STAT METHOD. Preparations for the pH-stat method were quite similar to those for the free-drift method, since the same spinning basket stirrer and the same particle size of limestone was used. The reactor was filled again with 1.25 liters of distilled water. In contrast to the free-drift method, the pH was maintained at a constant value during these experiments, by means of the pH controller.

A beaker filled with hydrochloric acid of known concentration was put on a balance and the pH controller was set to the desired pH and activated. Simultaneously, the spinning basket reactor was moved into the liquid contents of the reactor and stirrer speed was adjusted. From the amount of hydrochloric acid titrated per time, the dissolution rate was determined according to Equation 9.73. The symbol S refers to the slope of the titrated acid versus time curve which for all experiments could be approximated by a linear function.

Equation 9.73

$$r = \frac{1}{2} \cdot \frac{S \cdot c}{A \cdot \rho}$$

In Equation 9.73, c denotes the concentration, r represents the density of titrated hydrochloric acid, and A stands for the total area of the limestone surface in the experiments. The factor $1/2$ results from the assumption for the pH-stat method that two hydronium ions are required to dissolve a single ion of calcium or magnesium.

Experiments were carried out with pure Maxville limestone at 60°C and pH 5. Additional runs were performed at the same conditions using the three additives at various concentrations.

SOFTWARE. The free-drift method produced experimental data in the form of pH versus time curves which were plotted by a chart recorder. These graphical data had to be converted into dissolution rates of the limestones used. This conversion was achieved by the development of a BASIC program which is included in Reference 39. In the program, approximately 40-50 pH values versus time data points gained from the chart recorder serve as input values for each free-drift experiment. Additional input information like reactor volume and temperature or the presence and amount of an additive are required.

In the absence of an additive, limestone dissolution rates were calculated by dividing the hydronium ion change between two data points by their time difference and by the limestone surface area. The limestone surface area for the amount of limestone used was obtained according to the equation

Equation 9.74

$$A = \frac{6}{d_p} \cdot \frac{m}{\rho}$$

in which d_p is the average particle diameter, m is the mass of limestone used, and ρ is the bulk density of limestone. The calculated dissolution rates were finally divided by a calculated stoichiometric factor which accounted for the number of hydronium ions reacted per dissolved calcium ion. In free-drift experiments including additives, an additional factor had to be determined. This factor was the ratio of the number of hydronium ions reacted in the presence of the additive to the number of hydronium ions reacted in the absence of the additive.

Results and Discussion

INTRINSIC SOLUBILITY EXPERIMENTS. The solubilities of the calcium carbonate and magnesium carbonate, present in all five limestones examined, can be found in Reference 39. Typical results are given in Figure 9.26, From the combined results, a definite dependency of calcium and magnesium carbonate solubility on the solution pH can be observed. The calcium and magnesium ion concentrations increase with decreasing pH. This is in agreement with the chemical equilibria which describe the pH behavior of the carbonate/carbonic acid chain. Decreasing the solution pH to a low value and maintaining the low value results, according to Le Chatelier's principle, in a new equilibrium between the chemical compounds in the carbonic acid chain. At the new equilibrium, a high concentration of undissociated carbonic acid and a low concentration of carbonate ions will be present. As indicated by Equations 9.52 and 9.53 this leads to the conclusion that the concentrations of calcium and magnesium ions necessarily have to be increased in order to maintain the same solubility product in solution.

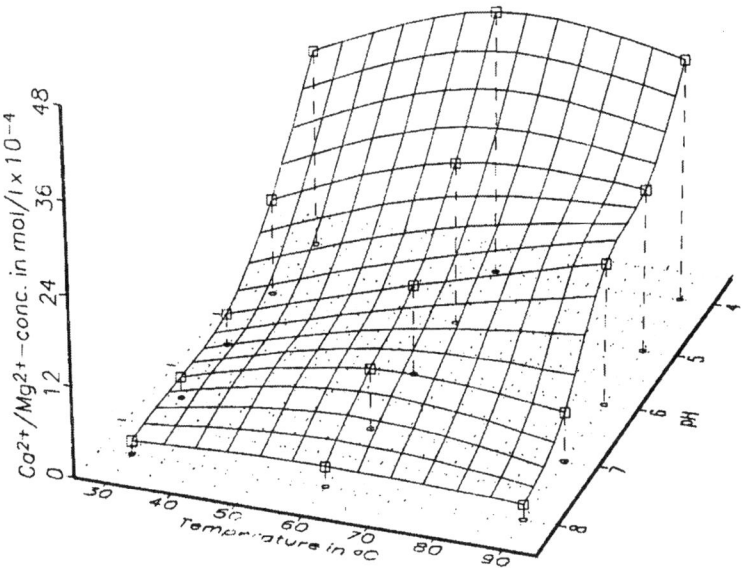

Figure 9.26 Experimental $Ca^{2+}+Mg^{2+}$ solubililty for Carey limestone.

The effect of the temperature on limestone solubility is far less pronounced than the pH effect. At a pH of 8, there seems to be a slight decrease in calcium and magnesium carbonate solubility with increasing temperature, but this is not consistent with the temperature effect at low pH values. The inconsistencies can be explained by the experimental error associated with solubility determinations. The standard deviation for these water chemistry determination was $\pm 2 \cdot 10^{-4}$ mol/l.

The influence of the limestone type on the solubility of calcium and magnesium ions is theoretically related to the amount of calcium and magnesium originally present in the mineral and to the amount of impurities in the limestone. Impurities in limestone lower the activity coefficients of calcium and magnesium ions in solution, thereby leading to an increase in their concentration.

To compare some of the results of this study with those reported by previous investigators, Table 9.5 contains solubilities. Weast [40] reported the solubility of natural calcite and aragonite in cold and hot water. It is assumed that these values are measured at the equilibrium pH, which should be around 8. Weast reported the solubility of natural calcite in cold water to be $1.39 \cdot 10^{-4}$ mol/l and in hot water to be $1.79 \cdot 10^{-4}$ mol/l. The solubility of natural aragonite was listed as $1.52 \cdot 10^{-4}$ mol/l in cold water and $1.89 \cdot 10^{-4}$ mol/l in hot water.

Table 9.5 Solubilities of Limestones Reported by Different Investigators

Investigator	Limestone	Experimental condition	Solubility in mol/l
Weast [40]	nat. Calcite	cold water	$1.39 \cdot 10^{-4}$
		hot water	$1.79 \cdot 10^{-4}$
	nat. Aragonite	cold water	$1.52 \cdot 10^{-4}$
		hot water	$1.89 \cdot 10^{-4}$
Baeckstroem [41]	Calcite	9°C	$129.8 \cdot 10^{-4}$
		35°C	$76.4 \cdot 10^{-4}$
	Aragonite	9°C	$145.8 \cdot 10^{-4}$
		35°C	$87.5 \cdot 10^{-4}$
This study	Mississippi	pH = 8, 30°C	$9.27 \cdot 10^{-4}$
		pH = 8, 90°C	$2.97 \cdot 10^{-4}$

To show the large difference in reported values from previous investigators, additional data can be obtained from Baeckstroem [41], who determined the solubility of calcite at 9°C to be $129.8 \cdot 10^{-4}$ mol/l and at 35°C to be $76.43 \cdot 10^{-4}$

mol/l. The solubility of aragonite was reported at the same temperatures of 9°C and 35°C to be $145.8 \cdot 10^{-4}$ mol/l and $87.5 \cdot 10^{-4}$ mol/l, respectively.

However, the high amount of calcium-containing Mississippi limestone tested in this study at a pH of 8, has a combined Ca^{2+}/Mg^{2+} solubility of $9.27 \cdot 10^{-4}$ mol/l at 30°C and $2.97 \cdot 10^{-4}$ mol/l at 90°C, which is in between the values reported by Weast [40] and Baeckstroem [41].

Boynton [42] lists the equilibrium pH of limestones to be in a range from 8 to 9, while dolomite has a value about 9 to 9.2. It was found in this study that Maxville limestone has its equilibrium value near 8.0, while the other four limestones, including the dolomitic limestones Carey and Bucyrus, reveal a lower equilibrium pH value. This value is near 7.5 and is therefore in disagreement with the pH value suggested by Boynton.

DISSOLUTION RATE EXPERIMENTS: FREE-DRIFT METHOD. A repeatability test for the free-drift method was performed at a stirring rate of 750 rpm. In this test Maxville limestone was dissolved using the spinning basket reactor at 90°C.

The experimental error for the free-drift method was calculated at three pH values. At pH 4.5, 5.0, and 5.5 the experimental error was determined to be $\pm 0.4 \cdot 10^{-5}$, $\pm 0.9 \times 10^{-6}$, and $\pm 0.19 \cdot 10^{-6}$ mmol•cm^{-2}•s^{-1}, respectively.

To estimate the stirring rate at which the dissolution rate is no longer liquid-film mass-transfer controlled, experiments were performed at varying stirrer speeds. The stirring rates ranged from 175 to 1400 rpm, while the solvent temperature was maintained at 90°C. A stirrer speed of 1000 rpm ensured that the dissolution of Maxville limestone was independent of stirrer speed. Figure 9.27 shows the combined results of these experiments.

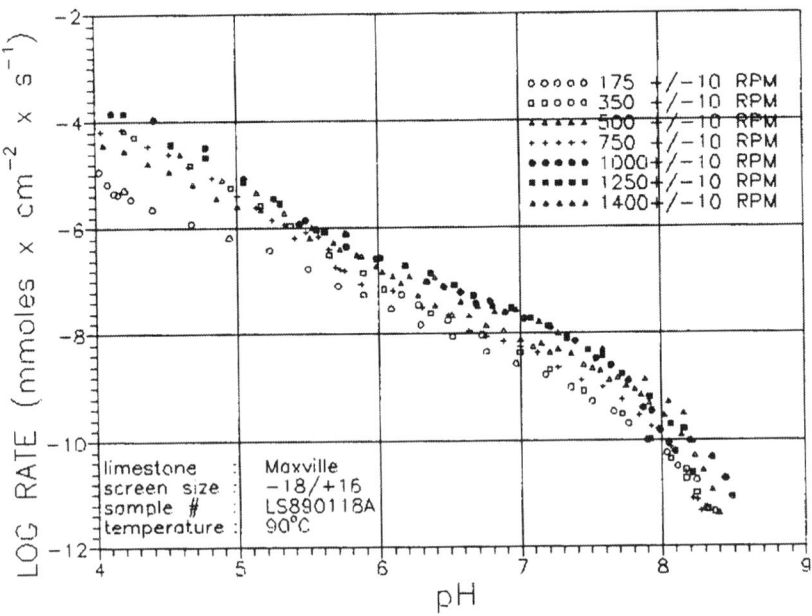

Figure 9.27 Effect of stirrer speed on observed dissolution rate.

The results of free-drift experiments for all limestones studied can be found in Reference 39. Figure 9.28 shows the dissolution rate of Carey limestone at 30, 60, and 90°C. Figure 9.29 gives the dissolution rates of all five limestones at 60°C. For all limestones examined in this study, a general dissolution rate behavior can be observed. The logarithm of the dissolution rate decreases in a linear fashion with an increasing pH value of the solution. This is equivalent to a diagram which uses linear axes and in which the dissolution rate would decrease exponentially with increasing pH.

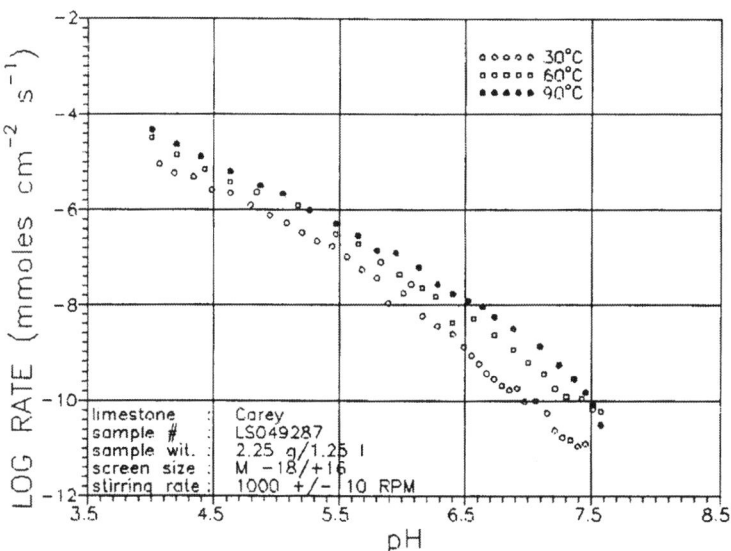

Figure 9.28 Free-drift method. Temperature dependency of Carey limestone dissoution rates.

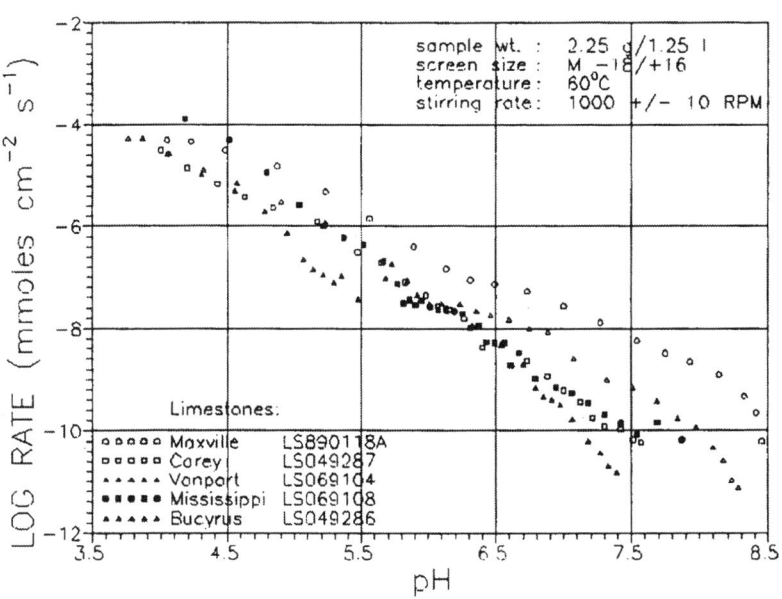

Figure 9.29 Free-drift method. Limestone dissolution rates at 60°C.

From the five limestones used in the free-drift method the two dolomitic limestones, Carey and Bucyrus, exhibit especially low dissolution rates at 30, 60, and 90°C, as is shown in Table 9.6.

Table 9.6 Dissolution Rates Evaluated at pH=5.5, Using the Free-Drift Method

Temperature	Limestone	Dissolution rate 10^{-6} mmol/(cm^2·s)
30°C	Maxville	0.10±0.19
	Carey	0.16±0.19
	Vanport	0.50±0.19
	Mississippi	5.01±0.19
	Bucyrus	0.18±0.19
60°C	Maxville	2.51±0.19
	Carey	0.32±0.19
	Vanport	0.50±0.19
	Mississippi	0.40±0.19
	Bucyrus	0.10±0.19
90°C	Maxville	1.41±0.19
	Carey	0.56±0.19
	Vanport	2.00±0.19
	Mississippi	2.51±0.19
	Bucyrus	1.00±0.19

With increasing temperature the spread between the dissolution rates of the five limestones narrows. Considering a single limestone, a significant temperature effect can be seen only for a temperature difference of at least 60°C.

Dissolution rates of limestones have been studied by many investigators at varying experimental conditions. Plummer et al. [26] used the free-drift method to study the dissolution rate of calcite at 25°C for a pH near equilibrium, between 5.5 and 7. At a pH of 6.5 and only a little CO_2 sparging, they found the logarithm of the dissolution rate to be about -6.6 (the non-logarithmic value was 2.51 · 10^{-7} mmol per cm^2 per sec). At similar experimental conditions in the present study, Mississippi limestone at 30°C and a pH of 6.5 was found to have a logarithmic dissolution rate value of -7.1 (the non-logarithmic value was $0.79 \cdot 10^{-7}$ mmol per cm^2 per sec). Compared to the non-logarithmic value reported by Plummer et al. the dissolution rate obtained in the present study agrees reasonably well. The slight difference between those values may be caused by differences in the concentrations of limestone impurities.

DISSOLUTION RATE EXPERIMENTS: PH-STAT METHOD. From each pH-stat experiment, the experimental data obtained were plotted as the amount of HCl titrated versus time. An example is shown in Figure 9.30 where the titration curves are basically linear. The slopes of the linear curves have been determined by fitting a straight line through the experimental data. From these slopes, the dissolution rates were calculated according to Equation 9.73.

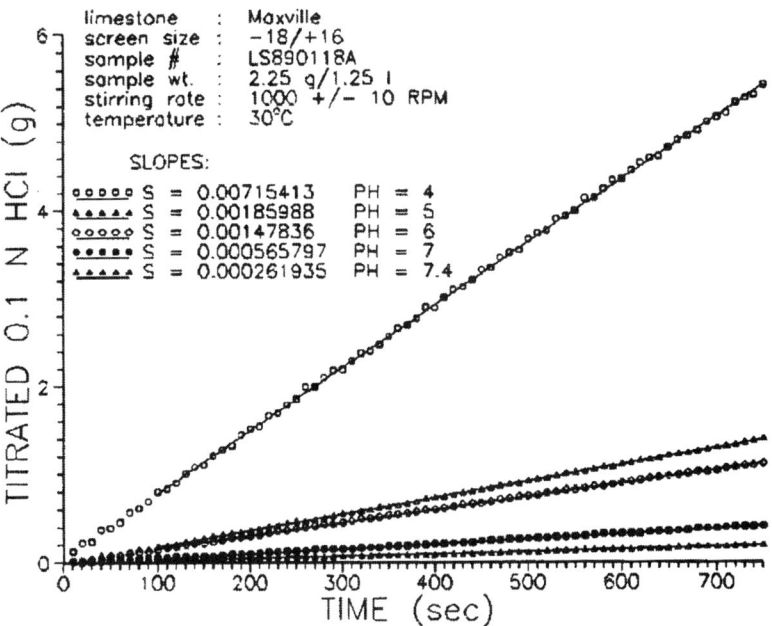

Figure 9.30 Amount of HCl titrated as a function of time.

The experimental error for the pH-stat method was obtained at three different pH values. At pH 4.0, 5.0, and 6.0, the experimental error was calculated to be $\pm 0.27 \cdot 10^{-5}$, $\pm 0.15 \cdot 10^{-5}$ and $\pm 0.03 \cdot 10^{-5}$ mmol·cm-2·s^{-1}, respectively.

The pH-stat method was utilized to find the dissolution rates of Maxville, Carey, Vanport, Mississippi, and Bucyrus limestones. As the name of the method implies, the pH was maintained constant at values of 4, 5, 6, 7. At pH values greater than 7, the equilibrium pH values, which are different for the five limestones, had to be considered. In most of those cases, a pH of close to 7.4 was maintained during the experiments.

Figure 9.31 gives representative dissolution rates for the five limestones tested at 60°C. The effect of temperature on limestone dissolution rate is found to be rather small, while the limestone type is the dominating factor. For example, the dissolution rates of Carey and Bucyrus limestone are lower than the dissolution rates of the other limestones at nearly all pH values investigated.

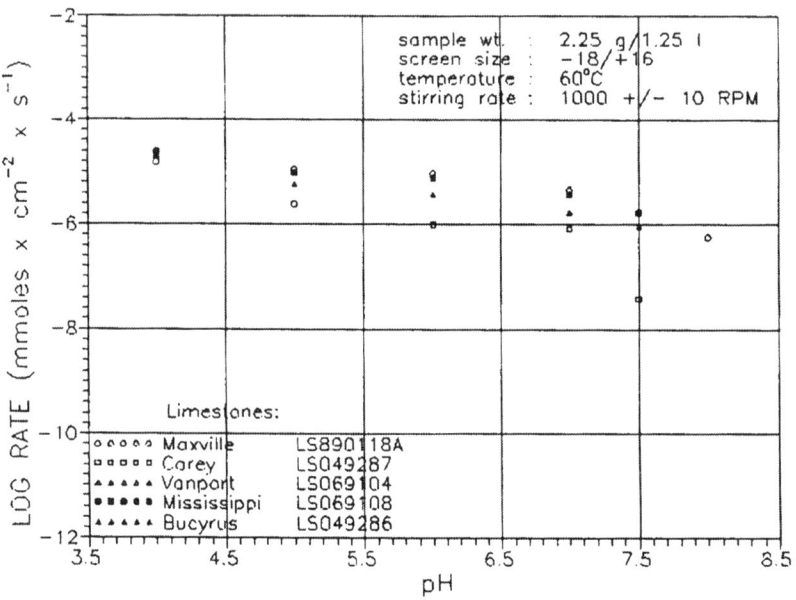

Figure 9.31 pH-stat method. Limestone dissolution rate at 60°.

Chan and Rochelle [16] measured the dissolution of reagent grade $CaCO_3$ with N_2 sparging in a solution containing 0.1 M $CaCl_2$. Using the pH-stat method they obtained dissolution rate constants at temperatures of 25 and 55°C. These curves have a similar shape to those in the present study. Nevertheless a direct comparison is not possible because these rate constants need to be related to the actual dissolution rate via the dissolved limestone concentration.

The study of Plummer et al. [26] included experiments using the pH-stat and free-drift methods. The free-drift method was applied near the equilibrium pH while the pH-stat method was used at pH values far from equilibrium. At the pH where both methods overlap, the slopes of the rate of dissolution curves do not match each other. In the present study this problem could not be observed. However, dissolution rates at the high pH values gained by the pH-stat method are higher than the rates gathered by the free-drift method (Table 7). This is due to the fact that the free-drift method was initiated at a pH of 4. When the pH finally reached a value around 7.5, the calcium ion concentration in solution led to a decrease in limestone dissolution rate.

Table 9.7 Dissolution Rates Obtained for Maxville Limestone at 60°C without Additives

	pH = 4	pH = 6	pH = 8
Free-drift method	$5.6 \cdot 10^{-5}$	$2.8 \cdot 10^{-7}$	$1.6 \cdot 10^{-9}$
pH-stat method	$2.5 \cdot 10^{-5}$	$1.0 \cdot 10^{-5}$	$6.3 \cdot 10^{-7}$

Conclusions and Recommendations: Solubilities and Rates of Dissolution

The solubility values acquired from the experiments indicate a relationship between the molar percentages of $CaCO_3$ and $MgCO_3$ in the parent limestone and the Ca^{2+} and Mg^{2+} ion concentrations in solution at equilibrium. A high molar percentage of $CaCO_3$ in the parent limestone yields a high solubility of Ca^{2+} ions in solution at equilibrium. The same statement is valid for magnesium. The amount of impurity in the parent limestone was found not to be related to the combined Ca^{2+} and Mg^{2+} solubilities in distilled water.

The effect of temperature on limestone solubility in the range of 30–90°C was identified to be rather small in comparison to the pH effect between pH 4 and pH 8. On the average, a temperature increase from 30°C to 90°C resulted in a 50% increase in limestone solubility while a pH decrease from 8 to 4 yielded an enhancement in solubility of 250%.

Previous investigators of limestone solubility have often reported values without reporting the pH value at which the solubilities were measured. In those cases it must be assumed that the pH at which the solubility was measured was the equilibrium pH of the solvent exposed to the atmosphere. Such a pH would not be representative of values found in a wet FGD system. This is because the pH in a wet scrubber is generally maintained at a value of about 5, while the equilibrium pH of limestone dissolved in distilled water is about 8.

Limestone solubility is not likely to play a major role in wet FGD because the amount of ions represented by the solubility value would be consumed rapidly in a scrubbing operation. More likely, the dissolution rate would be the rate-limiting factor for such processes.

The effect of temperature on limestone dissolution rate was investigated using the free-drift and pH-stat methods. Both methods showed that a difference in temperature from 30–90°C increases limestone dissolution rates by about a factor of 10. The effect of the limestone source on dissolution rate was found to be very important. The five limestones studied using the free-drift method yielded a variation in their dissolution rates of about 10 to 100 times at 30°C and about 10 to 50 times at 60° and 90°C, due to the effect of limestone type on the dissolution rate. The variations in dissolution rates observed by the pH-stat method showed differences of about 10 to 50 times only.

Dissolution rates obtained without additives using the pH-stat method were in agreement with those measured by the free-drift method at pH values less than four. For higher pH values, dissolution rates observed by the pH-stat method indicated higher dissolution rates than those obtained by the free-drift method at similar conditions. The reason for this difference in dissolution rates is the concentration of Ca^{2+} and Mg^{2+} ions in solution. This concentration keeps increasing throughout the free-drift experiment. This is a clear disadvantage of the free-drift method.

Further studies should be directed towards using specific ion electrodes for measuring limestone dissolution rates, instead of determining the calcium and magnesium ion concentration indirectly through the change in solution pH. Furthermore, it could be desirable to use a continuous sampling system, in combination with an ion chromatograph, to remove samples from the reactor in which the limestone dissolves.

Nomenclature

A	Bed cross-sectional area.
a_H	Specific area for heat transfer.
a^i	Activity of component i.
a^m	Specific area for mass transfer.
C_A	Concentration of SO_2 in the liquid phase.
C_B	Concentration of $CaCO_3$ in the liquid phase.
C_{Pair}	Heat capacity of air.
$C_{P_{H_2O}}$	Heat capacity of water.
C_{Ps}	Heat capacity of the solid.
C_i	Concentration of component i.
D_{Al}	Diffusivity of SO_2 in the liquid phase.
D_{Bl}	Diffusivity of $CaCO_3$ in the liquid phase.
D_{lime}	Diffusivity of lime in water.
d_p	Particle diameter.
f_s	Stoichiometric factor in Equation (62).
$\Delta G°$	Reaction free energy change.
H	Enthalpy in the gas phase.
H_i	Henry's law coefficient for component i.
$\Delta H°$	Reaction enthalpy change.
h_g	Air/water heat-transfer coefficient.
I	Ionic strength.
K_i	Equilibrium constant for component i.
k_{Ag}	Mass-transfer coefficient of SO_2 in the gas film.
k_{Bs}	Dissolution mass-transfer coefficient for $CaCO_3$.
k_{Wg}	Gas-side water/air mass-transfer coefficient.
L_i	Solubility product for component i.
m	Mass of limestone in Equation (74).
N_{CaCO_3}	Flux of $CaCO_3$.
N_{H_2O}	Flux of H_2O.

N_{SO_2} Flux of SO_2.
n   Adjustable parameter in Equation (1).
p_i Partial pressure of component i.
q   Heat flow.
R   Ideal gas law constant.
R_p Rate of lime dissolution.
S   Slope of curve in Equation (73).
$\Delta S°$ Reaction entropy change.
T_g Temperature of the gas phase.
T_l Temperature of the liquid/solid phase.
T_O Reference temperature.
t   Time.
v_g Gas-phase superficial velocity.
v_l Trickling water superficial velocity.
X   Moisture content.
X_e Equilibrium moisture content.
X_c Critical moisture content.
Y   Molal humidity.
z   Distance in direction of gas flow.
z_i Charge of component i.
δ Film thickness.
ε_g Bed voidage.
ε_{lT} Total liquid holdup.
γ_i Activity coefficient for component i.
λ Distance from the gas-liquid interface to the reaction plane.
λ_0 Latent heat of vaporization.
ρ_g Molal gas density.
ρ_l Density of the liquid phase.
ρ_s Density of the solid phase.
ψ_i Mole fraction of component i.

References

1. Statnick, R. M., F. P. Burke, B. J. Koch, D. C. McCoy and H. Yoon, "Status of Flue Gas Sorbent Injection Technologies," *Proceedings: Fourth Annual Pittsburgh Coal Conference*, Pittsburgh, PA (1987)
2. EER (Energy and Environmental Research Corp., Irvine, CA), *Global Approach for Enhanced Mass Transfer Effects in In-Duct Flue Gas Desulfurization Processes - Volume 1: Literature Review,* U.S.D.O.E. Contract No. DE-AC22-88PC88873 (1988)
3. Prudich, M. E., K. W. Appell, M. J. Visneski, J. D. McKenna, D. A. Furlong, J. C. Mycock, J. F. Szalay and J. E. Wright, *Small Pilot Plant Demonstration of ETS' Limestone Emission Control System,* Final Report, OCDO Grant No. CDO/R-86-24 (1988)
4. Prudich, M. E., K. W. Appell and J. D. McKenna, *Pilot-Scale Limestone Emission Control (LEC) Process: A Development Project,* Final Report OCDO Grant No. CDO/D-88-49 (1994)
5. Klingspor, J., H. T. Karlsson and I. Bjerle, "A Kinetic Study of the Dry SO_2-Limestone Reaction at Low Temperature," *Chem. Eng. Comun.*, 22, 81-103 (1983)
6. Jorgensen, C., J. C. S. Chang and T. G. Brna, "Evaluation of Sorbents and Additives for Dry SO_2 Removal," *Environmental Progress*, 6(2), 26-32 (1987)
7. Grading, M. and G. Svedberg, "Modeling of Dry Injection Flue Gas Desulfurization," *JAPCA*, 38(10), 1275-1280 (1988)
8. Rice, R. W. and G. A. Bond, "Flue Gas Desulfurization by In-Duct Dry Scrubbing Using Calcium Hydroxide," *AIChE J.*, 36(3), 473-477 (1990)
9. Damle, A. S., *Modeling of SO_2 Removal in Spray-Dryer Flue-Gas Desulfurization System,* U.S.E.P.A. Report No. EPA/600/7-85/038 (1985)
10. Harriott, P. and M. Kinzey, "Modeling the Gas and Liquid Phase Resistances in the Dry Scrubbing Process for SO_2 Removal," *Proceedings: Third Annual Pittsburgh Coal Conference*, Pittsburgh, PA (1986)
11. Maibodi, M. M., T. L. Pearson, R. M. Counce and W. T. Davis, "Simulation of Spray Dryer Adsorber for Removal of SO_2 from Flue Gases," *Proceedings: Tenth Symposium on FGD*, 1, EPRI CS-5167 (1987)
12. Newton, G. H., J. Kramlich and R. Payne, "Modeling the SO_2-Slurry Droplet Reaction," *AIChE J.*, 36(12), 1865-1872 (1990)
13. Appell, K. W., *A Mathematical Simulation of ETS= Limestone Emission Control Process Using the Method of Characteristics: Fixed-Bed Configuration/Gas-Phase Mass Transport Control,* M. S. Thesis, Ohio University (1989)
14. Visneski, M. J., *Modeling of the Low Temperature Reaction of Sulfur Dioxide and Limestone Using a Three Resistance Film Theory Instantaneous Reaction Model,* Ph.D. Dissertation, Ohio University (1991)
15. Chang, C. S., J. H. Dempsey, R. H. Borgwardt, A. J. Toprac and G. T. Rochelle, "Effect of Limestone Type and Grind on SO_2 Scrubber Performance," *Environmental Progress*, 1(1), 59-62 (1982)
16. Chan, P. K. and G. T. Rochelle, "Limestone Dissolution," in *Flue Gas Desulfurization*," Amer. Chem. Soc. Symp. Series 188, Amer. Chem. Soc., Washington, DC (1982)
17. Gage, C. L. and G. T. Rochelle, "Modeling of SO_2 Removal in Slurry Scrubbing as a Function of Limestone Type and Grind," poster paper presented at EPA/EPRI 1990 SO_2 Control Symposium, New Orleans, LA (1990)

18. Jarvis, J. B., F. B. Meserole, T. J. Selm, G. T. Rochelle, C. L. Gage and R. E. Moser, "Development of a Predictive Model for Limestone Dissolution in Wet FGD Systems," paper 5B-4, *Proceedings: First Combined Flue Gas Desulfurization and Dry SO_2 Control Symposium*, 1, EPRI GS-6307-V1 (1989)
19. Peterson, M. N. A., "Calcite: Rates of Dissolution in a Verical Profile in the Central Pacific," *Science*, 154, 1542-1544 (1966)
20. Berger, W. H., "Foraminiferal Ooze, Solutions at Depths," *Science*, 156, 383-385 (1967)
21. Weyl, P. K., "The Solution Kinetics of Calcite", *J. Geol.*, 66, 163-176 (1958)
22. Plummer, L. N. and T. M. L. Wigley, "The Dissolution of Calcite in CO_2-Saturated Solutions at 25EC and 1 Atmosphere Total Pressure," *Geochim. Cosmochim. Acta*, 40, 191-202 (1976)
23. Morse, J. W. and R. A. Berner, "Chemistry of Calcium Carbonate in the Deep Oceans," *Amer. Chem. Soc. Symp. Series 93*, 500-535, Amer. Chem. Soc., Washington, DC (1979)
24. Berner, R. A. and J. W. Morse, "Dissolution Kinetics of Calcium Carbonate in Sea Water. Part IV. Theory of Calcite Dissolution," *Amer. J. Sci.*, 274, 108-134 (1974)
25. Morse, J. W., "Dissolution Kinetics of Calcium Carbonate in Sea Water. Part III. A New Method for the Study of Carbonate Reaction Kinetics," *Amer. J. Sci.*, 274, 97-107 (1974)
26. Plummer, L. N., T. M. L. Wigley and D. L. Parkhurst, "The Kinetics of Calcite Dissolution in CO_2-Water Systems at 5EC to 60EC and 0.0 to 1.0 Atm CO_2," *Amer. J. Sci.*, 278, 179-216 (1978)
27. Toprac, A. J. and G. T. Rochelle, "Limestone Dissolution in Stack Gas Desulfurization," *Environmental Progress*, 1(1), 52-58 (1982)
28. Gage, C. L., *Limestone Dissolution in Modeling Slurry Scrubbing for Flue Gas Desulfurization,* Ph.D. Dissertation, University of Texas, Austin (1989)
29. Erga, O. and S. G. Terjesen, "Kinetics of the Heterogeneous Reaction of Calcium Bicarbonate Formation, with Special Reference to Copper Ion Inhibition," *Acta Chem. Scand.*, 10, 872-875 (1956)
30. Sjoberg, E. L., "A Fundamental Equation for Calcite Dissolution Kinetics," *Geochim. Cosmochim. Acta*, 40, 441-447 (1976)
31. Walter, L. M. and J. S. Hanor, "Effect of Orthophosphate on the Dissolution of Biogenic Magnesian Calcites," *Geochim. Cosmochim. Acta*, (1978)
32. Ingle, S. E., C. H. Culberson, J. Hawley and R. M. Pytkowicz, "The Solubility of Calcite in Seawater at Atmospheric Pressure and 35 Parts per Thousand Salinity," *Mar. Chem.*, 1, 295-307 (1973)
33. Berner, R. A., "The Solubility of Calcite and Aragonite in Seawater at Atmospheric Pressure and 34.5 Parts per Thousand Salinity," *Amer. J. Sci.*, 276, 713-730 (1976)
34. Garrels, R. M. and C. L. Christ, *Equilibre des Mineraux et de Leurs Solution Aqueuses*, Gathier-Villars, Paris (1967)
35. Jarvis, J. B., D. R. Owens and D. A. Stewart, "Comparison of the Effectiveness of FGD Additives for SO_2 Removal Enhancement and Additive Consumption," *Proceedings: Tenth Symposium on Flue Gas Desulfurization*, Atlanta, GA (1986)
36. Jarvis, J. B., F. B. Meserole, D. R. Owens and E. S. Roothaan, "Factors Involved in the Selection of Limestone Reagents for Use in Wet FGD Systems," *Proceedings: 1991 EPRI/EPA/DOE SO_2 Control Symposium*, Washington, DC (1991)
37. Reddy, S. N., *A Mathematical Simulation of ETS= Limestone Emission Control (LEC) Process Using a Moving-Bed Configuration*, M. S. Thesis, Ohio University (1991)
38. Jeffery, G. H., J. Bassett, J. Mendham and R. C. Denney, *Textbook of Quantitative Chemical Analysis, 5th ed.* (1989)

39. Maldei, M., *Low Temperature Dry Scrubbing Reaction Kinetics and Mechanisms: Limestone Dissolution and Solubility*, M. S. Thesis, Ohio University (1993)
40. Weast, W. C., *Handbook of Chemistry and Physics, 50th ed.*, CRC Press (1969)
41. Baeckstroem, H., *Z. Phys. Chem.*, 97, 179 (1921)
42. Boynton, R. S., *Chemistry and Technology of Lime and Limestone, 2nd ed.*, John Wiley & Sons, NY (1980)

CHAPTER 10   SIMULATION AND OPTIMIZATION OF A GRANULAR LIMESTONE FLUE GAS DESULFURIZATION PROCESS

D.W. Duespohl, K.J. Sampson, S. Chattopadhyay, M.E. Prudich
Department of Chemical Engineering
Ohio University
Athens, OH

Abstract

This chapter presents a process simulation model developed for the Limestone Emission Control process. The process involves contacting flue gas with a densely packed, wet bed of granular limestone. The process model includes the chemistry, mass transfer, and heat transfer associated with this system along with sorbent screening and recycle steps occurring outside the scrubber. Approximate cost correlations are applied to various system components and used to drive a design optimization for four different cases corresponding to two levels of sulfur capture and two levels of sulfur content in the coal. The results indicate that this technology is best suited for high-sulfur, small-scale applications. Future experimental work should evaluate the use of larger (1/4 inch) sorbent.

Introduction

The Limestone Emission Control (LEC) process is the basis for computer simulation and optimization considered in this study. The process is shown in Figure 10.1 and described in detail by Prudich et al. [1]. The process uses coarse wet limestone particles (approximately one-eighth inch in diameter) as the sorbent in a cross-flow moving-bed reactor configuration. It is a continuous process where the flue gas flows perpendicular to the direction of solids flow within the reactor. The limestone particles are wet, but may become dry as they move through the reactor bed. The sulfur dioxide in the flue gas reacts with the limestone to form a layer of calcium sulfite and calcium sulfate, referred to here as precipitate, on the outside of the sorbent. Upon exiting the reactor, the sorbent particles pass through a regeneration unit where the precipitate layer is removed and the unreacted limestone is recycled to the reactor feed. A pilot plant using this configuration was constructed and operated at Ohio University's physical plant in 1991. Prudich et al. [1] report that the plant achieved better than 90% removal in all of the experimental runs and in some cases achieved a 99% removal.

This work details the optimization of the design and operation of the LEC process. The optimum design (minimum cost) was found for four different cases. These cases were constructed using two power plant capacities (100 MW and 500 MW) and two coal sulfur contents (1.5% and 3.5%). A sulfur removal efficiency of 98% was used for all four cases. The optimum design is defined as that which meets the constraints and produces the minimum total cost.

The optimum was found with the aid of a computer simulation of the flue gas desulfurization plant. This simulation models the operation of the plant and generates both capital and operating cost estimates. The simulation is based on first principles type mathematical models that reflect the underlying transport and chemical reactions involved in the desulfurization process. The sulfur dioxide capture is described using a resistance in series model which accounts for the mass transfer resistances inherent to this reaction. Included are the resistance to diffusion from the bulk gas to the liquid layer, the resistance to diffusion within the liquid, and the dissolution of calcium from the solid stone into the liquid. In addition to the resistance in series model, the effects of dry capture and sorbent deactivation are modeled. As the limestone moves through the reactor, water present on the sorbent surface is evaporated, and the stone becomes dry. Once dry, the effectiveness for SO_2 capture is diminished, but not eliminated. Sorbent deactivation refers to the "blinding" that occurs on the surface of the stone. As the reaction proceeds, the product of the reaction tends to block the active surface of the stone, which reduces the mass transfer area available and therefore reduces the reaction rate.

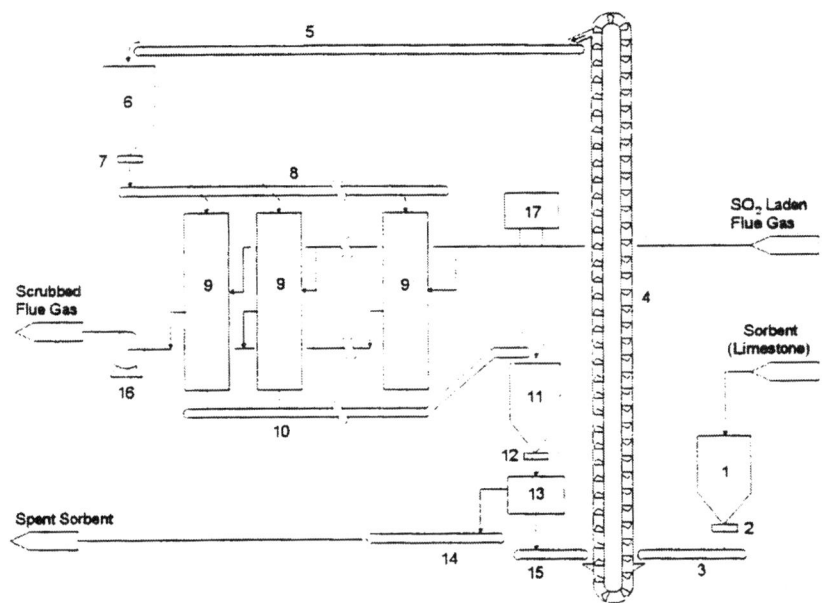

Figure 10.1 Schematic representation of the LEC process. Numbered equipment items are described in Table 10.1.

Because of the large number of calculations required, each simulation requires a significant amount of computation time. To make an optimization study practical, the time required for the evaluation of a single design had to be minimized. A considerable effort was applied to make the numerical algorithm efficient. Key ingredients include the use of an adaptive grid size and the use of vector processing on a Cray Y-MP8 supercomputer. The resulting code runs approximately 50 times faster than an early version run on a VAX/6440.

Table 10.1 Key to Figure 10.1

Identifier	Description
1	Feed hopper. (10 ton capacity).
2	Screw conveyor discharging the feed hopper.
3	Belt conveyor.
4	Bucket elevator.
5	Conveyor belt.
6	Sorbent storage tank (day tank) [one hour supply].
7	Screw conveyor discharging the day tank.
8	Conveyor belt.
9	Multiple reactors arranged in parallel.
10	Conveyor belt.
11	Holding bin (10 minutes holdup).
12	Screw feeder discharging the holding bin.
13	Regeneration units which separate sorbent by-products from unreacted limestone by means of a vibrating screen.
14	Conveyor belt.
15	Conveyor belt.
16	I.D. fan which compensates for the pressure drop across the sorbent bed.
17	Two-fluid atomizer which may be used to humidify the incoming flue gas.

Literature Review

Transport and Reaction

Flue gas desulfurization (FGD) systems can be classified as wet/wet, wet/dry or dry/dry, referring to whether the sorbent has liquid water present or not and to the condition of the waste product. The LEC process is a wet/dry process. The physical situation inside of the reactor bed can be described as flue gas flowing around relatively large wet limestone particles.

Several models for wet/dry scrubbing have appeared in the literature. Jozewicz and Rochelle [2] modeled a spray dryer sorption system. They assumed that the mass transfer was limited in the gas-phase, the liquid-phase ionic reaction rate was instantaneous, and that the dissolution rate of the lime was instantaneous. Harriot and Kinsey [3] used both gas- and liquid-phase mass transfer resistances, but also assumed an instantaneous lime dissolution rate. Karlsson and Klingspor [4] modeled a wet/dry spray dryer using gas-phase and dissolution-rate control.

Appell [5] modeled a fixed bed granular limestone reactor assuming a gas-phase reaction-rate control. He cited work by Gullett and Kramlich [6] which makes reference to experiments that show for SO_2 concentrations less than 800 ppm, the reaction in a single drop-drop tube reactor is gas-phase controlled. His model assumed an instantaneous reaction rate and did not include the effects of diffusion through the liquid layer, dissolution of the limestone, or sorbent blinding.

Visneski [7] presented a three-resistance-in-series model for sulfur dioxide capture with a granular limestone sorbent. His model used three rate equations: a gas-phase control equation, a liquid-phase control equation, and a dissolution-rate equation. Depending on what was limiting the removal of SO_2, the proper equation is selected. An instantaneous reaction rate was also assumed. This model was incorporated into a model for the fixed-bed granular-limestone LEC reactor.

Reddy [8] presented a model of the moving-bed granular-limestone LEC reactor. His model used the three-resistance-in-series transport model presented by Visneski [7] and also ignored sorbent blinding. He also assumed that the SO_2 capture rate went to zero in the absence of water. This model is the basis for the reactor model used in this work.

Klingspor, Karlsson and Bjerle [9] showed that in the absence of liquid water, the SO_2 capture does not stop, but is significantly reduced. Furthermore, the rate is shown to be a strong function of the gas relative humidity. At a relative humidity less than 24%, the reaction rate is not detectable, but the rate increases exponentially with the relative humidity for values higher than 24%.

Sorbent blinding is a potentially important effect. Jorgensen et al. [10] reported that when using hydrated lime as a sorbent, 60% to 90% of the Ca(OH)2 remained unreacted after the reaction stopped. This is attributed to surface blinding. Prudich et al. [1] determined the concentration of precipitate at which the reaction between limestone and sulfur dioxide stopped for various limestones.

Economics

Several economic analyses have been made for FGD systems. Most notably, Keeth, Ireland and Radcliffe [12] summarized the results of a study performed for the Electric Power Research Institute (EPRI) in which the economics of 28 FGD processes were analyzed. They found that in general, the cost of removal per ton of SO_2 is similar for many of the technologies. Also, dry injection systems generally have a lower capital cost, but have higher total annual cost (the sum of the annual operating cost and an annualized capital cost) due to the more expensive sorbent.

An economic analysis specifically for the LEC process has also been performed. Prudich et al. [11] conducted an economic comparison between the LEC process and a conventional lime scrubbing process. The study showed that the annualized cost of the combination of an LEC reactor and an ESP would be about 42% lower than the cost of a spray dryer/baghouse system. The report also gave the results of a pilot plant study where a cross-flow moving-bed reactor was constructed at Ohio University's physical plant to process a slip stream of 400 acfm of flue gas. The results of the study show that in some of the experimental runs, over 98% sulfur removal was achieved. This study also showed that the LEC process could provide a significant economic advantage over other FGD technologies on an industrial scale.

Chattopadhyay [13] compared the economics of two configurations for the LEC process. The two configurations studied were the cross-flow moving-bed reactor and the fixed-bed reactor. The study considered four different design cases. The cases are differentiated by two coal sulfur contents (1.5% and 3%) and two sulfur removal efficiencies (90% and 98%). For all cases a 500-MW power plant was used as the basis. For each of the four cases, the moving-bed configuration was determined to be the more economical option. A sensitivity study was also performed showing the effect changing one variable has on the cost. The variables studied were the approach to saturation temperature, the bed length, the particle size, the gas superficial velocity, and for the moving-bed reactor, the sorbent velocity.

Process Model

The process model consists of several components. The reactor is described using a first principles type mathematical model that reflects the chemical reactions and the transport occurring in the reactor bed. The results of the reactor model are used in the plant simulation to calculate the stream flow rates and to size all of the necessary plant equipment. These flow rates and equipment sizes are then used to generate the cost estimate for the plant. An optimization routine is used with this model to adjust the inputs to the simulation to achieve the lowest overall process cost.

Reactor Model

The reactor model consists of a set of partial differential equations that describe the material and energy balances in the reactor bed. Several physical property correlations and transport models are used to calculate the terms in the governing equations at each position in the reactor. The transport model used has its basis in the three-resistance-in-series model described by Visneski [7]. The three resistances are the transport of SO_2 through the gas film, diffusion of SO_3^- (or SO_4^-) and Ca^{++} in the liquid, and dissolution of calcium. Details of the transport equations are presented in Appendix C. The model is the same as that used by Chattopadhyay [13] with the additions of sorbent blinding and dry capture effects. The models used to describe these two effects are discussed below. The reactor model and the solution techniques are also described.

Sorbent Blinding

As the reaction of SO_2 and $CaCO_3$ proceeds, the reaction product covers a fraction of the sorbent surface and renders it nonreactive. The reaction product is $CaSO_3$ or $CaSO_4$ and is referred to here as precipitate. The sorbent blinding model calculates the fraction of the total surface area that has not been blinded by the precipitate. This unblocked area is the area available for mass transfer and is the area used in the three-resistance-in-series model to find the SO_2 transport rate. Modifications to the three-resistance-in-series model are required to account for the presence of the precipitate layer on the surface of the sorbent.

The surface of the sorbent is conceptually divided into blinded and nonblinded regions as shown in Figure 10.2. No mass transfer occurs in the blinded regions and the mass transfer in the nonblinded regions occurs as if the precipitate layer were not present. It is assumed that the width of each of the precipitate "piles" in the direction along the sorbent surface (the horizontal direction in Figure 10.2) is much smaller than the liquid thickness or the precipitate thickness. (Figure 10.2 is not drawn to scale.) This allows the transport in the liquid to be considered as a one-dimensional diffusion process with a step change in the mass transfer area at the top surface of the precipitate layer.

Simulation and Optimization of a Granular Limestone Flue Gas Desulfurization Process

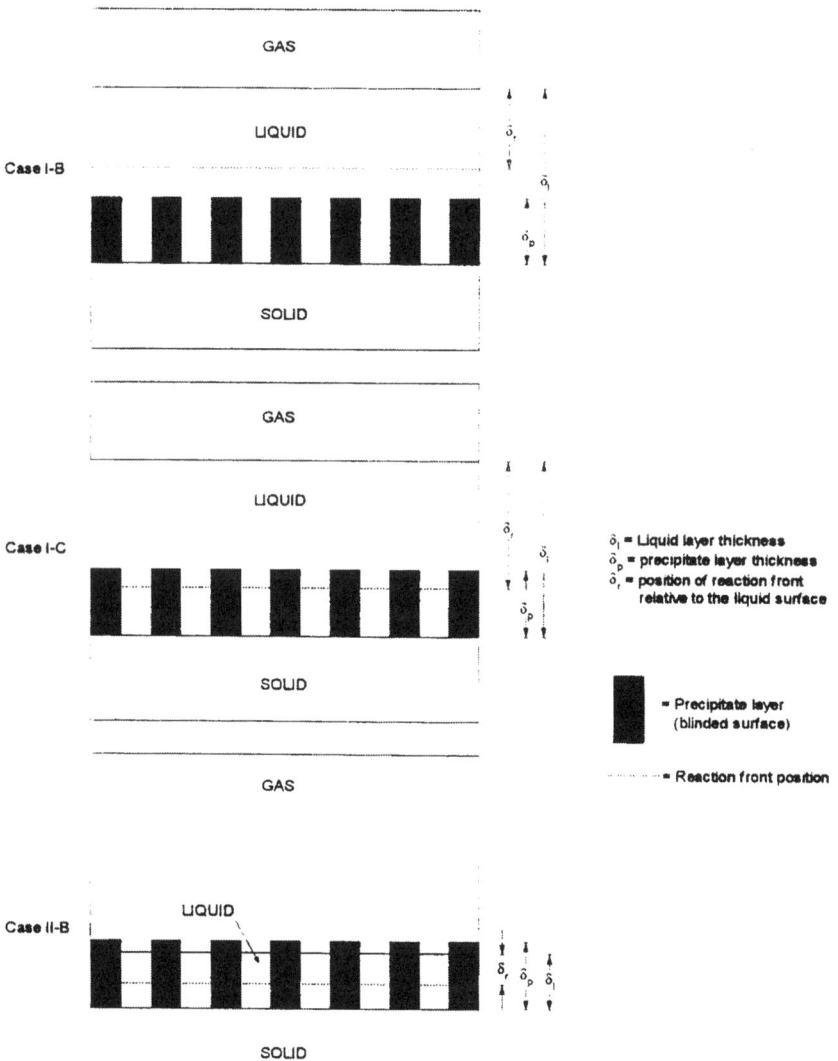

Figure 10.2 Schematic representation of the sorbent blinding model.

The liquid layer thickness must be adjusted to account for the precipitate layer that coexists with the liquid on the surface of the sorbent pellet. The liquid thickness is calculated by dividing the volume of the liquid by the unblinded surface area. If this thickness is greater than the precipitate thickness, then it is recalculated recognizing that the surface area above the precipitate layer is equal to the total surface area of the sorbent. The precipitate thickness is calculated using an empirical expression that relates the thickness to the concentration of precipitate.

The reaction rate is assumed to be instantaneous at the point where the reactants meet in the liquid. Expressions that give the transport rate through each of the resistances are solved simultaneously to give an expression in terms of the bulk gas phase concentration, C_{Sg}, and the solubility of calcium, C_{Cle}.

There are eight possible transport situations on the surface of the sorbent. These are determined by the thickness of the liquid layer with respect to the precipitate layer and the location of the reaction front. Three of these situations are depicted in Figure 10.2. The correct transport equation to use is dictated by which situation is present on the sorbent surface. The calculations proceed by determining the liquid layer thickness and then guessing in which region the reaction front lies. Using the appropriate transport equations for the guessed region, the position of the reaction front is calculated and compared to the assumption. If the calculated position is outside the guessed region, then the calculation must be repeated using the correct region. The detailed derivation of the transport equations for each of the eight cases is shown in Appendix C.

In order to employ these sulfur transport expressions, the precipitate thickness and mass transfer area within the precipitate layer is required. These are found using an empirical expression for the thickness. The empirical expression used has been developed by making several assumptions and observations.

The precipitate thickness is assumed to be a function of the concentration of the precipitate (moles/volume of sorbent). As the concentration increases, a greater portion of the sorbent is covered with the precipitate until the entire surface is covered. The concentration at which the entire surface is covered, C_{Sst}, is referred to as the terminal concentration and is the concentration at which the reaction stops. For concentrations less than C_{Sst} the blinded area will be less than the total area. If the assumption is made that the thickness will only increase with increasing concentration, then an upper limit is imposed on the thickness. This limit is the thickness when the amount of precipitate represented by C_{Sst} is spread evenly over the sorbent surface. Figure 10.3 is a qualitative graph that shows the region in which the thickness versus concentration curve must lie in order to satisfy these assumptions. Also shown is the general shape for the proposed empirical relation.

Simulation and Optimization of a Granular Limestone Flue Gas Desulfurization Process

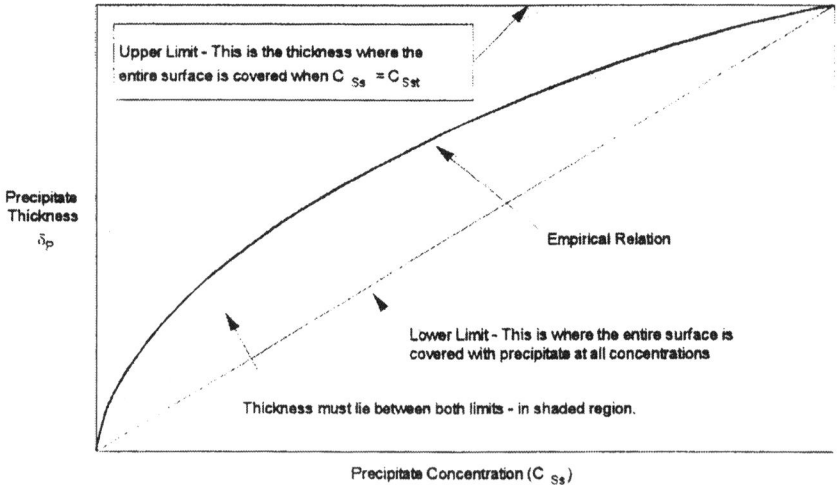

Figure 10.3 Qualitative behavior of precipitate layer thickness.

A proposed functional form for this relation is

Equation 10.1

$$\delta_p = k(C_{Ss})^{1/e}$$

In this equation, k and e are adjustable model parameters. The value of k can be determined by noting that the precipitate thickness at the terminal concentration is equal to the volume of precipitate divided by the total sorbent surface area.

Equation 10.2

$$\delta_{p-t} = \frac{C_{Sst}\varepsilon_s}{\rho_p a}$$

Substituting Equation 10.2 into Equation 10.1 and using $C_{Ss} = C_{Sst}$, the value of k is found to be

Equation 10.3

$$k = \frac{(C_{Sst})^{\frac{e-1}{e}} \varepsilon_s}{\rho_p a}$$

Substituting Equation 10.3 into Equation 10.1 gives the empirical expression for precipitate thickness.

Equation 10.4

$$\delta_p = \frac{(C_{Sst})^{\frac{e-1}{e}} (C_{Ss})^{\frac{1}{e}} \varepsilon}{\rho_p a}$$

To find the expression for the mass transfer area within the precipitate layer, a_{mp}, the blinded area is subtracted from the total. The blinded area is equal to the volume of precipitate divided by the thickness.

Equation 10.5

$$a_{mp} = a - a_p$$

and

Equation 10.6

$$a_p = \frac{C_{Ss} \varepsilon_s}{\rho_p \delta_p}$$

Substituting Equation 10.4 into Equation 10.6 and the result into Equation 10.5 gives the expression for the mass transfer area available within the precipitate layer.

Equation 10.7

$$a_{mp} = a\left(1 - \left(\frac{C_{Ss}}{C_{Sst}}\right)^{\frac{e-1}{e}}\right)$$

It is assumed that the value for C_{Sst} will be such that the precipitate thickness at this concentration will be independent of particle size. This implies that

Equation 10.8

$$C_{Sst} = \frac{\alpha}{S_p D_p}$$

where α is the primary parameter to be resolved from experiments. Towards this end, experiments have been conducted by Visneski [7] and by Prudich et al. [11]. The results of these experiments are shown in Table 10.2. Note that the maximum concentrations reported by Prudich et al. are in lbmol/ft^2 and those by Visneski are in lbmol/ft^3.

Table 10.2 Experimental Results for C_{Sst} Determination.

Reference	Limestone	Max Conc.	Particle Diam (ft)	α
Prudich	Maxville	6.26x10^{-2} lbmol/ft^2	2.10x10^{-4}	6.57x10^{-6}
Prudich	Mississippi	5.74x10^{-2} lbmol/ft^2	2.10x10^{-4}	6.03x10^{-6}
Prudich	Vanport	4.53x10^{-2} lbmol/ft^2	2.10x10^{-4}	4.76x10^{-6}
Visneski	Maxville	1.64x10^{-4} lbmol/ft^3	7.77x10^{-4}	9.83x10^{-4}
Visneski	Vanport	5.27x10^{-5} lbmol/ft^3	9.57x10^{-4}	3.16x10^{-4}

The wide variation in values of α are not easily explained. There is an apparent correlation with particle size, which suggests a higher value should be used when modeling desulfurization with granular limestone. The use of smaller values for C_{Sst} results in simulation results that are in strong disagreement with pilot scale experimental results [1]. A value of 9.83x10^{-4} has been used in this study.

Prudich et al. [11] also determined values for e (the exponent in Equation 10.7) experimentally. They report a value of 1.50 for Maxville limestone, 1.55 for Mississippi limestone, and 2.61 for Vanport limestone. An average value of 1.89 is used in this study.

Dry Capture Model

The limestone entering the top of the reactor may be wet or dry. Because the sorbent is cold relative to the flue gas, water condenses on the limestone after it enters the reactor. The limestone may become dry as it moves through the reactor due to the unsaturated gas flowing through it. Once dry, the reaction rate is

Simulation and Optimization of a Granular Limestone Flue Gas Desulfurization Process

greatly reduced and, as shown by Klingspor et al. [9], the reaction rate on dry limestone is a strong function of the relative gas humidity. This is assumed to be attributed to an increase in the amount of water adsorbed onto the surface of the stone as the humidity is increased. The model that is used to predict the dry capture rate here takes the form of an empirical equation that relates the effective fractional water coverage to the relative gas humidity. This equation is used with the three-resistance-in-series transport model to predict the SO_2 transport rate.

A correction factor has been defined and correlated to the relative humidity. The correction factor, f, is defined as

Equation 10.9

$$f = \frac{a_{ml}}{a_{mp}}$$

where a_{ml} is the area that is covered by water and a_{mp} is the total area available for mass transfer within the precipitate. This factor was correlated to the gas-phase relative humidity through a Langmuir isotherm.

Equation 10.10

$$f = \frac{(1-k)Y_r}{1-kY_r}$$

where k is the adjustable model parameter and Y_r is the fractional relative humidity. Experimental results obtained by Prudich et. al. [11] reveal that the average value for k is 0.89.

Differential Equations

The governing differential equations that describe the performance of the reactors are derived from energy and material balances. There are two energy balance equations, one for the gas phase and one for the combined liquid/solid phases (the condensed phases are assumed to be in thermal equilibrium). The mass balance equations for the gas phase account for the sulfur species, the water in the gas, and the noncondensables. The liquid phase is approximated as being composed entirely of water; therefore, only an overall material balance is required for the liquid. For the solid phase, a sulfur balance is maintained. The latter is needed because the concentration of the sulfur species in the solid is needed in the mass transport model. No other solid phase mass balances are maintained because it is assumed that the molar densities of calcium carbonate and calcium sulfite (or sulfate) are equal.

Appendix B contains the derivation of the partial differential equations. The seven partial differential equations used to describe the reactor along with the appropriate boundary conditions are shown below. (See the nomenclature section for variable definitions.)

Equation 10.11

$$\frac{\partial U_l}{\partial y} = \frac{g_w}{\rho_l}$$

Equation 10.12

$$(U_s \rho_s C_{pCs} + U_l \rho_l C_{pWl})\frac{\partial T_l}{\partial y} = g_g + g_w(H_{wg} - H_{wl})$$

Equation 10.13

$$U_s \frac{\partial C_{Ss}}{\partial y} = g_s$$

Equation 10.14

$$U_s(C_{Ng}C_{pNg} + C_{Wg}C_{pWg})\frac{\partial T_g}{\partial y} = -g_g$$

Equation 10.15

$$\frac{\partial U_g}{\partial x} = -\frac{g_w}{\rho_g} + \frac{U_g}{T_g}\frac{\partial T_g}{\partial x}$$

Equation 10.16

$$U_g \frac{\partial C_{Sg}}{\partial x} = -C_{Sg}\frac{\partial U_g}{\partial x} - g_s$$

and

Equation 10.17

$$U_g \frac{\partial C_{Wg}}{\partial x} = -C_{Wg}\frac{\partial U_g}{\partial x} - g_w$$

The boundary conditions used with these differential equations are listed below. At the top of the reactor (y=0) where the sorbent feed enters the boundary, conditions are

Equation 10.18
$$U_l = U_{l0}$$

Equation 10.19
$$T_l = T_{l0}$$

and

Equation 10.20
$$C_{Ss} = C_{Ss0}$$

At the side of the reactor (x=0) where the flue gas enters the boundary, conditions are

Equation 10.21
$$T_g = T_{g0}$$

Equation 10.22
$$U_g = U_{g0}$$

Equation 10.23
$$C_{Sg} = C_{Sg0}$$

and

Equation 10.24
$$C_{Wg} = C_{Wg0}$$

The simultaneous solution of this set of equations describes the SO_2 transport rate in the reactor which in turn dictates the overall performance and the required dimensions of the reactor. The coefficients in these equations are functions of physical properties and transport rates which are in turn functions of the dependent variables in the governing equations. The correlations for the physical properties and the models that predict the transport are described in detail in Appendix C.

Solution Algorithm for Partial Differential Equations

The solution to the governing equations is found using a modified Euler's method that employs a predictor-corrector algorithm and is extended to a two-dimensional problem.

All of the differential equations used to describe the operation of the reactor fit the form of

Equation 10.25

$$A\frac{\partial}{\partial x}\varphi + B\frac{\partial}{\partial y}\varphi = E$$

where φ represents any of the dependent variables and A, B, and E may be functions of any of the system variables or constants or zero. The predictor equation is

Equation 10.26

$$\varphi(y, x) = \varphi_0 + \Delta\varphi_1$$

In this equation, $\Delta\varphi_1$ is a step based on the slope at the known base point and φ_0 is the value of φ at the known base point. The value of φ_0 is taken to be a weighted average of two known points, $\varphi(y-\Delta y, x)$ and $\varphi(y, x-\Delta x)$.

Equation 10.27

$$\varphi_0 = \frac{A(y, x-\Delta x)\varphi(y, x-\Delta x)\Delta y + B(y-\Delta y, x)\varphi(y-\Delta y, x)\Delta x}{A(y, x-\Delta x)\Delta y + B(y-\Delta y, x)\Delta x}$$

$\Delta\varphi_1$ is a weighted average of the right-hand side of the differential equation multiplied by an appropriate step size.

Equation 10.28

$$\Delta\varphi_1 = \left(\frac{A(y, x-\Delta x)E(y, x-\Delta x) + B(y-\Delta x, y)E(y-\Delta y, x)}{A(y, x-\Delta x) + B(y-\Delta y, x)}\right)$$

$$\left(\frac{\Delta x \Delta y}{Ay, x - \Delta x \Delta y + By - \Delta y, x\Delta x}\right)$$

The predictor equation is explicit because all quantities are evaluated at known points.

The corrector equation uses the estimate of $\varphi(x,y)$ generated from the predictor equation to calculate an increment based on the derivative evaluated at the unknown point. The average of this increment and the increment used in the predictor equation is used to generate the new estimate for $\varphi(x,y)$.

Equation 10.29

$$\varphi(y, x) = \varphi_0 + \frac{(\Delta\varphi + \Delta\varphi_2)}{2}$$

In this equation, $\Delta\varphi_2$ is the increment evaluated at the unknown point.

Equation 10.30

$$\Delta\varphi_2 = E(y, x)\left(\frac{\Delta x \Delta y}{A(y, x)\Delta y + B(y, x)\Delta x}\right)$$

The forms reported above have been chosen based upon their behavior in certain limiting cases. For the case A=0 Equation 10.26 becomes

Equation 10.31

$$\varphi(y, x) = \varphi(y - \Delta y, x) + \left\{\frac{E(y-\Delta y, x)}{B(y-\Delta y, x)} + \frac{E(y, x)}{B(y, x)}\right\}\frac{\Delta y}{2}$$

which is the traditional modified Euler's method. Similarly for B=0 we have

Equation 10.32

$$\varphi(y, x) = \varphi(y, x - \Delta x) + \left\{ \frac{E(y, x - \Delta x)}{B(y, x - \Delta x)} + \frac{E(y, x)}{B(y, x)} \right\} \frac{\Delta x}{2}$$

As explained in the next section, the algorithm is applied with a single corrector step. No convergenge criterion is used.

Calculation Speed Enhancement

Two techniques were used to speed the solution of the differential equations. One is the use of an adaptive grid, which allows the numerical step size to be increased in regions of the reactor where the dependent variables do not change rapidly. The second is the use of vector processing on a Cray Y-MP8 supercomputer.

The adaptive grid is set up by first dividing the reactor space into several large grid elements. (See Figure 10.4.) Prior to completing the solution within each of the large grid elements, an appropriate numerical step size for that element must be selected.

Simulation and Optimization of a Granular Limestone Flue Gas Desulfurization Process

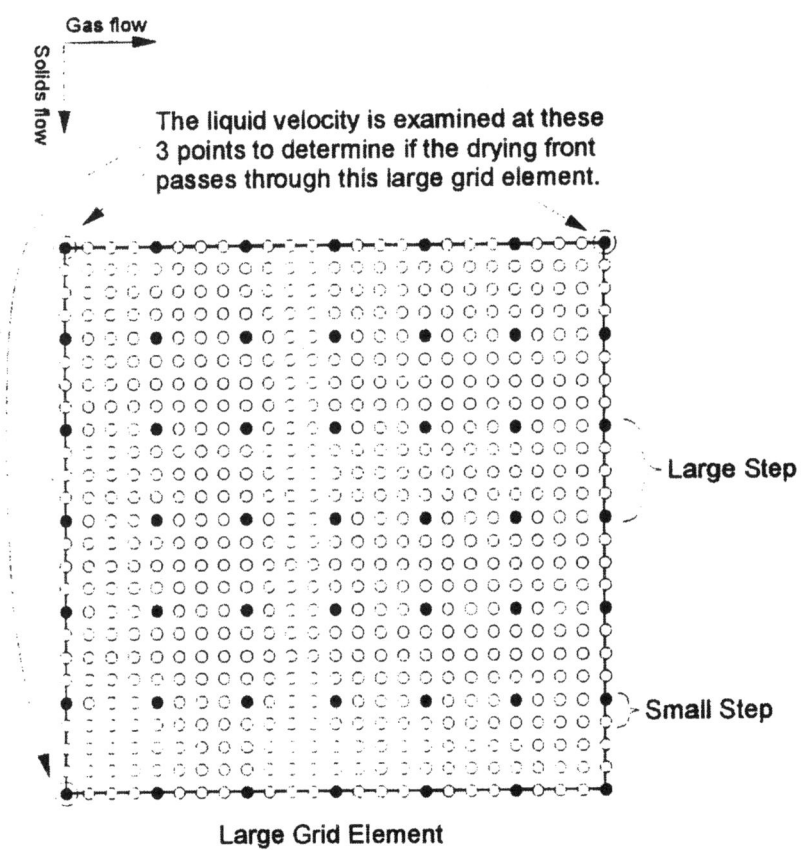

Figure 10.4 Arrangement of small grid elements within large grid element in adaptive technique.

From preliminary simulations it is known that the solution to the differential equations is fairly flat except in the region of the drying front. The drying front is the region in the reactor where the water is evaporating from the surface of the stone. It runs approximately diagonally from the corner of gas and solid inlet to the opposite side of the reactor. If the drying front does not pass through a particular large grid element, the step size for that element can be increased without risk of solution inaccuracy. In the region of the drying front, however, a very small step size is needed to provide accuracy and stability in the numerical algorithm.

To determine if the drying front passes through the large grid element, the liquid superficial velocity is examined at the top left, top right, and bottom left points of element. If the liquid superficial velocity is significantly different along the left edge or along the top edge of the large grid element, then the drying front passes through this element and the smaller step size is selected.

The precision and stability of the numerical procedure was evaluated by varying both the x and y step sizes in both the finely spaced and coarsely spaced large grid elements. These tests showed that the precision was insensitive to step size up until the point where the solution became unstable. The results presented below used a fine step size of 0.001 ft for both x and y and a coarse step size of 0.002 ft for both x and y.

Two special features in the numerical algorithm are required to enable the code to be vectorized. The first is that the number of corrector steps is fixed at one. This stipulation is required because conditional branch points are incompatible with vectorization. The specific number of corrector steps was established with only a cursory examination of the relationships between the number of corrector steps, the numerical step size, and the execution time.

The second is that the iteration procedure within a large grid element proceeds via successive diagonal rows of grid points. This approach is needed to eliminate data dependencies which occur when one element of the vector being processed depends on the value of other elements in the vector. The iteration formula represented by Equations 10.26-10.30 calls for every grid point (x_i, y_j) to be calculated from points (x_{i-1}, y_j) and (x_i, y_{j-1}). This structure causes data dependencies to exist regardless of whether the vector calculation proceeds via either rows or columns of the matrix of grid points. However, iterations for diagonal sets from the matrix of grid points can be vectorized because the iteration formulas for points (x_i, y_j), (x_{i+1}, y_{j-1}), (x_{i+2}, y_{j-2}), etc. depend only on points (x_{i-1}, y_j), (x_i, y_{j-1}), (x_{i+1}, y_{j-2}), etc.

Cost Model

The cost estimate is composed of both operating and capital costs. The operating costs are generated by multiplying a unit cost by an estimated usage for each of the operational expenses. These include power, water, limestone and waste disposal. The capital costs are estimated using formulas of the form

Equation 10.33

$$C_i = I_i E_i C^*_i$$

where C_i^* is the reference cost, E_i is an inflation escalator and I_i is an installation factor needed to convert purchased cost to installed cost. The total annual cost is the sum of the annual operating costs and the annualized fixed cost. Details of the cost equations are given in Appendix D.

Plant Simulation Model

In addition to the reactor simulation, a regeneration and recycle model is used to describe the separation of spent sorbent from reusable limestone. Pilot scale operating experience [1] indicates that the precipitate layer on the sorbent particles can be easily removed by mechanical agitation. Fine particles are formed by the agitation and are separated from the larger unreacted limestone particles by screening. The material balance equations and empirical relations used are described below.

Referring to Figure 10.5, the following material balance equations can be written

Equation 10.34

$$G_{14} = G_{10} - G_{15}$$

Equation 10.35

$$G1_5 = C_{Sso} G_6 / C_{Ssr}$$

Equation 10.36

$$G_1 = G_6 - G_{15}$$

Equation 10.37

$$G_{10} = G_6 + 20 Q_{go} C_{Sgo} F_{Sa}$$

and

Equation 10.38

$$C_{Ssw} = F_{Sa} Q_{go} C_{Sgo} \rho_l / G_{14}$$

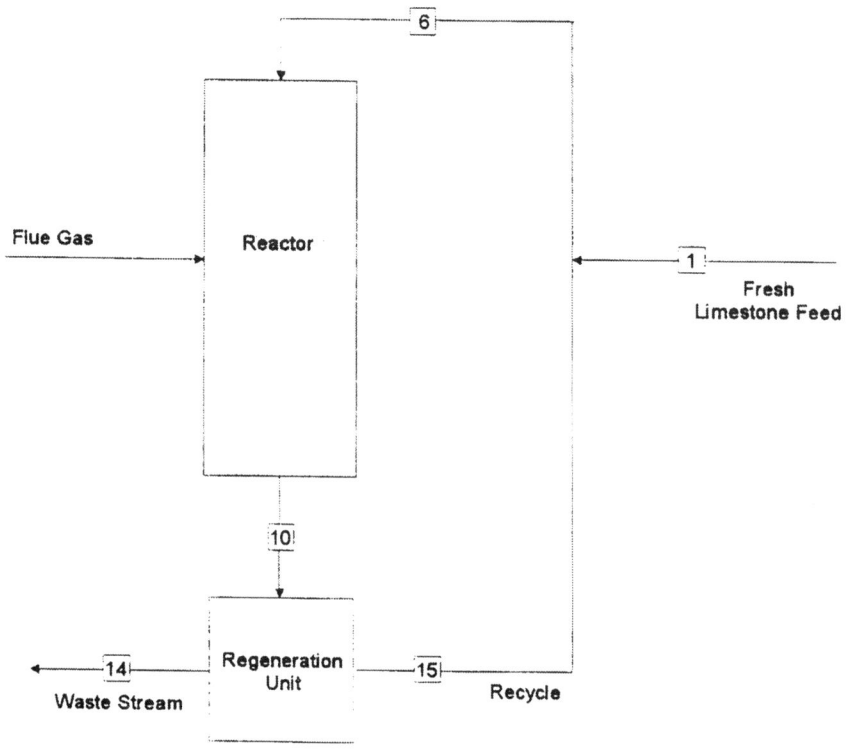

Figure 10.5 Regeneration and recycle model diagram.

Here the G's are the mass flow rates for the sorbent in the stream indicated by the subscript, C_{Ss0} is the molar concentration of sulfur in the sorbent feed to the reactor (stream number 6), C_{Ssw} is the concentration in the waste stream (stream number 14), and C_{Ssr} is the concentration in the recycled stream (stream number 15). Q_{g0} is the volumetric flow rate of the gas stream, C_{Sg0} is the molar concentration of SO_2 in the feed gas, F_{Sa} is the average fraction of sulfur dioxide removal, and ρ_i is the intrinsic density of the sorbent. These five equations have six unknowns (G_1, G_{10}, G_{14}, G_{15}, C_{Ssw}, and C_{Ssr}). C_{Ss0} is specified separately as a part of the optimization procedure and G_6 is determined from the specified sorbent velocity and reactor dimensions.

An empirical performance equation that describes the operation of the screen is added to this set to provide the sixth equation. This equation is

Equation 10.39

$$\left(\frac{C_{Ssw}}{C_{Sswm}}\right)^m = \frac{C_{Ssr}}{C_{Ssm}}$$

where C_{Sswm} is the maximum attainable sulfur concentration in the waste stream, and m and C_{Ssrm} are adjustable parameters. C_{Ssrm} is the maximum attainable concentration in the recycle and is taken to have the same value as C_{Sst} (the concentration at which the reaction stops). The value of C_{Sswm} is determined as follows.

The waste stream is comprised of the $CaSO_4$ formed in the reactor and unreacted limestone that passes through the screen. Since the operation is at steady state, the amount of $CaSO_4$ in the waste is equal to the amount of SO_2 captured (on a molar basis). The minimum amount of limestone in the waste is dictated by the size of the screen opening and the size of the feed particles. The outer surface of the stone is reacted to form $CaSO_4$, this layer is removed in the regeneration unit and the inner unreacted limestone is recycled back to the reactor. This process continues until the particles become small enough to fall through the screen, thus "contaminating" the waste with unreacted limestone. The minimum fraction of the unreacted limestone passing through the screen is equal to the ratio of the volume of a particle just small enough to pass through the screen to the volume of a feed particle, or

$$\gamma = \left(\frac{D_m}{D_p}\right)^3$$

Equation 10.40

Since the maximum concentration of sulfur in the waste stream corresponds to the minimum limestone loss, C_{Sswm} is found by dividing the moles of $CaSO_4$ produced by the volume of the waste stream (which is the volume of the $CaSO_4$ plus the volume of the limestone lost through the screen).

Equation 10.41

$$C_{Sswm} = \frac{Q_{go} C_{Sgo} F_{Sa}}{\left\{\frac{G_1 \gamma + 1200 Q_{go} C_{Sgo} F_{Sa}}{\rho_i}\right\}}$$

If the unreacted limestone is attrited in the process, the $CaSO_4$ concentration will be lower than C_{Sswm}. After setting C_{Ssw} equal to C_{Sswm} in Equations 10.34–

10.39 and after considerable algebraic manipulations the expression for C_{Sswm} can be simplified to

Equation 10.42

$$C_{Sswm} = \frac{\rho_i(1-\gamma)}{20(6-\gamma)}$$

No information is available at this time to establish an appropriate value for m in Equation 10.39. The value m=4 has been used in the simulations because it gives an intuitively reasonable relationship between the concentrations in the two product streams from the screen.

Making substitutions using Equations 10.35-10.39 to reduce the set to one equation with one unknown gives an implicit equation for C_{ssr}

Equation 10.43

$$C_{Sst}\left[\frac{F_{Sa}Q_{g0}C_{Sg0}\rho_i}{C_{Sswm}\left(G_{10} - \frac{C_{Ss0}G_6}{C_{Ssr}}\right)}\right] - C_{Ssr} = 0$$

The value of C_{Ssr} is found from this implicit equation by using an interval halving technique.

Optimization Algorithm

The algorithm used for finding the optimum plant design is a quasi-Newton, variable metric method with a line search. Specifically the Davidon-Fletcher-Powell method [14] is used to generate the search direction and a combination of the Davies-Swann-Campey and Powell unidimensional searches is employed to find the minimum in the specified direction [15]. This method was selected because it is generally robust and it converges relatively fast [14-16].

The algorithm starts by picking initial guesses for the independent variables and selecting a positive definite approximation to the inverse of the Hessian matrix. The identity matrix is used as the first approximation to the Hessian, which results in the first iteration selecting the steepest descent as the search direction. The gradient required for the method is approximated numerically using forward differences. Forward differences were selected over the more accurate central differences because they require fewer function evaluations (one less function evaluation per independent variable per gradient evaluation, or in this case, four fewer function evaluations per gradient calculation). The search direction is then calculated by pre-multiplying the gradient by the approximation of the inverse of the Hessian.

Equation 10.44

$$S(x^{(k)}) = -A^{(k)} \nabla f(x^{(k)})$$

Here, S is the search direction as a function of the independent variables, A is the approximation to the inverse of the Hessian, ∇f is the gradient of the function, and k indicates the iteration number. Once the search direction is calculated, a line search is used to find the minimum cost in this direction. The gradient is then calculated at the minimum point along the search direction. Using the values of the independent variables and the values of the gradients both at the starting point and minimum point, a new approximation to the inverse of the Hessian is calculated. This is then used to generate the next search direction.

The Davidon-Fletcher-Powell updating formula is used to approximate the inverse of the Hessian [14]. It is given by

Equation 10.45

$$A^{(k)} = A^{(k-1)} + \frac{\Delta x^{(k-1)} \Delta x^{(k-1)^T}}{\Delta x^{(k-1)^T} \Delta g^{(k-1)}} - \frac{A^{(k-1)} \Delta g^{(k-1)} \Delta g^{(k-1)^T} A^{(k-1)}}{\Delta g^{(k-1)^T} A^{(k-1)} \Delta g^{(k-1)}}$$

In this equation, the A's are the approximations to the inverse of the Hessian, the Δx's are the difference in the independent variable vectors between iterations, the Δg's are the difference in the gradient vectors between iterations, and k indicates the iteration number.

The line search is conducted by continually increasing the step size until the minimum is bracketed. At this point, a quadratic is fit to the three bracketing points and the minimum of the quadratic is used as the next guess for the minimum of the function along the selected search direction. This continues until the difference in the bracketing points is less than a preset tolerance.

The optimization algorithm is terminated when the difference between the starting point along a line search and the minimum found along the line is less than a selected tolerance.

There are eleven independent inputs required for the reactor simulation. Two of these are set by the design case being studied. (These are the flue gas sulfur concentration and the flue gas flow rate.) Of the remaining nine variables, three of these are set by practical limitations. These are the reactor width (the dimension perpendicular to both sorbent and gas flow) which is set to a practical upper limit of 20 feet, the gas humidity which is set equal to the value of the delivered flue gas (no humidification is performed) and the water addition rate which is set to zero. The humidification and water addition rates were set to zero to keep the waste product dry. Because of the length of the reactors in the direction of gas flow, humidifying the gas or adding water to the limestone resulted in a significant amount of water in the reactor solids outlet stream. This would cause operational problems in the regeneration step and would therefore be an unrealistic operating condition. This leaves six optimization search variables: the superficial gas velocity, the solids velocity, the inlet solid sulfur concentration, the sorbent particle size, the regeneration screen mesh size, and the reactor length (the dimension parallel to the gas flow). The reactor height (the dimension parallel to the solids flow) is set indirectly by the specification of the removal efficiency.

Four of the search variables are determined by the optimization algorithm. These are the concentration of sulfur in the solid phase, the limestone particle diameter, the superficial gas velocity, and the solids velocity. The reactor length is allowed to assume several values for each set of the four search variables and the screen mesh size is optimized at each reactor length. The cost for the plant is evaluated for successively deeper reactors until the minimum cost is bounded. This is done to make more efficient use of calculations. Since the same calculations used to simulate a three-foot-deep reactor would also be required to simulate a four-foot-deep reactor, both calculations are performed during one run thus making maximum use of each function evaluation. Once the optimum length has been bracketed, those variables that change with reactor length and have a significant impact on the cost are fit to a linear function of reactor length. These linear fits are then used in the calculation of the optimum length.

Since the screen mesh size does not affect the simulation of the reactor, it can be handled separately. For each reactor length evaluated, the optimum mesh size is determined. This is a constrained optimization limited by a practical lower limit on screen opening size and a natural upper limit. The lower limit is set to 100 mesh (0.147 mm) since screening operations are generally not economical below this limit [17]. The upper limit on the screen opening size is the particle diameter, since at this screen size all of the material would pass through the screen. To find the optimum, a region elimination technique (interval halving) is used. The trade-off that is occurring in the screen opening size optimization is that the screen capital and operating costs decrease as the screen opening gets larger, but the loss of limestone in the waste increases.

Design Cases

Four design cases were constructed by considering two power plant capacities and two coal qualities. The power plant capacities considered were 100 and 500 MW. This dictates the flow rate of the flue gas for a given coal quality. Table 10.3 shows the analysis for the two coals used in this study. The two coals were intended to match the compositions of Upper Freeport and Pittsburgh #8 but differ slightly in composition due a calculation error discovered after the optimization was completed. For all four cases studied, a 98% sulfur removal efficiency was specified for the FGD reactors.

Table 10.3 Coal Analysis

	Low Sulfur Coal		High Sulfur Coal	
Composition (wt%)	dry basis	wet basis	dry basis	wet basis
%C	72.89	71.05	75.11	73.20
%H_2	4.70	4.58	5.30	5.16
%H_2O		2.60		2.60
%N	1.28	1.33	1.32	1.28
%O_2	8.13	7.92	9.70	9.45
%S	1.58	1.54	3.69	3.59
%Ash	11.42	11.13	4.89	4.77
Heating Value (BTU/lb)	9,000		12,000	

The following equations are used to calculate the flue gas composition and flow rate for each of the four cases [18]. Complete combustion is assumed. The amount (lb/lb of coal) of theoretical dry air required is

Equation 10.46

$$T = 11.53 [C^*] + 34.34 ([H_2^*]-[O_2^*]/8) + 4.29 [S^*]$$

where T is the theoretical dry air required (lbs/lb coal), and $[C^*]$, $[H_2^*]$, $[O_2^*]$ and $[S^*]$ are the weight fractions of carbon, hydrogen, oxygen, and sulfur in the coal. The flow rates of the constituent compounds in the flue gas (lb/lb of coal) are given by

Equation 10.47

$$[CO_2] = 3.66\,[C^*]$$

Equation 10.48

$$[SO_2] = 2.00\,[S^*]$$

Equation 10.49

$$[N_2] = 0.7685T(1+e) + [N_2^*]$$

Equation 10.50

$$[O_2] = 0.2315Te,$$

and

Equation 10.51

$$[H_2O] = 8.94\,[H_2^*] + [H_2O^*] + [H_2O^=]$$

where e is the fraction of excess air, $[N_2^*]$ is the weight fraction of nitrogen in the coal, $[H_2O^*]$ is the weight fraction of water in the coal, and $[H_2O^=]$ is the moisture introduced with the air. The water introduced with the air is found by using the equation

Equation 10.52

$$[H_2O^=] = 18\left(\frac{T(1+e)}{28.9}\right)\left(\frac{HP_v}{1-HP_v}\right)$$

where H is the fractional relative humidity of the combustion air and P_v is the vapor pressure of water at the inlet air temperature.

The total flow rate of the flue gas is calculated from the power plant capacity. Dividing the capacity by the coal heating value of the coal gives the coal usage rate. Multiplying the coal usage by the specific flue gas flow gives the total flue gas flow rate. The volumetric flow rate of the flue gas is found using the ideal gas law and assuming a flue gas temperature (a pressure of 1 atmosphere is assumed). The assumptions used in the flue gas calculations for this study are shown in Table 10.4. Table 10.5 gives the values for composition and flow rate used in the simulation for the four cases studied. (Qg0 is the flow rate in cubic feet per second and x represents mole fractions of sulfur and water.)

Table 10.4 Flue Gas Calculation Assumptions

Combustion Air Humidity	80%
Excess Air	20%
Combustion air inlet temperature	70°F
Flue gas temperature	300°F
Power Plant efficiency	35%

Table 10.5 Flue Gas Flow Rate and Composition

	Low Sulfur Coal	High Sulfur Coal
100 MW Power Plant Cap.	Case 1 $Q_{g0} = 6991$ $x_S = 0.00115$ $x_W = 0.07729$	Case 2 $Q_{g0} = 5519$ $x_S = 0.00255$ $x_W = 0.08095$
500 MW Power Plant Cap.	Case 3 $Q_{g0} = 34956$ $x_S = 0.00115$ $x_W = 0.07729$	Case 4 $Q_{g0} = 27594$ $x_S = 0.00255$ $x_W = 0.08095$

Computer Code

Details of the computer program used to perform the process simulation and optimization are given in Duespohl [19].

Results

The optimum design and operating conditions for the four cases described in Table 10.5 are shown in Table 10.6. The optimum values for the six search variables are shown in Table 10.7.

Simulation and Optimization of a Granular Limestone Flue Gas Desulfurization Process

Table 10.6 Optimization Results

	Case 1	Case 2	Case 3	Case 4
X-Direction step siz.e (ft)	0.001	0.001	0.001	0.001
Y-Direction step size. (ft)	0.001	0.001	0.001	0.001
Reactor length (X-direction). (ft)	7.77	7.27	15.11	18.22
Reactor height (Y-direction). (ft)	42.0	59.3	78.6	106.2
Reactor width (Z-direction). (ft)	20	20	20	20
Percentage of SO_2 removed. (%)	98	98	98	98
Number of reactor units.	7.57	7.09	14.46	11.2
SO_2 gas conc. at reactor inlet. (ppm)	1150	2550	1150	2550
Initial solid sulfur conc. (lbmols/cuft)	0.0269	0.0268	0.0257	0.0264
Relative humidity at inlet. (%)	1.98	2.07	1.98	2.07
Inlet gas temperature. (R)	760	760	760	760
Inlet solid/liquid temperature. (R)	520	520	520	520
Inlet superficial gas velocity. (ft/s)	1.49	0.914	1.78	1.41
Inlet superficial liquid velocity. (ft/s)	0	0	0	0
Sorbent superficial velocity. (ft/h)	11.5	11.6	12.5	12.3
Sorbent particle diameter. (ft)	0.0205	0.0200	0.0246	0.0247
Input flue gas flow rate. (acf/s)	6990.	5520.	35000.	27600.
Press. drop. (inch. H_2O column)	5.42	1.99	12.4	9.41
Mesh size used in screen (mm)	3.33	2.84	3.71	3.26
Limestone mass flow rates:				
Flowrate thru Feed Hopper # 1. (lb/s)	24.0	40.1	112.	186.
Flowrate thru Elevator # 4. (lb/s)	659.	601.	2660.	2530.
Flowrate thru Storage Tank # 6. (lb/s)	637.	565.	2560.	2360.
Flowrate thru each Reactor # 9. (lb/s)	84.1	79.6	177.	211.
Flowrate thru Belt Conveyor # 10. (lb/s)	637.	565.	2560.	2370.
Flowrate thru Belt Conveyor # 14. (lb/s)	4.	6.	19.	28.
Flowrate thru Belt Conveyor # 15. (lb/s)	635.	561.	2550.	2340.

Table 10.6 Optimization Results

	Case 1	Case 2	Case 3	Case 4
Mass of sorbent:				
Sorbent in Feed Hopper # 1. (ton)	10.	10.	10.	10.
Sorbent in Storage Tank # 6. (ton)	1150.	1020.	4600.	4250.
Sorbent in Holding Bin # 11. (ton)	191.	170.	768.	710.
Power Requirements:				
Screw Feeder # 2. (KW)	0.7	1.1	3.0	5.1
Belt Conveyor # 3. (KW)	0.6	0.9	2.6	4.3
Bucket Elevator # 4. (KW)	177.0	161.0	714.0	680.0
Belt Conveyor # 5. (KW)	15.2	13.8	61.2	58.3
Screw Feeder # 7. (KW)	20.5	18.1	91.9	84.3
Belt Conveyor # 8. (KW)	10.9	9.1	102.5	89.8
Belt Conveyor # 10. (KW)	10.9	9.1	102.6	89.9
Screw Feeder # 12. (KW)	18.7	16.5	80.5	74.0
Belt Conveyor # 14. (KW)	0.1	0.1	0.4	0.7
Belt Conveyor # 15. (KW)	10.1	8.9	40.3	37.1
I.D. Fan # 16. (KW)	56.8	16.4	651.0	389.0
Humidification System # 17. (KW)	0.0	0.0	0.0	0.0
Total power requirement. (KWh)	427.0	365.0	2230.0	1910.0
Total Cost of Power. (K$/Year)	79.	67.	410.	352.
Total Cost of Water. (K$/Year)	0.	0.	0.	0.
Total Cost of Sorbent. (K$/Year)	531.	885.	2470.	4100.
Total Operating Cost. (K$/Year)	609.	952.	2880.	4450.
Cost of Feed Hopper # 1. (K$)	27.	27.	27.	27.
Cost of Screw Feeder # 2. (K$)	5.	5.	5.	5.
Cost of Belt Conveyor # 3. (K$)	113.	113.	113.	113.
Cost of Bucket Elevator # 4. (K$)	536.	489.	2162.	2058.
Cost of Belt Conveyor # 5. (K$)	113.	113.	113.	113.
Cost of Storage Tank # 6. (K$)	757.	695.	2002.	1894.
Cost of Screw Feeder # 7. (K$)	16.	15.	22.	22.
Cost of Belt Conveyor # 8. (K$)	71.	63.	234.	219.

Table 10.6 Optimization Results

	Case 1	Case 2	Case 3	Case 4
Cost of Reactor Units # 9. (K$)	2269.	2608.	8542.	8974.
Cost of Belt Conveyor # 10. (K$)	71.	63.	234.	219.
Cost of Holding Bin # 11. (K$)	216.	198.	571.	541.
Cost of Screw Feeder # 12. (K$)	10.	10.	14.	14.
Cost of Regenerator Units # 13. (K$)	1508.	1566	5412.	5692.
Cost of Belt Conveyor # 14. (K$)	113.	113.	113.	113.
Cost of Belt Conveyor # 15. (K$)	62.	62.	62.	62.
Cost of I.D. Fan # 16. (K$)	12.	7.	278.	165.
Cost of Humidification System # 17. (K$)	0.	0.	0.	0.
Total Fixed Cost. (K$)	5896.	6147.	19904.	20230.
System Cost (K$/Year)	1612.	1997.	6261.	7887.

Discussion of Results

The predictions in Tables 10.6 and 10.7 differ significantly from those expected based on experiments in the pilot-scale moving bed [11]. An optimum reactor length (in the direction of gas flow) of 7–18 ft is predicted while the pilot plant operated with only a 1.5 ft length. The expected increase in pressure drop in the larger bed is eliminated by the use of larger stone (approximately 1/4 here and approximately 1/8 in in the pilot experiments). The larger stone carries with it a larger cooling capacity because it has a higher ratio of volume to surface area. (Thermal mass scales with volume but total reactivity scales with surface area.) This effect depends on the assumption that sorbent is fed to the reactor at a cool 60°F. The increased cooling is enough to eliminate the need for humidification of the flue gas as well as the addition of water in the sorbent feed.

Table 10.7 Optimum Values for Search Variables

Case	CSS0 (lbmol/ft^3)	D_p (in)	U_{g0} (ft/s)	V_s (ft/h)	X (ft)	Mesh (mm)
1	.0269	0.246	1.49	11.5	7.8	3.33
2	.0268	0.240	0.91	11.6	7.3	2.84
3	.0257	0.295	1.79	12.5	15.1	3.71
4	.0264	0.296	1.41	12.3	18.2	3.26

Optimum values for other variables listed in Table 10.7 are very close to values used experimentally [11]. Superficial gas velocities of 1.0 to 1.5 ft/s used in the pilot plant correspond well to the range 0.9 to 1.8 ft/s predicted here. Bed velocities of 10–20 ft/h used in the pilot plant encompass the results in Table 7. Recycle concentrations were not evaluated in the pilot plant.

Comparisons between the four cases is less revealing. The inlet solid sulfur concentration (C_{SS0}) is fairly constant for all four cases. The particle diameter (δ_p) as well as the solids velocity (V_S) seem to be affected by only the power plant capacity and not the coal sulfur concentration. The gas velocity (U_{g0}) and the mesh size both decrease with increasing sulfur concentration and increase with increasing plant capacity. The reactor length (X) increases with power plant capacity. For the 100 MW cases (1 and 2) the length decreases with increasing sulfur concentration and for the 500 MW cases the length increases with increasing sulfur concentration. There is no apparent significance to these trends.

The total annualized cost increases with both the coal sulfur content and the power plant capacity. The economy-of-scale comparison between the 100-MW and 500-MW capacities indicates a negligible improvement with size for both the capital and operating costs. This suggests that this technology might best be applied to smaller boilers. The effect of variation in the sulfur level is also relatively small. Fixed costs are almost the same for the 1150 and 2550 ppm SO_2 flue gases. Operating cost increases with SO_2 content but much less than proportionally. Even the cost of sorbent increases less than proportionally to the sulfur content because a smaller screen size is suggested for the high-sulfur cases. The trend with sulfur content indicates that this technology is best suited for high-sulfur coals.

Conclusions

The simulation and optimization algorithm developed during this work produces a thorough and unbiased prediction of optimum operating conditions for a moving-bed granular limestone desulfurization process. The cost formulas have not been compared to those used in other studies so the absolute values of the cost estimates are not as significant as the comparison between different cases. The results indicate that the technology is best suited for high-sulfur, small-scale applications. Future experimental work should evaluate the use of larger (1/4 in) sorbent.

Simulation and Optimization of a Granular Limestone Flue Gas Desulfurization Process

Nomenclature

a	Geometric surface area of the sorbent. (ft^2/ft^3 of reactor)
A	Coefficient of dj/dx in the differential equations. Can be a function of the system variables. (Units specific to each equation)
a_{hg}	The heat transfer area for transfer between the gas and liquid phases. (ft^2/ft^3 of reactor)
a_{mg}	The mass transfer area in the gas phase boundary layer. (ft^2/ft^3 of reactor)
a_{ml}	The mass transfer area in the liquid layer. (ft^2/ft^3 of reactor)
a_{mp}	The mass transfer area in the precipitate layer. (ft^2/ft^3 of reactor)
A_{ri}	Surface area associated with a particular piece of equipment. (ft^2)
b	The fractional change in superficial liquid velocity over a large grid element (in either the x or y directions). Used to determine the location of the drying front.
B	Coefficient of dj/dy in the differential equations. Can be a function of the system variables. (Units specific to each equation)
C_i	Installed cost for unit(s) i. ($)
C_A	Annual system cost for the desulfurization plant. ($/yr)
C_{Cle}	The concentration of Ca++ in the liquid which is in equilibrium with the calcium in the unreacted limestone. The same as the solubility of the limestone. (lbmole/ft^3 of liquid)
C_{Ng}	Concentration of noncondensable in the gas phase. (lbmole/ft^3 of gas)
C_O	Operating cost for the desulfurization plant. ($/hr)
C_{pCs}	Heat capacity of sorbent. Includes sulfates, sulfites, and carbonates. (Btu/lbmole calcium ion/°R)
C_{pg}	Heat capacity of the gas phase. (Btu/lbmole/°R)

C_{pNg} Heat capacity of noncondensable. Includes SO_2, CO_2, N_2, and O_2. (Btu/lbmole/°R)

C_{pWg} Heat capacity of water vapor. (Btu/lbmole/°R)

C_{pWl} Heat capacity of liquid water. (Btu/lbmole/°R)

C_{Sg} The concentration of SO_2 in the gas. (lbmole/ft³ of gas)

C_{Sg0} The concentration of SO_2 in the gas at the reactor inlet. (lbmole/ft³ of gas)

C_{Ss} The concentration of sulfate or sulfite in the solid. (lbmole/ft³ of solid)

C_{Ss0} The concentration of sulfate or sulfite in the solid at the reactor inlet. (lbmole/ft³ of solid)

C_{Sse} The concentration of sulfate or sulfite in the solids in the reactor exit. (lbmole/ft³ of solid)

C_{Ssr} The concentration of sulfate or sulfite in the solids in the recycle stream. (lbmole/ft³ of solid)

C_{Sst} The terminal sulfate or sulfite concentration (the concentration when the reaction is assumed to stop). (lbmoles/ft³ of solid)

C_{Ssw} The sulfate or sulfite concentration in the solids in the waste stream. (lbmoles/ft³ of solid)

C_{Wg} The concentration of water in the gas. (lbmole/ft³ of gas).

CW_{g0} The concentration of water in the gas at the reactor inlet. (lbmole/ft³ of gas).

C_{Wge} The concentration of water in the gas at the reactor exit. (lbmole/ft³ of gas).

C_{Wgi} The concentration of water in the gas at the gas/liquid interface. (lbmole/ft³ of gas)

d	Power used in empirical relation for the concentration of sulfate or sulfite in the solids in the recycle stream.
D_{Cl}	Diffusivity of Ca++ in the liquid layer. (ft²/s)
D_m	Diameter of the hole in the screen mesh (ft)
δ_p	Sorbent particle diameter. (ft)
D_{Sg}	Diffusivity of SO_2 in the gas phase. (ft²/s)
D_{Sl}	Diffusivity of SO_3-- (SO_4--) in the liquid phase. (ft²/s)
D_{Wg}	Diffusivity of water in the gas phase. (ft²/s)
e	Exponent in empirical equation that predicts the blinding effect on the sorbent. (dimensionless)
E	Value of non-homogenous part in the differential equations. Can be a function of the system variables. (Units specific to each equation)
F_{Ei}	Inflation escalator for process unit i. (dimensionless)
F_{Ii}	Installed cost factor for process unit i. (dimensionless)
f_{Ng}	Flux of noncondensables in the gas phase in the direction of gas flow. Includes SO_2, CO_2, N_2, and O_2. (lbmole/s/ft² of reactor)
F_p	Packing factor of the reactor bed. (dimensionless)
f_S	Fraction of the sulfur removed from the flue gas at a given height in the reactor bed. (dimensionless)
f_{Sa}	Average sulfur removal for the reactor bed. (dimensionless)
f_{Sat}	Target average sulfur removal for the bed. (dimensionless)
f_{Wg0}	Flux of water in the gas phase in the direction of gas. (lbmole/s/ft² of reactor)
G_i	Mass flowrate through process unit i. (lb/s)

g_c	Conversion factor. (32 ft-lbm/lbf/s^2)
g_g	Rate of heat transfer from the gas phase to the liquid phase. (Btu/s/ft^3 of reactor)
g_S	Rate of transport of sulfur (dioxide) from the gas phase to the solid phase. (lbmole/s/ft^3 of reactor)
g_W	Rate of transport of water from the gas phase to the liquid phase. (lbmole/s/ft^3 of reactor)
H_i	Height of process unit i. (ft)
h_g	Overall heat transfer coefficient between the gas and liquid phases. (Btu/s/ft^2/°R)
H_S	Henry's law constant for SO$_2$. Based on concentration of SO$_3$-- (SO$_4$-) in liquid phase. Evaluated at sorbent temperature. (atm-ft^3/lbmole)
J_g	Factor for heat or mass transfer for gas phase. (dimensionless)
k	Empirical constant used in dry capture equation. (dimensionless)
k_{mC}	The pseudo mass transfer coefficient for Ca++ in the dissolution zone. Based on liquid phase concentrations. (ft/s)
k_{mS}	The mass transfer coefficient for SO$_2$ in flue gas for transport of SO$_2$ to the surface of the water layer. Based on gas phase concentrations. (ft/s)
k_{mW}	The mass transfer coefficient for water in flue gas for transport of water to or from the surface of the water layer. Based on gas phase concentrations. (ft/s)
k_{tg}	Thermal conductivity of the gas phase. (Btu/s/ft/°R)
L_i	Length of process unit i. (ft)
L_x	Length of bed in x direction. (ft)
L_y	Length of bed in y direction. (ft)

M_i	Mass of sorbent in process unit i or mass of process unit i. (lb)
M_g	Average molecular weight in the gas phase. (lb/lbmole)
N_i	Number of parallel units of type i. (dimensionless)
N_{Bi}	Biot Number. (dimensionless)
N_f	Number of calls to subroutine DEQTERMS.
N_{Fo}	Fourier Number. (dimensionless)
N_{Ga}	Galileo Number. (dimensionless)
N_{Prg}	Prandtl number for the gas phase. (dimensionless)
N_{Reg}	Reynolds number for the gas phase. (dimensionless)
N_{Rel}	Reynolds number for the liquid phase. (dimensionless)
N_{ScS}	Schmidt number for sulfur dioxide in the gas phase. (dimensionless)
N_{ScW}	Schmidt number for water in the gas phase. (dimensionless)
P_i	Power consumption for unit i. (kW)
P_W	Vapor pressure of water evaluated at sorbent temperature. (atm)
QA	The amount of compressed air required for humidification system. (lbmole/s)
$Qg0$	Total volumetric flowrate of flue gas in all parallel units at the gas inlet. (ft^3/s)
Q_W	The amount of water required for humidification system. (lbmole/s)
R	Ideal gas constant. (ft3atm/lbmole/°R)
S_p	Sphericity of sorbent particle. (dimensionless)
T_g	Temperature of the gas phase. (°R)
T_{g0}	Temperature of the gas phase at the reactor inlet. (°R)

T_{ge}	Temperature of the gas phase at the reactor exit. (°R)
T_{gi}	Temperature of the gas phase prior to being humidified. (°R)
T_l	Temperature of the liquid phase. (°R)
T_{l0}	Temperature of the liquid phase at the reactor inlet. (°R)
U_g	Superficial velocity of the gas phase. (ft^3 gas/ft^2 reactor/s)
U_{g0}	Superficial velocity of the gas phase at the reactor inlet. (ft^3 gas/ft^2 reactor/s)
U_l	Superficial velocity of the liquid phase. (ft^3 liquid/ft^2 reactor s)
U_{l0}	Superficial velocity of the liquid phase at the reactor inlet. (ft^3 liquid/ft^2 reactor /s)
U_s	Superficial velocity of the solid phase. (ft^3 gas/ft^2 reactor/s)
V_i	Volume of process unit i. (ft^3)
v_s	Velocity of the solid phase. (ft/s)
x	Horizontal direction. The value is zero at the point where the flue gas enters the bed.(ft)
y	Vertical direction. The value is zero at the point where the sorbent phase enters the bed. (ft)
Y_r	Relative humidity.
Y_{ri}	Relative humidity at the reactor inlet.
δ_l	Thickness of water layer. (ft)
δ_p	Thickness of precipitate layer. (ft)
δ_r	Thickness of SO3-- (SO4--) diffusion zone. Distance between gas/liquid interface and reaction front. (ft)
ΔP	Pressure drop across the bed. (atm)

Simulation and Optimization of a Granular Limestone Flue Gas Desulfurization Process

$\Delta\varphi_2$ — Intermediate quantity used in corrector equation for solving the differential equations (slope evaluated at the new point).

$\Delta\varphi_1$ — Intermediate quantity used in the predictor equation for solving the differential equations (slope evaluated at the base point).

γ — Fraction of limestone lost in waste stream.

ε_g — Fraction of bed occupied by gas. (dimensionless)

ε_{g0} — Fraction of bed occupied by gas at the reactor inlet. (dimensionless)

ε_l — Fraction of bed occupied by liquid. (dimensionless)

ε_{ls} — Maximum fraction of bed occupied by liquid when the liquid is not flowing relative to sorbent (static liquid volume fraction). (dimensionless)

ε_s — Fraction of bed occupied by solid. (dimensionless)

λ — Latent heat of vaporization of water. Evaluated at T l. (Btu/lbmole)

μ_g — Viscosity of the gas phase. (lbmole/ft/s)

μ_{g0} — Viscosity of the gas phase at the reactor inlet. (lbmole/ft/s)

μ_l — Viscosity of the liquid phase. (lbmole/ft/s)

φ_i — General dependent variable.

ρ_b — The bulk density of the sorbent. (lb/ft^3 of reactor)

ρ_g — The density of the gas phase. (lbmole/ft^3 of gas)

ρ_{g0} — The density of the gas phase at the reactor inlet. (lbmole/ft^3 of gas)

ρ_i — The intrinsic density of the sorbent. (lbs/ft^3 of sorbent)

ρ_l — The density of the liquid phase. Same as CWl. (lbmole/ft^3 of liquid)

ρ_p — The density of the precipitate layer. (lbmole calcium/ft^3 of precipitate)

ρ_s The density of the solid phase. Same as CCs. (lbmole calcium/ft³ of solid)

τ_i Holdup time for unit i. (s)

References

1. Prudich, M. E., K. W. Appell, J. D. McKenna, *Pilot-Scale Limestone Emission Control (LEC) Process: A Development Project,* Final Report OCDO Grant no. CDO/D-88-49, (1994)
2. Jozewicz, W. and G. Rochelle, "Modeling of SO2 Removal by Spray Dryers," *Proceedings; First Annual Pittsburgh Coal Conference*, Pittsburgh, PA, (1984)
3. Harriot, P. and M. Kinzey, "Modeling of the Gas and Liquid Phase Resistances in the Dry Scrubbing Process for SO_2 Removal," *Proceedings: Third Annual Pittsburgh Coal Conference*, Pittsburgh, PA (1986)
4. Karlsson, H. T. and J. Klingspor, "Tentative Modeling of Spray Drying Scrubbing of SO_2," *Chem. Eng. Technol.*, 10(2), p. 104 (1987)
5. Appell, K. W., *A Mathematical Simulation of ETS' Limestone Emission Control Process Using the Method of Characteristics: Fixed-Bed Configuration/Gas-Phase Mass Transport Control,* M. S. Thesis, Ohio University (1989)
6. Gullett, B. K. and J. C. Kramlich, "Fundamental Processes Involved in SO2 Capture by Calcium Based Adsorbents," *Proceedings: Fourth Annual Pittsburgh Coal Conference*, Pittsburgh, PA (1987)
7. Visneski, M. J., *Modeling of the Low Temperature Reaction of Sulfur Dioxide and Limestone Using a Three Resistance Film Theory Instantaneous Reaction Model,* Ph.D. Dissertation, Ohio University (1991)
8. Reddy, S. N., *A Mathematical Simulation of ETS' Limestone Emission Control (LEC) Process using a Moving Bed Configuration,* M. S. Thesis, Ohio University (1991)
9. Klingspor, J. and H. T. Karlsson, I. Bjerle, "A Kinetic Study of the Dry SO2-Limestone Reaction at Low Temperatures," *Chem. Eng. Commun.*, 22, 88 (1983)
10. Jorgensen, C., J.C.S. Chang and T.G. Brna, "Evaluation of Sorbents and Additives for Dry SO_2 Removal," *Environmental Progress*, 6, 26 (1987)
11. Prudich, M. E., L. Ben-Said, M. Maldei and K. J. Sampson, *Low Temperature Dry Scrubbing Reaction Kinetics and Mechanisms,* First Quarter Report OCDO Grant no. CDO/R-88-2C/B (1993)
12. Keeth, R. J., D.L. Baker, P. E. Tracy, G. E. Ogden and P. A. Ireland, *Economic Evaluation of Flue Gas Desulfurization Systems,* EPRI GS-7193, Volumes 1 and 2, EPRI Publications (1991)
13. Chattopadhyay, S., *Modeling and Evaluation of Granular Limestone Dry Scrubbing Processes,* M. S. Thesis, Ohio University (1992)
14. Reklaitis, G. V., A. Ravindran and K. M. Ragsdell, *Engineering Optimization,* Wiley, NY (1983)
15. Himmelblau, D. M., *Applied Nonlinear Programming,* McGraw-Hill, NY (1972)
16. Biegler, L. T., W. D. Seider and J. D. Seader, *Mathematical Modeling and Optimization,* AIChE Continuing Education Course Notes, American Institute of Chemical Engineers, NY (1988)
17. McCabe, W. L. and J. C. Smith, *Unit Operations of Chemical Engineering,* 3rd ed., McGraw-Hill (1976)
18. The Babcock & Wilcox Company, *Steam - Its generation and Use,* 38th ed., Babcock and Wilcox, NY (1972)
19. Duespohl, D.W., *Modeling and Optimization of a Cross-Flow, Moving-Bed, Flue Gas Desulfurization Reactor,* M.S. Thesis, Ohio University (1995)
20. Ozisik, M. N., *Heat Transfer, A Basic Approach,* McGraw-Hill, NY (1985)

21. Dwivedi, P. and N. Upadhyay, "Particle-Fluid Mass Transfer in Fixed and Fluidized Beds," *Ind. Eng. Chem. Process. Des. Dev.*, 16, 164, (1977)
22. Maldei, M. J., *Low Temperature Dry Scrubbing Reaction Kinetics and Mechanisms: Limestone Dissolution and Solubility*, M.S. Thesis, Ohio University (1993)
23. Foust, A. S., L. A. Wenzel, C. W. Clump, L. Maus and L. B. Ansersen, *Principles of Unit Operation*, 2nd ed., Wiley, NY (1980)
24. Reid, R. C. and T. K. Sherwood, *The Properties of Gasses and Liquids - Their Estimation and Correlation*, McGraw-Hill, NY (1958)
25. Chan, P. K.and G. T. Rochelle, "Limestone Dissolution-Effects of pH, CO2, and Buffers Modeled by Mass Transfer," *Amer. Chem. Soc. Symp. Series 188*, Washington, DC (1982)
26. Gullet, B. K.and K. R. Bruce, *Identification of $CaSO_4$ Formed by Reaction of CaO and SO_2*, Project Summary, EPA/600/S-7-88/024 (1989)
27. Rabe, A. E.and J. F. Harris, "Vapor Liquid Equilibrium Data for the Binary System, Sulfur Dioxide and Water," *J. of Chem. and En.g Data*, 8(3), 333-336 (1963)
28. Smith, J. M.and H. C. Van Ness, *Introduction to Chemical Engineering Thermodynamics*, 3rd ed., McGraw-Hill, NY (1975)
29. Specchia, V.and G. Baldi, "Pressure Drop for Two Phase Concurrent Flow in Packed Beds," *Chem. Eng. Sci.*, 32, 512-523 (1977)
30. Yu, H. C.and S. V. Sotirchos, *AIChE J.*, 33(3), 382 (1987)
31. Reid, R. C., J. M. Prausnitz and B. E. Poling, *The Properties of Gasses and Liquids*, 5th ed., McGraw-Hill, NY (1987)
32. 3Walas, S. M., *Chemical Process Equipment Selection and Design*, Butterworths Series in Chemical Engineering (1988)
33. Peters, M. S. and K. D. Timmerhaus, *Plant Design and Economics for Chemical Engineers*, 4th ed., McGraw-Hill, NY (1991)
34. Woods, Donald R., *Cost Estimation for the Process Industries*, McMaster University Bookstore Custom Courseware, Hamilton Ontario (1982)
35. Ulrich, Gael D., *A Guide to Chemical Engineering Process Design and Economics*, Wiley, NY (1984)
36. 3Stanley, W.M., *Chemical Process Equipment Selection and Design*, Butterworth (1988)
37. Perry, R. H. and C. H. Chilton, *Chemical Engineers' Handbook*, 5th ed., 9-6, McGraw Hill (1973)

Appendix A–Governing Differnetial Equations

Material and Enthalpy Balances

The following equations are a general starting point for development written in terms of component fluxes and inter-phase transport. The direction z is general and can stand for x or y depending on the direction of the flow.

Equation A.1
$$\frac{\partial}{\partial z} f_{sg} + g_s = 0$$

Equation A.2
$$\frac{\partial}{\partial z} f_{Wg} + g_W = 0$$

Equation A.3
$$\frac{\partial}{\partial z} f_{Ng} = 0$$

Equation A.4
$$\frac{\partial}{\partial z} f_g + g_W = 0$$

Equation 10.5
$$\frac{\partial}{\partial z} f_{Ss} - g_s = 0$$

Equation A.6
$$\frac{\partial}{\partial z} f_{Cs} = 0$$

and

Equation A.7
$$\frac{\partial}{\partial z} f_{Wl} - g_W = 0$$

The interphase transport rate for noncondensables and calcium is assumed to be zero. The interphase transport of SO_2 to the sorbent is assumed to be balanced by the interphase transport of CO_2 from the sorbent. The loss of O_2 from the gas phase due to adsorption (and reaction to convert sulfite to sulfate) is neglected. Material balances for the gas- and solid-phase sulfur and for the gas- and liquid-phase water are identified as "key" equations since these are the primary dependent variables of interest in the material balances.

The gas phase and solid/liquid phase enthalpy balances are

Equation A.8

$$\frac{\partial}{\partial z}(f_{Ng}H_{Ng} + f_{Wg}H_{Wg}) + g_g + g_W H_{Wg} = 0$$

and

Equation A.9

$$\frac{\partial}{\partial z}(f_{Cs}H_{Cs} + f_{W1}H_{W1}) - g_g - g_W H_{Wg} = 0$$

The heat effects associated with interphase transport of SO_2, CO_2 and O_2 are ignored. Noncondensables and sorbent species have averaged or representative enthalpies. Note in particular that H_{Cs} is the enthalpy of all sorbent species per lbmole of calcium. The liquid and solid phases are assumed to be at the same temperature. The latent heat of vaporization of water is assumed to be released to the liquid phase as condensation occurs. This corresponds to the assumption that the heat transfer resistance is in the gas boundary layer outside the liquid.

A critical assumption inherent in these equations is that the temperature in the solid phase is uniform across the particles. A rough estimate for the Biot number for this system gives a value of, $N_{Bi}=0.2$. The required values for the calculation are $D_p=0.02$ft, $k_{ts}=0.7$Btu/hr/ft/°F, $\rho_g=0.8$lb/ft^3, $U_g=1$ft/s, $\mu g=0.05$lb/ft/hr, $e_g=0.3$, $N_{Reg}=400$, $J_g=0.14$, $C_{pg}=0.24$Btu/lb/°F, $N_{Prg}=0.7$, and $h=14$Btu/hr/ft^2/°F. The given values imply that if one waits one minute following a step change in the gas temperature, the temperature in the center of a spherical solid pellet will fall (or rise) to within approximately 20 percent of the equilibrium value [20]. For this result the following values are also required: $\rho_s=15$lb/ft^2, $C_{ps}=0.21$Btu/lb/°F, and $N_{Fos}=3.3$. Since the bed residence time of sorbent particles is much longer than one minute, the assumption of a uniform temperature distribution in the solid phase seems to be adequate. However it is also true that inaccurate results will be generated near the solid feed point.

Gas Phase Material Balances

Before solving the four material balances they must be expressed in a set of four dependent variables. The key gas-phase balances are

Equation A.10

$$\frac{\partial}{\partial z} f_{Sg} + g_S = 0$$

and

Equation A.11

$$\frac{\partial}{\partial z} f_{Wg} + g_W = 0$$

For the sulfur and water fluxes, the following substitutions can be made.

Equation A.12

$$f_{Sg} = U_g C_{Sg}$$

and

Equation A.13

$$f_{Wg} = U_g C_{Wg}$$

Differentiating these expressions gives

Equation A.14

$$\frac{\partial}{\partial z} f_{Sg} = U_g \frac{\partial}{\partial z} C_{Sg} + C_{Sg} \frac{\partial}{\partial z} U_g$$

and

Equation A.15

$$\frac{\partial}{\partial z} f_{Wg} = U_g \frac{\partial}{\partial z} C_{Wg} + C_{Wg} \frac{\partial}{\partial z} U_g$$

This in turn requires an explicit expression for U_g and $\partial U_g/\partial z$. These quantities can be found from the general material balance for the total gas stream, Equation A.4. Using the ideal gas law we can substitute

Equation A.16

$$f_g = U_g r_g = U_g P/(RT_g)$$

into the general material balance. Note that the pressure is assumed to be constant. The result of the substitution is

Equation A.17

$$\frac{\partial}{\partial z}(U_g/T_g) = -\frac{g_w R}{P}$$

Taking the derivative of the products in this equation and rearranging the result gives

Equation A.18

$$\frac{\partial}{\partial z}U_g = \frac{U_g}{T_g}\frac{\partial}{\partial z}T_g - \frac{g_w}{\rho_g}$$

At any point in the bed, the r.h.s. of this equation can be calculated with the aid of the gas phase enthalpy balance which is yet to be derived. Values for Ug are found by integrating this equation from the gas feed point (z=0) to a particular point in the bed. Using this relationship plus Equations A.14 and A.15 in the key gas phase material balances, and substituting the expression for $\partial U_g/\partial z$ gives

Equation A.19

$$U_g \frac{\partial}{\partial z}C_{Sg} + C_{Sg}\left(\frac{U_g}{T_g}\frac{\partial}{\partial z}T_g - \frac{g_w}{\rho_g}\right) + g_s = 0$$

and

Equation A.20

$$U_g \frac{\partial}{\partial z}C_{Wg} + C_{Wg}\left(\frac{U_g}{T_g}\frac{\partial}{\partial z}T_g - \frac{g_w}{\rho_g}\right) + g_w = 0$$

Solid Phase Material Balances

For the solid phase we need to consider the key material balance, Equation 10.5. The following substitution can be made for the sulfur flux.

Equation A.21

$$f_{Ss} = U_s C_{Ss}$$

This result is

Equation A.22

$$\frac{\partial}{\partial z}(U_s C_{Ss}) - g_s = 0$$

Also,

Equation A.23

$$U_s = U_{S0}(\rho_{S0}/\rho_s)$$

and

Equation A.24

$$\frac{1}{\rho_s} = \frac{f_{Ss}/\rho_P + (f_{Ca} - f_{Ss})/\rho_C}{f_{Cs}} = \frac{1}{\rho_C} + \frac{f_{Ss}}{f_{Cs}}\left(\frac{1}{\rho_P} - \frac{1}{\rho_C}\right)$$

The expression for $1/\rho_s$ is a simple weighted average of the individual component molar volumes. Note that ρ_P and ρ_C are constants. An explicit form for ρ_s can be found after substituting

Equation A.25

$$f_{Ss} = U_s C_{Ss}$$

and

Equation A.26

$$f_{Cs} = U_s C_{Cs}$$

into the equation for $1/\rho_s$ and recognizing that

Equation A.27

$$C_{Cs} = \rho_s$$

This gives

Equation A.28

$$\frac{1}{\rho_s} = \frac{1}{\rho_C} + \frac{C_{Ss}}{\rho_s}\left(\frac{1}{\rho_p} - \frac{1}{\rho_C}\right)$$

which simplifies to

Equation A.29

$$\rho_s = \rho_C - C_{Ss}(\rho_C/\rho_p - 1)$$

Since the reaction is assumed to stop when C_{Ss} reaches C_{Sst}, this will be the maximum value for C_{Ss}, and will produce the minimum value for ρ_s. Using the values ρ_C=1.692 lbmoles/cuft, ρ_p=1.368 lbmoles/cuft, and C_{Sst} = 0.09873 lbmoles/cuft (calculated assuming D_p=0.02 ft), the resulting value for ρ_s will vary between

Equation A.30

$$1.669 < r_s < 1.692$$

The range of values for ρ_s is less than 2% of the maximum; therefore, ρ_s is assumed to be constant and equal to ρ_C. This means that U_s is a constant, and therefore the derivative of the solid sulfur flux is a function of concentration only. Using this result, Equation A.22 becomes

Equation A.31

$$U_s \frac{\partial}{\partial z} C_{Ss} - g_s = 0$$

The solid phase sulfur material balance is important because the reaction rate is assumed to be a function of the solid sulfur concentration.

Dry Scrubbing Technologies for Flue Gas Desulfurization

Simulation and Optimization of a Granular Limestone Flue Gas Desulfurization Process

Liquid Phase Material Balances

The liquid phase material balance is Equation A.7. The proper substitution for the liquid flux is

Equation A.32

$$f_{W1} = U_1 C_{W1} = U_1 \rho_1$$

Note that in Equation A.32 C_{Wl} is equal to ρ_l since the concentration of the dissolved species is neglected. Also, as described in Appendix B, the density of the liquid is taken to be a constant. Making these assumptions and substituting Equation A.32 into Equation A.7, the liquid phase material balance is

Equation A.33

$$\frac{\partial}{\partial z} U_1 - \frac{g_w}{\rho_1} = 0$$

Detailed Enthalpy Balances

The gas phase and solid/liquid phase enthalpy balances are Equations A.8 and A.9. Expanding these equations gives

Equation A.34

$$f_{Ng}\frac{\partial}{\partial z}H_{Ng} + H_{Ng}\frac{\partial}{\partial z}f_{Ng} + f_{Wg}\frac{\partial}{\partial z}H_{Wg} + H_{Wg}\frac{\partial}{\partial z}f_{Wg} + g_g + g_W H_{Wg} = 0$$

and

Equation A.35

$$f_{Cs}\frac{\partial}{\partial z}H_{Cs} + H_{Cs}\frac{\partial}{\partial z}f_{Cs} + f_{W1}\frac{\partial}{\partial z}H_{W1} + H_{W1}\frac{\partial}{\partial z}f_{W1} + g_g + g_W H_{Wg} = 0$$

The material balances (in their basic form) can be used to eliminate many of the terms from these equations. This leaves

Equation A.36

$$f_{Ng}\frac{\partial}{\partial z}H_{Ng} + f_{Wg}\frac{\partial}{\partial z}H_{Wg} + g_g = 0$$

and

Equation A.37

$$f_{Cs}\frac{\partial}{\partial z}H_{Cs} + f_{Wl}\frac{\partial}{\partial z}H_{Wl} - g_W(H_{Wg} - H_{Wl}) - g_g = 0$$

Using the general principal

Equation A.38

$$dH = C_p dT$$

in these equations gives

Equation A.39

$$(\langle f_{Ng}C_{P_{Ng}} + f_{Wg}C_{P_{Wg}}\rangle)\frac{\partial}{\partial z}T_g + g_g = 0$$

and

Equation A.40

$$(\langle f_{Ng}C_{P_{Ng}} + f_{Wg}C_{P_{Wg}}\rangle)\frac{\partial}{\partial z}T_l - g_W(H_{Wg} - H_{Wl}) - g_g = 0$$

The only other simplification needed is to substitute for the fluxes in terms of the explicit variables in the material balance. The following identities are used:

Equation A.41

$$f_{Ng} = U_g C_{Ng}$$

Equation A.42

$$C_{Ng} = r_g \cdot C_{Wg}$$

Equation A.43

$$f_{Wg} = U_g C_{Wg}$$

Equation A.44

$$f_{Cs} = U_{srs}$$

and

$$f_{Wl} = U_l C_{Wl} = U_l r_l$$

Equation A.45

The resulting two enthalpy balance equations are

Equation A.46

$$U_g((\rho_g - C_{Wg})C_{P_{Ng}} + C_{Wg}C_{P_{Wg}})\frac{\partial}{\partial z}T_g + g_g = 0$$

and

Equation A.47

$$(U_s \rho_s C_{P_{Ca}} + U_l \rho_l C_{P_{Wl}})\frac{\partial}{\partial z}T_1 - g_W(H_{Wg} - H_{Wl}) - g_g = 0$$

Appendix B Transport and Physical Property Correlations

Sulfur (Dioxide) Transport Rate (g_s)

A resistance-in-series model is used to estimate the reaction rate or equivalently the sulfur (dioxide) transport rate. The chemical reaction is assumed to occur instantaneously at the reaction front. Possible resistances include diffusion across the gas phase boundary layer, through the liquid layer, and the dissolution of solid calcium ions. The dissolution process is modeled using a mass transfer coefficient, k_{mC}. In order to determine the transport rate we must first assume a location for the reaction front, d_r, within either the liquid layer or precipitate layer and then calculate the position of the reaction front from a solution of the appropriate set of algebraic equations shown below. If the assumed position proves to be incorrect then the location may have to be recalculated assuming that the other layer controls. For example if the reaction front is assumed to be in the liquid layer and the calculated position is deeper than the liquid layer thickness then the position must be recalculated assuming the reaction front is within the precipitate layer. This is caused by the fact that there are two possible controlling resistance positions deeper than the liquid layer. Once the position of the reaction front is known the transport rate can be back-calculated from the equations below.

There are eight possible transport situations on the surface of the sorbent. (Figure 10.2 shows three of these cases.) Cases I-A, I-B, I-C, and I-D are for the situation where the liquid layer extends above the precipitate layer. Cases II-A, II-B, and II-C are for the situation where the liquid layer does not extend above the precipitate layer and case III considers dry capture. Cases I-A and II-A describe the situations where the reaction front is at the gas-liquid interface. Case I-B is for the reaction front within the liquid and above the precipitate layer. Cases I-C and II-B describe the situation where the reaction front is within the liquid and within the precipitate layer, and for cases I-D and II-C the reaction front coincides with the liquid-solid interface.

Case I-A corresponds to $\delta_l > \delta_p$ and $\delta_r = 0$. This is the case of gas phase control and the liquid layer extending above the precipitate layer. The final transport equation can be written down immediately. Note the interfacial concentration, C_{sgi}, is zero.

Equation B.1

$$g_S = k_{ms} a_{mg} C_{Sg}$$

Dry Scrubbing Technologies for Flue Gas Desulfurization

Case I-B corresponds to $\delta_l > \delta_p$ and $\delta_l-\delta_p > \delta_r > 0$. In this case five expressions for the SO_2 transport rate are solved simultaneously along with the gas-liquid equilibrium relationship. They are

Equation B.2

$$g_S = D_{Sl}a_{ml}C_{Sli}/\delta_r$$

Equation B.3

$$= D_{Cl}a_{ml}C_{Clp}/(\delta l-\delta_p-\delta_r)$$

Equation B.4

$$= D_{Cl}a_{mp}(C_{Cls}-C_{Clp})/\delta_p$$

Equation B.5

$$= k_{mC}a_{mp}(C_{Cle}-C_{Clsi})$$

and

Equation B.6

$$C_{sgi} = H_S C_{Sli} r_g/P = C_{Sli}(H_S/RT_g)$$

Simultaneous solution of these equations gives the transport rate.

Equation B.7

$$g_S = \frac{C_{Sg}}{\dfrac{1}{k_{ms}a_{mg}} + \dfrac{\delta_r}{D_{S1}a_{ml}(RT_g/H_S)}} = \frac{C_{Cle}}{\dfrac{\delta_1-\delta_p-\delta_r}{D_{Cl}a_{ml}} + \dfrac{\delta_p}{D_{Cl}a_{mp}} + \dfrac{1}{k_{mC}a_{mp}}}$$

and the location of the reaction front

Equation B.8

$$\delta_r = \frac{\dfrac{\delta_1-\delta_p}{D_{Cl}a_{ml}C_{Cle}} + \dfrac{\delta}{D_{Cl}a_{ml}C_{Cle}} + \dfrac{1}{k_{mC}a_{mp}C_{cle}} - \dfrac{1}{k_{mC}a_{mp}C_{Sg}}}{\dfrac{1}{D_{Sl}a_{ml}C_{Ssg}(RT_g/H_S)} + \dfrac{1}{D_{Cl}a_{ml}C_{Cle}}}$$

If the calculated value of δ_r is less than zero then the reaction is gas phase controlled and the equation listed under Case I-A must be used. If the calculated value of δ_r is greater than δ_l-δ_p then the reaction front is within the precipitate layer or the reaction is dissolution rate controlled and case I-C or I-D must be used.

Case I-C corresponds to $\delta_l > \delta_p$ and $\delta_l > \delta_r > \delta_l - \delta_p$. This case is the same as Case I-B except that the reaction front is within the precipitate layer. The equations are

Equation B.9
$$g_S = k_{mS}a_{mg}(C_{Sg}-C_{sgi})$$

Equation B.10
$$= D_{Sl}a_{ml}(C_{Sli}-C_{Slp})/(\delta_l-\delta_p)$$

Equation B.11
$$= D_{Sl}a_{mp}C_{Slp}/(\delta_r-\delta_l+\delta_p)$$

Equation B.12
$$= D_{Cl}a_{mp}C_{Cls}/(\delta_l-\delta_r)$$

Equation B.13
$$= k_{mC}a_{mp}(C_{Cle}-C_{Cls})$$

and

Equation B.14
$$C_{sgi} = H_S C_{Sli}\rho_g/P = C_{Sli}(H_S/R_{Tg})$$

Simultaneous solution of these equations gives the transport rate.

Dry Scrubbing Technologies for Flue Gas Desulfurization

Simulation and Optimization of a Granular Limestone Flue Gas Desulfurization Process

Equation B.15

$$g_s = \frac{C_{Sg}}{\frac{1}{k_{mS}a_{mg}} + \frac{\delta_1 - \delta_p}{D_{Sl}a_{ml}(RT_g/H_s)} + \frac{\delta_r - \delta_1 + \delta_p}{D_{Sl}a_{ml}(RT_g/H_s)}} = \frac{C_{Cle}}{\frac{\delta_1 - \delta_r}{D_{Cl}a_{ml}} + \frac{1}{k_{mC}a_{mg}}}$$

and the location of the reaction front

Equation B.16

$$\delta_r = \frac{\frac{\delta_l}{D_{Cl}a_{ml}C_{Cle}} + \frac{1}{k_{mC}a_{ml}C_{Cle}} - \frac{1}{k_{mS}a_{mg}C_{Sg}} - \frac{\delta_1 - \delta_p}{D_{Sl}a_{ml}C_{Sg}(RT_g/H_s)} + \frac{\delta_1 - \delta_p}{D_{Sl}a_{ml}C_{Sg}(RT_g/H_s)}}{\frac{1}{D_{Sl}a_{mp}C_{Sg}(RT_g/H_s)} + \frac{1}{D_{Cl}a_{ml}C_{Cle}}}$$

If the calculated value of δ_r is less than $\delta_1 - \delta_p$ then the reaction front is above the precipitate layer or the reaction is gas phase controlled and Case I-A or I-B must be used. If the calculated value of δ_r is greater than δ_1 then the reaction is dissolution rate controlled and Case I-D must be used.

Case I-D corresponds to $\delta_1 > \delta_p$ and $\delta_r = \delta_1$. This is the case for dissolution rate control. The transport expression can be written down immediately. Note that the concentration of calcium ions at the unreacted solid surface, C_{Cls}, is zero.

Equation B.17

$$g_s = k_{mC}a_{mp}C_{Cle}$$

Case II-A corresponds to $\delta_p > \delta_1$ and $\delta_r = 0$. This is the case for gas phase control and the precipitate layer extending above the liquid layer. The gas phase resistance is split into two parts: the film above the precipitate and the film within the precipitate. Note that the interfacial concentration, C_{sgi}, is zero. The equations are

Equation B.18

$$g_s = k_{mS}a_{mg}(C_{Sg} - C_{Sp})$$

Equation B.19

$$= D_{Sg}a_{mp}C_{Sp}/(\delta_p - \delta_l)$$

Simultaneous solution of these equations gives

$$g_s = \frac{C_{Sg}}{\dfrac{1}{k_{mS}a_{mg}} + \dfrac{\delta_p - \delta_1}{D_{Sg}a_{mp}}}$$

Equation B.20

Case II-B corresponds to $\delta_p > \delta_1$ and $\delta_1 > \delta_r > 0$. This case is similar to Case I-C. Five expressions for the SO_2 transport rate are solved simultaneously along with the gas-liquid equilibrium relationship. They are

Equation B.21

$$g_S = k_{mS}a_{mg}(C_{Sg} - C_{Sp})$$

Equation B.22

$$= D_{Sg}a_{mp}(C_{Sp} - C_{Sgi})/(\delta_p - \delta_1)$$

Equation B.23

$$= D_{Sl}a_{mp}C_{Sli}/\delta_r$$

Equation B.24

$$= D_{Cl}a_{mp}C_{Cls}/(\delta_l - \delta_r)$$

Equation B.25

$$= k_{mC}a_{mp}(C_{Cle} - C_{Cls})$$

and

Equation B.26

$$C_{sgi} = H_S C_{Sli} \rho_g / P = C_{Sli}(H_S/RTg)$$

Simultaneous solution of these equations gives the transport rate.

Simulation and Optimization of a Granular Limestone Flue Gas Desulfurization Process

Equation B.27

$$g_s = \frac{C_{Sg}}{\frac{1}{k_{mS}a_{mg}} + \frac{\delta_p - \delta_l}{D_{Sg}a_{mp}} + \frac{\delta_r}{D_{Sl}a_{mp}(RT_g/H_s)}} = \frac{C_{Cle}}{\frac{\delta_l - \delta_r}{D_{Cl}a_{mp}} + \frac{1}{k_{mS}a_{mg}}}$$

and the location of the reaction front

Equation B.28

$$\delta_r = \frac{\frac{\delta_l}{D_{Cl}a_{ml}C_{Cle}} + \frac{1}{k_{mc}a_{mp}C_{Cle}} - \frac{1}{k_{mS}a_{mg}C_{Sg}} - \frac{\delta_l - \delta_p}{D_{Sg}a_{mp}C_{Sg}}}{\frac{1}{D_{Sl}a_{mp}(RT_g/H_s)} + \frac{1}{D_{Cl}a_{ml}C_{Cle}}}$$

If the calculated value of δ_r is less than zero then the reaction is gas phase controlled and Case II-A must be used. If the calculated value of δ_r is greater than δ_l then the reaction is dissolution rate controlled and Case II-C must be used.

Case II-C corresponds to $\delta_p > \delta_l$ and $\delta_r = \delta_l$. This is the case for dissolution rate control. The transport rate can be written down immediately. Note that the concentration of calcium ions at the unreacted solid surface, C_{Cls}, is zero. This case is identical to Case I-D.

Equation B.29

$$g_s = k_{mc}a_{mp}C_{Cle}$$

Case III corresponds to $\delta_l = 0$. This is the case for the case of dry sorbent capture. The controlling resistance must be either the gas film or dissolution. This choice is made by calculating both rates and accepting the lower value. The rate expressions were shown previously. From Case II-A, the rate for gas phase control is

Equation B.30

$$g_s = \frac{C_{Sg}}{\frac{1}{k_{mS}a_{mg}} + \frac{\delta_p - \delta_l}{D_{Sg}fa_{mp}}}$$

From Case II-C the rate for dissolution control is

Equation B.31

$$g_S = k_m c f a_{mp} C_{Cle}$$

Note the incorporation of f. This is a dry capture correction factor for the mass transfer area. The value of f is found using an equation that relates f to the fractional relative humidity through a Langmuir isotherm form.

Equation B.32

$$f = \frac{(1-k)Y_r}{1-kY_r}$$

Values for the constant, k, has been report by Prudich, et al [11] for different limestones. The average value of k=0.89 is used here.

Water Transport Rate (g_W)

A mass transfer coefficient based upon the gas side driving force is used to correlate the water mass transfer rate.

Equation B.33

$$g_W = k_{mW} a_{mg}(C_{Wg} - C_{Wgi})$$

Gas/Liquid Heat Transport Rate (g_q)

Newton's law of cooling is used to correlate the heat transfer rate between the liquid and gas phases.

Equation B.34

$$g_g = h_g a_{hg}(T_g - T_l)$$

Gas/Liquid Heat Transfer Coefficient (h_g)

The vapor phase heat transfer coefficient is obtained using Colburn j-factor [21] as

Equation B.35

$$h_g = \frac{J_g C_{pg} \rho_g U_g}{N_{Prg}^{2/3}}$$

Simulation and Optimization of a Granular Limestone Flue Gas Desulfurization Process

The computational requirements for this correlation can be simplified by replacing $N_{Prg}^{2/3}$.

Equation B.36

$$N_{Prg}^{2/3} \approx 0.723427 N_{Prg} + 0.283067$$

The accuracy of this approximation has been checked for the values of N_{Prg} between 0.7 and 1.15. This range corresponds to a temperature range of 520 to 620°R. The simplified correlation is then

Equation B.37

$$h_g \approx \frac{J_g C_{pg} \rho_g U_g}{0.723427 N_{Prg} + 0.283067}$$

Calcium Pseudo-Mass Transfer Coefficient (k_{mC})

The mass transfer coefficient for limestone dissolution has been considered to be constant. The value used is an average value and is extracted from the data given by Maldei [22]. The assumed value is given as

Equation B.38

$$k_{mC} = 4.7 \cdot 10^{-4}$$

SULFUR DIOXIDE MASS TRANSFER COEFFICIENT (k_{mS}). The following form is reported [23].

Equation B.39

$$k_{mS} = \frac{J_g U_g}{N_{Scs}^{2/3}}$$

The computational requirements for this correlation can be simplified by replacing $N_{ScS}^{2/3}$.

Equation B.40

$$N_{Scs}^{2/3} \approx 0.603776 N_{Scs} + 0.404978$$

This approximation is accurate for values of N_{ScS} between 1.125 and 1.6. This range corresponds to a temperature range of 520–670°R. The simplified correlation is then

Equation B.41

$$K_{mS} \approx \frac{J_g U_g}{0.603776 N_{Scs} + 0.404978}$$

Water Vapor Mass Transfer Coefficient (k_{mS})

The following form is reported [23].

Equation B.42

$$k_{mW} = \frac{J_g U_g}{N_{ScW}^{2/3}}$$

The computational requirements for this correlation can be simplified by replacing $N_{ScW}^{2/3}$.

Equation B.43

$$N_{ScW}^{2/3} = 0.743154 N_{ScW} + 0.267317$$

This approximation is accurate for values of N_{ScW} between 0.60 and 0.85. This range corresponds to a temperature range of 520–670°R. The simplified correlation is then

Equation B.44

$$k_{mW} = \frac{J_g U_g}{0.743154 N_{ScW} + 0.267317}$$

Colburn j Factor for Heat and Mass Transfer in the Gas Phase (J_g)

The following correlation is valid over the range $10 < N_{reg} < 15000$. Dwivedi and Upadhyay [21] studied particle-fluid mass transfer in fixed beds and fluidized beds for different gases and liquids. They concluded after an iterative least-square study that the mass transfer factor is inversely proportional to the bed voidage and that gas and liquid phase data can be best presented as

Equation B.45

$$J_g = \frac{0.765/N_{Reg}^{0.82} + 0.365/N_{Reg}^{0.386}}{\varepsilon_g}$$

The computational requirements for this correlation can be simplified by replacing the numerator in the expression with a polynomial.

Equation B.46

$$j_g = \frac{4.15945 \cdot 10^{-7} N_{Reg}^2 - 0.00030226 N_{Reg} + 0.101048}{\varepsilon_g}$$

The accuracy of this approximation for values of N_{Reg} between 150 and 290 which corresponds to a temperature range of 520–770°R.

GAS PHASE PRANDTL NUMBER (N_{Prg}). By definition

Equation B.47

$$N_{Prg} = \frac{\mu_g C_{pg}}{k_{tg}}$$

GAS PHASE REYNOLDS NUMBER (N_{reg}). The gas phase Reynolds number used for packed beds is defined by

Equation B.48

$$N_{Reg} = \frac{D_p \rho_g U_g}{\mu_g \varepsilon_g}$$

LIQUID PHASE REYNOLDS NUMBER (N_{Rel}). By definition

Equation B.49

$$N_{Rel} = \frac{D_p \rho_l U_l}{\mu_l}$$

SCHMIDT NUMBER FOR SULFUR DIOXIDE (N_{ScS}). By definition

$$N_{ScS} = \frac{\mu_g}{\rho_g D_{Sg}}$$

Equation B.50

SCHMIDT NUMBER FOR WATER VAPOR (N_{ScW}). By definition

$$N_{ScW} = \frac{\mu_g}{\rho_g D_{Wg}}$$

Equation B.51

DIFFUSIVITY OF CALCIUM IONS IN THE LIQUID PHASE (D_{Cl}). At infinite dilution, ionic diffusivities are calculated as [21]

$$D_l = RT \frac{\lambda_0}{n(Fa)^2}$$

Equation B.52

where l_0 and n are the equivalent ionic conductivity and charge on the ion respectively, and Fa is the Faraday number. Liquid diffusivities at a particular temperature can be estimated using the Stokes-Einstein relationship [21]

$$D_l/T_l = D_{lR}/T_R$$

Equation B.53

Using a value of 85.035 ft²/s for the diffusivity of Ca^{++} at 298°K (Foust et al., 1980) in the Stokes-Einstein relationship gives

$$D_{Cl} = 1.5851976 \cdot 10^{-11} T_l$$

Equation B.54

DIFFUSIVITY OF SULFUR DIOXIDE IN THE GAS PHASE (D_{SG}). Reid and Sherwood [24] used the concept of rigid spherical molecules undergoing elastic collisions and empirically correlated the diffusion coefficient of a binary gas system at low pressure as

Equation B.55

$$D_g = \frac{1.1177 \cdot 10^{-5} T^{3/2}[(M_1 + M_2)/(M_1 M_2)]^{1/2}}{P(V_{b1}^{1/3} + V_{b2}^{1/3})^2}$$

where D_g is the diffusivity of binary gas system, M_1 and M_2 are molecular weights, and V_{b1} and V_{b2} are atomic volumes of the two species under consideration. Considering the pressure of the system to be constant, the equation can be written as

Equation B.56

$$D_g \alpha T^{3/2}$$

The diffusivity of a binary system evaluated at a particular temperature can be expressed, relative to a reference temperature and reference diffusion coefficient, as

Equation B.57

$$D_g/D_{gR} = (T/T_R)^{3/2}$$

In this correlation flue gas is modeled as air. Considering a diffusion coefficient for an SO_2/air system at 273°K and 1.0 atmosphere of 0.122 cm² /s [23] the diffusivity for the SO_2/gas system is

Equation B.58

$$D_{Sg} = 1.2055 \cdot 10^{-8} T^{3/2}$$

The simplified correlation is then

Equation B.59

$$D_{Sg} \approx 2.24678 \cdot 10^{-7}(T_g + T_1) - 9.2172 \cdot 10^{-5}$$

DIFFUSIVITY OF SULFUR DIOXIDE IN THE LIQUID PHASE (D_{sl}). Following the method used for D_{Cl} and using a value of 114.097 ft²/s at 298°K [21] in the Stokes-Einstein relationship gives

Equation B.60

$$D_{Sl} = 3.411633 \cdot 10^{-11} T_l$$

DIFFUSIVITY OF WATER VAPOR (D_{Wg}). The same method as used for D_{Sg} is used here. Considering a diffusion coefficient for a water/air system at 298°K and 1.0 atmosphere of 0.260 cm²/s [24], the diffusivity for the water/gas system is

Equation B.61

$$D_{Wg} = 2.2527 \cdot 10^{-8} T^{3/2}$$

The simplified correlation is then

Equation B.62

$$D_{Wg} \approx 4.18957 \cdot 10^{-7} (T_g + T_1) - 1.7873 \cdot 10^{-4}$$

DENSITY OF THE UNREACTED LIMESTONE LAYER (ρ_C). Chan and Rochelle [25] calculated the molar density of calcite to be 0.0217 gmmole/cm3 (2.172 g/cm3). This is equivalent to

Equation B.63

$$\rho_C = 1.692 \text{ lbmole of calcium ion}/ft^3$$

DENSITY OF THE GAS PHASE (ρg). Assuming ideal gas behavior we can write

Equation B.64

$$\rho_g = \frac{P}{RT}$$

At atmospheric pressure this correlation is

Equation B.65

$$\rho_g = \frac{1.3695}{T}$$

DENSITY OF LIMESTONE (ρ_i, ρ_b). The intrinsic mass density of limestone, ρ_i, is taken to be 169.3 lbs/ft^3. The bulk density is found by multiplying the intrinsic density by the solids volume fraction, es.

Equation B.66

$$\rho_b = \rho_i \varepsilon_s$$

DENSITY OF THE LIQUID PHASE (ρ_l). Over the range 492°R < T < 670°R the density of water is equal to 3.417 lbmole/ft^3 to within 2%.

DENSITY OF THE PRECIPITATE LAYER (ρ_p). The molar volume of $CaSO_4$ can be expressed in terms of its density. Gullet and Bruce [26] found $CaSO_4$ molar volume to be 45.64 cm^3/gmmole (2.98 g/cm3). This is equivalent to 1.368 lbmole/ ft^3.

GAS PHASE VISCOSITY (μ_g). The viscosity of a binary gas mixture (air/noncondensables) at low pressure is given by Wilke's equation as [25]

Equation B.67

$$\mu_g = \frac{\mu_{gW}}{1 + (y_{gN}/y_{gW})\phi_{WN}} + \frac{\mu_{gN}}{1 + (y_{gW}/y_{gN})\phi_{NW}}$$

where ϕ_{WN} and ϕ_{NW} are interaction parameters given by

Equation B.68

$$\phi_{WN} = \frac{(1 + (\mu_{gW}/\mu_{gN})^{1/2}(M_N/M_W)^{1/4})^2}{2\sqrt{2}(1 + M_N/M_W)^{1/2}}$$

and

Equation B.69

$$\phi_{NW} = \frac{(1 + (\mu_{gN}/\mu_{gW})^{1/2}(M_W/M_N)^{1/4})^2}{2\sqrt{2}(1 + M_W/M_N)^{1/2}}$$

μ_{gN} and μ_{gW} are the pure component viscosities, and noncondensables (N) and water (W) are the two components in the binary mixture. Replacing M_W and M_N with numerical values, we have the following expressions for ϕ_{WN} and ϕ_{NW}.

Equation B.70

$$\phi_{WN} = 0.2776(1 + 1.1261(\mu_{gW}/\mu_{gN})^{1/2})^2$$

Equation B.71

$$\phi_{NW} = 0.2189(1 + 0.8879(\mu_{gN}/\mu_{gW})^{1/2})^2$$

The pure component viscosities of water vapor and noncondensables are [6,24]

Equation B.72

$$\mu_{gW} = \frac{1}{60}e^{(0.0018334T - 11.7307)}$$

and

Equation B.73

$$\mu_{gN} = \frac{1}{60}e^{(0.00'0626T - 11.7274)}$$

Here we have modeled the noncondensables component as air. Considering that air and water vapor behave ideally at atmospheric pressure, the mole fractions of water and noncondensables can be written as

Equation B.74

$$y_{gW} = 0.7302 C_{Wg} T_g$$

and

Equation B.75

$$y_{gN} = 1 - y_{gW}$$

Over the range 520–620°R the interaction parameters can be approximated as linear functions of temperature.

Equation B.76

$$\varphi_{WN} \approx 4.73325 \cdot 10^{-4} T_g + 0.883549$$

and

Equation B.77

$$\varphi_{NW} \approx -3.05792 \cdot 10^{-4} T_g + 1.0178$$

In addition the pure component viscosities can be represented as linear functions of T_g.

Equation B.78

$$\mu_{gW} \approx 8.0629 \cdot 10^{-10} T_g - 7.85994 \cdot 10^{-8}$$

and

Equation B.79

$$\mu_{gN} \approx 5.1777 \cdot 10^{-10} (T_g + 1.53943 \cdot 10^{-7})$$

SOLUBILITY OF LIMESTONE (C_{Cle}). The equilibrium concentration of limestone in water is an average value extracted from the data given by Maldei [22]. The value is an average for several limestones, at ph=8, and a temperature of 140°F.

Equation B.80

$$C_{cle} = 2.5 \cdot 10^{-5} \; lbmole/ft^3$$

HENRY'S LAW CONSTANT FOR SULFUR DIOXIDE (H_S). The Henry's law constant for SO_2 in air (atm ft^3/lbmol) is given by Rabe and Harris [27] as

Equation B.81

$$H_s = \frac{15.8232}{e^{(5123/T_1^{-9.2379501})}}$$

The simplified correlation applicable for the temperature range 520–660°R is

Equation B.82

$$H_s \approx 0.00238961 T_1^2 - 2.39879 T_1 + 610.46$$

WATER VAPOR PRESSURE (P_W). The Clapeyron equation for the calculation of the latent heat of vaporization is (Foust et al., 1980)

Equation B.83

$$ln(P_W) = A - B/T$$

where A and B are constants. This correlation is valid over the range 492°R < T < 672°R. This equation is useful for many purposes, but it is not sufficiently precise to provide accurate values over a wide range of temperatures. The Antoine equation has the form

Equation B.84

$$ln(P_W) = A - B/(T+C)$$

where A, B and C are constants. This equation is more satisfactory and more widely used. Most vapor pressure estimation and equations stem from an integration of Clausius-Clapeyron [23] equation

Equation B.85

$$\frac{dP}{dT} = \frac{\Delta H_{vap}}{T \Delta V_{vap}}$$

where ΔH_{vap} is the enthalpy of vaporization and ΔV_{vap} is the volume change due to vaporization. The options one has in this integration are limited. However, results available in the literature vary widely as each researcher normally introduces correction terms to obtain more accuracy. One particular form that is commonly used [23] is

Equation B.86

$$IN(P_W) = A - \frac{B}{T+C} + DT + E ln(T)$$

where A, B, C, D and E are constants. This form is referred to here as a modified Antoine equation. The following specific form was found by regression.

Equation B.87

$$IN(P_W) = 4.0366 + \frac{3210.1855}{T - 210} + 0.006136T - 0.2256 In(T - 460)$$

The vapor pressure correlation can be approximated with

Equation B.88

$$P_W \approx 3.27778 \bullet 10^7 T_1^3 - 5.28814 \bullet 10^{-4} T_1^2 + 0.285635 T_1 - 51.6159$$

This approximation is valid for values of T_1 between 530 and 672°R.

GEOMETRIC SURFACE AREA (a). The geometric surface area is given by

Equation B.89

$$a = \frac{6\varepsilon_s}{D_p S_p}$$

The value of the shape factor, S_p, used is 0.5 (which is a typical value for non-uniform solids [28], and e_s is taken as 0.579.

GAS/LIQUID HEAT TRANSFER AREA (a_{hg}).

Equation B.90

$$a_{hg} = a.$$

LIQUID/SOLID HEAT TRANSFER AREA (a_{hl}).

Equation B.91

$$a_{hl} = a.$$

GAS PHASE MASS TRANSFER AREA (a_{mg}).

Equation B.92

$$a_{mg} = a.$$

LIQUID PHASE MASS TRANSFER AREA (a_{ml}).

Equation B.93

$$a_{ml} = a$$

PRECIPITATE LAYER MASS TRANSFER AREA (a_{mp}). The precipitate layer mass transfer area is found using an empirical relation that gives the precipitate layer thickness. The volume of the precipitate (which is known from the mass balance equations) divided by the thickness gives the area occupied by the precipitate. Subtracting this area from the total results in the mass transfer area available. The empirical relation used to calculate the precipitate thickness is

Equation B.94

$$\delta_p = \frac{(C_{Sst})^{\frac{e-1}{e}} (C_{Ss})^{\frac{1}{e}} \varepsilon_s}{\rho_p a}$$

The area occupied by the precipitate, a_p, is

Equation B.95

$$a_p = \frac{C_{Ss} \varepsilon_s}{\rho_p \delta_p}$$

Substituting the expression for d_p into the equation for a_p and subtracting the result from a gives a_{mp}.

Equation B.96

$$a_{mp} = a\left(1 - \left(\frac{C_{Ss}}{C_{Sst}}\right)^{\frac{e-1}{e}}\right)$$

It is assumed that the value for C_{Sst} will be such that the precipitate thickness at this concentration will be independent of particle size. This implies that

Equation B.97

$$C_{Sst} = \frac{\alpha}{S_p D_p}$$

Dry Scrubbing Technologies for Flue Gas Desulfurization

Prudich et al. [11] have determined values for the parameters: $a = 9.83 \times 10^{-4}$ lbmoles/ft^2 and $e = 1.89$.

PRECIPITATE LAYER THICKNESS (δ_p). The precipitate layer thickness is found using the empirical relation

Equation B.98

$$\delta_p = \frac{(C_{Sst})^{\frac{e-1}{e}} (C_{Ss})^{\frac{1}{e}} \varepsilon_s}{\rho_p a}$$

C_{Sst} is the concentration at which the reaction stops (see a_{mp} discussion above).

LIQUID LAYER THICKNESS (d_l). The dynamic holdup for a packed bed with trickling water has been correlated by Specchia and Baldi [29] as

Equation B.99

$$\varepsilon_d = 3.86(N_{Rel})^{0.545}(N_{Ga})^{-0.42}\left(\frac{aD_p}{1-\varepsilon_s}\right)^{0.65}$$

where N_{Rel} is the Reynolds number for the liquid phase, N_{Ga} is the Galileo number, and ε_s is the solid phase volume fraction. The liquid phase Reynolds number and the Galileo number are defined as

Equation B.100

$$N_{Rel} = D_p U_1 \rho_1 / \mu_1$$

and

Equation B.101

$$N_{Ga} = \rho_1^2 g D_p^3 / \mu_1^2$$

The total liquid holdup, ε_l, is the sum of dynamic holdup, and the static liquid holdup. Appel [5] measured the static holdup of water in a limestone bed to be 0.0432 pounds of water per pound of limestone. Multiplying this by the bulk density of limestone and dividing by the density of water gives the volume fraction of water. Using $\rho_i = 169.3$ lbs/ft^3, $\varepsilon_s = 0.543$, and $\rho_l = 3.417$ lbmoles/ft^3, ε_{ls} becomes

Equation B.102

$\varepsilon_{ls} = 0.0645$

The total liquid holdup is

Equation B.103

$\varepsilon_l = \varepsilon_{ls} + \varepsilon_d$

The liquid water thickness is tentatively calculated as

Equation B.104

$\delta_l = \varepsilon_l / a_{mp}$

If the calculated d_l is greater than the precipitate thickness, d_p, then the liquid thickness must be adjusted. The volume of liquid that exceeds the volume available within the precipitate layer is spread evenly over the surface of the sorbent. The liquid thickness is then the precipitate thickness plus the thickness of the liquid above the precipitate layer. The equation for d_l becomes

Equation B.105

$\delta_l = \delta_p + \dfrac{(\varepsilon_l - \delta_p a_{mp})}{a}$

SOLID PHASE VOLUME FRACTION (ε_s). Appel [5] experimentally determined the bulk density of limestone to be 91.83 lbs/ft^3. Dividing this by the intrinsic density, $r_i = 169.3$ lbs/ft^3, gives the solids volume fraction as

Equation B.106

$\varepsilon_s = 0.543$

Note that a value of $\varepsilon_s = 0.5$ has been reported elsewhere [30] for limestone.

HEAT CAPACITY OF SORBENT (C_{pCs}). The following correlation is valid over the range 536.4°R < T < 2160°R for $CaCO_3$ [28].

Equation B.107

$$C_{pCs} = 24.98 + 2.911 \cdot 10^{-3}T - 2.009 \cdot 10^6 T^{-2}$$

HEAT CAPACITY OF NONCONDENSABLES (C_{pNg}). The following correlation is valid over the range 491.4°R < T < 3240.0°R [15]. The heat capacity of air is being used to estimate the heat capacity of noncondensables present in the flue gas.

Equation B.108

$$C_{pNg} = 6.7135 + 2.6091 \cdot 10^{-4}T + 3.54 \cdot 10^{-7}T^2 - 8.0527 \cdot 10^{-11}T^3$$

HEAT CAPACITY OF WATER (C_{pNg}). The following correlation is valid over the range 491.4°R < T < 3240.0°R [28].

Equation B.109

$$C_{pWg} = 7.3 + 1.3667 \cdot 10^{-3}T$$

HEAT CAPACITY OF LIQUID WATER (C_{pWl}). The following correlation is valid over the range 491.4°R < T < 671.4°R [15].

Equation B.110

$$C_{pWl} = 4.3728 + 0.06268T - 8.7903 \cdot 10^{-5}T^2 + 5.3856 \cdot 10^{-8}T^3$$

Over the temperature range of interest, a constant value can be used.

Equation B.111

$$C_{pWl} \approx 18.0$$

ENTHALPY DIFFERENCE FOR WATER (H_{Wg}-H_{Wl}). By definition

Equation B.112

$$H_{Wg} - H_{Wl} = \lambda + \int_{T_l}^{T_g} C_{pWg} dT$$

LATENT HEAT OF VAPORIZATION OF WATER (λ). The following correlation is valid over the range 492°R < T < 672°R. The Watson correlation [28] estimates the latent heat of vaporization of a pure liquid at a particular temperature using a known value of the latent heat of vaporization at a reference temperature. The correlation has the form

Equation B.113

$$\frac{\lambda}{\lambda_R} = \left(\frac{T_{eW} - T}{T_{eW} - T_R}\right)^{0.38}$$

Using a value for the latent heat of vaporization of water at 590°R of 18400 Btu/lbmole [15] and a value for the critical temperature of water of 1165°R, the above equation becomes

Equation B.114

$$\lambda = 1640.77(1164.78-T)^{0.38}$$

A simple linear relationship can be used over the temperature range of interest.

Equation B.115

$$\lambda \approx -(12.2565 T_1 + 25602.8)$$

GAS PHASE THERMAL CONDUCTIVITY (k_{tg}). An empirical relation has been proposed [31] for thermal conductivity of gas mixtures.

Equation B.116

$$k_{tg} = \sum_{i=1}^{n} \frac{y_{g_i} k_{tg}}{\sum_{j=1}^{n} y_{g_i} A_{ij}}$$

where k_{tgi} is the thermal conductivity of species i, A_{ij} is an interaction parameter for species i and j, and y_{gi} is the mole fraction of species i in the gas phase. For a water/ noncondensables binary mixture the relation is

Equation B.117

$$k_{tg} = \frac{y_{gW} k_{tgW}}{y_{gW} + y_{gN} A_{WN}} + \frac{y_{gN} k_{tgN}}{y_{gW} A_{NW} + y_{gN}}$$

Considering that gas behaves ideally at atmospheric pressure we have

Simulation and Optimization of a Granular Limestone Flue Gas Desulfurization Process

Equation B.118

$$y_{gW} = C_{Wg}RT/P = 0.7302 C_{Wg}T$$

and

Equation B.119

$$y_{gN} = 1 - y_{gW}$$

The thermal conductivities of water and noncondensables (Appell, 1989) are

Equation B.120

$$k_{tgW} = 1.4702 \cdot 10^{-6} + 4.4643 \cdot 10^{-9} T$$

and

Equation B.121

$$k_{tgN} = 6.2407 \cdot 10^{-7} + 6.6358 \cdot 10^{-9} T$$

Here we have modeled the noncondensables component as air. It has been suggested [31] that A_{ij} can be expressed as

Equation B.122

$$A_{ij} = A_k \frac{[1 + (k_{tg_i}/k_{tg_j})^{1/2}(M_i/M_j)^{1/4}]^2}{[8(1 + M_i/M_j)]^{1/2}}$$

Using the suggested value of $A_k = 1.065$, the values of A_{ij} for the system under consideration are

Equation B.123

$$A_{WN} = 0.2958[1 + 1.266(k_{tgW}/k_{tgN})^{1/2}]^2$$

$$A_{NW} = 0.2331[1 + 1.261(k_{tgN}/k_{tgW})^{1/2}]^2$$

Over the range of interest, constant values for A_{WN} and A_{NW} can be assumed.

$$A_{WN} \approx 1.258102 \qquad \text{Equation B.124}$$

and

$$A_{NW} \approx 0.8782935 \qquad \text{Equation B.125}$$

Appendix C - Plant Simulation and Cost Model

Mass Balance Equations

In the following equations, the G's are the mass flow rates and the subscript refers to the equipment number. Refer to Figure 10.1 to associate the equipment item with the subscript. The mass balance equations start with the reactor. The flow rate through one reactor is the product of the volumetric flow rate and the density.

Equation C.1

$$G_9 = U_S W_9 L_9 \rho_i$$

The flow rates in the upstream equipment are equal to the combined flow rate into all of the reactors.

Equation C.2

$$G_6 = G_7 = G_g = N_9 G_9$$

The flow rate out of the reactors is equal to the flow rate in plus the mass of SO_2 scrubbed minus the mass of CO_2 released. This is the flow rate in the downstream equipment.

Equation C.3

$$G_{10} = G_{11} = G_{12} = G_{13} = G_6 + F_{Sa} C_{SGO} Q_{GO} (64 - 44)$$

Using the empirical relation described in section 3.2.2, C_{Sse} and C_{Ssw} are determined and used to calculate the waste and recycle rates.

Equation C.4

$$G_{14} = \frac{G_{10} C_{Sse} - G_6 C_{Ss0}}{C_{Ssw}}$$

and

Equation C.5

$$G_{15} = G_{13} - G_{14}$$

The fresh limestone flow rate from the feed hopper and associated equipment is found by assuming the hopper operates 10% of the time and, therefore, the instantaneous rate is 10 times the average. The average rate is equal to the difference between the reactor feed rate and the recycle rate.

Equation C.6

$$G_1 = G_2 = G_3 10(G_6 - G_{15})$$

The flow rate used to size the bucket elevator is the sum of the instantaneous fresh limestone rate and the recycle rate. This is also the rate for the belt conveyor downstream of the bucket elevator.

Equation C.7

$$G_4 = G_5 = G_3 + G_{15}$$

The flow rates of the gas streams are found using the ideal gas law with the volumetric flow rate. The volumetric flow rate for the humidifier (equipment ID 17) is the inlet flue gas flow rate.

Equation C.8

$$G_{17} = \frac{Q_{g0} M_W}{0.7302 T_{g0}}$$

The volumetric flow for the fan (equipment ID 16) is found by correcting the inlet flow for the amount of water that condensed from the gas. This is discussed along with the cost equation for fans in a later section.

Equipment Size Parameters

The retention time in tank number six is assumed to be one hour, and in tank number 11 it is assumed to be ten minutes. The volumes of these tanks are then found by multiplying the retention time by the flow rate from the tank. The volume of tank number one is found by assuming a ten-ton limestone capacity.

The length of the screw conveyors is assumed to be equal to the cube root of the volume of the tank with which they are associated.

Simulation and Optimization of a Granular Limestone Flue Gas Desulfurization Process

Equation C.9

$$L_2 = V_1^{1/3}$$

$$L_7 = V_6^{1/3}$$

$$L_{12} = V_{11}^{1/3}$$

The length of the belt conveyors number 3, 5, and 14 is assumed to be 200 feet. The length of belt conveyor number 15 is assumed to be 100 feet. The length of belt conveyors 8 and 10 is assumed to be equal to twice the length of the reactor units.

Equation C.10

$$L_g = L_{10} = 2N_9 L_9$$

The height of the bucket elevator is assumed to be 100 feet.

The surface area required for the regeneration units is denoted A_{13}. It is found by assuming the screen capacity is 85 lbs/s/sqft/ft of mesh size [17]. Dividing the flow rate through the screen by this capacity and dividing the result by the mesh size gives the required screen area.

Equation C.11

$$A_{13} = \frac{G_{13}}{85}\left(\frac{25.4}{\text{mesh}}\right)$$

G_{13} is the flow rate in lbs/s and mesh is the mesh size in millimeters. It is assumed that there are two regeneration units ($N_{13}=2$).

COST MODEL. The installed capital costs are estimated using a general form

Equation C.12

$$C_i = I_i E_i C^*_i$$

where C_i^* is the reference cost, E_i is the inflation escalator and I_i is the installed cost factor needed to convert purchased cost to installed cost.

Using a reference cost of $99,300 in 1986 for a 5650 ft³ carbon steel tank [12] and assuming a 0.7 capacity exponent, the tank cost formula is

Equation C.13

$$C_i = E_i I_i (99,300)(V_i/5650)^{0.7}$$

with $I_i = 2.3$ and $E_i = 372.5/318.0$.

Walas [32] gives costs for screw conveyors between 7 and 100 ft as

Equation C.14

$$C_i = E_i I_i (700 L_i^{0.78})$$

with $I_i = 1.4$ and $E_i = 372.5/325.0$.

Peters and Timmerhaus [33] reported costs for a 24-inch wide belt conveyor which handles up to 300 tons/hr as

Equation C.15

$$C_i = E_i I_i (7525 + 365 L_i)$$

with $I_i = 1.4$ and $E_i = 1.0$.

Woods [34] gives costs for bucket elevators as a function of capacity and lift.

Equation C.16

$$C_i = E_i I_i (1.3 \cdot 2(H_4(G_4 \cdot 60 \cdot 60/2000)))$$

with $I4 = 1.4$ and $E_4 = 372.5/300.0$.

The cost of the reactors is based on their weights. The total surface area of the rectangular reactors is calculated by

Equation C.17

$$(A_9 = 2(W_9 L_9 + W_9 H_9 + H_9 L_9))$$

Simulation and Optimization of a Granular Limestone Flue Gas Desulfurization Process

Using a bulk density of 483 lbs/ft^3 for iron, a $1/4$ inch wall thickness, and a multiplier of 2.0 to account for the associated plenums, the weight of the reactor vessel is given by

Equation C.18

$$M_9 = 2.0 \cdot 483 \cdot (1/4)(1/12)A_9 = 20.1 A_9$$

where A_9 is in square feet and M_9 is in pounds. This is then used in a cost formula for vertical vessels [32]

Equation C.19

$$C_9 = E_9 I_9 N_9 [17e^{(9.1 - 0.29 In(M_9) + 0.051 In(M_9)^2)} + 246\left(\frac{2M_9 L_9}{M_9 + L_9}\right)^{0.7} L_9^{0.7}]$$

with E_9=372.5/325.0 and I_9 = 1.5. A material factor of 1.7 is included in Equation C.19 for 304 stainless steel. N_9 is the number of reactor units. This is calculated by dividing the volumetric flow rate by the superficial gas velocity and area for one reactor. To ensure a reasonable minimum number, 2 is added to the result.

Equation C.20

$$N_9 = \frac{Q_{GO}}{U_{GO} W_9 H_9} + 2$$

The cost of a vibrating screen is given by Peters and Timmerhaus [33] as a function of screen area. Equation C.21 is adapted from this data.

Equation C.21

$$C_{13} = E_{13} I_{13} N_{13} (3200 + 840 A_{13})$$

with N_{13}= 2, I_{13} = 1.4 and E_{13} = 1.0.

Ulrich [35] presents purchased cost for fans as a function of gas flow rate, pressure rise and materials of construction for four different types of fans. A carbon steel centrifugal radial blade fan is assumed. Walas [32] fit the data given by Ulrich to an equation for cost.

Equation C.22

$$C_{16} = E_{16}I_{16}[2.2\exp(0.1522+01203\ \text{In}Q_{ge} + 0.0931(InQ_{ge})^2 + 0.2583\ \text{In}\ \Delta P - 0.0143(In\Delta P)^2]$$

with $I_{16} = 1.4$ and $E_{16} = 372.5/315$. Q_{ge} is the standard volumetric flow rate of gas exiting the reactors (see below) and DP is the pressure rise across the fan, which is assumed to be equal to the pressure drop across the reactor bed.

The pressure drop across the reactor bed is estimated using the Ergun equation [17].

Equation C.23

$$\Delta P = \frac{U_{g0}}{g_c \phi D_p}\left[\frac{1-\varepsilon_{g0}}{\varepsilon_{g0}^2}\right]\left[\frac{1501-\varepsilon_{g0}\mu_{g0}}{\phi D_p} + 1.75\rho_{g0}M_g U_{g0}\right]\Delta Z$$

where ϕ is the sphericity. The values for ε_{g0}, U_{g0}, and μ_{g0} are taken to be those at the entrance to the bed.

Since it is assumed that the reaction of one mole of SO_2 produces one mole of CO_2, the molar flow rate of gas from the inlet to the outlet only changes due to a difference in gaseous water flow rate. Therefore

Equation C.24

$$Q_{ge} = Q_{g0}\left(\frac{460(1-x_{WO})}{T_{g0}(1-x_{We})}\right)$$

where the x's are the water mole fractions. The effect of pressure on the volumetric flow rate has been ignored.

The electrical power required to operate the plant is estimated for each of the major pieces of equipment. The estimating equations used are detailed below. In the following equations, G_i is the flow rate through the equipment in lb/s, L_i is a characteristic length in feet, and P_i is the power requirement in KW. The sum of the power requirements is then used to calculate a power cost by multiplying by $0.03 per KWh of electricity.

The power requirement for screw conveyors is estimated using the equation

Equation C.25

$$P_i = 3.0(8.64 \cdot 10^{-3} G_i + 7.2 \cdot 10^{-5} G_i L_i)$$

This equation is adapted from the equation given by Stanley [36] for the power requirement of a belt conveyor, with a factor of 3.0 included to account for the added friction.

The equation used to estimate the power requirement for 24-inch belt conveyors with a running angle of repose of 38° considered for limestone is formulated as [36]

Equation C.26

$$P_i = 8.64 \cdot 10^{-3} G_i + 7.2 \times 10^{-5} G_i L_i$$

The power requirement for a bucket elevator is given by Keeth et. al. [21].

Equation C.27

$$P_i = 0.002 \left(0.746 \frac{G_i \cdot 60 \cdot 60}{2000} L_i \right)$$

Peters and Timmerhaus [33] give a graphical representation of the power required for a vibrating screen as a function of the screen area. The linear equation adapted from that graph is

Equation C.28

$$P_i = 0.072 + 0.077 A_i$$

where A_i is the screen area measured in square feet.

The fan power requirement can be expressed as [37]

Equation C.29

$$P_i = 1.5 \cdot 10^{-3} Q_g \Delta P$$

where Q_g is the volumetric flow rate in ft³/s and DP is the pressure rise in lbf/ft².

The power requirement for the humidification unit is estimated based upon that for an air compressor, which is given by [36]

Equation C.30

$$P_i = 6.9277 \cdot 10^{-3} \cdot 0.0283 \cdot 359 \cdot 3600 Q_A$$

where Q_A is the volumetric flow of air in lbmols/s. The air mass flow rate is estimated as 20% of the mass flow rate of water through the humidifier.

Water is (optionally) consumed in the humidification system. The rate of water consumption, Q_w, is calculated from the difference in the water concentration at the humidifier inlet and the concentration at the reaction inlet. The product of this difference times the gas flow rate gives the amount of water added. The cost for water is calculated based on a value of $1.50 per 1000 gallons.

The cost for limestone includes $10/ton for purchased stone and $10/ton for waste disposal. The purchased quantity is calculated using a mass balance; i.e., the flow rate into the reactor minus the recycle flow rate is the limestone feed rate.

The total annual costs include the operating cost and a multiple of the fixed costs. The total operating cost is the sum of the three operating costs listed above.

Equation C.31

$$C_O = C_p + C_W + C_S$$

The fixed cost is the sum of the installed costs for all of the pieces of equipment. The annual system cost is then expressed as

Equation C.32

$$C_A = 365 \cdot 0.7 C_O + 0.17 C_F$$

The factor 0.7 multiplied by the annual operating costs is an assumed plant operating factor (on-stream time). The factor of 0.17 that is multiplied by the fixed costs is the conversion factor from capital dollars to annual dollars. This factor includes estimated costs for depreciation (straight line method), interim replacement (@0.35% of the capital cost), taxes (@4% of the capital cost), insurance (@0.35% of the capital cost), and the price of capital (@9% of the capital cost).

Index

A
acid rain 2
  sulfur dioxide 422
additives 77, 83
  chemical effects 234
  effect on sorbent reactivity 647
  effects of on lime dissolution rate 634
  oxidizing abilities 652
  spray drying 229
  to enhance reactivity 214
  use of 77
air jet
  test results 375

C
calcination 12, 424
calcium 512
calcium based sorbents
  sulfation of 476
calcium silicate sorbents 139
calcium silicates 123
calcium-based sorbent 458
calcium-based sorbents
  calcination 460
calibration tests 368
capillary forces 261
carbonation reaction 561
circulating fluidized bed absorber 90
CO_2
  properties 116
  sorption studies 151

D
dry injection 29
dry/dry systems 6, 29
duct injection
  chemical additives 229

E
economizer zone injection 97
electrostatic forces 260
electrostatic precipitator (ESP) 51
entrained flow reactor 430, 536, 587
  construction and operation 537
  sulfation in 577
entrained flow reactors
  carbonation 561
ESP
collection efficiency 56
design considerations 58

F
fabric filters 60
  material selection 62
FGD
  processes 344
  transport processes 355
FGD applications
  new techniques 88
FGD process
  spray drying 37
FGD processes
  recovery systems 692
  throwaway systems 692
fixed-bed
  limestone emission control 705
fixed-bed reactors
  process model 723
flow
  particle-laden 400
  reversal 407
  single-phase 400
  spray-concurrent 401
flow patterns 87
flue gas desulfurization 88
  technology 4
fly ash resistivity 55
furnace sorbent injection 424

G
gas flow distribution 57

H
Hamaker constants 266
high temperature studies 72, 77
high-temperature reactors 435
high-temperature studies 208
hydrated lime injection 97

I
impeller fluidizer 92
in-furnace injection 10, 97
interparticle forces 266

L

LEC
  process theory 716
LEC
  model development 716
lime
  dissolution rate 630
  dissolution rate studies 623
  hydrated 29
limestone
  dissolution rates 713
  solubilities 713, 737
limestone emission control
  moving-bed 709
Limestone Emission Control (LEC) 691
Limestone Emission Control (LEC)
  process 95
Limestone Emission Control (LEC)
  system 91
Limestone/Lime Characterization
  Study 221
low-temperature studies 75, 79, 82
low-temperature studies 211

M

model
  dth-order deactivation 572
  kinetic 593
  powder dispersion 329
  spray dryer FGD 622
  sulfation 579
  third-order deactivation 569
model simulation 326
modeling
  calcination 493
  sintering 495
  sorbent contact geometries 262
  stochastic 317
  sulfation 497
models
  spray dryer 664
moisture
  effect on sorbents 302
moving-bed reactor
  process model 733

O

Ohio Coal Development Office
  (OCDO) 2

P

particle collection 451
particle resistivity 53
particle size 429
  effect on sulfation 483
particle-laden jet
  test results 375
particulate collection
  baghouses 615
  ESP 615
particulate control 50, 98
particulate control devices
  mechanical 65
pore structure 219
portland cement 182
powder characterization 257
  results 263
powder dispersion 303
powder feed system 433
powder testing 297
preparation 208
probe system 438

R

rapping 56
reagent preparation 37
reagents
  properties and preparations 611
reentrainment 57
repowering technologies 23

S

sintering 12, 219, 424
  role of temperature 430
slurry injection 29
SO2
  properties 116
  sorption studies 141,151
SO2 control
  dry 9
SO2 removal 617
  baghouse 67
  ESPs 67
sorbent
  characterization 224
  preparation 208

sorbent injection
  dry 96
  duct 532
  economizer 534
  facility design 355
  furnace 531
  test facility 358
  wet 96
sorbent reaction
  modeling 22
sorbent type
  role of 429
sorbent utilization 617
sorbents
  calcium based 71
  calcium-based 512
  sodium-based 117
spray dryer
  desulfurization processes 644
  experimental approach 641
  technologies 611
spray dryers 39
spray drying 34
  absorption 98
spray drying process 45
sulfation 17, 424
sulfation and carbonation reaction
  simultaneous 585
sulfation reaction 577
sulfur
  evolution 12
sulfur dioxide
  control 422
  removal 423

V
Van der Waals force 257

W
waste disposal 44
wet scrubbers 64
wet/dry systems 8, 31
wet/wet systems 6